CONSTRUCTION SAFETY MANAGEMENT AND ENGINEERING

Darryl C. Hill, CSP

EDITOR

AMERICAN SOCIETY OF SAFETY ENGINEERS ❖❖ Des Plaines, Illinois

Construction Safety Management and Engineering

Library of Congress Cataloging-In-Publication Data

Construction safety management and engineering / Darryl C. Hill, editor.
    p. cm.
  Includes bibliographical references and index.
  ISBN 1-885581-46-7 (alk. paper)
   1. Building–Safety measures. I. Hill, Darryl C.

  TH443.C6847 2003
  690'.22–dc22

                              2003057747

Copy Editor: Cathy Lombardi
Text Design, Composition, and Layout: Cathy Lombardi
Cover Design: Michael Burditt, ASSE
Managing Editor: Michael F. Burditt, ASSE

10 09   7 6 5 4

ISBN 1-885581-46-7

Cover photos courtesy of Hoffman Construction Company

Printed in the United States of America on acid-free, recycled paper.

Recycled

▪▪ Table of Contents ▪▪

▪▪ Preface ▪▪

Construction Safety: Engineering and Management Principles provides a comprehensive discussion in the field of construction safety for both new and veteran construction safety practitioners. The book describes how successful and effective owners, contractors, and subcontractors operate in an increasingly challenging, fast-paced, demanding, and perhaps revolutionary environment. Readers are offered an integrated view of the knowledge base, research, and practice of construction safety within a context of multiple perspectives and a wide range of thinking and experience. The book is a powerful resource for professional development focused on conquering the safety, health, and environmental challenges of the twenty-first century.

Each chapter facilitates and inspires the reader in the process of building a strong foundation on which to sustain his or her educational career. The focus is in proven principles and best practices intended to improve a company construction safety process. Chapters include learning objectives and questions to facilitate the academic environment. Questions are designed to model and encourage readers in the development of critical thinking and supported analysis of key issues. They also allow the reader to compare their experience to that of noted experts and professors in the field.

This book will be invaluable for anyone who wishes to take on greater responsibility in construction safety. The format is divided into five major areas. The first section explores what a practitioner must understand before the work commences at a job site. The second section looks at process components that are critical to the overall success of a project. These components parallel the zero-injury techniques that resulted from the Zero Accidents

Task Force research established by the Construction Industry Institute (CII). The task force used the BLS/OSHA lost-workday case incident rate (LWCIR) to measure performance on 25 construction projects that involved interviewing some 482 project personnel. The third section looks at legal aspects, including liability and regulatory requirements. The fourth section addresses the many technical aspects in construction, such as fall protection, steel erection, scaffolding, and much more. Finally, the fifth area examines the emerging issues and other key aspects in construction safety.

A number of nationally noted professors and safety professionals from leading institutions and progressive companies have provided original, expert reflections on critical topics within each chapter. The chapter authors have over 1000 years of combined construction safety experience. These authors are Professional Engineers, Certified Safety Professionals, Certified Industrial Hygienists, and Certified Hazardous Materials Managers. Many chapter contributors have advanced degrees.

The book content will be useful to a wide spectrum of people involved in or responsible for construction safety. Students will gain a solid understanding of the multiple facets of this dynamic discipline. Anyone interested in construction safety and its effects on students will find here a useful reference book and guide to understanding a very complex subject. Above all, this book will provide a template for how to design an effective safety process at a construction project. The aim is to provide a tool to address the many hazards in construction and to ensure that human suffering and harm is eliminated in construction throughout the world.

Darryl C. Hill, CSP
Editor

■■ Acknowledgments ■■

The publisher and editor are grateful for the support from the technical reviewers. Without their contributions, insight, and advice, this book would not be possible.

Richard S. Baldwin, CSP
Richard L. Barcum, CHMM, CIH, CSP
Patricia L. Bellm, CSP
Bruce W. Blakemore, MS, CMSP, CSHM, NREMTP
A. David Brayton, CPC, CSP
Richard Brayton
S. Renee Butler, CHST, ASP
Jeffrey C. Camplin, CPEA, CSP
Gregory D. Cantrell, CSP
Robert Capelli, ASP
Wayne C. Christensen, P.E., CSP
Ryan T. Corbin, CHST
Eileen Dare
Kathleen Dobson
Matthew Eisle
Don Garvey, CIH, CSP
Anthony Giovinazzo
Jack B. Hanna, CSP
Kara N. Herber
Jack M. Hill, CSHM, CSP

E. Tim Holden, CPEA
Donovan Jackson, CHMM, CSP
James H. Johnson, CHST, OHST, CSP
John H. Johnson, CSP
Paul J. Kendall, P.E., CSP
Mike D. Kinney, CSP
Leonard H. Kushner, P.E.
Sam Lybarger, Ed.S., CSP
Michael T. Maisey, CSP
Fred A. Manuele, P.E., CSP
Robert E. McClay, Jr., CSP
James McRoy, MA, CHCM
Robert T. Poole, CSP
Jim Quinn
Edward G. Ratzenberger, CSP
Larry Shaw Salazar, JD
Eugene Satrun, CIH, LIH, CSP
Donald L. Schmid, CSP
Steven C. Sobczak, CIH, CSP
Dan Souders, CSP
Kevin E. Stroup, CSP
Delmar E. Tally, P.E., CSP
T. Michael Toole, Ph.D., P.E.
Andy Whiton, CSP

▪▪ Foreword ▪▪

As Administrator of the American Society of Safety Engineers' (ASSE) Construction Practice Specialty (CPS), it is a great pleasure to introduce you to *Construction Safety: Engineering and Management Principles.* This textbook is the cumulative effort of selected construction safety experts among our 3500 CPS members.

Under the guidance of Editor Darryl C. Hill, CSP, the book has been organized into distinct sections highlighting performance objectives before the project commences, outlining key safety process components, and addressing technical issues, legal aspects, and a number of additional considerations for the construction safety practitioner.

The need for owners, contractors, and subcontractors to effectively manage the engineering/ construction process on any project has never been greater. As highlighted in this book, consideration for protection of people, property, and the environment has caused owners and contractors to expend additional resources for effective management of the construction process. While some of these issues addressed in the book focus on topics that have become important within the last few years (i.e., environment protection on construction sites), the facts remain that our greatest challenge must start with a reduction of fatal accidents in construction.

In researching construction-related fatalities from 1993–2001, the Bureau of Labor Statistics (BLS) reports an increase in every year except 2000, and the number of fatalities in 2001 were 33 percent higher than the totals recorded nine years ago in 1993. BLS reports 1225 construction fatalities in 2001 compared to 919 in 1993. By industry type, the construction industry had a fatality rate of 20.8 percent, the highest fatality rate of any industry, yet it employs only slightly more than 5 percent of our

nation's workforce. Amazingly, the Occupational Safety and Health Administration (OSHA) recordable injury/illness rate for the construction industry has decreased each year and shows a 35 percent reduction over this same period.

One could erroneously conclude that we are winning the war, but losing soldiers in the fight. The truth is, the loss of even one life during the construction process is the ultimate failure.

We can look back on two of the greatest construction projects America has ever known to illustrate this point. The construction of the Hoover Dam began in March 1931 and concluded on September 30, 1935. With peak manpower of over 5000 construction workers, the project was an engineering and construction marvel. The project was one of the first to require use of personal protective equipment in the form of hard hats, which saved countless lives, yet the Hoover Dam National Historic Landmark Society estimated 96 fatalities occurred during construction.

The construction of the San Francisco Golden Gate Bridge in the early 1930s established America's first designated "Hard Hat Area." The proect's Chief Engineer, Joseph B. Strauss, shared a vision with his grandfather that the workplace could be a safer environment for the worker. The project involved the most rigorous safety precautions in the history of bridge building. From the mandatory "Hard Hat Area" to the use of safety nets, Project Engineer Strauss was implementing safety precautions against the two greatest construction hazards (i.e., falls from elevations and struck-by incidents). The Golden Gate Bridge project was experiencing unprecedented safety success with the fall protection nets being credited with saving 19 lives. The project was well on its way to establishing a record for safety until an unforgettable day in February of

1937. With the scheduled opening only three months away, a section of scaffold collapsed, broke through the safety net, and plunged ten workers to their deaths. This terrible tragedy negated all the success of the past four years, as the scaffold came crashing down into the San Francisco Bay.

It is interesting that the construction safety hazards recognized by Strauss over 70 years ago continue to plague us today. According to recent OSHA data, falls from elevation and struck-by incidents resulted in over 50 percent of all construction-related fatalities in 2001. We must do a better job in managing these risks.

The devastation of a construction fatality is far reaching. The loss to a family is indescribable and the suffering is heart wrenching. The negative publicity associated with these events has increased dramatically and the legal ramifications can be staggering.

It is our intent that this book be utilized by safety practitioners, project managers, project engineers, academic professors, vocational counselors, students, workers, and anyone interested in the protection of our greatest asset—human life!

It is our desire to provide a tool that can be utilized to accomplish this goal.

R. Ronald Sokol, CSP
Administrator, ASSE Construction
Practice Specialty, 2000–2004

PART I

Before the Work Commences

■■ **About the Author** ■■

INTRODUCTION TO CONSTRUCTION SAFETY
by Tom Broderick

Tom Broderick has thirty years experience in construction. He has been a construction worker, co-owner of a small construction company, and a construction safety professional. In his safety career, he has managed safety programs at projects that include nuclear and fossil-fuel power plants, paper mills, and other large projects for Blount Construction, Stone and Webster Engineering Company, and the Rust Engineering Company. Tom holds a B.S. degree in speech-communications and an M.S. in safety management. He is currently Executive Director of the nonprofit Construction Safety Council and the Chicagoland Construction Safety Council. In that capacity he is the director of the annual Construction Safety Conference in Chicago—the largest educational gathering in the United States focused solely on construction safety, health, and environmental issues. Tom was nominated in 2001 by Secretary of Labor Elaine Chao to serve on the congressionally mandated OSHA Advisory Committee for Construction Safety and Health (ACCSH). He is a past president of Veterans of Safety International and a commissioner on the National Commission for the Certification of Crane Operators. He has authored numerous articles in the SH&E area and wrote a chapter titled The Role of the Safety Director in Injury Prevention in Susan Isernhagen's book, *The Comprehensive Guide to Work Injury Management.* Broderick also coauthored a chapter on overhead power line-hazards in Janine Reid's book, *Saving Lives.*

2

CHAPTER 1

Introduction to Construction Safety

LEARNING OBJECTIVES

- Explain the historical underpinnings for the current state of occupational safety and health.
- Discuss the ethical considerations that apply to the SH&E profession.
- Discuss the roles that construction SH&E professionals are expected to fill.
- Explain the role of the Occupational Safety and Health Administration (OSHA) in construction safety and health.
- List key components of successful SH&E programs in construction.
- Discuss the impact of corporate culture on safety and health in construction.

INTRODUCTION

Construction is the largest single-purpose industry in the United States. It has historically employed about 5 percent of the country's workforce, yet has accounted for a disproportionate number of occupationally related fatalities—from 17 to 20 percent. Table 1.1 outlines the steady increase for construction fatalities during the past 10 years. The causes of injuries and illnesses in construction have long been recognized and their persistence continues to frustrate construction safety and health practitioners. Traumatic injuries have historically been caused by falls, electrocution, being struck by objects, or being caught in or between objects. The work lives of many employed in the construction industry have been shortened by repeated physical insults from vibration, repetitive motion, overexertion, and the health hazards posed by exposure to lead, silica, solvents, heat and cold, and a plethora of other chemical and environmental challenges.

■ ■ **Table 1** ■ ■

Construction Fatalities, 1992–2001

Year	Annual Statistic
1992	963
1993	971
1994	1077
1995	1098
1996	1095
1997	1136
1998	1207
1999	1228
2000	1183
2001	1264

Source: U.S. Dept. of Labor, Bureau of Labor Statistics

The practice of occupational safety and health in construction holds difficulties not associated with other industries where the workforce is stable and the exposures to hazards are static. The types of construction vary, as do the firms that perform the work. Yet despite a high turnover among the construction workforce, the temporary nature of construction work, and the resulting travel requirements, construction safety and health practitioners seldom make career changes into other industries.

Broad categories of construction, including their Standard Industrial Classification (SIC) are:

- Building Construction General Contractors and Operative Builders (SIC 15)
- Heavy Construction Other Than Building Construction Contractors (SIC 16)
- Construction Special Trade Contractors (SIC 17)

[*Note:* the Standard Industrial Classification (SIC) system is changing to the North American Industry Classification System (NAICS) for government statistical purposes.]

Within these categories are all sorts of companies and projects that could employ safety and health professionals, from tunnels to skyscrapers and from single-family homes to 3000+-room hotel complexes. Each type of project can have unique exposures to health and/or safety hazards. Because the work site is ever-changing and heavy equipment and vehicles are often involved, careful planning, close coordination, and attention to safety are critical. Safety practitioners must assist in the creation of a system that can identify these hazards and intervene immediately. This means that all members of the project "team" will be responsible for identifying and correcting hazards. If adequate preplanning is done prior to mobilizing at the construction site, the number of hazards that will develop will be minimal.

HISTORICAL PERSPECTIVE

There is a rich history of the development of the safety, health, and environmental (SH&E) field. A short review of outstanding milestones will give the SH&E professional some appreciation for the field as we know it now. Some of the hazards that we confront today have been recognized for centuries.

The concept of building things with an eye toward safety has been around for quite some time. About 4000 years ago a Babylonian king named Hammurabi created a series of edicts that described certain "normal" parameters of social behavior and the consequences for straying from them. The issues ranged from theft of property to mistreatment of another's cattle. The punishments seem drastic by today's standards, as you can see in Hammurabi's building codes (Figure 1).

Although some of the remedies prescribed by Hammurabi may seem severe, note that there has been a recent increase in the number of U.S. employer prosecutions resulting in large fines and/or imprisonment for workplace safety hazards that lead to serious injury or death. While some states are more aggressive than others at this point, the practice has been established and is likely to expand.

Throughout antiquity we find historical references to occupational health and safety issues. Most were brought forth by early physicians. Their patients presented symptoms that seemed similar to other patients with the same occupation, and the diseases correlated with common exposures in their respective workplaces.

---

229. If a builder build a house for some one, and does not construct it properly, and the house which he built fall in and kill its owner, then that builder shall be put to death.

230. If it kill the son of the owner, the son of that builder shall be put to death.

231. If it kill a slave of the owner, then he shall pay slave for slave to the owner of the house.

232. If it ruin goods, he shall make compensation for all that has been ruined, and inasmuch as he did not construct properly this house which he built and it fell, he shall re-erect the house from his own means.

233. If a builder build a house for some one, even though he has not yet completed it; if then the walls seem toppling, the builder must make the walls solid from his own means.

■ ■ **Figure 1** ■ ■ **Hammurabi's building codes**

In the fourth century B.C., Hippocrates described the toxic nature of lead among miners exposed to airborne lead particulate and to workers smelting the lead ore and working with the metal. The handling of sulfur and zinc were noted to cause disease by Pliny the Elder, a Roman naturalist who created the first encyclopedia of natural science in the first century A.D. He was the first to document the use of respiratory protection to prevent airborne exposure to irritants or hazardous airborne contaminants such as cinnabar (red mercuric sulfide). A goat bladder was dried and worn over the worker's face to act as a filter, preventing particulate from entering the respiratory system. Fit testing would come much later. In the following century, the Greek physician Galen contributed to the literature when he described the toxic effects of lead exposure in detail.

An early system of compensation for injuries sustained in altercations was described in a series of statutes published in the seventh century by King Rothari of Lombardi. This early effort to legislate compensation for losses was followed by a more detailed system that the Nordic King Canute invoked in the eleventh century. This system was the first to differentiate the severity of loss through the assignment of values for different body parts. Higher value was placed on body parts deemed more essential for a person to function normally. This was the forerunner to the present-day workers' compensation system wherein anatomical charts are used to denote the relative value of body parts for the settlement of claims for their loss or loss of use.

In the late fifteenth century an Austrian physician named Ulrich Ellenbog authored the first publication devoted solely to the field of occupational injury and disease. He studied the gold mining process and the health of the workers engaged in the extraction of gold. The toxic effects of the heavy metals that the workers breathed were carefully documented. Ellenbog also described hygiene and other preventive methods that could reduce the toxic exposures.

Another milestone in the evolution of the field of health and safety came about in 1700, when Dr. Bernardo Ramazzini published *De Morbis Artificum Diatribor* (*The Diseases of Workers*). He described the toxic effects of materials found in the workplace and the ergonomic stressors that workers faced. He also described prevention strategies that could reduce or eliminate these harmful exposures. His early work lead to Ramazzinni earning the title of Father of Industrial Medicine in many circles. Unfortunately, many of the interventions described by Ramazzini were not assimilated into the burgeoning industrial infrastructure, as the Industrial Revolution rapidly changed society forever. The result was extensive carnage in the early factories, smelters, and other industrial facilities.

The detailed historical account of the construction of the Brooklyn Bridge in New York in the late nineteenth century provides a fascinating look into the dynamics of an early major American construction project. *The Great Bridge*, by David McCullough (Simon and Schuster, 1982) gives readers a candid view of a project that claimed a minimum of twenty bridge workers' lives. From the depths (caissons disease from getting the bends while working under compressed air in the confined spaces of the footings) to the heights (falls from the superstructure), and other areas in between, accidents were an accepted part of the construction process. Since there were no regulations requiring fatality reporting at the time, the *actual* number will never be known.

Throughout the nineteenth century there was widespread use of child labor. In fact, the 1900 U.S. census showed that nearly two million children were employed in factories, mines, and other hazardous workplaces.

The dawn of the twentieth century saw a number of reforms that had positive effects on worker health and safety and the compensation system for accident victims. Theodore Roosevelt did not advocate the practice of injured workers having to hire an attorney and file suit against their employer to receive medical coverage and indemnity payments for injuries suffered at work. In 1908 he saw the first workers' compensation system implemented at the federal level. By 1911, Wisconsin had created the first state workers' compensation program, and that was soon followed by other states.

The widespread problems of serious injuries and deaths in the workplace led to the first Cooperative Safety Congress in Milwaukee in 1912. Business leaders, insurance representatives, public officials, and early safety practitioners all came together to discuss solutions. A second meeting in 1913 in New York marked the creation of the National Council for Industrial Safety. The name for the organization was later changed to the National Safety Council to reflect the organization's broader scope to include health and safety at home, in schools, on the streets and roads, in the air, and anywhere else unintentional injuries and illnesses affected Americans. The National Safety Council received its charter from congress, indicating the considerable support the fledgling safety effort achieved.

One of the key topics discussed at these early safety conferences was the lack of a systematic collection of existing safety standards and codes and the lack of a framework for creating one in the future. In 1920, the American Engineering Standards Committee developed the National Safety Code Program to serve as a clearinghouse for standards developed by associations, societies, and various agencies. This program was later renamed and it became the American National Standards Institute or ANSI. Today ANSI is the preeminent standards-setting organization in the United States, and many of their safety standards form the basis for today's OSHA standards.

In the early 1900s, Dr. Alice Hamilton was a strong advocate for worker safety and health. She was considered the founder of modern occupational medicine, and became Harvard University's first female professor. Her 1943 autobiography, *Exploring the Dangerous Trades*, offers a candid look at the state of health and safety on the job in post–Industrial Revolution America. Her field investigations began soon after the turn of the century, and her writing and teaching pointed to the plethora of hazards facing stone masons, painters, boilermakers, and other common tradesmen. She described the health outcomes of those exposed workers, revealing widespread incidence of occupational disease. Her work hastened the proliferation of state workers' compensation programs throughout the United States. This reference is highly recommended as a primary resource that describes the evolution of the health and safety profession in America and in other countries.

Golden Gate Bridge

In contrast to the Brooklyn Bridge project only fifty years earlier, the construction of the Golden Gate Bridge in the early 1930s ushered in two important developments in the construction safety field. It marked the first use of a head protection device created by Edward D. Bullard, and the use of safety nets as fall protection. The headgear was the forerunner of the modern hard hat. Bullard's safety equipment firm continues to play a major role in protecting construction workers to this day.

The safety nets were placed under the areas where the iron workers plied their *high-wire* trade. A film produced by the Iron Workers Union documents this early bridge project and notes that nineteen men fell into the nets, saving them from certain death if they had fallen into the treacherous waters below. These workers were called the Halfway to Hell Club. The measures were effective; only one worker was killed between 1933 and 1937—quite an achievement in an era when it was predicted that one man would die for every million dollars of project cost. This was an early example of how a strong-willed construction manager who demands safety excellence can effect a successful safety record. Unfortunately, in February of 1937 a work platform holding twelve men broke loose and fell through the net system; ten men lost their lives. Despite this catastrophic loss, the Golden Gate Bridge project is still considered a significant milestone in the evolving field of risk control at construction sites.

The Government Takes Action

Although most early safety legislation came at the state level, in 1936 the Walsh-Healey Public Contracts Act was passed by congress. Any federal contract with a value of $10,000 or more required the contractor to follow all applicable safety requirements. Failure to do so could lead to disbarment or blacklisting from performing other government work. This act, along with the Longshoreman's and Harbor Worker's Compensation Act, were the principle federal regulations until the 1960s, when a flurry of social legislation was enacted. The Contract Work Hours and Safety Standards Act (Construction Safety Act), Federal Mine Safety and Health

Act, Service Contract Act, and several other bills became law.

By 1969, traumatic injuries in the workplace affected 2,200,000 workers per year, or about 8500 per day. Deaths on the job claimed 14,200 workers per year, or about 85 people per day. It is not clear how many of those workers practiced the construction trades, but there is no reason to think that construction didn't account for a significant portion of the injuries and deaths. This environment provided the impetus for the enactment of the most sweeping legislation aimed at protecting workers in the private sector. The Williams-Steiger Occupational Safety and Health Act of 1970 was passed and signed into law by President Nixon; it was fully implemented by April 28, 1971.

Turning the Tide with Owner Involvement

In 1982 a revolutionary series of reports was issued by the Business Roundtable. They addressed all facets of construction, including quality, productivity, availability of qualified personnel, and safety. The *A-3 Report* focused on the role of the owner in the construction safety process. Part of the foundation for the *A-3 Report* was an earlier study commissioned by the Roundtable and performed by Stanford University's Department of Civil Engineering. It examined, among other things, the ratio between direct and indirect costs of accidents, and the relationship between all parties to the construction process.

This report made it clear that owners who maintain a strict *hands-off* approach to construction safety miss an opportunity to save money, receive a better quality product, and improve public esteem in the bargain. The list of recommendations for owners is provided here:

Owners should:

1. Become familiar with the high cost of construction accidents to reinforce their moral commitments to provide a safe work environment.
2. Be prepared to financially support contractors' efforts to ensure an effective safety program.
3. Realize that merely adopting a safety program will not yield the desired results with-

out a serious and persistent management commitment.
4. Recognize that the principles of management control commonly applied to cost, schedule, quality, and productivity are equally applicable to safety and that, when used, they will improve safety performance.
5. Make safety an important consideration in the selection of contractors for bidding on their construction projects, including evaluation of contractors' past safety performance, safety attitude, and present programs and practices.
6. Explain to the contractor, prior to the bidding process, what is expected regarding safety performance.
7. Evaluate in the bid analysis the ability of the contractor to achieve expected safety performance and from this determine the degree of owner involvement required to meet safety objectives.
8. Become more directly involved in the safety activities of their construction projects and take proper measures to achieve better safety performance, such as:
   - Provide safety and health guidelines that the contractor must follow
   - Require a formal site safety program
   - Require the use of permit systems for potentially hazardous activities
   - Require the contractor to designate the responsible supervisor to coordinate safety on the site
   - Discuss safety at owner–contractor meetings
   - Conduct safety audits during construction
   - Require prompt reporting and full investigation of accidents.
9. Function with the contractor as a cohesive safety team during the planning and execution of a construction project.
10. Establish lines of communication at all levels with the contractor so that safe work practices are understood by both parties.

Even though this report was published over two decades ago, the information is still very relevant. This is evidenced by the fact that major players in the owner community still maintain a head-in-the-sand approach to construction safety, believing that contract documents and hold-harmless agreements will provide an acceptable level of protection from all ills created by commissioned construction work.

The *A-3 Report* is a technical reference that is highly recommended for review by all construction safety practitioners.

Eras of Safety Management

Dan Petersen captures the history of approaches to managing occupational safety in his book, *Safety Management: A Human Approach*. He describes the inspection era in the early 1900s, when safety efforts concentrated on inspecting the workplace for obvious physical hazards. Next, the safety movement focused on unsafe acts and unsafe conditions. Other historical periods include the hygiene era, followed by the noise era, the safety management era, the OSHA era, the accountability era, and, finally, today's human era, marked by the move toward behavior-based safety. Behavior-based safety uses peer-based observations and evaluation of task performance. Worker-to-worker relationships may provide an effective, nonthreatening environment for changing risk-taking behavior in the workplace. Favorable results have been reported in both general industry and construction. It should be noted that behavior-based safety is not embraced by all safety practitioners, and the concern has been raised that some employers may use it to shift the responsibility for a safe workplace onto the shoulders of the workers. They feel that management responsibility for all aspects of employee and site safety is the appropriate philosophy.

Modern contractor safety programs often combine some of the recent era's components, such as behavior-based safety with employee-to-employee observations, with vestiges of earlier era components. The OSHA era is still very much with us, as is the inspection and safety management era. This is not a bad thing. In fact, very effective contractor SH&E programs certainly do draw from the best of these historical eras.

Here is an example from the Construction Industry Safety Conference proceedings of October 2, 1963 (held at the Hotel Commodore, New York, NY):

Mr. Pintard, Turner Construction Company, outlined a general contractor's program maintained by his company. As the general contractor, Turner accepts full responsibility for accident prevention on their operation. Each superintendent outlines specifically and very early in the job what is expected of subcontractors. Representatives from each trade involved, management of subcontractors, frequently shop stewards or union rank and file, arrange for weekly job conferences. At these sessions unsatisfactory job conditions, conflicts with other trades, general contractor's requirements for certain protection, the use of personal protective devices and hard hats are examined and evaluated.

Although this construction safety conference took place years ago, the issues discussed then are still relevant, and this paragraph could have been excerpted from contemporary conference proceedings. It was reported that the general contractor essentially set the tone for the job by spelling out expectations for subcontractors, holding planning meetings that included worker input, and discussing the current safety issues at the site. If these principles were applied at all job sites since 1963, the injury and fatality statistics would have been positively impacted. The real challenge for the construction industry is to implement these practices on all projects going forward so that the *zero-injury goal* is realistic and achievable in the foreseeable future.

DISCIPLINES FALLING UNDER THE PURVIEW OF THE CONSTRUCTION SAFETY PRACTITIONER

Early construction safety practitioners generally had scanty formal education in safety and health. Many had been trained in first aid and had a dual role: inspect the jobs for physical hazards and provide first aid for those injured on the job. Education often consisted of short seminars provided by insurance carriers and trade associations.

The 1980s and 1990s saw a significant growth in the number of graduate and undergraduate educational programs in SH&E. This has provided an opportunity for the construction community to draw from a pool of prospective construction risk-control professionals better prepared educationally to enter the field.

As the range of educational opportunities increases, so too does the scope of the practitioners' responsibilities. Today's construction safety managers often have position descriptions that cover many more responsibilities than the "safety man" of the past. The following list of disciplines is not all inclusive, nor does any single practitioner necessarily perform or participate in all of these functions.

Marketing

The SH&E professional has become a significant participant on the construction firm's marketing, development, design, and management teams. Enlightened construction firms call on the SH&E professional to compile a package of information that describes the corporate safety policy and commitment, programmatic elements, safety awards and recognition for the program, OSHA compliance data, injury and illness rates, insurance EMRs, testimonials, and any other positive information that promotes the firm's proactive approach to managing projects with safety as a core value. While some owners expect—even demand—such information, others will view it as a value-added benefit in the contractor selection process.

Some construction firms develop comprehensive marketing presentations with safety as a module, often produced in Power Point or other presentation software. Whether or not this is the case, it is a good idea for the SH&E professional to develop this package both in hardcopy for inclusion in proposals and also in presentation format that can be used to "pitch" a proposal in an interview setting.

It should be noted that this information should be included in an effective crisis management plan. In the face of a crisis, it is essential to have positive information about the firm that can be readily disseminated to customers, media groups, and the public. This information is also useful in the development of a project's safety orientation program. It emphasizes the firm's environmental, safety, and health policy and sets the tone for safety excellence and the resultant safety culture on the site.

Design/Engineering

The 1990s saw a marked increase in the utilization of SH&E professionals in the design and planning phase of construction. This may involve both professional and peer participation in the design process, specifications development, constructability reviews, contract document development, value engineering, and other activities depending on the role of the employing firm in the construction process.

The development of a detailed project safety plan is a major responsibility and performance expectation of the SH&E professional serving in the site's safety-coordinator role. This plan should be developed well before construction begins, and it should be functional, flexible, and comprehensive. The more significant the project design and management team's involvement is in this safety planning effort, the greater the likelihood of positive outcomes for the project's risk-control program.

Environmental

Since the arrival of the OSH Act of 1970, many other regulations have been promulgated, including the environmental laws enforced by the Environmental Protection Agency (EPA) as well as various state environmental agencies and departments of natural resources. Other environmental requirements are enforced by the U.S. Department of Transportation. The contractor's SH&E professional is often responsible for administering the environmental compliance program for the business. This can be a challenge for those companies with multistate operations, as individual states may have particular requirements that differ from other states or even from federal regulations.

Insurance/Risk Management

Managing the insurance portfolio for a construction business has become an increasingly complex task. Many older construction firms, such as second- and third-generation family-owned companies, still use the same insurance agent that the founder used when the firm was first established. Where the agent responded to the dynamic nature of the process by keeping abreast of all of the risk-management tools needed to support customers' evolving insurance needs, this is often a trusted and continuing relationship. The number of insurance brokerages that handle construction business exclusively has grown in the last twenty years. Other brokerage firms handle multiple industries, but have developed specialized departments that concentrate on their construction book of business.

Very large construction firms often employ a risk manager whose primary function is management of the insurance portfolio. More often, though, we see the SH&E professional assuming the risk-management functions as part of a growing position description for SH&E professionals in construction.

At the project level, the SH&E department is often responsible for many of the insurance-related administrative duties. This may mean communicating and coordinating activities with the medical provider, supervisors, and injured employees on the workers' compensation side. Maintaining an aggressive fleet safety policy will help to keep auto liability and collision claims in check. All subcontractors must have the proper insurance coverage in place with current certificates of insurance on file with the general or managing contractor. It is common practice to require subcontractors to add the general or managing contractor as a named insured to their policies, and to indemnify and hold that party harmless if and when losses occur.

On projects covered by owner-controlled insurance programs (OCIPs) or contractor-controlled insurance programs (CCIPs) this role will be more significant for the construction management firm's or general contractor's SH&E department. Tight administration of the OCIP or CCIP is the key to realizing financial rewards from these programs. Many owners and their representatives have reaped the benefits of these comprehensive insurance packages through loss reductions from effective risk-control efforts and aggressive management of loss claims.

The support role that insurance brokers and carriers can provide to construction safety practitioners is substantial. The training and "real world" experience that many of today's insurance risk-control personnel possess can inure to the benefit of the insured's corporation and their individual projects. The insight that the insurance provider's underwriting information offers, including detailed loss runs, should not be underestimated. The following list of support activities and resources that may be provided is not necessarily inclusive:

- management-level loss-control review
- design, development, and delivery of a strategic risk-control management system
- setting goals and objectives
- safety claims/review meeting
- broker oversight and coordination of the insurance carrier's risk-control services
- review of the carrier's service plan for OCIP or CCIP coverage
- monitoring insurance-related regulatory issues
- training and education assessment and delivery where appropriate
- statistical safety performance analysis
- referral of safety and health professionals to assist the contractor with staffing needs.

Medical Coverage/Injury Management

The SH&E department's direction of the medical responder/provider comes from a long tradition where a project's "safety man" was frequently placed in that position by virtue of his knowledge of first aid, often from experience in the military service as a medic. Medical coverage on today's construction project can range from a first responder with basic first-aid and CPR training, to a large project with a fully staffed clinic that could include an on-site physician, physician's assistant, nurses, paramedics, or any combination of these medical professionals. In some situations these services are contracted while these professionals hold staff positions elsewhere. The senior safety manager often has overall responsibility for the direction of the on-site medical team and provides an interface with any outside medical facilities.

An extension of the medical and insurance management roles places the SH&E department in the position of work-injury management also. The OSHA standards relating to occupational health-related hazards mandate medical surveillance activities such as respirator qualification, blood tests for lead, asbestos physicals, and other procedures. Also, the SH&E department monitors work-related injuries to determine OSHA recordability, often carried out in collaboration with the medical personnel, if appropriate. Aggressive, supportive management of work-injury diagnosis and treatment can markedly improve outcomes, keep the morale of injured workers strong, and ensure that the best possible care is rendered for employees.

The construction SH&E professional is familiar with the work tasks performed by each of the trades working for his or her company, or on the site. It is critical to avoid prolonged periods of lost time for injured workers, if they can safely perform some of the essential tasks of their craft. A number of studies have demonstrated that the longer the disability period away from work the less likely it is that the worker will ever return to work. This reality, combined with encouragement from insurance carriers to promote enlightened return-to-work policies, has tempered the old "absolutely no light duty on this job" mentality that was universally accepted in the past. The ability for the SH&E professional to work with the injured worker, his treating physician, and his supervisor to fashion modified work validates the practice of merging these disciplines within the SH&E department.

Security

Oversight of corporate and project security protection is usually placed with the SH&E professionals. This field continues to evolve as technology and social conditions change. The increasing dependence on computers for everything from scheduling and cost accounting to payroll and material control, has created a challenge for companies to preserve the proprietary nature of internal data. Removing information surreptitiously can now be accomplished through copying thousands of files onto discs or CDs and removing them, whereas in the past such theft would require physically removing quantities of paper files.

Another security challenge that SH&E professionals must contend with falls in the broad category of *homeland security*, a highly publicized term used frequently in the wake of the events of September 11, 2001. Although passenger aircraft had been the target of terrorist hijackings in the past, never had they been intentionally crashed into buildings, therefore becoming weapons of mass destruction (WMD). Also, a widespread anthrax scare that sickened and killed several innocent citizens and postal workers pointed to our vulnerability to harm from WMDs. Building departments in many major U.S. cities sounded the alert to owners of large high-rise and infrastructure projects to ratchet up security measures at these projects.

Most construction firms hire contract security forces for major projects. Most large security service providers have had experience with protecting construction operations and are familiar with standard procedures such as badging in and out, lunch box checks, on-site surveillance, vehicle searches, and other measures. Diligent SH&E professionals will interview prospective security providers to ensure that the firm has experience with construction operations. All project security personnel should have a thorough indoctrination in the particulars of the site security plan.

Industrial Hygiene

While the construction industry recognizes that physical hazards claim the lives of over 1000 workers each year, the death toll from health hazards continues to remain under the radar screen of most construction professionals. This is unfortunate since the number of deaths attributable to disease caused by exposure to health hazards in the construction environment is much larger than those caused by traumatic injuries. The extent of the problem is difficult to quantify. Death certificates of tradespersons afflicted with work-related respiratory exposures often state COPD, or chronic obstructive pulmonary disease, without indicating its etiology. Therefore, these death certificates are indistinguishable from those in the general population whose COPD is caused by smoking, nonoccupational exposure to airborne contaminants, and/or other underlying pathologies unrelated to construction exposures.

Other health hazards lead to occupational disease with potentially fatal outcomes as well. Some health hazards will not cause *fatal* outcomes, but may still disable or permanently diminish the quality of life for affected workers. Coatings and adhesives containing isocyanates can cause respiratory distress, sometimes fatal, to those sensitized to them. Long-term exposure to other materials considered carcinogenic can lead to cancers of target organs.

Exposure to temperature extremes can harm worker health. Vibration-related injuries disable workers, as do other ergonomic hazards such as lifting heavy and/or awkward loads and performing overhead work for extended periods. Noise exposure on construction sites has led to widespread noise-induced hearing loss among the construction community.

Does this mean that every construction site needs an industrial hygienist to identify these potential health hazards? Not necessarily. However, all construction safety practitioners should acquire a good educational foundation in industrial hygiene principles.

This may well suffice where the projects have few, uncomplicated exposures to health hazards. On larger projects with more complex health hazards, such as asbestos and or carcinogen exposures, the safety practitioner will be an educated consumer of industrial hygiene consulting services.

Crisis Management

This subject will receive extensive coverage later in this book. The author of the chapter on crisis management will define the roles of each of the members of the crisis management team, among them the SH&E professionals play a critical role in the process. In fact, the SH&E manager will generally author the site-specific crisis management plan, facilitate recruitment of the team, provide training, and play an important role in handling situations that require utilization of the plan.

Fire Protection and Suppression

Each year fires strike hundreds of construction sites, threatening the well-being of the workers and destroying property. While a handful of *megaprojects* may have dedicated fire-protection personnel, the responsibility for fire protection/suppression usually falls on the SH&E manager. Plans for fire protection need to begin well before construction. A plan should specify that an adequate number of fire extinguishers are available to distribute as the project develops, and they will be continually checked and serviced as needed. Actively involving the local fire department that will provide first response to a fire/rescue emergency is a good idea. Inviting mem-

bers of that agency to periodically tour the site will allow them to optimize their resources and target their response if a fire or other emergency situation requires their help. An ancillary benefit of establishing such a relationship can be the agency's assistance with fire-extinguisher training, first-aid and CPR training, OSHA-mandated confined space and excavation rescue response, and a number of other support services.

Training

A construction business that treats training as an essential, ongoing, multidimensional process has a necessary ingredient for an effective risk-control program. Each level in the organizational hierarchy has needs for training, and those needs are quite dynamic. Although safety and health training is a critical part of this training regimen, other topics include cost and scheduling, EEOC and other human resources issues, quality assurance, and technical developments and changes. Although the SH&E department will generally only facilitate the training related to risk control, senior management may well enlist the SH&E professional to assist in coordinating other training also.

A significant challenge faces today's safety and health trainers: language. The influx of non-English-speaking workers to the construction industry has produced a sharp increase in the incidence of fatalities within this sector, primarily the Latino worker population. OSHA recognizes this trend and has dedicated resources to address the issue. Language barriers and cultural issues will be addressed in depth in a later chapter.

Another training-related issue for SH&E professionals is one of portability (or lack thereof) of training credentials. Because of the temporary nature of construction projects and the high rate of turnover in the industry, workers often receive duplicative training when moving from one employer to another. Regional initiatives, such as Safe2Work in the Midwest, have created systems to reduce the redundancy of training workers. In this program, workers' training records along with drug-testing documentation are maintained, and cards are issued with this information imbedded so that

---

Code of Ethics and Professional Conduct

This code sets forth the code of ethics and professional standards to be observed by holders of documents of certification conferred by the Board of Certified Safety Professionals. Certificants shall, in their professional safety activities, sustain and advance the integrity, honor, and prestige of the safety profession by adherence to these standards.

Standards

1. Hold paramount the safety and health of people, the protection of the environment, and protection of property in the performance of professional duties, and exercise their obligation to advise employers, clients, employees, the public, and appropriate authorities of danger and unacceptable risks to people, the environment, or property.

2. Be honest, fair, and impartial; act with responsibility and integrity. Adhere to high standards of ethical conduct with balanced care for the interests of the public, employers, clients, employees, colleagues, and the profession. Avoid all conduct or practice which is likely to discredit the profession or deceive the public.

3. Issue public statements only in an objective and truthful manner and only when founded upon knowledge of the facts and competence in the subject matter.

4. Undertake assignments only when qualified by education or experience in the specific technical fields involved. Accept responsibility for their continued professional development by acquiring and maintaining competence through continuing education, experience, and professional training.

5. Avoid deceptive acts which falsify or misrepresent their academic or professional qualifications. Not misrepresent or exaggerate their degree of responsibility in or for the subject matter of prior assignments. Presentations incident to the solicitation of employment shall not misrepresent pertinent facts concerning employers, employees, associates, or past accomplishments with the intent and purpose of enhancing their qualifications and their work.

6. Conduct their professional relations by the highest standards of integrity and avoid compromise of their professional judgment by conflicts of interest.

7. Act in a manner free of bias with regard to religion, ethnicity, gender, age, national origin, or disability.

8. Seek opportunities to be of constructive service in civic affairs and work for the advancement of the safety, health, and well-being of their community and their profession by sharing their knowledge and skills.

■ ■ **Figure 2** ■ ■ **Ethical standards of the Board of Certified Safety Professionals**

electronic card readers can produce current transcripts for prospective employees. A number of contractor safety councils have been formed, primarily in Texas and other Gulf States, to alleviate the problem of redundant training. The OSHA Process Safety Management (PSM) Standard has been a driving force there, and owners have been the catalyst for most of the councils. The PSM standard's training requirements require owners to ensure that contract employees are trained and understand the training. The safety councils provide the generic hazard information to workers, test them, and issue cards that are recognized by other councils through reciprocity agreements, thus eliminating the need for one worker to attend multiple orientation classes in a single year. Site-specific training is then provided either by the host employer or by proxy at the local safety council under the owner's direction.

Summary

That SH&E positions have become so enmeshed in the corporate fabric of construction firms is a good sign. This indicates that enlightened executives view SH&E professionals as multidimensional, multitalented individuals with the ability to enhance the bottom line through the reduction of loss. We will discuss corporate cultures and subcultures later in this chapter, but suffice it to say that a company that involves its safety and health professionals in multiple business functions probably understands that the safety effort can be a true profit center.

PROFESSIONAL ETHICS

Various professional societies and associations have published lists of ethical canons for practitioners in the disciplines encompassed by the SH&E field. Since the majority of SH&E practitioners in the construction workplace are trained in traditional safety education programs, the professional society most aligned with them is the American Society of Safety Engineers. The preeminent certification for safety professionals is the CSP—Certified Safety Professional issued by the Board of Certified Safety Professionals (BCSP). A Construction Specialty is available to safety professionals seeking a construction-specific certification. The award of the CSP designation is predicated on successful challenge of two examinations. Passing the first exam qualifies the practitioner as an Associate Safety Professional, or ASP. When the second specialty exam is successfully completed, the practitioner holds the CSP designation.

The BCSP created the Council on Certification of Health, Environmental, and Safety Technologists (CCHEST). The CCHEST offers two other designations, achieved through examination. The Safety Trained Supervisor (STS), as the name implies, attests to a supervisor's knowledge of important health and safety principles in the workplace. Since it is not expected that supervisors will have the same level of safety training as the SH&E professional, another designation has been developed for construction SH&E practitioners that may lack the level of education held by CSPs. This is the Certified Health and Safety Technician (CHST), conferred on those successfully challenging the CHST examination.

The BCSP has produced an ethical code (Figure 2), which contains principles that all construction SH&E professionals, whether holding one of these certifications or not, should follow. This document is provided here in its entirety.

PROFESSIONAL STATUS

Although the reporting order in the corporate hierarchy may differ among construction businesses, the trend has been to have the director of the SH&E function report to the highest-ranking executive at their respective level—whether at the corporate office or at a field site. In the case of field sites, the site safety manager will generally send a dotted-line report back to the corporate SH&E official.

By requiring the senior SH&E position to report to the top executive within the organization, a clear message is sent to the entire company, "We are serious about safety and that commitment starts at the top."

Relationship with Top Management

Regardless of the reporting order, it is critical that the relationship between SH&E professionals and top management demonstrate a high level of mutual respect and support, with a commonness of purpose. By participating in staff meetings that address planning issues, cost and scheduling, purchasing, and other critical management considerations, the senior SH&E professional will reinforce the concept that safety and health is an integral part of the construction process, rather than an afterthought.

The relationship between the purchasing department and the SH&E department is very important. The purchasing agent can assist the SH&E effort by ensuring that all materials regulated by the Hazard Communication Standard are delivered with at least one copy of the material safety data sheet (MSDS). That department can also make sure that a copy of the sheet accompanies the product to the site and that it is furnished to the custodian of the on-site material safety data sheet book or file. Also, the purchasing agent can specify that all equipment arrive with all OSHA-required safeguards. By working closely with the purchasing department, SH&E professionals can provide training that will help purchasing personnel look for equipment that is safer and emits less noise or dust, and consumable materials that are less noxious.

Another practice, with both practical and symbolic value, involves the senior project executive taking regular tours of the job site(s) and making safety- and health-related observations during a significant part of the walk. Verbalizing these observations to the field superintendent(s) and other key project staff will help underscore the priority of SH&E issues.

Relationship with Employees

The relationship between SH&E professionals and the construction workforce is very important. It has often been said that if you want to find out what's really going on at a construction site, ask one of the workers. A high level of mutual respect and two-way communication between the workforce and safety practitioners is a key to achieving safety excellence at the project level. By building mechanisms to facilitate good communication in the SH&E program, this can be achieved in any setting. Safety committees, newsletters, safety suggestion or *idea* boxes, and other information-sharing components will go a long way toward promoting management's concern for worker health and safety.

This doesn't mean that the SH&E professional will always make judgment calls that will be universally popular. At times the SH&E professional will make decisions that don't seem to please either management or the workers. Here is where solid knowledge of the field with its standards and practices, combined with a keen sense of need to protect both the company and its workers, comes into play in making effective decisions. The most effective practitioners make decisions daily that are firm, fair, and consistent, with the protection of all assets the goal. Of course our *workers* are our *most important asset.*

Interpersonal communication in the field can help set the tone for a site. It helps to define the *culture* at the site, and will be dealt with later. The SH&E professional must be able to effectively communicate with the rank and file with the same ease as he or she does with project management. A colleague once described this as *the moves*—what make the safety program tick—with dynamic communication flowing both up and down the chain of command. Here are some tips from seasoned SH&E professionals with years of field experience:

- Know about the project: what you are building and the processes that will be used to construct the project. Acquiring this knowledge should begin well before the ground is broken during the planning phase. This will help the SH&E professional to anticipate potential hazards. Don't hesitate to ask questions if you are unfamiliar with a process, the schedule, engineering details, or other information that is unclear. Asking well-thought-out questions does not reflect badly,

rather it speaks to your interest in getting on top of the job. Ultimately, this will allow you to speak with authority to the project team based on the best information available.

- Work with the senior project manager to set goals for the project. This relationship should be cemented as soon as possible, preferably during the planning process before the first dirt is turned. Work together on how to most effectively communicate the project safety expectations.

- Your words and your actions help define the safety tone, or culture, for the site. Your expectation of firm adherence to the safety rules doesn't mean that a "take no prisoners" demeanor is necessary, or desirable. Get to know the names of the workers. Friendly and approachable behavior can help set the tone for the health and safety agenda. However, this may have to be temporarily interrupted in situations where other approaches have failed and firm, fair enforcement actions are necessary. Remember, consistency should always be maintained in discipline situations.

- Know the project schedule.

- Know the site—every nook and cranny. Failure to do so can cause a problem area to be missed, and can also cause some embarrassing moments. Don't let an unfamiliarity of the site catch you offguard—the stakes can be high.

- When correcting a behavior that compromises safety, do so with respect and discretion. None of us want to be openly criticized in front of our peers. To do so can temporarily change the unwanted behavior, but the embarrassment this will cause and the resulting lack of respect it will engender are not worth it. Also, the recidivism rate is higher.

- Schedule some time with each supervisory person to walk the site. This will foster a better mutual understanding of your respective roles. It gives you the experience of "walking in the other person's moccasins" as the old adage goes. It also will serve as a mutually beneficial learning opportunity that can help achieve some of the project safety goals.

- Attend project meetings and don't hesitate to recognize positive contributions to the safety effort. This positive stroking can return large dividends. As stated above, correcting safety problems can most effectively be accomplished with respect and discretion—at every level.
- Listen.
- Never, never take yourself too seriously.

Discipline—Tough Choices

In a perfect world, every worker would receive an orientation that covered every potential hazard that could be encountered on site, learn how to remove the hazard, and, if necessary, avoid harmful contact with the hazard while others abate it. A part of this avoidance training would include the use of personal protective equipment as a last defense if other intervention strategies are impossible or have not yet been implemented. Of course the use of standard protective gear may always be required; it could include items such as hard hats, a form of eye protection, work shoes, high-visibility garments, and any other appropriate garb.

Following the safety requirements will be a condition of employment, and this message will definitely be communicated in the site orientation, with employees affirmatively recognizing and agreeing to this prior to beginning work at the site. Even with an effective safety orientation, followed up with appropriate safety training for specific hazards and continual reminders about the safety rules, there will be occasional instances when employees will break the rules. This must be anticipated in the project safety plan and the corporate safety program with clearly defined procedures for documenting noncompliance and clearly stated consequences. Some infractions may be so serious that immediate removal from the site and termination will be required. These infractions and the hazards to which they relate should be highlighted in the orientation, and any additional information should be provided to workers upon hire and during ongoing safety training and awareness activities. Most, however, will be addressed with a progressive discipline system. Typically the first infraction will be a warning, with

following infractions carrying more serious penalties and culminating in discharge for cause.

Firm, fair, and consistent administration of the safety requirements is incumbent on all levels of the supervisory and management hierarchy. In an environment that lacks a true safety culture, supervisors call on the SH&E professionals to discipline others for safety infractions. At a project with a proper health and safety program, supervisors will treat safety infractions with the same disciplinary procedures that are used for any other case of employee misconduct.

Documentation of disciplinary measures is very important. First, it provides evidence of a proper progressive discipline system that is applied to all infractions of requirements that are conditions of employment, including those relating to health and safety. This documentation can help to protect a company in cases of retaliatory discharge and other types of litigation.

Certain OSHA citations issued to an employer for an employee violating an OSHA standard may be challenged with an affirmative defense. The thrust of the defense is that the OSHA inspector observed an isolated incident of employee misconduct (i.e., an infraction of the company safety requirements). The following elements must be present to support this challenge:

- There must be a written requirement that applies to the work in progress and that addresses the hazard observed by the OSHA inspector.
- This requirement must be clearly communicated to the employee prior to observation by OSHA. There must be documentation of that communication, such as a signed training attendance sheet.
- No supervisor may be in the immediate area, allowing the unsafe act to take place.
- A disciplinary system must be in place, with evidence that it is used. This is where documentation of warnings or other disciplinary actions for safety infractions is very important. Simply having a paper disciplinary policy without documentation that it is actually used will likely cause this "isolated incident" or

"idiosyncratic employee behavior" defense to fail.

When a project safety program and safety requirements are clearly communicated to all participants in the construction process, and the rules are firmly, fairly, and consistently applied, the need to use the progressive discipline should be minimal. In those situations where it is appropriate to discipline employees, it should be done with respect and with precise adherence to procedural protocol, and followed up with careful documentation.

CORPORATE CULTURE

Occupational Cultures and Subcultures

The practice of developing a *safety culture* in the workplace is an interesting and positive trend in the health and safety arena. To fully appreciate this concept, an examination of the term culture and how a culture affects behavior will be helpful.

Culture is the social framework in which we function. It affects the way we think, the choices we make, and the way we relate to others. Humans are not confined to a single culture, and we frequently move through separate but overlapping cultures in a single day. Ethnicity, religion, and political persuasion are examples of cultural identity. Two ingredients of a culture are *ideology* and *cultural form*. Ideologies, or strongly held beliefs, are often allied in a single culture. The second component is cultural form, manifested in shared slang, myths and legends, even gestures. Commonality of these elements helps to define a particular culture.

This definition of culture is very broad. Social scientists have observed that this phenomenon also takes place in microcosm in the workplace. In fact, workplaces can have both cultures and subcultures. A construction project is a good example. Each one is a social unit made up of one or more management hierarchies; a project-controls group that keeps track of cost, scheduling, and engineering matters; and workers plying different trades. Each project has its own character, or culture. Although normal operating procedures of the company in charge of construction can influence the culture on the project, the individual directing the project for that company has the most influence. If his or her direction on the job favors production over quality and safety, that will be the general perception on the site. Similarly, if safety is promoted as a core value by the person directing the project, an environment favorably disposed to work safely will be created.

The terms *corporate culture* and *safety culture* are phrases that are frequently used in contemporary SH&E-related literature and discourse among professionals. The American Society of Safety Engineers published a significant study of this phenomenon in a recent (July 2002) issue of *Professional Safety* (Molenaar, K. et. al, "Corporate Culture—A study of firms with outstanding construction safety," pp. 18–27). Although the study included a relatively small cohort of three construction firms, the findings were significant and pointed to the need for further study in this area. The article should be studied by construction safety practitioners, as the conclusions are reached through analysis of responses to a highly insightful questionnaire which explored construction process elements that included management and field components, subcontractor relationships, the safety plan, training and education, safety values, behavior-based safety, and incentives and disincentives.

The study concluded that, ". . . corporate safety culture had an integral effect on construction safety performance in the three firms studied." The authors state, "The hypothesis that corporate culture affects construction safety was proven correct through a comparison of responses between three companies' upper management, middle management, and field personnel." The limitations of the study, including the small size of the cohort, are also discussed.

SAFETY PROGRAM ELEMENTS

Effective risk-control programs may contain many different elements. Although there is no *cookie cutter*, one-size-fits-all construction safety program that will achieve *zero* accident results, the Construction Industry Institute (CII) has devoted considerable resources to the study of this subject. CII is a consortium of leading owners, contractors, suppliers, and academia interested in raising the safety bar in the construction industry by improving the constructed project and the capital investment process. CII developed and published a list of elements that have been implemented on construction sites, and research has demonstrated that they are effective in lowering injury rates. As such, they can be considered essential elements of a construction safety program that may yield zero-injury results. It should be noted that the validation of these techniques was derived from detailed examination of 38 projects that ranged in size from $50 million to $600 million. Also, it is not a coincidence that these key elements track with those mentioned earlier in this chapter associated with the Business Roundtable's *A-3 Report*. CII is an outgrowth of Business Roundtable, and to some extent this is an extension of the earlier work.

The nine key elements include:

- demonstrated management commitment
- staffing for safety
- safety planning
- safety training and education
- worker participation and involvement
- subcontractor management
- recognition and rewards
- accident/incident reporting and investigations
- drug and alcohol testing.

The following list is a compilation of specific program elements that have been implemented on construction projects. All of them are consistent with and support the nine general elements advocated and tested by the CII. Their sources are as diverse as the backgrounds of the entities that use them— owners, insurance carriers providing wrap-ups or owner-controlled insurance programs (OCIPs), construction managers, general contractors, and sub-tier contractors. This list, although extensive, is not complete.

- SH&E policy
- accountability system and matrix
- prebid qualifications
- specifications for SH&E
- SH&E marketing
- prework orientation—worker
- prework orientation—managers and supervisors
- prework orientation—subcontractors
- prework orientation—public and vendors
- prejob orientation checklist
- safety manual
- training
  - orientation
  - hazard-specific
  - new task
  - hazard communication
  - contractor safety forum

- prejob safety analysis/planning
- job/task hazard analysis
- JHA audit team
- work permit system
- safety permit observation team
- safety meetings
  - prework
  - toolbox talks
  - supervisor
  - project
  - huddles

- hazard elimination process
- incident analysis investigation
- safety suggestion program
- near-miss reporting and analysis
- worker safety recognition
- contractor safety recognition
- probationary status program
- alcohol and substance-abuse policy
- drug and alcohol testing
- employee assistance program
- SH&E site manager/coordinator.

SAFETY PRACTITIONER PROFESSIONAL DEVELOPMENT

Undergraduate and Graduate Programs

The number of educational opportunities for prospective SH&E professionals continues to increase, as does ease in access to them. Undergraduate options include both two-year associate's degree programs and four-year bachelor's degree programs. The proliferation of distance learning has made access to SH&E educational programs even easier.

Graduate-level programs have also flourished. The number of master's-level programs in environmental science, industrial hygiene, occupational safety, and the fire sciences has grown exponentially over the last twenty years. Recognizing the plight of the working professional seeking an advanced degree, some institutions have tailored graduate programs to accommodate work schedules of these candidates.

While the number of quality programs continues to rise at the associate, bachelor, and master level, doctoral programs are still somewhat limited and tend to focus on industrial hygiene and environmental engineering. As the SH&E community grows and the demand for quality higher education becomes more vocal, the academic opportunities at the post-graduate level should also increase.

There is a hole in the educational platform, however. While a large number of SH&E students will enter the construction field working for contractors, owners, and insurance-related companies, few programs provide a construction-specific focus. Another educational shortcoming involves the curriculum for construction technology and construction management programs, which are often bereft of safety and health information. One notable exception to this is the University of Wisconsin, Stout (UWS), in Menomonie, Wisconsin. Until 1996, UWS maintained the separation between risk control and construction management that is found at most schools. Then, working with the St. Paul Fire and Marine Insurance Company's Construction Group, the Construction Safety Council, and some dedicated volunteers, UWS connected the two programs in several significant ways.

First, each of the classes in the construction management program was dissected and appropriate construction risk-control information was integrated with the other technical information. Too often construction managers and supervisors experience a disconnect with site SH&E professionals. For example, in a soil mechanics class that teaches construction management students to prepare a site for foundation work, the soil composition is described in geologic nomenclature. Sandy loam and glacial till are typical geological terms that might be used to describe a site's soil. In a class on risk control, addressing the hazards of excavation work and the various worker protections available, students learn to classify soil into types A, B, and C. The program now provides the knowledge necessary for a construction manager to properly prepare the site to receive the foundation, and to do so in a manner that keeps the construction workers safe. In addition, UWS offers a minor in Construction Risk Control. This gives students the opportunity to enter the workplace with a sound educational underpinning in both technical and risk-control aspects of the construction process.

The integration of construction technology and construction management curricula with SH&E curricula offers both challenges and opportunities for the future. Since both disciplines have coexisted for a very long time as discrete entities, it is incumbent on both the educational and construction communities to push for bilateral integration. Graduate and undergraduate programs in the SH&E disciplines, including industrial hygiene and occupational health, should have some consideration of construction exposures throughout. Similarly, construction programs should prepare students in risk-control elements of the construction process throughout.

Short Courses and Seminars

Many SH&E-related educational and training programs are available to the construction industry. Many offer topical information about specific hazards, while others take a broad-brush approach, such as the OSHA 10- and 30-hour classes. Hazard-specific courses assist workers and supervisors to meet the knowledge component of OSHA's *competent person* requirements for standards such as fall protection, trenching, scaffolding, and others. Employers must realize that these classes do not create competent persons under the OSHA standards, however. The definition of the term requires that competent persons have both the *knowledge* to recognize and correct or control the hazards and the *authority* to do so.

Other courses offer more complete SH&E training, intended to provide SH&E construction professionals with the essential tools to work at a site's SH&E professional level, but that fall short of the level of education needed for a four-year bachelor's degree. This includes programs that award certificates upon completion of a series of short, topical SH&E-related classes.

Sources for these courses continue to increase as the market expands. A partial list of categories of construction-related SH&E course providers follows:

- trade associations
- building and construction trade unions
- consultants
- OSHA Training Institute
- official OSHA Training Institute Education Centers
- universities and community colleges
- nonprofit organizations, including safety councils and professional societies.

OSHA'S ROLE IN CONSTRUCTION

Structure of OSHA

The Occupational Safety and Health Administration was created by Congress in 1970. The Occupational Safety and Health (OSH) Act established OSHA as the agency principally responsible for the health and safety of private-sector workers through the promulgation of safety and health standards, enforcement of those standards, and the training of employers and workers. The training is aimed at helping employers and workers to recognize hazards in the workplace, eliminate those hazards where possible, and create safeguards to reduce the risk posed by those hazards that cannot be eliminated.

The OSH Act also created the National Institute for Occupational Safety and Health (NIOSH). As OSHA's sister agency, NIOSH is also charged with protecting the health and safety of American workers. The role of that agency is not enforcement, however. NIOSH performs and supports research in occupational health and safety, supports education of health and safety professionals as well as nurses and doctors in the field of occupational medicine, and supports intervention projects. While the OSH Act placed OSHA under the U.S. Department of Labor, NIOSH was established under the Department of Health, Education, and Welfare, now known as the Department of Health and Human Services.

The United States is divided into ten OSHA regions, each with an office headquarters directed by the Regional Administrator. Each region is divided into areas, with Area Directors administering the agency's activities in that district. It is the Area Office that employs and dispatches compliance officers, including industrial hygienists, to inspect workplaces. Each state has the option of administering its own OSHA program. Currently 26 states exercise that option. Further, states may create occupational health and safety standards that differ from the federal regulations found in Title 29, Part 1926 of the Code of Federal Regulations (CFR). The OSH Act does mandate that these standards provide a level of protection that meets or exceeds the federal regulations. Although some "state plan" states exercise this option, others elect to allow the federal requirements to prevail.

OSHA has several Technical Centers with laboratory capabilities for testing various materials such as asbestos, silica, benzene, and any other substances that are hazardous to workers' health. The centers support the efforts of compliance officers and industrial hygienists by assisting with laboratory analysis of samples, assistance with evidence from accident investigations, and other scientific/technical support required by the agency.

The OSHA Office of Training and Education and the OSHA Training Institute are located in Arlington Heights, Illinois. This branch of the agency is responsible for providing health and safety training and education for OSHA personnel, employees from other federal agencies, and the private sector. It is here that we find the headquarters for the extensive network of *outreach instructors* located throughout the United States. These instructors are the front line of OSHA's educational outreach program. They are authorized to teach both OSHA 10- and 30-hour hazard recognition classes in either general industry or construction. In construction, the OSHA 10-hour class has become a benchmark for safety training, with an ever-increasing number of projects requiring the 10-hour card for supervisors, and for all workers in some cases. Some contractors are requiring it of supervisory personnel within their own ranks, even asking subcontractor supervision to acquire the OSHA 30-hour card.

To assist OSHA with its training mission, Education Centers have been authorized in each region. Each center is administered by a university or community college, often in partnership with another training agency such as a nonprofit organization, trade association, or labor organization. The development of OTI's Education Centers allowed for an exponential increase in the amount of training that OSHA was able to deliver to all parts of the United States.

In addition to direct assistance from OSHA, the OSH Act provides for each state to receive support for the delivery of safety consultation and training to the private sector. These state programs typically target the smaller employers, those less likely to have an SH&E practitioner on site. Refer to the OSHA Web site at www.OSHA.gov for more information and a list of state consultation programs.

The Directorate of Construction is, as its name implies, responsible for providing technical expertise as a resource to the agency in all matters relating to construction. This dedicated group has assisted with major accident investigations; analyzed data relating to injuries, illnesses, and fatalities in construction; developed policy and standards affecting construction; and conducted outreach and assistance to the construction industry. Departments within the Directorate include the Office of Construction Services, the Office of Construction Standards and Guidance, and the Office of Engineering Services.

The OSH Act created a panel known as the Advisory Committee for Construction Safety and Health (ACCSH) to provide a liaison between the Secretary and the construction industry. The committee is comprised of five labor representatives, five management representatives, two public representatives, one NIOSH representative and one representative of the state-plan states. Work groups develop recommendations for new standards and review existing ones, direct OSHA's attention to emerging issues in construction health and safety, and perform other work at the direction of the ACCSH member(s) who chair the respective work groups. These work groups are comprised of industry volunteers, other affected parties, and an OSHA staff person.

In the late 1990s, OSHA invoked a process known as negotiated rulemaking. The concept involves bringing all parties affected by a particular standard together with a professional facilitator to *negotiate* the content and language for a new or revised standard. The first standard selected for updating was the steel erection standard, subpart R. The crane standard, subpart N, is currently in the process of revision through negotiated rulemaking.

OSHA's role as an enforcement agency is probably the most recognized. The agency has the duty to inspect workplaces for violations of OSHA standards. Although a few categories of business that have very low exposure to risk will seldom, if ever, see an OSHA inspection, the construction industry does not fall into that category. The relatively high incidence of serious injuries and deaths in the industry makes it a target for enforcement activity.

There are several triggers for a visit from OSHA. The first is a planned or scheduled inspection. The agency has had a longstanding relationship with the University of Tennessee, requesting the university to prepare a schedule of prospective construction sites to visit in each region. The university utilizes the *Dodge Reports* to glean the location and scheduling information on each site. These inspections are typically *wall-to-wall-type audits* that will include a review of programmatic elements and records, as well as a walk-through inspection of the job site.

A variation of that inspection is called a *focused inspection.* This type of inspection concentrates on the four major fatality-producing hazards: falls, electrocution, being struck by objects, or being caught in or between objects. Minor hazards may not draw a penalty if they are abated immediately. This is afforded to sites that meet the following criteria:

- have a safety and health program in place
- have a competent person on staff to administer the program, make inspections, and take corrective actions where necessary.

It should be noted that the overall intent of the focused-inspection initiative is to provide abbreviated inspections to construction sites that are making a good-faith effort to comply with the intent of the OSHA Act. This effectively frees up OSHA compliance resources to seek out the "bad actors" and deliver more stringent enforcement actions where workers are routinely placed in harm's way.

Other site inspections are spurred by complaints, referrals from other agencies, and fatality/catastrophe investigations. It should be noted that employers do have a right to refuse entry to their site. In these cases, OSHA has the authority to seek a search warrant, and return to the site with the warrant—often accompanied by local law-enforcement officers.

Another legal issue relating to OSHA involves the employer's handling of worker concerns about health and safety on the job. Section 11C of the OSH Act provides for serious penalties against employers that discriminate against employees perceived by the employer to be *whistle blowers* or who are raising issues about safety to the detriment of the company. OSHA offices have special agents who follow up and prosecute these cases. Penalties can include fines and reinstatement of workers with back pay and benefits. This is not to be confused with the employer's obligation to discipline supervisors or workers for documented infractions of the safety rules.

OSHA has developed several programs that provide opportunities for *stakeholders* to work with the agency to improve health and safety on the job. These programs are consistent with OSHA's philosophy of supporting employers committed to providing workers with a safe and healthy workplace, while freeing up resources to pursue employers who choose to take the "low road" and ignore the welfare of workers. The programs include strategic partnerships, alliances, and a voluntary protection program (VPP).

Finally, one of the finest resources available to construction SH&E professionals is the OSHA Web site at www.OSHA.gov. The site is rich with tools to help create and support a corporate and/or a site safety program. In addition to the resources found on this site, links to other SH&E-related sites are also provided.

SUMMARY

Construction is an industry that has long been considered *dangerous*. The relatively high number of injuries and deaths associated with the industry validates this perception. The nature of the work will always present some hazards, particularly falls, electrocutions, cave-ins, and falling objects, as well as a plethora of others. What can be changed through effective risk-control strategies is the *degree* of danger. The construction industry has demonstrated that it can effectively manage these hazardous elements at a level that allows workers to perform in a manner that renders them less susceptible to harm than those in most other industrial occupations. This book is dedicated to those workers and that principle.

REVIEW EXERCISES

1. Will complying with OSHA standards ensure zero injuries at a job site? Explain your answer.
2. Explain the role of owners in relationship to contractors' safety programs and liability.
3. Explain why termination is the best course of action for a worker who continually presents safety issues to his supervisor.
4. Discuss the impact that *modified or light duty* may have relative to potential liability and other issues at a job site.
5. Does *corporate safety culture* have a significant impact on worker behavior in the field? Explain your answer.

■ ■ **About the Author** ■ ■

CONSTRUCTION SAFETY PROGRAM ESSENTIALS
by Charlotte A. Garner, MSS, CSP

Charlotte A. Garner, MSS, CSP, is Corporate Safety Administrator for Webb, Murray & Associates, Houston, Texas, where she manages the companywide safety, health, and environmental compliance activities. In addition, she provides safety engineering and consulting for Webb, Murray customers, many of whom are VPP participants. Previously she was Manager of Safety Services for the American Iron and Steel Association, Washington, DC, and a Field Safety Coordinator and Assistant to the Corporate Director of Safety for Kaiser Steel Corporation, Oakland, California. Ms. Garner is an ASSE Professional Member and a past president of the ASSE Gulf Coast chapter, and a board member and former president of the Houston Area Contractors Safety Council. She holds a Master of Safety Sciences from the University of Southern California.

24

CHAPTER 2

Construction Safety Program Essentials

LEARNING OBJECTIVES

■ Discuss the essential elements of a model construction safety, health, and environmental (SH&E) management system.

■ Describe the criteria that are being universally accepted as the model benchmark SH&E management system.

■ List *additional* elements to consider that will enhance the success of the construction SH&E management system.

■ Explain how to organize the construction SH&E elements into an orderly, systematic, all-inclusive process.

PACKAGING THE COMPLETE PROGRAM

Putting all of the essential elements into a workable package can be a challenge for the owner and the construction SH&E manager. The majority of the parts fit into any effective SH&E system, but there are some parts that fit only into the construction and/or contractor system. The challenge is to enfold all of the essential pieces into a systematic, orderly, all-inclusive management process. You want to be sure that you have complied with regulatory requirements, have a program that can score high when compared to a benchmark management system, and that, most importantly, provides the utmost of SH&E protection for the workers. Just meeting the regulatory requirements is not enough for construction SH&E since this is the industry that leads all others in the number of fatalities and disabling injuries. Your program must go beyond the minimum, but where is the benchmark that sets the bar that you must clear? There are additional essential elements that should be added to those traditionally accepted in the construction industry. These elements augment and include the traditional essential elements of the construction SH&E program under structured criteria that both contractor and owner can accept and use as a standard against which to evaluate the SH&E process.

These criteria recognize that the construction industry has specific differences from general industry, are flexible, and can include any hazard and its control(s). This is taking into account that construction involves several phases during its complete cycle. Although all construction projects do not include all of the life-cycle phases, they can at times include: preplanning, planning, design, construction, operation/maintenance/repair, and completion or disposal. The construction process does not just involve the craftspeople and workers at the construction site. It involves other people with a variety of skills and other functions for each phase before and after the actual construction begins and ends. This means that each phase has its own SH&E requirements that must be systematically applied, monitored, and verified.

The model system is at the top level of planning a comprehensive and complex project such as a construction job. The system safety philosophy and techniques that are implemented during each phase fit comfortably into the model program. So long as the systems and subsystems of each phase are recognized and planned for, the criteria of the model program applies as well to each phase as it does to the complete life cycle of the construction process.

There is evidence that these criteria are being universally accepted. (See "OHS Management Systems: A Survey" by Peggy Farabaugh in *Occupational Health & Safety Magazine*, June 2000.) Since construction is an international industry, it bodes well for the companies who work internationally to have an exemplary and flexible SH&E management system in place.

THE MODEL PROGRAM

The basic model program is the OSHA-approved, integrated SH&E process that incorporates the fundamental principles and practices of risk assessment, OSHA Voluntary Protection Programs (VPP) safety and health criteria (refer to the Federal Register 54 FO 3904, January 26, 1989, *Safety and Health Program Management Guidelines*), and administrative requirements to create an umbrella process under which all of the essential elements can be systematically and logically organized. The VPP as it is being practiced in the construction industry will be discussed further in a later chapter. In this chapter, we will concentrate on the fundamental elements of the criteria and link the essential and traditionally accepted elements of the construction industry with the model system. You will find the nine key elements and the specific program elements of construction are easily accommodated within these criteria.

Essential Elements of a Benchmark Construction SH&E Management System

The current, effective construction SH&E programs emphasize management and employee ownership and active involvement in implementing and sustaining the management system. No more "Safety is not my job" attitude. The model SH&E system intentionally involves *everyone* in the company—top management, middle management, line supervisors, and general employees. Incorporating one set of flexible, performance-based criteria, the model system contains management accountability for worker safety and health, continuous identification and elimination of hazards, and active involvement of employees in their own protection. These criteria are successful for the construction industry and any other industry, union and nonunion, and for employers large and small, private and public.

The four critical elements that must be part of the construction company's SH&E management system are:

- management leadership and employee involvement
- work-site analysis
- hazard prevention and control
- safety and health training.

Management Leadership and Employee Involvement

Enlightened management leadership provides the motivating force and the resources for organizing and controlling any activity within the organization. In the model SH&E system, leaders regard worker safety and health as a fundamental and valuable activity of the organization and apply energy and attention to worker safety and health protection with as much vigor as to the other organizational purposes and goals.

Employee involvement will provide the means through which workers develop and express their own commitment to safety and health protection for themselves and for their fellow workers. Employees effectively participating in this SH&E process become stakeholders. They then develop a sense of ownership for the success of the venture and support it.

Table 1 presents the functions in the model program for management commitment and employee involvement, responsibility, and accountability.

■■ Table 1 ■■
Critical Element 1: Management Commitment and Employee Involvement

Management Commitment

- Commitment to safety and health protection
- Policy
- Goal and objectives
- Commitment to maintaining the standards of the model program
- Planning for SH&E as a part of the overall management process
- Written SH&E program containing all of the critical elements of the model program appropriate to the size of the work and the construction project and phases: management leadership and employee involvement; work-site analysis; hazard prevention and control; and safety and health training

Visible Management Leadership Includes

- Establishing a clear line of communication with employees
- Setting an example of safe and healthful behavior
- Inaugurating an environment that allows for reasonable employee access to top management
- Ensuring that all workers at the site are provided equally high-quality SH&E protection
- Written clear definition of responsibility in all areas, ensuring no unassigned area
- Assignment of commensurate authority to those who have responsibility
- Holding managers, supervisors, and nonsupervisory employees accountable for meeting their responsibilities
- Ensuring that all subcontractors have in place SH&E management systems equal to or exceeding the general contractor's system
- Implementing an annual self-evaluation system to examine at top level the overall processes for strengths, weaknesses, improvements, achievements of meeting or not meeting goal and objectives. Conducting an evaluation immediately prior to completion of the construction to provide improvement indicators for future projects or deficiencies to correct prior to completion sign-off.

Employee Involvement

- Enable and encourage effective employee involvement in the planning and operation of the SH&E program and in decisions that affect employees' safety and health. Examples of active and meaningful participation are:
  - Participating in ad hoc SH&E problem-solving programs
  - Participating in accident and incident investigations
  - Developing or participating in employee-involved suggestion programs
  - Training other employees in SH&E
  - Analyzing job or process hazards
  - Acting as safety observers
  - Serving on SH&E committees (in conformance to the National Labor Relations Act)

OSHA *Revisions to the Voluntary Protection Programs to Provide Safe and Healthful Working Conditions*, Federal Register, July 20, 2000.

A managerial commitment to worker safety and health protection

Managers must personally and actively perform their SH&E responsibilities to such a degree that all employees recognize and accept that management's personal commitment to the goal of excellence in the SH&E system is just as dedicated as commitment to the profitability of the company. In its preamble to *The Voluntary Protection Programs Management Guidelines* (Federal Register, 1989) OSHA states:

> . . . Actions speak louder than words. If top management gives high priority to safety and health protection *in practice*, others will see and follow. If not, a written or spoken policy of high priority for safety and health will have little credibility, and others will not follow it.
>
> Plant managers who wear required personal protective equipment in work areas, perform periodic "housekeeping" inspections, and personally track performance in safety and health protection demonstrate such involvement. . . .

Top site management's personal involvement

Section III of OSHA's *Revisions to the Voluntary Protection Programs to Provide Safe and Healthful Working Conditions* (July 24, 2000) addresses personal involvement.

> Each applicant must be able to demonstrate top-level management leadership in the site's safety and health program. . . . Managers must provide visible leadership in implementing the program. This must include:
>
>> Establishing clear lines of communication with employees;
>> Setting an example of safe and healthful behavior;
>> Creating an environment that allows for reasonable employee access to top site management.

A system in place to address safety and health issues/concerns during overall management planning/purchasing/contracting

Management systems for corporatewide planning must address protection of worker safety and health. That is, are expenditures for SH&E included in the project budget planning? Including the SH&E needs for each phase of the construction schedule will synchronize these monetary outlays with the project progress. For example, if demolition is required in the preconstruction phase, the demolition and preparatory work may carry more risk and intensive preventive activities, thereby requiring more SH&E expenditures than the hazards that are present during the completion phase.

Safety and health management integrated with your general day-to-day management system

Authority and responsibility for employee safety and health must be integrated with the overall management system of the organization and must involve employees. This is the part of the management system that can incorporate the communication link between management and worker that is available from day-to-day, the process for handling such communications, and who is responsible for receiving, investigating, and then responding to the reporting worker. The appraisal and discipline program can be a part of this commitment section. Specific SH&E assignments for workers can be included as well. There are others, but the objective of this criterion is to ensure that employee participation is a part of everyone's daily activities and that SH&E is integrated into their daily responsibilities.

A written safety and health management system that addresses the criteria elements with policy and procedures specific to the company's project, appropriate for the company's size, construction characteristics, and specialty

The four critical elements of the basic SH&E management system must be a part of the written program. The purpose is to ensure that *all* of the factors relating to an effective SH&E management system are addressed.

A safety and health policy communicated to and understood by employees

The top-level safety policy specific to the company values must clearly demonstrate commitment to meeting and maintaining the requirements of the model SH&E management system.

The policy emphasizes the value that management places on SH&E requirements and purpose. If this value is internalized by all of the employees in the company, it becomes the basic point of reference for all decisions affecting SH&E. It also becomes the criterion by which the adequacy of all protective actions is measured.

During the management system's evaluation, validation of the policy is so well stated that all employees not only understand the priority of SH&E in their personal tasks and responsibilities, and in relation to other organizational values, but they are living it day to day and nothing interferes with that first priority.

Safety and health management system goals and results-oriented objectives for meeting those goals

The goal should be broadcast throughout the construction site, along with the objectives to attain the goal. Although there may be an underlying goal stated in the SH&E management document that addresses zero accidents, injuries, and illnesses, the goal for project purposes is the specific goal or goals to accomplish during the active life cycle of the construction company's involvement in the project. For example, "Our goal is to prevent fall incidents for the duration of this project."

Along with the goal, develop the objectives to achieve the goal, such as:

> During this project, we will use bulletins, training, hands-on practice, observations, and job safety analyses during each phase of the project. Suggestions from the workers are encouraged. Mechanical means will be used as much as possible for elevated work. Our focus will be to eliminate fall incidents from this project. Activities relating to fall prevention and safety awareness of hazards relating to falls will be documented. During the project, employees will be regularly advised of our progress toward the goal.

There may be different goals with related objectives for different phases of the project. It depends on the scope of the project, how many goals are considered essential relative to the resources that are available. If yours is a small or medium-size company, you can have as few as can be supported financially or accomplished during the life of the project.

Look for evidence that the goal and the objectives are sufficiently specific to direct the affected employees to the desired results and to the methods for achieving them. Communicating the goal and objectives directs the employees toward success. However, follow-up is necessary to ensure the correct route to achievement has been taken. During the project-end evaluation, consider the methods used and examine how close you and the employees came to achieving the goal and what can be changed next time to get closer to or stay on the goal.

Clearly assigned safety and health responsibilities with documentation of authority and accountability from top management to line supervisors to site employees

Ownership is established by investing in the venture, whether it is buying a home, starting a business, working at a job. The same scenario applies to establishing ownership in the industrial SH&E processes. In the prior version of the VPP guidelines, OSHA noted that when responsibility for safety and health protection is assigned to a single staff member or even a small group, other staff members and employees develop the attitude that someone else is taking care of safety and health problems—"that isn't my problem." *Everyone*—all levels of employees—has some responsibility for safety and health in the model SH&E management system. At each level, each employee has his or her individual and functional safety and health-related activities, and each one at each level should know what those are. Not only should they know what their specific duties are, it is essential that each one understands them, is trained in how to perform them, and is clearly assigned the authority to act upon their relevant requirements.

It is counterproductive to assign responsibility without providing commensurate authority and resources to get the job done. To do so sets the employee up for frustration and disappointment and sets the SH&E management system up for failure. A person with responsibility for the safe operation of a piece of machinery should have the authority to shut it down and get it repaired when it malfunctions.

All employees must know what is expected of them. Ensure that the responsibilities for specific

SH&E functions are assigned, that the authority and resource provisions are specified, are known, and are drawn upon by all levels of employees from management to craftspeople to administrative people. Describe the performance appraisal system for all levels. Be specific—whether it is job performance evaluations, management by objectives, warning notices, contract language, or progressive penalties ending in termination for repeated violations.

A declaration that certain performances are expected from managers, supervisors, and employees means little if management is not serious enough to track performance, to reward it when it is well done, and to correct it when it is not.

It is standard operating procedure that management holds everyone who works for the company accountable for meeting their responsibilities on the job. That is the essence of any effective and profitable company. Holding them accountable for their safety and health responsibilities and duties is another integral component of the worker's job and company's plans for success and stability. The system of accountability must be applied to everyone from senior and corporate management to hourly employees. If some are held firmly to expected performance and others are not, the system will lose its worth. Those held to the expectations will be resentful and those allowed to neglect expectations may continue to downgrade the quality of their work. The result can increase the chances of serious injury and illness. Especially in construction work, this particular procedure is not an acceptable option.

Leadership personally must ensure that the accountability system is fair, reasonable, and consistently applied no matter who the offender may be or what status in the company he or she may have. This requires checking the system closely during the evaluation process to verify that such a system is in place and that the rules are impartially administered, no matter who is the offender.

Necessary resources to meet responsibilities, including access to certified safety and health professionals, other licensed heatlh-care professionals, and other experts, as needed

Needed resources include operational and capital expenditure of funds as well as responsible, well-trained, and equipped personnel. The model SH&E management system documents assignment of authority and funding for access to occupational safety and health specialists. Professional industrial hygienists, healthcare providers, and safety personnel must be available and accessible when their services are needed. Also, identify any internal or external resources that may be available, such as hazard monitors, fire safety personnel, SH&E instructors, emergency responders, and others.

Selection and oversight of subcontractors to ensure effective safety and health protection for all workers at the site

The model SH&E process includes the requirement that the general contractor's SH&E management system include the provision that the subcontractors for the construction project are ensured safe and healthful working conditions consistent with and equal to general contractor's employees and that their safety performance is tracked by the same system and procedures that are applied to the general contractor's employees.

The subcontractors must follow work-site SH&E rules and procedures applicable to their activities while on the project. Ensure that contractors are informed and understand that SH&E performance below the company criteria can cause termination of the contract and removal from the construction site. Do not omit any contractor. The management system includes all work-related activities of the construction project or the subcontractor employees. They are expected to develop and operate an effective safety and health program management system.

At least three ways employees are meaningfully involved in activities and decision making that impact their safety and health

Since an effective program depends on commitment by employees as well as managers, it is important that the system reflects their concerns. As mentioned earlier, to be effective, the model SH&E management system must include all personnel in the organization—managers, supervisors, and field employees.

By this all-inclusive requirement, OSHA in no way intends to transfer responsibility for compliance to employees. The OSH Act clearly places responsibility for safety and health protection on the employer. In order to attain the optimum SH&E performance, however, OSHA has found that everyone in the company must be involved. Important to attaining the best performance is the employees' intimate knowledge of the jobs they perform and the special concerns that they bring to the job. This gives them a unique perspective that can be used to improve the program.

Employees should have an impact on the decision process through methods such as hazard assessments, hazard analyses, inspections, SH&E training, evaluation of the SH&E management system, and any other SH&E activities particular to the company's construction specialties or industry. OSHA does not classify the following as "meaningful" involvement:

- expecting employees to work safely
- wearing PPE
- participation on investigation teams, except as the injured or affected employee
- attending SH&E training sessions
- other basic SH&E activities that are part of a regular program.

The number or percentage of employees involved should be stated. This provides an estimate of how many employees are actually involved. The same ones should not be participating constantly. *All* employees should be involved from time to time.

During the closing evaluation of the project, discover if these methods are available to the employees, if a work environment has been created that welcomes, encourages, and supports employee participation, and whether it is happening. Also evaluate whether the participative activities are resulting in improved safety and health program performance. Ensure that there are in place the verification and validation procedures to measure the activities and to yield an analysis of the results.

Annual and/or project conclusion safety and health management system evaluations of the

critical element functions in a narrative format, containing recommendations for improvements, and documented follow-up

The results of this evaluation establish a baseline from which the SH&E planners can discover the weaknesses and the strengths of the company's and/or project's systems. Under the model SH&E requirements, a self-evaluation should be conducted annually. With particular reference to construction projects, Section III of OSHA's *Revisions to the Voluntary Protection Programs to Provide Safe and Healthful Work Conditions* (July 24, 2000) states:

In construction, the evaluation must be conducted annually and immediately prior to completion of construction. The final evaluation is to determine what has been learned about safety and health activities that can be used to improve the contractor's safety and health program at other sites.

It is a tool to assist in ensuring the success and stability of the company's SH&E management system. Again, OSHA's 2000 VPP revisions say that the "comprehensive program audit evaluates the whole set of safety and health management means, methods, and processes, to ensure that they are adequate to protect against the potential hazards at the specific work site." This is the one VPP program element that OSHA has found to be generally lacking or misunderstood on sites applying for VPP approval. That is unfortunate because, when the annual evaluation of the management system is consistently administered, VPP participants agree that all of the processes are constantly improving.

Work-site Analysis

BASELINE HAZARD ANALYSIS

A baseline hazard analysis identifies and documents common hazards associated with the project and the site, such as those found in OSHA regulations, building codes, national fire codes, other recognized industry standards, and those for which existing work practices and controls are well established, and the many other certifications and mandatory requirements under which the construction industry regularly operates.

■ ■ **Table 2** ■ ■

Critical Element 2: Work-site Analysis

- Procedure to ensure analysis of all newly acquired or altered facilities, processes, materials, equipment, and/or phases before use or work begins to identify hazards and preventive or control measures.

- Comprehensive safety and health surveys at appropriate intervals for the project to include:
  - Identification of hazards during the initial baseline survey then subsequent surveys as needed to ensure prevention or control measures are in place
  - Identification of health hazards and employee exposure levels through industrial hygiene monitoring and planning
  - Use nationally recognized procedures for sampling, testing, and analysis with written records of results

- Routine examination and analysis of safety and health hazards with:
  - Individual jobs, processes, or phases
  - Inclusion of the results in training and hazard control programs
  - Emphasis on special safety and health hazards of each craft and each phase of work

- A system in place for conducting routine self-inspections that follow written procedures, result in written findings, and tracking of hazard elimination or control to completion

- On construction sites, inspections of the entire work-site should occur weekly due to the volatile and rapid change in the work-site conditions

- A process that:
  - Allows employees to notify management in writing of unsafe or hazardous conditions
  - Notified the employees of timely, appropriate response
  - Tracks the responses and hazard elimination or control to completion

- An accident/incident investigation system that includes:
  - Written procedures or guidance
  - Written reports of findings and hazard elimination or control tracking to completion
  - Identifying the root causes of the accident or incident
  - Near-miss incidents

- A system to analyze trends through review of the inspections, employee reports, accidents, and other sources to identify common causes to be eliminated or controlled.

OSHA *Revisions to the Voluntary Protection Programs to Provide Safe and Healthful Working Conditions*, Federal Register, July 20, 2000.

Procedures should also be developed for all newly acquired or altered facilities, processes, materials, equipment, and phases before use or implementation to identify hazards and their prevention or control. This is part of the system safety methods to conduct preliminary hazard analyses for all phases of the construction project and the systems and subsystems of the phases.

Within each phase of the project, the appropriate hazard analyses should be conducted regularly:

1. Describe the system used to examine and evaluate the SH&E hazards associated with routine jobs, tasks, and processes, which identifies uncontrolled hazards and leads to hazard elimination or control. Emphasize special SH&E hazards of each craft and phase of work. Describe the forms used to record the findings and how discrepancies are tracked to completion, such as job safety analysis, prework permits, sign-off on task safety cards.

2. Describe the procedure to identify uncontrolled hazards prior to an activity or use, which leads to hazard elimination or control of significant changes, including nonroutine

tasks, new processes, materials, equipment, and facilities. Ensure that preliminary hazard analyses are conducted prior to integrating these tasks or processes into the company's SH&E system and that the tasks involved are considered. Identify the occasions when and under what circumstances you use safety or health professionals to assist with these types of hazard analysis.

Do not confuse this requirement with self-inspections covered under the self-inspection section. What is being done here concerns the specific task hazard analyses and precautionary measures for the day-to-day jobs—installing pipe, assembling a scaffold, putting down concrete for foundations, building frames, excavating a trench and preparing it for entry—tasks that have inherent hazards and particular precautions in the many diverse jobs performed on a construction site.

Include in the analysis:

- how the results of these analyses are used to train employees to do their jobs safely
- how the analyses contribute to planning and implementation of hazard correction and control programs
- the analytical methods—job hazard analysis, job safety analysis, prejob hazard assessment, or others—used to identify the task hazards of the job.

Explain how new/modified equipment, materials, processes, and nonroutine tasks are assessed and analyzed for potential or existing hazards prior to implementation.

SAMPLES, TESTS, AND ANALYSES

Samples, tests, and analyses to identify health hazards and employee exposure levels that follow nationally recognized procedures must be conducted regularly and as needed. Describe the methods used to record a baseline analysis to identify occupational health and environmental hazards associated with your specific work environment, such as air contaminants, noise, lead, asbestos. Identify the SH&E professionals who were or will be engaged in these activities. Be specific about the industrial hygiene surveys that concern sampling rationale and strate-

gies. The occupational health and environmental management systems should address:

- sampling strategies and priorities that are preventive, including exposures or spills
- initial screenings, and the circumstances under which they are necessary
- how sampling strategies apply to any possible health hazards in the work environment and accurately assess employee exposures (i.e., duration, route, frequency, number of workers exposed)
- how the sampling results compare to OSHA's permissible exposure limits (PELs) to adequately assess employee exposure
- the qualifications of the individual(s) conducting the sampling and industrial health (IH) assessment
- the system used to document and track the IH monitoring and the information included in the documentation.

Documentation of the sampling strategy is necessary to identify health hazards and accurately assess employees' exposure, including duration, route, frequency of exposure, and the number of exposed employees.

SELF-INSPECTIONS

Self-inspections should be regularly conducted by trained staff to cover the entire site weekly during the construction work. Deficiencies will be recorded along with the tracking of hazard corrections to completion. If the construction is extensive and consists of several facilities and structures at one site, the self-inspections can be scheduled so that the entire site is inspected during each quarter.

Describe the procedures for conducting SH&E inspections. Include specific information on:

- what you consider to be a comprehensive inspection
- hazard-recognition training
- inspection schedules
- IH sampling and monitoring
- who conducts the inspections
- the system that tracks corrective actions to completion

- where applicable to IH hazards, summarize the testing and analysis procedures used and qualifications of those conducting them
- how you verify that you are performing scheduled inspections
- if you use a matrix of inspections to ensure that all required inspections are conducted, describe the information contained in the matrix.

WRITTEN HAZARD REPORTING SYSTEM

Crucial to successful implementation of the model process, a written hazard reporting system must be in effect that enables employees to report their observations or concerns to management without fear of reprisal and to receive timely responses. A mechanism must be in place for employees to report anonymously and there must also be a mechanism to provide a response. Responses can be handled through newsletters, postings on bulletin boards, or by electronic mail.

ACCIDENT/INCIDENT INVESTIGATIONS

Sites should have a written procedure for investigation of accidents, near-misses, first-aid cases, and other incidents. The procedure should include:

- investigators' training
- what incidents warrant investigation, but should include investigating first-aid and near-miss cases
- accident/incident investigations conducted by trained staff with written findings that aim to identify all contributing factors, corrective actions required, tracking actions to completion, and a summation of what action was taken to prevent similar events in the future
- when someone other than the supervisor conducts the investigation
- how the root cause is derived
- considerations other than employee fault.

SYSTEM TO IDENTIFY AND ANALYZE INCIDENT DATA

A system should be in place to analyze injury, illness, and related data—including inspection results; ob-

servations; near-miss and incident reporting; first-aid, injury, and illness records; and employee reports of hazards—to identify common causes and corrections in systems, equipment, or programs.

Include how the data is collected and analyzed:

- what, if any, trends were noted for the last 12 months, or for the duration of the construction if less than 12 months
- a description of the system to implement corrective actions after identification of the trends
- the tracking system used to monitor corrections indicated by a trend.

Hazard Prevention and Control

SYSTEM FOR ELIMINATING OR CONTROLLING HAZARDS

An effective system for eliminating or controlling hazards must be in place. This system emphasizes engineering solutions to provide the most reliable and effective protection. It may also utilize, in preferred order, administrative controls that limit daily exposure, such as job rotation or work-practice controls (i.e., rules and work practices that govern how a job is done safely and healthfully), and personal protective equipment. All affected employees must understand and follow the system.

Engineering controls:

- Provide examples of engineering controls to eliminate or control hazards by reducing severity, likelihood of occurrence, or both.
- Describe measures such as:
  - reduction of pressure or amount of hazardous material
  - substitution with less hazardous material
  - reduction of noise levels
  - fail-safe design
  - ergonomic changes
  - guards, barriers, interlocks
  - grounding and bonding
  - pressure relief valves
  - caution and warning devices (i.e., detectors and alarms in conjunction with the above).

Administrative controls: Describe how daily exposures to hazards are limited by adjusting work schedules or work tasks, such as job rotation.

Work-practice controls: Identify the workplace rules, work practices, and procedures for specific operations to reduce employee exposure through:

- changing work habits
- behavior-based training
- improving sanitation and hygiene practices
- other appropriate changes in how the job is performed.

Personal protective equipment: Describe the written programs for the use of PPE, including res-

pirators, hearing protection, ergonomic equipment, fall protection, and other types of PPE used on the work site.

Hazard control program: Identify the major technical programs and regulations that pertain to the construction work, such as hazard communication, hearing conservation, respiratory protection, fall protection, lockout/tagout, confined space entry, excavation and trenching, and other particular ones. Include safe work practices for each one.

TRACKING HAZARD CORRECTION

A system for tracking hazard correction should be operational. It should include documentation on

---

■ ■ **Table 3** ■ ■

Critical Element 3: Hazard Prevention and Control

Site hazards identified during the hazard analysis process must be eliminated or controlled by developing and implementing the following systems by the means that assure the most control.

- The hazard controls must be:
  - Understood and followed by all affected parties
  - Appropriate to the hazards of the site
  - Equitably enforced through a clearly communicated written disciplinary system that applies to all levels of employees—managers, supervisors, and nonsupervisory employees
  - Written, implemented, and updated by management as needed and used by the employees
  - Incorporated into training, positive reinforcement, and correction programs

- The systems of hazard prevention and control are:
  - Initiating and tracking hazard elimination or control in a timely manner
  - Written for and ongoing documentation of the monitoring and maintenance of workplace equipment such as preventive maintenance
  - An occupational health-care program that uses licensed health-care professionals to assess employee health status for prevention of and early recognition and treatment of illness and injury
  - Procedure for response to emergencies on all shifts. The procedure must be written and communicated to all employees, must list personal protective equipment, first aid, medical care, and emergency egress

- The following order should govern actions to eliminate or control hazards, with engineering controls preferred wherever possible:
  - Engineering controls that design changes to eliminate or limit the severity and/or likelihood of hazard exposure
  - Administrative controls that limit daily exposure by change of the work schedule
  - Administrative work-practice controls such as workplace rules, safe and healthful work practices, and procedures for specific operations
  - Personal protective equipment

OSHA *Revisions to the Voluntary Protection Programs to Provide Safe and Healthful Working Conditions*, Federal Register, July 20, 2000.

how and when hazards are identified, controlled or eliminated, and communicated to employees.

Ensure that the hazards identified during the following activities are included in the tracking hazard-correction system:

- accident/incident investigations
- inspections
- employee hazard reporting
- industrial hygiene surveys
- annual or prior projects' SH&E evaluations.

The system should track the identified items to ensure completion of abatement and the use of interim protective measures when necessary.

WRITTEN PREVENTIVE MAINTENANCE SYSTEM

Describe the preventive/predictive maintenance system that reduces safety-critical equipment failures and schedules routine maintenance and monitoring. For each phase of the project, include a summary or list of the safety-critical equipment to be maintained.

OCCUPATIONAL HEALTHCARE PROGRAM

Develop an occupational healthcare program appropriate for the project and/or the construction site, as well as a corporate policy for companywide occupational health issues. It should include, at a minimum, nearby medical services, staff trained in first aid and CPR, and hazard analysis by licensed healthcare professionals, as needed. Include:

- on-site and off-site health services and/or the availability of qualified healthcare professionals
- coverage provided by employees trained in CPR and first aid, in which one they are trained, and their current certification
- how healthcare services are made available to employees on more than one shift
- how the company addresses hearing conservation and other health issues
- how occupational healthcare professionals provide their services (i.e., design/implementation of health surveillance and monitoring programs.

CONSISTENT DISCIPLINARY SYSTEM

A consistent disciplinary system must be enforced and applied to all employees—including supervisors, managers, and contractors—who disregard the rules. The system should include:

- the company's general safety and health rules
- the written disciplinary system for enforcing the rules
- how the system equitably enforces the disciplinary system for managers, supervisors, and employees.

ERGONOMICS

Describe the company's ergonomics program, including assessment of work areas, the system used to identify and track hazards and corrections, who is responsible, and who monitors the reports and the system.

WRITTEN EMERGENCY RESPONSE PROGRAM

Include written plans to cover emergency situations, including emergency and evacuation drills for all shifts. Also include emergency response personnel organization, equipment, and procedures for responding to the emergency, containing the emergency situation, decontamination of the affected area, and other relevant aspects of a well-organized emergency response program.

PROCESS SAFETY MANAGEMENT

For construction work on sites subject to the process safety management (PSM) standard, describe the company's PSM system and compliance methods compared to the OSHA standard's requirements, including:

- employee participation
- process safety information
- process hazard analysis
- operating procedures
- training
- contractor
- prestartup safety review
- mechanical integrity

- hot-work permits
- management of change
- incident investigations
- emergency planning and response
- compliance audits.

Of course, several of these have already been addressed in your model SH&E program management system. But, for a PSM work site, the ones not addressed in your SH&E system should also be included.

Safety and Health Training

Program description: Include both the formal and informal SH&E training provided for managers, supervisors, and employees.

- the system used to ensure that all compliance and other required training is completed
- how the effectiveness of the training program is verified.

Supervisors: Describe the verification process that supervisors:

- understand their responsibilities
- meet their SH&E responsibilities effectively
- understand the hazards associated with the jobs performed by their employees
- understand their role to ensure that their workers also understand and follow the rules and practices designed to protect them.

Employees: Describe:

- records that verify employees understand the hazards associated with their jobs and the importance of following rules that apply to their jobs
- how employees are made aware of the hazards associated with their work and in their work area
- methods used to teach employees to recognize hazardous conditions and signs and symptoms of work-related illnesses.

Emergencies: Describe the procedure for verifying that supervisors, employees, and visitors know what to do in an emergency.

Personal protective equipment: Describe the training or other methods to ensure that employees understand:

- why PPE is necessary
- PPE limitations
- how to maintain it
- how to use it properly.

Administer and document training for:

- managers and supervisors, emphasizing safety and health leadership responsibilities
- all employees in the site's safety and health management system, the hazards, and the hazard controls in place

▪▪ Table 4 ▪▪
Critical Element 4: Safety and Health Training

- Managers and supervisors understand their safety and health responsibilities and are able to carry them out effectively.
- Managers, supervisors, and nonsupervisory employees (including subcontractor employees) learn the safe work procedures through being taught the work procedures and through reinforcement.
- Managers, supervisors, and nonsupervisory employees (including subcontractor employees), and visitors understand what to do in emergency situations.
- Where personal protective equipment must be used, employees understand that it is required, why it is required, its limitations, how to use it, and how to maintain it and use it properly.

OSHA *Revisions to the Voluntary Protection Programs to Provide Safe and Healthful Working Conditions*, Federal Register, July 20, 2000.

■ ■ Table 5 ■ ■

Documents and Records for the Model Safety, Health, and Environmental Management System

1. Written safety, health, and environmental program
2. Management statement of commitment to SH&E
3. OSHA Form 300 for the site and for all subcontractors
4. Safety, health, and environmental manuals
5. SH&E rules and emergency procedures
6. System for enforcing SH&E rules
7. Reports from employees of SH&E problems and documentation of management's response
8. Self-inspection procedures, reports, and correction tracking
9. Accident investigation reports and analyses
10. SH&E committee minutes (if there is a committee)
11. Employee orientation and SH&E training programs and attendance records
12. Baseline safety and industrial hygiene programs and attendance records
13. Industrial hygiene monitoring records, results, exposure calculations, analyses, and summary reports
14. Annual SH&E program evaluations, site audits, and other audit documents
15. Preventive maintenance program and records
16. Accountability and responsibility documentation (e.g., performance standards and appraisals)
17. Occupational health-care programs and records
18. Available resources devoted to SH&E
19. Hazard and process analyses
20. Process Safety Management (PSM) documentation, if applicable
21. Employee involvement activities
22. Other records that provide relevant documentation

OSHA *Revisions to the Voluntary Protection Programs to Provide Safe and Healthful Working Conditions*, Federal Register, July 20, 2000.

■ how all employees recognize hazardous conditions and understand safe work procedures applicable to their work environments
■ assessment of employee comprehension and training effectiveness.

Documentation Assessment

Included in the annual evaluation and the construction project evaluation prior to conclusion should be a review of the SH&E documentation that is maintained by site management. The model SH&E management system expects the documentation listed in Table 5 to be part of the SH&E management system. It should be relevant to the scope and purpose of the overall company SH&E management

system and to the SH&E management system of the particular construction project.

LAST THOUGHTS

This chapter has presented the fundamental and essential elements of a model safety, health, and environmental management system for the construction industry. The elements were defined and their subparts were described. The extent of the definitions and descriptions was limited by the confines of one chapter as part of an entire book. However, other chapters in this book elaborate on some of the elements. More information on any elements *not* covered in this book can best be found in some of the references that follow. The construction industry is becoming more and more involved in the SH&E management system presented here. Companies are finding it to be a comprehensive and all-inclusive system that yields reduced injuries, increased profit, and safe and healthy workers. The system is a win–win deal for all of those involved—construction company owners, OSHA, and, foremost, the workers.

REFERENCES

American Society of Safety Engineers, Des Plaines, IL.; Web site: www.asse.org.

Christensen, Wayne C. and Fred A.Manuele. *Safety Through Design.* Chicago, IL: NSC Press, 1999.

Construction Safety and Health Management. Richard J. Coble, Theo C. Haupt, and Jimmie Hinze (eds.). Upper Saddle River, NJ: Prentice Hall, Inc., 2000.

Garner, Charlotte A. and Patricia O. Horn. *How Smart Managers Improve Their Safety and Health Systems: Benchmarking with OSHA VPP Criteria.* Des Plaines, IL: American Society of Safety Engineers, 1999.

Haltenhoff, C. Edwin. *The CM Contracting System: Fundamentals and Practices.* Upper Saddle River, NJ: Prentice Hall, Inc., 1999.

Hinze, Jimmie W. *Construction Safety.* Upper Saddle River, NJ: Prentice Hall, Inc., 1997.

Hislop, Richard D. *Construction Site Safety, A Guide for Managing Contractors.* Boca Raton, FL: CRC Press LLC, 1999.

Levy, Sidney M. *Subcontractor's Operations Manual: Forms, Processes, and Techniques.* New York: McGraw-Hill, 1999.

Occupational Safety and Health Administration, U.S. Department of Labor, Washington, DC, Web site: www.osha.gov/.

Richardson, Margaret R. *Preparing for the Voluntary Protection Programs: Building Your Star Program.* New York: John Wiley & Sons, 1999.

Stephenson, Joe. *System Safety 2000, A Practical Guide for Planning, Managing, and Conducting System Safety Programs.* New York: Van Nostrand Reinhold, 1991.

System Safety Workshop, presented by NASA Safety Training Center/Johnson Space Center Learning Center, Houston, TX, 1993.

Vincolli, Jeffrey W. *Basic Guide to System Safety.* New York: Van Nostrand Reinhold, 1993.

Voluntary Protection Programs Participants Association, Falls Church, VA., www.vpppa.org.

Wilkinson, Bruce S. *A Comprehensive Guide to Contractor Safety & Health in the Construction Industry.* Rosslyn, VA: Construction Education Foundation, 1994.

REVIEW EXERCISES

1. What are the four critical elements of a model SH&E management system?
2. Which critical element has the most influence on achieving positive results from the model SH&E management system? In your own words, why?
3. Which employees should be accountable for their safety? Why?
4. List three meaningful methods to involve employees in the decision-making aspects of the SH&E management system.
5. List three safety activities that are *not* considered "meaningful."
6. At what period or periods during a construction project should an SH&E evaluation be conducted. Why?
7. What is the purpose of the work-site hazard analysis?
8. What technique would you use to identify the hazards of daily tasks? Describe the process in two or three sentences.
9. Name at least four parts that an accident/incident investigation should include. What is their importance?
10. Identify four engineering controls that should be considered to control or eliminate hazards.
11. Who in the company should receive SH&E training? Why?
12. Identify at least four documentation records that must be maintained.

■■ **About the Author** ■■

INCIDENT CAUSATION
by Darryl C. Hill, CSP

Darryl C. Hill, CSP, is Director of Safety & Health for ABB Inc. He earned a B.S. in Occupational Safety from Iowa State University and an M.S. in Hazardous Waste Management from Wayne State University (Detroit). Mr. Hill is currently pursuing a Ph.D. in Educational Leadership at Oakland University where he is an adjunct instructor teaching Construction Safety & Environmental Standards. He also is an Oakland University Safety & Health Program Advisory Committee member. He is on the ASSE Board of Directors, serving as Region VII Vice-President, 2001–2005. Mr. Hill is an elected officer for the Michigan Safety Conference and Safety Council for Southeast Michigan. He has served as (ASSE) Construction Practice Specialty Administrator and was named Construction Practice Specialty Safety Professional of the Year in 1997 and 2003. In 1997 Darryl Hill was also named the Edgar Monsanto Queeny Safety Professional of the Year.

Incident Causation

INCIDENTS ARE CAUSED BY SYSTEM DEFICIENCIES

Tragedy #1

A painter at a new stadium project was killed when a hydraulic lift he was on fell into the first tier of seats. Workers exclaimed it was just a freak accident, primarily due to the fact that the job had been accident-free to date. Another painter in the area refused to go up on the lift prior to the accident, fearing it was unsafe. The machine involved in the accident was inoperable the previous day due to a faulty hydraulic pump. Both lifts being used at the stadium project at the time were old. On the day of the accident the second lift was not working. Several workers complained that the job was being rushed because the football season was scheduled to begin within a month. The painting contractor was given two unrelated serious violations prior to the accident. A subdued friend of the fallen painter stated he could speculate on what caused the tragic accident, but he could not be certain what was the root cause.

Tragedy #2

A laborer fell three stories to his death when a scaffold reportedly gave way. Preliminary investigation findings included:

■ The scaffold components were not capable of supporting, without failure, the scaffold's weight plus at least four times the maximum intended or transmitted load.

- Footings were not level, sound, and rigid, and capable of supporting the loaded scaffold without settling or displacement.
- Employees were not instructed to recognize and avoid unsafe conditions relating to outrigger and fabricated frame-type scaffolds.

Despite these findings, project management stated during the subsequent inspection that the fatality was an aberration and cited their low incident rate as corroboration.

Tragedy #3

A Texas construction firm faced a $238,000 fine following a trenching fatality that claimed the life of a worker. The company was cited for nine alleged safety and health violations, including three willful violations, for not protecting employees involved in excavation work from cave-in hazards. The fatal accident occurred when the wall of an approximately 20-foot-deep trench collapsed, causing fatal blunt-force injuries and asphyxia to a 17-year-old pipe layer. Investigators stated it would take several weeks to determine what caused the fatality.

These incidents are examples of a disturbing trend in the construction industry. During 2001, construction fatalities increased to their highest level (1210 deaths) since 1992, which was the first year the fatality census was taken (T. Nighswonger). This figure represented a 6 percent increase. One of every five U.S. workplace fatalities is a construction worker (OSHA Web site). To effectively address this disturbing trend, one must have a thorough understanding of incident causation. To do that, let's first examine the definition of an accident.

Funk and Wagnalls New Encyclopedia defines the word *accident* as "an unintended and unforseen event, usually resulting in personal injury or property damage." In law, the term is usually limited to events not involving negligence. Negligence involves the carelessness or misconduct of an involved party. In popular usage, the term accident designates an unexpected event, especially if it causes injury or damage, without reference to the negligence or fault of an individual.

In construction, we normally think of an accident as causing unwanted effects such as the loss of a system or part of a system; the injury to, or fatality of, operators or personnel in near proximity; and property damage to related equipment or hardware. An accident is usually a dynamic event since it results from the activation of a hazard and culminates in a flow of sequential and concurrent events until the system is out of control and a loss is produced. While we may think in terms of the events proceeding logically, it must be understood that environmental influences are part of these logical relationships. Accident events may include explosion, high-energy release, and separation of parts of the system. The goal should be to focus on the set of events that occurs and leads to the accident which results in a loss. To render a consistent meaning and universally accepted concept, an operational definition must be established. Ed Adams, in his book *Total Quality Safety Management*, used this operational definition: "An accident is a process event within a system. It is an unwanted transfer or flow of energy that, due to barriers and/or controls that are less than adequate, result in harm to the persons or objects in the path or exposed to unwanted transfer."

The medical profession and many injury-prevention specialists are opposed to focusing *only* on accidents. The term itself is limited in scope when evaluating causation models. By focusing only on *accidents*, we exclude occupational illness, hearing loss, fires, and near-misses. A preferred term is *loss incident* or simply *incident*, which includes a near-miss.

As a modern industrialized society, there is considerable injury and fatality data collected and published. This data is tracked, the consequences are measured, and governmental agencies publish the information. The construction profession is one of many industries that tracks its own data on errors. There is an abundance of numbers, and the Internet has made the information immediately available. One must realize the difference between information and knowledge. Information is provided by print and electronic media, and can provide a brief overview of a subject. Information on incidents and their causes will not lead to the elimination of the hazards that contribute to these negative occurrences. Only from knowledge—usually as a result of theory gained from education and application—can incidents be significantly reduced.

Organized efforts for the prevention of incidents began in the nineteenth century with the adoption of factory inspection laws, first in Great Britain, then in the United States and other countries. Fire and accident insurance companies made efforts to enforce safety rules and educate the public. Factory inspectors and inspectors from other industries carried on the campaign against hazardous environments. Preventing incidents is difficult in the absence of an understanding of the causes of incidents. Many attempts have been made to develop a theory of incident causation, but most have not been universally accepted. Researchers from several fields of science and engineering have been trying to develop a theory of incident causation that will help to identify, isolate, and, ultimately, remove the factors that contribute to or cause incidents.

Since the inception of the safety profession, a satisfactory explanation for incident causation has been lacking, probably because incident causation requires a thorough understanding of factors that are normally complex, multifaceted, and difficult to identify. Multiple incident-causation theories have been introduced to the safety profession. They range from the simple to the very complex. Some theories focus on employees and how their actions, or lack thereof, contribute to incidents. Other theories look at management roles and responsibilities for incident prevention. Theories are not facts, and must be viewed as tools that assist in predicting outcomes. Incidents have specific causes that will allow someone to predict the possibility of an outcome. Nevertheless, considerable research must still be done to determine actual incident causes—as evident by the disturbing construction fatality trend.

The construction safety profession has evolved to a point where incidents are no longer accepted as part of doing business at a job site. Companies are recognizing the business of safety and how understanding the causes of incidents, and loss-avoidance can be realized, contributing positively to the bottom line. Contractors recognize that reducing incidents can make them more competitive by lowering their insurance cost and improving productivity. Contractors are incorporating proactive management tools to effectively minimize or eliminate factors that contribute to incidents. This approach takes on greater meaning in times of escalating workers' compensation costs, liability resulting from common-law negligence, OSHA penalties, costs associated with damaged raw and finished products, and project delays.

CONSTRUCTION INJURY FACTORS

Design

One often overlooked but critical aspect of preventing job-site injuries is the design process. Design concepts are sometimes ignored due to perceived cost implications and feasibility. During the planning stage of a project many functions are responsible for a particular segment, but little consideration may be given to determining how, when assembled together, hazards may be created. These hazards ultimately contribute to job-site injuries. Safety, included in the design specifications, will ensure that guidelines and parameters are in place to minimize or eliminate hazards.

Many times, the basic design has hidden flaws that lead the worker or equipment to perform in a manner that may result in a loss-producing situation. These designs provoke, invite, or mislead the worker into performing these acts. The concerns may not be obvious at first, but after repeated exposure a loss incident may occur. Safety should be considered and built in at the earliest stage of the design. Construction safety professionals must interface with designers and engineers and provide information in clear, concise terms. Engineering and process design activities have considerable positive influence on ultimate project costs. In *Construction Safety Planning* (David MacCollum), the following example illustrates the importance of design on a construction project.

A worker fell twenty feet to his death during construction of an office complex as he was welding hangers of a precast concrete cladding panel to the building's steel framework. Investigation revealed that the design of the hangers was such that the panels could not have been welded into place from inside the building where the worker would have had effective fall protection by standing on the floor, eliminating the need for him to be suspended outside the building to make welds. The hangers were anchored in the panels in such a way that they would not hang plumb, requiring the crane from which they were suspended for installation to exert a side-pull so the hangers could be properly attached to the building frame. Thus, the panels were defective because attachment to the building had to be made from the outside, and they could not hang plumb when being lifted into place by the crane. The question that must be asked is why wasn't safety involved in the design process? The defective hanger attachment system could have been identified, the hazard eliminated, and a death prevented. In this case, a professional safety engineer familiar with construction practices was not included on the design team. The company did not understand the benefits of a close relationship between the design and erection processes.

Those contractors who coordinate the design with erection and other installation processes usually have superior safety results. Safety, integrated into the design phase, allows evaluation of the construction methods to be used and can eliminate hazards before they are created. In "Time for Change in Construction Safety" (*Professional Safety*), it is noted that a hazard is an unsafe condition that exists in three modes. The three modes include:

1. *Dormant:* A hazard created on the drawing board, either from omission or inclusion. It lies in wait, undetected, in proposed materials or construction methods to be used.
2. *Armed:* A dormant hazard that might not have been detected during construction planning and is ready to be activated. Interaction between equipment on the job and those operating it or working around it might also trigger a dormant hazard into the armed mode.
3. *Active:* What happens when the right combination of factors triggers a dormant or armed hazard into action.

As noted in the definitions, the design phase is critical to eliminating hazards that contribute to construction incidents. Safety professionals must also understand that construction methods, including design, are effective problem-solvers at job sites. Safety professionals, and others charged with preventing incidents, should use these valuable resources from the design stage through final installation tasks. The greatest opportunity to influence design occurs during the early concept stages. Engineering and process design activities have the greatest influence on construction costs. Unfortunately, many contractors address hazards once they have mobilized on site. Safety professionals have a tremendous opportunity to interact with designers/engineers during the project bid stages and integrate safety solutions that will prevent human suffering.

A preliminary safety analysis or survey should be conducted during the proposal or design phase. To minimize the potential for undesired incidents, safety must be considered during all phases of a product's life cycle: design, engineering, fabrication, testing, installation, and startup. Safety professionals who are involved in design reviews can have a significant (positive) impact on overall project success. It is a given that projects are understaffed, underbudgeted, and aggressively scheduled. As a result, there aren't sufficient resources or time to do the job "right," but there is always a way to *rework* and have a punch-list when the project is near completion. Usually, it is near project completion that workers are susceptible to an incident.

A useful tool is *Design for Safety Toolbox*, a computerized design aid that enables designers to conduct a comprehensive and systematic construction safety analysis during the design phases of a project. The University of Texas at Austin developed the theoretical framework, program design, and prototype for this interactive, multimedia software. The system includes the following capabilities:

- identification of a construction hazard
- ability to summarize and prioritize hazards
- incorporation of lessons learned for hazard and design solutions
- documentation of project hazard-related decisions
- report generation, which includes hazard tracking, owner acceptance, and residual risk.

Schedule

Schedule is a critical aspect of a project that could have an impact on job-site incidents. Once a project gets behind schedule, there is a tendency to increase the production effort to get the schedule aligned with the construction contract. When this happens there is an increased probability that an incident may occur. Therefore, it is imperative for a construction project to progress according to schedule in a systematic and logical manner. Coordination among multiple contractors is important to ensure schedules are met, thus reducing the potential for job-site incidents.

Depending on the scope of the project, the construction manager or general contractor will prepare a schedule. The schedule will outline the multiple activities that will occur for the duration of the project. Time parameters are identified to show when these activities must be completed. It is important that each contractor understands how each task is related to the others and considers the duration for each task. Construction workers are task-oriented, and they want to complete their duties in a timely manner.

A project that is properly managed and adheres to the schedule reduces the potential for rework. Rework creates the opportunity for workers to be exposed to unnecessary hazards. A properly executed master schedule and the detailed preparation of short-interval schedules will assist in the overall project success. Project schedules should be continuously reviewed and updated, along with consideration for a safe job site, throughout the duration of the project.

Overtime

Construction workers tend to work overtime more than workers in other industries. Research conducted by Sue Dong (CPWR newsletter) has demonstrated a high degree of risk for construction workers. This is especially true for laborers, who have the highest degree of risk among the trades. The research covered responses from over 12,000 men and women in yearly interviews from 1979–1994 and in 1996. There were 550 to 700 construction workers interviewed each year. Initial findings showed construction employees work about 30 to 45 minutes longer than other blue-collar workers each day. Approximately one-third of construction workers put in overtime, compared to 25 percent of employees in other industries. Based upon worker responses, the risk of injuries for employees who usually worked overtime is about 1.57 times higher than those who worked regular hours—and about 2.6 times higher for construction laborers.

The impact of scheduling for a safe job site is well documented. There are numerous occasions when a project schedule can get behind, the owner makes demands, and sometimes safety is sacrificed. Safety doesn't "pay the bills"; therefore, the focus is getting the job completed, even if it means safety is no longer a priority. Although we all know safety is important, project completion is paramount. Overtime due to an upset project schedule is a serious issue that can have a major impact on construction safety.

Job-site Safety Culture

Many times at a construction site, incidents are followed by an investigation to determine the sole cause of the incident. Incidents are blamed on human error, mechanical failure, or another single cause. As noted previously, often there are multiple contributing causes in an interrelated web of events. The root causes get to the foundation of the ultimate cause of the accident. In essence, root causes are factors that, if changed, could prevent many other incidents from occurring.

One prevalent root cause is a flaw in the safety culture of the organization. Safety culture is the general attitude and approach to safety reflected by those who participate in an industry or organization, including management and workers. In the construction industry that may include senior management at the corporate office, project management, superintendents, foremen, and other key project personnel in the field. Also, there may be varying cultures relative to the many contractors and subcontractors working side by side. Safety cultures may contribute to complacency because the term *safety* itself is a difficult term to measure. The reason for this difficulty is that a proactive safety process leads to a lack of incidents. Many successful safety processes with a proven track record could be viewed as irrelevant.

Events with a low probability of occurrence, but high consequences, tend to be ignored as situations that couldn't happen at all. Similarly, complacent safety efforts signal that risk somehow decreases over time. The potential for an incident is ignored with the assumption that everything is under control (thus lacking an understanding of variation), and that an incident couldn't happen.

As a subset component of the safety culture in construction, ineffective organization structure influences incident causation. It can adversely impact safety efforts.

- Safety efforts are hampered if safety personnel lack independence. When safety professionals report to a function that is directly responsible for the project budget and schedule considerations, there is a high probability that schedule and budgetary pressures will override safety.
- Diffusion of responsibility and authority can leave the burden of ensuring safety on individuals who do not have the ability to carry out their tasks. This is often the case when safety professionals have a low-level status.
- When safety is not included during the bidding process, the responsibility may fall, as a collateral responsibility, on someone who may not have knowledge or experience to successfully execute the safety function. Also, a contractor whose bid does not include the cost of safety may be awarded work because the bid appears more attractive.

Owner Influence

As discussed earlier, the demand to meet a project schedule can influence job-site safety. There is a trend toward owners rewarding accelerated construction schedules by offering bonuses for early completion. Owners can influence project conditions by allocating space to contractors for staging. For example, during highway construction, is the contractor allowed to close a road, or is moving traffic immediately next to the crew acceptable? Owners are recognizing their role in promoting safety on their projects. Owners can have a particularly strong impact on job-site safety during the bidding process. Utilizing low-bid subcontractors who fail to build the cost of safety into their bids can have a negative impact on the safety process. Subcontractors may introduce hazards to other contractors on the job site long after the awarding contractor is gone.

Proactive owners are taking construction safety seriously and are addressing this area during the selection process.

EXPLORING CAUSES AND EFFECTS

A review of causes and effects is needed to evaluate incident-causation models. To be considered a *cause*, the event must precede the incident in time. The event (factor or condition) may be a necessary or sufficient condition for the effect; however, rarely is one event *both* the necessary and sufficient condition. In most cases, it is enough to identify the events/factors/conditions that increase the probability of the effect. Furthermore, an argument can be made that causes and effects are one and the same. The difference is primarily how we perceive them in time. When we start with an effect of consequence, we want to prevent it from occurring. When we ask "why" it occurred, we find a cause; but if we ask "why" again, what was just a cause becomes an effect. Look at the analogy below relative to construction safety.

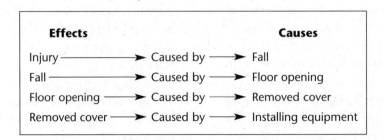

Effects		Causes
Injury	Caused by	Fall
Fall	Caused by	Floor opening
Floor opening	Caused by	Removed cover
Removed cover	Caused by	Installing equipment

■ ■ Figure 1 ■ ■ Construction cause and effect

In the example, a *fall* is a cause when viewed as the precursor to the *injury*. However, a fall could also be an effect of *floor opening* when viewed as the result of slipping. We could also add *slipped* as the *cause* of fall and the *effect* of the floor opening if we wanted to include more causes in the example. As we will see later, there are always causes in between causes, and logic can help us find the optimum sequence. In essence, we have our own perceptions of a correct alignment based on our individual knowledge of the specific causal relationships.

Our perspectives are important, but others may perceive a cause or effect differently or more deeply if they have a greater understanding of the causal relationships. For example, we know we have a cold when we ache and cough, whereas a doctor knows we have a cold when he or she can observe a virus on a microscope slide. The situation of the effect is the same, but the knowledge of the causes is significantly different depending on perception and knowledge.

By understanding that a cause and effect are one-and-the-same—merely viewed from different perspectives—we begin to see how they are part of a continuous set of causes that has no beginning or end. As we observe the structure of the cause chain created by asking "why," we are drawn to a linear path of causes. When we keep asking "why," the chain of causes seems eternal. The answer to our initial "why" is a function of our perspective. If we are the people responsible for valve maintenance in this example, we may choose to look at the leaky valve, or possibly the seal failure. If we are the safety professional, our primary interest would be preventing the injury, so we would probably focus on the injury when looking for causes.

Cause-and-Effect Principles

- All undesirable events are caused to happen. These events are the result of design deficiencies, human errors, equipment malfunctions, and other elements.
- The root cause(s) of an event can be determined by analyzing cause-and-effect relationships.
- Because undesirable events are caused to happen, they are actually effects created by additional causes(s).

TERMINOLOGY

Causal factor – a factor that shaped the outcome of the situation.

Contributing causes – causes that by themselves would not have caused the problem, but include secondary reasons for the event's occurrence.

Potential causes – conditions that appear to have caused the event, but need verification.

Presumptive causes – causes that may be apparent at the beginning of the investigation, or that emerge during the data-collection process. These include hypotheses that explain the effects of the problem, but need validation.

Primary effects – undesirable events (i.e., errors or incidents), the occurrence of which was critical for the activity of the situation being evaluated.

Root causes – the basic reasons for an incident.

■ ■ Figure 2 ■ ■ Cause-and-effect definitions

Determining the root cause is a process for systematically detecting and analyzing the possible causes of an accident. Root-cause determination is based on internal logic and reasoning skills to derive conclusions.

ROOT-CAUSE ANALYSIS

To effectively have a complete understanding of incident causation, one must evaluate the event, define the problem, and identify the cause. This three-step process will allow the safety professional to look at the system, including activities and processes, and determine hazards present in the system and the resulting causes for nonconformance and errors.

The safety professional must clearly identify and describe the loss incident he or she is attempting to solve. This will provide focus for a root-cause analysis by determining the who, what, when, where, and how specific to the undesirable event, thus defining the loss incident. During a systematic attempt to define the loss, a safety professional, or anyone responsible for determining the causes of an incident, may reveal multiple problems that can be addressed to prevent reoccurrence. To identify potential incident causes one must:

- explore the undesirable events and situations inherent in the system
- determine the deviation(s) from any one requirement or expectation
- evaluate the primary effect necessary for the situation to occur.

It is tempting to prejudge a probable cause or sequence based upon first observations or identified factors. This often results in forming an incomplete or erroneous conclusion. Root causes are those which, when corrected, would bring about positive results. Therefore, it is very important that management understands its role specific to root causes of incidents.

FAILURE MODE AND EFFECT ANALYSIS

A failure mode and effect analysis (FMEA) is an engineering technique used to define, identify, and eliminate known and/or potential problems, failures, and errors from a system, design, and/or process before they reach the ultimate user. The analysis of the evaluation may take the following courses of action: (1) historical data compared to similar data for like products and/or services and other appropriate available information will define failures; (2) mathematical modeling, inferential statistics, and reliability engineering may be used to identify and define failures.

An FMEA will provide the safety professional with useful information that may reduce the risk load in the system, design, or process. The FMEA is a logical and progressive potential failure analysis technique that allows the task to be performed more efficiently. FMEA is effective in identifying and eliminating causes in a system, design, or process which will prevent failures and errors from occurring. This will allow a methodical way of studying and evaluating causes and effects of failures before the design and process is completed. Basically, the FMEA provides a systematic method of exploring the many ways a failure (incident) might occur.

ACCIDENT-CAUSATION THEORIES

Paradigms and Theories

Paradigms are derived from the traditional sciences; they are the framework of thought that relies on a belief system and some scientific data and facts. Simply, a *paradigm* is the set of rules by which something operates, while a *theory* is contained in the sphere of abstract knowledge or speculative thought. A *model* is a simplified description of a system to assist calculations and predictions. The changing of company rules, theories, or paradigms is the earliest sign of organizational change and a cultural shift.

Scientists see unfounded explanations as ideologies. Ideologies will influence decisions that are based on personal interpretation and not on rational inquiry and analysis and scientific methods. Many in the scientific and engineering community do not hold ideology in high regard because it is often based on speculation or a body of unproven doctrine. Paradigms in the social sciences are highly likely to emerge from ideologies rather than scientific theories because they cannot be tested against objective realities (F. English). Scientists suggest that paradigms based on ideology are much less challengeable, and thus less provable, than those that are based on theory (J. A. Barker).

Accident-Proneness Theory

While a *model* is a simplified description of a system to assist in calculations and predictions, a *theory* is a system of ideas based on general principles independent of the particular things to be explained. Perhaps the best-known and controversial theory of accident causation is the *accident-proneness theory*. This theory focuses on the worker's relationship to accident causation. It maintains that, within a given set of workers there exists a subset of workers who, because of their personal characteristics, are more likely to be involved in accidents. Many in management support this theory. Advocates of the theory state that accidents are not randomly distributed or that sustaining an injury is not simply a chance occurrence. These individuals assert that some workers have permanent characteristics that predispose them to a greater probability of being injured.

Several research studies have been conducted to determine the validity of the accident-proneness theory, although recently there has not been any validated research in this area. The studies focus on examining a given population and assessing the distribution of injuries in that population. Obviously, the theory allows supporters to place the blame on the injured worker. However, this concept which implies blame for repeat accidents on an individual's lack of awareness, skills, or experience, "has no empirical foundations" (R. Sass and G. Crook). Researchers have not been able to conclusively prove this theory because most of the research work has been poorly conducted, and most of the findings are contradictory and inconclusive. This theory is not generally accepted because different jobs present different hazard exposures, and employees who are injured are not exempt from another incident. However, many safety professionals believe incident repetition (a series of near-misses, reinjury, or otherwise unrelated accidents experienced by the same person) is a real phenomenon. Some suggest a Pareto distribution is demonstrating that 20 percent of employees account for 80 percent of all accidents. Anecdotal information demonstrates that, while a small number of workers experience a chain of accidents for a period of time, it appears they move out of this pattern, and others may replace them to account for multiple accidents. "Evidence indicates that when the period of observation is sufficiently long, the small group of persons responsible for most of the accidents is essentially a shifting group of individuals falling in and out of the group" (M. Schulzinger). Also, a survey of 27,000 industrial and 8000 nonindustrial accidents indicated that the accident-repeater was involved in only 0.5 percent of them; whereas, 75 percent involved the relatively infrequent experiences of a large number of people. Morris Schulzinger concluded:

- Irresponsible and maladjusted individuals are significantly more likely to have accidents than individuals who are responsible.
- Men are significantly more likely to have accidents than women with a 2:1 ratio in nonindustrial studies and an even higher ratio in industrial studies.

- Most accidents involve young workers. Of the nonindustrial accidents studied, 70 percent involved workers under the age of 35, and nearly 50 percent involved workers under 24.
- Most accidents are due to the relatively infrequent, solitary experiences of large numbers of individuals (86%).

In a contrasting argument, Mintz and Blum attacked the accident-proneness theory by asserting the following:

- Accidents are considered to be "rare" events; therefore, the proper statistical model to be used in interpreting the occurrence of accidents and whether these happen with greater-than-chance frequency in a given situation is the statistical frequency distribution known as the Poisson Distribution.
- According to the Poisson Distribution, 9 percent of a given population should have 39 percent of the accidents, and 39.5 percent should have 100 percent of the accidents.
- Therefore, if accident-proneness is to be considered a reasonable explanation for any given distribution of accidents, then the distribution should be more extreme than these (i.e., significantly fewer than 9% should have 39% of the accidents, or 9% of the population should have significantly more than 39% of the accidents).

It has been concluded that, if any empirical evidence supports this theory at all, it probably accounts for only a very low proportion of accidents without any statistical significance. Finally, the personal factors contributing to accident-proneness are not consistent traits within workers, but vary over time and during different situations. Therefore, this theory is invalid and should not be used.

The Domino Theory

According to W. H. Heinrich who developed the so-called *domino theory*, 88 percent of all accidents are caused by unsafe acts by people, 10 percent by unsafe conditions, and 2 percent by "acts of God."

He proposed a *five-factor accident sequence* in which each factor would actuate the next step, like toppling dominoes lined up in a row. The sequence of the accident factors is:

1. ancestry and social environment
2. worker fault
3. unsafe act together with mechanical and physical hazard
4. accident
5. damage or injury.

In the same way that the removal of a single domino in the row would interrupt the sequence of toppling, Heinrich suggested that removal of one of the factors would prevent the accident and resultant injury; the key domino to remove would be number 3. Heinrich provided no data for his theory. Logically, it is unlikely that accidents occur

in a unilinear fashion with all causal factors dependent only on one preceding event. Most accident investigations reveal two or more independent chains of events leading to the loss incident. Too often the narrow interpretation of the domino theory has led only to accident *symptoms* (Dan Petersen).

Figure 3 illustrates a traditional accident-causation flowchart. Using this approach has not resulted in significant advances in preventing accident recurrence. The act and/or condition only identifies symptoms, not the causes. By addressing and removing only symptoms, the root causes remain, allowing another accident to occur. The act or condition should be viewed as the *proximate cause* of an accident; neither addresses root causes. Figure 4 illustrates some direct and proximate accident causes. A full accident analysis will usually demonstrate the importance of design and management action in causation.

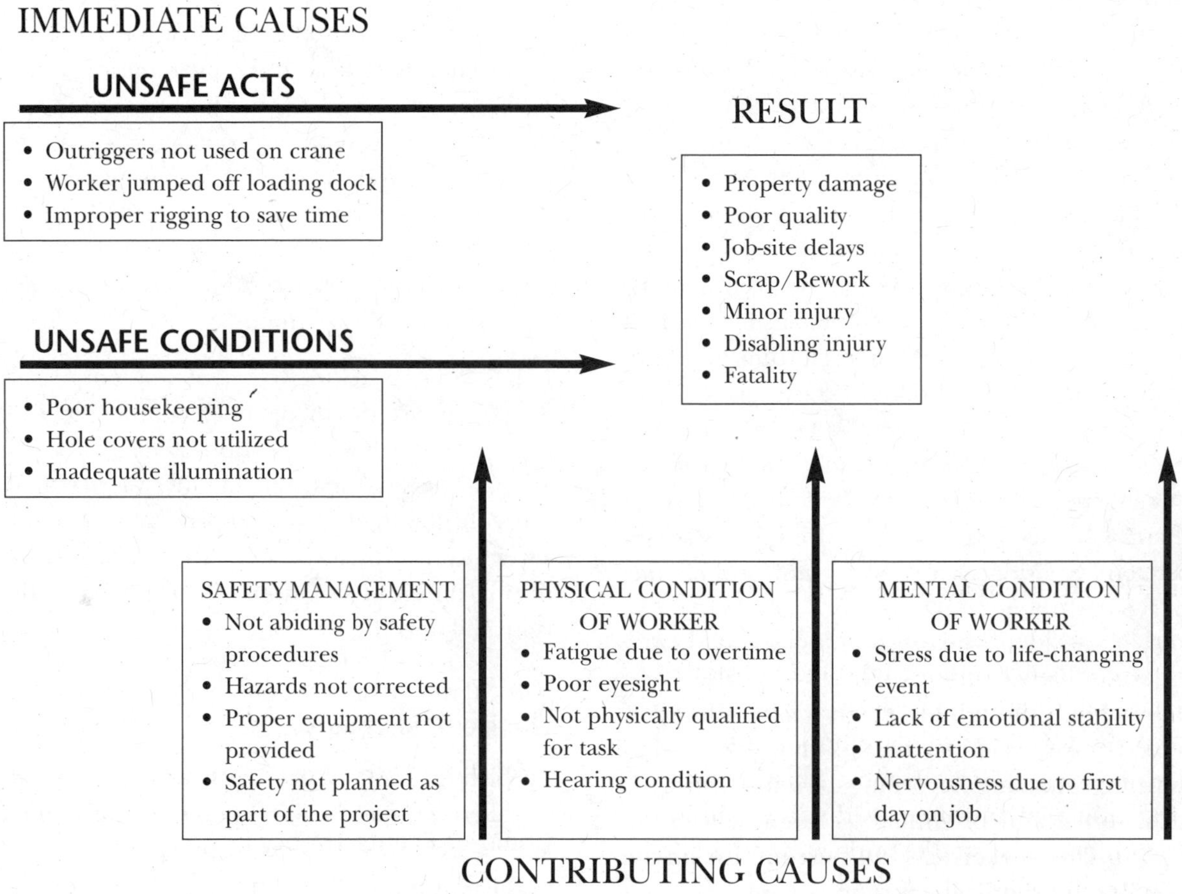

IMMEDIATE CAUSES

UNSAFE ACTS

- Outriggers not used on crane
- Worker jumped off loading dock
- Improper rigging to save time

RESULT

- Property damage
- Poor quality
- Job-site delays
- Scrap/Rework
- Minor injury
- Disabling injury
- Fatality

UNSAFE CONDITIONS

- Poor housekeeping
- Hole covers not utilized
- Inadequate illumination

SAFETY MANAGEMENT
- Not abiding by safety procedures
- Hazards not corrected
- Proper equipment not provided
- Safety not planned as part of the project

PHYSICAL CONDITION OF WORKER
- Fatigue due to overtime
- Poor eyesight
- Not physically qualified for task
- Hearing condition

MENTAL CONDITION OF WORKER
- Stress due to life-changing event
- Lack of emotional stability
- Inattention
- Nervousness due to first day on job

CONTRIBUTING CAUSES

■ ■ **Figure 3** ■ ■ **Traditional accident causation flowchart**

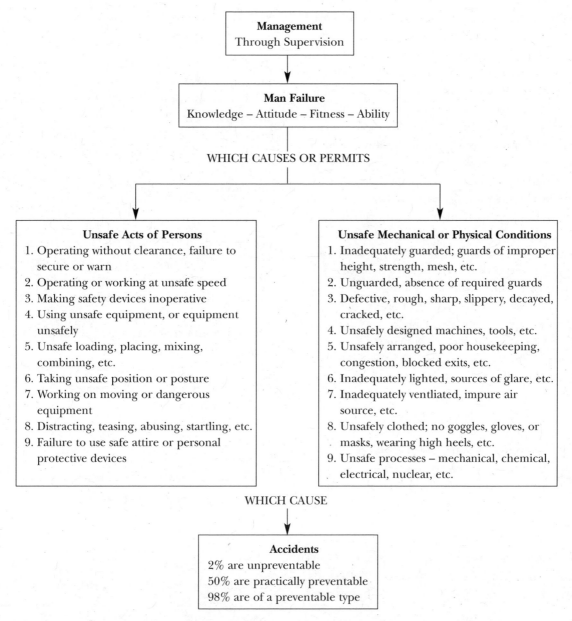

■ ■ **Figure 4** ■ ■ **Flowchart of direct and proximate accident causes** (Source: *Heinrich Revisited: Truisms or Myths* by F. A. Manuele, NSC Press, 2002; reprinted with permission)

Heinrich's Accident-Causation Model

Heinrich's accident-causation model has had a tremendous influence on the SH&E profession for several decades. Although his teachings are still widely accepted, recent evidence has shown a gradual shift away from his axioms. According to Heinrich, 88 percent of direct and proximate accident causes are due to unsafe worker acts, 10 percent are unsafe physical or mechanical conditions, and 2 percent of accidents are unpreventable. Basically, man failure

is the primary problem, and the field of psychology is an integral element in addressing this issue.

The Heinrich model, behavioral safety programs, and other incident-causation theories have traditionally focused on worker issues and not on the root causes of hazards.

Chain-of-Events Theory

Very similar to the domino theory is the *chain-of-events theory*. Many in the safety profession have argued that the two theories are basically the same. Incidents are occasionally characterized as occurrences that are the result of a series of events. The events in a series are all linked in that event and followed by yet another event. This portrayal of incident occurrence is referred to as the chain of events. The occurrence of all events in the series, or chain, ultimately results in the incident. Furthermore, if any event in the chain had not occurred, the incident might have been averted.

The last event preceding many worker injuries is some action performed by the injured worker. Thus, it is common for many injuries to be blamed on worker behavior, if for no other reason than that the worker is often the last party involved in the chain of events. However, stopping the occurrence of any event in the chain will break the chain; that is, breaking any link will break the chain. If safety is to be promoted, it is important to consider all events in the chain, not just the final action of the worker who becomes injured. To simply blame injured workers for many of their own injuries is to ignore the roles that other parties play in influencing worker behavior.

Because every link in the chain is vital to incident prevention, each is therefore a potential target for incident prevention. The links may relate to physical working conditions, first-line supervisors, various levels of management, company policies, or other factors beyond the control of the injured worker. The parties associated with the various links in the chain have the opportunity to alter the course of events, and thereby prevent incidents—that is the premise for all models.

The chain-of-events theory, consistent with other models, is best used as a means of preventing incidents. It is not intended to place blame. Serious consideration of the events in the chain and their roles in an incident can be helpful in incident prevention. Such consideration addresses factors other than simply the action or actions of the party that is ultimately injured. Focusing on the chain of events focuses attention on the physical working conditions that might contribute to the risk of an incident. Furthermore, other means of averting incidents can be identified, including worker training, supervisory instructions, project planning, corporate policies, and improved project designs.

Adjustment Stress Theory

The *adjustment stress theory* states that safe performance is compromised by a climate that diverts the attention of workers. This theory contends that "unusual, negative, distracting stress" placed on workers increases their "liability to accident or other low-quality behavior." The theory also includes the premise that negative factors in the workers' environment create diversions of attention, and that the lack of attention caused by the diversions can be detrimental to safety.

This theory requires a safety professional to recognize that a worker can bring stress onto the construction site. The worker may be experiencing financial problems, illness of a child, fatigue, and even bodily pain. These stresses may affect the worker's ability to remain safe at the site. Table 1 identifies life events that may negatively impact a worker's day at a job site during any given day.

Everyone has a certain degree of stress. It is an inappropriate level of stress that leads to error. Not enough stress leads to inattention, fatigue, and mindless behavior (J. C. Miller). Too much stress is likely to lead to attention-narrowing, reduced working memory capacity, shifting from conservative to risky performance, shifting to quicker but more error-prone performance, regression to well-learned habits, considering fewer alternatives, and persevering with inappropriate strategies (J. E. Driskell & E. Salas).

Another stress-related factor is depression. Each year, between 15 and 20 million adult Americans, or as much as 10 percent of the adult population, experience a depressive illness that results in more than 200 million lost workdays (T. Nighswonger). The number of Americans treated for depression increased from 1.7 million in 1987 to 6.3 million in 1997, researchers reported in the *Journal of the American Medical Association*. Yet, only one-third of those with depression seek treatment.

Accident-causation models only provide a simplistic view of why workers get injured. There are multiple factors that lead to an injury, many of them complex in nature.

Other Theories

When an assumption is made based on common factors present in incidents that can be discerned

■ ■ ■ **Table 1** ■ ■ ■

Negative Impact of Life Events on Worker Performance

Life Event	Value	Life Event	Value
Death of a spouse	100	Son or daughter leaving home	29
Divorce	73	Trouble with in-laws	29
Marital separation	65	Outstanding personal achievement	28
Jail term	63	Spouse begins or stops work	26
Death of close family member	63	Begin or end school	26
Personal injury or illness	53	Change in living conditions	25
Marriage	50	Revision of personal habits	24
Fired at work	47	Trouble with boss	23
Marital reconciliation	45	Change in work hours or conditions	20
Retirement	45	Change in residence	20
Change in health of family member	44	Change in schools	20
Pregnancy	40	Change in recreation	19
Sex difficulties	39	Change in church activities	19
Gain of new family member	39	Change in social activities	18
Business readjustment	39	Mortgage or loan less than $10,000	17
Change in financial state	38	Change in sleeping habits	16
Death of close friend	37	Change in number of family get-togethers	15
Change to different line of work	36	Change in eating habits	15
Change in number of arguments with spouse	35	Vacation	13
Mortgage over $10,000	31	Christmas	12
Foreclosure of mortgage or loan	30	Minor violations of the law	11
Change in responsibilities at work	29		

Source: Holmes and Rahe.

with statistical analysis of the *right* data from incident investigations, this is called *factorial theory*. This theory assumes hypotheses about determinant variables can only be identified by *secondary* examination of facts. Criteria are scope, data, and outputs dictated by hypothesis rather than direct observations from incidents. This requires extensive exercise of investigators' judgments, often from the use of data reporting forms. It also requires the occurrence of sufficient incidents to build a database. In practice, this results in differentiation between fact-gathering during field investigation and the secondary data-analysis function.

The *logic tree theory* assumes that accidental events are predictable, and structures a predictive search for alternative event pathways leading to a selected undesired event through speculations by knowledgeable systems analysts. This theory follows rules of procedure for structuring speculations and assigning probabilities in branched event-chain displays, and demands ordering events in incident sequences. Displays facilitate communication, discovery, constructive criticism, and technical inputs. The theory provides a basis for identifying data needed during operations to update probability estimates. Displays can provide guidance during investigation of actual incidents, and incidents can be used to upgrade predictions. It does not provide for incorporation of event time relationships and duration; criteria for undesired event choices are unspecified.

The *process theory* assumes incidents are a transient segment of a continuum of activities, and views an incident as a transformation process by which a momeostatic activity is interrupted with an accompanying unintended harmed state. Interacting actors describe the process with actions, each acting in a sequential order, with each sequence related to each other interacting sequence in specific temporal and spatial precede–follow logic. Investigative tasks call for identification of the actors, their actions and interactions, and resultant changes of state from the initiating perturbation through the last sequential harm to the actors. It uses prescribed criteria for beginning and end of accident, and for data search, selection, recording, organization, and testing. The display provides a *time coordinate* to discipline events' timing relationships and the hypothesis generation method, in addition to the benefits of the logic-tree displays described earlier.

HUMAN ERROR

Operator error is frequently cited as an accident cause, though a complete understanding of human error is still evolving. Griffion-Fucco and Ghertman define human error as: a human action revealing a deviation from the one that would have averted the event or reduced its seriousness (*New Technology and Human Error*, pp. 193–207). Human error is any member of a set of human actions that exceeds some limits of acceptability. One might at least view human error as a deviation from a specified norm.

Often, operator error is cited as the *sole* accident cause. However, data about operator effects on incident rates can be biased and incomplete. Positive actions by operators are rarely identified. For example, when a crane capsizes, often the operator is blamed for the incident. However, there are many known scenarios of operators averting potential incidents, such as adjusting A-frames on lattice-work boom cranes to accommodate low underpasses. These operators are described as doing what they were taught. The term *human error* has a negative connotation and implies that the worker is at fault for the failure in what is usually a complex system. One must understand that there are many factors that lead to mistakes in a system through no fault of the worker. Learning from human-error incidents, such as the Concorde crash in 2000 and the deadly gas leak at Bhopal, India, is important to advancing the profession relative to causes of incidents.

Separating human error from design error is difficult at best. Employees must work with the system interface provided by management. Even with a global shift to automation, human error is not eliminated. Also, automation does not remove workers from systems. Automation puts a human in different roles within the system. Human operators are adaptable and flexible. Therefore, workers may use problem-solving skills and creativity to address unusual circumstances, and human operators can exercise judgment. Human error is the inevitable outcome of this adaptability and flexibility. Therefore, we can say that the qualities that lead to operator error are the same ones that make operators valuable.

Workers are inquisitive. They form mental models about what a system is doing. They usually work within a system and attempt to validate or invalidate those models and make the necessary adjustments. A worker's mental model may actually mirror the functioning of the system. Systems at construction sites are dynamic—in a constant state of change. After installation, startup, and service, the actual functioning of the system many times is different from the designer's model.

The study of human reliability analysis, including the man–machine interface, is critical in the understanding of human error and risk. The analysis examines the effectiveness and layout of the design of the workplace and its functioning. The study of dials, switches, and knobs on construction equipment; the messaging and alarms; the workflow during installation; and the value and type of procedures all have their importance for human-error and incident causes. The analysis will thus allow for quantification of the human-error probability (HEP). The HEP is a function of the actual tasks undertaken, the stress and time available, the quality of training and procedures, degree and extent of the management support structure, and effectiveness of team decisions and actions. Thus, all actions have to be viewed in what is called the *context*, meaning the circumstances or environment in which they occur. Human-centered design (HCD) considerations are gaining increased importance in an attempt to reduce the possibilities of human error during work activities.

ROLE OF MANAGEMENT IN ACCIDENTS

"The sign leading to the construction site's entrance, YOU ARE RESPONSIBLE FOR JOB-SITE SAFETY, *was prominently displayed. When I ascended the stairs to get to another level, I nearly fell sev-*

eral feet, the handrails were so loosely put in place." (Statement contributed by a laborer who began to question management's safety commitment.)

A prevailing view in the SH&E profession is that management plays a critical role in the prevention of accidents. Many safety professionals and others charged with accident prevention view accidents as a failure in the system—or management ineffectiveness. Since management (project management) oversees schedules, project coordination, capital purchases, and other project-execution activities, it is obvious they have a major influence on the safety process. As noted in Figure 4, management controls human parameters—knowledge, attitude, fitness, ability—hence unsafe acts are a result of supervision and the direction it provides.

The Navy Surface Weapons Center conducted a study that focused exclusively on management's role in injury accidents; it was based on the premise that "all accidents are indicators of management failures" (B. Fine). The hypothesis for the study included the following: During the incident investigation, there can always be found some degree of management involvement or activity that might in some way have prevented the incident. Therefore, it can be reasonably assumed that management will be responsible for the causes of every incident. The Navy study also investigated incidents that showed clearcut causes. It became evident that management opportunities were present that could have minimized or eliminated hazards that led to incidents.

Goals-Freedom-Alertness Theory

The *goals-freedom-alertness theory* relies on the premise that safe work performance is the result of a psychologically rewarding work environment. According to this theory, accidents are viewed as low-quality work behavior, occurring in an unrewarding psycho-logical climate that does not contribute to a high state of alertness. The theory states that management should provide a well-defined goal and should give the worker the freedom to pursue that goal.

Management must allow workers an avenue for achieving organizational goals. One researcher used the following analogy to describe the status of economic goals: "They need profits in the same way as any living being needs oxygen. It is a necessity to stay alive, but it is not the purpose of life (K. Thomas)." While profits and market share remain very important, this theory demonstrates the need for workers to achieve goals in order to maintain their focus.

UNIVERSAL CAUSATION MODEL

There are over twenty incident-causation models referenced in textbooks, journals, and other safety literature, in addition to those already mentioned in this chapter. Several safety professionals have expressed a desire for a universally accepted model that would have relevance over a specified range of desired application. Robert E. McClay addressed this issue in his article, "Toward a more Universal Model of Loss Incident Causation" (*Professional Safety*, 1989). His paper focused on a model that would strengthen the safety discipline and, more importantly, lead to improved strategies for assessing risks and controlling hazards.

The *universal model* is one of the more contemporary and comprehensive models of loss incident causation. It is so named because it can be used to better understand loss incidents of all types across the full spectrum of adverse human experience. Losses include events such as accidents, structural collapses, explosions, occupational illnesses, and hazardous material spills. There is no model that captures all incidents. McClay's model uses a generalized sequence of states and events leading up to the undesirable final effects (see Figure 5).

| Distal causal factors | → | Proximal causal factors | → | Point of irreversibility | → | Mitigating and aggravating factors | → | Loss incidents | → | Final effects |

■ ■ **Figure 5** ■ ■ **Loss sequence**

The *arrows* shown in the figure represent causation and the relationships are determined through the logical examination of experience, investigation, and analysis.

Final effects are a measure of the harm created by the loss incidents preceding these effects. Final effects can be expressed as injuries, number of fatalities, lost time, or cost. The severity of the incident is directly measured by the magnitude of the final effects.

A *loss incident* is an event, which, without any subsequent events, has the capacity for producing *adverse* final effects. Loss incidents can occur alone or in groups. One loss incident can lead to other loss incidents. For example, a "crane starts tipping over" example (a loss incident), together with other factors, eventually results in other loss incidents: "load drops onto truck" and "crane tips over." Each subsequent loss incident that occurs increases the severity of the final effects. When there are no adverse final effects resulting from a given loss incident, we classify that incident as a *near-miss*. No injury or damage may have resulted, but if the potential for damage and/or injury was there, it is still a loss incident.

Other factors that can affect the severity of the final effects are called *aggravating factors* or *mitigating factors*. Aggravating factors are states or events that make the final effects more severe than they otherwise would be. Mitigating effects keep the final effects from being more severe—an employee wearing personal protective equipment is a mitigating factor. It will not prevent the loss incident, but it does keep the final effects from being worse than they otherwise would be. Since these factors occur after the *point of irreversibility*, they are not called causal factors, but they are important nonetheless. Three types of states or events constitute mitigating and aggravating factors:

- human actions or inactions (an event)
- exceeding the functional limitations of system elements (an event)
- physical, chemical, or biological conditions (a state)

If the incident sequence reaches the point of irreversibility (POI), then a loss incident or series of loss incidents will, by definition, be unavoidable.

This is important for the loss-incident sequence as it separates causal factors from their effects. States and events occurring prior to the POI are regarded as causal factors; none of those states or events occurring after the POI is regarded as a causal factor. The POI has been reached if it is not possible to return the situation to its original condition without a major effort or expenditure. If the sequence can be stopped, and the situation returned to normal without the possibility of a loss incident, then the POI has not been reached.

Proximal causal factors are the hazards that are present at the time and location of the loss incident. These causal factors consist of the same three types of states and events that were identified under the aggravating and mitigating factors.

Physical, chemical, or biological conditions are the structures, equipment, and environment in which the employee functions. Everything that can be seen, smelled, or touched (states) fits into this category. Weather conditions are another example. People are included in this category, but not the actions they perform. The crane, load, operator, and rigging are examples of physical, chemical, or biological conditions. The position or location of these conditions is frequently either an important causal factor or an aggravating/mitigating factor in a loss-incident sequence (i.e., truck 75 feet away from the crane).

Human actions and inactions are proximal causal events that can cause a change in states and help bring the sequence closer to the POI. Examples of human action causal factors from the crane scenario are: operator cuts the load loose, delivery truck driver parks the truck 75 feet away from the crane.

The last proximal causal factor is the exceeding of any functional limitation. Every element in any system has one or more functions to perform as dictated by the system design and operation. Associated with each function are one or more limitations. If a functional limitation is exceeded, this constitutes a hazard, and is therefore a proximal causal factor. This factor has not been included in other incident-causation models, but is recognized by other professions. Reliability Engineering is one field that studies the limitations of nonhuman elements, and ergonomics is also an area that studies human limitations.

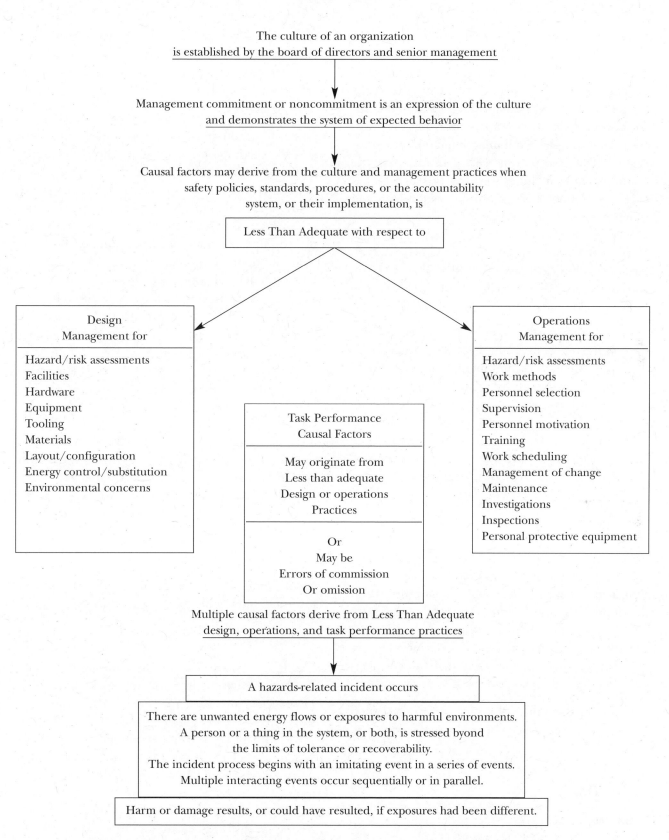

The culture of an organization
is established by the board of directors and senior management

Management commitment or noncommitment is an expression of the culture
and demonstrates the system of expected behavior

Causal factors may derive from the culture and management practices when
safety policies, standards, procedures, or the accountability
system, or their implementation, is

Less Than Adequate with respect to

Design
Management for

Hazard/risk assessments
Facilities
Hardware
Equipment
Tooling
Materials
Layout/configuration
Energy control/substitution
Environmental concerns

Task Performance
Causal Factors

May originate from
Less than adequate
Design or operations
Practices

Or
May be
Errors of commission
Or omission

Operations
Management for

Hazard/risk assessments
Work methods
Personnel selection
Supervision
Personnel motivation
Training
Work scheduling
Management of change
Maintenance
Investigations
Inspections
Personal protective equipment

Multiple causal factors derive from Less Than Adequate
design, operations, and task performance practices

A hazards-related incident occurs

There are unwanted energy flows or exposures to harmful environments.
A person or a thing in the system, or both, is stressed byond
the limits of tolerance or recoverability.
The incident process begins with an imitating event in a series of events.
Multiple interacting events occur sequentially or in parallel.

Harm or damage results, or could have resulted, if exposures had been different.

■ ■ **Figure 6** ■ ■ **A systematic causation model for hazards-related incidents** (Source: *On the Practice of Safety* by F. A. Manuele, Van Nostrand Reinhold, 1997; reprinted with permission)

Humans have several functional limitations because they have so many functions in system operation. Reaction time, grip strength, and the number of activities a person can perform at any given time are all examples of human functional limitations. *Knowledge of hazard exceeded* is probably the most common example of a human limitation being exceeded. Examples of this causal factor from the crane example include: strength of truck bed is exceeded, balance of crane is exceeded, and soil stability of trench is exceeded.

In *On the Practice of Safety*, F. A. Manuele proposed that a new name be created to encompass all hazard-related incidents—HAZRIN. The HAZRIN causation model (see Figure 6) includes the following elements:

1. An organization's culture is the primary influence concerning hazard development.
2. Management commitment is recognized as the extension of the organization's culture, and is the source of decision making that affects the elimination and control of hazards.
3. There must be a balance of considerations for causal factors that are derived from inadequate policies, standards, or procedures.

HAZRIN better defines the accident phenomenon, and it is better than terms such as "undesired events" or "accidents." These terms are used in the SH&E vernacular, but are not precise and could result in several interpretations. Manuele further states that a sound causation model for hazards-related incidents must identify and stress the significance of the design management aspects, the operations management aspects, and the task performance aspects of the causal factors, and that those aspects are interdependent and mutually inclusive.

Let's explore the following scenario and then determine the cause(s) of an unfortunate incident.

CASE STUDY

What Went Wrong?

The trumpet fanfare as the crane arrives on site is the car horns from the motorists behind the trac-

tor trailer carrying the crane, as it has difficulty getting from the street to the job site. Someone forgot to widen the access gate and grade the access ramp!

Serious Work

Once on the job site, it is time for some serious work and the rigging crew is installing the lattice boom sections. While the crew prepares, the crane oiler helps the operator set the counterweights on the crane. The last counterweight for the front bumper is being set. The operator booms up so he can get the counterweight closer to the bumper. Both the oiler and operator are watching the counterweight. Unfortunately, the operator does not see the overhead temporary electrical lines. The boom hits the lines and breaks them. Not only is the crane damaged, but also the job site is shut down because there is no power.

On day two, despite the crane crew's mistakes and a little rain, they are ready to go to work. The general contractor has shown where the crane is set up, and he tells the operator that there will be a crew pouring concrete next to the crane. He also points out several open utility trenches in the adjacent area. The superintendent's parting words are: "Now that you have delayed my job, you have to accelerate your work so I can get back on schedule—at no additional cost to me!"

Having an immense amount of pride, the crane operator and rigging crew are going to show the superintendent that they cannot only make up for lost time, but get ahead of the contractor!

Setting Steel

The lattice boom is set up near the building so it can easily reach the working areas. The first task is to set structural steel for the steel erectors. The structure is 120 feet high to ensure that the 250-ton crane with 170 feet of boom should not have any problems. The crane is set up so that the extended outriggers are within 10 feet of a 14-foot-deep utility trench.

As the day progresses, the crane operator is having no difficulty picking and setting the beams and columns. To impress his co-workers, he occasionally picks up several bundles of metal decking at once, and sets them on the structural frame-work. The superintendent, still fuming about the delay of the previous day, decides to remind the crane operator that the job is behind schedule be-cause of him. The crane operator passes the infor-mation along to his rigging crew. As a result, everyone is upset, and working faster than they should.

Foreman's Decision

The rigging foreman now decides to hook four steel beams to the load using four chokers. Now, the crane operator can pick four beams at a time, speed-ing up the ironworkers' work rate. Lifting the 5000-pound steel beams is no problem for the crane, even with four connected at a time. About 30 min-utes before quitting time, a delivery truck arrives with equipment that must be set in the building. The operation will take two hours, but the superintend-ent wants it done before the end of the day. After some persuasion, the operator agrees to make the lift. The only problem is that they do not have a spreader beam for the task. The rigging foreman says they can wrap the chokers with cardboard to act as softeners.

The delivery truck can only get within 75 feet of the building because the utility trench is be-tween him and the crane. Each piece of equipment weighs about 40,000 pounds, and the location in the building where the equipment will sit is 110 feet from the crane. As the crane rigs the first piece of equipment, the superintendent tells them it will be dark soon and that there are no lights for nighttime work.

Things Happen

The first piece of equipment is lifted from the truck. The crane operator is hurrying and raises the load about 35 feet above the bed of the truck. He is peeking over the side of the crane and starts to bring the boom back toward the crane to reduce the radius. All of a sudden, things start to happen:
the crane starts tipping
the utility trench collapses
the crane tips more rapidly

the crane operator cuts the load loose
the load (expensive equipment) falls on the bed of the truck and is damaged beyond repair
the truck bed collapses as the load and crane boom fall on it
the crane tips over.

What happened? What could have occurred, and what action could have prevented this unfor-tunate incident? What occurred in this scenario is common at many construction projects: many contractors working diligently to meet schedule demands, tempers flaring, safe work practices not followed, and many other critical factors.

Based on good safety and construction prac-tices, both the general contractor and the crane company should have analyzed the conditions and agreed on the correct working method. The fol-lowing describes what they should have been do-ing. The accidents described here can result, and have resulted, in the injury and death of workers. Through diligent preplanning, appreciating the operational procedures for the crane, and cooper-ating with contractors, accidents can be prevented and quality improved in the construction industry. Finally, the accident causes are complex, and the result of several interwoven variables.

The following four lists of "should haves" are taken from ASSE's *Construction Practice Specialty Newsletter* (Summer, 1994) and are reprinted here with permission.

THE CONTRACTOR'S COMPANY SHOULD HAVE:

- widened the gate opening for the crane
- placed the flagmen on the street to control the traffic while the crane entered the site
- determined the heaviest load to be lifted and the maximum distance for the crane to reach (radius)
- warned the crane crew of overhead electrical lines and any other potential hazards
- coordinated the work so the crane would not have to operate around open trenches or excavations or near any work where lifts

would have to be made over the heads of the workers
- graded or compacted the ground so the crane could set up on firm, solid ground.

THE CRANE COMPANY SHOULD HAVE:

- informed the contractor of minimum gate width needed
- asked for traffic control when they first walked the project
- reviewed the crane load charts to determine the proper crane for the weight and radius of the lifts
- looked for any hazards that could affect the setting up and operation of the crane
- reviewed the work so the crane could be set up where loads would not have to be lifted over the workers' heads
- taken into account the soil conditions, including the muddy conditions created by the rains. Use of large cribbing mats could have helped distribute the outrigger loads.

THE CONTRACTOR'S SUPERINTENDENT SHOULD HAVE:

- discussed with the crane crew how, if possible, they are planning to make up for the down time. An open statement about delay problems does not foster good working relationships.
- reviewed the crane company's proposed method for lifting and setting loads. When the crane company started lifting four beams one below the other in a manner that is called "treeing," the superintendent should have stopped work and discussed the potential safety problems.
- instructed the crane operator that no loads surrounding the crane manufacturer's load charts will be lifted.
- asked the crane operator (when the special equipment arrived) if he had the time and proper equipment for setting the equipment. Bullying the crane operator does not create good working relationships.

PRINCIPLES OF VARIATION

There are two kinds of accidents in the context of system operation. The distinction between them is in the type of cause involved.

- Type 1. The outcome is from common causes of variation.
- Type 2. The outcome is from a special cause.

This distinction is important to gaining a solid understanding of incident causation, and focusing on the wrong causes will result in system outcome recurrences.

It is frequently important to locate, estimate, and control major sources of variation in incident prevention. Of all the devices for analysis of data, perhaps the most valuable is the simple graph. Since understanding comes as a result of information properly communicated, the mere existence of information is not enough. Whether the objective is the control of a manufacturing plant, care of a patient, or a construction-site incident evaluation, it is not only important that appropriate information be collected, but also that this information be fed back into a readily understood form to those responsible for taking action.

To be informative, data must be displayed so that present and past experience can be readily compared, and concomitant variation in two or more impinging responses can be simultaneously considered. Appropriate plotting of data is never a futile effort as formal analysis can be conducted to prevent incidents. Evaluating data also frequently reveals unexpected characteristics that might otherwise be overlooked. The run chart is a simple statistical tool that involves plotting over a period of time. With the results of a statistically significant number of trial runs in hand, the result of each trail is plotted on a chart. This produces the basic run chart, a plot of test results over a series of tests. Walter Shewhart, a mathematician and scientist, studied variations for the future of the telephone industry during the 1920s. Shewhart studied variations and determined the average number of faults (variations) for the entire group of trials. The value was charted as a line around which the individual plots would be located. By differentiating between desirable and

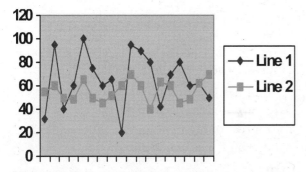

■ ■ **Figure 7** ■ ■ **Run chart with faults per 1000 inputs (average = 60)**

undesirable variations, one could improve the system. The objective would be to identify the factor that caused a condition above the average for the system. Next, the factor(s) would be eliminated or minimized to improve performance. The goal would be continuous improvement. Then, looking at Figure 7, the farther spread points would be closer to line 2, indicating increased optimal performance.

The dispersion of data utilizing the average, or $\bar{x}$ line, is the basis of deviation. Using three standard deviations from either side of $\bar{x}$ will include 99.7 percent of the values of a normal population. This became Shewhart's upper and lower control limits, limits that result from the system itself. A system operating within those limits is in normal operation for the condition it is in. The random variations occurring within these limits are normal, and are known as *common causes*. Accidents on the highway arise mostly from common causes; for example, driving under the influence of alcohol. Other common causes of highway accidents include, range of speed on the same road and unintelligible road signs.

When points are outside the control limits there is an indication that an abnormal event has occurred. The causes of these events are unusual and usually easy to identify. Initially they were called *assignable causes*, but later Deming referred to them as *special causes*. A process control chart with a normal variation is shown in Figure 8. Figure 9 shows that a special cause variation in the system has occurred.

The construction safety professional can also use data and these charts to measure the frequency of variation occurrences. Eighty-five percent or more of the variations that occur in systems that are operating smoothly are due to common causes

that arise from the system. Therefore, in order to prevent and/or minimize incidents, one must improve the system by minimizing undesirable variations. Only project management can effect changes to control the system.

It is easier to address special causes than common causes. Special causes are usually readily identifiable (e.g., malfunction of a crane; new, untrained pipefitter; heater-box explosion). Controlling common causes—those causes of variation built into the system or process—is a more complicated issue. In those cases, things only change if the system itself is changed. Unlike special causes, where the worker can institute changes to correct a hazardous environment, common causes require project management intervention to correct things like poor scheduling, high worker turnover, inadequate training, or bad tools. A system operating with no special causes (any points outside control limits) is said to be operating under statistical control.

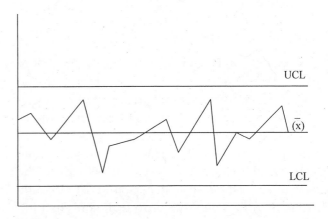

■ ■ **Figure 8** ■ ■ **Normal operating system**

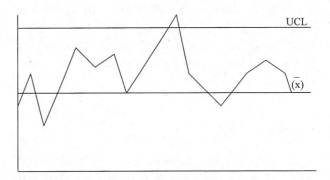

■ ■ **Figure 9** ■ ■ **System with a special cause incident**

In *Total Quality Safety Management* (E. Adams), five significant variation facts were outlined:

1. Variation is normal in every system. (Therefore, one must question the statement, "All accidents are preventable.")
2. The causes of variation lie either within the system (common causes), or outside the system (special causes).
3. Common causes arise out of the characteristics of the system, which are determined by management and can only be corrected by management action to improve the system. Workers have no control over common causes.
4. When a system that is running consistently within its upper and lower control limits under statistical control is left untouched, the variations that occur are due to common causes.
5. The 85–15 rule of system variation applies: In a normal system, 85 percent or more of variations are due to common causes; 15 percent or less are due to special causes.

As stated in Deming's *Out of the Crisis*, no matter what effort is put into a system, whether in manufacturing, construction, maintenance, or service, it will not be free of accidents. Figures on accidents do nothing to reduce the frequency of accidents. The first step in their reduction is to determine whether the cause of an accident belongs to the system or to some specific person or set of conditions. Statistical methods provide the only method of analysis to serve as a guide to the understanding of accidents and their reduction. The system guarantees an average frequency of accidents will occur at unpredictable places and times. Accidents that arise from common causes will continue to happen with their expected frequency and variations until the system is corrected.

SEVERE INCIDENTS

Up to this point, the chapter has primarily addressed total incidents. The construction safety professional must evaluate the severe injury potential separate from the gross of all incidents (including near-miss incidents). Studies in recent years suggest that severe injuries are fairly predictable in certain situations (Manuele). Those situations include:

- unusual, nonroutine work
- nonproduction activities
- sources of high energy
- certain construction situations.

All are examples of conditions that are prevalent on construction projects; each different in size and scope. When a system is operating within its upper and lower control limits, the potential for a severe incident is minimized. When equipment fails or an explosion occurs, severity potential is greatly increased. Equipment failure and an explosion are examples of special causes.

For the most part, hazards and risks that provide severe injury potential are identifiable. Individuals charged with construction safety responsibility must address these hazards/risks in the design processes for buildings and in the design of the work methods. Too many times someone given safety responsibility assumes that if efforts are focused on frequently occurring incidents, then severe injury potential will also be addressed. This approach will actually result in severe injury potential being overlooked since the types of incidents that often produce severe injuries or illnesses are rarely equated to incidents that occur frequently.

CONCLUSION

Several incident-causation theories have marked the safety landscape for a number of decades. Thorough analysis usually demonstrates the need for several incident-causation models and causal tracing techniques. A single model or theory cannot cover all potential causal categories. While safety professionals, academia, and regulatory agencies have supported the traditional theories, there is a growing need for a universally accepted model. The need for such a model has been widely supported in recent years. However, theories alone cannot explain the causes of construction incidents. Principles of variation, cause-and-effect models, FMEA, design, scheduling, and overtime each play an important role in the understanding and interaction of construction incidents. It is also important to understand the importance of management controls relative to actions and decision making at job sites. Finally, construction professionals must recognize their role during the design stage, and how they can prevent incidents during construction.

REFERENCES

Adams, E. *Total Quality Safety Management*. Des Plaines, IL: American Society of Safety Engineers, 1995.

Ammerman, M. *The Root Cause Analysis Handbook*. Portland, OR.: Productivity, Inc., 1998.

Barker, J. A., *Paradigms: The Business of Discovering the Future*. New York: Harper Business Publishers, Inc., 1993.

Cohen, H. and D. Cohen. "Human error: Myths about mistakes," *Professional Safety*, Oct. 1991: 32–35.

Deming, W. E. *Out of the Crisis*. Cambridge, MA: MIT Press, 2000.

_____. *The New Economics: for Industry, Government, Education*, 2nd ed. Cambridge, MA: The MIT Press, 1994.

Driskell, J. E. and E. Salas. *Stress and Human Performance*. Mawah, NJ: Lawrence Erlbaum, 1996.

English, F. "A post-structural view of the grand narratives in educational administration." Organizational theory dialogues. Bloomington, IN: Organizational Theory SIG (AERA), Indiana University, 1993.

Fine, B. Technical Report 75-104. *A Management Approach in Accident Prevention*. Naval Surface Warfare Center, July 1975.

Friend, M. and J. Kohn. *Fundamentals of Occupational Safety and Health*. Rockville, MD: Government Institutes, 2001.

Funk & Wagnalls New Encyclopedia, Volume I. New York: Funk & Wagnalls Corporation, 1996: 98–99.

Griffion-Fucco, M. and F. Ghertman. "Data Collection on Human Factors," *New Technology and Human Error*. New York: John Wiley & Sons, 1987.

Harrell, R. "What Can Go Wrong?" *ASSE Construction Division Newsletter*, Summer 1994.

Hill, D. "Time to Transform? Assessing the Future of the SH&E Profession," *Professional Safety*, Nov. 2002: 18–26.

Hinze, J. *Construction Safety*. Upper Saddle River, NJ: Prentice-Hall, Inc., 1997.

Holmes, T. and R. Rahe. "The Social Readjustment Rating Scale," *Journal of Psychosomatic Research*, no. 2, 1967: 21318.

Lack, Richard. *The Dictionary of Terms Used in the Safety Profession*, 4th ed. Des Plaines, IL: American Society of Safety Engineers, 2001.

MacCollum, D. *Construction Safety Planning*. New York: Van Nostrand Reinhold, 1995.

_____. "Time for change in Construction Safety," *Professional Safety*. Feb. 1990: 17–23.

Manuele, F. A. *Heinrich Revisited: Truisms or Myths*. Itaska, IL: National Safety Council, 2002.

_____. *On the Practice of Safety*. New York: Van Nostrand Reinhold, 1997.

_____. "Severe Injury Potential: Addressing an Overlooked Safety Management Element," *Professional Safety*. Feb. 2003: 26–31.

McClay, R. "Toward a More Universal Model of Loss Accident Causation," *Professional Safety*. Jan./Feb. 1989.

Miller, J. C. *Controlling Pilot Error: Fatigue*. New York: McGraw-Hill, 2001.

Mintz, A. and M. Blum. "A re-examination of the accident proneness concept," *Journal of Applied Psychology*, 33: 195–211, 1949.

Nighswonger, T. "Depression: The Unseen Safety Risk," *Occupational Hazards*. April 2002: 38–42.

Center to Protect Workers' Rights. *On Center* newsletter, "Overtime May Be Tied to Construction Injuries," vol. 2, no. 1, Nov. 2002.

Petersen, D. *Techniques of Safety Management: A Systems Approach*, 4th ed. Des Plaines, IL: American Society of Safety Engineers, 2003.

Saas, R. and G. Crook. "Accident Proneness: Science or Non-Science?" *International Journal of Health Sciences*, Vol. 11, No. 2, 1981.

Stamatis, D. H. *Failure Mode and Effect Analysis: FMEA from Theory to Execution*. Milwaukee, WI: ASQ Quality Press, 1995.

Schulzinger, M. *Accident Syndrome*. Springfield, IL: C. Thomas, 1956.

Thomas, K. *Intrinsic Motivation at Work*. San Francisco: Berrett-Koehler Publishers Inc., 2000.

REVIEW EXERCISES

1. How does the term *incident* differ from *accident*? Explain the importance for differentiating these two terms when addressing job-site safety.
2. What is the difference between a theory and a model? Considering the many traditional incident-causation theories/models, does the safety discipline need a universal causation model? Support your answer.
3. What are the causes of human error? Discuss practical applications to implement at a construction site to minimize these losses.
4. What is the role of project management in incident causation? Discuss who has greater control in eliminating construction safety incidents—project manager or construction worker.
5. Discuss how an understanding of the principles of variation can support or disprove the argument that "all accidents are preventable."
6. Root-cause analysis and FMEA have been tools used in the quality discipline for many years. Explain how these tools can assist the construction safety professional.

■ ■ **About the Author** ■ ■

HIGH COSTS OF WORKER INJURIES
IN CONSTRUCTION
by Darryl C. Hill, CSP

Darryl C. Hill, CSP, is Director of Safety & Health for ABB Inc. He earned a B.S. in Occupational Safety from Iowa State University and an M.S. in Hazardous Waste Management from Wayne State University (Detroit). Mr. Hill is currently pursuing a Ph.D. in Educational Leadership at Oakland University where he is an adjunct instructor teaching Construction Safety & Environmental Standards. He also is an Oakland University Safety & Health Program Advisory Committee member. He is on the ASSE Board of Directors, serving as Region VII Vice-President, 2001–2005. Mr. Hill is an elected officer for the Michigan Safety Conference and Safety Council for Southeast Michigan. He has served as (ASSE) Construction Practice Specialty Administrator and was named Construction Practice Specialty Safety Professional of the Year in 1997 and 2003. In 1997 Darryl Hill was also named the Edgar Monsanto Queeny Safety Professional of the Year.

CHAPTER **4**

High Costs of Worker Injuries in Construction

LEARNING OBJECTIVES

- Discuss the five components that most impact workers' compensation costs.
- Explain direct and indirect costs and the relationship these costs have to construction project profitability.
- Discuss the workers' compensation system and its relationship to overall injury costs.
- Explain how the premiums for workers' compensation are calculated.
- Discuss fraud in workers' compensation and any applicable warning signs.

INTRODUCTION

A shift is appearing on the safety landscape. Safety professionals are realizing that communicating the need for a safety process solely for humanitarian and regulatory reasons is no longer sufficient. There is a growing trend toward integrating safety into the financial goals of an organization. A safety professional should realize that understanding financial principles is as important as having thorough knowledge of OSHA regulations. The reason is simple: Businesses are looking for every strategic advantage to stay competitive. Company executives are exploring every facet of the organization and determining which functions add value and which should be eliminated or outsourced.

According to a Liberty Mutual Study (*Safety + Health*), the cost of disabling workplace injuries and illnesses grew faster than inflation between 1998 and 2000, while direct costs of workers' compensation claims were concentrated in a relatively small number of injury cases. The Liberty Mutual Workplace Index found that direct costs of claims from disabling work-related injuries grew 8.3 percent (2.5 percent after inflation adjustment) between 1998 and 2000. This equates to $42.5 billion in direct claim costs. The index found that a small percentage of workers' compensation claims are responsible for the majority of direct costs. During 2000, disabling workplace accidents accounted for 18 percent of workers' compensation claims, but 93 percent of direct costs.

The $42.5 billion in direct costs is only a fraction of the total cost of work-related injuries. Indirect costs (discussed later) may significantly increase this amount. The study suggests that the total financial impact of disabling injuries is between $170 and $255 billion.

CONSTRUCTION INJURY COSTS

Costs are incurred whenever an accident occurs at a construction site. Injuries can result in substantial costs and have a significant impact to the "bottom line" of a financial ledger. Figure 1 shows employer spending on workers' compensation by industry, as a percentage of payroll (U.S. Bureau of Labor). Construction/mining is the highest at $5.17 billion.

There is little documented information on the costs of implementing safety processes in construction. Estimates of administering a construction safety and health process range from 2 to 4 percent of direct labor costs. Business Roundtable's data for the cost of such a program for the construction industry is approximately $2 billion.

Figure 2 illustrates the impact of construction injuries. Workers' compensation and liability costs, business, society, and the worker are all affected.

Workers' compensation costs are a result of job-site injuries that significantly impact project profitability. Injuries may be medical only or necessitate indemnity payments to the worker. Liability may be incurred based upon the nature of the incident. There is a growing concern that an employee can "go around" the workers' compensation system and seek *uncapped* benefits under the owner or general contractor's general liability policy. Construction

companies and other interested parties are paying closer attention to insurance and contract terms, such as risk transfer, additional insureds, indemnification, and hold-harmless obligations (see Figure 3). Based upon the number of workers' compensation claims and liability from lawsuits plus penalties, a business may be adversely affected. Profit margins are impacted, public image is damaged, and, if severe enough, jobs may be lost. Finally, the worker has a decrease in wages, which lessens the buying power for the consumer. The more than 194,000 annual injury and illness cases with days away from work in construction mean losses not only to workers, but also to their families, employers, and society (Center to Protect Workers' Rights). Some of the costs are in wage replacement and medical treatment [known as direct (billable) costs] that can be measured. However, these workers' compensation payments are only a small amount of injury- or illness-related expenses. A family member may have reduced income as a result of having to stay home to watch an ill or injured family member. Thus, there are several elements that contribute significantly to construction costs due to injuries and illnesses. Figure 2 also illustrates how the five components impact one another. A conclusion can be made that the final cost of one injury can be significant based on these five major components.

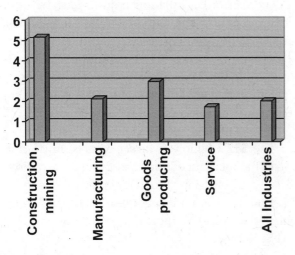

■■ **Figure 1** ■■ **Workers' compensation as a percentage of payroll** (Source: U.S. Bureau of Labor Statistics)

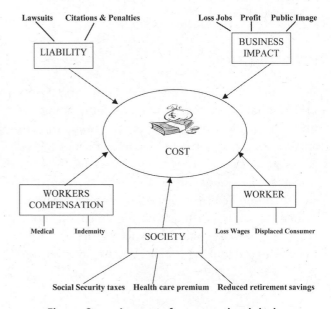

■■ **Figure 2** ■■ **Impact of construction injuries**

INSURANCE/CONTRACT TERMINOLOGY

Additional insured – a person or entity that can make claims against an insurance policy that was procured and paid for by someone else.

General liability insurance policy – a type of indemnity contract that obligates one party, the "insurer," to pay any sums that other parties, the "insureds," become legally obligated to pay as damages to third parties who suffer bodily injury or property damage, usually because the damage was caused by an insured or someone for whom the insured is responsible.

Indemnify – to save harmless; to pay for injury or damage; to fully reimburse an entity that experiences a loss.

Named insured – the person or entity to whom an insurance policy is issued, generally referred to in policy as "you" or "your," and who is the primary indemnitee under the insurance contract with reference to whom all other insureds are identified.

Risk transfer – an agreement that the financial burden of certain accidental losses that have not yet occurred, but which may occur in the future, will be shifted from one party or its insurer to another party or its insurer.

Third-party-over lawsuit – a lawsuit by an employee or an independent contractor for bodily injuries that occurred within the scope of employment, against a third party other than the employer.

Waiver of subrogation – contracting terms in which parties relinquish any right they would otherwise have to make claims against each other for damages that are covered by insurance.

■■ **Figure 3** ■■ **Insurance and contract definitions for construction safety professionals**

While a bad public image and displaced consumers may be difficult to quantify, profits, lawsuits, and penalties are easily understood in boardrooms. It is incumbent upon project leadership to realize that construction injuries can have catastrophic consequences.

Direct and Indirect Costs

Injury costs can be categorized as either direct or indirect. *Direct costs* are attributed primarily to indemnity and medical costs. Indemnity is also known as loss wages for injuries arising out of the scope of employment. In essence, direct costs are covered by workers' compensation insurance. These costs are well understood and can be quantified with accuracy. Profit margins are impacted as illustrated in Table 1. The construction safety professional will be able to demonstrate the amount of extra sales or revenue needed to offset the direct costs of an injury.

Published estimates of the total cost of nonfatal injuries for all injuries in the United States range from $131.2 to $145 billion per year (National Safety Council, Archives of Internal Medicine). The insurance industry and those who underwrite workers' compensation insurance have collected data that demonstrate the significant financial burden resulting from injuries. Liberty Mutual has compiled a list of the ten causes of injuries and illnesses that cost the most in wage replacement and medical payments. Table 2 illustrates several leading causes related to the cost for loss wages and any medical payments, also known as direct costs.

■■ **Table 1** ■■

Sales Needed to Cover Direct Costs of Injury—Based on Profit Margins

Direct Costs of Injury	Profit Margin				
	1%	2%	3%	4%	5%
$1000	$100,000	$50,000	$33,000	$25,000	$20,000
$5000	$500,000	$250,000	$167,000	$125,000	$100,000
$10,000	$1,000,000	$500,000	$333,000	$250,000	$200,000
$25,000	$2,500,000	$1,250,000	$833,000	$625,000	$500,000
$100,000	$10,000,000	$5,000,000	$3,333,000	$2,500,000	$2,000,000

For a $25,000 injury, a company with a 4% profit margin will need $625,000 in additional sales to offset the injury costs.

■ ■ **Table 2** ■ ■

Causes of Injury/Illness and Associated Costs

Injury/Illness	Estimated Direct Costs ($ billions)
Overexertion	10.3
Fall on same level	4.6
Bodily reaction	3.8
Fall to lower level	3.7
Struck by	3.4
Repetitive motion	2.7
Highway accident	2.4
Struck against	1.7
Caught in/compressed	1.6
Temperature extremes	0.4
Other	5.5

Source: Center to Protect Workers' Rights, *The Construction Chart Book*, 3d ed., 2002.

The ten leading causes listed were reportedly responsible for $34.5 billion, or 86%, of the total $40.1 billion paid by employers in 1999. While there is no breakdown available for construction, the items listed match those of greatest concern in construction, and 32% of the total estimated costs involved ergonomics (repetitive motion and overexertion).

Indirect costs (see Figure 4) to an employer for a construction-site injury can be enormous and have a significant financial impact. Usually indirect costs are associated with the injured construction worker. When the worker is injured, it is common practice for that worker to continue to be compensated (per state requirement) while receiving medical treatment. First, there is the cost for transporting the injured worker to receive treatment. Unless the injury is severe, a project worker will usually accompany the injured worker until treatment is received, adding the cost of (1) wages paid to the transporting worker and (2) the company vehicle being used. A worker with a minor injury may return to work within one to three hours, depending on the distance to the medical facility, traffic, and the number of people at the medical facility. Once the worker returns, he or she will probably have to assist in completing an injury report. Co-workers may want to discuss the circumstances regarding

---

INDIRECT COSTS

■ Fines
■ Lawsuits
■ Damaged equipment and its repair or replacement costs
■ Production delays
■ Loss of productivity
  • The injured worker at time of injury
  • Job shutdown at time of injury
  • Injured worker's reduced capacity upon return to work
  • Co-workers' involvement, including assisting the injured worker and training a replacement worker
  • Co-workers who are shorthanded following an injury
  • Administrative personnel completing and submitting required paperwork
  • Supervisor/management time hiring or retraining a replacement worker
  • Management time investigating and reporting incident to government, insurance, and news media representatives

■ ■ **Figure 4** ■ ■   **Indirect costs of construction injury**

the incident once the worker returns. If the worker does not return to work the day of the injury, productivity is hampered by work assignments being rearranged. If the injury is severe, workers may feel uneasy about the sequence of events, which will affect productivity further.

WORKERS' COMPENSATION

To appreciate today's workers' compensation system, look at recent history. Not long ago, the serious injury of an employee could bankrupt a business or an employee. Until 1911, the law did not mandate workers' compensation coverage in the United States. Injured employees had to take legal action against their employers to collect compensation. The difficulty of this procedure and the high risk of lawsuits for employers created a general dissatisfaction with the system. Beginning in the first decade of the twentieth century, each of the fifty states and the

District of Columbia began passage of workers' compensation legislation that varies widely in coverage and benefits but that, in general, provides employers with immunity from lawsuits in exchange for payments to the affected worker or dependent survivor. In 1911, Wisconsin enacted the first workers' compensation laws. Immunity can be waived in cases in which the employer is not in compliance with relevant OSHA standards, or where gross negligence or criminal activity can be proven.

Workers' compensation is a no-fault insurance for occupational injuries and illnesses. The idea of providing workers with this kind of special treatment for workplace injuries and illnesses spread. Today, all fifty states have their own workers' compensation laws in effect. This compensation pays the benefits of covered employees for job-related illnesses, injuries, and deaths. Benefits include medical expenses, death benefits, lost wages, and vocational rehabilitation. Failure to carry workers' compensation coverage leaves an employer out of compliance with the law and vulnerable to paying all of the benefits in addition to possible fines.

Claims arising from work-related injuries are usually the result of one-time events where neither the existence of the injury nor the fact that it is work-related is doubted (e.g., a 30-foot fall from a fire escape that collapsed). Other cases may be less obvious: A laborer who continually must carry heavy containers and dispose of trash alleges low back pain. The employer questions the worker about what activities he did over the weekend. In addition, low back pain cannot be substantiated with objective medical evidence.

Occupational illnesses are more difficult to assess, and only a small number of cases of work-related diseases are thought to be covered by workers' compensation. It is estimated that occupational diseases claim the lives of an estimated 50,000 to 60,000 workers each year (AFL-CIO *Fact Sheet*). Occupational illnesses are rarely the result of a single identifiable incident, and many, like cancer, may be diagnosed many years after exposure to the causative agent. Therefore, many costs are not compensated, partly because they are difficult to correlate to specific work exposures. Construction workers may have several employers in a specified period. Musculo-

skeletal disorders arising out of the scope of employment can be costly in expenses and human suffering. The disorder is a result of repetition over several months or years. Another example of an illness that can be hard to substantiate is carpal tunnel syndrome, characterized by a gradual onset that may leave doubt about whether a case is definitely occupational in nature. Finally, it is important to understand that musculoskeletal disorders for record-keeping purposes may be classified as an injury or "other illness," making it a challenge from a statistical and data-gathering perspective.

The National Academy of Social Insurance (Mont) estimates that approximately $40 million in benefits was paid out annually from 1993 to 1996, although other estimates are as much as $70 billion to $100 billion (Leone and O'Hara). A number of states suggest that workers' compensation is underutilized, however, with as much as 30 to 60 percent of work-related fatalities not found in workers' compensation records (Cone, Stout and Bell, Leigh). Reasons why worker fatalities may not be properly reported are outlined in Figure 5.

Parker (1994) discovered that 67 percent of eligible injuries were not reported to the workers' compensation system. Reasons presented for the underutilization include workers' desire to be part of the "team," workers' not aware of their rights, and fear of retaliation from an employer considering the financial benefit of minimizing claims.

Assigned risk plans are pooling mechanisms, mandated by some states, which ensure that all employers needing workers' compensation insurance

---

Why worker fatalities may not be reported:

- Family-owned business
- Incident occurred on a small farm
- Employer didn't provide the appropriate insurance
- Employment of independent contractors
- Ignorance of (or no regard for) reporting requirements

■ ■ **Figure 5** ■ ■ **Reasons construction fatalities go unreported**

Monopoly States	Option States
A business must use the monopoly state fund for workers' compensation if it is doing business in: • North Dakota • Ohio • Washington • West Virginia • Wyoming • Puerto Rico & U.S. Virgin Islands	A business has the option of using the state fund or a private insurance company if it is doing business in: • Arizona • California • Colorado • Idaho • Maryland • Michigan • Minnesota • Montana • Nevada • New York • Oklahoma • Oregon • Pennsylvania • Utah

■ ■ **Figure 6** ■ ■ **Workers' compensation funding by state**

can get it, even if private insurers turn them down as an undesirable risk (see Figure 6). In an assigned risk plan, the employer still gets a workers' compensation insurance policy, and coverage for the employees is identical to coverage from a private insurance policy. The individual insurance company pays the claims out of a pool, instead of insurance funds. State government assessments on insurance companies doing business in the state as well as the premiums paid by those companies in the plans typically fund this pool.

Definitions

Understanding key workers' compensation terms is important when managing safety at a construction site. Even though somebody may handle this responsibility at the corporate office or another location, the construction safety professional should have a basic understanding of the following terms.

Assigned risk pool: the insurance pool where policies are generally written for companies that are perceived to be too high of a risk for the voluntary market.

Audited premium: the premium the insurance company comes up with after determining actual payrolls for the policy period.

Credits/Debits: the sometimes subjective process that affects premiums after classifications rates and experience rating are applied.

Declaration page: page near the front of the company's workers' compensation policy that details all factors that go into the calculation of the company's premium.

Estimated premium: the premium developed using estimated payrolls, which applies at the policy's beginning.

Experience modification factor (EMF): an adjustment to the manual premium, based on prior years' losses and the payrolls of a particular insured company compared with the average claims paid out in that industry.

Manual premium: the premium that results from multiplying payrolls by the rates for the various classifications.

Modified premium: the premium that results from applying the experience modification factor to the manual premium.

Paid versus incurred losses: paid losses refer to the actual dollars of claim payments made by an insurer to policyholders in a specified time frame. Paid claims may be related to policies from any number of earlier periods. *Incurred losses* are paid claims plus changes in loss reserves during a given period. Incurred losses better reflect the loss experience associated with exposures during a given period because of changes in loss reserves, which may have been established from earlier periods.

Premium discount: a legal discount that is state determined; it is based on the size of the standard premium.

Remuneration: payment of salary or equivalent for work performed. It may include gross wages and salaries, commissions, holiday pay, vacation pay, bonuses, substitutes for money payments, and overtime pay. (There are exclusions for overtime.)

Schedule credit/debit: a discretionary adjustment to the modified premium, based on underwriting decisions by the insurance company.

CLASSIFICATION CAN IMPACT PREMIUMS

- More than one business entity may be insured on a policy; one of the businesses may be eligible for a separate and lower classification.
- The insurance company may be using the wrong classification if neither it nor the rating bureau has inspected the business.
- The description of the company's operations in the inspection report may be inaccurate.
- If a part of the company operations are separate and distinct from the rest of the operations and not a normal part of the primary business, the separate part may be eligible for a lower rating classification.
- There may be an unusual exception to an NCCI classification rule in a state that has been overlooked by either an insurance company and/or an agent.
- Even if the rating bureau has inspected a business, classification definitions may have changed.

▨▨ **Figure 7** ▨▨ **Classification factors that impact a premium**

Standard premium: the premium that results when an insurance company applies schedule credits and debits to the modified premium.

Voluntary market: all workers' compensation policies written outside of the assigned risk pool. An insurer writes policies voluntarily.

Rating Bureau

The majority of states have their classification system implemented by the National Council on Compensation Insurance (NCCI). The NCCI is a centralized rating bureau funded by member insurance companies. It compiles information used to price workers' compensation insurance. The NCCI has developed an elaborate system of classifications for businesses, which it continually updates to reflect changes in business practices. In most states the NCCI is responsible for determining which classifications should be used on the workers' compensation policy for a particular insured company. A review can be requested if the member insurance company does not agree with the NCCI's decision. The insurance company must abide by the NCCI's final decision. Although a business may be located in a non-NCCI state, independent rating bureaus often use the NCCI system as the basis for their own system.

There are, however, many instances where a company's classification may be wrong. Items that affect premiums and that a contractor should be aware of are outlined in Figure 7.

Loss Ratios

The loss ratio is the employer cost measure most frequently utilized in empirical analyses of the impact of insurance regulation. The term has several alternative definitions. Figure 8 provides a general guide to workers' compensation underwriting experience. The list is general because it starts with *premium* and does not specify whether the losses are paid or incurred. The alternatives in the measure of premiums, in the method of reporting data, and between paid and incurred benefits can be significant. A general understanding of ratios used to analyze underwriting results in workers' compensation insurance will allow the construction safety professional to effectively communicate the ramifications of job-site injuries. Loss ratios will be an indicator of employers' costs.

WORKERS' COMPENSATION UNDERWRITING EXPERIENCE

1. Premium
2. − Losses
3. − Loss adjustment expenses
4. − Underwriting expenses (commissions and brokerage expenses; state and local insurance taxes; licenses and fees; general expenses; and other operating expenses)
5. − Dividends
6. = Underwriting results
7. +/− Net investment gain/loss and other income
8. = Overall operating results prior to state and federal income taxes

▨▨ **Figure 8** ▨▨ **Determining underwriting experience**

Using the terms in Figure 8, one can determine underwriting expenses as follows:

- Loss (sometimes called pure loss) ratio = line 2 ÷ line 1
- Loss plus loss adjustment expenses ratio = (line 2 + line 3) ÷ line 1
- Combined ratio = line 6 ÷ line 1
- Overall operating ratio = line 8 ÷ line 1

Losses normally are the largest cost component of net premiums. Therefore, the loss ratio is sometimes used as a crude proxy for the inverse of insurer profits. Basically, the higher the loss ratio, the smaller the portion of premium that is left for other expenses and profits. Also, the inverse of the loss ratio is sometimes used as a measure of workers' compensation costs in a state. For example, if losses per $100 of payroll are the same in two states, then a relatively higher inverse loss ratio in one state means that premiums per $100 of payroll (employer costs) are also higher in that state.

Danzon and Harrington used average losses per $100 of payroll and its growth over time as workers' compensation cost measures. The data is based on NCCI estimates of incurred losses (cash, medical, and total benefits). In essence, they are projections of the ultimate cost of claims with accident dates in a policy year, using data from the "first report" of loss experience that, in turn, are extrapolated us-

ing historical trends to the "ultimate report" basis. Projections of the eventual (ultimate) cost of a claim will be modified as updated loss data become available. Furthermore, the data does not include all of the costs of workers' compensation insurance to employers, such as underwriting expenses and carrier profits.

Premium Calculation

Generally, workers' compensation premiums are calculated using a three-step process:

1. The manual premium is determined.
2. The experience rating adjusts the manual premium.
3. Miscellaneous credits and debits are applied.

An example is provided in Table 3.

Workers' Compensation Fraud

The workers' compensation system provides medical and income benefits for injured workers and helps protect employers from the potentially devastating liabilities of medical costs and lawsuits. When used properly, the system provides benefits to the interested parties involved. But unfortunately, there are employees, employers, medical providers, and those in the legal profession and the insurance industry who abuse the system to the detriment of

■■ Table 3 ■■
Workers' Compensation Premium Calculation

Classification	Payroll	Rate (per $100)	Premium
Clerical	$100,000	$0.28	$280
Laborer	$250,000	$2.08	$5200
Ironworker	$250,000	$4.34	$10,850

Manual Premium	$16,330
EMR	0.84
Modified Premium	$13,717
Premium Discount	0%
Association Credit	10%
Scheduled Credits	15%
Total Credits	25%
Net Premium	$10,228

those it was meant to protect. Their actions can add millions of dollars to the cost of the system—dollars that could be used to lower premiums or improve a company's safety process. Florida investigated workers' compensation fraud extensively and discovered that 63.5 percent of all contractors surveyed had experienced fraudulent workers' compensation claims (Coble). Thus, owners, contractors, and construction firms are giving increased attention to workers' compensation fraud. Because fraud raises everyone's costs, insurance providers, contractors, and regulatory agencies have increased their efforts to find and prosecute those responsible for fraud.

Medical providers, employees, employers, and insurance agents can all be guilty of workers' compensation fraud, which can take on a variety of forms. Whenever, someone intentionally makes a false statement to an insurer to receive benefits or coverage they are not legally entitled to, or knowingly misrepresents information necessary for the insurer to calculate an accurate premium, workers' compensation fraud has been committed.

Many construction companies have established formal training programs to help reduce fraudulent claims. These training programs may include educating project managers, superintendents, and foremen in how to recognize suspicious activities or circumstances related to job-site incidents. Many large contractors provide information related to the signals of fraudulent claim activity to their employees. These include: project newsletters, periodic toolbox meetings, and paycheck inserts. Warning signs for workers' compensation fraud include:

The worker

- claimant has a history of reporting subjective injuries (back pain, CTDs)
- injured worker often changes jobs
- injured worker has been recently terminated or demoted
- injured worker continuously makes demands
- injured worker calls soon after injury and asks for quick settlement of the case
- injured worker relocates after injury and/or is difficult to contact
- claimant is extremely uncooperative

- the worker asserts he/she cannot work while medical provider indicates a full recovery

The injury

- the incident occurs just prior to work termination, layoff, or near the end of the worker's probationary period
- there were no witnesses to the incident
- the incident is not reported in a timely manner and/or per company's reporting policy
- medical providers have vastly differing opinions regarding the injured worker's disability
- injured worker is claiming disability exceeding that which is normally consistent with such an injury/illness
- the incident occurs in an area where the injured employee would not normally be
- the incident occurs late Friday afternoon or shortly after the employee reports to work on Monday
- the task that caused the incident is not the type that the employee should be involved in
- details of the incident are contradictory and/or vague

Medical status

- claimant frequently changes medical providers/physicians or changes medical provider when a release for work has been issued
- claimant frequently misses physician's appointments

Legal issues

- the attorney representation letter is dated the day of the reported incident
- claimant is unusually familiar with workers' compensation claim-handling procedures and laws
- claimant's attorney threatens further legal action unless a quick settlement is made
- there is an unusually high number of applications from a specific firm

■ claimant's attorney inquires about a settlement during the early stages of the claim

External factors

■ Co-workers allege the injured worker is either working elsewhere or is active in recreational activities

■ injured worker is exaggerating an injury in order to get time off to work on personal interests

■ injured worker is seasonally employed, making it attractive to be "injured" during the off-season

■ injured worker leaves different daytime and evening telephone numbers

■ injured worker is never home or is always busy

CONCLUSION

As safety professionals, we must understand the financial directives presented to senior management to achieve economic goals and demonstrate a profit to shareholders. Demands placed on organizations will only increase, and a safety professional must function successfully within the financial framework. While the construction safety professional confronts difficult issues daily, they must be perceived as problem solvers who also assist the company in meeting its financial obligations. Successful construction safety professionals will not only be problem solvers in areas such as fall protection, scaffolding, and crisis management, but also be knowledgeable about financial measures specific to their company.

REFERENCES

AFL-CIO. "Facts about Worker Safety and Health." *Fact Sheet.* Washington DC: AFL-CIO, Dept. of Occupational Safety and Health, April 2001.

Burton, J., Jr. and F. Blum. "Workers' Compensation Costs in 2000: Regional, Industrial, and Other Variations," *Workers' Compensation Policy Review,* 3–11 (based on U.S. Bureau of Labor Statistics data). Brentwood, TN: M. Lee Smith Publishers, LLC, May/June 2001.

Business Roundtable. "Improving Construction Safety Performance," *Report A-3.* Houston, TX: Business Roundtable, January 1982.

Center to Protect Workers' Rights. "Workers' Compensation and Other Costs of Injuries and Illnesses in Construction," *Construction Chart Book,* 3rd ed. Silver Springs, MD: Center to Protect Workers' Rights, 2002.

Coble, R. *A Study of Fraudulent Workers' Compensation Claims.* Building Construction Industry Advisory Committee (BCIAC), Technical Publication no. 92. Gainesville, FL: University of Florida, 1994.

Cone, J. E., A. Dapone, D. Makofsky, R. Reiter, C. Becker, R. J. Harrison, and J. Balmes. "Fatal injuries at work in California," *Journal of Occupational Medicine,* 33:813–817, 1991.

Danzon, P. and S. Harrington. *Rate Regulation of Workers' Compensation Insurance: How Price Controls Increase Costs.* Washington DC.: American Enterprise Institute, 1998.

Leigh, J., M. Fahs, P. Landrigan, S. Markowitz, and C. Shin. *Occupational Injury and Illnesses in the United States.* Washington, DC: Archives of Internal Medicine, July 1997.

Leone, F. H. and K. O'Hara. "The market for occupational medicine managed care," *Occupational Medicine: State of the Art Reviews,* 13:869–879, 1998.

Liberty Mutual. "Liberty Mutual Study: Workplace Injury Costs Grow," *Safety & Health,* June 2003.

Mont, D., J. Burton, Jr., and V. Reno, *Workers' Compensation: Benefits, Coverage, and Costs, 1996.* Washington, DC: National Academy of Social Insurance, 1999.

_____. *Workers' Compensation: Benefits, Coverage, and Costs, 1997–1998: New Estimates.* Washington DC: National Academy of Social Insurance, 2000.

National Safety Council. *Injury Facts,* 2001 edition. Itaska, IL: NSC, 2001.

Parker, D. L., R. Carl, L. French, and F. Martin. "Characteristics of adolescent work injuries reported to the Minnesota Department of Labor and Industry," *American Journal of Public Health,* 84:606–611, 1994.

Stout, N. A. and C. Bell. "Effectiveness of source documents at identifying fatal occupational injuries: a synthesis of studies," *American Journal of Public Health,* 81:725–728, 1991.

REVIEW EXERCISES

1. Discuss the factors that influence employer spending on workers' compensation. What makes it the highest expenditure (as a percentage of payroll) in the construction industry?
2. Discuss both the advantages and disadvantages that could apply to a company that is included in an assigned risk plan.
3. What is the National Council on Compensation Insurance? Explain how NCCI can help a construction company achieve cost savings.
4. How can an understanding of profit margins assist the construction safety practitioner in achieving safety excellence on a project?
5. List and discuss strategies a contractor or subcontractor can use to proactively address the growing problem of workers' compensation fraud.

CONSOLIDATED INSURANCE PROGRAMS
by Patricia Ennis, ARM, CSP

Trish Ennis, ARM, CSP, has 14 years experience in safety and loss control. Her employment includes work as Safety Manager for a multistate utility contractor, Loss Control Specialist for an insurance carrier, and, most recently, Vice President, Safety and Loss Control Manager for Hilb, Rogal & Hamilton Company of Colorado.

Trish has provided safety services for a number of municipalities, public and private entities, and construction firms nationwide. She has been directly involved in the evaluation, development, and implementation of safety and health programs for individual companies as well as several large owner-controlled insurance programs. She has developed and conducted numerous safety training programs for construction and industry and has worked closely with OSHA on strategic partnership agreements.

Trish is a frequent guest trainer for the Denver Associated Builders and Contractors (ABC), the Associated General Contractors (AGC), the Colorado Safety Association, and the Denver Metro Home Builders Association. She is an active member on the AGC Safety Committee and is a past president of the American Society of Safety Engineers (ASSE), Colorado chapter.

Trish is a Certified Safety Professional (CSP) and has earned her Associate in Risk Management (ARM). She holds a B.A. from Evergreen State College in Olympia, Washington.

CHAPTER 5

Consolidated Insurance Programs

LEARNING OBJECTIVES

- Explain what a consolidated insurance program (CIP) is and how it differs from a more conventional approach to insurance.
- Identify and explain some of the potential advantages and disadvantages experienced by contractors in consolidated insurance programs.
- Describe the key lines of insurance coverage included in a CIP.
- Explain the role the experience modification factor plays in rating workers' compensation policies.
- Identify and explain the key elements in a safety program that contribute to a successful CIP program.

INTRODUCTION

The popularity of *consolidated insurance programs* in the construction industry has grown in recent years. For large-scale construction projects, CIPs are becoming the standard rather than the exception.

This chapter will provide the safety professional with an objective overview of CIPs, also known as *wrap-ups*, and will answer some of the frequently asked questions from contractors who participate in these programs. It will highlight some of the elements that make wrap-ups successful and identify some of the potential pitfalls. CIPs include many lines of insurance that contractors typically need on a construction project, whether the project is covered by a consolidated program or not, so the information in this chapter will provide a general insurance overview that will be helpful to those who are not involved in a CIP as well as those who are or may be in the future.

WHAT IS A CONSOLIDATED INSURANCE PROGRAM?

A CIP is an insurance program in which the CIP sponsor purchases specific lines of insurance coverage designed to cover multiple contractors working on a defined project. This differs from the more traditional arrangement in which each contractor purchases his own insurance and passes those costs on to the owner, or general contractor, by building the costs into the bid.

The optimum size of a construction CIP program is one that will have hard construction costs of at least $100 million and labor costs that equal 25–30 percent of the total project costs. CIPs have, however, been successfully implemented for smaller projects.

Control of the insurance program is a key component of a wrap-up, including the related risk-management functions of claims management, loss control, and safety. The parties involved in a wrap-up typically include the owner, construction manager, general contractor, and all subcontractors on site. Other involved parties include the insurance broker, who often will be contracted to provide wrap-up administration services including safety and claims oversight, and the insurance carrier. The roles of each of these players will be discussed.

The CIP sponsor is frequently the owner of the project; however, in some cases, the general contractor will sponsor the program. Common types of CIPs are outlined below.

- ■ *Owner-controlled insurance program (OCIP):* The owner of the development or project purchases the insurance coverage for all parties working on the defined construction site. This is the most common type of wrap-up.
- ■ *Rolling owner-controlled insurance program (ROCIP):* The owner purchases the insurance as in an OCIP, but the program is designed to "roll" multiple project sites into one existing CIP program. Also known as a rolling wrap-up, this type of program is commonly used in situations where one owner may have multiple construction projects in different locations, such as a residential contractor, a school district, or a university. The insurance program is developed and new work is added as it occurs.
- ■ *Contractor-controlled insurance program (CCIP):* The contractor purchases, or sponsors, the program and covers all subcontractors working on the project.
- ■ *Partner-controlled insurance program (PCIP):* This is a more recently employed program in which several partners collaborate to sponsor the insurance. Partners may be owners, municipalities, contractors, or other combinations of players.

PRIMARY REASONS FOR IMPLEMENTING A CIP

There are many different reasons to implement a wrap-up on a construction project. One of the most common is to take advantage of the economies of scale that can be achieved by purchasing a large volume of coverage. In addition to realizing savings by volume-purchasing, the CIP sponsor may save on premium taxes by implementing a loss-sensitive program with a large self-insured retention. Purchasing one large program can result in reduced administrative, loss-control, and related costs. It is important to note that positive safety results must be present to allow economies of scale to be realized.

Cost savings are important, but are not the only motivating factor in the decision to employ a wrap-up. Other reasons include providing broader insurance limits for all of the subcontractors working on the project; reducing cross-litigation; and increasing uniformity of safety efforts, compliance, and oversight.

Subcontractor Compliance with Insurance

On large construction projects that are not covered by a CIP, the general contractor must audit the insurance compliance of all the subcontractors working on site. The project bid specifications will define the coverage and limits required for the project, as well as additional insured and waiver-of-subrogation requirements. The general contractor must then monitor all subcontractors to ensure that these provisions are met. Certificate tracking is a complex and time-consuming process.

In a CIP, this process is significantly reduced, although not entirely eliminated. Certificates of insurance may still be required to show evidence of required coverage outside the CIP; however, this process is simplified substantially by having to communicate with fewer insurance companies and brokers. One broker will typically administer the program, and the number of insurance carriers is usually only two or three.

Broader Insurance Limits

In most cases, a CIP will offer broad policy limits and terms that may not be available to a smaller contractor. Recent changes in the insurance market have resulted in coverage exclusions for areas of risk that previously were available to contractors. Examples include broad mold exclusions for all industries, exterior insulation finish systems (EIFS), and construction defect and subsidence exclusions in the residential construction market.

Purchasing one large policy for a project can provide the coverage for exposures that a smaller company may not be able to obtain through their core policies. In addition, the limits may be higher than those the contractors have on their own policies.

Contractors who are considering working under a CIP should inquire if the program limits are shared with the owner and other subcontractors on the project, and if the limits are for the life of the project or reinstated annually. This will help the contractor determine if the limits are adequate to cover a catastrophic event. The CIP sponsor may elect to issue separate limits to each contractor enrolled in the CIP. These limits, whether individual or shared, can be supplemented by an umbrella policy. A number of excellent publications are available for inexperienced contractors to guide them through the process of evaluating and enrolling in a CIP. *The Wrap-Up Guide* (third edition), by Gary E. Bird, CPCU, ARM, published in 2000 by the International Risk Management Institute, Inc.® (IRMI®), is a valuable publication for all parties considering entering or implementing a CIP. Another resource for contractors is the publication developed by the Associated General Contractors of America titled *OCIPs Look Before You Leap! A Contractor's Guide To Owner Controlled Insurance Programs*, published May 18, 2001.

Project Safety

The success or failure of a wrap-up is strongly influenced by the quality of the project safety program. A project safety/loss-control manual is a critical tool, and should be developed in time to be incorporated into the contract documents in order to obtain approval and buy-in from the participating contractors. It is not unusual for CIPs to require safety practices that exceed the standards set by OSHA, such as requiring a six-foot fall protection level for all trades or implementing a mandatory, projectwide hard-hat requirement. By identifying project-specific safety and safety administration requirements prior to the bid phase, contractors can anticipate and plan for costs in the bidding process.

Contractor prequalification is an important part of any construction project. The general contractor or owner can establish guidelines for prequalifying contractors who will work on their project. Standards for acceptance can include requiring an injury incident rate at or below the national average, as published by the Bureau of Labor Statistics (BLS), and evaluation of a written safety program. The injury incident rate can be calculated using the formula laid out in Figure 1.

Both rates are calculated by multiplying the number of work-related injuries by the standard 200,000, which represents 100 employees working 2000 hours per year. Divide the product by the total number of man-hours worked by all employees, including overtime hours. The rate should be calculated using the number of cases recorded on the OSHA 300 Log, as opposed to the loss run provided by the workers' compensation insurance carrier. Not all losses reported to the insurance carrier are considered OSHA-reportable.

Incident rate for OSHA recordable injuries:

$$IR = \frac{\text{Number of Recorded Injuries} \times 200{,}000}{\text{Total Number of Hours Worked}}$$

Incident rate for lost-time or restricted-duty cases:

$$IR = \frac{\substack{\text{Number of Injuries with Lost} \\ \text{or Restricted Workdays} \times 200{,}000}}{\text{Total Number of Hours Worked}}$$

■ ■ **Figure 1** ■ ■ **Calculating the incident rate**

A contractor's OSHA citation history can be a valuable tool for evaluating contractor safety, although the absence of citations does not necessarily reflect good safety—it may mean the contractor has never had an OSHA inspection. Another popular tool for evaluating contractor safety is the experience modification factor (EMF) issued by the National Council on Compensation Insurance (NCCI). For more on the EMF, see the section titled "Workers' Compensation Insurance" later in this chapter.

POTENTIAL DISADVANTAGES OF A CIP

Contractors should carefully review CIP requirements before agreeing to participate. Some of the more common concerns voiced by contractors relate to the structure of an insurance program and its safety requirements. Common objections to participating in a CIP include:

- Concern that the CIP does not provide adequate limits.
- Concern that the CIP does not provide the same dividends and credits as the contractor's core program.
- Increased administrative requirements associated with having more than one insurance program.
- Concern that claims management will be less proactive than the contractor's current practices.
- Compliance with substance-abuse programs.
- Associated potential costs if CIP safety requirements exceed OSHA guidelines.
- Difficulty in getting second- and third-tier subcontractors to comply with project safety requirements.

These concerns are valid and should be evaluated individually. Contractors should consult with their own brokers, inviting them to attend prebid meetings in order to get assistance in evaluating the CIP, and possibly adjusting existing insurance programs to ensure adequate coverage. Smaller subcontractors typically are not invited to the preconstruction meetings and subsequently do not get site-specific safety information prior to bidding the job. The CIP program should be designed to ensure accountability is built into every level of the project, guaranteeing compliance by even the smallest subcontractor.

ENROLLMENT AND COST DEDUCTION

Cost Deduction

There are wide variations in how enrollment and cost deductions are handled on wrap-ups. Theoretically, contractors enrolling in a wrap-up should see insurance costs that are the same as their existing coverage—no more, no less. The sponsor will *carve out* or *back charge* enrolled contractors for their insurance costs. The insurance costs can become an issue when the CIP sponsor implements an insurance program that has higher limits than those the typical contractor carries on his own policies. For example, most contractors carry general liability coverage with a limit of $1 million, with an umbrella policy that provides an additional $1 million. The CIP sponsor might purchase a program that provides $2 million per occurrence on the general liability policy with a $6 million aggregate and a $25 million umbrella policy. Contractors should only pay the premium for insurance limits that they would be required to carry if the project were not structured as a wrap-up ($1 million general liability, $1 million umbrella). The CIP sponsor can determine contractors' insurance costs by requiring them to submit their bids both with and without the insurance costs included.

Enrollment

The CIP administrator should answer contractor questions during the enrollment process. Some questions that arise might relate to how to identify administrative costs associated with the wrap-up, and whether contractors will be reimbursed for any of these expenses. These costs can include tracking and reporting payroll to the CIP, time for employee orientation and drug testing, and requirements for implementing a return-to-work program. Additional concerns might include:

- Identifying excluded classes of work or workers, such as temporary labor, truckers, or high-risk work such as blasting.
- Defining procedures that contractors must follow for tracking subcontractor enrollment and providing subcontractor certificates to the sponsor.
- Identifying deductibles that contractors may be subject to under the program.
- Outlining the kinds of safety services that will be provided under the program.
- Identifying any other requirements that are beyond the scope of a typical construction project.

CIP INSURANCE COVERAGE

Lines of Insurance Most Often Included in a CIP

The typical wrap-up will include general liability, builders' risk, and workers' compensation insurance. Many programs also include professional liability, pollution liability, and an umbrella liability policy.

GENERAL LIABILTY

This type of coverage protects owners of businesses or contractors from a variety of exposures. These can include liabilities that result from operations, contractual obligations, and accidents.

General liability coverage can be structured in more than one way. The policy may provide for aggregate per-occurrence limits in which all of the insured parties are covered under one limit for all claims arising from one accident. The policy will have an annual aggregate limit that can be two to three times the per-occurrence limit. Another option is to structure the program to extend an individual per-occurrence limit to each contractor. In this case each insured is covered individually up to the policy limits. General liability insurance will contain the following:

1. *Contractual Liability:* Protects the insured for a loss for which the insured has accepted liability under a written contract.

2. *Property Damage Liability:* Provides protection from liability for damage to the property of another.

3. *Contractor's Protective Liability:* Provides coverage for claims that can arise from the insured parties' supervision of work.

4. *Completed Operations Liability:* Provides coverage for claims arising out of operations that have been completed and turned over to the owner. The typical policy will provide coverage for three years beyond the end of the project. Contractors may need to purchase supplemental insurance through their own insurer to cover a longer period of time. In some jurisdictions the warranty period can extend seven to ten years past the project completion date.

5. *Premises and Operations:* Provides coverage for claims arising out of an insured's premises and business operations.

6. *X,C,U Coverage:* Provides coverage for claims arising out of explosion, collapse, and damage to underground property. Explosion coverage insures against property damage that results from blasting or other explosions. Collapse coverage is triggered when property damage is caused by demolition or excavation activities, and underground coverage applies when property damage results in damage to utilities from digging, pile driving, or related activities.

BUILDERS' RISK

Builders risk is insurance specifically designed to cover buildings or structures in the course of construction. The policy should provide coverage for the following risks:

- earthquake/flood
- boiler and machinery
- damage to existing or adjoining property
- collapse
- business interruption.

If the structure is damaged during the course of construction, the builders' risk policy will help pay for the costs of reconstruction. Builders' risk insurance is a requirement on many projects whether they are associated with a CIP or not. It is common for the owner to pass the deductible requirements on to the general contractor, or for the general contractor to pass them on to the subcontractors. Deductibles on construction builders' risk policies can range from thousands to hundreds of thousands of dollars or more, depending on the contractor's appetite for risk and ability to self-insure. Passing on the deductible can be a powerful incentive for contractors to practice good risk management! In cases where the deductible is passed on, the requirements should be clearly communicated in bid and contract documents.

In a CIP, the builders' risk policy may be included in the program or added to the program as a separate coverage. It is not unusual for the builders' risk policy to be provided by a different insurance carrier than the other lines of coverage.

WORKERS' COMPENSATION INSURANCE

Workers' compensation insurance has been around in some form since the early 1900s. By 1934, all states had some form of workers' compensation law in place. The laws vary from state to state, so this section will provide a general overview of the coverage and how the premium is calculated.

Workers' compensation insurance can have the greatest impact on the success of the CIP because of the volume of the premium and the high degree of control over losses. The majority of employers are statutorily required to provide workers' compensation insurance, whether they are involved with a CIP or not. The purpose of workers' compensation insurance is to provide medical and disability benefits to workers who incur an occupational injury or illness.

MEDICAL BENEFITS

Medical benefits are payment for the medical treatment of an injured employee. In many states these benefits may be unlimited in both dollar amounts and duration.

DISABILITY BENEFITS

Disability benefits are designed to replace an injured worker's loss of income or future earning capacity that results from a work-related injury or illness. Disability benefits are set by state statute and expressed as a percentage of the worker's average weekly wage (usually 66.66%). Like many other insurance benefits, the disability wages are not subject to income tax deductions. There are four categories of disability benefits:

- Temporary total disability (TTD)
- Temporary partial disability (TPD)
- Permanent partial disability (PPD)
- Permanent total disability (PTD).

Temporary total disability means that an injured worker is expected to recover and return to work, but is temporarily unable to work due to the nature of the injury or illness. This category promotes the modified or light-duty program. In some cases TTD can be reduced or even eliminated by bringing an injured worker back to work in a reduced or modified capacity.

Temporary partial disability means that the injured worker can perform some sort of work during the recovery process. At this point the employer has the opportunity to bring the injured worker back in some sort of modified or light-duty position. When an employee is working, even part time and at partial wages, the recovery time can be faster and the cost of the claim can be significantly reduced. If the employer elects to pay the injured worker less than the preinjury wage, workers' compensation will make up the difference at the limits established by the local insurance division. For instance, if the injured worker was paid $18 per hour preinjury and the employer decides to pay $12 for modified duty, the workers' compensation insurance carrier will pay $4 (66.66%) of the wage difference.

Permanent partial disability occurs when a worker suffers some permanent effects from an injury or illness. At some point during treatment of an injury or illness, a physician may determine that the worker is not going to improve more with further treatment. At that point, what was considered a temporary partial disability may become a permanent partial disability. For example, a construction worker breaks several fingers when he gets his hand caught in a sling used to lift a piece of pipe. The fingers heal but have a reduced range of motion, resulting in permanent partial disability. The disability benefits are designed to compensate for potential reduced future earning capacity. These benefits can vary widely from one state to another.

Permanent total disability results when an injury is so severe that the employee is unable to perform any kind of work for the remainder of his or her lifetime. Benefits are calculated based on the average weekly wage. In many states PTD benefits are provided for the rest of the injured worker's life.

WORKERS' COMPENSATION PREMIUM CALCULATION

Workers' compensation insurance premiums are based on the classification codes of covered employees, payroll, and the experience modification factor.

Rates are based on the industrial classifications of the employers and classification of the covered employees. There are two types of codes used to arrive at the proper classification. The first is the Standard Industrial Classification (SIC) code, and the second is the National Council on Compensation Insurance (NCCI), or similar rating-bureau classification code.

SIC codes and the North American Industry Classification System (NAICS) are the U.S. government's standardized systems for classifying businesses by the operations and activities in which they are engaged. Why are these systems important to the safety professional? The SIC code and NAICS system provide the information used in statistical analysis to develop the accident and incident rates published by the BLS and used by OSHA for measuring safety performance and tracking trends by industry. Insurance carriers also use these systems to classify businesses and develop underwriting criteria.

SIC

The four-digit SIC codes are used to facilitate the collection, tabulation, presentation, and analysis of data relating to establishments. The codes are divided into hierarchical tiers:

- Division
- Major group
- Industry group.

Here is an example SIC for construction:

Division C: Construction.
Major group 15: Building Construction, General Contractors & Operative Builders.
Industry group 154: General Building Contractors, nonresidential.

A fourth digit further classifies the operation to a specific group—1542:

General Contractor, nonresidential buildings other than industrial.

NAICS

While the SIC system is a four-digit system that presents limitations in coding, the NAICS is a six-digit system. This system was established jointly by the United States, Canada, and Mexico in 1997 in order to improve statistical comparison between countries on the North American continent. The Office of Management and Budget (OMB) and federal agencies adopted the NAICS in 2002. *It is important to note that the 2002 NAICS differs significantly from the 1997 NAICS.* The first five digits are standardized. The sixth digit allows each country to tailor the codes to industries that are unique to that country. Like the SIC, the NAICS is broken into hierarchical tiers:

- Major sector (first two digits)
- Subsector (third digit)
- Industry group (fourth digit)
- NAICS international industry (fifth digit)
- National industry (fifth and sixth digits).

The following is an example of the 2002 NAICS for Construction:

Major sector 23: Construction.
Subsector 236: Construction of Building.
Industry group 2362: Nonresidential Building Construction.

NAICS international industry 23622: Nonresidential buildings, other than industrial buildings and warehouses, general contractors.

For more information about SIC and NAICS, visit the Occupational Safety and Health Administration Web page at www.osha.gov and follow the "SIC/NAICS Search" link under the Statistics heading. The *2002 NAICS United States Manual* is available for purchase from the National Technical Information Service of the U.S. Department of Commerce, or visit the Census Bureau at www.census.gov for information online.

The SIC and NAICS systems help carriers and governmental agencies classify business operations, but for rating workers' compensation policies the carrier relies on the classification codes established by a rating bureau such as NCCI.

NCCI Classification Codes

Once the operation of the employer has been determined, it is necessary to determine the classification of the employees. Classifications are established to group employees who have common exposure to hazards in the workplace. In general, the classification will cover the majority of the workers. For example, a large bridge-building contractor may have up to 90 percent of payroll classified under the 5222 class code—concrete bridge construction. This includes everyone performing physical work at the project site, from equipment operators to carpenters. Essentially, anyone performing physical labor in the field will fall under the same class code. The remaining payroll can fall under 5606—executive supervisors, for project management, and possibly 8810—employees who solely perform clerical work. The premium is calculated by applying a rate by class code for every $100 of payroll reported for that class.

As you can imagine, the rates for various class codes will vary widely from one another due to the amount of risk that applies to the operation. Clerical workers are employed in a relatively nonhazardous occupation, and may have a rate of $0.40 to $0.50 per $100 of payroll. Roofers, on the other hand, are employed in a more hazardous occupation, and those rates may be more than $20 per $100 of payroll.

The National Council on Compensation Insurance, in operation as a nonprofit organization since 1923, develops advisory rates for each class code and provides those rates to insurers. Now a for-profit company, NCCI, Inc., is the oldest and largest of the rating bureaus. Some states may use an alternate rating bureau, but NCCI is by far the most recognized authority.

NCCI collects, manages, and distributes information based on statistical information provided by insurance carriers nationwide. The claims' data that NCCI collects enables them to identify and project trends, which in turn enables them to recommend advisory rates, known as *loss-cost rates*, based on expected losses for specific classes of work. Insurers refer to these rates as a benchmark, as they apply their own multiplier to the loss-cost to arrive at a final rate. The rate is applied to the estimated payroll to generate a manual premium. This premium is then subjected to a variety of multipliers and discounts, including the employer's experience modifier, to arrive at an annual premium.

The rating process is complicated and varies widely from state to state. For the purposes of this chapter, it is important to understand that workers' compensation rates are generated based on statistical data that projects the "expected" costs of losses for a specified amount of payroll by trade. When generating these rates, the rating bureau considers many factors, including medical costs, wages, and other economic factors in geographic regions. The advisory rates in Georgia for example may not be the same as the rates in Colorado.

Payroll

Rates are quoted per $100 of payroll for each classification covered. The insurance carrier will need an estimated payroll in order to calculate the workers' compensation premium. The estimated payroll for the current policy period can be based on historical records plus projected growth. At the end of the policy period, the contractor's payroll will be audited to determine the final premium. The insured will pay an additional premium if the payroll estimations were low. Conversely, if the payroll is less than estimated, the insured may get a refund or credit.

Experience Modification Factor

In addition to generating advisory rates for classifications, NCCI compiles statistical loss data on individual employers. The EMF is an employer-specific multiplier that compares the company's loss experience to that of other employers in its state in the same industry.

The rating bureau generates the EMF based on three years of loss experience, excluding the most recent year. For example, the 2002 EMF will be based on experience from 1998, 1999, and 2000. The 2001 losses will be allowed to develop or mature for a year and will not affect the EMF until 2003. The EMF is revised every year prior to the policy renewal date, dropping the oldest year of experience and adding the more recent year. In essence, an employer can continue to pay for claims' experience long after the carrier has closed the claim, as the costs can continue to impact the EMF for three of the next four years. On the other hand, an employer with less than average loss experience can realize the benefits of that experience for as many years.

An average EMF of 1.0 should reflect the average loss experience of the state classification. Less than 1.0 reflects better-than-average experience, greater than 1.0, worse-than-average experience. The modifier is applied to the manual premium to arrive at a modified premium. A company with a modifier below 1.0 will have a credit mod, which results in a credit on the workers' compensation premium calculation. A contractor with an experience mod over 1.0 will have a debit mod, which translates into a debit on the premium calculation. It is easy to see who has the advantage. Since the EMF is calculated based on loss experience, it is widely used as a benchmark to measure the effectiveness of an employer's safety program. However, the EMF is an actuarial factor that is designed to help insurance carriers rate companies based on their loss experience and was never intended to be used as a measure of safety.

The experience modification factor calculation is very complex. Information factored into the calculation is:

- expected loss rates for class of business by payroll
- frequency of losses
- severity of losses
- deductibles for net reporting states.

When insurance rates are reduced in a state, the expected loss rates (ELR) go down as well. These factors are multiplied against payroll to determine *expected losses*. If a company's actual losses are greater than its expected losses (the losses expected for the company based on size and industry) then there is an inflationary impact on the mod factor. These factors, or *multipliers*, are then applied against loss experience that is two to three years old for losses that are not going down and could even be increasing as claim costs continue to develop. This can cause the EMF to creep up from year to year even as rates appear to be going down. This increase is independent of the actual safety efforts implemented by employers. It is not unusual for an owner or general contractor to refuse to accept bids from contractors who have EMFs over 1.0. As the modification factor increases, it impacts the premiums paid by employers and may result in contractors being excluded from bid lists.

Let's see how an EMF can impact a contractor in the bidding process by looking at three theoretical paving contractors:

▪▪ Table 1 ▪▪

Costs

	Contractor A (Good)	Contractor B (Average)	Contractor C (Poor)
Manual workers' comp premium	$250,000.00	$250,000.00	$250,000.00
EMF multiplier	.75	1.0	1.25
Modified workers' comp premium	$187,500.00	$250,000.00	$312,500.00

Contractor A has the clear advantage of paying $125,000 less for workers' compensation insurance than Contractor C, making him more competitive.

Contractors working on a CIP will be impacted by loss experience just as they are on other projects. Most CIPs place strong emphasis on safety as a measure of controlling losses, and it is not unusual for a contractor to improve an EMF as a result of working under a CIP.

Insurance carriers use the experience modification factor to rate insurance policies based on an employer's past loss history when compared to peers in the industry. It is an indication of the effectiveness of accident prevention efforts taken by employers; however, there are a number of factors that can impact the EMF that have nothing to do with the safety program. If the EMF is going to be used as a benchmark for measuring safety performance, it should be used in conjunction with other methods, such as OSHA incident rates and OSHA compliance history.

PROFESSIONAL LIABILITY

Some CIPs will include professional liability insurance for architects, engineers, or consultants such as brokers. Professional liability insurance responds to claims that arise from liability associated with the acts, errors, or omissions resulting from the rendering of, or failure to render, professional services. This insurance is especially important when the project includes design–build components. Contractors should be familiar with the insurance provisions included in the CIP if they will be assuming any of the responsibility for the design as well as the construction of the project. This coverage is typically not included under general liability.

POLLUTION LIABILITY

Pollution liability may be included in the CIP, depending on the scope of the project. Typically not covered under general liability, pollution coverage is triggered by claims for bodily injury or property damage arising from pollution that originates from the insureds premises or operations. In a tight insurance marketplace, this coverage can be difficult for smaller contractors to obtain. Participating in a CIP can provide increased limits and coverage to the contractors for work that takes place on the projects designated by the CIP.

Lines of Insurance Excluded from a CIP

A wrap-up is designed to cover only on-site exposures. This means that contractors need to provide coverage for all operations outside of the CIP: a few of the key areas of risk include automobile liability, contractor's equipment coverage, and surety bonds. Contractors will be required to show proof of workers' compensation insurance for operations outside of the CIP before being awarded a contract.

Typically, the CIP will only cover materials and equipment on site that are to be incorporated into the work. This means that tools, equipment, mobile offices, trailers, and related items will not be covered. This requirement should be clearly specified in the CIP contract documents and presented during the prebid phase to prevent confusion.

Surety bonds are not commonly included in a wrap-up. The relationship between a contractor and surety bond provider is confidential due to the nature of the financial information disclosed. Bonding requirements for the CIP should be clearly spelled out in bid packages and contract documents.

CIP SAFETY MANAGEMENT

There are several parties involved in a wrap-up, and all of them need to be aware of their roles and responsibilities in order for the project to be successful. This is especially important for effective project loss control and safety management. Loss-control and safety responsibilities should be defined in prebid meetings and outlined in contract documents to preserve the chain of command and make it enforceable. Most CIP sponsors require contractors to have an effective safety program in place prior to starting work.

Roles and Responsibilities

CIP SPONSOR

The CIP sponsor, in many cases the owner, is responsible for hiring a broker to market the insurance and provide CIP administration services. The sponsor will work with the broker to prepare and implement the project's loss-control and safety programs administered by the broker and the construction manager or general contractor.

BROKER

In addition to marketing the insurance program and developing the CIP enrollment and procedures manual, the broker usually provides a variety of loss-control services for the CIP. These services may include:

- Development of the loss-control and project safety guidelines.
- Providing project safety audits.
- Assisting with claims, handling and coordinating insurance-carrier services for the program.
- Establishing a substance-abuse prevention program.
- Establishing a designated medical provider for all CIP-related injuries.

The broker's loss-control oversight on a CIP will range from assigning a full-time CIP safety manager to providing part-time service on a fee basis.

CONSTRUCTION MANAGER/GENERAL CONTRACTOR (CM/GC)

Ultimate responsibility for the CIP safety and loss-control program should rest with the construction manager/general contractor. The CM/GC may be required to:

- Provide safety and loss-control personnel to review subcontractors' written safety programs.
- Develop contractor prequalification guidelines.
- Develop a site-specific orientation training session for contractor and subcontractor personnel.
- Promote and enforce the project's loss-control program.
- Establish a project safety committee composed of contractor safety representatives. The committee should meet regularly to identify and resolve safety concerns.
- Conduct regular project safety audits and notify contractors of violations or problems.
- Review accident reports and ensure contractors conduct accident investigations.
- Develop and implement a project emergency evacuation plan.

CONTRACTORS AND SUBCONTRACTORS

Contractors and subcontractors are required to follow the guidelines set forth in the CIP loss-control and safety manual. Each may be required to:

- Submit a written safety program to the CM/GC to be accepted prior to starting work.
- Ensure that all workers attend the site safety orientation, including subcontractors' workers.
- Attend any preconstruction meetings to discuss safety-related issues and concerns.
- Provide and identify competent persons and/or on-site safety coordinators who will be responsible for safety on the project site.
- Conduct regular toolbox safety meetings.
- Participate in the project's safety committee meetings.
- Comply with project loss-control and safety requirements.
- Develop safe work plans or job safety analysis (JSA) for critical work or high-risk exposures.
- Comply with CIP substance-abuse prevention policies.
- Develop a return-to-work program for injured workers.
- Investigate accidents.
- Follow CIP reporting guidelines for all claims.

INSURANCE CARRIER

The insurance carrier plays a key role in a successful CIP. The carrier's loss-control personnel provide valuable services to supplement the CIP loss-control program. Carrier loss-control personnel may:

- Conduct regular project safety and loss-control audits.
- Provide industrial hygiene services, such as monitoring for airborne contaminants or noise.
- Recommend corrective actions based on identified hazards.
- Attend project safety committee meetings.
- Assist with accident investigations and claim management activities.

■ Conduct safety-training classes for contractors working on site.

Loss-Control and Safety Procedures Manual

The CIP loss-control and safety manual should be designed to enhance the contractor's existing safety manual. A CM/GC's existing safety manual may not be comprehensive or specific enough to apply to the multiple trades on a large construction project and may need to be revised before being used on a CIP project. The CIP loss-control and safety manual should be designed to enhance the contractor's existing program, not replace it with a set of entirely new rules. Items that should be defined in the safety/loss-control manual include:

1. *Requirements for safety staff trained by subcontractors.* This is the place to let subcontractors know if full- or part-time safety personnel will be required on site, and identify the minimum training requirements for these individuals (i.e., OSHA 10-hour training, competent-person status, etc.).

2. *Safety compliance requirements.* If the CIP sponsor expects the contractors to exceed OSHA standards, the specific requirements should be defined in the manual. For example, it is not uncommon for a CIP to require 100% fall protection at 6 feet for all trades on site, including steel erection. The chain of authority should be clearly spelled out, with each contractor tier notified of its safety responsibilities.

3. *Substance-abuse programs.* If the CIP sponsor is going to require participation in a project-specific substance-abuse program, the program details should be included here. Prior to bidding the work, the contractor should understand who will administer the program and how the owner or program sponsor plans to reimburse contractors for any costs they incur by participating in the program. Additionally, the contractor should understand how, or if, the project's substance-abuse program will impact existing union agreements. Procedures for protecting confidentiality of test results must be clearly defined.

4. *Return-to-work programs.* Some CIPs require that contractors return injured workers to modified or light duty. This can be a positive program for both the CIP sponsor and the contractors if it is administered properly. Contractors should understand how the program will impact them in the event the injured employee is returned to a light-duty position that is outside the CIP and what kind of assistance the CIP can provide to contractors trying to identify modified-duty options.

CONCLUSION

A wrap-up program can be an advantageous arrangement for everyone involved, from project owners to the smallest subcontractor. Advantages include uniform insurance coverage for all participants, increased limits, and potential savings through volume purchasing and aggressive safety management. Many small contractors can benefit from the sophisticated loss-control and safety resources available through the CIP, such as industrial hygiene services, safety training, and technology that may otherwise be inaccessible to them. As insurance markets continue to evolve, contractors can expect CIPs to become more and more common. If the CIP is carefully evaluated and questions are addressed, contractors will find that the CIP is a winning arrangement.

REVIEW EXERCISES

1. Explain what a consolidated insurance program is and how it differs from a more traditional approach to insurance for a construction project.
2. List four of the more common types of CIPs.
3. Explain two primary advantages for implementing or participating in a CIP program.
4. Discuss SIC and NAICS and why these systems are important to safety professionals.
5. What is an experience modification factor? Explain why the EMF should or should not be used as a measure of contractor safety.
6. List four safety activities or standards that can help contain losses on a construction project.

■ ■ **About the Authors** ■ ■

CONTRACTOR SELECTION
by Everett A. Beaujon, CHMM, CPEA, CSP,
and Greg L. Smith, CSP

Everett A. Beaujon, CHMM, CPEA, CSP. Until his death in March
2003, Everett A. Beaujon was the Director of Safety and Health for Austin
Industrial, Inc., in Houston, Texas. Austin Industrial, Inc., is a maintenance and construction contractor that provides a variety of services for
the petrochemical, refining, power generation, and manufacturing industries in the southern and southeastern United States. Everett held various safety and health positions with Austin Industrial during his 14 years
with the company. He received a B.S. in Mechanical Engineering from
the University of Cincinnati, and was a Certified Safety Professional and
a Certified Hazardous Materials Manager. Prior to joining Austin Industrial, Everett was the Safety and Hygiene Administrator at a large overseas
refinery complex where he was employed for 19 years.

Greg L. Smith, CSP, is president of Construction Safety & Health Inc.
in Austin, Texas. Mr. Smith has over 20 years of experience in the safety
field, working in both general industry and construction. He holds a
B.S. in Industrial Hygiene and Safety from the University of Houston,
Clear Lake, and is completing an M.S. in that discipline from the same
institution. He was named the Gulf Coast chapter's Safety Professional
of the Year in 2001, and is a member of the Advisory Council for the
University of Houston's Safety Engineering program.

CHAPTER 6

Contractor Selection

LEARNING OBJECTIVES

- Discuss economic and regulatory factors that may influence contractor selection.
- Discuss the selection criterion used to ensure that a contractor meets basic qualification requirements.
- List key SH&E program components that assist in a proactive contractor safety process.

INTRODUCTION

While the construction industry was 6.6 percent of the total workforce in 2000, it accounted for 20.8 percent of all workplace fatalities in the United States according to the United States Bureau of Labor Statistics. The BLS report also found that the fatality rate in construction fell significantly from 1999 to 2000, but then showed a dramatic increase in 2001 (see BLS Figure 1).

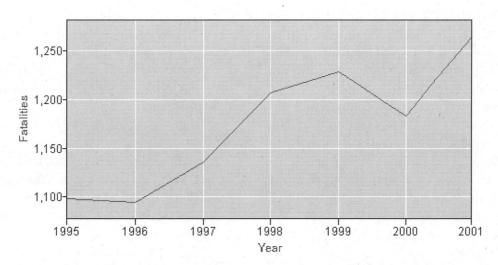

■■ **Figure 1** ■■ **Construction fatalities, 1995–2001** (Source: Bureau of Labor Statistics)

The Occupational Safety and Health Administration says that construction safety will be "a continuing focus area" for the agency and, if companies are to achieve the goal of reducing workplace injuries and deaths, they must take proactive measures to reduce the accident, and most importantly, the fatality numbers.

Increasingly, as safety knowledge becomes more widespread and available, OSHA is citing both contractors and owner companies for violations if it is believed the owner company knows or should have known about risks to which workers were exposed. Using contractors does not absolve an owner company of its responsibility for safety, though it may transfer a limited degree and amount of liability.

The intent for the responsibility is clearly stated in OSHA's Construction Industry Standard 29 CFR Part 1926.16(c), and more clearly defined in its General Industry Standard 29 CFR 1910.119. [*Note:* 29 CFR 1926.16, as part of Subpart B (General Interpretation), has pertinence only to the application of section 107 of the Contract Work Hours and Safety Standards Act (The Construction Safety Act) and has no direct significance in the enforcement of the OSH Act.]

However, irrespective of 29 CFR 1926.16, OSHA *does have* a policy on multiemployer work sites, as enunciated in the OSHA *Field Operations Manual,* Chapter V, Section F.

Poor safety performance influences not only the cost of the construction work, but also can be the cause of other serious problems such as lawsuits as a result of accidents, work delays, inferior quality work, and negative media reports.

This chapter will review the economic and regulatory factors that impact construction work and the risks and liabilities confronted by both an owner and a contractor. It will further offer the guidelines an owner can use to select a contractor to minimize and control the risks and liabilities encountered during the execution of a construction project.

UNDERLYING ECONOMIC AND REGULATORY FACTORS

Selecting the right contractor is crucial in today's competitive business environment. The safety performance of an outside contract workforce can influence not only the cost and quality of the outsourced project, but many other areas of business operations as well—from regulatory compliance to community relations.

A construction project, small or large scale, must have a comprehensive, well-defined management control process that is implemented consistently during all phases of the project to ensure that the project is executed and completed according to the planned time and cost estimates.

The safety, health, and environmental program at a project should be implemented and executed through a management control process that is an integral part of the overall management process of the project.

In the absence of a comprehensive management process for the execution of the work, including engineering, the following typical factors are negatively affected, either directly or indirectly.

a. *Completion Schedules*
Delays and subsequent late completion of phases of the project and of the overall project can and will be caused by: inadequate planning, poor engineering, late delivery of material, accidents, material damage, inferior quality of work, inadequately skilled workers, work delays, etc.

b. *Project Cost*
The same factors that cause the project completion date to be delayed will cause the cost of a project to overrun. The overrun is generally not merely a cause of the cost of labor, but also material damage, overstocking, etc.

c. *Incidents*
Injuries, accidents, material damage or failure, lawsuits, and insurance costs will add additional operational overhead to the project.

d. *OSHA Inspections and Citations*
Construction work sites inspected by the OSHA area office are seldom scheduled for programmed inspections (in certain cases the contractor may have entered into a partnership agreement with OSHA, which requires an annual verification audit by an OSHA representative), unless the area director is pursuing a special-emphasis program. Inspections may be the result of

complaints by disgruntled employees, public complaint to the area OSHA office, or a random inspection initiated by a compliance officer due to issues noted as they drove by a site. Depending on the nature of the inspection, it can take from one day to several days or weeks. The delays, interruptions, and resources required to attend to the OSHA representatives can add a substantial cost to the project. In addition, penalties due to any resulting citations and/or the cost to contest citations will also be an additional burden and affect the overall cost of the project.

Regulatory Factors

The safety and health management control process implemented at a construction project should have as its principal purpose the protection of employees, other workers, the public, and company property. To successfully accomplish this objective, a safety and health management control process must include the various elements and components that are required by OSHA's 29 CFR 1910 and 1926, general industry and construction standards.

The following are some examples of a few specific areas covered by standards: hazard communication, personal protective equipment (PPE), ladders, scaffolding, electrical equipment, and welding.

OSHA standards may also require specific procedures and training based on the type of work activities being performed. Some examples include forklift safety, crane safety, lockout tagout, confined space entry, excavation, and fall protection.

OSHA 29 CFR 1926.20 specifically requires that construction companies should implement workplace safety and health programs. Additionally, for certain hazardous processes, the Process Safety Management Standard, generally known as OSHA PSM, 29 CFR 1910.119 par. (h), mandates that when selecting a contractor, the owner "shall obtain and evaluate information regarding the contract employer's safety performance and programs."

Under the OSHA *Field Inspection Reference Manual* (CPL 2.103 Multi-Employer Work Site Citation Policy), the owner, general contractor, and subcontractor can all be cited for violation of an OSHA standard. Either entity on a multiemployer work

site may have created the violation or hazard (see the chapter on multiemployer work sites for more details).

The responsibilities for compliance with the applicable OSHA standards, specifically with the construction standard is clearly stated in OSHA 1926:

- 29 CFR Part 1926.16(a) "In no case shall the prime contractor be relieved of overall responsibility for compliance with the requirements of this part for all work to be performed under the contract."
- 29 CFR Part 1926.16(c) "With respect to subcontracted work the prime contractor and any subcontractor or subcontractors shall be deemed to have joint responsibility."

This language is born out of long-established industry practice as relates to the hierarchy of controls on construction work sites.

Owner and Contractor Liabilities

RISK MANAGEMENT

Owners have a direct economic stake in the safety performance of their contractors because accident costs are an expense to the contractor and are subsequently passed on, one way or another, to the owner. This is obvious in cost-reimbursable contracts where incurred costs are passed on to the owner by contractual agreement, but it also applies to fixed-price contracts where the fee to the contractor will not change unless a change order is issued by the owner.

Owners must consider three kinds of costs when dealing with safety in construction.

- direct costs of accidents and insurance
- indirect costs of accidents
- costs of safety programs and compliance.

Direct Costs of Accidents and Insurance
Workers' Compensation

Most contractors buy insurance for their workers' compensation (WC) exposure. Those that do not are either self-insured or are covered by insurance carried by their client. The cost of insurance coverage varies with a contractor's accident record, which is based on that contractor's injury costs in a given geographical area, as expressed in an experience modification rate (EMR). The other part of the premium formula involves the workers' compensation insurance rate set for a particular type of work (i.e., carpentry, plumbing, steel erection), by a state rating bureau, expressed in dollars per $100 of payroll and based on the injury experience for that type of work in the rating state. The contractor's insurance premium is the product of state rates multiplied by the contractor's EMR for that state. It is important to note that the work classification rates are constant for all contractors doing similar work in a specific state, so it is a contractor's EMR in conjunction with other company information and insurance carrier policies that determines the cost.

Liability—Third Party

Owners should take particular note of the magnitude of third-party liability costs as one of the indirect costs of accidents. Litigation against a third party has become more common in recent years, and dollar losses in some jurisdictions can be significant for the owner when an employee of a contractor sustains an injury or illness. Agreements of indemnification (hold-harmless clauses) sometimes tend to be ineffectual in protecting owners from either dollar loss or adverse publicity. However, when carefully drafted, such clauses can provide some protection to owners by establishing intent, and should be considered in all contracts.

Accidents resulting in injury to anyone other than one's own employees or damage to the property of others constitute liability exposures. Coverage for general liability, automobile liability, and completed operations is considered to be a part of liability costs. In addition to the actual insurance premium, other expenses such as deductibles and legal fees should be considered in compiling total costs to the contractor for liability. Published rates exist for liability insurance coverage; however, the cost is relatively insignificant compared to workers' compensation rates. There are certain high-risk operations that may command insurance premiums in the range of 15 percent of direct labor payroll; however, most contractors pay about 1 percent for liability coverage.

Property

Real property, such as the facility under construction and construction equipment, provides a potential for accidents and resulting losses. The costs of such insurance as a builder's risk, equipment floaters, and installation floaters must be considered. In addition to the insurance premiums, deductibles and the possibility of losses affecting uninsured property should be considered. Accidents in construction work that is in progress may bring catastrophic losses, either because of the severity of the accident itself, resulting long delays in completing the work, or both.

Indirect Costs of Accidents

The insurance costs discussed so far are readily identifiable as a specific cost of doing business and may therefore be projected. However, when an accident occurs, not only direct but also indirect costs are involved. Indirect costs include:

- loss of productivity
- disrupted schedules and overtime wage costs
- administrative time for investigations and reports
- training of replacement personnel
- wages paid to the injured worker(s) and other workers for time not worked
- clean up and repair
- adverse publicity
- third-party liability claims against the owner
- equipment damage.

Estimates of the ratio between indirect and direct costs have varied from 4 to 1, up to 175 to 1 for high-profile multiple-fatality incidents. This ratio varies with the magnitude of the accident; however, it is not necessarily linked to the severity of the injury. In other words, an extremely serious and costly accident may occur without any person sustaining injury.

Costs of Safety Programs

Insurance costs, costs of injuries, and the expense of liability suits are easily documented and rather readily available. The cost of establishing and administering a construction safety and health program is somewhat less tangible, but can be estimated with reasonable accuracy. Data collected from a significant sample of contractors working at various construction sites in 1980 indicate the cost of administering a construction safety and health program usually amounts to about 2.5 percent of direct labor costs. These costs include:

- salaries for safety, medical, and clerical personnel
- safety meetings
- inspections of tools and equipment
- orientation sessions
- site inspections
- personal protective equipment
- health programs such as respirator fit tests and medical monitoring
- sampling/monitoring equipment and laboratory fees
- miscellaneous supplies and equipment.

RISK MINIMIZATION

Owners can, and do, successfully influence construction job safety. The degree to which owners should involve themselves in this process should be based on the costs, benefits, and risks involved. All owners have a legal and moral responsibility to use reasonable care to correct or warn contractors of any non-apparent hazards present on the site that could affect the safe performance of the construction, and to use reasonable care to prevent contractors from injuring others on the site. Owners must ensure that contractors recognize and conform to their contractual responsibility to perform safely.

The owner has considerable flexibility to adjust the degree of involvement and control in each situation. The incentives for increased involvement are lower costs, quality work, improved productivity, adherence to schedule, reduced exposure to bad publicity, and minimal disruption of the owner's employees and facilities.

On the other hand, increased owner involvement, if not handled adroitly, can interfere with the contractor's productivity, cause ill will between an owner and the contractor, and could transfer a heavy degree of liability to the owner. Management should consider each situation separately, and a decision should be made regarding the appropriate degree of involvement. Once this decision is made, the success of the program will depend on good owner–contractor communications. These commu-

nications should include the owner's safety expectations, an understanding of the contractor's safety program, and effective dialogue among all levels throughout the life of a project.

Owners can be successful in their efforts to improve job safety on construction projects. Comments from contractors indicate positive support for proactive owner programs. Proper management by owners in this phase of their business activities can make a significant contribution to a reduction of injuries and a reduction of construction costs.

Selection Criteria

GENERAL AND FINANCIAL

The selection and eventual hiring of a contractor to provide a service or construct a major facility should be managed with a process that ensures the selected contractor meets the basic qualification criterion set by the owner's management.

A contractor selection or screening committee created by the owner should set the qualification specification. The committee or task force should be comprised of health, safety, environmental, accounting, purchasing, engineering, and legal/insurance department representatives.

The qualification criterion should include basic information about the contractor which enables the owner to verify that the contractor is financially sound, has a history of performing quality work, and has the technical qualifications to complete the work. The criterion should also include provisions to verify that the contractor has a successful safety and environmental compliance record.

From a legal perspective, the owner should require the contractor to provide documentation relative to the various types of insurance required for the type of service the contractor provides. These include coverage for workers' compensation, general liability, auto insurance, bonding, and builder's risk.

Figure 2 contains the general and financial portion of a sample "Prequalification Questionnaire" used to collect and verify data on a prospective contractor. These questionnaires vary in length and complexity but, in general, ask for substantially similar information.

GENERAL INFORMATION		
1. Company Name:	Telephone:	Fax:
Street Address:	Mailing Address:	
e-mail:		
2. Officers: President:		Years with Company:
Vice President:		
Controller:		
3. How many years has your organization been in business under your present firm name?		
4. Parent Company Name:		
City:	State:	Zip:
Subsidiaries:		
5. Under Current Management Since (Date):		
6. Contact for Insurance Information:		
Title:	Telephone:	Fax:
7. Insurance Carrier(s):		

Name	Type of Coverage	Telephone

8. Are you self-insured for Workers' Compensation Insurance? Yes ❑ No ❑	
9. Contact for Requesting Bids:	Title:
e-mail: Telephone:	Fax:
10. PQF Completed by:	Title:
e-mail: Telephone:	Fax:

■ ■ **Figure 2** ■ ■ **A general and financial prequalification questionnaire** (Source: Houston Business Roundtable)

ORGANIZATION

11. Form of Business:	Sole Owner ❑ Partnership ❑ Corporation ❑

12. Percent Minority/Female Owned:	EEO Category:

13. A. Describe Services Performed:

❑ Construction
❑ Construction Design
❑ Original Equipment Manufacturer and Installer
❑ Project Maintenance
❑ Maintenance

SIC Code:

❑ Original Equipment Manufacturer and Maintenance
❑ Service Work (e.g., janitorial, clerical, etc.)
❑ Manpower and Resource

B. Work Categories:

Check the categories in which you are interested in bidding and in which you are qualified to perform work.
Feel free to attach additional information clarifying your capabilities and specialties.

(C) denotes work done by company employees   (S) denotes work done by subcontractors

C	S		C	S	
		1. Air Conditioning/Refrigeration			11. Field Maintenance
❑	❑	Comfort Cooling/HVAC	❑	❑	General
❑	❑	Process Refrigeration	❑	❑	Hot Tape/Line Stops
			❑	❑	Leak Sealing (online)
		2. Buildings	❑	❑	Field Machining
❑	❑	Remodeling	❑	❑	Tank/Vessel Code
		New (steel, brick, block, other)	❑	❑	Boiler Code
			❑	❑	Exchanger Retubing
		3. Cleaning	❑	❑	Rotating Equipment
❑	❑	Industrial	❑	❑	Valve
		Janitorial	❑	❑	Cooling Tower
			❑	❑	High Alloy Welding (list type)
		4. Civil	❑	❑	Lead Lining
❑	❑	Concrete	❑	❑	Glass Lining
❑	❑	Excavation/Grading	❑	❑	Heat Treating
❑	❑	Paving	❑	❑	Nonmetallic materials
❑	❑	- Asphalt	❑	❑	Pipe Fabrication
❑	❑	- Concrete	❑	❑	Mobile Equipment Repair
❑	❑	5. Demolition/Dismantling	❑	❑	12. New Construction
		6. Electrical	❑	❑	13. Painting
❑	❑	General			
❑	❑	High-voltage/High-line	❑	❑	14. Refractory/Acid Brick
❑	❑	Heat Tracing			
❑	❑	Cathodic Protection	❑	❑	15. Rigging/Equipment Erection
❑	❑	Grounding Systems			
			❑	❑	16. Scaffolding
		7. Inspection & Testing			
❑	❑	General NDT	❑	❑	17. Scale Maintenance
❑	❑	Infrared Scanning			
❑	❑	Eddy Current Testing	❑	❑	18. Structural Steel Fab/Erection
❑	❑	Acoustic Emission			
❑	❑	Column Scanning	❑	❑	19. Tanks – Field Erection
❑	❑	Civil/Soils			
❑	❑	High Voltage Electrical	❑	❑	20. Other _____
❑	❑	Electrical Ground Inspection	❑	❑	_____
❑	❑	Fiberglass Inspection			
❑	❑	Other			

■ ■ **Figure 2** ■ ■ (cont.)

❏ ❏	8. Instrumentation General	❏ ❏	21. Consulting - Mechanical		
❏ ❏	DCS Control Systems	❏ ❏	- Electrical		
		❏ ❏	- Chemical		
	9. Insulation	❏ ❏	- Metallurgical		
❏ ❏	General	❏ ❏	- Controls		
❏ ❏	Asbestos Abatement	❏ ❏	- Other _____		
		❏ ❏	_____		
	10. Linings/Coatings for:	❏ ❏	_____		
❏ ❏	Metal	❏ ❏	_____		
❏ ❏	Concrete	❏ ❏	_____		
		❏ ❏	_____		
		❏ ❏	_____		

14. Describe Additional Services Performed:

15. List other types of work within the services you normally perform that you subcontract to others:

16. Attach a list of major equipment (e.g., cranes, man lifts, forklifts) your company has available for work and the methods of establishing competency to operate.

17. A. Do you normally employ?   Union Personnel ❏ Non-Union Personnel ❏ Leased Personnel ❏

    B. Average number of employees for last 3 years _____

COMPANY WORK HISTORY

18. Annual Dollar Volume for the Past Three Years:

19. Largest Job During the Last Three Years:

20. Your Firm's Desired Project Size: Maximum: Minimum:

21. D & B Financial Rating: Annual Sales: Net Worth:

22. Major jobs in progress:

Customer/Location	Type of Work	Size $M	Customer Contact	Telephone

23. Major jobs completed in the past three years:

Customer/Location	Type of Work	Size $M	Customer Contact	Telephone

24. Are there any judgments, claims, or suits pending or outstanding against your company?   Yes ❏   No ❏

    If yes, please attach details.

25. Are you now or have you ever been involved in any bankruptcy or re-organization proceedings?   Yes ❏   No ❏

    If yes, please attach details.

■ ■ **Figure 2** ■ ■ (cont.)

HEALTH, SAFETY, AND ENVIRONMENTAL PERFORMANCE

Owners must recognize and honor the moral and legal obligations to provide a safe work environment to minimize injuries. Owners have, in addition to their moral and legal obligations, an *economic* incentive to help reduce the number of accidents that occur on their construction projects. The high cost of accidents gives owners, as users of contractors, good reason to concern themselves with the safety efforts of the contractors they hire. Past research has shown that accidents are, to some extent, controllable by all levels of construction management. Studies also show that reasonable reductions in the frequency and severity of accidents would lower the cost of accidents by as much as 8 percent of direct construction labor payroll each year. There is ample economic incentive, in addition to humanitarian concerns, for owners to play an important role in construction safety.

One way that an owner can carry out this responsibility is to hire contractors who have a record of good safety, health, and environmental performance. A prospective contractor with a history of good SH&E performance is more likely to perform safely in the future than a contractor with a poor, or less-than-average, SH&E record. Contractors who hold their management accountable for accidents, as well as productivity, costs, scheduling issues, and quality generally have the best SH&E records. Several objective measures of past safety performance are available, notably in the experience modification rate that is applied to workers' compensation insurance premiums and OSHA-recordable injury and illness incident rates. Both may be obtained from contractors in the prequalification process, or prior to contract execution.

Owners cannot maintain a completely hands-off policy toward construction activity on the owner's property. The owner is charged with the legal duty to use reasonable care to correct or warn against nonapparent site hazards that the construction contractor might encounter in the course of his performance. Owners could, and commonly do, face third-party lawsuits in states that do not have workers' compensation as the sole remedy. These suits and actions are brought by the contractor's employees for injuries caused by the owner's potential breach of this duty; even if the independent contractor status of the construction contractors has been maintained and hold-harmless clauses are included in the contract language. The owner's duty often extends to unsafe activities by contractors, who create dangers for others on the site. Thus, the owner could be liable for injuries to individuals on the site that are caused by apparent unsafe practices of the construction contractor. (See Chapter 12—Multiemployer Work-Site Issues.)

Owners should recognize that the principles of management control commonly applied to costs, schedules, quality, and productivity are equally applicable to safety and that, if used properly, will improve safety performance. By showing more concern for, and focusing more attention on, construction safety, owners can help reduce injuries and loss of life—and the billions of dollars needlessly wasted by construction accidents.

Owners ultimately pay for a contractor's workers' compensation insurance and the other costs of worker injury. In fact, all monies spent by a contractor come from owners. As a contractor's safety focus and execution decreases, the owner's costs increase. If an owner chooses to work with select, safer contractors, then the owner's costs will be significantly less. Owners who pay attention to contractor safety records experience fewer third-party lawsuits and get more efficient execution of their work.

Three sources of information offer ways for owners to evaluate the probable safety performance of prospective contractors:

- experience modification rates for workers' compensation insurance
- OSHA total incidence rates
- contractor safety attitudes and practices.

Experience Modification Rates for Workers' Compensation Insurance

The insurance industry has developed experience rating systems as an equitable means of determining premiums for workers' compensation insurance. These rating systems consider the average workers' compensation losses for a given firm's type of work and amount of payroll, and predict the dollar amount of expected losses to be paid by that employer in a designated rating period, usually three years. Rating is based on comparison of firms doing similar types of work, and the employer is rated against the average expected performance in each work classification. Losses incurred by the employer for the rating period are then compared to the expected losses to develop an experience rating.

Workers' compensation insurance premiums for a contractor are adjusted by the experience modification rate. Lower rates, meaning that fewer or less severe accidents had occurred than were expected, result in lower insurance costs. A contractor's EMR is adjusted annually by using the rate for a percentage of a set number of previous years. Owners often use this information in the bid–cost analysis for contract awards, but it is also a valuable tool for assessing trends of a contractor's safety performance.

OSHA Incident Rates

The Occupational Safety and Health Act (1970) requires employers to record and report accident information on an Occupational Injuries and Illness Annual Survey Form known as the OSHA 300 Log (formerly the 200 Log). The employer must retain completed forms for five years. OSHA promulgated major revisions to the original record-keeping standard 29 CFR 1904 and 1952 in 2001. The revised record-keeping standard now requires employers to maintain, record, and report the accident information on an OSHA 300 Log Form. The 200 Log statistics will still have an impact until, over time, they are averaged out by the 300 Logs.

Information available from a contractor's OSHA log includes:

- number of fatalities
- number of injuries and illnesses involving lost work days
- number of injuries and illnesses involving restricted work days
- number of days away from work
- number of days of restricted work activity
- number of injuries and illnesses without lost work days.

A contractor, who has the number of hours his employees worked during the year, can compute incident rates for any or all of the items just listed using the following formula:

$$\frac{\text{No. of incidents} \times 200{,}000 \text{ hours}}{\text{No. of hours worked}} = \text{Incident Rate}$$

(The 200,000 hours in the formula represents the equivalent of 100 employees working 40 hours per week, 50 weeks per year, and is the standard base for incident rates.)

In calculating the OSHA-recordable incident rate, the number of incidents in the formula are the total number of fatalities, injuries, and illnesses involving lost and restricted work days, and injuries and illness without lost work days. The Bureau of Labor compiles construction industry incidence-rate averages each year for fourteen separate classifications of construction work and a variety of employee size groupings. Since review of the contractor's OSHA logs is common (and sometimes mandatory) prior to selection, these statistics may lead to further inquiries into a contractor's safety performance. Keep in mind that a significant change in numbers both here and in EMRs does not always indicate parallels in safety performance. Totals of hours worked and number of employees, varying from year to year, may also affect the numbers.

Contractor Safety Attitudes and Practices

Management accountability for health, safety, and environmental performance is a very important factor in determining a company's safety record. Companies that hold project management accountable for safety along with productivity, schedules, quality, and other areas will usually have the best safety records.

Management plays a major role in the safety culture of a company. The influence of management on the field operations is considerable and constant. The importance of working safely will be emphasized when management makes its commitment to safety known to personnel in the field. Whether it is through field safety inspections conducted by top managers, upper-level participation in safety initiatives such as training and orientation, or accident and incident investigations by company leaders, upper management can contribute considerably to the overall safety performance of the project.

The accountability and support of a contractor's program can be measured by the effectiveness of the following components of the contractor's SH&E program:

- frequency and distribution of accident reports (field superintendents, vice-president, president)
- frequency of project safety inspections and the degree to which they include project and field superintendents
- frequency of SH&E meetings for field supervisors.

PRACTICES OF THE SH&E PROGRAM

OSHA compliance is often expressed, incorrectly, as the objective of safety program implementation. The true objectives of a sound and effective safety program are protecting workers and eliminating losses due to poor working practices that could impact workforce well-being, project cost, and schedule, or that could adversely impact the environment. OSHA regulations define minimum legal requirements. Compliance does not guarantee safety, and in some instances does not even meet OSHA requirements (such as in performance-based standards where evaluation and reevaluation are required) The following sample questions are those an owner might use to evaluate a contractor's SH&E program.

Comprehensive Written Safety Programs
Does the contractor cover essential areas such as hazard communication, fall protection, and incident investigations, especially when those areas apply to the owner company's scope of work for the contractor? Are the programs/procedures adequate, and how is adequacy determined? Is a program "purchased," shared with prospective clients, then put away and not used to administer work? Does the contractor have evidence of essential programs such as incident investigations, audits, injury management, ongoing training, daily safety briefings, disciplinary policy, and substance-abuse testing?

OSHA Competent Workers and Supervisors
Does the contractor have documentation of OSHA regulatory-requirement competency? Does the contractor have a sufficient number of competent employees?

Qualified Supervision
Have supervisors been trained to be leaders, especially in safety? Do supervisors have copies of safety rules? Are they held accountable for safe performance by their teams? By management?

Capable/Competent Safety Staff
Has the contractor just given someone a safety title, or is the person truly competent/capable by degree, certification, experience, training, or some combination of these? A key determining question here is "Does the individual have the authority to take corrective actions?" If the answer is no, then the individual, regardless of expertise, is not truly the *competent person* on the site.

PREQUALIFICATION QUESTIONNAIRE

Figure 3 is an example of an owner's prequalification questionnaire used to collect and verify the safety data of a prospective contractor. (*Note:* this form is usually combined as one form with the general and financial questionnaire found earlier in the chapter.)

Control—Audits and Inspections

The contractual agreements and the SH&E plans for construction projects should contain necessary language and provisions for responsibilities that address multiemployer work-site issues (refer to Part III—Legal Aspects).

Owners should ensure that the SH&E program or plan for the construction project is implemented at the onset of construction activities.

The responsibilities for implementing and monitoring the SH&E plan should have been reviewed for compliance with the project contract terms and objectives and approved before a contractor is ever authorized to conduct on-site activities.

The following are SH&E controls that owners can implement on construction projects.

SAFETY & HEALTH PERFORMANCE						
26. Workers Compensation Experience Modification Rate (EMR) Data						

26. Workers Compensation Experience Modification Rate (EMR) Data	
a. EMR is ❑ Interstate rate ❑ Intrastate rate ❑ Monopolistic State rate ❑ Dual rate	b. EMR for three last years:
c. State of Origin:	d. EMR anniversary date:

27. Injury and Illness Data:

a. employee hours worked last three years (excluding subcontractors)	Hours/Year			
	Field			
	Total			

b. Provide the following data (excluding subcontractor) using your OSHA 200 and/or 300 Forms from the past three years:

	No.	Rate	No.	Rate	No.	Rate
Injury related fatality Rate = $\dfrac{\text{Total Co. 1} \times 200{,}000}{\text{Total Employee Hours}}$						
Lost workday case injuries involving days away from work, or days of restricted work activity, or both Rate = $\dfrac{\text{Total Co. 2} \times 200{,}000}{\text{Total Employee Hours}}$						
Lost workday case injuries involving days away from work Rate = $\dfrac{\text{Total Co. 3} \times 200{,}000}{\text{Total Employee Hours}}$						
Injuries involving medical treatment Rate = $\dfrac{\text{Total Co. 6} \times 200{,}000}{\text{Total Employee Hours}}$						
Total OSHA Recordable Injury Rate Rate = $\dfrac{(\text{Total Co. } 1 + 2 + 6) \times 200{,}000}{\text{Total Employee Hours}}$						
Illness related fatality Rate = $\dfrac{\text{Total Co. 8} \times 200{,}000}{\text{Total Employee Hours}}$						
Lost workday case illnesses involving days away from work, or days of restricted work activity, or both Rate = $\dfrac{\text{Total Co. 9} \times 200{,}000}{\text{Total Employee Hours}}$						
Lost workday case illnesses involving days away from work Rate = $\dfrac{\text{Total Co. 10} \times 200{,}000}{\text{Total Employee Hours}}$						
Illnesses not involving lost workdays or restricted workdays Rate = $\dfrac{\text{Total Co. 13} \times 200{,}000}{\text{Total Employee Hours}}$						
Total OSHA Recordable Illness Rate Rate = $\dfrac{(\text{Total Co. } 8 + 9 + 13) \times 200{,}000}{\text{Total Employee Hours}}$						
Total OSHA Recordable Injury/Illness Rate Rate = $\dfrac{(\text{Total Co. } 1 + 2 + 6 + 8 + 9 + 13) \times 200{,}000}{\text{Total Employee Hours}}$						

Notes: 1) Data should be the best available data application to the work in this region or area.
2) If your company is not required to maintain OSHA 200 or 300 forms, please provide information from your Worker's Compensation insurance carrier itemizing all claims for the last three years.
3) Note: OSHA 300 has column letters instead of numbers

■ ■ ■ **Figure 3** ■ ■ ■ **An owner's prequalification questionnaire for contractor safety and health performance** (Source: Houston Business Roundtable)

28. Have you received any regulatory (EPA, OSHA, etc.) citations in the last three years: If yes, please attach copies.	Yes ❑ No ❑

SAFETY & HEALTH MANAGEMENT

29. Highest ranking safety/health professional in the company:

Title:	Telephone:	Fax:

30. Do you have or provide:				
a. Full time Safety/Health Director	Yes	❑	No	❑
b. Full time Safety/Health Supervisor	Yes	❑	No	❑
c. Full time Job Safety/Health Coordinator	Yes	❑	No	❑

31. Do you have or provide:				
a. Safety/Health incentive program	Yes	❑	No	❑
b. Does the program address the following key elements?				
1. Management commitment and expectations	Yes	❑	No	❑
2. Employee participation	Yes	❑	No	❑
3. Accountabilities and responsibilities for managers, supervisors, and employees	Yes	❑	No	❑
4. Resources for meeting safety & health requirements	Yes	❑	No	❑
5. Periodic safety and health performance appraisals for all employees	Yes	❑	No	❑
6. Hazard recognition and control	Yes	❑	No	❑
c. Does the program satisfy your responsibility under law for:				
1. Ensuring your employees follow the safety rules of the facility?	Yes	❑	No	❑
2. Advising owner of any unique hazards presented by the contractor's work, and of any hazards found by the contractor?	Yes	❑	No	❑

32. Does the program include work practices and procedures such as:				
a. Equipment Lock out and Tag Out (LOTO)	Yes	❑	No	❑
b. Confined Space Entry	Yes	❑	No	❑
c. Injury & Illness Recording	Yes	❑	No	❑
d. Fall Protection	Yes	❑	No	❑
e. Personal Protective Equipment	Yes	❑	No	❑
f. Portable Electrical/Power Tools	Yes	❑	No	❑
g. Vehicle Safety	Yes	❑	No	❑
h. Compressed Gas Cylinders	Yes	❑	No	❑
i. Electrical Equipment Grounding Assurance	Yes	❑	No	❑
j. Powered Industrial Vehicles (Cranes, Forklifts, Manlifts, etc.)	Yes	❑	No	❑
k. Housekeeping	Yes	❑	No	❑
l. Accident/Incident Investigation	Yes	❑	No	❑
m. Unsafe Condition Reporting	Yes	❑	No	❑
n. Emergency Preparedness, including evacuation plan	Yes	❑	No	❑
o. Waste Disposal	Yes	❑	No	❑
p. Back Injury Prevention	Yes	❑	No	❑

33. Do you have written programs for the following:						
a. Hearing Conservation		Yes	❑	No	❑	
b. Respiratory Protection	Yes ❑	No	❑	N/A	❑	
Where applicable, have employees been:						
Trained		Yes	❑	No	❑	
Fit Tested		Yes	❑	No	❑	
Medically approved		Yes	❑	No	❑	
c. Hazard Communication		Yes	❑	No	❑	
Have employees been trained		Yes	❑	No	❑	

34. Do you have a substance abuse program?	Yes	❑	No	❑
If yes, does it include the following:				
- Pre-placement Testing	Yes	❑	No	❑
- Random Testing	Yes	❑	No	❑
- Testing for Cause	Yes	❑	No	❑
- DOT Testing	Yes	❑	No	❑

35. Medical						
a. Do you conduct medical examinations for:						
- Pre-placement	Yes ❑	No	❑	N/A	❑	
- Pre-placement Job Capability	Yes ❑	No	❑	N/A	❑	
- Hearing Function (Audiograms)	Yes ❑	No	❑	N/A	❑	

■ ■ **Figure 3** ■ ■ **(cont.)**

	Yes		No		N/A	
- Pulmonary	Yes	☐	No	☐	N/A	☐
- Respiratory	Yes	☐	No	☐	N/A	☐

b. Describe how you will provide first aid and other medical services for your employees while on site. Specify who will provide this service:

c. Do you have personnel trained to perform first aid and CPR?			Yes	☐	No	☐

36. Do you hold site safety and health meetings for:

	Yes		No			
Field Supervisors	Yes	☐	No	☐	Frequency	
Employees	Yes	☐	No	☐	Frequency	
New Hires	Yes	☐	No	☐	Frequency	
Subcontractors	Yes	☐	No	☐	Frequency	
Are the safety and health meetings documented?			Yes	☐	No	☐

37. Personal Protective Equipment (PPE)

	Yes		No	
a. Is applicable PPE provided for employees:	Yes	☐	No	☐
b. Do you have a program to assure that PPE is inspected and maintained?	Yes	☐	No	☐

38. Do you have a corrective action process for addressing individual safety and health performance deficiencies?

	Yes		No	
	Yes	☐	No	☐

39. Equipment and Materials

	Yes		No		N/A	
a. Do you have a system for establishing applicable health, safety and Environmental specifications for acquisition of materials and equipment?	Yes	☐	No	☐	N/A	☐
b. Do you conduct inspections on operating equipment (e.g., cranes, forklifts, manlifts) in compliance with regulatory requirements?	Yes	☐	No	☐	N/A	☐
c. Do you maintain operating equipment in compliance with regulatory requirements?	Yes	☐	No	☐	N/A	☐
d. Do you maintain the applicable inspection and maintenance certification records for operating equipment?	Yes	☐	No	☐	N/A	☐

40. Subcontractors
Do you use subcontractors? (If no, skip to question 42)

	Yes		No		N/A	
	Yes	☐	No	☐		
	Yes	☐	No	☐	N/A	☐
a. Do you use safety and health performance criteria in selection of contractors?						
b. Do you evaluate the ability of subcontractors to comply with applicable health and safety requirements	Yes	☐	No	☐	N/A	☐
c. Do your subcontractors have a written Safety & Health Program?	Yes	☐	No	☐	N/A	☐
d. Do you include your subcontractors in:	Yes	☐	No	☐	N/A	☐
- Safety & Health Orientation	Yes	☐	No	☐	N/A	☐
- Safety & Health Meeting	Yes	☐	No	☐	N/A	☐
- Inspections	Yes	☐	No	☐	N/A	☐
- Audits	Yes	☐	No	☐	N/A	☐

41. Inspections and Audits

	Yes		No	
a. Do you conduct safety and health inspections?	Yes	☐	No	☐
b. Do you conduct safety and health program audits?	Yes	☐	No	☐
c. Are corrections of deficiencies documented?	Yes	☐	No	☐

SAFETY & HEALTH TRAINING

42. Craft Training

	Yes		No	
a. Have employees been trained in appropriate job skills?	Yes	☐	No	☐
b. Are employees job skills certified where required by regulatory or industry consensus standards?	Yes	☐	No	☐

c. List crafts which have been certified:
Examples: Carpenter, Pipefitter, Equipment Operator etc.

43. Safety & Health Orientation

	New Hires				Supervisors			
a. Do you have a Safety & Health Orientation Program for new hires and newly hired or promoted supervisors?	Yes	☐	No	☐	Yes	☐	No	☐
Does program provide instruction on the following:	Yes	☐	No	☐	Yes	☐	No	☐
- New Worker Orientation	Yes	☐	No	☐	Yes	☐	No	☐
- Safe Work Practices	Yes	☐	No	☐	Yes	☐	No	☐
- Safety Supervision	Yes	☐	No	☐	Yes	☐	No	☐
- Toolbox Meetings	Yes	☐	No	☐	Yes	☐	No	☐
- Emergency Procedures	Yes	☐	No	☐	Yes	☐	No	☐
- First Aid Procedures	Yes	☐	No	☐	Yes	☐	No	☐
- Incident Investigation	Yes	☐	No	☐	Yes	☐	No	☐
- Fire Protection and Prevention	Yes	☐	No	☐	Yes	☐	No	☐
- Safety Intervention	Yes	☐	No	☐	Yes	☐	No	☐
- Hazard Communication	Yes	☐	No	☐	Yes	☐	No	☐

■ ■ **Figure 3** ■ ■ (cont.)

c. How long is the orientation program?					
d. Are written exams given? If no, how do you verify comprehension?	Yes	❑	No	❑	
44. Safety and Health Training					
a. Do you know the regulatory safety and health training requirements of your employees?	Yes	❑	No	❑	
b. Have your employees received the required safety and health training and retraining and is it documented?	Yes	❑	No	❑	
c. Do you have a specific safety and health training program for supervisors?	Yes	❑	No	❑	
d. Are all employees trained in the work practices needed to safely perform his/her job?	Yes	❑	No	❑	
e. Is each employee instructed in the known potential of fire, explosion, or toxic release hazards Related to his/her job, the process and the applicable provisions of the emergency action plan?	Yes	❑	No	❑	
45. Training Records					
a. Do you have safety & health and crafts training requirements for your employees?	Yes	❑	No	❑	
b. Do the training records include the following:					
Employee identification	Yes	❑	No	❑	
Date of training	Yes	❑	No	❑	
Name of trainer	Yes	❑	No	❑	
Method used to verify understanding	Yes	❑	No	❑	
b. How do you verify understanding of training? (Check all that apply)					

❑ Written Test	❑	Job monitoring
❑ Oral Test	❑	Other (List) _____
❑ Performance Test		

INFORMATION SUBMITTAL

Please provide copies of checked (✓) item with the completed PQF:

_____ Safety Commitment Letter from Executive Officer
_____ EMR documentation from your insurance carrier
_____ Insurance Certificate(s)
_____ OSHA 200 and/or 300 Logs (Past 3 years)
_____ Safety & Health Program (Table of Contents)
_____ Safety & Health Incentive Program
_____ Substance Abuse program
_____ Hazard Communication Program
_____ Respiratory Protection Program
_____ Housekeeping Policy
_____ Accident/Incident Investigation Procedure
_____ Unsafe Conditions Reporting Procedure
_____ Safety & Health Inspection Form
_____ Safety & Health Audit Procedure or Form
_____ Safety & Health Orientation (Outline)
_____ Example of Employee Safety & Health Training Records
_____ Safety & Health Training Schedule (Sample)
_____ Safety & Health Training for Supervisors (Outline)
Note: Owner checks items to be provided with evaluation questionnaire

■ ■ **Figure 3** ■ ■ **(cont.)**

Manager–Contractor Meetings

Safety should be the first order of business in all meetings, whether they are schedule or progress meetings. This provides a regular forum for management to reemphasize its commitment to safety. Starting meetings with a review of safety issues and a discussion of accidents also reinforces the owner's safety interests. Additionally, beginning with a "Safety Moment" (a brief current-interest topic that everyone can relate to regardless of job function) may be a good idea. For example, in the middle of the summer a brief "Safety Moment" concerning heat stress might be appropriate and can enhance the participants' perception of the owner's interest in the safety of everyone concerned with the project. Other subjects to cover periodically include recent safety-related incidents, accident statistics, identification of the contractor with the fewest incidents, and new safety objectives.

These meetings should continue (with contractor participation) throughout the entire construction phase. Contractors may initially view these meetings as a burden and show a reluctance to participate; however, benefits become apparent as they receive additional input to work-planning considerations that may have been overlooked. These meetings provide a forum for intracontractor and client discussions of support.

Safety inspections and audits

On a regularly scheduled basis, all site contractors should participate in site safety inspections. Management participation in site safety walks is a highly visible demonstration of a company's commitment to safety. The return on the time invested by management is well worth the involvement. Managers visiting the job site gain an increased level of contact with employees, which personalizes the project from both sides. Scheduled management tours usually supplement daily audits, allowing additional opportunities for hazard assessment and resolution.

Regular inspections conducted by site safety representatives and contractor supervisors enable prompt feedback to contractors of identified shortcomings. Substandard conditions should be corrected immediately. If necessary, suspend work until proper safety precautions are taken. Audits also allow employees to bring unsafe conditions and practices to the attention of supervision while the situations are taking place.

Significant safety observations made during the walks and other safety-related issues should be discussed at production meetings. Document this information and make it available for daily or weekly toolbox meetings. This demonstrates management's concern with job safety and shows that safety is managed with the same interest, and viewed with the same importance, as schedule and production, sending a positive message to the workforce.

Summary

The success of a construction SH&E program depends to a great extent on management involvement. One cannot place a dollar value on the humanitarian aspects of a safety program. It is also impossible to place a dollar value on the negative effects accidents have on labor relations and publicity. Without serious and persistent management commitment, merely adopting a safety program does not yield desired results.

Once having a good program becomes a priority, implementation depends on good communications between the owner and the contractor, as well as active owner participation. Human nature places emphasis upon that by which one will be evaluated. If safety is given due importance by project management, it will receive proportionate attention from the rest of the organization.

REFERENCES

Bennett, Brian T. "Orientation Program for Casual Contractors in the Chemical Industry," *Professional Safety*, August 2000.

Bureau of Labor Statistics. *Workplace Injuries and Illnesses in 2000*. Washington DC: United States Department of Labor, Government Printing Office, 2000.

Construction Industry Institute. "Zero Injury Economics," *Construction Industry Institute Special Publication 32-2*, September 1993.

Fagan, John L., Tyrone Monte, Darlene A. Powell, and Charles J. Croncini. "Contractor Review Committee," *Professional Safety*, May 1998.

Hislop, Richard D. "A Construction Safety Program," *Professional Safety*, September 1991.

Houston Business Roundtable, "Standardized Pre-Qualification Form (PQF)," www.houbrt.com.

Krizan, William G. "Owner Demands Craft Training," *Engineering News Record*, November 23, 1998.

Krzywicki, Robert S. "Guidelines for Contractor Safety," *Professional Safety*, June 2000.

_____. "Guidelines for Hiring Safe Contractors," *Blueprints*, American Society of Safety Engineers, Fall 2001, Volume 1, Number 1.

"Making Zero Accidents a Reality" presented at the Construction Industry Institute Annual Construction Project Improvement Conference, Austin, Texas, September 2001 (http//www.construction-institute.org/cpi.htm).

Occupational Safety and Health Administration, (OSHA Regulations – 29 CFR). Washington DC: United States Department of Labor, Government Printing Office.

1910.119, "Process Safety Management of Highly Hazardous Chemicals."

OSHA Directive CPL 2-0.124, "Multi-Employer Citation Policy 12/10/1999."

1926.16, "Rules of Construction."

1903.1, "Purpose and Scope," Williams–Steiger Occupational Safety and Health Act of 1970.

29 CFR Parts 1904 and 1952, "Recording and Reporting Occupational Injuries and Illnesses."

The Business Roundtable, "Improving Construction Safety Performance," *Report A-3*. Houston, TX: Business Roundtable, January 1982.

Tulaz, Gary J. "Construction Leads in Deaths Despite Low Fatality Rate," *Engineering News Record*, August 27, 2000, http://www.enr.com.

REVIEW EXERCISES

1. List and discuss three economic factors that can impact the financial stability of any construction project.
2. Explain why a construction safety professional should have a basic understanding of third-party liability at the job site.
3. Explain how a company can design a pre-qualification form that will better address the leading causes of death in the construction industry.
4. What role should job-site owners have prior to the start of a construction project?

.. PART II ..

Safety Process: Key Components

■■ **About the Authors** ■■

SUBSTANCE-ABUSE PROGRAMS
FOR CONSTRUCTION SITES
by Arnold R. Braver, M.B.A., and Gordon E. Wall

Arnold R. Braver, M.B.A., is currently the Human Resource Manager for a large transportation construction company headquartered in Michigan. Mr. Braver has been involved in substance abuse and behavioral health administraton for over twenty years. He obtained a Bachelor's Degree in Psychology from the University of Michigan and an M.B.A. from Oakland Univeristy, Rochester Hills, Michigan.

Gordon E. Wall is the Corporate Safety Manager for a large transportation construction company headquartered in Michigan. He obtained his Bachelor of Science Degree in Occupational Safety and Health in 1979 from Ferris State University. Mr. Wall is a member of the Detroit chapter of the ASSE. He is the former chairman of the GLCA Safety Committee and is currently a member of the MRBA Safety Committee.

7

Substance-Abuse Programs for Construction Sites

LEARNING OBJECTIVES

- Discuss the history of drug-testing programs in the workplace.
- Explain the reasons for and against drug testing.
- Discuss the impact of drug testing on the construction industry.
- Identify best practices for implementing a drug-testing program.
- Identify other elements of a comprehensive safety program.

INTRODUCTION

Substance abuse in the workplace has prompted many companies to develop drug-testing programs intended to reduce safety incidents and improve productivity and morale while at the same time making workers and the public feel safer. This chapter will examine the history of drug testing, arguments for and against its implementation, its current prevalence throughout industry, and the results of drug testing generally in the workplace population as well as specifically in the construction industry. Model programs and their impacts are described. An analysis of drug-testing results and its impact on other safety measures concludes that drug testing, as an element of a comprehensive safety program, is highly successful in helping companies improve their safety records.

HISTORY OF DRUG-TESTING PROGRAMS

The earliest industry efforts to impact employee performance problems based on drug and/or alcohol use were initiated in 1914 when the Ford Motor Company introduced a profit-sharing plan that was supplemented by visits from the company's sociological department to employees' homes to evaluate aspects of their lives (fidelity, thrift, sobriety, etc.). Ford's management showed foresight in recognizing that there is a close relationship between home life and work productivity.

In the prewar and wartime shipyards of California, Washington, and Oregon, as well as in mining towns in Michigan, high turnover rates, accidents, and absenteeism were epidemic and caused many employers, most notably Kaiser, to develop in-house healthcare delivery systems—again in recognition of the relationship between employee health (in general, but including substance abuse) and productivity.

Beginning in the 1960s, many companies developed policies—often included in collective bargaining agreements—prohibiting drinking on the job, acknowledging that sober workers were the most productive. These agreements and policies were the precursors of today's employee assistance programs, found in many modern companies. These programs provide prevention, early identification, and treatment for alcohol and other drug users, often at company cost. They create a rehabilitative rather than a punitive environment for employees with substance abuse or other personal problems that may be impacting their work performance.

Beginning in the early 1980s, some employers began collecting urine or blood specimens from prospective or current employees in an attempt to identify those who might present productivity, attitude, or performance problems. Although the rationale for drug testing reflects some highly controversial elements, it does seem to successfully impact safety records, insurance rates, and other productivity issues.

In 1986, Executive Order 12564 established the Drug-Free Federal Workplace, which defined the federal government's goal of a drug-free work environment and made refraining from the use of illegal drugs, both on and off duty, a condition of federal employment. In 1988, the federal government passed the Drug-Free Workplace Act. This act requires that federal contractors earning over $100,000 participate by providing:

- Active, visible leadership and support of the program.
- Clearly written drug-free workplace policies and procedures that are applied uniformly.
- Employee and union involvement in program development.
- Management, supervisors, union representatives, and employees who are knowledgeable about their rights and responsibilities.
- Access to treatment and follow-up by employees who are experiencing substance-abuse problems.
- Methods of identifying alcohol and drug abusers, including drug testing, for the purpose of offering the opportunity for treatment, recovery, and the return to work

As a result of several highly publicized railroad accidents and many over-the-road, freight-hauler accidents, in 1995 the federal government mandated substance-abuse testing for employees in certain safety-sensitive positions. These included federal highway workers, drivers, airline pilots, and traffic control professionals.

For more details on the history of drug-testing programs, go to the Web site for the Substance Abuse and Mental Health Services Administration (SAMHSA), Division of Workplace Programs (www.drugfreeworkplace.gov).

PROS AND CONS OF DRUG TESTING

Opponents' Arguments

Many opponents of drug testing cite the fact that drug tests measure use of illegal drugs rather than measuring *impairment* leading to job-performance issues. In fact, drug-testing methodologies cannot determine *when* a drug was used, only its presence in the bloodstream. The quantity used cannot be determined either.

Opponents also cite the varying dissolution times of drugs; for example, marijuana can be detected in urine drug screens for up to 45 days while alcohol can only be detected for a period of hours. Therefore, according to researcher Ethan Nadelmann, "It [drug testing] also creates a bizarre incentive: If one wants to get inebriated on Friday night and still pass a urine test Monday, smoking a joint would be foolish. Cocaine and alcohol would represent the 'safer' choices of intoxicants because alcohol is 'legal' and cocaine cannot be detected in the body as long." ("Drawing the Line on Drug Testing," Intellectual Capital.com, October 14, 1999, p. 2)

In fact, because of both the prevalence of marijuana use and its lengthier absorption time, most drug tests giving a positive result are positive for marijuana. Opponents of drug testing argue that marijuana users do not represent severe safety or performance risks in the workplace.

Another argument presented by opponents is reliability of the drug-testing procedures themselves. According to the American Civil Liberties Union

(ACLU), false positives are reported from 10–30 percent of the time. Many legal and common substances are believed to create false-positive test results (ACLU Briefing Paper #5).

The final and most far-reaching argument posed against drug testing is legal invasiveness into employees' private lives. The ACLU, and others, consider the process demeaning, stating, "It is unfair to force workers who are not even suspected of using drugs, and whose performance is satisfactory, to prove their innocence through a degrading and uncertain procedure that violates personal privacy."

Proponents' Arguments

According to the 2001 National Household Survey on Drug Abuse, commissioned by the federal government's Substance Abuse and Mental Health Services Administration, an estimated 15.9 million Americans age 12 and older used an illicit drug within the past month. Also in 2001, 25.1 million Americans reported driving under the influence of alcohol within the past twelve months. (Results of this survey are available on the SAMHSA Web site.)

Proponents of workplace drug-testing programs also cite statistics such as the following:

- 75 percent of illegal drug users are employed in either full- or part-time positions.
- An estimated 6–17 percent of the workforce uses drugs.
- Estimates indicate that drug abusers cost companies between $7,000–10,000 per employee annually.
- Drug users incur higher medical costs (as much as 300 percent over non-drug-using employees).
- Drug users present other performance problems, notably excessive absenteeism (up to 16 times that of nonusing employees).

These statistics and others are available on the Web site PreEmploymentdrugtests.com.

Proponents also argue that testing will reduce the number and severity of accidents, workers' compensation claims, injury rates, and experience modification rates. In addition, some companies point to customer relations as a reason to implement drug-testing programs.

In fact, companies that perform postaccident drug testing on employees may, in some jurisdictions, be able to successfully dispute workers' compensation claims for employees who test positive. However, in many states the employer must demonstrate that the positive drug test was the *proximate cause of the accident*. The Ohio Supreme Court recently struck down a statute *requiring* postaccident drug testing and requiring denial of workers' compensation benefits after a positive drug test, citing that the drug-testing requirement interfered with employees' fourth amendment rights to undue search and seizure (State ex rel. *Ohio AFL-CIO v. Ohio Bur. of Workers' Comp.*, 97 Ohio St.3d 504, 2002-Ohio-6717).

COMPONENTS OF DRUG-TESTING PROGRAMS

Drug-testing programs can test potential employees prior to hiring (postoffer testing is recommended to avoid potential Americans with Disabilities Act legal issues). Many employers limit their testing to potential employees only, attempting to avoid hiring "bad" employees.

Posthire drug testing can include any or all of the following categories: random, postaccident, annual, return to work, and for cause.

Most companies engaging in drug testing use urinalysis collected and analyzed at a laboratory as the methodology for checking specimens. Urine "dipstick" testing gives a quicker result and is less costly, but is more prone to error. Hair-follicle testing and other similar methodologies are more accurate, less invasive, and more costly. Blood testing is quite accurate, but highly invasive, costly, and creates other safety issues.

Most companies conducting drug testing today utilize a medical review officer (MRO) to review any preliminarily positive results and rule out false positives, as well as positives due to valid prescription use. This medical professional interviews the (prospective) employee and reviews any prescriptions or other information that may result in a false-positive report prior to reporting the result to the employer.

While the federal government requires a five-panel drug screen for its safety-sensitive employees (Department of Transportation, Department of Defense), the ten-panel drug screen is more popular with most employers. The five-panel drug screen tests for marijuana, amphetamine, cocaine, opiates, and phencyclidine, while the ten-panel test adds screening for barbiturate, benzodiazapene, methadone, methaqualone, and propoxyphene to those drugs tested in the five panels.

Finally, there seems to be a constant battle between users and labs to develop or detect *adulterants.* The Internet is full of Web sites offering drug users potential ways to *fool* the system, while laboratories are continually offering more sophisticated methods of detecting adulterants, or maskers.

The federal government also establishes criteria for laboratories, requiring consistent, documented, and monitored handling of drug-testing specimens; cut-off levels for reporting positive results; the use of medical review officers; and many other safeguards to ensure consistent treatment of samples.

PREVALENCE OF DRUG-TESTING PROGRAMS

In 1983, less than 1 percent of employees were subject to drug-testing programs. In 1991, 52 percent of companies surveyed reported utilizing drug-testing programs. By the year 2000, this had leveled off to 47 percent. (American Management Association, "A 2000 AMA Survey: Workplace Testing: Medical Testing: Summary of Key Findings.) In fact, the Bureau of Labor Statistics noted that almost 33 percent of the businesses they surveyed had dropped drug testing between 1988 and 1990. Most of the drop had been in small companies (less than 50 employees) who cited cost as the main reason they had terminated their programs ("Anti-Drug Progams in the Workplace: Are They Here to Stay?" *Monthly Labor Review,* US Bureau of Labor Statistics, April 1991, pp. 26–28).

According to the 2000 AMA survey, approximately two-thirds of all companies tested employees, with over 60 percent testing new-hires, and approximately 50 percent also testing existing employees. The survey showed that new-hire and regular employee testing is most common in manufacturing industries and least common in the financial services, as illustrated in Table 1.

One reason some companies give for discontinuing drug-testing programs is an excessive cost per positive result. In the early 1990s, two studies reported that the cost per positive result was between $20,000 and $77,000. These costs were based on positive result rates of only 0.5 percent, as recorded in the Cornell/Smithers Report ("Workplace Substance Abuse Testing, Drug Testing: Cost and Effect," Utica, NY: Cornell University, January 1992).

In 2001, Quest Diagnostics Incorporated reported positive results of 4.6 percent (based on 6.3 million tests performed nationally, among both federally mandated and nonmandated populations). This rate is the lowest since tracking positive result rates began in 1988. In general, rates of

▪▪ Table 1 ▪▪

Testing Prevalence by Industry Type

Industry	Preemployment drug testing*	Postemployment drug testing*
Manufacturing	78.9%	38.3%
Financial services	33.6%	6.9%
Wholesale/retail	56.2%	30.9%
Business/professional services	42.8%	19.8%
Other services	60.0%	39.5%

*Percent of companies conducting drug testing

positive results in the workforce are going down over time both for federal workers and the general population. In 2001, mandated testing of federal workers recorded a 2.9 percent positive rate (The Drug Testing Index, Quest Diagnostics Incorporated, 2002.) It can be postulated that, as drug-testing programs become more common in each industry, positive rates will go down because workers using drugs will leave that industry.

In the "testing reason" category, Quest's results for 2001 indicate the following for the general U.S. workforce:

▪▪ Table 2 ▪▪

Positive Drug Tests by Testing Reason

Testing reason	Percent positive
For cause	26.1
Random	7.0
Postaccident	6.0
Return to duty	5.3
New-hire	4.4
Periodic	3.4

Quest also reports that approximately 60 percent of their positive test results are for marijuana, approximately 14 percent for cocaine, and 6 percent each for amphetamines and opiates (based on 2001 results for the combined U.S. workforce).

CONSTRUCTION INDUSTRY TRENDS

There appears to be a paucity of published research relating to construction-industry-specific drug use, testing, or result data. However, Jonathan Gerber and George Yacoubian's article, "Evaluation of Drug Testing In the Workplace: Study of the Construction Industry," published in the *Journal of Construction Engineering and Management* (127:6), November/ December 2001, is one article that thoroughly evaluates this phenomena.

According to the article, the U.S. Department of Health and Human Services estimated that, in 1997, the construction industry had nearly double the rate of employees using illicit drugs or heavy alcohol when compared to all industries. Despite this, the percentage of companies performing pre-hire or postemployment drug screening was extremely low when compared to other industries. These statistics are particularly compelling given the safety-sensitive nature of construction work and the heavy machinery used, as well as the potential danger to the public.

In this study, data was collected from 69 construction firms (out of 405 randomly selected and sent questionnaires). The data measured perceptions of substance-abuse problems in the industry, existence and features of drug-testing programs, and impact of the drug-testing programs on accident rates, experience modification ratings, and perceived results. Of the 69 responding companies, 49 (or 71%) had drug-testing programs in place.

As a safety-sensitive industry, construction companies typically have high workers' compensation insurance premiums. The workers' compensation insurance industry compares companies' loss experiences by the establishment of the experience modification rating (EMR). Basically, an EMR *less* than 1.00 means that the company's claim experience is *better* than the industry average, whereas an EMR *greater* than 1.00 suggests that the company's claim experience is *worse* than the industry average. Also, premium rates are impacted by the EMR (EMRs lower than 1.00 result in a premium reduction; EMRs higher than 1.00 pay a premium penalty).

Currently, commercial and some public projects *require* EMRs lower than 1.00 in order to submit bids for new work.

Gerber and Yacoubian noted that those construction companies with drug-testing programs in place had lower EMRs every year from 1995 through 2000 than did their non-drug-testing counterparts, as indicated in Table 3.

Generally, companies giving drug tests improved their EMRs over time after the institutionalization of a drug-testing program, whereas EMRs of companies that did not drug test remained constant over time. Thus, the percentage of improvements noted increased over time. This suggests that drug-testing programs are successful at reducing the incidence and severity of workers' compensation cases, and therefore workers' compensation costs for construction companies.

▪ ▪ Table 3 ▪ ▪
Construction EMR Trends

Year	EMR of companies giving drug tests	EMR of companies not giving drug tests	Percent improvement
1995	0.923	0.935	1.3
1996	0.936	0.955	2.0
1997	0.895	0.957	6.5
1998	0.898	0.982	8.5
1999	0.833	0.958	13.0
2000	0.842	0.950	11.4

Source: Gerber and Yacoubian. Reprinted with permission of ASCE.

This study also identified the reasons for implementing drug-testing programs, as well as the barriers to implementation. As identified by the questionnaire respondents, the top three implementation reasons were:

1. The promotion of the safety of their workers and those who use their products and services.
2. The belief that drug testing contributes positively to a company's image.
3. The recognition that screening is an effective deterrent.

Identified as the three most important barriers to implementing drug-testing programs were:

1. A concern for increased legal liability.
2. Concerns that testing could be too costly.
3. The existence of state regulations prohibiting or limiting a company's right to test.

This landmark study also demonstrated an impact of drug-testing programs on incidence rates. Incidence rates, as defined by OHSA, are the recordable occupational injuries and illnesses per 200,000 work hours (100 workers) per year. In the construction industry as a whole, the incident rate decreased from 14.6 to 8.8 between 1988 and 1998. It is postulated that this was due to a number of workplace safety programs that were implemented during this time period.

Among the respondent companies that have implemented drug-testing programs, the Gerber–Yacoubian article reports average reductions of .51 percent in the incident rates two years after implementation of drug testing.

Although this study reviewed only a small sample of construction companies, and only those willing to respond to a questionnaire, the data reported suggest that drug-testing programs are very effective mechanisms for objectively improving companies' safety records. It appears that drug testing is particularly effective when (a) a thorough, consistent drug-testing regimen is practiced, and (b) other safety enforcement programs are also introduced.

GREAT LAKES CONSTRUCTION ALLIANCE (GLCA) PROGRAM

In 1998, in response to the safety-sensitive nature of construction work, the (perceived) substance-abuse problem in the industry, and the high rate of safety incidents, the Great Lakes Construction Alliance developed the Management and Unions Serving Together (MUST) program. This program operates in Michigan.

MUST (currently managed by Safe2Work™) is a comprehensive drug and alcohol screening program that also offers CD-ROM training on a variety of safety modules.

Construction owners subscribing to the program (including many large employers such as the Big Three automobile companies, DTE Energy, etc.) have agreed to register their construction sites as drug free. In order to work on these sites, craftspeople must agree to register in the program, which means that they have been drug-tested and have submitted to the minimum of a site safety orientation. They then receive a membership card they can provide to new employers at job sites and other valid locations. Currently, owners are adding safety-module completion requirements as well.

One unique feature of the MUST/Safe2Work program is *portability*. Given the sometimes transient nature of construction employment, an individual [or his or her (prospective) employer], once registered, may pull up his or her "report card" on a Web site. This will verify their status for any interested party, including new employers or construction-site owners or managers.

Safe2Work requires preemployment (preregistration), annual, for-cause, and random drug testing of all "members." Twenty-five percent of all enrollees are tested annually on a random basis. This random testing includes breath alcohol testing. Safe2Work uses a medical review officer and has strict prohibitions against adulteration. The penalty for a first-time positive drug test is 30-day removal from all Safe2Work job sites, with a 90-day removal for the second positive, and one-year removal for the third (and any subsequent) positive results. Again, management and union representatives have jointly developed this program.

Safe2Work currently has over 63,000 registered employees and over 1400 registered employers. Safe2Work conducted almost 28,000 drug and alcohol tests in 2001 (with a 5.25% positive rate) and over 22,000 through August 2002 (with a 4.69% positive rate) (MUST, "Safe2Work Statistical Information," September 1, 2002). No data is available regarding the incident rate or any other safety impact of the program.

According to Donald O'Connell, the Executive Director of MUST, "It is difficult to isolate the effect of drug testing on objective safety measures. In most cases, numbers are reduced due to a positive corporate safety attitude. Drug testing, safety training, and reinforcement of policies all influence the end results" (personal e-mail correspondence, October 4, 2002). While this is no doubt true for all drug-testing programs, it appears that drug testing is one critical element in the development of a corporate safety culture that will reduce incident rates and EMRs.

MODEL PROGRAM RESULTS

A model program was instituted by a large, multi-faceted road-building company with headquarters in a Midwest state and operations throughout its home state and Florida. Currently, the company has over 500 employees (including laborers, operators, teamsters, and cement masons). The company has completed over 600,000 work hours per year in each of the last three years.

In early 2001, the company became more aggressive and comprehensive in its drug-testing program, although new-hire and mandated DOT drug testing were in place prior to that time. The company's modified drug policy, adopted and implemented April 1, 2001, included the following elements:

1. *The MUST program was adopted companywide.* Prior to that time, the company applied MUST testing policies on MUST job sites, utilized the company drug-testing policy on other sites, and implemented DOT testing for its drivers. However, given a high level of mobility of the workforce, the company found that employees who had not been MUST-certified would suddenly be required on a MUST job site. Therefore, the decision was made to have all employees, field and office, participate in MUST.
2. Along with the required MUST testing (prehire, annual, for cause, and random), the company added testing for:
   a. **postincident:** a requirement that *any* employee involved in *any* safety incident (including vehicular or equipment accident, injury, utility hit, speeding ticket, etc.) would be required to have an immediate drug screen and breath alcohol test.
   b. **return to work:** a requirement that any employee returning after a 60-day layoff obtain a drug test prior to returning to work.
3. In addition, the company initiated very strict prohibitions against refusal or adulteration, requiring immediate termination.
4. Finally, the company added a strict company response to positive results:
   a. New-hire positives result in immediate termination.
   b. The first positive result for an existing employee will normally result in immediate

discharge as well. However, a last-chance agreement *may* be offered to an employee who has demonstrated a strong history of performance, appropriate circumstances and attitude surrounding this result, and a willingness to:
1) Seek evaluation by a designated substance-abuse professional (SAP) at the employee's cost.
2) Follow the treatment and/or education recommendations of that SAP.
3) Obtain clearance to return to work from the SAP.
4) Undergo follow-up testing on demand for a period of 12 months (normally 6–12 tests are done during this time period).
c. If there is a second positive result at any time during the employment, immediate discharge results.

Table 4 shows the results this comprehensive drug-testing program, combined with other changes from instituting a corporate safety culture, has had in the first two years after implementation.

DISCUSSION

The company described has experienced significant improvement in all indicators measured, similar to improvements noted in the study of the construction industry by Gerber and Yacoubian. Also, drug-testing results for 2001 and 2002 are better than the Quest National Data, the data from the construction industry, or the GLCA results.

Possible reasons for this are that, in addition to the implementation of a comprehensive and consistent drug-testing program, the company has taken several other steps toward intensifying the development of a safety culture, including:

■ ■ **Table 4** ■ ■

Model Company Results*

Year	Work hours	EMR (1)	Incident rate (2)	Lost-day rate (3)	Drug tests	Positive rate
1999	600,000	0.93	28.90	6.50	no data	no data
2000	583,000	1.00	17.53	3.08	403	4.2%
2001 (4)	1,507,869	0.92	9.42	1.86	934	2.9%
2002 (5)	976,865	0.78 (6)	6.14	1.43	**777**	**2.9%**

*Source: internal company data

Table Notes
(1) **Experience Modification Rating:** EMRs are based on claim experience of the three years prior, excluding the most recent year. In other words, the EMR for 2001 is based on claim experience from 1997–1999.
(2) **Incident rate:** The number of recordable occupational injuries and illnesses per 200,000 work hours (100 workers) per year. Incident rates of 8.00 or below are desirable in construction work.
(3) **Lost-day rate:** The number of recordable occupational injuries and illnesses per 200,000 work hours (100 workers) per year that *result in lost time.* Lost-time incident rates of 6.00 or lower are generally acceptable in the construction industry.
(4) **2001 work-hour data** include all operations. Prior work-hour data reflect Midwest operations only.
(5) **2002 data** are through September 30.
(6) **2003 EMR** is projected to be approximately 0.5, based on claim experience from 1999–2001.

1. The development and filling of a Corporate Safety Manager position with appropriate authority and total support from upper management.
2. The implementation of a corporate safety committee, consisting of upper management, which regularly reviews accident trends, etc.
3. The implementation of safety incentive programs.
4. Careful monitoring of toolbox talk meetings.
5. Regular meetings with field management to review safety issues.
6. The implementation of disciplinary processes for noncompliance with safety guidelines.
7. Focusing attention on personal protective equipment.

It is estimated that drug testing will cost this model company approximately $55,000 this year. However, the claim, premium, and productivity savings make it worth that price tag.

It is this development of a comprehensive attention to, and value of, safety—*including drug testing*—that has helped this company seriously improve its safety record.

Each of these corporate actions enables employees to understand that the company takes safety seriously. Over time, these kinds of priorities help safety-oriented companies attract and retain safety-oriented employees. *Good* employees know that the company is looking out for them. Similarly, drug-using employees tend to self-select out of companies where they know drug testing (and the overall safety culture) is taken seriously.

Among model companies within the construction field, and across industries in general, it appears that comprehensive drug-testing programs are effective in impacting positive-result trends, EMRs, and safety-incident trends, and certainly are a critical element in successful safety programs.

REFERENCES

Gerber, Jonathan and George S. Yacoubian, "Evaluation of Drug Testing in the Workplace: Study of the Construction Industry," *Journal of Construction Engineering and Management*, 127:6, November/December 2001:435–44.

REVIEW EXERCISES

1. Describe some of the precursors of drug testing. Why were early employers interested in the home life of their employees?
2. Describe opponents' arguments against the use of drug testing.
3. In what ways has drug testing impacted safety records in the construction industry?
4. Identify the components of a model drug-testing program.
5. What other elements of a safety program are required to make it comprehensive and successful?

■■ **About the Author** ■■

WHAT'S WRONG (OR RIGHT)
ABOUT INCENTIVES
by John Gambatese, Ph.D., P.E.

John Gambatese, Ph.D., P.E., is an Assistant Professor in the Department of Civil, Construction and Environmental Engineering at Oregon State University. Dr. Gambatese's educational background includes Bachelor and Master of Science degrees in Civil Engineering from the University of California at Berkeley, and a Ph.D. in Civil Engineering from the University of Washington, focusing on Construction Engineering and Management. He is a member of the American Society of Safety Engineers (ASSE), American Society of Civil Engineers (ASCE), and the American Institute of Constructors (AIC), and actively participates on ASCE's Construction Site Safety Committee, Constructability Committee, and Construction Research Council. Dr. Gambatese has taught courses on a variety of construction and engineering topics, including construction safety and productivity improvement, construction contracts and specifications, and construction planning and scheduling. He has published numerous articles on construction safety and other topics, and performed research on construction worker safety, project constructability, construction automation, construction contracting, and life-cycle properties of constructed facilities. He is a licensed Professional Civil Engineer in California.

CHAPTER 8

What's Wrong (or Right) About Incentives

LEARNING OBJECTIVES

- List the characteristics of incentives and the various types of incentives used to motivate workers.
- Discuss the types of incentives commonly used in the construction industry and how incentive programs are implemented by construction companies.
- Discuss issues of concern associated with the implementation of incentive programs in construction companies.
- Explain the impact incentives impose on safety performance in the construction industry.
- Develop a number of practices to assist in crafting and implementing incentive programs in construction companies.

THE NATURE OF INCENTIVES

There are many ways to motivate workers to improve performance. Worker motivation is often linked to a person's desire to fulfill his or her needs. Personal needs from the highest down to the most basic level include: self-fulfillment, self-esteem, social belonging, safety, and survival. Once a need has been satisfied, that need no longer acts as a motivator, and, in the absence of any needs, there is no motivation. Motivation is also believed to stem from a worker feeling that he or she is performing meaningful work. A sense of achievement, along with recognition, responsibility, advancement, and growth, are factors on which employees base their motivation to work. Worker performance is generally high when these motivators exist within the workplace.

One way employers have attempted to motivate workers to improve safety performance is through incentives. An incentive is a benefit that is offered as a result of an accomplishment or an exhibited behavior. The benefit can be anything of value, either real or perceived, to the person striving to accomplish the associated objective. It can be monetary, such as a weekly allowance, increased salary, or an end-of-year bonus, or it can be an asset that contains some value, commonly some type of personal possession. Benefits may be nonmonetary but still retain some value. Examples of nonmonetary benefits are an extra day off from work, a promotion, or a conveniently located parking spot. Prestige, fame, and importance are also commonly considered benefits. While these may not result in immediate benefit, their anticipated potential for future monetary or material benefit is the primary value.

An incentive involves both a benefit and a performance objective. A performance objective is an established goal which, when met, provides some value to the person or entity offering the benefit and also, possibly, to the person receiving the benefit. A person is motivated to successfully meet the stated performance objective with the expectation of receiving the benefit. The person receives the benefit when the objective is successfully met. If the stated performance is not met, the benefit is not received. Specific programs are set up to implement and administer incentives. Such programs outline the nature and structure of the incentives and can incorporate a variety of different benefits for meeting a number of different objectives.

Typically, worker incentives are created to acknowledge and promote a specified worker performance or behavior. When a level of performance is set, workers must attain the specified level in order to receive the benefit. A set number of days without sustaining an injury is an example of a commonly used performance criteria related to safety. Production level, project duration, and project cost are other criteria often used as the basis for a level of performance that is to be met. Incentives may also be implemented to establish or shape certain behaviors. An employee, for example, who takes extra time out of his or her day to teach a newly hired worker how to safely perform a task might be rewarded with some type of benefit. In this case, the desired result is not the achievement of a specified level of performance, but a behavior needed to attain the level of performance. It is anticipated that, by accomplishing the stated performance objective, the worker will behave in a desired manner. While the behavior is not the central focus of the incentive, it is the desired outcome. Incentives that are inherently designed to alter behavior as opposed to just achieving a desired level of performance will generally lead to long-term improvements in worker performance. This is especially true for construction worker safety. A successful safety incentive program will raise awareness of safety issues and instill proactive behaviors that create a long-lasting, safe working culture.

Incentives may also be implemented to curb or reduce *negative behavior*. Negative reinforcement of a behavior can be provided in the form of a disincentive—any form of punishment or penalty designed to inhibit unwanted levels of performance or types of behavior. For example, if workers do not wear the required personal protective equipment for a work activity, they are punished or penalized. The performance objective is not met, and, instead of a benefit, the workers are punished or assessed a penalty. The consequences might be a verbal reprimand, a written letter of negative performance, garnished wages, or suspension from work. For severe or repeated actions, the consequence might be termination from the job.

SAFETY INCENTIVES IN CONSTRUCTION

Incentives are often used in construction to help improve worker safety performance. Implementing incentives is frequently the first, and sadly sometimes the only, step that construction companies take to improve safety. Because of their prominent use and the lack of safety management knowledge within companies, often the incentive program is considered to be a company's safety program. This is unfortunate. Incentives should be considered as just one part of an overall safety management program. Studies show that construction companies generally have better safety records when their incentive program does not stand alone, but is one part of a comprehensive program (Liska, 1993).

There are essentially three different types of safety incentives that are commonly implemented by construction companies: outcome based, behavior based, and activity based.

The Outcome-based Safety Incentive

The traditional form of safety incentives that construction companies use focuses on outcomes. The objective of this type of incentive is to meet a specified outcome or level of performance. Typical outcome measures are the number of days and number of labor-hours worked without sustaining an injury. A company, for example, may offer an incentive to an employee or crew for working 100 days without an injury. The employee or crew receives a gift or reward if 100 days without an injury is attained. At the end of the 100 days, the day-count is restarted. If an injury occurs within the stated length of time, no benefit is received and the day-count is restarted. The period of time to be injury-free might also be stated as a month, calendar quarter, year, or the duration of a project. Generally the value of the benefit increases as the duration increases. A small gift such as a baseball cap or T-shirt might be offered for a month without an injury, while a television or $1000 bonus is given for a year of injury-free work.

This type of incentive is relatively easy to implement and therefore common among construction companies. The employer simply needs to establish the performance objective (e.g., number of labor-hours without an injury) and the benefit (e.g., a baseball cap), and then monitor when an injury occurs and the length of time between injuries. The simple format helps when communicating the incentive to employees. A clear understanding of the incentive minimizes confusion and discouragement regarding employee participation and motivation to attain the level of performance.

To implement this type of incentive, the employer must define the type of injury that triggers a restart of the day or hour count and loss of the benefit. It must be established whether all injuries are counted, including minor cuts and bruises, or whether only other, more severe injuries are of interest. Some companies, for instance, limit it to only OSHA-recordable injuries. Whatever type of injury is included, it should be consistent with the duration over which the performance will be measured. If all injuries are included, no matter how minor they may be, shorter time periods should be used because minor cuts and bruises can be a common occurrence. Severe injuries typically occur less frequently and, therefore, the corresponding time period should be longer. If only severe, uncommon injuries are included and the time period is short, the incentive benefit will be too easy to achieve. On the other hand, if a common, minor injury triggers a loss of the benefit and the time period is long, workers will get discouraged and not try as hard to achieve the desired level of performance. In the latter case, workers may even forego reporting injuries,

either willingly or as a result of peer pressure, in order to receive the incentive benefit.

The underlying premise of this type of incentive is that, in order to attain the specified number of days or hours without an injury, the employees must work in a safe manner. By doing so, it is assumed that safe worker behavior is established. Thus, this type of incentive is indirectly related to worker behavior. While a relationship between a lack of injuries and safe worker behavior can be assumed, it should be realized that not having injuries does not necessarily mean safe worker behavior. Unsafe work practices, in some cases, may not result in an injury or may result in a near-miss. Therefore workers who receive gifts for not having an injury may actually be benefiting while still performing their work in an unsafe manner. If this is the case, then the incentive does not promote safe worker behavior.

The Behavior-based Safety Incentive

With this type of incentive, workers receive a benefit for exhibiting certain behaviors. The performance objective is not a measurable outcome, as is the case with an outcome-based incentive, but rather is a type of behavior that is assumed will result in a desired outcome. Employees are rewarded for demonstrating good behavior or practices during their work. For example, a worker might be rewarded with a gift if he or she is observed performing a work task in a particularly safe manner. Employees who take personal time to learn about new safety procedures or who take extra steps to create safe working conditions might also be rewarded. While the incentives might be restricted to behaviors observed on a particular project, there is generally no time period associated with the incentives. The behaviors could take place at any time and with any frequency during the course of the project.

This type of incentive is typically more difficult to implement than outcome-based safety incentives. Instead of defining an outcome (e.g., number of labor-hours without an injury) and the type of injuries considered, the employer must establish the types of behaviors that are deserving of a reward.

This is often a difficult task since many different safe behaviors might be exhibited. Behaviors such as wearing gloves for a particular task, using a hammer in an appropriate manner, and remembering to turn on an exhaust fan while cutting wood might all qualify. So that workers can get a sense of what are desired behaviors, the types of behaviors that will lead to a reward should be established when the incentive is implemented.

It is important as well that rewards be *consistently* provided for safe behavior. When a reward is given to one employee for an observed behavior, it should be given to all employees who exhibit the same behavior. Workers will consider the incentive program to be unfair if some are rewarded for their actions while others who exhibit the same behavior are not rewarded. This requirement can be a drawback for employers because they must commit resources to observing all worker behavior. If the necessary resources are lacking, safe work behaviors may go unnoticed and therefore not be rewarded.

For this type of incentive, it is assumed that safe behavior will lead to improved safety performance in terms of reduced injuries. For example, if a worker uses the correct procedures when using a mobile scaffold, it is assumed that injuries related to use of the mobile scaffold will not occur or be less frequent. This connection may be clear and direct for some behaviors. For other behaviors and practices, though, the relationship to good safety performance might not be as direct. Therefore, the desired outcomes might not be achieved when certain behaviors are rewarded. When giving the reward, the employer should consider the behavior's potential impact on safety performance.

The Activity-based Safety Incentive

Activity-based incentives are similar to behavior-based incentives, but focus on participation in specified activities rather than behaviors. Employees are rewarded when they participate in sanctioned activities that relate to safety, such as safety toolbox meetings, safety training classes, and safety and health conferences. The activities might be periodically scheduled as part of the construction project or within established programs for the public. The more activities the employee takes part in, the more the employee is rewarded.

This type of incentive is generally easier to implement than behavior-based incentives. Performance with respect to the incentive is essentially measured by whether the employee participated in the activity. This is usually verified with a review of an attendance sheet or certificate of completion. With its focus on activities, it is essential that the employer define the activities that are acceptable and the level of reward that will be provided with each type of activity. This type of incentive is difficult to implement without establishing the sanctioned activities, especially if employees do not see the rewards being fairly distributed for seemingly equivalent activities.

Similar to behavior-based incentives, it is assumed that there is a relationship between participation in certain activities and improved safety performance. If an employee attends a safety conference, for example, it is assumed that the employee will bring that knowledge and training back to the job site upon his or her return. This in turn will result in improved safety performance for the worker, and possibly other employees. To ensure that this actually takes place, the employer should consider the resulting impacts of different activities. The employer might also incorporate a mechanism to ensure that the potential impacts are actually realized. Lacking a connection between the activity and improved safety performance, the reward is given for outcomes that do not materialize.

Types of Incentive Rewards

The gift or benefit received by those who fulfill the performance objective can take many forms. Gifts such as baseball caps, belt buckles, jackets, and duffle bags with the company logo are common. More expensive gifts, such as a CD-player, television, tickets to a sporting event, or a new car, are also awarded for meeting more challenging performance objectives. These gifts may be given directly to the workers, or the workers might obtain them through *safety bucks* earned during the project. A safety buck might be given to a worker or crew who works a specified number of hours without an injury or is observed performing a task safely. Safety bucks are accumulated by each worker and then redeemed for a prize. Higher-valued prizes require more safety bucks to obtain. Other awards that a company might provide include monetary awards or extra vacation days. One study of large construction firms and large construction projects (Hinze, 2002) revealed that 26 percent of incentive rewards were financial in nature, 38 percent were small gifts such as T-shirts and jackets, and the remainder consisted of both gifts and money. Another study of 150 general contractors in the southeastern part of the United States (Banik, 2002, see Table 1) investigated the type of reward more closely. It found that the most common incentive rewards given were special hard hats and mugs, dinner or lunch, and a written letter of appreciation or recognition.

In addition to offering different types of rewards, distribution of the awards may also be handled in different ways. One method commonly used when the performance objective is outcome-based, such as a specified number of days without an injury, is to make the reward's value progressive in nature. With this method, the value of the reward increases as the length of injury-free time increases. It is often the case, although not necessarily recommended, that the reward's value grows disproportionately with time.

▪ ▪ Table 1 ▪ ▪

Types of Incentive Rewards Offered

Types of incentive reward	Percent of companies that offer a reward
Speical hard hat/mug	92
Dinner/lunch	90
Written appreciation/recognition	88
Cash	62
Tickets to sporting events	62
Training/education	62
Bonus	56
Television/stereo	44
Freedom/special assignments	40
Promotion/advancement	37
Paid vacation days	33
Stock ownership	25
Denim jackets	15
Pick-up trucks	13
Other	27

Source: Banik, 2002

Another method used is to make the incentive cumulative. Cumulative incentives involve the accumulation of rewards or benefits, such as safety bucks, that can be redeemed for gifts. If an injury does not occur during a specified time period, the benefit is earned. No benefit is earned during the period if the workers do not go injury-free. In the study involving large construction projects (Hinze, 2002), most projects had progressive incentives (64 percent), while approximately 32 percent of the projects had cumulative incentives, and the remaining 4 percent had both progressive and cumulative incentives.

The incidence of injuries, as opposed to safe behavior, is the primary focus of incentives in many construction companies. This is revealed in the study of large construction firms which found that 64 percent of the companies surveyed base their incentive rewards on incidence of injuries (Hinze, 2002). An injury incidence rate is the traditional performance objective used in construction and is easier to implement than an incentive based on safe behavior. The use of incidence of injuries as the primary focus for incentives should not necessarily be associated with improved safety performance. This is discussed further later in the chapter.

EXAMPLE SAFETY INCENTIVE PROGRAMS IN CONSTRUCTION

Construction companies have developed many different types of safety incentive programs and implement them in a variety of ways. Described below are examples of actual incentive programs found within construction companies. Examples are given for small-, medium-, and large-sized companies, and for incentive programs that range from simple to complex. The incentive programs described here are only one part of a comprehensive safety program that each company implemented. Each company developed its incentive program to complement and interact with other features of its overall safety program. Therefore, while improvements in a company's safety performance occurred after the incentive program was implemented, the improvements were most likely not directly related to the implementation of an incentive program, but were a result of a combination of factors.

Small-sized Companies

Company A: This small-sized company installs coatings and linings for industrial construction projects. The company is 30 years old and employs approximately 60 workers. Its annual work volume ranges from $3–5 million.

The incentive program employed by this firm is based on recognizing good safety performance and positive reinforcement. The company recognizes superior safety performance in a company newsletter that describes ongoing projects. The employee's name and a description of the safe acts are published in the newsletter, which is distributed to all employees. Employees who go "above and beyond" that which is expected are recognized in a Safety Hall of Fame. Each month an employee is added to the Safety Hall of Fame for performance during the month. The monthly winner is heralded in the company newsletter. Additionally, each year one employee is recognized as the top safety performer for the year. The employee is selected for this award by his or her fellow employees. The yearly award winner is recognized at the annual Christmas holiday lunch, has his or her name added to a Safety Hall of Fame plaque, and is given a savings bond.

Company B: An electrical subcontractor with approximately 50 employees, this small-sized company has an annual work volume averaging about $1.5 million. It is a family-owned business with a 25-year history, specializing in electrical construction and maintenance services for the residential and commercial building industry.

The company's incentive program is multifaceted, based on the accumulation of points. Workers who accumulate points are able to exchange the points for gifts from a product catalog. The catalog includes gifts such as jackets, electronics, household items, and tools. More points are required for more expensive gifts. When ordered, the gifts are brought to the job site and given out by the supervisors. Points are accumulated in the following ways:

1. *Hours worked.* Each employee receives points for every hour worked as follows: craft employee: 0.25 points/hour, foremen: 0.35 points/hour, general foremen: 0.45 points/hour, and superintendents: 0.55 points/hour.

2. *Training classes.* Participation and completion of training classes earns 150 points. The classes could be those offered by the company or an approved outside organization (e.g., CPR class, first-aid class). Before points can be earned, the classes must be approved by the employee's supervisor.

3. *Safety performance.* An employee who exhibits outstanding safety performance receives 15 points. Outstanding safety performance is defined as the completion of a project without any accidents or OSHA-recordable injuries.

4. *Safety recommendations.* Employees who identify potential safety hazards and find ways to mitigate those hazards are awarded points as well. The number of points awarded depends on the situation. The greater the impact of the suggestion, the more points are awarded.

5. *Attendance.* Employees can earn points for perfect attendance.

In addition to rewarding employees with points for positive behavior and outcomes, the company also *deducts* points in an effort to deter poor safety performance. Points are deducted for injuries sustained, near-misses, and safety violations. A poor score on a safety inspection and an unexcused absence will also result in deducted points. Generally, more points are deducted for more serious incidents. The same amount of points that are deducted from an employee's total are also deducted from that of the foreman, general foreman, and superintendent who oversee that employee.

Company C: This company has been in business for 30 years and employs about 35 workers. It is an electrical subcontracting firm, performing work primarily on industrial projects.

The company has several monetary safety incentives that it has developed over the years. Workers are offered a $0.10-per-hour pay increase for safe performance. This increase accumulates each pay period if the employee does not sustain an injury during the period. No incentive pay is received for the period if an injury occurs.

Two other parts of the incentive program are structured as games. At the end of each week, employees who did not sustain an injury during the week draw a card from a deck of cards. The cards are kept by the employees until the end of the month. If the month only has four pay periods, an extra card is drawn so that all participants have five cards. At the end of the month, a round of poker is played with the cards. The employee with the high hand and the employee with the low hand each receives a $25 gift certificate.

The company also awards an incentive using the letters from a Scrabble game. At the end of each week, employees who did not sustain an injury during the week randomly draw a letter. At the end of the month, the employees make four- or five-letter words based on the Scrabble game rules. The employee who makes a word with the most points receives a $50 gift certificate.

Lastly, when the company receives any monetary award from its clients for safety performance, it passes the award on to its employees. The award is split between the employees based on the number of hours each employee worked on the project.

Medium-sized Companies

Company D: This mechanical contractor has been in business for approximately 18 years, performing industrial construction projects, plant maintenance, and shutdown work. The company has an annual work volume of $40 million and close to 900 workers.

Based on its belief that training and employee involvement are essential, along with employee buy-in and realistic goals, the company has several elements in its incentive program. Foremen and general foremen accumulate points for every hour that their crew(s) works safely. Cash awards are distributed based on the crew(s) injury incidence rate. If a crew's injury incidence rate is less than 10, the foreman or general foreman receives cash as follows: $200 for 3000 to 6999 injury-free labor-hours; $300 for 7000 to 12,499 injury-free labor-hours; $400 for 12,500 to 19,999 injury-free labor-hours; and $500 for 20,000 or more injury-free labor-hours. No award is given if the incidence rate is greater than 10.

An incentive is also offered to superintendents. For each project, superintendents are rated according to their performance with respect to specified criteria. The project manager and the corporate safety manager rate each superintendent in each of the following nine areas:

1. Availability of the necessary safety equipment at the job site.
2. Proper and consistent use of the necessary safety equipment by the crew.
3. The safety example set and adhered to by the superintendent.
4. Reduction of the number of repeat hazards discovered on the job site.
5. Timely reporting of injuries and near-misses.
6. Submission of paperwork related to safety on the job site.
7. Handling of potential hazards on the job site.
8. Injury incidence rate of the superintendent's crew.
9. Enforcement of the company's safety program among its subcontractors.

Superintendents with high rates in the categories are given a safety award at the end of the project. The superintendents receiving an award also receive a gift, commonly a weekend trip, tickets to sporting events, or an electronic appliance.

All employees are eligible for a monetary award. Each is evaluated on a quarterly basis and accumulates bonus pay at a rate of $0.02 per hour worked. At the end of each quarter, if an employee has not sustained an injury, bonus pay is awarded in the form of a gift certificate. If the worker is injured, the bonus pay for that quarter is lost.

Company E: This company is a general contractor that has been in business for 12 years and has approximately $50 million in annual revenue.

The company has a two-part incentive program, one for its craft employees and another for the craft supervisors. The incentive for craft employees is designed to help establish and maintain a safe job site for all employees. For each safe quarter of work (3 months of work without an injury or accident), a craftsperson receives an extra $0.25 per hour of pay. The extra pay is not awarded if the employee sustains an OSHA-recordable accident, receives discipline for an unsafe act, or fails to report an injury in a timely manner. As an additional incentive, if an entire work crew does not experience an

OSHA-recordable injury during a quarter, each crew member receives an additional $0.15 per hour.

Craft supervisors (foremen and general foremen) are offered an incentive as well. For a safe quarter of work, supervisors are given $1.00 extra for every hour of work. A safe quarter is defined as a 3-month period of time during which: no employee assigned to the supervisor has an OSHA-recordable injury or is not properly supervised with regard to safety performance and practices; and the supervisor does not receive a poor safety evaluation or formal safety-related discipline.

The company also establishes safety incentives for other companies with which it subcontracts work. An incentive is written into the subcontractual agreement that awards the subcontractor for working safely on the job.

Company F: This company is a painting and insulation subcontractor that has been in business for 30 years. The company has over 1000 field personnel and 150 foremen in seven regional offices. It maintains a $10-million equipment inventory and earns approximately $45 million in revenue each year.

The company provides both monetary and non-monetary incentives. Based on a weekly safety audit conducted by the company's safety manager with assistance from various on-site employees, the employees receive a $30 gift certificate for achieving an audit score above a specified level. If the specified level is not achieved, no certificate is awarded. The audit results and awards are presented at weekly safety meetings.

Superintendents can also receive cash bonuses at the end of the year based on safety performance. Performance is measured with respect to the company's self-insurance program. The amount of award equals the percentage of savings realized in the company's workers' compensation program on the project compared to the premium cost that the company would otherwise pay an outside carrier if it were not self-insured. In the course of a typical year, this can amount to almost $10,000.

Nonmonetary awards are also distributed. The company rewards employees, including salaried employees, who have worked 1000 hours or more without any type of work-related injury. The rewards are given at the end of the year and consist of jackets, shirts, and tools, depending on the number of injury-free hours attained. Safety goals are also set periodically during a project. When these periodic goals are met, employees receive hard-hat stickers, baseball caps, lunch, T-shirts, and other small gifts. Also, a drawing is held periodically for select individuals who have taken additional steps toward safety.

Lastly, the company holds a safety slogan contest for every work season. Employees are asked to submit safety slogans that are then judged for their impact and ingenuity. The employee who submits the winning slogan receives a cash prize. The winning slogan is then displayed on a banner in the home office, attached to the bulletin board in all job-site offices, and printed on weekly paychecks.

Company G: This company provides general contracting and construction management services for major commercial, industrial, and healthcare projects. The company brings in approximately $80 million in revenue and employs almost 250 workers.

The company's incentive program has three parts. On a quarterly basis, all employees are offered a choice of one of several small gifts if they do not have an injury that requires a doctor's visit. The gifts are generally small in value and bear the name of the company. At the crew level, any crew that makes it through a quarter or the entire project without an injury to one of its members is rewarded. The reward is a free lunch for the entire crew. Foremen are offered an additional quarterly safety incentive. They receive bonus pay if their crew works a specified number of hours without a recordable injury .

Large-sized Companies
Company H: This company is a general contracting firm that has been owned and operated by one family for more than 120 years. The company specializes in carpentry work on commercial building projects and construction management. It employs more than 1000 people, located in offices in 20 different cities. Its annual volume of work earns approximately $2 billion.

The incentive program within the company contains a variety of elements. Safety milestones are established for each project undertaken. A typical milestone is to reach 250,000 labor-hours without sustaining a lost-time accident on the project. If this is achieved, a barbeque lunch and baseball caps are given to everyone on the project. Rewards of greater value are given for reaching more difficult milestones. When 1.25 million labor-hours are reached without a lost-time accident, the employees are treated to a special dinner and given shirts with the company's logo. The employees and their families are invited to a Family Day celebration when no lost-time accident is sustained over 2 million labor-hours.

A safety banner contest is held every quarter during the project. Safety slogans submitted by employees are judged and the winning slogan is placed on the banner. The winning employee receives a $50 gift certificate, and the banner is prominently displayed. A similar contest is also held for a safety insignia for the project. The employee who submits the winning insignia receives a $100 gift certificate. The insignia is placed on all program gifts on the project and displayed throughout the project site.

Monthly drawings for cash awards are also held to recognize injury-free work. Approximately one out of four eligible employees receives a $100 award each month. Additionally, all employees plus subcontractors, salaried staff, and nonsalaried workers, are randomly assigned to teams each month. If members of the team do not suffer an OSHA-recordable injury during the month, one out of every four members of the team receives at least $100, based on a random drawing. A third drawing, based on attendance, is held only for hourly employees. A lump sum of $1000–3000, depending on the size of the workforce, is distributed equally among those employees whose names are randomly drawn from the pool of workers who have a perfect attendance record. These drawings are held at periodic safety luncheons during the course of the project.

Company I: This company specializes in integrated design, engineering, and construction of industrial facilities, process plants, and distribution centers. The company employs 325 workers and takes in approximately $200 million in revenue each year.

The company's incentive program is based on positive reinforcement for taking extra steps to en-

sure safety. A "Stop Work For Safety" award is given to recognize those employees who have done more in the area of safety than expected. Employees who receive the award are sent a personal letter of commendation from the company president and receive a $100 gift certificate.

Incorporated into the incentive program are also disincentives for unsafe acts or hazardous site conditions exhibited by subcontractors. When an unsafe act has been committed or a hazardous site condition is left without proper mitigation by a subcontractor's employee, the subcontractor is fined. Fines range from $50 for working without a hard hat to $200 for a trenching or confined space violation. The money collected by the company is given to other employees who have exhibited safe work habits or to a local charity on behalf of all of the workers. The fine doubles and the employee must leave the job site if the violation is repeated. The employee can no longer work on the project if a third offense occurs.

Company J: This large-sized company employs over 2500 workers on site and in the home office. The company's annual revenue is approximately $800 million. It provides construction services and engineering on petro-chemical projects.

Safety awards are given out periodically during a project when safety milestones are met. If no injuries occur over a span of one month, a special luncheon is held for all employees at the job site. Other gifts, such as T-shirts and baseball caps, are given to employees when other safety milestones are achieved. Additionally, dividends received from the company's workers' compensation policy are distributed to all employees as a reward for safe job performance.

Incentive Issues of Concern

While there are many different ways in which safety incentives are structured and implemented by construction companies, their use is not entirely widespread throughout the industry. In fact, it is not uncommon for some companies to terminate their safety incentive programs after an initial trial period because of a perceived lack of results, or unwanted results. Recent studies have shown that safety incentives are used by approximately 80 percent of all construction companies (Banik, 2002; Hinze, 2002). Consequently, incentive programs have become somewhat controversial in their use in construction. Those who favor incentive programs feel they are successful in helping to deliver long-lasting, superior safety performance. The main argument in favor of incentives is that they change workers' behavior, leading them to work more safely. Some feel that employees who receive a reward for recognizing and eliminating unsafe behavior eventually will *police each other* to use the proper procedures.

On the other hand, many feel that safety incentives have limited value and are not appropriate for various reasons. Incentives are often regarded as "toxic" to a company's safety culture on a long-term basis because employees eventually consider them entitlements. Some other reasons for not incorporating safety incentives into a comprehensive safety program include: the reward is too easy to earn, they may actually reward unsafe work practices if unsafe activities occur without an injury, and they do not necessarily alter behavior because they do not sufficiently motivate workers to change their behavior. Table 2 presents the common reasons construction firms cite for not implementing a safety incentive program.

Various labor organizations have voiced an opinion against incentives that reward workers when a prescribed length of time has been worked without sustaining an injury. They contend that this type of incentive promotes underreporting of injuries. This view is also held by OSHA, and employers approved for participation in its Voluntary Protection Programs (VPPs) are discouraged from using injury frequency as the basis for incentive awards because of a fear of underreporting.

There are several issues that should be taken into consideration when a construction company considers implementing safety incentives. When developing the incentive structure, a company should consider whether to reward safe behavior or a lack of injuries. As discussed previously, rewarding a worker for having no injuries over a period of time assumes that no unsafe behaviors or near-misses have occurred. This may not actually be the case,

even if no injuries occur. If unsafe behavior actually occurs but there are no injuries, then rewards are being given for unsafe behavior. On the other hand, rewarding safe behavior requires added effort, but is a proactive approach. Rewarding safe behavior focuses on the process of doing the work safely rather than merely striving for an end result, often at almost any cost if the value of the reward is high.

Another issue for consideration is whether to reward a worker for expected behavior as opposed to extraordinary behavior. Some companies take the position that it is appropriate to give workers an extra benefit for just working safely. This might be considered positive recognition of performing their job as required or as part of the standard benefits package. Others feel that this is rewarding workers for something that they should be doing every day as part of their job. By rewarding the workers for doing what is expected of them, there is no incentive to improve. If it is expected, for example, that employees wear hard hats while on the job site, there should be no reason to provide an extra benefit for doing so. The motivation they receive for wearing their hard hats may be better provided through a disincentive. Therefore, some companies establish their incentives to only reward employees if they go "above and beyond" what is required. If workers exhibit extraordinarily safe behavior or work an exceptional amount of time without an injury, then a reward is warranted.

■■ Table 2 ■■

The Case *Against* Safety Incentive Programs

Reasons for not having a safety incentive program	Percent of companies
Does not improve safety performance	86
Difficult to administer regularly and fairly	79
Busy with other aspects of construction	43
Management does not believe in it	36
Not an owner requirement	28
Too expensive	21
Other	28

Source: Banik, 2002

The structure of an incentive program will vary depending on whether the company considers the gift as a "reward" or an "award." A *reward* is generally given after a person performs a specified act, or acts in a certain way. If an incentive is offered for attaining a specified number of days without an injury, workers who achieve this level of safety performance receive a reward. Any number of workers can receive the reward as long as they attain the specified level of safety performance. An award generally relates to some type of competition in which those involved compete for a prize. Giving a pick-up truck to the worker who works the most hours without an OSHA-recordable injury would be considered an award. All qualified workers are in competition for the truck, but only one worker can win the truck.

Some companies offer gifts in their incentive programs while other companies primarily give recognition. In some companies, both gifts and recognition are given for workers who achieve the stated performance objective. Many safety professionals argue that the monetary value of the rewards is not as important as the recognition that goes along with the award. The perceived value of a belt buckle, for example, may only last a short time, while recognition by one's employer or co-workers stays with the employee much longer. Recognition from an employer and co-workers is typically held in high regard. It helps boost the worker's morale and sense of worth to the company. Some feel that the best incentive is to recognize that employees are important people who deserve to be respected and listened to, and that employees should be publicly and verbally recognized at company events whenever circumstances merit such recognition. It should be remembered, though, that some gifts may be quite highly regarded—more than recognition. A belt buckle with an especially desired insignia may bring status and prestige to a worker within his or her peer group that recognition cannot match.

Regardless of whether the benefit received is a gift or recognition, the benefit should be matched to the level of effort required to reach the performance objective. If it is difficult to meet the stated objective, workers will expect a benefit that is relatively high in value. The value of the benefit should increase as the difficulty in achieving the objective increases. If the value of the benefit is perceived to be low compared to the level of effort required to obtain the benefit, workers may not strive to meet the objective. The *rate of return* on their investment of time and energy is not desirable. If the value of the benefit is perceived to be very low, workers might not strive to meet the objective regardless of how easy it is to meet. On the other hand, if the value of the benefit is perceived as being great and relatively little effort is required to obtain the benefit, the reward is too easy to earn. In this case little improvement in safety performance would be expected. Whatever the value of the reward, the employer should make sure that it is appropriate for the effort required to receive it.

The Impact of Incentives on Safety in Construction

The impact of incentives on safety performance has received much attention. Based on personal experience and anecdotal evidence, many construction professionals believe that incentives positively affect worker safety. In the survey of general contractors (Banik, 2002), 44 percent of the contractors felt that incentives improved safety performance. In the same study, 21 percent felt that safety incentives do not improve safety, while 35 percent were unsure of their impact on safety.

Many companies view the gifts and recognition given for meeting safety objectives as a way of thanking the employees for their efforts. The effect this has on employees goes only as far as the employees recognize the company's efforts as *giving thanks*. One study (Molenaar et al., 2002) addressed this issue in its survey of the safety culture within construction firms. The study focused on the viewpoints of upper and middle management and field personnel on several topics related to safety. Regarding safety incentives, upper and middle management believed that they were showing thanks to the field personnel for their safety performance by offering the rewards. The field personnel, though, did not feel as strongly that this was the case.

One study of large construction firms and large construction projects (Hinze, 2002) investigated the impact of safety incentives to a greater extent. In the study, the relationships between various safety incentive characteristics and injury incidence rates were examined. The results of the study are summarized below:

- Higher injury incidence rates are experienced on construction projects and by construction firms that employ safety incentives. Thus, good safety performance can be achieved without a safety incentive program.
- Identifying a specific injury incidence rate as a performance objective for a project, as opposed to not specifying any rate at all, results in a better safety record for the project. However, better safety performance results when the rewards are tied to safe worker behavior as opposed to an incidence of injuries.
- Construction firms that do not utilize progressive incentives generally have better safety records.
- If an injury occurs during a given period, better safety performance generally results when all workers begin the new period on an equal basis.
- The reward value should not grow exponentially or geometrically when no injuries are sustained, and it is not good to offer gifts with an extremely high value when outcome-based objectives are used.
- Better safety performance occurs when the time period between distribution of the rewards is short. Weekly distribution is better than monthly; monthly distribution is better than quarterly. More frequent offering of rewards allows the company to keep its commitment to safety fresh in the minds of the workers and remind them often of the importance of safety.

- Better safety performance results when incentives are offered to other employees in addition to craft workers (e.g., to foremen, general foremen, superintendents, project managers).
- Better safety performance exists in firms that sponsor safety dinners. Even better safety performance results if the company president attends the safety dinners, and family members are invited as well. Companies that do not invite field workers to the safety dinners generally had poorer safety records.
- Better safety performance results when workers are evaluated, at least in part, on their safety performance on a project.
- Better safety performance results when rewards are given to the entire crew for their performance instead of just one worker in the crew.

Implementing Safety Incentive Programs in Construction

A safety incentive program should not be implemented on a project or within a construction firm without giving consideration to the positive and negative aspects of incentives. Poor implementation of a safety incentive program could cause harm to the safety culture of a company. Additionally, safety incentives should not be a substitute for poor management. The following questions should be considered when deciding whether to introudce a safety incentive program:

1. ***Does the project or company culture provide the foundation for an incentive program?***

    Some projects and companies maintain safety cultures that are very positive, in which the behavior of the workers is based on the intrinsic value of working safely. If this is the case, introduction of a safety incentive program might create a more materialistic, selfish view toward improving safety. If incentives are not part of the project or company culture, or have caused harmful distractions in the past, they should not be implemented. On the other hand, if incentives have been used successfully on other project or company goals (e.g., cost, schedule, quality, etc.), there may be some merit in their use.

2. *Are sufficient resources available to implement a safety program?*

If the safety culture can support an incentive program, a company should address whether it has the resources needed to establish and administer the program. Time and effort are required for personnel to monitor the worker performance, decide on whether a reward is warranted, and distribute the rewards. This can often be quite time-consuming if the workforce is large and the observance of worker behavior is required. If insufficient resources are provided to effectively implement the program, an inability to administer the program and fairly award deserving workers might arise.

3. *Whose performance should be rewarded?*

Consideration should be given to which employees are able to participate in the incentive program. Limiting it to one group of employees not only generally leads to poorer safety performance, but also may create tension between different levels of employees. Better safety performance usually results when incentives are offered to foremen, general foremen, superintendents, and project managers in addition to craft workers.

4. *With what frequency will the rewards be given?*

A decision should be made regarding the frequency with which the incentive rewards will be distributed. It is beneficial to keep the program in the minds of the workers. Therefore, offering rewards more frequently generally leads to better safety performance.

A combination of frequencies might be used in which, for example, some rewards are given out weekly, while others with greater value are given out quarterly.

5. *What kind or level of performance deserves to be rewarded?*

Answering this question is often difficult. The performance objective should not be set too high so that it is perceived as not attainable, yet it should not be set too low, making the reward too easy to earn. The goals should be kept reasonable and attainable. The company's past safety performance could be used as an initial basis for the goal. For example, the incentive performance level might be set at 50 percent higher than the average performance level. If the goal involves not sustaining an injury, the type of injury that constitutes loss of the reward needs to be defined as well. It is recommended that the day-count should not be restarted for minor scrapes and bruises.

It is generally accepted that the reward should be given for performance that is *above and beyond* that which is expected. Rewarding workers for the minimum expected of them does not give the workers any motivation to improve. Additionally, the reward should be tied to positive and proactive performance.

6. *What will be the value of the reward?*

The value of the reward, actual and perceived, is also a concern. It is not desirable for the reward to be considered more important than the positive changes in safety-related actions and the underlying attitudes regarding those actions. Additionally, workers will not be motivated to achieve the incentive goal if the reward value is too low. It is suggested that incentives with lower value be used and given out more often. This helps keep the costs of the rewards low and enables continual feedback. Employees themselves might be asked to suggest rewards in order to gain an understanding of what is valued.

7. ***With what frequency will the rewards be given?***

The manner in which the rewards are distributed is important as well. Consideration should be given to who distributes the rewards and at what occasions or events the rewards are distributed. Awarding the gifts and giving recognition in front of co-workers and family members is generally a good practice.

8. ***How will the program be monitored and modified?***

Safety incentive programs require monitoring to ensure that they are fairly implemented, the rules are consistently followed, and communication about the program remains open and effective. In addition, the programs should be modified periodically so that they do not become stale in the minds of the employees. Indicators of an incentive program that is not working effectively include: workers not being open, honest, and engaged to improve safety; workers thinking that the incentive program is a gimmick; and workers talking only about the program rewards rather than working safely. Sufficient resources should be set aside to ensure that these tasks are met. Employee involvement in the incentive program also benefits its effectiveness. Safety committees composed of both craft workers and management staff provide an opportunity to establish worker involvement that will lend assistance and validity to the program.

The following are some additional suggestions to consider when designing and implementing an incentive program:

- Hold a kick-off celebration when the incentive program is introduced to the employees.

This will set up an opportunity to inform the employees about the program, describe why the program is being implemented, and help to increase employee involvement and buy-in to the program.

- Investigate all hazardous site conditions and near-misses reported by workers as part of the incentive program. The workers may become discouraged if they do not see action to correct the hazards.
- Encourage proper reporting of injuries. Workers should not *hide* injuries suffered on the job in order to win the reward. Additionally, workers should not *be afraid* to report injuries due to pressure from co-workers who, as a result, will not receive a reward.

REFERENCES

Banik, Gouranga C. "Criteria for Construction Safety Incentive Programs—A Perspective," *The American Professional Constructor*, American Institute of Constructors (AIC), 2002: Vol. 26, No. 1, pp. 33–38.

Hinze, Jimmie W. "Safety Incentives: Do They Reduce Injuries?" *Practice Periodical on Structural Design and Construction*, American Society of Civil Engineers (ASCE), 2002: Vol. 7, No. 2, pp. 81–84.

_____. *Construction Safety*. Upper Saddle River, NJ: Prentice-Hall, Inc., 1997.

Levitt, R. E. and N. W. Samelson. *Construction Safety Management*. New York: John Wiley & Sons, Inc., 1993.

Liska, R. W., D. Goodloe, and R. Sen. "Zero Accidents," Source Document 86. Austin, TX: Construction Industry Institute (CII), 1993.

Molenaar, K., H. Brown, S. Caile, and R. Smith. "Corporate Culture: A study of firms with outstanding construction safety," *Professional Safety*, American Society of Safety Engineers (ASSE), July 2002, pp. 18–27.

REVIEW EXERCISES

1. What are the different components that make up all types of incentives, and how do the components impact worker performance?

2. How does an outcome-based safety incentive differ from a behavior-based safety incentive? Describe the circumstances in which one might be favored over the other.

3. Describe the different types of rewards commonly used as incentives by construction companies. Suggest how the rewards should be distributed in relation to potential recipients and the frequency and severity of injury.

4. Concern is often raised with regards to the implementation and effectiveness of incentive programs. Describe any issues of concern and then how a construction company might overcome those issues.

5. Provide examples of how a construction company might integrate the workers into the effort to implement an incentive program.

6. Describe how incentive programs might differ among small-, medium-, and large-sized construction companies.

7. How might a construction company increase the positive impact of an incentive program on safety performance?

■ ■ **About the Author** ■ ■

PREJOB/PRETASK PLANNING
by John A. Gleichman, CSP

John A. Gleichman, CSP, is Director of Safety and Loss Control at
Barton Malow Company, a design/construction firm based in Southfield,
Michigan. He is certified by the BCSP as a Construction Safety Specialist
and has more than 30 years' experience in construction safety. During
this period, he held numerous positions in professional organizations.
Highlights include ASSE Region X Vice-President, National Construction
Division chairman, and Detroit chapter president; National Safety Council
Board of Directors and Industrial Division chairman; Michigan Safety
Conference president; and American National Standards A-10 Committee
for Construction and Demolition Operations member. He has been
ASSE Detroit chapter Safety Professional of the Year and a speaker at
the ASSE Professional Development Conference and the National
Safety Congress.

CHAPTER **9**

Prejob/Pretask Planning

LEARNING OBJECTIVES

■ Discuss the importance of forming a multidisciplinary safety planning team.
■ Discuss how to incorporate project safety goals into trades selection and management.
■ Explain how to prepare for possible crisis situations.
■ Explain how to maintain site security.
■ Identify the tools (job hazard analysis and task safety assignment plans) for assessing and defusing risks inherent in upcoming work.

INTRODUCTION

If you are a program manager (PM) or construction manager (CM), safety planning starts when the management plan is developed. During this time, project owners are made aware of the legal safety requirements, PM/CM safety standards, and the benefits of a safety-focused project: few injuries, cost savings, high morale, and positive press.

Sophisticated owners in high-profile industries usually realize that accidents can result in huge losses of revenue from claims, lost time, and bad publicity; therefore, they take a leading role in mon-

itoring project safety. Such clients usually write very specific procedures into their contracts to ensure that best practices are used to prevent incidents. Most, though, need more guidance on structuring the project to limit risk. On both human and financial levels, it is less expensive to set up a comprehensive safety program than it is to incur injury.

To be relieved of liability, owners must leave the construction process completely in the hands of a CM or general contractor (GC). If they try to contract the work directly, they become program managers, construction managers, or general contractors as viewed through the eyes of expert witnesses in litigation issues. Recently, for example, an injury occurred during construction of an auto plant. In the course of the project, the owner issued several CM contracts and was subsequently held as the controlling contractor.

CORPORATE CULTURE

While this chapter focuses on the creation of client programs, it is necessary to take a step back and examine your own corporate culture. Is safety considered to be of greater importance than any other aspect of project operations or administration? Are your processes and procedures continuously refined and implemented to reduce the likelihood of accidents and job-related illnesses? Is safety training mandatory for employees and trades? Buy-in at the CM and PM levels—reflecting strongly defined values—is essential to delivering safe projects.

Remember that on the project site, what's necessary is more than a set of technical guidelines. *The overall goal is to create a culture that actively pursues safe practices and emphasizes individual as well as team responsibility for creating an injury-free workplace.*

GENERAL PRECAUTIONS

While a comprehensive safety program and a motivated staff go a very long way toward reducing site risk, every company should still prepare a detailed crisis management plan so field personnel do not have to make near-impossible decisions in times of great stress. The plan must include the following:

A crisis management handbook. This handbook should spell out procedures for dealing with emergencies, designated contacts within your company, designated client contacts, responsibilities of each member of the field team in dealing with the situation, phone numbers for contact personnel (office, home, cell), and the appropriate way to deal with the media. The caller will receive immediate support from the appropriate specialists within the company.

A "crisis card" that every crew member can carry. It should list emergency numbers (fire, police, ambulance, utilities, as well as principal client and corporate contacts); a checklist of information to be gathered, if at all possible (What happened? Who was involved? When and where did it happen?); and a sample media statement. A reminder such as "Evacuate injured immediately; give exact location to ambulance" is helpful as well.

Training in, and practice of, emergency procedures. Knowing what to do and whom to call can mitigate extreme situations and save lives.

CONSTRUCTION PLANNING

At the macro level, a safety plan will help to define responsibilities (see Figure 1).

While many of the individual tasks shown in the flowchart belong to the CM, the two most important words are found in the very first box: "PROJECT TEAM." Often, safety planning is considered the sole responsibility of the constructor, but involving others produces a more successful outcome. When we say "The plan must be unique to the project," we mean more than a particular set of field conditions; we mean encompassing all the knowledge available to this particular team—your client's safety officer and facility staff, the project architect, subcontractors. Regular meetings, keeping everyone informed and welcoming input, are essential to site safety. If you truly want to look beyond the most obvious safety considerations, then you must have a multidisciplinary staff.

Work with the architect and engineer right from program inception: Review drawings for the safest site setup and decide overall construction methods and strategies before work is bid in order to minimize any construction-erection hazards. Get rid of potentially risky situations as early as possible rather than having to deal with them later. It is a truism in the construction industry: the earlier in the project such issues are addressed, the greater the alternatives and, in general, the cheaper and more effective the solution. Establish a plan that allows for proper sequencing and an adequate time frame to ensure that all operations can be carried out safely and with assurance of quality workmanship.

Material choice frequently carries safety implications; weather, soil conditions, and surroundings are all major factors in determining the safest type of construction. Safety concerns may well drive the choice between caissons and spread footings, precast and cast-in-place concrete, or concrete and structural steel, for example. Phasing, and therefore the order in which design documents are finalized, must also consider the number and location of workers on site. These are tasks best considered from diverse perspectives.

Multidisciplinary teams involving a CM's technical and field staff, the architectural engineers, and the client, came up with the following building solutions early in project planning. In all cases, worker safety (and the well-being of those already using the site) was consistent with building quality, budget and schedule goals, and design intent.

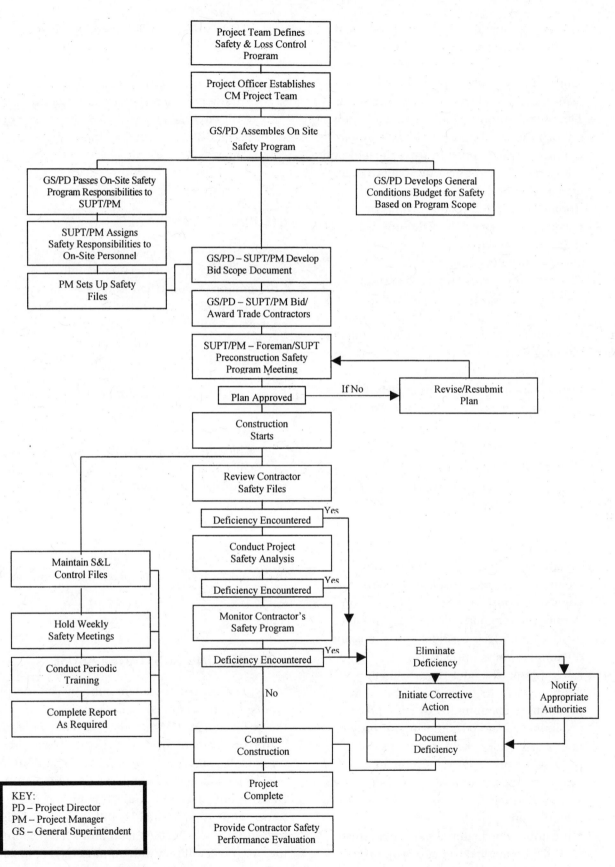

■ ■ **Figure 1** ■ ■ **Safety plan overview**

■ In keeping with an owner's set of values, the architect/constructor/client team decided on precast brick panels, leaving the concrete rough; this not only improved the phasing plan, but also cut down on airborne silica.

■ The use of vibroflotation to compact sandy soil prior to building a new 750,000-square-foot hospital enhanced bearing capacity and reduced settlement; measurements of stiffness, hydraulic conductivity, and shear strength were all dramatically improved. Further, this methodology did not induce shock waves (potentially damaging to adjacent structures).

■ To best comply with upgraded earthquake-resistance standards, meet load requirements as high as 300 lb/ft^2 in the finished structure, minimize vibration to an adjoining building to the greatest degree possible, and phase the work to ensure trade safety, one team came up with an unusual hybrid structure: concrete on the lower levels and steel above.

Activity Checklist

At this stage of a project's development, these activities are recommended:

■ meeting with the owner to define specific project safety scope

■ mobilizing internal staff (administrative, technical); meeting with owner, architect, engineers, consultants

■ developing reimbursable and general conditions budgets for safety programs and obtaining owner approval

■ assigning specific safety responsibilities to a field team

■ preliminary planning for training program

■ preliminary material-gathering for project's on-site safety file system—contents will vary by jurisdiction, owner requirements, and project complexity, but will most likely include

• any applicable local, state, and federal regulations

• the applicable material safety data sheets (MSDS) published by manufacturers to help workers and emergency personnel manage hazardous materials

• corporate policy and procedure manuals: safety, quality, crisis management

• trade contractor materials: organizing files to hold all contractor safety programs, safety meeting files, accident investigation files.

INSURANCE COVERAGE

Insurance type will depend on the project's size and complexity. Large projects—generally those over $100 million in construction value—may apply for owner-controlled insurance. Owner-controlled or contractor-controlled programs combine workers' compensation and usually general liability coverage under one policy for all labor, including contractors on the project. In most cases, this type of coverage is only for those contracts with employees who primarily work at the site location.

Insurance costs are either subtracted from contractor bids, or bids are submitted minus insurance coverage. One policy is then purchased by the owner for all construction labor performing within the physical perimeters of the project. Deliveries to and from the project are usually not included. One appealing feature is that the same insurance carrier covers third-party claims. A good safety record on the project can result in substantial savings to the project owner.

CONTROLLING SUBSTANCE ABUSE

Many owners have excellent substance-abuse programs, but others need help formulating one for the project. PM/CMs should start by suggesting a full program of preemployment, postaccident, reasonable cause, and—the real key—random testing. All tests should be administered by an approved Department of Health and Human Services Substance Abuse and Mental Health Services Administration (HHS/SAMHSA) laboratory. The extent of the programs will vary with an owner's relationship to the community and the labor contracts. Someone who tests positive cannot be allowed to work on site; the potential danger is too great.

SITE SECURITY

Site security is planned before construction starts. Some projects will need to be completely fenced to keep the public off the site. Others, usually those in less-trafficked areas, may require only signs posted around the site perimeter.

If possible, position the office trailer at the entrance gate so that the supervisor or someone in the office can readily observe suspicious traffic on site. Get to know the local police and fire departments; determine their capability to help in an emergency; recontact them and ask them to visit the site as the project progresses.

Provisions must be made to prevent an unauthorized person, visitor, or the general public from gaining access to the project site. Special care should be taken where there are changing conditions between an existing occupied facility and the project. In all cases, there should be signs placed in strategic locations stating that visitors are to check into the CM office before entering the job site.

Everyone on site must wear an approved hard hat and hard-soled shoes with cut-resistant leather uppers. Depending on site conditions and the type of construction activity, visitors may be required to wear safety glasses, ear protection, and other safety equipment. In general, running shoes are not permitted, but there are some OSHA-approved tennis shoes now on the market that meet ANSI requirements.

Site Security Checklist
Site Evaluation

- Review neighborhood characteristics.
- Meet with local police to discuss security risk in the area.
- Review site access plan.
- Determine average response time for the security alarm.
- Identify time of day for highest security risk.
- Identify phase of construction that poses highest security risk.

- Determine visitor access requirements.
- Identify site lighting requirements.
- Establish normal working hours.
- Determine emergency-service access requirements. Where an existing facility needs emergency egress through the construction site, there needs to be clear coordination and communication with the owner. The path of egress must remain clear of obstruction or hazards.
- Define construction-site boundaries.

Site Security Components

- Define contractor responsibility in bid scope documents.
- Define site operations in bid scopes (working hours, vehicle access, location of fence, trailers allowed on site, material laydown space).
- Construction warning signs, site access, and site-access restrictions.
- Type and location of perimeter fence.
- Locking of gates and handling of keys.
- Determine whether a security-guard service should be used and, if so, include this cost in the budget.
- Establish key personnel, phone numbers, and notification procedures for security violations.
- Coordinate security program with contractors during preconstruction/safety meeting.
- Review entire security program with your Director of Safety and your client before commencing on-site work.

Terrorism

Terrorist threats must be taken seriously; they should be addressed in the crisis management handbook and, without creating panic, procedures should be reviewed with field staff. A firm's reputation and project location, type, and sensitivity are all considerations when deciding how much to emphasize this aspect of site security.

If a threat is in writing, make certain the original note is saved for future investigation. If it is issued over the phone and the caller warns of immediate action or gives the site staffer any reason to believe that danger is imminent, evacuate the premises. If the danger is not immediate, the person receiving such a call should try to gather as much information as possible without being confrontational.

- Listen to the tone and characteristics of the voice for future identification.
- Ask the time, nature, and location of the attack.
- Ask who is calling (name of person or group).
- Ask the reason for the action.
- Ask what the caller wants—money, a public statement, surrender of a person on site?
- The person taking the call should have a co-worker contact police, the phone company, and your corporate safety director immediately. Dialing *69 after hanging up usually reveals the caller's number, but it is not difficult for someone to block that information, so remaining on the line is important.

Make every reasonable effort to alert all personnel in or near the area under threat, including but not limited to site staff, and evacuate the premises.

CONSTRUCTION PROCESS REQUIREMENTS

Construction process safety requirements need to be spelled out in the bid documents. Consider everything from excavations, scaffolds, and erection of large formwork and steel work, to critical lift safety requirements, guardrails and floor covers; material or personnel hoists, and trash chutes; the location of cranes, storage materials, and offices; and the proximity of walkways to work areas.

For example,

- *Hazardous materials removal.* No toxic materials should be permitted to leave the site without knowing where they will be transported. Some materials must be sent to specifically rated landfills that are lined to control leaching.
- *Clean-up/rubbish removal procedures.* Include procedures in the contract to control the buildup of waste materials on site. Clean-up should be required immediately upon completion of the task and at least at

the end of each shift. Specify if each contractor is responsible for all clean-up or only for identifiable materials, with general clean-up provided by a different entity and the cost allocated to the contractor.

- *Permitting, such as scaffolds.* A colored tag should be required on all scaffolds. RED designates that the scaffold is not complete and no one is permitted on it. YELLOW designates that the scaffold is structurally sound, but fall protection is required while working on the platform. GREEN designates that the scaffold is structurally sound, has standard guardrail protection on all exposed sides, and that proper access is provided to the work platform. Scaffolds should be checked and the tag signed and dated by a competent person at the beginning of each shift.
- *Excavations.* A permit should be required for all excavations. No excavation should be dug until there has been a thorough check for underground and overhead utilities. All states now have a one-call system that needs to be notified prior to digging. This service will call all affected utility companies, and each company will come to the project to stake out the location and depth of their lines. Hand-digging must be done within certain specified distances of these lines until they are located.

  This permit should also note soil type, method of excavation, and whether the banks will be sloped, benched, or shored. Only cohesive soils may be benched. Noncohesive soils must be sloped or shored. Shoring is to be done according to OSHA regulations, as directed by the shoring manufacturer, or consistent with the stamped and signed drawings of a professional engineer. A competent person should perform daily inspection of the excavation.
- *Lifts.* Permits should be required for any critical lift: any lift more than 75 percent of the capacity of the crane at the furthest horizontal distance from the center pin of the crane. Capacity includes the weight of the load, all rigging, the headache ball and/or block, and all the cable.

- **Confined spaces.** Permits are required for entry into confined spaces (vaults, manholes, vessels, tanks, etc.). The OSHA general industry standard for confined spaces and rescue requirements (Regulation 29 CFR 1910.146) governs selected situations.
- **MSDS program.** In compliance with the OSHA Hazard Communication Standard (29 CFR 1910.1200), procedures should be set up to keep workers informed about potentially dangerous chemicals created on or brought onto the project site. This includes substance labeling, education, and readily accessible material safety data sheets (MSDS). Everyone must receive special instruction before beginning work with such substances.

THE ROLE OF THE SAFETY MANAGER

The owner and the CM will decide on the need for a dedicated project safety manager. This will depend on the size and scope of the project, but is determined before the first phase of work. This manager does not assume the safety responsibility of the contractors on site, but will monitor contractor compliance with their own programs and the project safety requirements.

The safety manager, or employee charged with site safety, has absolute authority to stop work that is deemed hazardous. Upon observing an unsafe act or condition, he or she will immediately approach the tradesperson, listen to an explanation, and then offer reasons why the situation is dangerous. If the violation is minor and the tradesperson receptive, the discussion is viewed as a teaching opportunity. It is not confrontational. A severe violation, however, may mean that the very next incident results in dismissal; the situation is immediately brought to the attention of the safety representative, who is reminded of the responsibility for maintaining safe conditions.

By contrast, a safety manager who sees extraordinary safety performance should immediately and publicly commend the crew and later mention their performance to others, including an announcement at the job progress meeting. This provides positive reinforcement.

The safety manager should file a daily report, with observations, concerns, actions taken, and tasks performed.

INCENTIVES

Positive reinforcement is also provided through an incentive program; one that addresses both leading and trailing indicators should be established for the project. Instant recognition should be given to those who of their own volition exert effort to fix an unsafe condition, remind a worker to use the proper personal protective equipment when he or she momentarily forgets to do so, or make some significant contribution to improve a safety plan.

We are all more familiar with the trailing indicators of a person or a crew going a certain length of time without having an accident. These incentives should be given in increments of time—the more time without an injury, the larger the incentive: A crew that goes a month without an injury may be treated to a pizza and soda lunch, for example, while a project that goes six months without an injury is rewarded with a fully catered meal (choose something a little unusual so it will be memorable) and a drawing for prizes.

Top management and the owner should attend all such occasions; they should address the recipients, giving recognition for the achievement. Logo clothing—baseball caps, T-shirts, sweaters, and jackets—printed with the reason for the award are worn proudly off site, providing visible evidence of a program that promotes worker well-being.

HIRING AND MANAGING TRADES

Prebid Qualifications

The safety record should be the key to the selection of all contractors. Historically, if a firm has performed safely (or even better) on work similar to your project if the specific tradespeople assigned to your project have performed safely, it indicates a serious awareness of reducing risk. Even in tight job markets, do *not* consider firms that ignore good safety practices.

Prebid Conference

A meeting is held with all potential bidders. Here, owner and construction manager requirements for general safety, sequencing of work, and material-handling requirements are detailed. All contractors must have an acceptable safety program, and the tradespeople proposed for the work must have acceptable training and experience in the tasks they are to perform.

Safety issues specific to the project are brought to the attention of the bidders; for example, limitations on the amount of material that may be stored on site, the proximity of occupied campus areas to the proposed job site, and the serious consequences that follow disregard of safety standards. Remind everyone that this will be a drug-free site. OSHA training requirements for tasks involved in the contractor's phase of work must be emphasized at the prebid meeting.

Postbid Conference

After all bids have been evaluated, the PM/CM will meet with each prospective project participant. At this time, the general safety requirements and those specific to the project should again be reemphasized. Documentation of required training should be included in the bid. If training has not been completed, then the contractor must submit a plan of how it will be accomplished before contract award. Only then should the bid be awarded.

Contractors and subcontractors are responsible for the safety and security of employees and work areas under their control. They must provide a written site-specific safety program, and submit it to the CM. It must be at least as stringent as the CM's safety program included in the bidding documents. Maintain copies of their individual programs as part of your on-site documentation.

Preconstruction Meeting

The CM/GC must meet with key management from each contractor before starting work to reinforce general safety requirements and those specific to this project. In this way, safety policies and practices are communicated directly to those who will be on site. Each contractor is then expected to teach employees and subcontractors, in a formal session, about the safety and health requirements of the project, and enforce adherence to safe work procedures.

Contractors assign an individual to act as *safety representative*. This is a competent person capable of identifying existing and predictable hazards, who will have the responsibility of resolving safety matters and act as a liaison among the other trade contractors, the CM, and the owner. This individual must be on site, have the authority necessary to immediately correct unsafe practices or hazardous conditions, and be available for regularly scheduled periodic safety meetings as directed by the CM.

All trade contractors and subcontractors should have their site supervisor and/or safety representative at the preconstruction safety meeting to review and agree to the following:

- safety procedures at the project
- safety orientation and meetings for all trades (schedule and methods to be used)
- record-keeping requirements for inspections, violations, and variances
- employee complaint and discipline
- accident-report and emergency procedures
- sanitation and water-supply system
- tagging and lockout system procedures
- confined space procedures
- risk assessment and problem solving.

RISK ASSESSMENT AND PROBLEM SOLVING

This meeting should include explanations of job hazard analysis and task safety assignments (described later) that actively involve trades in safety planning at every step of the project. This lets project participants know that their knowledge is valued—that practical job plans, from those who are literally in the trenches, will be incorporated into the day-to-day management and performance of the work. Supervisors and staff are expected to gain proficiency in these vital skills.

TRAINING

Review contractors' documentation for completion of training before they mobilize on site. Be clear that training requirements will be checked *before* a task is started. In the absence of documentation, personnel will not be permitted to perform work.

Among other tasks, special training is required for:

- working on a scaffold
- working in confined spaces
- using respiratory protection
- using personal fall protection equipment
- becoming a competent person for scaffolds and excavations
- handling hazardous materials
- trenching
- hazard communication (identification, documentation, labeling).

This meeting goes a long way toward eliminating misunderstanding of requirements and procedures as well as improving safety awareness and teamwork.

Construction Phase

ORIENTATION

At a minimum, every employee on the project must attend the project safety orientation before starting work. This reinforces the general safety rules for the project and emphasizes that workers must be willing to abide by these standards if they wish to continue on this assignment.

The owner and CM should determine the content of the general safety orientation given to every trade-contractor employee who will work on the project. Attendance is mandatory. Depending on the project, orientation may include the following:

- the scope of the project and the benefit of the project to the community
- top management's commitment to safety
- basic safety rules and procedures as set out by the CM and client
- schedule of safety meetings
- record-keeping requirements for inspections, violations, and variances
- accident-report and emergency procedures.

IDs are issued only after orientation is completed and acceptable results of substance-abuse tests have been received. Ideally, IDs are in the form of large,

bright stickers for hard hats. In this way, it is quick and easy to ascertain who has been through the training program's preproject drug testing.

JOB PROGRESS MEETINGS

Safety is the first item discussed at every job progress meeting. General safety requirements (pertinent to the phase of work) are reiterated, safety coordination problems are resolved, and the group looks ahead to upcoming issues. This underscores the construction manager's commitment to safety, concern for those on site, responsiveness to issues raised by workers, and openness to worker suggestions.

TOOLBOX TALKS

The foreman of each crew (two or more people performing the work) should conduct toolbox talks pertinent to their work, and submit a copy of the toolbox form to the CM/GC weekly. The safety manager on the project periodically attends the talks to observe the foreman's depth of understanding and commitment. Keep in mind, though, that if a foreman emphasizes production over safety *after* the talk, the crew is far less likely to place safety first in accomplishing their tasks. Consistency is the key.

MEASUREMENTS

Each contractor should be measured monthly on the degree of seriousness of written warnings issued, overall safety management attitude, and frequency and severity of their injuries. (It's important, too, to record and review the positives: the number of safety suggestions, number of safe practices observed, etc.). Those companies that have experienced injuries are given their measurement of injury as compared to the whole project and to the Bureau of Labor Statistics for their category of work.

Project Completion

Each contractor is formally rated on safety performance. Contractors with a high severity of injuries, and who had a greater amount of written warnings, should be placed on a list of contractors who will not be permitted to bid future work.

Task	Hazard	Preventive Measures
Concrete placement	Concrete burns	Gloves will be worn when working with wet concrete. Tradespeople will be required to wash wet concrete off within a half-hour of exposure.
		Rubber boots (knee high) will be worn.
		If clothing becomes wet from splashes, tradespeople will be required to change within a reasonable amount of time—but within the hour.
	Elevated concrete floor placement	Standard guardrail protection will be placed around perimeter of floors, floor openings, and wall openings prior to placement of concrete.
	Falls through resteel mat openings	Wire mesh (4" x 4") will be installed across top mat of resteel before concrete placement.
	Walls	A platform with standard guardrail protection will be used to place concrete in wall forms. A ladder will be provided to gain access to the work platform.
	Columns	A platform with standard guardrail protection will be used to place concrete in columns;
		or personal fall protection with a group of rebars protruding above the formwork as anchorage will be used.
		In both instances a ladder or mechanical lift will be provided as access to the top of the column.
	Vibrators	Will be on ground-fault circuit interrupters.
		Operators and those in the vicinity will wear required eye protection.
	Suspended concrete pump hoses at point of distribution	Hose operators will wear full face shields.
	Movement of ground-placed concrete pump hoses	Concrete pump hose hooks or rope will be used by crew moving a concrete pump hose, so they are all standing with their backs 45 degrees or straighter.
		The lead hose operator will be the signal person when the hose is moved.
		Where practical, slides will be placed under hose connections.
	Concrete finishing	Only nonconductive bull float handles will be used.
		Gloves will be worn when handling wet concrete.
		Wet gloves will be changed within a reasonable amount of time.
	Concrete troweling machines	Self-propelled, riding troweling machines shall not be operated closer than 10 feet to an elevated edge.
	Level lasers	Warning signs shall be posted in areas where lasers are used. Only up to Class II lasers are permitted.
	Other physical exposures	Eye protection will be worn at all times.
		Rubber boots, knee high, will be worn when walking or standing in wet concrete.
		Hard hat will be worn by all crew members.

■ ■ **Figure 2** ■ ■ **Job hazard analysis (excerpt): Placing formwork for a concrete column**

Task	Hazard	Preventive Measures
Concrete formwork	Fall protection	See WORKING FROM HEIGHTS
	Scaffolds	See SCAFFOLDS
	Design failure	Form systems are built as designed by the manufacturer. Deviations from formwork systems are designed by registered professional engineers.
	Stripping Simmons panels	Slide panels down from scaffold platform by using a plank or other "ramp system" to steady and control panels.
	Back strains from lowering forms with rope	Stand in vertical position while lowering panels.
Use of Simmons panels to build short (10-ft) walls	Fall protection (6 ft)	Set first tier of Simmons panels. Tie panels on both sides together with wall ties. Set scaffold brackets and place scaffold planks on brackets.
		Secure planks to brackets. Install standard guardrail protection.
		Provide ladder access.
Resteel installation	Falls	Use of horse scaffolds under 4 feet high for placement of bars in 10-foot wall forms.
	Falls	Place wood runways in general paths of egress across raised resteel mats.
Confined space	Atmosphere	A competent employee shall check the content of the atmosphere for oxygen, explosives, and potential toxic gases before entry.
		A permit from the site safety coordinator will be required to enter any confined space.
Cranes	Maintenance	A copy of the annual crane inspection report must be located in the cab of the crane.
		Daily visual inspection will be made and a record logged and a copy placed in the cab of the crane.
		An ABC-type fire extinguisher shall be located on the crane.
		OSHA-required placards will be placed on the crane and will include danger swing radius signs placed on each side of the counterweight.
		Lifting lines should have torque converters or their equivalent.
	Critical lifts	All lifts of more than 75% of capacity are considered critical lifts, and a specific written procedure will be discussed with parties involved prior to the lift.
	Miscommunication	A competent signal person will be used to direct all lifts.
		The crane operator will follow only the signal of the designated competent signal person.

■ ■ **Figure 2** ■ ■ **(cont.)**

JOB HAZARD ANALYSIS *AND* TASK SAFETY ASSIGNMENT PLAN: THE KEYS TO SUCCESS

Everyone on site should know how to create and follow a job hazard analysis and a task safety assignment plan. *They are the heart of a safe construction project.*

A *job hazard analysis* examines a work process, such as building a large wall that requires the performance of multiple tasks. By contrast, a *task safety assignment plan* is a subset of the job analysis, and is created for each type of work each crew must accomplish. For example, one crew will place the rebar for the wall and another will place the gang form. If either crew moves on to a different task relating to the wall, a new task safety assignment plan must be created. The job hazard analysis encompasses the entire process and would not be altered, unless scope changes or new hazards are introduced.

Job Hazard Analysis

A job hazard analysis should be written for all potentially injurious work operations and major work processes. It involves listing each step of the process, evaluating each step for risks, and then coming up with a strategy by which each risk will be eliminated or minimized.

In addition to the list of anticipated problems and solutions, the finished job hazard analysis should include equipment; materials; any required electrical and hand tools; and the qualified, trained staff required to perform the work. Figure 2 is an excerpt from a job hazard analysis for concrete column formwork.

Those responsible for the process write the analysis. In this example, the project superintendent for the contractor placing the concrete column, the general superintendent, the foreman, and the crane operator placing the gang forms would meet to plan the work together. Each person has a different area of expertise and can contribute unique insights.

A thorough job hazard analysis increases the likelihood that the job will be completed safely. Moreover, such planning will result in greater efficiency and saved time. Thinking through the process step by step helps ensure that everyone has the right tools, equipment, and materials on site and in the right locations before they begin work. And the team will check to see that only trained people familiar with the process are assigned to the task. Also, check to see if there are crew members performing the work for the first time; they should be assigned less hazardous tasks in the process or paired with more experienced trades.

Without instruction, however, there is a tendency to rush through the job hazard analysis just to get the paperwork done, instead of using it as a safety tool. Job hazard analysis training is usually accomplished in a small enough classroom setting so that everyone can participate.

The instructor, with input from the class, should demonstrate how an analysis is written. Then the students should be divided into small groups, each with a work process that requires an analysis. A spokesperson from each group should explain their work to the other groups. Everyone can ask questions. Later, the class should discuss a well-thought-out analysis brought to class by the instructor and try to improve upon it. This process covers the key learning elements of seeing, hearing, and then practicing the skill.

Task Safety Assignment Plans

A task safety assignment plan refers to a type of work within the job hazard analysis. Before beginning work, the crew, under the direction of the foreman, should think through the steps of performing the tasks—what the risks are, and how they, the team, will control, minimize, or eliminate them. This includes deciding on the correct tools and having the right materials in sufficient quantity. Above all other considerations, everyone who will be performing the work must know how to complete it safely.

The foreman should complete the plan form, and the other team members should sign it. The form should include a standard checklist of items that require consideration. The crew leader should carry a copy of the form for reference at any time, and it should be available for the supervisor to check at any time. Include a place on the form where the team, upon completion, can offer suggestions on how to improve workflow and safety practices. [See Figure 3 for a sample form (two-sided trifold).]

General Information:

1. What are the special hazards associated with the task?

_____

Have they been explained to the employees?
☐ Yes ☐ No

2. What weather conditions could effect the safety performance of this task?

Tools & Equipment:

3. Inspection is required on all tools, ladders, electrical cords, rigging and safety equipment. Has this been completed? ☐ Yes ☐ No

Material Storage:

4. Has a material storage area been identified and approved? ☐ Yes ☐ No

5. **MSDS** checked ☐ Yes ☐ No

Scaffolds:

6. Inspect all scaffolds/ladders before use. Has the scaffold tag been signed? ☐ Yes ☐ No

Emergency Equipment:

7. Identify below the location of the nearest phone.

_____

Emergency # _____

Housekeeping:

8. Are trash containers available in the work area?
Location: _____

Fall Protection:

9. Have areas been identified as requiring fall protection systems? Have they been installed? (static lines, barricades, hole covers, etc.)
☐ Yes ☐ No

Explain: _____

Fire Protection:

10. Is a fire watch required? ☐ Yes ☐ No
Name: _____

11. Are flammable/combustible materials stored, separated and secured per procedure?
☐ Yes ☐ No

Procedures/Permits Required:

	Yes	No
Confined Space	☐	☐
Crane Lift	☐	☐
Excavation	☐	☐
Hot Work	☐	☐
Line Break/Hot Tapping	☐	☐
Lock, Tag, Try	☐	☐
Scaffolds	☐	☐
Signs/Barricades	☐	☐
Other (specify)_____	☐	☐

Employee Authorization Required:

	Yes	No
Crane Operator	☐	☐
Aerial Lift	☐	☐
Forklift Operator	☐	☐
Mobile Equipment Operator	☐	☐
Powder-Actuated Tool User	☐	☐
Competent Person (lead, asbestos, excavations, confined space, hazardous material, scaffolds)	☐	☐
Other (specify)_____	☐	☐

SAFETY TASK ASSIGNMENT

The STA should be completed daily for each task. Each crewmember involved with the task should sign this STA. At the end of the task, give this STA to the Project Management. If deviation from known safe work practice/procedure occurs, work must be stopped.

Supervisor: _____ Date: _____

Location of Task: _____

Task Description: _____

Does task require special training?
☐ Yes ☐ No

If yes, what type? _____

Personal Protective Equipment Required:

	Yes	No	Type
Fall Protection	☐	☐	_____
Eye/Face	☐	☐	_____
Respirator	☐	☐	_____
Foot	☐	☐	_____
Hand	☐	☐	_____
Hearing	☐	☐	_____
Coveralls	☐	☐	_____

PPE: Monogoggles, face shield, sandblasting hood, welding (goggles, shield, sleeves), ear protection, gloves (leather, chemical resistant, gauntlets), shin/foot protection, boots (rubber, hip), rain suit, life vest, safety harness, fall protection equipment, breathing air assembly.

■ ■ ■ **Figure 3** ■ ■ ■ **Safety task assignment form**

GENERAL FOREMAN: _____

◆

SAFETY TASK ASSIGNMENT PROCESS

◆

ZERO ACCIDENTS!

SAFETY TASK ASSIGNMENT

SUPERVISOR: _____

DATE: _____

1. WAS ANYONE INJURED OR DID AN UNPLANNED INCIDENT OCCUR TODAY? IF YES, EXPLAIN.

    YES _____ NO _____

    _____

2. WAS IT REPORTED TO THE SAFETY DEPARTMENT?

    YES _____ NO _____

3. WHAT PROBLEMS DID YOU HAVE WITH TODAY'S WORK ASSIGNMENT?

    _____

4. WHAT CAN WE DO TOMORROW TO IMPROVE PERFORMANCE?

    _____

5. MISCELLANEOUS CONCERNS:

    _____

REVIEWED BY:

SUPERVISOR: _____

Name   **Badge #**

_____   _____

_____   _____

_____   _____

_____   _____

_____   _____

_____   _____

My Safety Principles:

• *Plan Every Task*

• *Anticipate Unexpected Events*

• *Use The Right Tool For The Job*

• *Use the planned procedures*

• *Team members are trained to perform the task*

• *Correct hazards immediately*

• *Watch out for each other*

• *Analyze how task can be improved*

TURN FORM IN TO MANAGEMENT WITHIN 24 HOURS.

■ **Figure 1** ■■ (cont.)

INSTRUCTIONS: SAFETY TASK ASSIGNMENT FORM

This set is for all supervisors who assign work to employees. It takes into consideration all aspects of the task to be performed while emphasizing safety.

General

1) A safety task assignment (STA) involves showing or explaining to each employee the safety application that pertains to the job he/she is to do.
2) Management down through foreman is responsible for giving STA briefings to all employees (individually or in a group) before they actually begin any assigned task, making sure they are clear about what they must do.

Procedure

1) Each foreman must analyze each job or task for specific hazards before work begins, in order to give accurate instructions for each job his/her employees will be engaged in during that shift.
2) Each foreman is responsible for giving STAs every day to every group of employees. The STA should include any specific hazard that group may encounter, safety equipment, and any personal protective devices that may be needed.
3) The magnitude of the task will generally determine the extent of the STA—anything from a few words identifying the hazard to an actual demonstration of how the job can be done safely.
4) All employees involved must be checked to ensure they understand the safety requirements of their task.
5) Each foreman will direct employees to be on alert for hazards that may be encountered during the course of assigned work. They will report such hazards to the foreman for correction.
   • Following notification by an employee, the foreman must initiate corrective action to achieve hazard abatement.
6) The STA form should be posted in a conspicuous place near the work area. This will enable employees to review it during the day.
   • The STA form should be turned in at the end of each shift so the area manager can review it.

CONCLUSION

Companies with long-term records of safety excellence achieve their results by specifically incorporating safety into the design of the production system, from project inception through completion. In planning and in construction, many builders place safety on a par with quality, schedule, and cost. But truly, it is first. The lack of a strong safety program will ultimately jeopardize the achievement of all other project goals.

Results are most fully realized when safety is the basis of the overall construction plan. Companies that design safety into their total production system are much more likely to have satisfied clients, happy and healthy workers, and a good reputation. Everyone wants to work a lifetime and remain vigorous and injury free. Planning safety into each task and procedure can make this possible.

REFERENCES

Earnest, R.E. "Making Safety a Basic Value," *Professional Safety,* August 2000.

Hansen, Larry. "Safety Management: A call for (R)evolution," *Professional Safety,* March 1993.

MacCollum, David. *Construction Safety Planning.* New York: John Wiley & Sons, 1997.

Occupational Safety and Health Administration. *OSHA Standards for the Construction Industry, 29 CFR 1910.* Davenport: American Safety Training/Mangan Communications, 2001.

_____. *OSHA Standards for the Construction Industry, 29 CFR 1926.* Davenport: American Safety Training/Mangan Communications, 2001.

Reid, Janine. *Crisis Management: Planning and Media Relations for the Design and Construction Industry.* New York: John Wiley & Sons, 2000.

Saving Lives: Proven Methods to Eliminate Job-Site Fatalities. Janine Reid (ed.). Denver: Janine Reid Group, 2001.

REVIEW EXERCISES

1. Specifically, what are the benefits of having the architect, owner, and builder work collaboratively to plan a safety program?
2. What steps should be taken before hiring subcontractors to help assure their support of a stringent site safety program?
3. In what specific ways can a constructor help trades "think safety" to promote their concern for and awareness of safety issues?
4. List some specific steps field personnel can take to assure site security.
5. With co-workers, prepare a portion of a job hazard analysis and a task safety assignment plan to get a sense of the techniques required.

■■ About the Author ■■

ACCIDENT INVESTIGATION/RECONSTRUCTION
by Ronald B. Cox, Ph.D., M.B.A., P.E.

Ronald B. Cox, Ph.D., M.B.A., P.E., holds the Burkett Miller Chair of Excellence in Management and Technology at the University of Tennessee, Chattanooga (UTC). He has senior-level management experience in higher education, government, and industry, including service as Dean of the UTC College of Engineering and Computer Science and as Executive Director of the UT Research Corporation. He holds the Ph.D. degree in Mechanical Engineering from Rice University, the M.B.A. degree from Vanderbilt University, and both B.S. and M.S. degrees in Mechanical Engineering from the University of Tennessee.

Accident Investigation/ Reconstruction

THE ROLE OF SENIOR MANAGEMENT

Accidents occur all too often. According to the Bureau of Labor Statistics, there were more than 497,000 nonfatal accidents in the U.S. construc- tion industry alone during the year 2000 (U.S. Department of Labor, December 2001).

Needless to say, the prudent manager/owner should take all reasonable steps to eliminate accidents—*all* accidents. There should be no ac- ceptable level of accident frequency. The goal for all projects should be zero accidents and zero injuries. Achieving this goal requires active involve- ment of senior management. The top management must set the standard by cultivating a "zero acci- dents, zero injuries" attitude. This philosophy must become a part of the culture of the business. It must be so prevalent that it permeates all levels of management and all activity by all employees. Quite simply, the philosophy must become a part of the thought process of every employee as they go about their respective tasks throughout the day. This behavior must recur day after day.

The cost of accidents is enormous. If one con- siders occupational accidents in general, the annual cost to society in 1982 was estimated to be greater than $30 billion, and workplace hazards posed seri- ous risks to more than 100 million workers. In 1982 the BLS reported 4000 work-related fatalities and 4.9 million occupation-related injuries in the United States (*U.S. Chamber of Commerce Annual Report, 1982*). These numbers have climbed in recent years. For the construction industry alone, the number of nonfatal injuries climbed from 493,000 in 1999 to 497,200 in 2000 (BLS, Dept. of Labor, December 2001). Fatal injuries in the construction industry have risen from 1155 in 2000 to 1225 in 2001 (BLS, *Census of Fatal Occupational Injuries*, Dept. of Labor, December 2001). These data are cited simply to impress upon owners and managers of construction companies, and their employees, the enormity of the costs of accidents (both human and financial). Again, construction managers at all levels must recognize that the prevention of accidents and injuries makes good business sense (short term and long term). The results of achieving, or even approaching, the "zero accidents, zero injuries" goal will translate into:

1. Improved safety
2. Improved profits
3. Improved morale
4. Improved success in obtaining new contracts
5. Improved worker retention and lower insurance costs.

THE INQUIRY PROCESS

The purpose of discussing accidents, accident investigation, and accident reconstruction is to inform the reader about the process of inquiry that typically follows an accident that is serious enough to warrant litigation or the threat of litigation. Additionally, the discussion will suggest to construction managers a list of actions to take that will lessen (hopefully eliminate) the risk of liability resulting from accidents. Whether or not litigation results, all accidents should be investigated. The fundamental outcome of the investigation should be the determination of steps to take in the future that will eliminate accidents. Such steps could lead to changed hiring practices, improved training programs, better supervision, or a host of other measures.

Consider a rather simple, hypothetical situation in which a crane operator picks up a load of materials on a construction site. While maneuvering the load, a cable breaks. The load drops toward the ground and strikes a worker below, causing severe, disabling injuries.

Not so long ago, the investigative report on this accident might have stated that the cable broke, the load fell, and the worker was struck. It might state further that the worker contributed to the injury (i.e., he committed an unsafe act)—he put himself in a dangerous position. Today, the accident described above is going to result in much more complex analysis and many different groups becoming involved in the assessment. Why? Basically, for two reasons:

1. To find out what went wrong in order to correct the problem, learn from it, and eliminate any future occurrence.
2. To determine who was at fault (totally or partially) and, if relevant, to derive economic recovery from the responsible parties for whatever loss is determined to have occurred with regard to the injured worker, with regard to property loss, and with regard to other losses (such as clean up, failure to meet contractual time lines, etc.) that may have occurred as a result of the accident.

The following types of activity and inquiry, referred to as *accident reconstruction*, will likely occur once the injured worker has been tended to.

The general contractor will begin to collect information, asking: "What happened? Who was involved? Were there witnesses? What did the witnesses observe? Were the involved parties employed by us or subcontractors?" This early stage of the investigation is extremely important and should be done by a skilled person who has some degree of independence from the event. A safety director, for example, would be a good person to direct and carry out the investigation for the contractor. The immediate supervisor of the injured worker might not be the best investigator since he may have been directly involved in the worker being where he was and might not collect unbiased information.

The collection of information and data at this stage might make use of an approach called *root-cause analysis*. This involves a logical step-by-step evaluation of the various actions involved in the work that was being performed. The collection of information leads to an understanding of what was being done and how that behavior deviated, if at all, from what should have been done. As a result, corrective actions may be identified and undertaken. Even though the immediate supervisor of an injured employee may not have been directly involved in the collection of the data, he and all other levels of supervision must be involved in the company's review of the incident and the planning for any changes in practices or procedures.

Along with collecting the basic written statements regarding the accident, it is also important to preserve the evidence. In the case above, the crane should be preserved along with the cables, the load, the cable attachments, and the pallet or structure supporting the load. Additionally, photographs should be taken showing the accident site, the path of travel of the crane (if it moved), the crane, the load, the cable, and anything that may have been related to the accident.

Since there are undoubtedly various insurance carriers involved, they will have been notified and will have representatives coming to the scene (the location of the accident where all equipment and objects are still positioned where they were immediately following the accident), or the site (the location where the accident occurred, but normally after some or all objects have been moved). These insurance representatives will begin to collect their own information by way of photographs and interviews.

Since the injured worker probably has dependents or at least some entity (perhaps a union) looking out for his rights, there may be (and most probably will be) an attorney representing the injured worker. Now, the inquiry may take on increased complexity. The attorney will probably hire experts to look into the matter and to perform an accident-reconstruction analysis. The attorney and the experts will begin to inquire about the following:

The history of the injured worker
- When was he hired?
- Who hired him?
- What work task was he hired to perform?
- How long was he employed?

- Was he new to the company? Was he new to the trade?
- What training did he receive? How did he receive the training?
- Who provided the training?
- Were videotapes used? Were written materials used?
- Was some testing or certification required?
- What was the injured worker's record as an employee?
- What was his work schedule?
- Was he fatigued?

The circumstances surrounding the accident
- Why was his client there?
- Who was his supervisor?
- Who directed his client to be near and under the load?
- Who should have been aware of his client's position under the load?
- Did the operator of the crane see his client (or should he have seen him)?
- Should someone have had his client move away from the crane?
- Did the employer conduct daily safety meetings?
- When was the last safety meeting?
- Who attended the meetings?
- What was the content of the meetings?

The company (contractor and/or subcontractor) history and practice
- Had there been other accidents?
- When?
- Where?
- What was their nature?
- Who was involved?
- What was the outcome?
- Who are the managers, owners, supervisors?
- What is each one's background?
- Is there an established safety program? What is it?
- Are documents (training manuals, films, etc.) used in the training?
- What is the content?
- What is the evidence (if any) of this company being concerned about the safety of employees?

- What contractual documents exist? What do they say?
- Do they refer to compliance with any particular codes and standards?

The site

- What was the condition of the site?
- Was the work site cluttered?
- Were there organized work crews?
- What was the condition of the equipment?
- What was the morale of the employees?
- What was the degree of planning, organization, and control?
- Who was in charge? Who gave the orders for getting tasks accomplished?
- What happened on the day of the accident?
- What was the overall environment (at the site) at the time of (and just before) the accident occurred?
- What time of day was it?
- Was it wet or dry?
- What was the temperature?
- What was the wind velocity?

The product or products—the manufacturers of the various products will probably be brought into the matter as defendants. The integrity and applicability of the design, application, service, and maintenance of the products will be questioned. Typical questions will be:

- Why did the cable fail? Who manufactured the cable and when?
- Did the cable fail due to an overload? How did the cable come to be overloaded (i.e., who made such a decision)?
- Was the cable being used improperly (i.e., not according to its intended use and purpose)?
- How did that come about?
- What was the load?
- Should it have been lifted at all? Was it properly packaged?
- Was it properly identified as to its weight?
- Were there warnings about its being lifted?
- Did the crane operator behave properly?
- Did he know the condition of the cable?
- Was the cable damaged in some way prior to the accident in question?

- How was it damaged? When and how are cables inspected?
- What was the known condition of this cable?
- Who knew it and how did they come to know it?
- What about the crane? Did the controls work properly?
- Did the controls malfunction causing the load to be "jerked"?
- What about the crane operator, was he alert and attentive?
- Did he operate the crane properly?
- Did he drive into a hole or over an object?

Some of these questions can only be answered through engineering analysis and experimentation. Engineers may remove some or all of the evidence from the site for evaluation. Records and specification data may be requested from the contractor. Such records may include original specifications for products and equipment (i.e., the crane, the load, etc.) Also, maintenance records may be requested along with inspection records (such as cable and crane inspections and certifications).

The use of codes and standards

- What codes and standards were applicable?
- Had any OSHA standards been violated in the past?
- Were there any violations of any code or any standard at the time of the accident?
- Did employees understand that there were codes and standards relating to the work site, the equipment, the use of warnings, etc.?
- Did the violation of any code or standard result in a situation that caused or contributed to the cause of the accident?

At this point, it is clear that there are (or could be) several defendants identified as possibly playing some role in the causation of this accident. Yes, the injured worker was where he should not have been; however, his presence did not cause the load to fall. Also, it will be asked whether this worker, with his level of training and experience, should have and could have appreciated the risk. Further, it will be asked whether the overall operating environment (the time constraints, the rush of getting a task completed—all imposed by the company) served as an intervening cause of the worker being where he was. Each of the various parties drawn into the matter now begin their own inquiry and reconstruction with the goal of demonstrating that either their actions or their product did not cause or contribute to the cause of the accident. Nevertheless, at some point, as a result of the reconstruction evaluation, testing, and analysis, opinions are expressed and causation is presented by one or more of the parties. There may be general agreement among the parties as to where the responsibility rests, in which case an economic agreement is reached and the matter is settled. Such a resolution may be reached with the aid of an arbitrator, or simply reached among the parties themselves. On the other hand, if the dispute is not resolved, the matter will be litigated in a court of law and a jury will make the decision regarding where fault lies and what the compensation level will be, based on the evidence presented by the various stakeholders.

ACTIONS TO BE TAKEN

With the prospect that such incidents like the one described above might occur, it is noted once again that the owner/manager/stakeholders should strive to avoid all accidents and all injuries. Policies and practices must be put in place and a culture must be created to accomplish this. Further, if such policies and practices are in place, accidents will be less likely to occur, and, if one does occur, the company is in the best possible position to defend itself in the event of litigation. The following actions, although presented in a simplified format here, are recommended to the chief executives of construction companies:

1. Recognize that a "zero accidents, zero injuries" culture begins with the corporate leadership.
2. Strive to create a "zero accidents, zero injuries" culture because it is the right thing to do with regard to all stakeholders (owners, stockholders, employees, customers, suppliers, partners). Further, it makes good sense economically.
3. Build *safety* into the strategy of the enterprise. Make safety just as important as quality, planning, scheduling, operating, execution, cost control, and marketing.
4. State and reinforce the behavior you seek by establishing practices and policies that are consistently and repetitiously reinforced over time.
5. Hold officers and managers accountable for safety performance.
6. Assure that the following actions are carried out:
   a. Hire qualified individuals at the appropriate levels in the organization. This implies that there is a commitment to safety and an understanding of safe practices on the part of all managers. In addition, the managers should be qualified for the particular functional areas for which they are responsible.
   b. Recognize the existence and importance of codes and standards to the conduct of business. Assure that managers, supervisors, and all employees adhere to established codes and standards.
   c. Establish a formal training and education program involving all employees. Such training should position safety (i.e., zero accidents and zero injuries) as an essential and high-priority goal of the company. Training should include written materials and videos, if possible.
   d. Through contracts and agreements, assure that safety issues are addressed with partners, associates, and subcontractors. Be careful about who has what responsibilities (therefore, who may be held liable).
   e. Understand the environment in which a project will occur (the complexity, the level of skill of employees, etc.)
   f. Recognize that *new employees* are often involved in accidents. Focus attention on their training and supervision.
   g. Create a culture with open communication among all employees, encouraging a willingness to identify and eliminate situations and practices that are not safe (i.e., attempt to foresee what might happen).

h. Reward success in safe practices just as you reward success in other areas (remember, a safe project produces improved profits).

i. Document! Document employee records, training programs, safety requirements, work-order changes, equipment specifications, maintenance records, design requirements and changes, and purchases. Remember, after an accident occurs, experts will be requesting a variety of records. Even though you may not be required to furnish every record requested, your ability to provide complete and comprehensive documentation in response to the types of questions posed will enhance your opportunity to minimize the threats posed to the company as a result of accidents.

Other chapters of this book present excellent guidance to those who are decision makers in the construction industry. Heightened awareness and aggressive attention to these matters of safety will reduce both the number and severity of accidents.

REFERENCES

Bureau of Labor Statistics. *Census of Fatal Occupational Injuries.* Washington, DC: U.S. Department of Labor, Government Printing Office, December 2001.

_____. *U.S. Chamber of Commerce Annual Report, 1982.* Washington, DC: U.S. Department of Labor, Government Printing Office, 1982.

Handbook of Human Factors (Section 7, "Human Factors in Occupational Injury, Evaluation, and Control"). Gavriel Salvendy (ed.). New York: JohnWiley, 1987.

Standard Handbook for Civil Engineers, 4th ed. (Section 4, "Construction Management"). Frederick S. Merritt, M. Kent Loftin, and Jonathan T. Ricketts (eds.). New York: McGraw Hill, 1995.

REVIEW EXERCISES

1. What should be the manager's goal with regard to reducing (and eliminating) accidents?
2. In what ways will accident reduction benefit the company? Comment on both financial benefits and personnel benefits.
3. What types of information will be collected by the accident reconstructionist following an accident?
4. What information will likely need to be provided to the reconstructionist by the company?
5. What actions should be taken by the senior management of a company to minimize the number of accidents and reduce the threat of litigation?

■■ **About the Author** ■■

TRAINING

by Christine Fiori, Ph.D., P.E.

Christine Fiori, Ph.D., P.E., joined the faculty at the Del E. Webb School of Construction (Arizona State University, Tempe, AZ) as an assistant professor in January 2002. She received her Doctorate of Philosophy in Civil Engineering from Drexel University in 1997, where she also completed her master's and bachelor's degrees in Civil Engineering. She is a Registered Professional Engineer in the state of Pennsylvania, and is currently conducting courses in construction management and safety that include project delivery, management styles, and leadership within project teams.

She is a certified Occupational Safety and Health Administration Construction Outreach Trainer, authorized to conduct 10- and 30-hour construction outreach training in accordance with guidelines provided by the OSHA Training Institute.

Dr. Fiori is one of two academics serving on the Construction Industry Institute's Educational Committee: EM 160-21, Making Zero Incidents a Reality.

The focus of her efforts is to develop safety training material that can be used by construction companies in their safety training programs.

CHAPTER 11

Training

- Discuss the importance and benefits of developing, implementing, enforcing, and assessing an employee safety training program.
- List guidelines, regulations, and requirements for safety training as detailed by OSHA and MSHA.
- Develop a procedure for quantifying safety training needs, requirements, and implementation.
- Identify the key elements of a successful safety training program and how they can be successfully employed.
- Discuss the different tools available for the creation, implementation, enforcement, and assessment of a safety program.

INTRODUCTION

Proper training is important in all aspects of life. Can you imagine taking a driver's examination without ever sitting behind the wheel of a vehicle, or playing a sport without learning the rules and preparing physically? You can't, right? Then why do so many employers minimize the importance of proper safety training since it affects the lives and well-being of all of their employees? Safety training tends to be brushed aside or given minimal thought and financial support. After all, performing your work safely is all common sense, right? Then why does the construction industry still have such a high accident rate? The work performed is dangerous and conditions can often be hazardous. Without proper training, individuals can seriously injure themselves or even be killed on any job site on any given day.

Knowledge of how to perform work safely is crucial to the welfare of every worker. How do employees gain this knowledge? Formal safety training has proven to be the most effective and successful way to ensure that workers possess the knowledge required to perform tasks in a safe manner. Safety training is another form of insurance. It gives owners, managers, and supervisors peace of mind, knowing that the workforce on their job is trained in safe work practices and is aware of the dangers present on a job site.

In the construction industry, an employee cannot simply pick up tools and begin work as a journeymen carpenter. It is no more likely that someone will jump into the cab of an excavator and immediately operate the equipment. There are apprenticeship programs and equipment training programs required prior to performing certain work or operating heavy equipment. Additionally, a young graduate engineer or construction manager is not employed by firms without the appropriate education, certifications, or experience. Most professionals have some level of formal training required for practice in their chosen field. The same should be true with respect to safety. Safety training must receive the same respect as apprenticeship programs and other professional licensing procedures. Formal safety training programs need to be developed for every company and every job site.

Still not convinced that safety training is imperative? Consider the goal of safety training: To create a hazard-free work environment by training all employees in the recognition of unsafe conditions and empowering them to correct those conditions. These actions develop a work environment that is safe for everyone. In simple terms, a safety training program serves to protect people from physical harm, illness, and possibly death. If nothing else convinces you of the importance of a formal safety training program, the possibility of death should. Consider the unpleasant task of investigating a fatal accident in which the employee was unaware of certain unsafe job-site conditions due to lack of training.

Developing a safety training program is the right thing to do! It is not only the right thing to do, it is the law. In 1970, OSHA was created with the passage of the Occupational Safety and Health (OSH) Act. Of the numerous standards promulgated by OSHA, the regulations that pertain specifically to the construction industry are contained within 29 CFR 1926. While a majority of the regulations focus on the creation of safe work environments, some regulations concentrate on the responsibility of management to maintain safe work sites by ensuring that appropriate safety training is received by their employees. Subsection 1926.21, entitled "Safety Training and Education," outlines the requirements of employers to provide employee training; however, it does not describe the details of how to meet these requirements.

In addition to the training standards implemented by OSHA, a construction firm may also have to comply with the Federal Mine Safety and Health Act of 1977. This act requires that each mine operator have a health and safety training program for its miners, and it created the regulations contained within 30 CFR Parts 1–199. Part 46, "Training and Retraining of Miners Engaged in Shell Dredging or Employed at Sand, Gravel, Surface Stone, Surface Clay, Colloidal Phosphate, or Surface Limestone Mines," and Part 48, "Training and Retraining of Miners within Subchapter H," outline the training requirements for different types of mining operations.

At a minimum, all training programs must meet the guidelines and regulations promulgated by OSHA. These requirements are that all employees in the construction industry are to be trained in the recognition, avoidance, and prevention of unsafe conditions.

Despite the fact that training is a requirement, in the period from October 2001 to September 2002, there were 1610 citations relating to the education and training standard that resulted in 1576 investigations by OSHA officials. Of these incidences, the fines assessed for violation of §1926.21 amounted to over $1.5 million.

What are the key components of a safety training program? What should be included in a safety training plan? What are the drivers of a successful safety program? The goal of this chapter is to answer these questions and to develop tools for use as guidelines in the creation, implementation, enforcement, and assessment of a safety training program. Remember, you are protecting both your employees and yourself by developing, implementing, and enforcing a strict safety policy.

DEVELOPING A SAFETY TRAINING PROGRAM

There are many predetermined laws and regulations regarding the minimum standards required in an effective safety program to help guide the development of a safety training program. But such a sterile and almost boiler-plate approach to safety will not win your company safety awards or move your firm to a new level of safe working. Having a written safety policy does not ensure stellar performance from your workers. Remember the keys to a successful program are to develop, implement, and then enforce the program. Regardless of how well written a plan is, without a firm commitment to safety from the corporate office, it may not achieve the intended goal. Demonstrated commitment to safety is of paramount importance to an effective training program.

Defining Company Safety Philosophy

At the outset of developing a successful safety training program, the corporate philosophy toward safety should be firmly established. The company attitude and commitment to safety should be reflected within mission statements or formal declarations of company policies. This management culture must be clearly communicated to field personnel at all job sites to achieve maximum effectiveness. Every worker needs to know that upper management is concerned for their well-being, and that unsafe work practices will not be tolerated. Developing a strong safety culture within a company requires more than having a written policy in place. Management must lead by example and become involved in safety training. Without demonstrated commitment, even the most well-written and well-intentioned safety training programs are destined to fall short of their goals.

Management needs to take an active role in the enforcement of any developed program. There are many ways to send clear messages to all personnel regarding corporate dedication to safe work practices. Most of these methods are inexpensive and simply require a corporate presence on the job site and during training sessions. Some effective measures for exerting strong influence on the safety culture of a project are:

- Have the company president or other top management address every safety orientation class, reiterating the company safety policy. This speaks volumes to employees. If the company president invested the time to address the session, safety must be a serious matter.
- Have home-office representation at job-site safety inspections on a regular basis. If man-

agement presence is felt on site during safety inspections, workers will clearly see the importance placed on safety.

- Ensure corporate review of all safety performance reports. Knowing that the corporate office is scrutinizing safety performance on every job site, supervisors and employees alike will pay closer attention to safe work practices.
- Establish corporate participation in accident investigations. Employees clearly see the level of importance placed on their welfare when management invests the time to determine the root cause of unsafe actions.

Quantify Training Needs

Each employer will be conducting different types of work requiring various levels of safety training to meet regulatory guidelines. The first step in the development of an effective company training program is to determine the regulatory requirements.

Training and education requirements are outlined in general terms by OSHA in §1926.21. Other subsections address specific training requirements for the safe operation of various equipment and tools. These specific requirements and their respective subsections are illustrated in Figure 1.

§1926.21 Safety Training and Education

(a) General requirements. The Secretary shall, pursuant to section 107(f) of the Act, establish and supervise programs for the education and training of employers and employees in the recognition, avoidance, and prevention of unsafe conditions in employments covered by the act.

(b) Employer responsibility.

(1) The employer should avail himself of the safety and health training programs the Secretary provides.

(2) The employer shall instruct each employee in the recognition and avoidance of unsafe conditions and the regulations applicable to his work environment to control or eliminate any hazards or other exposure to illness or injury.

29 CFR 1926 Section	Applicable Safety Training Requirements	29 CFR 1926 Section	Applicable Safety Training Requirements
Subpart C **General Safety and Health Provisions**	General Safety and Health Provisions Safety Training and Education Employee Emergency Action Plans	**Subpart P** **Excavations**	General Protection Requirements
Subpart D **Occupational Health and Environmental Controls**	Medical Services and First Aid Ionizing Radiation Nonionizing Radiation Gases, Vapors, Fumes, Dusts, and Mists Hazard Communication Methylenedianiline Lead in Construction Process, Safety Management of Highly Hazardous Chemicals Hazardous Waste Operations and Emergency Response	**Subpart Q** **Concrete and Masonry Construction**	Concrete and Masonry Construction
		Subpart R **Steel Erection**	Bolting, Riveting, Fitting-up, and Plumbing-up
		Subpart S **Underground Construction, Caissons, Cofferdams, and Compressed Air**	Underground Construction Compressed Air
Subpart E **Personal Protective and Life-Saving Equipment**	Hearing Protection Respiratory Protection	**Subpart T** **Demolition**	Prepatory Operations Chutes Mechanical Demolition
Subpart F **Fire Protection and Prevention**	Fire Protection	**Subpart U** **Blasting and Use of Explosives**	General Provisions Blaster Qualifications Surface Transportation of Explosives Firing the Blast
Subpart G **Signs, Signals, and Barricades**	Signaling		
Subpart I **Tools—Hand and Power**	Power-operated Hand Tools Woodworking Tools	**Subpart V** **Power Transmission and Distribution**	General Requirements Overhead Lines Underground Lines Construction in Energized Substations
Subpart J **Welding and Cutting**	Gas Welding and Cutting Arc Welding and Cutting Fire Prevention Welding, Cutting, and Heating in Way of Preservative Coatings	**Subpart X** **Stairways and Ladders**	Ladders Training Requirements
		Subpart Y **Diving**	Commercial Diving Operations
Subpart K **Electrical**	Ground-fault Protection	**Subpart Z** **Toxic and Hazardous Substances**	Asbestos 13 Carcinogens Vinyl Chloride Inorganic Arsenic Cadmium Benzene Coke Oven Emissions 1,2-Diborom-3-Chloropropane Acrylonitrile Ethylene Oxide Formaldehyde Methylene Chloride
Subpart L **Scaffolding**	Scaffolding—Training Requirements		
Subpart M **Fall Protection**	Fall Protection—Training Requirements		
Subpart N **Cranes, Derricks, Hoists, Elevators, and Conveyors**	Cranes and Derricks Material Hoists, Personnel Hoists, and Elevators		
Subpart O **Motor Vehicles, Mechanized Equipment, and Marine Operations**	Material-handling Equipment Site Clearing		

■■ Figure 1 ■■ OSHA training requirements

(3) Employees required to handle or use poisons, caustics, and other harmful substances shall be instructed regarding the safe handling and use, and be made aware of the potential hazards, personal hygiene, and personal protective measures required.

(4) In job-site areas where harmful plants or animals are present, employees who may be exposed shall be instructed regarding the potential hazards, and how to avoid injury, and the first-aid procedures to be used in the event of injury.

(5) Employees required to handle or use flammable liquids, gases, or toxic materials shall be instructed in the safe handling and use of these materials and made aware of

the specific requirements contained in Subparts D, F, and other applicable subparts of this part.

(6) (i) All employees required to enter into confined or enclosed spaces shall be instructed as to the nature of the hazards involved, the necessary precautions to be taken, and in the use of protective and emergency equipment required. The employer shall comply with any specific regulations that apply to work in dangerous or potentially dangerous areas. (ii) For purposes of paragraph (b)(6)(i) of this section, "confined or enclosed space" means any space having a limited means of egress, which is subject to the accumulation of toxic or flammable contaminants or has an oxygen deficient atmosphere. Confined or enclosed spaces include, but are not limited to, storage tanks, process vessels, bins, boilers, ventilation or exhaust ducts, sewers, underground utility vaults, tunnels, pipelines, and open top spaces more than 4 feet in depth such as pits, tubs, vaults, and vessels.

In addition to OSHA standards, the construction industry may be faced with complying with the training requirements of the Mine Safety and Health Administration (MSHA), depending upon the type of operations performed. The definition of miner contained within 30 CFR Part 46.2 (g)(1)(ii) includes any construction worker who is exposed to the hazards of mining operations. The provisions outlined in 30 CFR 48 are more selective. For work in underground mines and surface mines or surface areas of underground mines, if the construction work is not of a "major addition," or if the mine is producing material, or if a regular maintenance shift is ongoing, then training is required.

MSHA defines training requirements according to the minimum courses of instruction and the minimum number of hours required and depends on the classification of the miner. Both Parts 46 and 48 classify miners as (1) new miners, (2) experienced miners, or (3) miners assigned to a task in which they have had no previous experience.

Before a new miner begins work at the mine, they must receive no less than 4 hours of training in the subject areas listed.	An introduction to the work environment, including a visit and tour of the mine, or portions of the mine that are representative of the entire mine (walk-around training). The method of mining or operation utilized must be explained and observed.
	Instruction on the recognition and avoidance of electrical hazards and other hazards present at the mine, such as traffic patterns and control, mobile equipment (e.g., haul trucks and front-end loaders), and loose or unstable ground conditions.
	A review of the emergency medical procedures, escape and emergency evacuation plans in effect at the mine, and instruction on the fire-warning signals and firefighting procedures.
	Instruction on the health and safety aspects of the tasks to be assigned, including the safe work procedures of such tasks, the mandatory health and safety standards pertinent to such tasks, information about the physical and health hazards of chemicals in the miners' work area, the protective measures a miner can take against these hazards, and the contents of the mine's HazCom program.
	Instruction on the statutory rights of miners and their representatives under the Act.
	A review and description of the line of authority of supervisors and miners' representatives and the responsibilities of such supervisors and miners' representatives.
	An introduction to your rules and procedures for reporting hazards.
No later than 60 calendar days after a new miner begins work at the mine, they must be provided with training in the listed subjects	Instruction and demonstration on the use, care, and maintenance of self-rescue and respiratory devices, if used at the mine.
	A review of first-aid methods.
No later than 90 calendar days after a new miner begins work at the mine, they must be provided with	The balance, if any, of the 24 hours of required training on any other subjects that promote occupational health and safety for miners at the mine.

■ ■ **Figure 2** ■ ■ **Applicable MSHA training standards for new miners**

Before a newly hired, experienced miner begins work at the mine they must be provided with training in the following subjects, which must also address site-specific hazards:	An introduction to the work environment, including a visit and tour of the mine, or portions of the mine that are representative of the entire mine (walk-around training). The method of mining or operation utilized must be explained and observed.
	Instruction on the recognition and avoidance of electrical hazards and other hazards present at the mine, such as traffic patterns and control, mobile equipment (e.g., haul trucks and front-end loaders), and loose or unstable ground conditions.
	A review of the emergency medical procedures, escape and emergency evacuation plans in effect at the mine, and instruction on the fire-warning signals and firefighting procedures.
	Instruction on the health and safety aspects of the tasks to be assigned, including the safe work procedures of such tasks, the mandatory health and safety standards pertinent to such tasks, information about the physical and health hazards of chemicals in the miners' work area, the protective measures a miner can take against these hazards, and the contents of the mine's HazCom program.
	Instruction on the statutory rights of miners and their representatives under the Act.
	A review and description of the line of authority of supervisors and miners' representatives and the responsibilities of such supervisors and miners' representatives.
	An introduction to your rules and procedures for reporting hazards.
No later than 60 calendar days after a newly hired, experienced miner begins work at the mine, they must be provided with	Instruction and demonstration on the use, care, and maintenance of self-rescue and respiratory devices, if used at the mine.

■■ Figure 3 ■■ Applicable MSHA training standards for newly hired, experienced miners

The training requirements of Part 46 most directly relate to the construction industry. These standards are outlined in Figures 2–4, according to the various classifications of miners. The MSHA regulations also dictate that annual refresher training be conducted for personnel and key safety issues; these standards are summarized in Figure 5.

New miners must be provided with no less than 24 hours of training as noted in Figure 2. Miners who have not yet received the full 24 hours of new-miner training must work where an experienced miner can observe that they are performing their work in a safe and healthful manner.

Training for newly hired, experienced miners must be conducted in the timeframe indicated in Figure 3 and also include the subjects listed.

Additional training requirements pertaining to the construction industry are contained in the MSHA regulations that are summarized in Figure 4. These regulations are focused on the actual operations being performed and not on the level of experience of the miner.

The most extensive section regarding training and education of employees in the OSHA standards can be found in §1926.59, Hazard Communication. MSHA regulations also contain a Hazard Communication standard found in 30 CFR §47.1-92. There was a time when employees had no legal means to force their employers to provide them with information about the hazardous substances they used at work. They demanded a "right-to-know"—the right to have access to such information. The Federal OSHA Standards detailed within 29 CFR 1910.12 and 1926.59, better known as the "Hazard Communications" or "Right-to-Know" standards require employers to do just that. This section of the

Regulation	Applicable Safety Training Requirements
30 CFR §46.7	New task training
30 CFR §46.8	Annual refresher training
30 CFR §46.11	Site-specific hazard training

■■ Figure 4 ■■ Applicable MSHA safety training references

standards requires employers to develop and implement employee training programs that inform them of the hazards of chemicals used in the workplace and the appropriate protective measures available for hazard avoidance. The MSHA HAZCOM standards were adopted more recently with full compliance by all sizes of mines to be in effect as of March 21, 2003. The standards are similar to those promulgated by OSHA, but focus on the specific hazards of the mining industry.

Each employer must have a written hazardous communication program for the workplace and all employees must have access to it. Employees must be trained regarding the hazardous chemicals present in the workplace and must possess knowledge of where to locate the list of these chemicals and the material safety data sheets for each chemical they might come in contact with. Employees should know what they are working with, where they can find information about the substance—either in the corporate office or on the job site—and what precautions they should be taking regarding exposure and accidents.

The purpose of this section is clearly stated as:

§1926.59 Safety Training and Education
(a) Purpose

(1) The purpose of this section is to ensure that the hazards of all chemicals produced or imported are evaluated, and that information concerning their hazards is transmitted to employers and employees. This transmittal of information is to be accomplished by means of comprehensive hazard communication programs, which are to include container labeling and other forms of warning.

Every company will have specific training needs and requirements, depending upon the performed work, the type of contract, and the owner. Upon assessing the overall regulatory requirements applicable to general construction work, the next step in safety-program development is to determine the work primarily performed by your employees. Although not mandatory on all projects, to achieve safety success, the training program should be unique to every project. Assessment of needs is crucial for each project, as each is different and presents its

own challenges. The importance of site-specific training will be addressed later in this chapter. When identifying the needs of a program, determine the type of work to be performed and then review the applicable OSHA standards. Different types of work and the use of equipment associated with that work have specific training requirements as listed in the OSHA regulations. The education of your employees regarding these standards is required by law, so be sure the training program includes these topics.

In addition to the regulatory and specific jobsite requirements, a determination must be made regarding the solution of any safety problem via training. Do safety problems currently exist? Will training solve the unsafe practices of employees or are there other actions that must be taken? Not every problem can be solved through training—shoddy equipment or unavailability of materials, for instance. Problems that can be addressed effectively by training include:

- lack of knowledge of a work process
- unfamiliarity with equipment
- incorrect execution of a task.

In determining what training is needed, the employer must identify the work the employee is expected to do and if the employee's performance is deficient. To obtain this information, an analysis of what an employee needs to know in order to perform a job in a safe manner must be conducted. With this information, specific training requirements will be identified.

These requirements determine the minimum that should be included in a safety training program. They are a place to start—guidelines to help program development. The goal of any safety training program is to educate both employers and employees in the recognition of unsafe or hazardous work conditions. The training program should raise awareness of these conditions so that avoidance and prevention of unsafe work practices is second nature to all involved. Once the kind of training that is needed has been determined, employees must be made aware of all the steps involved in a task or procedure. Training should focus on those steps so performance will be improved and *unnecessary* training will be avoided.

Determine the Goals and Objectives of the Training Program

In the twenty-first century, safety has taken a more active role in the culture of the workplace. To accompany the increased level of importance placed upon a safe work environment, employers must develop safety training programs for all employees. The employer is responsible for instructing each employee in the dangers associated with the work he or she will be performing. The employee must be able to recognize unsafe conditions, be aware of the regulations governing the work performed, and be instructed in measures to minimize, control, or eliminate hazardous exposures to injury or illness.

To develop a program that will have a strong impact on the work environment, goals are invaluable. When you know what it is you want to achieve, it is easier to reach the bar. By setting goals and tracking progress, the rewards of safe practices will be easier to quantify and benchmark, and training is most effective when designed in relation to the goals of the employer's total safety and health program.

Therefore the first step is to define the requirements of a successful training program. The goals should reflect the importance the company places on safe workers, safe working environments, and the minimization of workplace hazards, illnesses, and injuries caused by unsafe practices. When determining the goals ask questions that focus on results, such as:

- What are the benefits of a safety training program?
- How safe is the current workplace?
- How has safety improved since implementing the training program?

When defining the general goals of your program, consider the following areas where adequate training and education can really affect the bottom line.

1. Regulatory or performance-based goals (i.e., meet OSHA/MSHA standards).
2. Reduction in accidents, injuries, and illnesses.
3. Reduction in lost work days, hours, etc.

4. Improved EMR rating.
5. Strong safety records.

All of the above goals work toward development of a strong safety culture. There are several benefits to stellar safety performance. Reduced insurance premiums, better morale in the workplace, a good reputation in the industry are just a few. Each of these factors contributes to the increased possibility of consideration for various types of employment. With the rising popularity of alternate delivery methods, where the low bidder is not always awarded the job, safety is typically considered one of the factors to be evaluated for potential contract awards.

In addition to overreaching company goals and improving the bottom line, goals should also be set that involve the level of training, the frequency of training, and the quality of training. Consider goals such as:

1. Corporate managers and top executives will receive annual safety training.
2. Supervisors, project managers, and superintendents will receive project-specific training.
3. Every employee receives an initial company safety orientation within one to three days of starting work.
4. Every employee receives refresher training at regular intervals.
5. Every employee is instructed in safe practices and the hazards associated with an activity prior to the commencement of that activity.

In order to effectively administer a safety training program, guidelines and objectives must be determined for the training. The goal of instructional objectives is to clearly state what employers want their employees to do, to do better, or to stop doing as a result of the training. These objectives must be definitive and measurable, precisely indicating the skills or knowledge that must be demonstrated by an employee upon completion of the training. The learning objectives should also state the conditions for acceptable employee performance and the level of competency required to satisfactorily complete training. The OSHA training guidelines suggest using the following when defining effective learning objectives:

- specific, action-oriented language, that describes the preferred practice or skill and its observable behavior
- sufficient detail in the wording so that other qualified persons can recognize when the desired behavior is exhibited.

Ideas to consider include:

1. Rather than using the statement "The employee will understand how to use hearing protection," instead, say, "The employee will be able to describe when hearing protection should be used, why it is needed, and where it can be found in the workplace."

2. Instead of stating, "The employee will understand the Hazard Communication Plan," say, "The employee knows the location of the written Hazard Communication Plan, the location of the Material Safety Data Sheets, and the dangers associated with the chemicals used in the work."

By establishing predetermined safety goals and objectives at various levels, your company will set milestones to achieve. Keeping these goals in mind will also aid in developing the key components of the safety training program. In Figure 5, a sample outline of each level of goals and objectives is illustrated. The sample is modeled after an existing Hazard Communication Plan developed by Oklahoma State University.

Identifying Critical Elements and Key Players

The critical elements of a training program are the content of the program, the people delivering the training, and the employees receiving the training. After development of safety goals and the objectives of the training program, employers can then begin to define the elements of the program. The content of the program should be outlined with precise activities that the employee will take part in during training. These activities should simulate the actual job environment as closely as possible to best ensure that the employee will transfer the skills learned in training to the job. By arranging the course material to mimic the sequence of events that would be used on site, a specific safe practice can be developed.

Program Goals	
Goals of hazard:	To help reduce the risks involved in working with hazardous materials
Communication:	To transmit vital information to employees about real and potential hazards of substances in the workplace
Program:	To reduce the incidence and cost of illness and injury resulting from hazardous substances
	To promote public employer's need and right to know
	To encourage a reduction in the volume and toxicity of hazardous substances

Training Goals	
Training shall be conducted:	Within 30 days of inital employment or assignment to a new job
	Whenever new hazards are introduced into the workplace
	Annually (as a review)
Training must cover:	Method to detect presence of release Physical and health hazards
	Measures or personal protection
	Details of company/department plan

Training Objectives	
Employees must be informed of:	Any operation in their area where hazardous chemicals are used
	Requirements of regulations
	Location and availability of MSDS and plan

■■ **Figure 5** ■■ **Sample training goals and objectives**

For example, if an employee is learning the process of safe operation of a power tool, the sequence might be:

1. Ensure the power-source cord is free of defects.
2. Check that the power source is connected.
3. Identify and acknowledge that safety devices are in place and operating.
4. Understand the manufacturer's recommendations for safe operation.
5. Identify the power switch and its operation.
6. Determine and use the appropriate personal protective equipment, etc.

When determining the actual content of the training program, examine the needs that were previously identified. Use Figures 1–4 as guides to determine the actual content of the program, or consider the following as a starting point.

Every year OSHA lists the most violated standards in the construction industry. For the past decade, the following areas appear somewhere in the top ten.

- hazard communication
- electrical
- scaffolding
- fall protection
- ladders/stairways
- fire protection.

Each of these areas has specific requirements with respect to training that can be cited. In the period from October 2001 to September 2002, there were 919 citations and 788 investigations leading to over $376,000 in fines for violation of the scaffolding training requirements outlined in §1926.454. Falls are one of the leading causes of deaths on construction sites. During the same time period, there were 1191 citations and 1106 investigations for failures to comply with §1926.503, fall protection training. These citations yielded over $598,000 in fines. Table 1 highlights the top three most violated OSHA training standards. Is it just a coincidence that falls from heights are the leading cause of deaths in the construction industry, and that the most violated OSHA standard is §1926.451, General Requirements for all Types of Scaffolding? Based on the above facts, every safety training program should begin the development of its content by focusing on those standards.

■ ■ Table 1 ■ ■

Most Cited OSHA Training Violations

Standard	Description	Number cited	Fine
§1926.503	Fall protection training requirements	1191	$598,583.50
§1926.021	Construction safety training and education	1029	$963,478.25
§1926.454	Training requirements for all types of scaffolding	919	$376,312.31

In addition to the content of the course, other things employers must consider when developing training are:

- knowledgeable instructors with appropriate certifications
- adequate and up-to-date training material
- appropriate facilities in which to conduct the training
- effective evaluation tools
- instructor-to-student ratios.

Adequacy and appropriateness of the training program's curriculum development, instructor training, distribution of course materials, and direct student training should also be considered.

When developing the training curriculum, include appropriate technical input from outside reviewers, trainers, and industry representatives. The program should be pretested and evaluated by representative members of the target audience. Careful attention must be given to the following items:

- duration of the training
- course schedules and agendas
- different training requirements of the different target audiences
- adequate amount of hands-on demonstration and instructional methods
- monitoring of employee safety, progress, and performance during training.

The commitment to safety does not end when an employer completes the development of the training program. The program must be delivered in an effective manner in order for it to be successful in promoting safe work habits. The performance of the safety staff and the delivery effectiveness of the training program need to be given the same attention as the course content. The training should be conducted by a trainer or staff member who has demonstrated competency and leadership and who possesses the ability to deliver high-quality health and safety training.

Budgeting for a Comprehensive Training Program

If a company is serious about promoting safe work practices, there is a financial investment that must be made. There are costs associated with hiring an adequate number of safety personnel on staff, training time, and investments in appropriate and ample supplies of personal protective equipment. To adequately complete a safety training program, training materials need to be developed and distributed to employees, and time must be invested to train employees. The employer must identify and train an appropriate number of competent safety personnel or hire a certified safety professional to provide training to the employees. Research conducted by the Construction Industry Institute in 2002, has shown that there is an increase in safe performance on job sites if the safety representatives on the job site are actual company employees whose primary responsibility is the safety of fellow employees. Dedicating in-house staff to the promotion of safety illustrates the safety commitment of the employer to workers. Additionally, some incentive rewards for safe performance may also be planned into the operating budget of a company or a particular project. Based on the above discussion, safety needs to be a line item in both the company budget and the project budget.

KEY COMPONENTS OF A SUCCESSFUL PROGRAM—BEYOND REGULATORY REQUIREMENTS

In addition to the regulatory requirements that employers must meet in relation to training programs, thought must be given to other important components that contribute to the success of a training program. Research completed on safe workplaces indicates that certain actions can influence the overall success or failure of a safety program. These components are easy to implement; most only take time and planning prior to completion of work. However, what makes implementation difficult is that these steps attempt to change individual attitudes and habits. Although the steps are tangible methods that can be outlined on paper, the real effect and delivery is somewhat intangible. The focus of these components is making the shift from written policy to action at all levels.

Establishing the Safety Foundation

Safety demands a constant effort by all parties involved. The degree of success of any training program is contingent upon field implementation of the knowledge acquired during the instruction. Practicing the guidelines detailed within the training on a daily basis must be expected from all employees. When establishing a strong safety foundation, a message must be sent to employees indicating that not only are they expected to follow the safety regulations, but that they should immediately take care of any safety problems they encounter. If they are unable to correct the problem by themselves, they should contact their immediate supervisor to get the appropriate help.

All employees need to know that they are empowered to correct any unsafe conditions on a site, and that they can stop working if they feel they are in an unsafe environment. With this concept in mind, and knowing that it is supported by the employer, employees will begin to act safely. Employees must see that compliance with safe work practices is for their benefit, and that violation of these practices will not be tolerated. To enforce that belief, it must be clear that noncompliance with the safety regulations will result in immediate action. For example, some employers have adopted a three-strikes system that deals with safety violations in stages. The employee is issued a warning regarding his or her unsafe behavior for the first infraction. Next the employee is given time off without pay. Finally, if behavior fails to improve, the employee is terminated. This system, if followed consistently, will send a clear message regarding the company's policy toward safety compliance.

Improving the Overall Attitude

In order to increase the success of a training program, the attitude of the worker must be changed, and apathetic behavior must be modified. People can be trained so that acting safely and performing work in a safe manner is second nature. The atmosphere that needs to be created is one where unsafe acts are not tolerated, and that atmosphere needs to permeate every level of employment. Regardless of the level, whether top management, project managers, field personnel, or office administrative staff, the environment created must be one in which unsafe practices do not exist. If safety is the first consideration given to every task by every employee, then the "safety first" sign at the project entrance is more than just a sign—it has real significance.

The attitude toward safety is even more important during the actual training sessions. The focus of the training should not be a negative one. Safety should not be viewed as something that has to be done to avoid a punishment. It should be addressed in a favorable light, with enthusiasm and a positive attitude. Training should also reflect a level of seriousness since the goal of the program is to ultimately save lives and reduce injuries. Make sure employees understand what could happen if the practices are not followed. Part of a safety training program should include images of what can happen as a result of unsafe practices. Although sometimes graphic, these images will leave a lasting impression. Additionally, both OSHA and MSHA publish documents easily accessible on their Web sites that highlight fatal accidents. Employees can learn from others' mistakes.

Employees need to know that the company cares about their safety and well-being. Safety training should take center stage regularly throughout the month. It should be viewed as an investment and not as an expenditure. As the old saying goes, "People don't care how much you know, until they know how much you care." Keep that phrase in mind as training is conducted. Focus on issues that are important to workers and listen to their feedback. Try something novel with a safety twist, such as a safety essay or poster contest for employees' children. Lastly, remember that during training the attitude of trainees is often a reflection of the trainer's attitude.

Promotion of the Safety Culture

People who practice safe work habits develop a mindset that is in synch with other employees of the same mindset; together they achieve a new safety culture. Employees need to be an active and integral part of the safety team. To help promote this sense of inclusion and responsibility, have your employees sign a personal commitment to safety and positive interaction. This document is a powerful tool that noted safety professional Art Fettig developed several years ago. People need to learn to interact with each other in positive ways to promote safe work environments. Part of the training program must address this issue. On how many job sites where accidents occur do you hear co-workers say, "I knew they were going to hurt themselves or someone else, they were always taking chances."? Why don't co-workers get involved? A positive interaction program can only work if you first have the participants' signed permission. It should state that they give their permission for other employees to point out to them when they are acting unsafely. Workers need to admit they are human and that they do make mistakes. Injuries and accidents most often occur when individuals become complacent with their work and let their guards down. If part of the training program emphasizes that everyone is an equal player in the game of workplace safety, everyone will benefit.

IDENTIFYING WHO NEEDS SAFETY TRAINING

Who has the greatest need for safety training on a job site? Despite the fact that all employees are entitled to the knowledge of the safety and health hazards to which they are exposed, employers may not have the resources to train everyone to such a high degree. Employers need to determine which employees must receive which level of training. They must also access those on their job sites in the greatest need of information and instruction.

Employees at Risk

The OSHA training guidelines suggest that one way to differentiate between employees who have priority needs for training and those who do not is to identify employee populations that are at higher levels of risk. Having completed an assessment of training needs at the beginning of the training program development, the nature of work performed by certain employees will provide an indication of who should receive priority based on occupational safety and health risks.

To prioritize who should receive training, review these criteria:

- Pinpoint hazardous occupations.
- Examine the incidence of accidents and injuries, both within the company and within the industry.
- Complete thorough accident investigations.
- Identify both specific employees who could benefit from training and companywide training needs as a result of accident investigations.

According to OSHA sources, research has identified the following variables as being related to a disproportionate share of injuries and illnesses at the work site on the part of employees and should be considered when identifying employee groups for training in occupational safety and health.

1. The age of the employee (younger employees have higher incidence rates).
2. The length of time on the job (new employees have higher incidence rates).
3. The size of the firm (in general terms, medium-sized firms have higher incidence rates than smaller or larger firms).
4. The type of work performed (incidence and severity rates vary significantly by SIC Code).
5. The use of hazardous substances (by SIC Code).

It is imperative that high-risk employees receive top training priority. In addition to high-risk employees, there are several other levels of employees that require safety training, but the contents of the training program should be targeted to their job-performance duties.

Training Based Upon Employment Level

Another significant factor that influences the success of a training program is its appropriateness. The level and detail of training required is dependent upon the intended audience. The training for project managers and top executives will be distinctly different from the specific topics covered in the training of field-level personnel. Employers need to determine what level of safety training and material coverage is appropriate for each level of employees. The following are suggested topics for inclusion in safety training programs for various levels of employees.

OWNERS AND CORPORATE MANAGEMENT

These are the people who will set the overall tone of the safety environment. In addition to the list of topics presented below, be sure to illustrate how safety affects the bottom line of the company. Fewer injuries on the job significantly reduce the costs associated with insurance, workers' compensation claims, and litigation over these issues. Competitiveness is increased by lowering incident rates, allowing jobs to be delivered within time and budget constraints. Management must understand that training is an investment that pays handsome dividends. The statistics presented by J. Diether and G. Loos in the *Occupational Health and Safety Journal* (2000) indicate that corporate America provides nearly two billion hours of training to 60 million employees at a cost of $55 billion annually. Is the investment worth it? Some companies have reported a $10 to $20 return for each dollar spent on training, according to data published by M. Kleiman in the same journal. Training works; but if management does not appreciate the impact, that attitude will be reflected in work performance. Safety training at this level should highlight the following issues:

- benefits of a safety training program
- importance of top-management commitment
- leading by example
- promoting the safety culture
- direct involvment in the safety process
- active participation in accident investigations
- reviewing safety performance on individual projects.

PROJECT MANAGERS

These are the individuals who will ensure that the safety culture developed by upper management is actually implemented on individual projects. Budgeting, scheduling, and planning for safety is their responsibility, and these must be highlighted in their training. Project manager training should focus on site-specific safety and should be scheduled prior to the commencement of a new project. Suggested elements include:

- knowledge of the on-site dangers
- responsibility for communicating the hazards on the job
- complete job hazard analysis
- pretask safety planning
- evaluating the safety training required for each job
- understanding that the functioning of subcontractors also affects the level of safety experienced on the job site.

FIELD PERSONNEL

These are the individuals who will enforce the safety culture on a daily basis. Their training should equip them with the necessary tools to implement the job-site safety plan. Without adequate preparation, equipment, and support, these vital players in the safety program will not be able to complete their important job. Training should address the following issues:

- safety training of field personnel
- the ability to determine hazardous situations
- being the competent person who identifies hazards and has the authority to make necessary changes.
- responsiblity for daily job-site safety.

LABOR FORCE

The labor force is made up of the individuals who must adhere to the safety standards promoted from above. Their training must be focused, timely, and related to the particular work they are completing in order to be effective. Additionally, they must feel empowered to correct any safety problems without repercussions. Their training should incorporate the following issues:

- awareness of the hazards that will be encountered during the work day
- training focused on tasks that will be completed that day

- training in safe actions and recognition of unsafe work practices.

LEVEL OF TRAINING

Each safety training program is unique. There is no one-size-fits-all for every project or every company. To adequately address the needs on each project, safety training must reflect the actual work that is to take place, not just the general guidelines for safe work practices. Individuals performing project tasks must be aware of the dangers associated with each new task being performed. This goes beyond the initial safety orientation. Workers must continue to receive task-specific safety briefings throughout the life of a project, preferably at the beginning of each shift or when new hazards or conditions are introduced into the workplace. By relating the necessary safety precautions and safe work practices to workers prior to the commencement of each day or new task, workers are more likely to perform the work safely than workers who are not reminded of the dangers of their job. Most construction workers accept the fact that their work has some inherent risks and dangers. Acceptance sometimes turns into complacency that can ultimately lead to unsafe practices. Construction workers do not think that the work they perform on a daily basis is dangerous or that a particular shift could be their last. By reiterating the dangers, it brings safety to the forefront of their minds and, with that raised consciousness, they begin their tasks. The saying "safety first" needs to be taken seriously and not viewed as some trite cliché mouthed by industry. It must be adopted as policy and practiced on a daily basis.

General Safety Training

Training should be presented in a manner that clearly conveys its organization and meaning to the employees. To achieve this goal, employers should strive to provide a concise overview of the materials being presented and how this new knowledge relates to the employees' experience. Workers must also be aware of what they are expected to learn in the program. Make sure to summarize the program's objectives and the key points of information that they are responsible for learning. Employees must be convinced of the importance of the material they are learning and how it is relevant to their work.

A key element of an effective training program is allowing employees to participate in the training process and to practice their skills or knowledge. Not only will the employees feel they are part of the safety process, but it is a means to ensure that they are learning the required knowledge or skills. By having activities built into the training sessions, employers can also correct any unsafe practices that may be exhibited during the hands-on portion of the training. Make sure to involve employees in training. This can be done by encouraging participation in group discussions, asking questions of an audience, and allowing participants to contribute their knowledge and expertise on the issues discussed.

Project-specific Training

Every new project will have distinct and different tasks that must be completed. Safety programs need to be tailored to meet the needs of different job sites. It is important not only to have a general safety orientation for all employees, but each new project should also develop its own safety training program. Aspects of jobs will be different, the types of work performed and the employees may change. Most general contractors not only have permanent staff, but also employ additional local labor forces for each new contract landed. In addition to the employees on their own payroll, several subcontractors are also on site. Will safety training be conducted by the general contractor or will the training be left up to the individual subcontractor's staff? Making sure that subcontractors' employees are trained before they are allowed to work on the job site is the best way that the general contractor has of ensuring that all employees on the site have received at least the minimum training required.

When developing a project-specific safety training program, a job hazard analysis (JHA) must be conducted to determine the needs of the training. JHA is a procedure that was developed by OSHA for completing the following:

- studying and recording each step of a job
- identifying existing or potential hazards
- determining the best way to perform the job in order to reduce or eliminate the risks.

OSHA suggests that if an employer's safety training needs can be met by revising an existing company training program rather than developing a new one, or if the employer already has some knowledge of the process or system to be used, appropriate training content can be developed through such means as:

- Using company accident and injury records to identify how accidents occur and what can be done to prevent them from recurring.
- Requesting employees to provide in writing, and in their own words, descriptions of their jobs, including the tasks performed and the tools, materials, and equipment used.
- Observing employees at the work site as they perform tasks, asking about the work, and recording their answers.
- Examining similar training programs offered by other companies in the same industry.
- Obtaining suggestions from such organizations as

  the National Safety Council
  the Bureau of Labor Statistics
  OSHA-approved state programs
  OSHA full-service area offices
  OSHA-funded state consultation programs
  the OSHA Office of Training and
    Education.

IMPLEMENTATION TOOLS

Once the content of the training program is established and the initial orientation training complete, the real challenge to employers is the implementation of the program on site. Several methods can be utilized to help in the implementation phase, including checklists, demonstrated policy enforcement, and routine safety meetings. Also, the establishment of a job-site safety committee comprised of employees and supervisors alike is an effective method for implementing the goals of a developed program.

Checklists

Checklists are a quick, concise way to ensure that the proper safe practices are being employed on a job site. Checklists can be developed for key safety requirements when performing any job-site task. They should be written and tailored to the type of operation performed. The content of a checklist should include applicable regulatory requirements, personal protective equipment required, and a step-by-step procedure that will ensure safe operations and task performance.

Several contractors have condensed key safety points and OSHA requirements for each subsection of the construction codes onto a laminated card that is issued to all employees. The job-site routine is for each employee to review the procedures on the card prior to commencing a particular type of work. This is a simple, inexpensive way to implement a safety program.

Policy Enforcement

Regardless of the content of a safety training program, the quality of the instructor, and the demonstrated management commitment, no safety program will be taken seriously by employees unless it is enforced. Enforcement must be consistent and applied to everyone and every accident equally. All employees should be required to sign a statement indicating that they have received safety training and understand the policy of the employer regarding unsafe work practices. Employees must complete the training program with the following elements clearly embedded in their minds:

- They are empowered to stop working if unsafe conditions are present or if someone around them is acting in an unsafe manner.
- They must immediately correct unsafe conditions or contact a supervisor to do so.
- If they witness unsafe practices by a fellow employee, they must express their concerns to the other employee.

- They must believe that everyone is responsible for safety.
- They realize that accidents do not just happen.
- If they do not comply with the above listed safety guidelines, they will be subject to immediate disciplinary action.

This is a hard line to take and may seem harsh to some employers. Some feel that everyone deserves a second chance, but in the safety world that attitude will lead to accidents. In some cases a second chance is impossible, since the unsafe action of an individual may have caused the death or work-ending injury of that individual or others. This situation should not be tolerated; unsafe actions must result in swift enforcement of the company safety policy.

Safety Meetings

To further enforce and bolster the safety commitment and culture on a job site, regular safety meetings must be held. The frequency of these meetings is determined by the employer, but a good rule of thumb is to hold meetings once a week, preferably on a Monday morning. Why Monday morning? Injury data published has shown that most accidents take place on Mondays and then taper off throughout the week. Also, making safety the first thing that is addressed at the beginning of the work week continually reinforces the importance of safety in the workplace.

Scheduling the meeting is not the only thing that must be considered. There are several components of a meeting that must be carefully planned to achieve the desired results of the program. Who is required to attend the meetings? Who will conduct the meetings? Where will the meetings be held? What topics will be addressed in each meeting? What kind of a record will be kept of the meetings? Each of these questions raises valid points that must be addressed prior to conducting any meeting. Without good delivery, pertinent topics, and the proper audience, a safety meeting will not meet the goal of enforcing company safety policy on site. In fact, it may become a joke to workers, living up to the unflattering stereotypes of these meetings which permeate the industry.

Art Fettig wrote a book entitled *Winning the Safety Commitment*. In his book, he offers numerous suggestions and methods for delivering effective, motivating, and creative safety meetings that really convey the importance of safety. It is highly recommended that the session presenter use as many resources as possible to create the safety meetings.

One of the key components of an effective safety meeting is the participation of the employees. There are several ways to actively engage a crew during a meeting that will encourage continued and sustained participation in future meetings. Listed below are some suggestions for doing this.

- Assign crew members topics that they must prepare and present at a meeting.
- Ask questions of the workers to begin the meeting. Use the answers to direct the discussion of the meeting.
- Avoid lecturing.
- Ask about personal experiences.
- Don't allow one person to dominate the discussion.
- Keep meetings short and to the point.
- Ensure the topic discussed is relevant to the work that will be performed.

Safety meetings do not need to be formal, or even called safety meetings. A supervisor can simply incorporate the safety message in daily discussion of the work that will be conducted during that shift. The National Safety Council has referred to informal gatherings to discuss the work being performed where safety is built into the discussion, as the "best kind of safety meeting" (NSC, 1985). These five-minute, informal, pretask meetings are important from both a productivity and a safety standpoint. Employers should develop the habit of holding such meetings.

ASSESSMENT TOOLS

To make sure that the training program is accomplishing its goals, an evaluation of the training must be completed. Critical components of every training program are assessment tools used to measure the effectiveness of the training. The method used to meet this goal should be developed in conjunction with the course objectives and content. Evaluation of the program will help employers benchmark the learning achieved and determine if an employee's performance has improved based upon the training.

Measuring the Success of the Program

There are several tools and methods that can be used to evaluate the effectiveness of training. Regardless of the tools used, be sure that they target all of the levels of participation: students, instructors, managers, and corporate officers. Each stakeholder in the program must receive feedback regarding the effectiveness of the program. Recommended methods of evaluating training are:

1. *Participant opinion.* Surveys and informal discussions with employees can help employers determine the relevance and appropriateness of the training program. This can also lead to suggestions for improvements by employees, making them feel that they are a part of the safety team.
2. *Supervisors' observations.* Supervisors observe their workers on a daily basis and possess a vantage point like no other person on the job site. They can observe an employee's performance both before and after the training and note the improvements or changes in an employee's attitude or work practices.
3. *Workplace improvements.* The ultimate success of a training program may bring about changes throughout the workplace that result in reduced injury or accident rates. These may be improved housekeeping standards or a raised awareness of safety importance.

Regardless of how the assessment is conducted, it will provide employers with the knowledge of how well the program was received, that its message is having an impact on job-site safety, and how employees feel about the training. Equipped with this information, employers can make necessary changes to the training program itself.

Incentive Programs

Find a way to keep score of safe practices and promote healthy competition within the workplace. Nothing motivates people more than when they have a personal buy-in to the success of a project. By rewarding people for performing safely, others will be motivated to do the same. These incentives need not be costly or even monetary. Simple rewards like a sticker for a hard hat, a new hard hat, a T-shirt, or a feature in the company newsletter can really make a positive difference. Also, mentioning employees' safe work practices in weekly toolbox meetings is another way to positively influence the safety culture. Remember the rule, "Praise in public, scold in private." Do not discuss negative or poor safety performances in groups; make sure to individually council unsafe workers, and do it in a timely manner. If punishment or reprimand is not immediate, it may only lead to continued unsafe actions.

Innovative incentive programs practiced by some companies include the following:

- Evenly dividing the annual rebates received from the insurance provider for a good safety record among employees.
- Issuing specialty credit cards on which employees can earn dollars for safe performance. Employees spend their dollars by ordering from a catalog of items.
- Hosting family dinners with a safety focus.
- Providing on-site lunches for the safest crew every month.
- Awarding special hats or T-shirts quarterly to the safest crew.

Employee Feedback

The employees themselves can provide valuable information on the training they need. Safety and health hazards can be identified by questioning employees regarding their attitude and feelings toward the level of safety on the job site. Some questions to ask employees that are suggested by OSHA include:

- Does anything about your jobs frighten you?
- Have you experienced any near-miss incidents?
- Do you feel you are taking risks in your work?
- Do you believe that your job involves hazardous operations or substances?

Employee feedback is one of the best methods for determining what training needs are required and if the current training program is actually achieving its goals. By listening to employees, employers can gain a real sense of how the safety culture is being promoted and reinforce the safety team concept.

SUMMARY

A safety training program is an education tool for both employers and employees. Development of a safety training program involves several steps: include regulatory requirements and company objectives of the program, incorporate the corporate safety philosophy, determine the program needs, identify the key players, and budget for safety. Once the program content is developed, training must be focused to specific audiences and contain pertinent information for each target group and each project. To successfully launch a safety training program, there are several available means to assist with enforcement of the developed policy. Both OSHA and MSHA offer extensive literature, guidelines, videos, and other resources to employers to help them develop safety training programs. Most of these items can be purchased for a nominal fee, and a large number of the items are free.

A key component to the safety training process is the assessment of the program. Methods to determine the effectiveness of the program must consider input and feedback from all stakeholders in the process. Safety training is the most effective way to increase the performance and efficiency of all employees and should be a priority of every company.

REFERENCES

"Contractors Take Safety Seriously," *Concrete Construction*, February 2001, Vol. 46, No. 2, pp. 42–44.

Diether, J. and G. Loos. "Advancing Safety and Health Training," *Occupational Health and Safety*, 2002, Vol. 69, pp. 28, 34.

Fettig, A. *Winning the Safety Commitment*. Battle Creek, MI: Growth Unlimited Inc., 1998.

Goetsch, D. L. *Construction Safety and Health*. Upper Saddle River, NJ: Prentice Hall, 2003.

Hinze, J. W. "Safety Plus: Making Zero Accidents a Reality," Construction Industry Institute Research Report 160-11, 2003.

_____. *Construction Safety*. Upper Saddle River, NJ: Prentice Hall, 1997.

Kleiman, M. "What Happens If You Don't Train Them and They Stay?" *Occpational Health and Safety*, 2000, Vol. 69, pp. 18, 70.

Mine Safety and Health Administration (MSHA), 30 CFR 46–48. Washington, DC: U.S. Department of Labor, 1999.

Occupational Safety and Health Administration (OSHA). *OSHA Standards for the Construction Industry*, 29 CFR 1926. Washington, DC: U.S. Department of Labor, 2001.

Sarvadi, D. "The Importance of Safety Training," *Compliance Magazine*, February 2000, pp. 16–17.

REVIEW EXERCISES

1. What are the general requirements of employers in the construction industry relating to safety training of their employees?
2. What is the purpose behind the HAZCOM standards?
3. What safety problems can be effectively addressed through training?
4. What are the critical elements of a safety training program?
5. What are the most cited OSHA training violations?
6. In addition to presenting safety training with a positive attitude, what other methods can be utilized to improve an employee's attitude toward safety?
7. Name/list criteria used to determine which employees receive safety training.
8. List suggested elements of a safety training program for the following groups:

    owners and corporate management
    project managers
    field personnel
    labor force
9. Why is it important to include employee input in the safety training process?
10. What is job hazard analysis and why is it important in developing safety training?
11. List three sources for suggestions and guidelines for safety-training materials.
12. Give some examples of effective safety-program implementation tools.
13. Why is it important to define the safety culture?
14. List the benefits associated with conducting pretask safety meetings.
15. Why is swift and equal enforcement of the company safety policy important to the program's success?

.. PART III ..

Legal Aspects

■■ **About the Authors** ■■

MULTIEMPLOYER WORK-SITE ISSUES
by Everett A. Beaujon, CHMM, CPEA, CSP,
and Greg L. Smith, CSP

Everett A. Beaujon, CHMM, CPEA, CSP. Until his death in March 2003, Everett A. Beaujon was the Director of Safety and Health for Austin Industrial, Inc., in Houston, Texas. Austin Industrial, Inc., is a maintenance and construction contractor that provides a variety of services for the petrochemical, refining, power-generation, and manufacturing industries in the southern and southeastern United States. Everett held various safety and health positions with Austin Industrial during his fourteen years with the company. He received a B.S. in Mechanical Engineering from the University of Cincinnati, and was a Certified Safety Professional and a Certified Hazardous Materials Manager. Prior to joining Austin Industrial, Everett was the Safety and Hygiene Administrator at a large overseas refinery complex where he was employed for nineteen years.

Greg L. Smith, CSP, is President of Construction Safety & Health Inc. in Austin, Texas. Mr. Smith has over twenty years of experience in the safety field, working in both general industry and construction. He holds a B.S. in Industrial Hygiene and Safety from the University of Houston, Clear Lake, and is completing an M.S. in that discipline from the same institution. He was named the Gulf Coast chapter's Safety Professional of the Year in 2001, and is a member of the Advisory Council for the University of Houston's Safety Engineering program.

CHAPTER 12

Multiemployer Work-Site Issues

LEARNING OBJECTIVES

- Explain the citation doctrine that OSHA follows for inspections and citation issuance to employers at multiemployer work sites.
- Discuss the roles and responsibilities of the various employers, owners, and contractors at a multiemployer work site as they relate to compliance with the applicable OSHA standards.
- Discuss the various safeguards owners and contractors can implement to minimize the exposure to risk and liabilities in execution of a project.
- List the elements and components of a site safety and health process that owners and contractors can implement to protect employees and consequently avoid OSHA citations.

INTRODUCTION

A single construction company contractor seldom executes major construction projects without subcontractor assistance. The construction of houses, highways, bridges, major buildings, structures, and manufacturing facilities is commonly accomplished by a number of employers [i.e., owners, engineers, general and specialty (or sub) contractors].

The Occupational Safety and Health Act, Section 5(a)(2) states that each employer has a responsibility to comply with the Occupational Safety and Health standards promulgated under the Act. Employers are subject to citations by OSHA for violations of its standards. The Occupational Safety and Health Administration has provided clear examples of how its compliance officers may issue citations under its multiemployer citation policy to both the general contractor and various subcontractors that operate within a single work site. OSHA's *Field Inspection Reference Manual* (FIRM) [which replaced the *Field Operations Manual* (FOM) in September 1994], provides OSHA field offices with a reference document for identifying the responsibilities associated with the majority of their inspection duties at all multiemployer work sites.

OSHA's *Field Inspection Reference Manual*, Directive CPL 2.103 of September 1994, Chapter III, Section C.6, covers multiemployer work sites and specifically references the issuance of citations to employers at multiemployer work sites, both in construction and general-industry settings.

The multiemployer citation policy has been the target of employer groups, particularly in recent years, with many companies arguing that they should not be held responsible for protecting employees who are managed by other companies or subcontractors. The policy has been upheld in several important court decisions, but has also been determined by some courts (most notably the 5th Circuit Court of Appeals) to be unconstitutional and hence unenforceable. Nevertheless, on December 10, 1999, OSHA issued a revised Directive CPL 2-0.124 to clarify the agency's original policy of 1994.

The new directive, which can be found at http://www.osha.gov, replaces paragraph C.6 of Chapter III in the agency's *Field Inspection Reference Manual* and has been in effect since December 10, 1999. Generally, this latest directive ". . . continues OSHA's existing policy for issuing citations . . ." on multiemployer sites, but gives more guidance on how the compliance officer determines whether and to what degree an employer should be held responsible.

General contractors have often borne the brunt of enforcement actions at multiemployer work sites, triggering complaints that they have been held accountable for protecting employees who are by contract beyond their control.

The new directive categorizes various types of employers based on the extent to which employers are exposing employees, creating or correcting hazards, or the degree of control they have over the work site.

This chapter explores the principal issues related to violations of OSHA standards that employers confront at a multiemployer work site, regardless of whether or not they created the hazard(s).

Discussions with the safety managers of a number of general contractors in the petrochemical industry, Gulf Coast Area, confirms that OSHA applies the multiemployer citation doctrine for citations of violative conditions to both the general contractor and the owner of the facility when the compliance officer inspects the work sites of the general contractor at an owner's facility (one of the author's employers is a general contractor in the Texas Gulf Coast Area).

OSHA does apply the doctrine in all industry sectors—manufacturing, construction, and others—whenever it encounters a multiemployer work site.

Several examples of typical situations that may be encountered or created on a multiemployer site are described with the related action steps that employers can take to protect their employees from the hazards, and consequently avoid OSHA citations. A listing of the most frequently cited OSHA standards for construction and other SIC major groups, can be found at: (http//www.osha.gov/cig-bin/std-stdser1?).

MULTIEMPLOYER WORK SITES

Who are the employers and other entities involved on a multiemployer work site, whether it is a construction project or a general-industry facility? On such work sites there are usually two types of employers: those who exercise some control over the work site and others who have very little control over activities on the site. Control—as far as the execution of the work—is proportional to the responsibility each employer has for executing the work.

Principal Employers

It is not unusual to have more than one owner of a project. Owners are classified as those who provide the capital funds and, at times, have physical control of the site. General contractors (usually one) are responsible for the overall coordination, planning, scheduling, and executing of the work. Subcontractors, on the other hand, are usually contracted to the general contractor and are responsible for planning, scheduling, and execution of specific sections of the overall project. Subcontractors are generally specialty contractors dealing with one subsystem of the project (i.e., concrete work). The construction of projects of major proportions, both in size as well as capital outlay, are not managed by owners due to a scarcity of the resources required to self-manage major construction projects. Instead, owners hire separate professional firms or general contractors to manage the execution of the project.

In most cases, owners are the principal employers involved in the construction process through project managers, the general contractor, and the subcontractors on multiemployer work sites.

Other Employers or Entities

Facilities, whether they are commercial buildings or manufacturing plants, producing chemicals or other products, are conceived, designed, and then transferred to construction plans and drawings (blueprints) by professional architects and engineers. These firms, hired by the owner, may be considered employers, depending on their direct or indirect involvement at the work site. Alleged hazards stemming from the engineering design of equipment might be one reason for these entities to be involved.

The U.S. Department of Labor's Occupational Safety and Health Administration is also considered an "other entity" when involved at a multiemployer work site. OSHA's compliance officers are the parties that conduct the work-site compliance inspections and recommend citations be issued to employers at the work site.

WORK-SITE ISSUES

The OSH Act and Employers' Duties

The Williams–Steiger Occupational Safety and Health Act of 1970 (commonly known as the Occupational Safety and Health Act of 1970) has several requirements that employers and employees must meet. First, in Section 5(a)(1) it states: ". . . each employer shall furnish to each of his employees employment and a place of employment which are free from recognized hazards that are causing or are likely to cause death or serious physical harm to his employees." Section 5(a)(2) continues with: ". . . shall comply with occupational safety and health standards promulgated under this Act." Section 5(b) states "Each employee shall comply with occupational safety and health standards and all rules, regulations, and orders issued pursuant to this Act which are applicable to his own actions and conduct."

Section 5(a)(1) of the William–Steiger Occupational Safety and Health Act of 1970 has become known as the general duty clause. It is a "catch-all" used for citations if OSHA identifies unsafe conditions for which a regulation does not exist or may not reflect *current best practices*.

In practice, OSHA court precedent, and the Occupational Safety and Health Review Commission (a review panel created under the 1970 OSH Act) have established that if the following elements are present, a general duty clause citation may be issued.

1. The employers failed to keep the workplace free of a hazard to which employees of that employer were exposed.
2. The hazard was recognized. (Examples might include: through your safety personnel, employees, organization, trade organization, or industry customs.)
3. The hazard was causing or was likely to cause death or serious physical harm.
4. There was a feasible, alternative means to correct the hazard.

Multiemployer Citation Policy

BACKGROUND

The revision to OSHA's policy CPL 2-0.124 continues OSHA's existing policy for issuing citations on multiemployer work sites. However, it gives clearer and more detailed guidance than did the earlier description of the policy in the FIRM, including new examples explaining when citations should and should not be issued to exposing, creating, correcting, and controlling employers. These examples, which address common situations and provide general policy guidance, are not intended to be exhaustive. In all cases, the decision on whether to issue citations should be based on all of the relevant facts revealed by the inspection or investigation.

CITATIONS

On multiemployer work sites in all industry sectors, more than one employer may be citable for a hazardous condition that violates an OSHA standard. The Compliance Safety and Health Officer (CSHO) must follow a two-step process in determining whether more than one employer is to be cited. (Refer to the quick guide in Figure 1.)

The first step the CSHO must follow is to determine whether or not the employer is a "creating," "exposing," "correcting," or "controlling" employer. Employers may have more than one role on a work site. Only the *exposing* employer may be cited under the general duty clause violations.

If an employer falls into one or more of the categories, he or she has a responsibility to comply with the Occupational Safety and Health standards promulgated under the OSH Act.

The second step the CSHO must follow is to determine if the employer's actions were sufficient to meet those obligations under the Act. The extent of the actions required of employers varies in accordance with which category applies to the employer. The extent of the measures that, for instance, a controlling employer must take to satisfy the duty to exercise responsible care to detect and prevent violations is less than what is required of an employer with respect to protecting his own employees.

STEP 1: DEFINING THE TYPE OF EMPLOYER

Determining the category of an employer in order to cite the employer is the first step a CSHO must follow. During an OSHA work-site inspection at a multiemployer site, hazardous conditions that violate OSHA standards discovered by a CSHO will be documented, investigated, and analyzed to determine which employer is citable for the discovered violative conditions. The following are the categories of employers.

The Creating Employer is the employer that caused a hazardous condition that violates an OSHA standard. The directive states: ". . . Employers must not create violative conditions. An employer that does so is citable even if the only employees exposed are those of other employers at the site."

The Exposing Employer is the employer whose own employees are exposed to the hazard. Here the explanation is more complex: ". . . If the exposing employer created the violation, it is citable for the violation as a creating employer. If another employer created the violation, the exposing employer is citable if it (1) knew of the hazardous condition or failed to exercise reasonable diligence to discover the condition, and (2) failed to take steps consistent with its authority to protect its employees. If the exposing employer has authority to correct the hazard, it must do so. If the exposing employer lacks the authority to correct the hazard, it is citable if it fails to do each of the following: (1) ask the creating and/or controlling employer to correct the hazard; (2) inform its employees of the hazard; and (3) take reasonable alternative protective measures. In extreme circumstances (e.g., imminent danger situations), the exposing employer is citable for failing to remove its employees from the job to avoid the hazard. . . ."

The Correcting Employer is the employer who is engaged in a common undertaking on the same work site as the exposing employer and is responsible for correcting a hazard. This usually occurs where an employer is given the responsibility of installing and/or maintaining particular safety/health equipment or devices. Now some subjectivity starts to develop in the directive which can lead to confusion ". . . The correcting employer must exercise reasonable care in preventing and discovering violations and meet its obligations of correcting the hazard. . . ." The subjectivity of course is in the definition of "reasonable care," which can refer back to the mandate to the employer to be knowledgeable of, and apply, regulations that are applicable to their types of work activities.

The Controlling Employer is the employer who has general supervisory authority over the work site, including the power to directly correct safety and health violations or require others to correct them. Control can be established by contract or, in the absence of explicit contractual provisions, by the exercise of control in practice. Per the directive ". . . A controlling employer must exercise reasonable care to prevent and detect violations on the site. The extent of the measures that a controlling employer

OSHA CITATIONS—COMPLIANCE RESPONSIBILITIES ON MULTIEMPLOYER WORK SITES

QUESTION

If a subcontractor created a hazard or is responsible for correcting it, when is the General Contractor or other employer subject to citation?

Two Stage Analysis

Step (1): Does the company have (or exercise) sufficient control over the sub?
Step (2): If the company does have control, did it exercise reasonable diligence to prevent or discover the violation?

■ Step (1): DOES THE TRADITIONAL GENERAL CONTRACTOR HAVE CONTROL OVER THE SUBCONTRACTOR?

The Traditional General Contractor

- Has full contractual responsibility and authority over subs.
- Has full authority to compel sub to fulfill its safety responsibilities by virtue of the General's supervisor role.
- Control established contractually
- Subject to a citation if, and only if, the General failed to exercise reasonable diligence.

■ Step (2) DID THE GENERAL EXERCISE REASONABLE DILIGENCE?

- Standard of Care: General must exercise reasonable diligence in preventing or discovering violation. (What would a reasonable person do?)
- No legal basis for holding a company strictly liable.
- General is not required to inspect as frequently or thoroughly as the sub.
- Not required to have the same expertise as specialized sub.

SUBCONTRACTOR

If a subcontractor did not create the hazard, and is not responsible for correcting it, the subcontractor must:

- Protect its employees to the extent possible
- Try to get the general contractor (or whoever is responsible) to correct the problem.

OSHA's VIEW

The policy is not supposed to result in strict liability. A General's duty of care is not the equivalent of the sub's. Safety and Health Program quality is relevant.

We do not want to create safety disincentives.

We are looking at ways to give more clarity to the policy and how it is enforced.

■■ Figure 1 ■■ Quick guide to compliance responsibilities

must implement to satisfy this duty of reasonable care is less than what is required of an employer with respect to protecting its own employees. This means that the controlling employer is not normally required to inspect for hazards as frequently or to have the same level of knowledge of the applicable standards or of trade expertise as the employer it has hired. . . ." They do however need to be aware of any regulations applicable to the project type and scope.

STEP 2: DETERMINING ACTIONS TAKEN

Analysis

The second step an OSHA CSHO must follow is to analyze each discovered hazardous condition that violates an OSHA standard to determine which employer is citable as the creating, exposing, correcting, or controlling employer. Since employers can be categorized as having more than one definition, they can be citable under any one of them.

Violation Examples

The following are descriptions of hypothetical situations and conditions created and allowed to exist at a multiemployer work site that violate an OSHA standard. The CSHO must analyze each violative situation or condition to determine which employer(s) is citable under the OSH Act. In other words, which one created the hazard, which one has employees exposed to the hazard, which one could or should have corrected the hazard, and which one is the controlling employer.

The descriptions of hypothetical situations and conditions can be found at http://www.osha.gov/pls/oshaweb/owadisp.show_document?p_table=DIRECTIVES&p_i

CREATING EMPLOYER

Example 1: Employer Host operates a factory and contracts with Company S to service machinery. Host fails to cover drums of a chemical despite S's repeated requests that it do so. This results in the creation of airborne levels of the chemical that exceed the permissible exposure limit (PEL).

Analysis

Step 1: Host is a creating employer because it caused employees of S to be exposed to the air contaminant above the PEL.

Step 2: Host failed to implement measures to prevent the accumulation of the air contaminant. It could have met its OSHA obligation by implementing the simple engineering control of covering the drums. Having failed to implement a feasible engineering control to meet the PEL, Host is citable for the hazard.

Example 2: Employer M hoists materials onto floor 8, damaging perimeter guardrails. Neither its own employees nor employees of other employers are exposed to the hazard. It takes effective steps to keep all employees, including those of other employers, away from the unprotected edge and informs the controlling employer of the problem. Employer M lacks authority to fix the guardrails.

Analysis

Step 1: Employer M is a creating employer because it caused a hazardous condition by damaging the guardrails.

Step 2: While it lacked the authority to fix the guardrails, it took immediate and effective steps to see that the hazard was corrected. Employer M is not citable since it took effective measures to prevent employee exposure to the fall hazard: It took effective steps to keep all employees away from the hazard and notified the controlling employer of the hazard.

EXPOSING EMPLOYER

If the exposing employer created the violation, it is citable for the violation as a creating employer. If another employer created the violation, the exposing employer is citable if it (1) knew of the hazardous condition, and (2) failed to take steps consistent with its authority to protect its employees. If the exposing employer has authority to correct the hazard, it must do so. If the exposing employer lacks the authority to correct the hazard, it is citable if it fails to do each of the following: (1) ask the creating and/or controlling employer to correct the hazard;

(2) inform its employees of the hazard; and (3) take reasonable, alternative protective measures. In extreme circumstances (e.g., imminent danger situations), the exposing employer is citable for failing to remove its employees from the job to avoid the hazard.

Example 1: Employer Sub S is responsible for inspecting and cleaning a work area in Plant P around a large, permanent hole at the end of each day. An OSHA standard requires guardrails. There are no guardrails around the hole and Sub S employees do not use personal fall protection, although it would be feasible to do so. Sub S has no authority to install guardrails. However, it did ask Employer P, which operates the plant, to install them. P refused to install guardrails.

Analysis
Step 1: Sub S is an exposing employer because its employees are exposed to the fall hazard.
Step 2: While Sub S has no authority to install guardrails, it is required to comply with OSHA requirements to the extent feasible. It must take steps to protect its employees and ask the employer that controls the hazard—Employer P—to correct it. Although Sub S asked for guardrails, since the hazard was not corrected, Sub S was responsible for taking reasonable alternative protective steps, such as providing personal fall protection. Because that was not done, Sub S is citable for the violation.

Example 2: Unprotected rebar on either side of an access ramp presents an impalement hazard. Sub E, an electrical subcontractor, does not have the authority to cover the rebar. However, several times Sub E asked the general contractor, Employer GC, to cover the rebar. In the meantime, Sub E instructed its employees to use a different access route that avoided most of the uncovered rebar and required them to keep as far from the rebar as possible.

Analysis
Step 1: Since Sub E employees were still exposed to some unprotected rebar, Sub E is an exposing employer.
Step 2: Sub E made a good-faith effort to get the general contractor to correct the hazard and took

feasible measures within its control to protect its employees. Sub E is not citable for the rebar hazard.

CORRECTING EMPLOYER

The correcting employer must exercise reasonable care in preventing and discovering violations and meet its obligations of correcting the hazard.

Example 1: Employer C, a carpentry contractor, is hired to erect and maintain guardrails throughout a large, 15-story project. Work is proceeding on all floors. C inspects all floors in the morning and again in the afternoon each day. It also inspects areas where material is delivered to the perimeter once the material vendor is finished delivering to that area. Other subcontractors are required to report damaged or missing guardrails to the general contractor, who forwards those reports to C. C repairs damaged guardrails immediately after they are reported. On this project, few instances of damaged guardrails have occurred, except where material has been delivered. Shortly after the afternoon inspection of floor 6, workers moving equipment accidentally damage a guardrail in one area. No one tells C of the damage and C has not seen it. An OSHA inspection occurs at the beginning of the next day, prior to the morning inspection of floor 6. None of C's own employees are exposed to the hazard, but other employees are exposed.

Analysis
Step 1: C is a correcting employer since it is responsible for erecting and maintaining fall protection equipment.
Step 2: The steps C implemented to discover and correct damaged guardrails were reasonable in light of the amount of activity and size of the project. It exercised reasonable care in preventing and discovering violations; it is not citable for the damaged guardrail since it could not reasonably have known of the violation.

CONTROLLING EMPLOYER

A controlling employer must exercise reasonable care to prevent and detect violations on the site. The extent of the measures that a controlling employer must implement to satisfy this duty of reasonable care is less than what is required of an employer with respect to protecting its own employees. This means that the controlling employer is not normally required to inspect for hazards as frequently or to have the same level of knowledge of the applicable standards or of trade expertise as the employer it has hired.

Factors Relating to the Reasonable-Care Standard
Factors that affect how frequently and closely a controlling employer must inspect to meet its standard of reasonable care include:

a. The scale of the project.
b. The nature and pace of the work, including the frequency with which the number or types of hazards change as the work progresses.
c. How much the controlling employer knows both about the safety history and safety practices of the employer it controls and about that employer's level of expertise.
d. More frequent inspections are normally needed if the controlling employer knows that the other employer has a history of non-compliance. Greater inspection frequency may also be required, especially at the beginning of the project, if the controlling employer had never worked with the other employer or does not know its compliance history.
e. Less frequent inspections may be appropriate where the controlling employer sees strong indications that the other employer has implemented effective safety and health efforts. The most important indicator of an effective safety and health effort by the other employer is a companywide commitment to safety that has resulted in a consistently high level of safety performance. Other indicators include the use of an effective, graduated system of enforcement for noncompliance with safety and health requirements coupled with regular job-site safety meetings and safety training.

Evaluating Reasonable Care
In evaluating whether a controlling employer has exercised reasonable care in preventing and discovering violations, the CSHO will consider questions such as whether the controlling employer:

a. Conducted periodic inspections of appropriate degree and frequency (frequency should be based on all the factors listed in the Factors Relating to Reasonable Care section).
b. Implemented an effective system for promptly correcting hazards.
c. Enforced the other employer's compliance with safety and health requirements with an effective, graduated system of enforcement and follow-up inspections.

Types of Controlling Employers
Control Established by Contract. In this case, *the employer has a specific contract right to control safety.* To be a controlling employer, the employer must itself be able to prevent or correct a violation or to require another employer to prevent or correct the violation. One source of this ability is explicit contract authority. This can take the form of a specific contract right to require another employer to adhere to safety and health requirements and to correct violations that the controlling employer discovers.

Example 1. Employer GH contracts with Employer S to do sandblasting at GH's plant. Some of the work is regularly scheduled maintenance and is therefore considered general-industry work; other parts of the project involve new work and are considered construction. Respiratory protection is required. Further, the contract explicitly requires S to comply with safety and health requirements. Under the contract, GH has the right to take various actions against S for failing to meet contract requirements, including the right to have noncompliance corrected by using other workers and back-charging for that work. S is one of two employers under contract with GH at the work site, where a total of five employees work. All work is done within an existing building. The number and types of hazards involved in S's work do not significantly change as the work progresses. Further, GH has worked with S over the course of several years. S provides periodic and other safety and health training and uses a graduated system of enforcement of safety and health rules. S has consistently had a high level of compliance at its previous jobs and at this site. GH monitors S through a combination of weekly inspections, telephone discussions, and a weekly review of S's own inspection reports. GH has a system of graduated enforcement that it has applied to S for the few safety and health violations that have been committed by S in the past few years. Further, due to respirator equipment problems, S violates respiratory protection requirements two days before GH's next scheduled inspection of S. The next day there is an OSHA inspection. There is no notation of the equipment problems in S's inspection reports to GH, and S made no mention of it in its telephone discussions.

Analysis
Step 1: GH is a controlling employer because it has general supervisory authority over the work site, including contractual authority to correct safety and health violations.
Step 2: GH has taken reasonable steps to try to make sure that S meets safety and health requirements. Its inspection frequency is appropriate in light of the low number of workers at the site, lack of significant changes in the nature of the work and types of hazards involved, GH's knowledge of S's history of compliance, and its effective safety and health efforts on this job. GH has exercised reasonable care and is not citable for this condition.

Example 2: Employer GC contracts with Employer P to do painting work. GC has the same contract authority over P as Employer GH had in the previous example. GC has never before worked with P. GC conducts inspections that are sufficiently frequent in light of the factors listed above. Further, during a number of its inspections, GC finds that P has violated fall protection requirements. It points the violations out to P during each inspection, but takes no further action.

Analysis
Step1: GC is a controlling employer since it has general supervisory authority over the site, including a contractual right of control over P.

Step 2: GC took adequate steps to meet its obligation to discover violations. However, it failed to take reasonable steps to require P to correct hazards since it lacked a graduated system of enforcement. A citation to GC for the fall protection violations is appropriate.

Example 3: Employer GC contracts with Sub E, an electrical subcontractor. GC has full contract authority over Sub E. Sub E installs an electric panel box exposed to the weather and implements an assured equipment grounding conductor program, as required under the contract. It fails to connect a grounding wire inside the box to one of the outlets. This incomplete ground is not apparent from a visual inspection. Further, GC inspects the site with a frequency appropriate for the site. It saw the panel box, but did not test the outlets to determine if they were all grounded because Sub E represents that it is doing all of the required tests on all receptacles. GC knows that Sub E has implemented an effective safety and health program. From previous experience it also knows that Sub E is familiar with the applicable safety requirements and is technically competent. GC had asked Sub E if the electrical equipment was OK for use and was assured that it was.

Analysis
Step 1: GC is a controlling employer since it has general supervisory authority over the site, including a contractual right of control over Sub E.
Step 2: GC exercised reasonable care. It had determined that Sub E had technical expertise, safety knowledge, and had implemented safe work practices. It conducted inspections with appropriate frequency. It also made some basic inquiries into the safety of the electrical equipment. Under these circumstances, GC was not obligated to test the outlets itself to determine if they were all grounded. It is not citable for the grounding violation.

Control Established by a Combination of Other Contract Rights
Where there is no explicit contract provision granting the right to control safety, or where the contract says the employer does not have such a right, an em-

ployer may still be a *controlling* employer. The ability of an employer to control safety in this circumstance can result from a combination of contractual rights that, together, give it broad responsibility at the site, involving almost all aspects of the job. Its responsibility is broad enough so that its contractual authority necessarily involves safety. The authority to resolve disputes between subcontractors, set schedules, and determine construction sequencing is particularly significant because these activities will affect safety. (*Note:* citations should only be issued in this type of case after the CSHO consults with the Regional Solicitor's Office).

Example 1: Construction manager M is contractually obligated to: set schedules and construction sequencing, require subcontractors to meet contract specifications, negotiate with trades, resolve disputes between subcontractors, direct work, and make purchasing decisions that affect safety. However, the contract states that M does *not* have a right to require compliance with safety and health requirements. Further, Subcontractor S asks M to alter the schedule so that S would not have to start work until Subcontractor G has completed installing the guardrails. M is contractually responsible for deciding whether to approve S's request.

Analysis
Step 1: Even though its contract states that M does not have authority over safety, the combination of rights actually given in the contract provides broad responsibility over the site and results in the ability of M to direct actions that necessarily affect safety. For example, M's contractual obligation to determine whether to approve S's request to alter the schedule has direct safety implications. M's decision relates directly to whether S's employees will be protected from a fall hazard. M is therefore a controlling employer.
Step 2: In this example, if M refused to alter the schedule, it would be citable for the fall-hazard violation.

Example 2: Employer ML's contractual authority is limited to reporting on subcontractor's contract com-

pliance to owner/developer O and making contract payments. Although it reports on the extent to which the subcontractors are complying with safety and health infractions to O, ML does not exercise any control over safety at the site.

Analysis
Step 1: ML is not a controlling employer because these contractual rights are insufficient to confer control over the subcontractors, and ML did not exercise control over safety. Reporting safety and health infractions to another entity does not, by itself (or in combination with these very limited contract rights), constitute an exercise of control over safety.
Step 2: Since it is not a controlling employer, it had no duty under the OSH Act to exercise reasonable care with respect to enforcing the subcontractors' compliance with safety; thus there is no need to go to Step 2.

Architects and Engineers
Architects, engineers, and other entities are controlling employers only if the breadth of their involvement in a construction project is sufficient to bring them within the parameters discussed above.

Example 1: Architect A contracts with owner O to prepare contract drawings and specifications, inspect the work, report to O on contract compliance, and certify completion of work. A has no authority or means to enforce compliance, no authority to approve or reject work, and does not exercise any other authority at the site, although it does call the general contractor's attention to observed hazards noted during its inspections.

Analysis
Step 1: A's responsibilities are very limited in light of the numerous other administrative responsibilities necessary to complete the project. It is little more than a supplier of architectural services and conduit of information to O. The responsibilities it does have are insufficient to make it a controlling employer. Merely pointing out safety violations did not make it a controlling employer. (*Note:* In a cir-

cumstance such as this, it is likely that broad control over the project rests with another entity.)

Step 2: Since A is not a controlling employer it had no duty under the OSH Act to exercise reasonable care with respect to enforcing the subcontractors' compliance with safety; there is therefore no need to go to Step 2.

Control Without Explicit Contractual Authority

Even where an employer has no explicit contract rights with respect to safety, an employer can still be a controlling employer if, in actual practice, it exercises broad control over subcontractors at the site. (*Note:* Citations should only be issued in this type of case after consulting with the Regional Solicitor's Office.)

Example 1: Construction manager MM does not have explicit contractual authority to require subcontractors to comply with safety requirements, nor does it explicitly have broad contractual authority at the site. However, it exercises control over most aspects of the subcontractors' work anyway, including aspects that relate to safety.

Analysis

Step 1: MM would be considered a controlling employer since it exercises control over most aspects of the subcontractor's work, including safety aspects.

Step 2: The same type of analysis on reasonable care that was described in the examples in "Control Established by Contract" would apply to determine whether a citation should be issued to this type of controlling employer.

Multiple Roles

a. A creating, correcting, or controlling employer also is often an exposing employer. The CSHO will consider whether the employer is an exposing employer before evaluating its status with respect to these other roles.

b. Exposing, creating, and controlling employers can also be correcting employers if they are authorized to correct the hazard.

OSHA STANDARDS' INTERPRETATIONS

Interpretations of responsibility for compliance on multiemployer work sites for specific performance-type OSHA standards, such as the Hazard Communication Standard, that require specific documentation and training can be found at http://osha.gov by researching the specific standard in question.

Liabilities

The relationships formed through the contractual language used between the *host employer* (usually the customer or facility owner), the *prime controlling contractor* (usually the general contractor), and the *subcontractor(s)* can delegate certain responsibilities in any number of different ways.

However, a typical problem within the contractual language used by most contractors is that it does not adequately take into account the requirements of specific OSHA standards as they apply to construction. A lack of knowledge concerning both the applicable OSHA standards and the requirements of the multiemployer work-site-compliance enforcement guidelines can leave a general contractor or the subcontractors in situations that result in unnecessary and excessive liability risk exposures.

To understand this dilemma, all employers, including contractors and owners, need to thoroughly understand the OSHA 29 CFR 1926.16 policy, Rules of Construction, (a), (b), (c), and (d) and OSHA Directive CPL 2.103 Multi-Employer Work Site Citation Policy. (*Note:* Subparts A and B in 29 CFR 1926, which include 1926.16, do not apply to projects that are not federally funded).

This standard and OSHA policy in general designates the specific roles and responsibilities that the prime or controlling contractor and the subcontractor(s) have to ensure workplace compliance with applicable OSHA standards.

A clarification by OSHA of 29 CFR 1926.16 as it relates to work sites can be found in the reply letter of the interpretations of OSHA to Ms. Ellen Zielinski, Editor, American Society of Safety Engineers, August 31, 1990. An additional example is found in the reply letter to Mr. Jack Fees, Safety Education

Director, Construction Industry Service Program, June 6, 1978.

Applying the Guidelines

These relationships and responsibilities are key to understanding the potential application of the compliance enforcement guidelines for multiemployer sites. Because these standards clearly state that the *prime* or *controlling* contractor is ultimately responsible for compliance by all personnel, including those of the subcontractor(s)—even if the prime contractor attempts to subrogate these responsibilities through a contractual assignment—this necessitates much more stringent control of subcontractors and their activities at the job site.

Consequently, the general contractor can be held responsible and liable for violations incurred by the action of employees of any subcontractor on the job.

This is true from the perspective of civil actions such as OSHA citations and fines, civil litigation if such violations result in injury or death, and also potentially from a criminal perspective in cases where gross negligence is found to be the cause of such violations.

Key personnel and ownership of the general and subcontractor entities are most susceptible to the application of criminal prosecution in cases where negligence is established.

As an example, in the case of *Universal Construction Co. v. OSHRC* (18 OSHC 1769), the Tenth Circuit joined five other circuits in upholding the multiemployer citation policy, under which Universal Construction Co. was cited for several alleged violations in October 1997.

In this case, an OSHA compliance officer allegedly witnessed an employee of A. Zahner Sheet Metal Co., a subcontractor of Universal Construction, in violation of two OSHA construction standards. The employee, working in an aerial lift, failed to wear and attach a safety belt to the lift basket, and subsequently climbed out of the lift basket onto a building roof.

OSHA argued that Universal's field manager and foreman at the site had the authority to correct the hazards—or direct Zahner's foreman to correct them—but failed to do so. OSHA then cited Universal for a serious violation, which was upheld both by an administrative law judge in March 1998 and by the appeals court the following year.

Another example is the case in which the Texas Supreme Court found a general contractor liable for the death of a subcontractor's employee. The following is an excerpt from the *Construction Law Bulletin* of Austin Industries, Inc., Company Law Department.

Lee Lewis Construction was hired as the general contractor by Methodist Hospital in Lubbock, Texas, to remodel the eighth floor and add a ninth and tenth floor of a hospital tower. The contract between Methodist Hospital and Lee Lewis contained standard form provisions published by the American Institute of Architects (AIA) that required Lee Lewis to be "solely responsible" for every aspect of the work, including "initiating, maintaining, and supervising all of the safety precautions and programs," and assigning one of its employees the duty of preventing accidents.

Lee Lewis subcontracted the glass-glazing work to KK Glass Co., who assumed toward Lee Lewis all of the duties and obligations Lee Lewis had agreed to with Methodist Hospital. KK Glass was to abide by all applicable governmental safety rules as well as Lee Lewis's safety rules. Lee Lewis was authorized to remove from the project any KK Glass employee who failed to comply with the safety requirements.

Harrison, a KK Glass employee, was installing thermal insulation and caulking between the window frames on the tower's tenth floor when he fell to his death. Although there is disputed evidence as to what type of safety system was in use, the evidence is undisputed that Harrison, an employee of the subcontractor, was not using an independent lifeline that would have prevented his fall.

KK Glass, as a workers' compensation subscriber, settled the claim for gross negligence prior to trial. Lee Lewis was sued for negligence and gross negligence and tried by a jury. The jury rendered a verdict for Harrison, finding that Lee Lewis had retained the right to control safety at the construction site, that Lee Lewis was both negligent and grossly negligent, and that Lee Lewis was 90 percent responsible for the accident. The jury awarded $7.0 million in compensatory damages, and $5 million in punitive damages. Including interest, the total award was approximately $18 million.

During trial, Lee Lewis admitted that the job superintendent routinely inspected the job site to see that subcontractors properly utilized fall protection equipment, that he personally witnessed and approved of the fall protection systems used by KK Glass, and did not object to the KK Glass employees' use of a bosun's chair (a wooden board suspended from the roof by a rope) without an independent lifeline.

The Supreme Court of Texas upheld the finding of negligence, stating that the above facts constitute enough evidence to support that Lee Lewis retained the right to control fall protection systems on the job site. Therefore, Lee Lewis owed a duty of care to Harrison, an employee of a subcontractor.

Gross negligence was upheld because: (1) working on the tenth floor of a building without an independent lifeline creates an extreme risk of a fatal fall and (2) Lee Lewis knew that KK Glass was not using an independent lifeline, did nothing to remedy it, and even required their own employees to use independent lifelines. The resulting damages assessed against Lee Lewis were almost $13 million.

Another case with a different resolution occurred when a worker employed with a construction company that was performing work for the DOW Corporation was injured on a DOW site in Texas. Since workers' compensation is not an exclusive remedy in the state of Texas, DOW was named in the ensuing litigation for not protecting the contractor's employee. The court ruled DOW had no liability in the case and released them from the action.

Minimizing Risk and Loss

Employers can minimize risk and loss on multiemployer work sites by negotiating contractual terms (language) and implementing controls for work-site hazard identification and correction.

CONTROLLING EXECUTION OF THE PROJECT

To minimize liability risk exposure, the following are recommended steps employers should take.

1. Review the contractual language with regard to: specific task and responsibility for subcontractors as well as the methods of recourse for indemnification in the event they are to be held responsible for violations or other legal actions precipitated by subcontractors' actions.
2. Conduct prebid reviews with subcontractors to evaluate their intended course of action with regard to *high-risk* job activities and to inform them of the requirements for compliance-related issues involved with such activities.
3. Monitor the job-site activities of subcontractors and consistently enforce standards that are applicable to that job site whether they are OSHA's, the general contractor's, or customer-specific standards.
4. Educate key staff on the standards and guidelines and ensure full understanding of the implications and liability of noncompliance.
5. Design and implement a comprehensive safety and health program to address the employer's entire scope of operations, including those completed by subcontractors.
6. Establish and maintain thorough and accurate documentation at the job site (especially with regard to communications between the employer's staff and that of the subcontractor(s), as well as the host employer (or customer), as these communications relate to compliance and safety issues.
7. Ensure that all required and necessary training has been provided for the employer's personnel and those of all subcontractors(s) prior to commencement of work activities. Require documentation be submitted prior to commencement of work.
8. Regularly monitor the training efforts of the subcontractors to ensure adequacy. Include the provisions in 3 to monitor and enforce training requirements.

9. Select the best-qualified general contractor and subcontractors—those who have demonstrated special efforts to perform at the highest levels of quality.
10. Select companies that are prequalified through a validated certification process to ensure that they meet the industry's highest standards of quality and sound business practices.

CONTRACTUAL CONTROLS

The goal in this instance is to minimize liability risk by negotiating and ensuring the inclusion of specific contractual safeguards. The following are some guidelines as they pertain to specific participants in a project.

Owners. (1) Attempt to shift responsibility for safety and health prevention, identification, and corrections to the general contractor by contract; (2) acknowledge in the contract that construction safety is not within your expertise and that you are relying on the expertise of the general and subcontractors; (3) ensure the contract includes an unambiguous indemnification clause; (4) maintain a minimal employee presence on site and (5) do not shift responsibility for safety and health contractually and then act to control the contractors' safety programs.

General Contractor. (1) Avoid assuming absolute responsibility for work-site safety and health by contract; (2) include acknowledgements that trade-specific safety and health is not your expertise, and that you will rely on your subcontractors to provide expertise within their scope of work; (3) maintain a minimal employee presence on site; (4) if you contractually assume safety and health responsibility, be diligent in monitoring and enforcing it by making it a value on the site: holding meetings, inviting complaints to be addressed, and stopping work (if necessary) until a hazard is corrected; (5) attempt to shift the responsibility for safety and health to the subcontractors by contract; (6) acknowledge in the contract that trade-specific safety and health is not your expertise and that you will rely on sub-

contractors; (7) require the sharing of safety and health program rules and practices among subcontractors, with copies provided to you if you are responsible for safety under your contract with the owner; (8) require immediate reporting to you by subcontractors of safety and health violations or problems not within their control; and (9) ensure the contract includes an unambiguous indemnification clause.

Subcontractors: (1) Do *not* attempt to assume safety and health responsibility for any work-site locations other than your own by contract; (2) include acknowledgements that you have no safety or health expertise outside of your trade and that you will rely on the general and other subcontractors for their expertise; (3) negotiate a provision that you will not be responsible for any safety or health violation that you did not create, control directly, or did not know about; (4) include language stating that the general contractor will not hold you responsible (a hold-harmless agreement) for the safety and health wrong-doings and miscalculations of their subcontractors; (5) instruct your employees to work and travel only in designated areas that are essential to their work activities; and (6) be diligent about safety and health in your work areas and areas where your employees go.

Subcontractors should also report all problems that they cannot control, in writing, to the contractor who created or controlled the hazard and to the general contractor. If no action is taken, or the issue is not fully resolved, take realistic measures to protect your employees, including walking off the job. Hobbs said that ". . . it is best to first try to negotiate to avoid later litigation. . . ." (See *Contract and Tort Liability for Employers Multi-Work Site Safety and Health*, by Eric E. Hobbs, Esq.)

REFERENCES

American National Standards Institute. *American National Standards for Construction and Demolition Operations* (ANSI A10.18–1996). Itasca, IL: National Safety Council, 1996.

_____. *Safety and Health Program requirements for Multi-Employer Projects* (ANSI A10.33–1992). Itasca, IL: National Safety Council, 1992.

Beaujon, Everett A. "A Project Safety Execution Plan," *Professional Safety*, American Society of Safety Engineers, August 1990.

Hedin, Melvin. "Contingent Liabilities of the Multi-Employer Work Site," *ABC Today*, Associated Builders and Contractors, Inc., March 1998.

Hobbs, Eric E. "Employer Liability for Safety and Health on Multi-Employer Worksites," *Proceedings of the 1999 ASSE Professional Development Conference*. American Society of Safety Engineers, 1999.

Occupational Safety and Health Administration, OSHA Regulations – 29 CFR. Washington, DC: U.S. Department of Labor, Government Printing Office.

    1910.119, *Process Safety Management of Highly Hazardous Chemicals*, 1996.

    OSHA Directive CPL 2-0.124, *Multi-Employer Citation Policy*, 12/10/1999.

    1926.16, *Rules of Construction*.

    1903.1, *Purpose and Scope—Williams–Steiger Occupational Safety and Health Act of 1970*, 29 CFR Parts 1904 and 1952, "Recording and Reporting Occupational Injuries and Illnesses," 1970.

_____. Standard Interpretations, Construction, Multi-Employer Work Sites, http://osha.gov/pls/oshaweb/owadisp.show_document?p_table=INTERPRETATIONS&p_i.

REVIEW EXERCISES

1. What is the philosophy of OSHA as it relates to the citation for violative conditions at a multiemployer work site?
2. What is the argument most often used by employers when contesting citations?
3. Can any employer on a multiemployer work site be cited under the general duty clause?
4. What is OSHA's position relative to applying the general duty clause as opposed to utilizing a specific standard?
5. List the four principal categories of employers that OSHA considers to be present at a multiemployer work site, and give a description of their classification.
6. Which employer is expected to exercise "reasonable care"?
7. What are the main factors considered in the determination of "reasonable care"?
8. What is a typical liability problem associated with the contractual language used in most contracts? Explain.
9. Can risk and loss be minimized at multiemployer work sites? Explain some of the steps employers can take, both contractually and on the site, to accomplish this.
10. Can risk and loss and, consequently, liabilities be minimized by contractual language? Discuss recommended guidelines for the inclusion of contractual safeguards.

■■ **About the Author** ■■

MANAGING SUBCONTRACTOR LIABILITY
by by Adele L. Abrams, Esq., CMSP, P.C.

Adele L. Abrams, is an attorney and president of the Law Office of Adele L. Abrams, P.C., in Beltsville, MD. Ms. Abrams is a Certified Mine Safety Professional and MSHA-approved trainer. She is a professional member of the American Society of Safety Engineers' Construction and Mining practice specialty groups (2000 "SPY" Award recipient, Mining practice specialty). She is an active member of the International Society of Mine Safety Professionals, the National Safety Council (past chairman of the Cement, Quarry & Mineral Aggregates section), the Energy and Mineral Law Foundation, Women in Mining (past president, DC chapter), the Holmes Safety Association, and the Washington Metropolitan Area Construction Safety Association. She earned a Juris Doctor degree from the George Washington University National Law Center in Washington, DC, and a Bachelor of Science degree from the University of Maryland, College Park. She is admitted to the Maryland and DC Bars, as well as to practice before the U.S. District Courts of Maryland and DC, and the U.S. Court of Appeals, DC Circuit.

Managing Subcontractor Liability

OSHA'S VIEW OF SUBCONTRACTOR COMPLIANCE OBLIGATIONS

Since its establishment in 1970,[1] the federal Occupational Safety and Health Administration has implemented an enforcement approach that creates tension between employers, general contractors, and subcontractors. Too often companies give little thought to such critical issues as which employer has primary responsibility for compliance with mandatory standards and which controls hazards at a work site and directs the activities of the workforce until after an inspection has occurred and citations have been issued.

Often, the agency issues parallel citations to general contractors and their subcontractors for the subcontractor's alleged violations. OSHA may assess higher civil penalties against the employer or general contractor than against the company directly involved in the violative action because OSHA deems the primary employer or general contractor to have a higher level of culpability or a better understanding of OSHA compliance responsibilities. In addition to costing companies significant money in civil penalties—currently, OSHA civil penalties can reach as high as $70,000 per violation—OSHA's dual-citation policy often pits the general contractor and its subcontractors against each other in litigation.

Significantly, OSHA citations can be introduced in some state tort actions to prove "negligence *per se*."[2] This legal theory simplifies a plaintiff's burden of proof on the "breach" element of a prima facie[3] negligence case. Plaintiffs in negligence cases ordinarily have to produce detailed factual evidence regarding the precise manner in which the defendant breached the duty of reasonable care. *Negligence per se* allows the plaintiff to prove "breach"[4] by showing simply that the defendant violated a statute or regulation that (a) covers the class of activities giving rise to the plaintiff's injuries, and (b) was designed to protect the class of persons to which the plaintiff belongs.[5] The plaintiff is still required to prove that the statutory violation was both the legal and proximate cause of his injuries; therefore, *negligence per se* is not the same as a theory of strict liability. The jury must still determine the issue of proximate cause (whether the defendant's negligence in violating the statute or regulation caused or contributed to the plaintiff's injury).[6]

Therefore, it is imperative that general contractors take appropriate action to manage subcontractor risks from the outset and ensure that the companies who will provide services pursuant to a contractual agreement are aware of OSHA compliance obligations, have employees who are trained and supervisors who are competent, and demonstrate a commitment to safety and health through appropriate safety and health management programs.

OSHA Standards That Create Contractor Coordination Responsibilities

Although a complete review of OSHA standards and regulations that contain unique provisions requiring multicontractor coordination is outside the scope of this chapter, any standards that require development of written programs, certifications, or specific employee training or competency should be carefully reviewed as they may contain requirements that place additional importance on the prequalification of contractors and subcontractors.

As a general starting point, in order to comply with OSHA Standard 1926.20(a)(1), all employers who perform any part of a construction project must ensure that no employees must work in surroundings or under conditions that are unsanitary, hazardous, or dangerous to health or safety. This is not a duty that can be delegated to another party or contracted away. In addition, specific safety and health standards contain mandates to share infor-

mation with contractors and/or nonemployees who may be present at the work site. Finally, the contracting party, the general contractor, and subcontractors may all be cited for failure to provide a safe and healthful work environment, as mandated under OSHA's "General Duty Clause."

In addition to including specific and general safety requirements in contracts used between prime contractors and subcontractors, it is critical to communicate this information to each worker at the job site in a consistent and uniform manner. The information communicated should cover basic safety and health rules and regulations, as well as specific requirements for the job site. The communication is most effective when verbal instructions are supplemented by a written contractor/employee handbook that includes both safety and health information and other site-specific material (such as contact telephone numbers and addresses for key personnel, PPE requirements, rules on alcohol/drugs/smoking/eating at the work site, and emergency procedures). It is recommended that the prime contractor obtains written documentation of the receipt of the handbook from the subcontractor and/or each individual at the work site. It is important to note, however, that such a handbook is not a substitute for the formal written programs that each employer must maintain to comply with OSHA requirements.

DEFINITIONS AND APPLICABLE STANDARDS

The OSH Act, codified at 29 U.S.C. §651 et seq., defines *employer* as "a person engaged in a business affecting commerce who has employees, but does not include the United States (not including the United States Postal Service) or any State or political subdivision of a State." The term *employee* means "an employee of an employer who is employed in a business of his employer which affects commerce."[7] OSHA's so-called "General Duty Clause," which was Section 5(a) of the OSH Act, requires each employer to "furnish to each of his employees employment and a place of employment which are free from recognized hazards that are causing or are likely to cause death or serious physical harm to his employees." 29 U.S.C. §654 (a)(1).

In addition, employers must comply with the occupational safety and health standards promulgated by OSHA and codified in 29 CFR Parts 1900–1926. The standards of most relevance to the construction industry are found in 29 CFR Part 1926; however, there are sections of OSHA's "general industry" standards (found in 29 CFR Part 1910) that must also be adhered to by construction employers and contractors. Some standards have specific requirements that place burdens on construction companies engaged in operations at multiemployer work sites to coordinate their safety and health activities with the other contractors and subcontractors, to ensure the safety of all workers present.

Normally, a prime contractor assumes the entire responsibility under the contract and the subcontractor assumes responsibility with respect to his portion of the work. With respect to subcontracted work, the prime contractor and any subcontractor or subcontractors often are deemed to have joint responsibility. This implies that the general (or "prime") contractor has a nondelegable duty to ensure compliance on the work site with all applicable standards and regulations.

UNIQUE CONTRACTOR COORDINATION ISSUES

Some of the most critical standards requiring coordination of contractor programs and activities include: OSHA's hazard communication standard (29 CFR §1910.1200)[8]; confined space standard (29 CFR §1910.146); lockout/tagout standard (29 CFR §1910.147); process safety management (29 CFR §1910.119 and §1926.64); asbestos standards (29 CFR §1926.1101)[9]; hazardous waste operations and emergency response (29 CFR §1926.65);[10] steel erection (29 CFR Part 1926, Subpart R); and rules governing hazardous air contaminants such as lead (29 CFR §1926.62). Table 1 illustrates the risks and requirements associated with selected mandatory requirements.

In addition, 29 CFR §1926.16, provides that the prime contractor and its subcontractors:

. . . may make their own arrangements with respect to obligations which might be more appropriately treated on a jobsite basis rather than individually. Thus, for example, the prime contractor and his subcontractors may wish to make an express agreement that the prime contractor or one of the subcontractors will provide all required first-aid or toilet facilities, thus relieving the subcontractors from the actual, but not any legal, responsibility (or, as the case may be, relieving the other subcontractors from this responsibility). In no case shall the prime contractor be relieved of overall responsibility for compliance with the requirements of this part for all work to be performed under the contract. . . . To the extent that a subcontractor of any tier agrees to perform any part of the contract, he also assumes responsibility for complying with the standards in this part with respect to that part. Thus, the prime contractor assumes the entire responsibility under the contract and the subcontractor assumes responsibility with respect to his portion of the work. With respect to subcontracted work, the prime contractor and any subcontractor or subcontractors shall be deemed to have joint responsibility . . . . Where joint responsibility exists, both the prime contractor and his subcontractor or subcontractors, regardless of tier, shall be considered subject to the enforcement provisions of the Act.

This standard applies to work performed pursuant to Section 107 of the Contract Work Hours and Safety Standards Act (Construction Safety Act) dealing with federally funded projects. Section 107 requires, as a condition of each contract that is entered into under legislation subject to Reorganization Plan Number 14 of 1950 (64 Stat. 1267), and which is for construction, alteration, and/or repair, including painting and decorating, that no contractor or subcontractor contracting for any part of the contract work shall require any laborer or mechanic employed in the performance of the contract to work in surroundings or under working conditions that are unsanitary, hazardous, or dangerous to his health or safety, as determined under construction safety and health standards promulgated by the Secretary by regulation. Section 1926.16 does not have direct significance with respect to enforcement of the OSH Act.[11]

■■ **Table 1** ■■

Risks and Requirements of Mandatory Standards

Standard	Subject	Risk	Requirement
1910.119 and 1926.64	Process safety management	Process-related incident; fire; explosion; environmental releases	**Owner:** Must maintain contractor work-related injury/illness data. **Contractor:** Must assure that each employee follows facility safety rules and work practices.
1926.65	Hazardous waste emergency responder	Exposure to hazardous chemicals	Inform contractors, subcontractors and/or their representatives of the site emergency response procedures and any potential fire, explosion, health, safety, or other hazards of the operation.
1926 Subpart R	Steel erection	Falls from heights, may be struck by falling objects	Controlling contractor must inspect decking and guardrail systems, designate areas for overhead lifting, enforce usage of PPE.
1910.146	Confined space entry	Exposure to uncontrolled confined space hazards (dangerous atmospheres, engulfment, etc.).	Inform host employer of the permit space program that the contractor will follow and of hazards confronted or created in permit spaces, through debriefing or during entry operation.
1910.147	Lockout/tagout	Exposure to uncontrolled hazardous energy sources	Prime contractor and subcontractors must inform each other of their respective lockout/tagout procedures. Should ensure that no persons are in a "danger zone" when equipment will be reenergized.
1910.1101	Asbestos	Exposure to asbestos fibers (health hazard)	Building and facility owners must inform employers of potential hazards, and employers must inform workers who perform activities in areas containing asbestos, or asbestos-containing materials, of the presence and location of ACM and/or PACM that may be contacted during worker activities.
1910.1200	Hazard communication	Chemical hazards created by lack of information about potential health risks, methods of handling hazardous chemicals, and appropriate PPE	Employer must have written hazard communication program, train workers, provide them with access to Material Safety Data Sheets for each hazardous chemical at the work site, and maintain chemical inventory list. Each employer must inform other employers/contractors/subcontractors at the work site about hazardous chemicals to which individuals may be exposed and make "hazcom" materials available to all workers upon request.

OSHA Enforcement Actions Involving Subcontractors

OSHA's multiemployer work-site policy states that more than one employer may be citable for a hazardous condition that violates an OSHA standard, except for those citations issued under OSHA's General Duty Clause.[12] A two-step process is used to determine whether more than one employer shall be cited: (1) determine whether the employer is a "creating, exposing, correcting, or controlling employer"; and (2) conclude that, if the employer falls into one of these categories, it has obligations with respect to OSHA requirements. The extent of the actions required of employers varies based upon which category applies.

The *controlling employer* is the one with "general supervisory authority over the work site, including the power to correct safety and health violations itself or require others to correct them. Control can be established by contract or, in the absence of explicit contractual provisions, by the exercise of control in practice." OSHA explains that the controlling employer "must exercise reasonable care to prevent and detect violations on the site."

The *correcting employer* is one who is "engaged in a common undertaking on the same work site, as the exposing employer and is responsible for correcting a hazard." This usually occurs where an employer is given the responsibility of installing and/or maintaining particular safety/health equipment or devices. Often, the controlling and correcting employers are a single entity. The applicability of the "controlling employer" designation may be linked to the contractual arrangements between employers and their contractors. The controlling employer's level of "reasonable care" will depend in part on whether it "enforces the other employer's compliance with safety and health requirements with an effective graduated system of enforcement and follow-up inspections."

OSHA policy defines a *creating employer* as the party that "caused a hazardous condition that violates an OSHA standard." Such employers are citable even if the only exposed employees are those of other employers at the work site. The specific example used in the policy is a host employer whose workplace has airborne chemical levels that exceed the PEL because the host fails to take adequate control measures.

By contrast, an exposing employer is one whose own employees are exposed to the hazard. This category can include both the primary employer and contractor–employers. If an exposing employer did not create the condition, it can still be cited if (1) it knew of the hazardous condition or failed to exercise reasonable diligence to discover the condition; and (2) it failed to take steps consistent with its authority to protect its employees. Even if the exposing employer lacks authority to correct the hazard, it can still be cited if it fails to (1) ask the creating/controlling employer to correct the hazard; (2) inform its employees of the hazard; and/or (3) take reasonable alternative protective measures and remove its employees from the hazardous area.

At some point, most contractors will fall within the controlling, creating, exposing, or correcting category and face possible prosecution by OSHA for the sins of another employer. Therefore, advance

---

Do the OSHA Construction Standards Apply to Me?

1. The standards apply to:
   - All contractors who enter into contracts that are for construction, alteration, and/or repair, including painting and decorating
   - All subcontractors who agree to perform any part of the labor or material requirements of a contract
   - All suppliers who furnish any supplies or materials, if the work involved is performed on or near a construction site, or if the supplier fabricates the goods or materials specifically for the construction project, and the work can be said to be a construction activity.
2. The controlling contractor assumes all obligations under the standards, whether or not he subcontracts any of the work.
3. To the extent that a subcontractor agrees to perform any part of the contract, he assumes responsibility for complying with the standards with respect to that part.
4. With respect to subcontracted work, the controlling contractor and any subcontractors are deemed to have joint responsibility.

Source: Occupational Safety and Health Administration, U.S. Department of Labor (2003).

■ ■ **Figure 1** ■ ■ **Responsible parties**

---

planning is critical in order to demonstrate that due diligence was applied in selecting contractors/subcontractors for a particular project.

TORT LIABILITY AND LABOR LAW CONSIDERATIONS

Analysis of Agency Principles

Basic agency principles state: "Since the relation of master and servant is dependent upon the right of the master to control the conduct of the servant in the performance of the service, giving service to two masters at the same time normally involves a breach of duty by the servant to one or both of them. Therefore, an individual cannot be a "servant of two masters" in doing an act as to which an intent to serve one necessarily excludes an intent to serve the other." [Restatement (Second) of Agency Section(s)

226, Comment a, p. 499 (1957).] However, a "person may be the servant of two masters . . . at one time as to one act, if the service to one does not involve abandonment of the service to the other." [Restatement (Second) of Agency Section(s) 226, at 498 (emphasis added).]

Because of these distinct theories of agency, and in light of the doctrine of *respondeat superior* (which holds an employer responsible for the negligent acts of his/her agents), it is imperative for companies to determine clearly which individuals at a work site will be legally considered their employees, and which are clearly contractors as a matter of law.

Assumption of Risk

The assumption-of-risk defense arises where the plaintiff voluntarily enters into a situation knowing fully the risks, and fully appreciating the dangers. This is a defense that must be pleaded separately from contributory or comparative negligence.[13] As with these other affirmative defenses, assumption of risk generally does not apply to employees in the course of their employment. Whether assumption of risk is applicable is a question for the trier of fact—the jury—and the defendant bears the burden of proof.

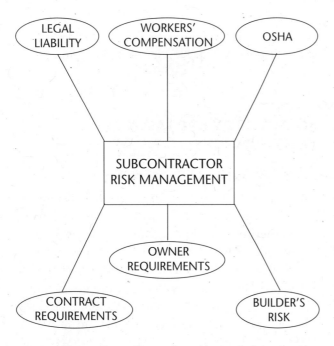

■■ **Figure 2** ■■ **General contractor/subcontractor risk management structure**

Contributory Negligence and Comparative Negligence

Contributory negligence is applicable to mitigate damages when an individual who is not covered by workers' compensation is injured where "alternative courses of action are available to the injured party, and he chooses the unreasonable course."[14] In essence, contributory negligence is measured by what a reasonable person would have done under similar circumstances.[15] It is generally understood that the affirmative defense of contributory negligence is, in all cases, a question of fact left to the jury.

Significantly, the defense of contributory negligence generally is not available against employees in actions based on another employer's breach of the statutory duty to provide reasonably safe working conditions for wage earners, except where the injured worker's actions were taken with "willful, wanton, or reckless disregard" for his or her own safety.[16]

Some states use a *comparative negligence* standard when determining if complaints seeking compensation for personal injuries can be sustained even though the claimant's negligence played a role in the event causing the injuries. The underlying theory in comparative negligence is that the plaintiff's contributory negligence shall not bar a recovery by the plaintiff or his legal representative where such negligence was not greater than the causal negligence of the defendant or defendants against whom recovery is sought, but any damages sustained by the plaintiff shall be diminished in proportion to the amount of negligence attributed to the plaintiff. However, these standards vary from state to state (some deny any compensation where the claimant was more than 50 percent responsible, while others have a pro-rata approach to damage apportionment), so each state's rule of law should be examined carefully—especially where a choice of forum may be available to the parties.

Some states have adopted either the Uniform Contribution Among Tort-feasors Act or the Uniform Apportionment of Tort Responsibility Act.[17] These model laws initially were promulgated when contributory negligence on the part of a claimant in a negligent tort action was an absolute defense against liability and a small contribution of fault by the plaintiff was enough to bar compensation for injury. The Acts address the questions of multiple liability when there is more than one tort-feasor responsible for injury, and provide for joint and several liability, with right of contribution in the event one tort-feasor pays another tort-feasor's share—even when there is no action in concert. This permits an injured person to receive full compensation even if some tort-feasors do not actually provide compensation.[18] The 2002 act, the new Uniform Act, adopts a partial or modified form of comparative fault. This concept is especially critical where more than one company may be involved in environmental pollution of land at an Environmental Protection Agency-designated "Superfund" site.[19]

Regardless of the apportionment of blame in a particular state's law, before beginning any such analysis, potential defendants must remember that comparative negligence principles do not apply to a nonnegligent plaintiff. This is because "the raison d'etre[20] and rationale of comparative negligence are tied, hand-and-foot, to the narrow parameters of a blameworthy plaintiff's claim."[21]

IRS Distinctions Between Employees and Contractors

As more employers hire outside contractors to perform tasks that were previously performed by direct employees, the federal Internal Revenue Service has begun to closely scrutinize these relationships to ensure that they are not an attempt to evade reporting and withholding tax laws. Although the tests established by the IRS have little to do with occupational safety and health, in practice the so-called "common law" and "IRS tests" are often used by OSHA and by the courts to determine liability for the acts of workers.

There are two general tests for determining whether a person is a contractor or an employee. The first is the common law "right to control" test. The courts consider whether the employer retains the right to control the manner and means by which the result is to be accomplished, or whether control is reserved only as to the result sought.

The second approach is the "economic realities" test. The courts consider the following factors:

1. The hiring firm's right to control the means and manner of the individual's work.
2. The worker's investment in equipment, tools, and facilities.
3. The worker's opportunity for profit or loss.
4. The permanency of the relationship between the worker and the employer.
5. The skill of the worker required to perform the work.
6. Whether the service rendered is an integral part of the employer's business.

Legal Liability Checklist

✔ Is workers' compensation shield available?

✔ Is prime contractor the "statutory employer" of its subcontractor's workers?

✔ Are there "joint tort-feasors" responsible for the injury?

✔ Did the plaintiff's negligence contribute to the injury?

■ ■ **Figure 3** ■ ■ Who's liable?

In addition, the IRS's "20-factor" test for distinguishing between contractors and employees should also be part of the general contractor/employer's analysis to determine the status of individuals at the work site. These factors are:

1. Instructions
2. Training
3. Integration
4. Services rendered personally
5. Hiring, supervising, and paying assistants
6. Continuing relationship
7. Set hours of work
8. Full time required
9. Doing work on employer's premises
10. Order or sequence set
11. Oral or written reports
12. Payment by hour, week, month
13. Payment of business and/or traveling expenses
14. Furnishing of tools and materials
15. Significant investment
16. Realization of profit or loss
17. Working for more than one firm at a time
18. Making service available to general public
19. Firm's right to discharge
20. Worker's right to terminate.[22]

If the employer "guesses wrong" and fails to withhold tax or provide other benefits (e.g., overtime pay) mandated by federal and state labor laws, and the IRS determines that a purported contract worker is, in fact, deemed to be an employee of the company or general contractor, significant monetary penalties could ensue. Criminal penalties are also available where willful conduct to evade taxes can be proven. Since 1988, the IRS has assessed nearly a billion dollars in penalties as a result of audits involving contract employees. However, §530 of the Revenue Act of 1978[23] does provide a "safe harbor" for companies who demonstrate that there was a reasonable basis for calling an individual an "independent contractor" rather than a company "employee." This is a further reason supporting reducing all agreements to writing so the terms of engagement are completely delineated prior to commencement of work on a multiemployer project.

Workers' Compensation, Statutory Employer, and Borrowed/Loaned Employee Doctrines

Prior to the establishment of state workers' compensation laws, a third party whose negligence was partially responsible for a nonemployee's injury could join an employer as an additional defendant on the basis of joint liability. If it was determined that both the employer and the third party were responsible for the injury, the employee could obtain full recovery from the third party, but the third party could also obtain a contribution or indemnity from the employer to the extent of that employer's statutory liability under the old Workmen's Compensation Act.[24]

Today, in addition to workers' compensation providing an exclusive remedy (and shield from tort litigation) for direct employers, immunity from joinder[25] has been extended to statutory employers (e.g., a work-site general contractor), as well as to the plaintiff's principal employer, in many states.[26]

Although these rules are state-specific,[27] and determinations are made on a case-by-case basis, there are some fairly universal factors that are applied across the United States. The basis for such laws is a prevailing legislative interest in substituting "a limited but certain remedy" for a more uncertain, potential remedy in tort.[28]

The ordinary dictionary definition of *employee* includes any "person who works for another in return for financial or other compensation."[29] These so-called "statutory employer" laws or the "borrowed employee" doctrines minimize tort exposure of employers, facilitate subrogation of workers' compensation claims, and ensure that injured or deceased workers are compensated for their losses regardless of their direct employer's insurance coverage. When a claim is subrogated, one party is substituted in the place of another with respect to liability. For example, an insurance company can "step into the shoes" of the party whom it has compensated, and the insurer can sue any party whom the compensated party could have sued. The flip side of this concept is indemnification, where one party agrees to reimburse another upon the occurrence of an anticipated loss. In essence, through a contractual agreement, all or part of a loss shifts from the individual or company who is only technically or passively at fault to another who is primarily or actively responsible.[30] In such instances, the presence or absence of an OSHA citation against the prime contractor and/or its subcontractors can be critical in determining whether indemnification provisions are triggered.

Sometimes the issue becomes more complicated because the law of a state recognizing this doctrine may govern companies' contracts, while the work is performed in a different state that lacks such statutes. It is generally held that, if a damage suit is brought in the forum state by the employee against the employer or statutory employer, the forum state will enforce the bar . . . of a state that is liable for workers' compensation as the state of employment relation, contract, or injury.[31] Thus, an understanding of "choice of law" principles can be pivotal in determining a construction contractor's liability for workplace injuries and fatalities.

WORKERS' COMPENSATION AND TORT ISSUES

Today, all states provide a system of workers' compensation coverage that requires employers to provide insurance protection for the injuries and illnesses suffered by their employees, arising out of the conditions and activities of their employment. Generally, two tests are used to determine whether the required employer–employee relationship exists: (1) the "right to control" and (2) the "relative nature of the work."

The first test concentrates on direct evidence of the right to control the worker, the method of payment, the furnishing of equipment, and the right to terminate the worker. The second test is more subjective, and examines whether the work performed is integral to the business of the employer. Another critical consideration is whether the worker separately provides independent business or professional services to other employers. Where such an employment relationship is found to exist, workers' compensation provides an exclusive remedy for occupational injuries and illnesses suffered by the employee—at least with respect to his/her employer.

The majority of states prohibit the joinder of employers by third parties for any purpose.[32] This is an exception from the general right to contribution from joint tort-feasors,[33] and it prevents a third party from seeking contribution or indemnity from a victim's employer, even though the employer's own negligence may have been the primary cause of the worker's injury.

STATUTORY EMPLOYER LAWS

The statutory employer laws have been enacted by many state legislatures to ensure payment of medical expenses and lost wages (often without the need for prolonged litigation) while shielding the general contractor from third-party tort liability. In many states with such doctrines the exclusive remedy against statutory employers is workers' compensation insurance.[34]

The Virginia Supreme Court has stated that, when deciding whether an owner of a company is the statutory employer of a general contractor, or of a subcontractor's injured employee, the focus is on whether the general contractor or subcontractor's employee is performing work that is a part of the owner's trade, business, or occupation.[35]

Among other tests, the courts review whether the subcontractor's employee is performing work that is not part of the owner's trade, business, or occupation, but is part of the work that the general contractor agreed to perform for the owner.

The complexities of such litigation are illustrated by the U.S. Court of Appeals decision in *Stephens v. Witco Corp.*, 198 F.3d 539 (5th Cir. 1999). In that case, Plaintiff Stephens was seriously injured in an explosion and fire while he was supervising a crew engaged in replacing a steel bar joist at a Louisiana chemical plant owned and operated by the defendant. At the time of the accident, the victim was employed by a contractor who provided maintenance services at the defendant's plant for "construction, maintenance, and plant services" under a contract between the two companies.

Stephens sued Witco Corporation for personal injuries under tort law, while the victim's direct employer intervened to recover workers' compensation benefits paid to Stephens. At the U.S. District Court level, Witco successfully moved to dismiss, arguing that it was Stephens's "statutory employer" under Louisiana state law,[36] and was therefore immune from tort liability. In the alternative, Witco argued that Stephens was its "borrowed employee," also entitling it to immunity under Louisiana's workers' compensation law. The district court granted summary judgment for Witco based on the statutory employer argument.

On appeal, the central question was whether the contract work performed by Mundy was part of Witco's "trade, business, or occupation," and the court reversed the holding, pending an evaluation of all relevant factors. Louisiana, like many other jurisdictions, applies a "totality of the circumstances test," requiring a fact-intensive consideration of all pertinent factors.[37]

Among the factors often considered in determining whether a statutory employment relationship exists are:

1. The nature of the business of the alleged principal
2. Whether the work was specialized
3. Whether the contract work was routine, customary, ordinary, or usual
4. Whether the alleged principal customarily used his own employees to perform the work, or whether he contracted out all or most of such work
5. Whether the alleged principal had the equipment and personnel capable of performing the contract work
6. Whether those in similar businesses normally contract out this type of work or whether they have their own employees perform the work
7. Whether the direct employer of the claimant was an independent business enterprise who insured his own workers and included that cost in the contract
8. Whether the principal was engaged in the contract work at the time of the incident.[38]

The specific task being performed by the individual employee at the time of the accident is not controlling; rather, the entire scope of the contract work must be considered. In some instances, this analysis is further broken down into a "normal work test" and a "subcontracted fraction exception" under state statutes and prevailing case law.

In addition, where contracts between general contractors and subcontractors are vague as to statutory employee requirements, courts will often find[39] that such an arrangement should be weighed in favor of the subcontractor employee's position that he is not the general contractor's statutory employee—in other words, permitting him to proceed against the general contractor for personal injury or wrongful death damages under the state's applicable tort laws.

BORROWED, LOANED, AND LEASED EMPLOYEE ISSUES

Under the *borrowed* employee or *loaned* employee doctrine, an employee of one company may become the "servant" of another if the former transfers him to the employ of the latter. Under Utah law, for example, the loaned employee doctrine provides that, if a labor service loans an employee to a special employer for the performance of work, then the employee, with respect to that work, is the employee of the special employer for whom the work or service is performed.[40] A *special employer* is the business to which the employee is assigned, while the labor service providing the employee is the general employer. For purposes of workers' compensation, the special employer is the loaned employee's employer, and workers' compensation is the employee's exclusive remedy, provided that the special employer pays workers' compensation insurance for the loaned employee.

The loaned employee doctrine may be triggered when three requirements are met:

a. The employee has made a contract of hire, express or implied, with the special employer
b. The work being done is essentially that of the special employer
c. The special employer has the right to control the details of the work.[41]

In some states there is a presumption that a general employer retains control of his employees.[42] Although there is no fixed test, "borrowed employee" case law often is decided based on the following factors:

1. Right of control
2. Selection
3. Payment of wages
4. Power of dismissal
5. Relinquishment of control by general employer
6. Which employer's work was being performed at the time in question
7. Agreement between the borrowing and lending employer
8. Furnishing of tools and place of performance of work in question
9. Length of employment
10. Acquiescence by the employee in the new work arrangement.[43]

Often, these determinations rest upon the persuasiveness of circumstantial evidence presented by the parties, as borrowed employee language is seldom utilized in contracts between construction companies and their subcontractors.

Employee *leasing*—the use of temporary services, contract management, day laborers, and job shops to increase an employer's workforce on a temporary basis—is also becoming more common. This approach to staffing brings with it new issues when determining liability within the workers' compensation framework.[44] These include whether both entities in the shared/leased employee context will enjoy the statutory immunity from common law tort claims because of workers' compensation protections, and which entity (or both) should be insured for the risk of such occupational injuries and illnesses.

As with the statutory employer analysis above, the criteria include examining the contractual documents between the individual and employer, the nature of the work being performed, and the amount of control that the leasing employer has over the details of the individual's work. If all of these criteria are satisfied, generally both the lessor employer and the lessee employer will be immune from common lawsuit.[45]

It should be noted that, in some jurisdictions, the loaned or borrowed employee doctrine is falling into disfavor. For example, at one time, the rule in the Virgin Islands was that the exclusivity of the workers' compensation remedy prohibited suit against a secondary employer.[46] More recently, the Legislature altered its position and enacted 24 V.I.C. S 263a, which states in pertinent part: "It shall not be a defense to any action brought by or on behalf of an employee, that the employee at the time of his injury or death, was the borrowed, loaned, or rented employee of another employer."

Thus, the statute abolished the borrowed servant doctrine and clarified that an employer of an independent contractor is not immune from suit simply because the contractor is protected by the exclusivity provision of the workers' compensation laws.[47] Because this is a constantly evolving area of law, employers are advised to research the applicability of the statutory employer and borrowed or loaned employee doctrines each time they enter into a new contractual agreement, and to pay specific attention to both the law governing the contract and the applicable law in the state(s) where the actual construction work will be performed.

NEGLIGENT TRAINING AND NEGLIGENT SUPERVISION TORT CLAIMS

Many states increasingly make damage awards based on the tort of negligent supervision. This theory holds that an employer may be subject to liability for negligent supervision of its workers where it knows, or should have known, that an employee's conduct would subject third parties to an unreasonable risk of harm.[48] This may subject employers to tort action for the negligent actions of their employees that result in injury to a third party (including a subcontractor's employee).

For the doctrine of *respondeat superior* to apply, an employee must be liable for a tort committed in the scope of his employment. Likewise, an underlying requirement in actions for negligent supervision and/or negligent training is that the employee is individually liable for a tort or guilty of a claimed wrong against a third person, which then seeks recovery against the employer.[49] Thus, in order to maintain an action against the employer, a third party must allege that one of the employees is individually liable for a tort.[50] Moreover, most jurisdictions require the plaintiff to prove two elements to hold an employer liable for negligent supervision or training: (1) that an incompetent or untrained employee committed a tortious[51] act resulting in injury to the plaintiff; and (2) that prior to the act, the employer knew or had reason to know of the employee's incompetence or lack of training.[52]

Because of the increasing frequency of these tort actions, it is critical for all employers to document the training given to their employees and the training directions provided to outside contractors and subcontractors, and to determine the line of supervision appropriate to each part of a particular project and each area of the work site. Many OSHA standards have specific training requirements, often involving operation of specific types of equipment. First-aid training is mandated under 29 CFR §1926.50,53 while §1926.21 requires that all construction industry employers (both prime and subcontractors) provide the following training to their workers:

> The employer shall instruct each employee in the recognition and avoidance of unsafe conditions and the regulations applicable to his work environment to control or eliminate any hazards or other exposure to illness or injury. . . . Employees required to handle or use poisons, caustics, and other harmful substances shall be instructed regarding the safe handling and use, and be made aware of the potential hazards, personal hygiene, and personal protective measures required. . . . In job site areas where harmful plants or animals are present, employees who may be exposed shall be instructed regarding the potential hazards, and how to avoid injury, and the first aid procedures to be used in the event of injury. . . . Employees required to handle or use flammable liquids, gases, or toxic materials shall be instructed in the safe handling and use of these materials and made aware of the specific requirements contained in . . . other applicable subparts of this part. . . . All employees required to enter into confined or enclosed spaces shall be instructed as to the nature of the hazards involved, the necessary precautions to be taken, and in the use of protective and emergency equipment required. The employer shall comply with any specific regulations that apply to work in dangerous or potentially dangerous areas.

Training documentation to comply with §1926.21 or other OSHA standards can be as simple as a signup sheet for toolbox talks or as complex as a written examination on the information presented.[54] Prime contractors should require all subcontractors to provide OSHA-mandated training to its workers and should spot check such training, including review of training documentation and questioning of workers to verify that the requirements are being satisfied. Wherever possible, however, prime contractors should refrain from training nonemployees (absent an indemnification agreement) because of the risk of negligent training litigation.

SUBCONTRACTOR SELECTION CONSIDERATIONS

As noted in the discussion above, the use of contractors and subcontractors involves legal risk—both from OSHA enforcement actions arising from the misconduct of other employers at a multicontractor work site, and from third-party tort claims where nonemployees are injured or killed based on negligent conduct at the work site. Therefore, unless a company prequalifies and selects its contractors and subcontractors wisely, it can be exposed to (a) government investigations, civil penalties, and criminal proceedings; (b) tort claims and liability; and (c) extensive defense and litigation costs arising from the misconduct or safety/health/environmental violations of contractors and subcontractors.

Selection criteria of contractors and subcontractors should include safety and health considerations of the outside company's ability to conduct the anticipated services and work in a manner consistent with the safety and health practices of the contracting party and meeting safety and health regulatory requirements. The degree to which selection criteria are set should be commensurate with the level of risk that the contractor's expected services and work will involve.

The creation and implementation of a contractor liability prevention program requires recognition of the legal theories outlined above, which give rise to these risks. Therefore, the following steps are critical to ensuring a safe multiemployer work site:

a. Knowledge of the particular work site's practices and procedures.
b. A diligent evaluation of the safety programs, practices and track record of the general contractor and all subcontractors involved in a project.
c. A companywide policy for contractor utilization, written contracts that implement the policies and provide for enforcement.
d. Procedural guidelines for use in contractor selection, orientation, auditing, and contract enforcement.

Written Program Requirements

Facility owners and general contractors should consider having written policies mandating environmental, safety, and health protections for their workers and for individuals employed by subcontractors. All personnel, regardless of employment status, MUST be protected.

All safety- and health-related requirements (as well as insurance and indemnification requirements) must be spelled out in the applicable contracts in order for an employer or general contractor to raise this as a defense. Therefore, if a subcontractor will be required to have OSHA-mandated written programs and proof of employee training/certifications, or furnish equipment that satisfies the regulatory requirements, this should be set forth in writing before a job commences. Too often these requirements are verbally communicated, leaving the primary employer without recourse after an accident or OSHA inspection occurs.

Due Diligence Considerations

Risk assessment of planned contractor activities must be conducted to determine the degree of risk to which contractor activities will expose the contracting party's employees and property. It should be company policy to require proof of the subcontractor's EMR, OSHA 300 Log (or MSHA 7000-1 Forms), citation history, and record of employee claims for discrimination (under both Civil Rights Acts and Section 11(c) of the OSH Act). Other considerations in contractor/subcontractor selection, depending upon the nature and duration of the job, may include: technical/professional certifications of contractor/subcontractor personnel; corporate involvement in trade and professional safety organizations, licenses, permits and bonding, and verification of references.

This type of due diligence analysis will quickly reveal which contractors are falling short in their willingness or ability to ensure regulatory compliance and provide a safe and healthful work environment.

Elements of a Safety Partnership Program

Admittedly, it is difficult for an employer or a general contractor to control all hazards at a work site all of the time because construction sites are complex, dynamic, work environments. When you add subcontractors and others into the mix, it further complicates issues—particularly if the subcontractor does not have the same level of safety management experience (or knowledge of OSHA requirements) as the employer or general contractor does.

The contracting party should determine its managing role with other contractors and adjust its safety and health training and oversight obligations accordingly. In order to minimize liability, the contracting party must establish a system of monitoring contractor activities on its premises. However, to best control risk and minimize legal liability, all companies active at a work site must form a partnership of safety.

ELEMENTS OF A PRIME/SUBCONTRACTOR SAFETY PARTNERSHIP

- Providing all necessary information about OSHA requirements (and working cooperatively to guide a contractor/subcontractors toward sources of information).
- Prequalifying contractors/subcontractors to ensure that they have sound safety programs and a culture of safety.
- Providing appropriate site-specific training to contractors/subcontractors and ensuring that the contractor–employer has complied with any OSHA training and programmatic requirements.
- Informing contract workers of health and safety hazards to which they may be

exposed (as well as sampling results, where appropriate).

■ Ensuring that a communications system is in place to alert all companies at a multiemployer work site of new hazards, equipment, changes in conditions, or other information that is critical to safe job performance.

■ Coordinating emergency procedures before commencement of work, including evacuation procedures and determination of who shall provide emergency services for contractor/subcontractor employees when required.

■ Routinely checking to ensure that contractors/subcontractors are not exposing other employees to hazards.

CONCLUSION

Construction industry prime contractors have increasing liability exposure as the use of specialty subcontractors becomes more prevalent at work sites. If a subcontractor violates OSHA standards, the general contractor can be held liable if it failed to oversee activities and enforce compliance. If an injury or property damage results from the violative actions, the general contractor and its subcontractor(s) may find themselves named as joint tort-feasors in litigation brought by, or on behalf of, the injured party. Moreover, general contractors have tort exposure for injuries to subcontractor employees, absent a finding of "statutory employer" status, and may also find themselves charged with negligent supervision or negligent training.

Because of the significant monetary risks involved, prequalification of subcontractors is a critical step prior to commencing any construction project. The general or prime contractor must ensure that the companies with which it contracts have similar proactive cultures with respect to safety and training and are committed to full compliance with all applicable laws and regulations. It is also essential that all employers at a work site work cooperatively to ensure program integration, where this is required by law or is otherwise desirable to ensure the safety and health of all individuals at a work site, as well as the general public.

By forging a partnership in safety among employers, general contractors, and subcontractors on construction projects, accidents, illnesses, tort liability, and unwarranted OSHA-enforcement actions can be avoided by everyone concerned.

REVIEW EXERCISES

1. What are the indicia that distinguish "contractors" from "employees" under the Internal Revenue Service criteria?
2. How can a prime contractor protect itself from tort liability exposure, in situations where a subcontractor fails to carry workers' compensation coverage?
3. Identify three OSHA standards that require coordination of written program information between contractors and subcontractors.
4. Outline a plan for implementing a work-site safety partnership.
5. What are the distinctions between contributory negligence, comparative negligence, and assumption of risk?
6. What are basic steps that a prime contractor should take to prequalify subcontractors with respect to safety and health?
7. What are the differences between the controlling, creating, correcting, and exposing employer?

Chapter Notes

[1]OSHA was created through the Occupational Safety and Health Act, Public Law 91-596, 91st Congress, S.2193, December 29, 1970 (hereinafter, "OSH Act").

[2]For example, Pennsylvania's rule is: "a party who fails to observe a safety regulation has the burden of showing not merely that [its] fault might not have been one of the causes [of the loss], or that it probably was not, but that it could not have been . . . . *United States v. Nassau Marine Corp.*, 778 F.2d 1111, 1116 (5th Cir. 1985).

[3]A "prima facie" case is one that will prevail until contradicted and overcome by other evidence. A prima facie case consists of sufficient evidence to get a plaintiff past a motion for directed verdict or a motion to dismiss, which forces the defendant to proceed with its case. *Black's Law Dictionary* 825 (Abridged 6th ed., 1991).

[4]The generic legal meaning of "breach" is to break or violate a law, right, obligation, engagement, or duty, either by commission or omission. *See Black's Law Dictionary* 130 (Abridged 6th ed., 1991). In the sense that the term is used in this section, it refers to a party's duty to provide a safe environment, which was "breached because of a regulatory violation that resulted in injury to the plaintiff."

[5]See *Scott v. Ford Motor Company*, 229 F.3d 1143, 229 F.3d 1143 (4th Cir. 2000) (finding negligence where defendant furnished a ladder that did not comply with OSHA requirements).

[6]*Duty v. East Coast Tender Serv., Inc.*, 660 F.2d 933, 947 (4th Cir. 1981) (en banc) (per curiam). Compare *Horne v. Owens Corning Fiberglass*, 4 F.3d 276, 284 (4th Cir. 1993) (addressing OSHA as evidence of the standard of care rather than as a basis for *negligence per se*), and *MacCoy v. Colony House Builders*, 239 Va. 64 (1990) (addressing independent contractor problems in the *negligence per se* context, rather than proximate cause issues).

[7]29 U.S.C. §652 (5) and (6). Some OSHA standards have more specific definitions that must be observed. For example, OSHA's general safety and health provisions for construction define employee as: "every laborer or mechanic under the Act regardless of the contractual relationship which may be alleged to exist between the laborer and mechanic and the contractor or subcontractor who engaged him." 29 CFR §1926.32(j).

[8]Standard 1910.1200(e) provides that "Employers who produce, use, or store hazardous chemicals at a workplace in such a way that the employees of other employer(s) may be exposed (for example, employees of a construction contractor working on-site) shall additionally ensure that the hazard communication programs developed and implemented under this paragraph (e) include [all mandatory elements]. . . ."

[9]This standard, which is frequently cited by OSHA, provides, in relevant part: "Asbestos hazards at a multi-employer work site shall be abated by the contractor who created or controls the source of asbestos contamination. For example, if there is a significant breach of

an enclosure containing Class I work, the employer responsible for erecting the enclosure shall repair the breach immediately." 29 CFR §1926.1101(d)(2).

[10]This standard requires, among other things, that the employer's or general contractor's written safety and health program shall be made available to any contractor or subcontractor or their representative who will be involved with the hazardous waste operation, as well as being accessible by employees, their representatives, OSHA, and other government agencies with regulatory authority over the work site.

[11]Letter from OSHA Assistant Secretary Gerard F. Scannell to Ellen Zielinski, American Society of Safety Engineers, August 31, 1990.

[12]The General Duty Clause is Section 5(a)(1) of the Occupational Safety and Health Act of 1970. It permits OSHA to issue a citation in the absence of a specific standard where there is a serious risk of injury or illness to the employee that is known to the employer. Only employer(s) whose own employees are exposed to the hazard may be cited.

[13]*Kanelos v. Kettler*, 406 F.2d 951 (DC Cir. 1968).

[14]See *Socony Vacuum Oil Co. v. Smith*, 305 U.S. 424, 432-33 (1939) (finding of contributory negligence proper where the individual "knowingly failed to choose an available safe method of doing his [or her] work," such as making "use of a defective appliance knowing that a safe one is available").

[15]See *Am. President Lines, LTD v. Welch*, 377 F.2d 501, 504-05 (9th Cir. 1967).

[16]*Martin v. George Hyman Construction Co.*, 395 A.2d 63 (D.C. 1978).

[17]Uniform Law Commissioners, 2002.

[18]See, e.g., *Howell Construction Co. v. Luckey*, 205 W.Va. 445, 518 S.E.2d 873 (W.Va. 1999).

[19]See, e.g., *New York v. Solvent Chemical Co.*, 984 F.Supp. 160 (W.D., N.Y. 1997).

[20]Something that gives meaning or purpose, or the justification for something's existence. See, e.g., *Merriam-Webster OnLine Dictionary*, http://www.m-w.com/cgi-bin/dictionary (March 2003).

[21]*Berry v. Empire Indemnity Ins. Co.*, 634 P.2d 718, 719 (Okla. 1981).

[22]Internal Revenue Service, Rev. Rul. 87-41, 1987-1, CB 296.

[23]Pub.L. 95-600, *as amended by* Pub.L. 96-167, § 9(d), and Pub.L. 104-188, §1122(b).

[24]See, e.g., *Winters v. Herdt*, 400 Pa. 452, 162 A.2d 392 (1960).

[25]This is the concept of joining two or more parties in some legal step or proceeding. In the sense that this term relates to construction litigation, it means that two or more defendants may be charged in the same case "if they are alleged to have participated in the same act or transaction or in the same series of acts or transactions constituting an offense or offenses." *Black's Law Dictionary* 581-82 (Abridged 6th ed., 1991). The "joined" defendants may be charged together or

separately and all of the defendants need not be charged in each count. *See* Fed. R. Crim.Pro. 8(b). In some situations, "joinder" of parties may be compulsory if complete relief cannot be afforded without joinder, or if grave injustices will be done if some plaintiffs or defendants are omitted. *See* Fed.R.Civ. P. 19(a).

[26]See, e.g., *Anskis v. Fischer,* 326 Pa. Super. 374, 474 A.2d 287 (1984) (explaining Pennsylvania's statutory employer doctrine).

[27]For example, Wyoming and New Jersey currently do not appear to recognize the "statutory employer" defense to tort litigation.

[28]See *Garcia v. Am. Airlines, Inc.,* 12 F.3d 308, 312 (1st Cir. 1993) (citing 4 Arthur Larson, *Workmen's Compensation Law* §88.13, at 16-187 (1992)).

[29]*American Heritage Dictionary* 604 (3d ed. 1992). *See also Black's Law Dictionary* 363 (Abridged 6th ed., 1991) (an employee is a "person in the service of another under any contract of hire, express or implied, oral or written, where the employer has the power or right to control and direct the employee in the material details of how the work is to be performed").

[30]*See Black's Law Dictionary* 529 (Abridged 6th ed., 1991).

[31]See *Stuart v. Colorado Interstate Gas Co.,* 271 F.3d 1221 (10th Cir. 2001).

[32]See *Heckendorn v. Consolidated Rail Corp.,* 293 Pa. Super. 474, 439 A.2d 674 (1981).

[33]"Joint tort-feasors" refers to two or more persons who are jointly or severally liable in tort for the same injury to person or property, including those who act in concert in their tortuous conduct as well as those whose independent actions result in causing a single injury. *See Black's Law Dictionary* 584 (Abridged 6th ed., 1991).

[34]See, e.g., Colo. Rev. Stat. §§ 8-41-102, 401 (2001).

[35]See *Evans v. B.F. Perkins Co.,* 166 F.3d 642 (4th Cir. 1999); *Cinnamon v. IBM Corp.,* 384 S.E.2d 618, 621 (Va. 1989).

[36]Under Louisiana law, a principal who hires a contractor to perform work that is part of its trade, business, or occupation is a statutory employer of the contractor's employees. *See* La. Rev. Stat. Ann. §23:1061 (1990). The statutory employer is legally required to pay workers' compensation benefits, but is immune from tort liability. *See* La. Rev. Stat. Ann. §23:1032 (1989).

[37]See *Kirkland v. Riverwood Intern. USA, Inc.,* 681 So.2d 329, 336 (La. 1996).

[38]*Kirkland,* 681 So.2d at 336-37.

[39]See, e.g., *Kirkland* at 337.

[40]*Ghersi v. Salazar,* 883 P.2d 1352, 1356 (Utah 1994).

[41]1B Arthur Larson, Workers' Compensation Law §48.00, at 8-434 (1992).

[42]See *Marzula v. White,* 488 So.2d 1092, 1095 (La. App. 2 Cir. 1986).

[43]See, e.g., *Green v. Popeye's Inc.,* 619 So.2d 69 (La. App. 3 Cir. 1993).

[44]See, generally, G. L. Hammond, *Flexible Staffing Trends,* 10 The Labor Lawyer 161 (American Bar Association, 1994).

[45]*Id.* at 176-177.

[46]See *Vanterpool v. Hess Oil Virgin Islands Corp.,* 766 F.2d 117 (3d Cir. 1985).

[47]To further drive home this point, the Virgin Islands also amended its workers' compensation laws to provide: "[A] contractor shall be deemed the employer of a subcontractor's employees only if the subcontractor fails to comply with the provisions of this chapter with respect to being an insured employer. The "statutory employer and borrowed servant" doctrine are not recognized in this jurisdiction, and an injured employee may sue any person responsible for his injuries other than the employer named in a certificate of insurance. . . . " *See also Gass v. Virgin Islands Telephone Corp., Raco, Inc.,* ___ F.3d ____ (3d Cir. Nov. 18, 2002).

[48]See *Pascouau v. Martin Marietta Corp.,* 185 F.3d 874 (10th Cir. 1999).

[49]*Strock v. Pressnell,* 527 N.E.2d 1235, 1244 (Ohio 1988).

[50]*Greenberg v. Life Insurance Co.,* 177 F.3d 507 (6th Cir. 1999).

[51]The term "tortious" means wrongful, and is used throughout the Restatement, Second, Tort (the lawyer's "bible" on tort law principles) to denote the fact that conduct is of such character as to subject the actor to liability under the principles of tort law. To establish a "tortuous act," a plaintiff must establish both the existence of actionable wrong and damages resulting form the wrong. *See Black's Law Dictionary* 1036 (Abridged 6th ed., 1991).

[52]See e.g., *Smith v. First Union National Bank,* 202 F.3d 234 (4th Cir. 2000).

[53]This standard provides that, the absence of an infirmary, clinic, hospital, or physician that is reasonably accessible in terms of time and distance to the work site, a person who has a valid certificate in first-aid training from the U.S. Bureau of Mines, the American Red Cross, or equivalent training that can be verified by documentary evidence, must be available at the work site to render first aid. First-aid supplies must also be readily accessible. Where a medical facility *is* near the workplace, OSHA requires the employer to insure that, in areas where accidents resulting in suffocation, severe bleeding, or life-threatening injury/illness can be expected, a 3-to-4-minute response time is required. In other circumstances, i.e., where a life-threatening injury is an unlikely outcome of an accident, a longer response time of up to 15 minutes is acceptable; and if employees work in areas where emergency transportation is not available, the employer must make provisions for acceptable emergency transportation.

[54]It is recommended that employers maintain training information as long as possible, as this can be extremely useful in demonstrating compliance or in supporting an "employee misconduct" affirmative defense to a citation issued several years after the training initially was provided.

■■ **About the Author** ■■

REGULATORY REQUIREMENTS
by T. Michael Toole, Ph.D., P.E.

T. Michael Toole, Ph.D., P.E., is an Assistant Professor in the Department of Civil & Environmental Engineering at Bucknell University in Lewisburg, Pennsylvania. He received his Bachelor of Science in Civil Engineering from Bucknell University and his Master of Science in Civil Engineering and Ph.D. in Technology Strategy from the Massachusetts Institute of Technology. Dr. Toole is a professional civil engineer registered in four states and an authorized OSHA instructor for the construction industry. He is a member of the American Society of Safety Engineers and the American Society of Civil Engineers. Within the latter, he is a member of the Construction Site Safety Committee and the Chairman of the Engineering Management Group of the Central Pennsylvania Section.

Regulatory Requirements

INTRODUCTION

This chapter summarizes the content, applicability, and revision process of the federal safety standards for construction. Key duties of employers are also discussed. Related documents, such as referenced ANSI standards and OSHA's manual for its inspectors, are briefly covered. Specific regulatory para-graphs that require the involvement of a competent person, qualified person, or an engineer are identified. Web locations (URLs) are provided to allow the reader to access the actual text being discussed when not quoted in this chapter.

Why Safety Requirements Are Important

This chapter may be one of the most important in this book. Although the actual OSHA text is rather tedious, the better a safety manager knows the standards and how they are enforced, the more effective the manager can be. Safety and operational managers need to be intimately familiar with the general safety duties ascribed to their companies and how to quickly locate the OSHA text applicable to a specific task or hazard. Companies that ignore or do not meet safety standards can be fined by OSHA, prohibited from bidding on some projects, and will likely find their workers' compensation modification factors increasing as valuable employees incur injuries.

Conversely, companies need to know when they are exceeding mandated requirements. There is nothing wrong with having a stricter safety program than OSHA requires—indeed, the safety programs of the best-managed companies go well beyond minimum standards. But safety practices that exceed mandated standards should be the result of intentional management decisions, not inefficient practices unintentionally carried over from a previous project or client who had unique safety needs.

Another reason for safety and operational managers to be familiar with the OSHA requirements is that there is no shared understanding of respective safety roles within the construction industry. A recent article published in the American Society of Civil Engineers' *Journal of Construction Engineering and Management* (Toole, 2002A) presented the results of a survey of general contractors, subcontractors, and civil engineers that indicated widespread disagreement on which party on a construction site was responsible for various aspects of safety management. Many construction industry professionals ascribe all or most safety management responsibilities to the general contractor, which is in accordance with the General Conditions documents commonly used on building construction projects. As discussed in this chapter, the federal safety standards conflict with this commonly held belief.

KEY FACTS ABOUT THE FEDERAL OSHA STANDARDS

Introduction to 29 CFR 1926

The Occupational Safety and Health Administration and federal safety standards were first created in 1970 with the passage of the Occupational Safety and Health Act (the OSH Act) and Contract Work Hours and Safety Standards Act passed in 1969. Written safety and health standards established by OSHA are promulgated as part of the Code of Federal Regulations. OSHA standards apply to the private sector only and exclude self-employed workers. Specific standards for the construction industry are promulgated as Title 29, Part 1926 of the Code of Federal Regulations (29 CFR 1926), which covers all construction work, including maintenance and repairs. Table 1 lists the subparts that comprise 29 CFR 1926. The majority of this chapter will discuss key portions of this text.

■■ **Table 1** ■■

Subparts in 29 CFR 1926

Subpart	Subpart Title
A	General
B	General Interpretations
C	General Safety and Health Provisions
D	Occupational Health and Environmental Controls
E	Personal Protective and Life Saving Equipment
F	Fire Protection and Prevention
G	Signs, Signals, and Barricades
H	Materials Handling, Storage, Use, and Disposal
I	Tools—Hand and Power
J	Welding and Cutting
K	Electrical
L	Scaffolds
M	Fall Protection
N	Cranes, Derricks, Hoists, Elevators, and Conveyors
O	Motor Vehicles, Mechanized Equipment, and Marine Operations
P	Excavations
Q	Concrete and Masonry Construction
R	Steel Erection
S	Underground Construction, Caissons, Cofferdams, and Compressed Air
T	Demolition
U	Blasting and the Use of Explosives
V	Power Transmission and Distribution
W	Rollover Protective Structures; Overhead Protection
X	Stairways and Ladders
Y	Commercial Diving Operations
Z	Toxic and Hazardous Substances

Note: The contents of these subparts can be accessed by going to http://www.osha.gov/pls/oshaweb/owastand.display_standard_group?p_toc_level=1&p_part_number=1926&p_text_version=FALSE and clicking on the hyperlink for a specific subpart.

Construction industry firms are also subject to safety standards and procedures required of firms in other industries. 29 CFR 1903 addresses the processes governing OSHA inspections, citations, and proposed penalties. 29 CFR 1904 addresses requirements for recording and reporting occupational injuries. Portions of 29 CFR 1910, safety standards for general industry, may also be applicable to construction-industry firms (see Table 2). Where 29 CFR

■ ■ **Table 2** ■ ■

Subparts in 29 CFR 1910

Occupational Safety and Health Standards for General Industry

Subpart	Subpart Title
A	General
B	Adoption and Extension of Established Federal Standards
C	[Removed and Reserved]
D	Walking and Working Surfaces
E	Means of Egress
F	Powered Platforms, Man lifts, and Vehicle-Mounted Work Platforms
G	Occupational Health and Environmental Control
H	Hazardous Materials
I	Personal Protective Equipment
J	General Environmental Controls
K	Medical and First Aid
L	Fire Protection
M	Compressed Gas and Compressed Air Equipment
N	Materials Handling and Storage
O	Machinery and Machine Guarding
P	Hand and Portable Powered Tools and Other Handheld Equipment
Q	Welding, Cutting, and Brazing
R	Special Industries
S	Electrical
T	Commercial Diving Operations
U	Y [Reserved]
Z	Toxic and Hazardous Substances

Note: The contents of these subparts can be accessed by going to http://www.osha.gov/pls/oshaweb/owastand.display_standard_group?p_toc_level=1&p_part_number=1910&p_text_version=FALSE and clicking on the hyperlink for a specific subpart.

1926 explicitly addresses a subject, the text supersedes text on the same subject in 29 CFR 1910, as set forth in 1926.20(d)(1):

> If a particular standard is specifically applicable to a condition, practice, means, method, operation, or process, it shall prevail over any different general standard which might otherwise be applicable to the same condition, practice, means, method, operation, or process.

Where 29 CFR 1926 does not address a subject, firms performing construction work must also comply with applicable paragraphs in 29 CFR 1910, as set forth in 1926.20(d)(2):

> On the other hand, any standard shall apply according to its terms to any employment and place of employment in any industry, even though particular standards are also prescribed for the industry to the extent that none of such particular standards applies.

An excerpt from an OSHA interpretation letter (explained below) dated December 3, 1993 helps clarify this issue:

> With regard to the application of 1926.20(d)(2), please be advised that the plain reading of this paragraph would indicate that any standards, including Part 1910 standards, would apply to an activity at a construction site if no specific construction standard exists. However, OSHA's policy is to apply only those Part 1910 standards to construction that have been identified in notices such as the June 30 Federal Register as being applicable to construction. (See http://www.osha.gov/pls/oshaweb/owadisp.show_document?p_table=INTERPRETATIONS&p_id=21334&p_text_version=FALSE)

Although the OSH Act covers all employers in the 50 states and territories and jurisdictions under federal authority, including the District of Columbia, Puerto Rico, and the Virgin Islands, 29 CFR 1902 allows states to establish their own construction safety regulations. State standards must be at least as strict as those contained in 29 CFR 1926. State standards can apply only to employees of state and local governments or to both public- and private-sector employees. Currently, 23 states and 2 territories (see Table 3) have their own safety plans and 3 states (Connecticut, New Jersey, and New York) have plans that cover only government employees. Some states, such as California and Michigan, have safety standards that substantially differ from federal standards (see Figure 1). Other states have standards that represent only minor deviations from the federal standards. When federal OSHA promulgates new or revised standards, states must establish comparable standards within 6 months.

The federal OSHA standards for the construction industry are revised frequently, but the vast majority of requirements do not change from one year to the next. Revised editions of the standards are printed annually (and sold) by the Government

■ ■ **Table 3** ■ ■

States with OSHA-approved Plans

State Agency	URL
Alaska Department of Labor	http://www.labor.state.ak.us/lss/lss.htm
Hawaii Department of Labor	http://www.state.hi.us/dlir/hiosh/
Industrial Commission of Arizona	http://www.ica.state.az.us/ADOSH/oshatop.htm
Indiana Department of Labor	http://www.in.gov/labor/iosha/
California Department of Industrial Relations	http://www.dir.ca.gov/occupational_safety.html
Iowa Division of Labor Services	http://www.state.ia.us/iwd/labor/index.html
Connecticut Department of Labor	http://www.ctdol.state.ct.us/osha/osha.htm
Kentucky Labor Cabinet	http://www.kylabor.net/kyosh/index.htm
Maryland Division of Labor and Industry	http://www.dllr.state.md.us/labor/mosh.html
Michigan Department of Consumer and Industry Services	http://www.cis.state.mi.us/bsr/
New Jersey Department of Labor	http://www.state.nj.us/labor/lsse/lspeosh.html
North Carolina Department of Labor	http://www.nclabor.com/osha/osh.htm
Minnesota Department of Labor and Industry	http://www.doli.state.mn.us/mnosha.html
Oregon Department of Consumer & Business Services	http://www.orosha.org/
Nevada Division of Industrial Relations	http://dirweb.state.nv.us
Puerto Rico Secretary of Labor and Human Resources	http://dtrh.prstar.net
New Mexico Environment Department	http://www.nmenv.state.nm.us/OHSB_web site/ohsb _home.htm
South Carolina Department of Labor Licensing and Regulation	http://www.llr.state.sc.us/osha.asp
New York Department of Labor	http://www.labor.state.ny.us/working_ny/worker_rights/safety_ health.html
Tennessee Department of Labor	http://www.state.tn.us/labor-wfd
Labor Commission of Utah	http://www.labor.state.ut.us/Utah_Occupational_Safety___Hea/ utah_occupational_safety___hea.html
Vermont Department of Labor and Industry	http://www.state.vt.us/labind/vosha.htm
Wyoming Department of Employment	http://wydoe.state.wy.us/doe.asp?ID=7
Virginia Department of Labor and Industry	http://www.doli.state.va.us/
Virgin Islands Department of Labor	
Washington Department of Labor and Industries	http://www.lni.wa.gov/wisha/default.htm

See http://www.osha.gov/oshdir/states.html or the URLs shown above for addresses and phone numbers.

Printing Office (GPO). Employers must comply with the revised standards when they are enacted. That is, employers must comply with standards that have been enacted but which are not yet included in the most recent printing of the standards. Private publishers also publish bound copies of the OSHA text and often charge less than the GPO does. The GPO also sells compact disks containing 29 CFR 1926 and provides free access to the latest standards via the OSHA Web site (www.osha.gov).

Revisions to the standards may involve new sections on recently recognized hazards or revisions to existing sections reflecting updated knowledge of previously recognized hazards. Revisions may be initiated within OSHA itself or in response to suggestions from other government organizations, national standards organizations, or organizations representing employers or labor. A list of all new or revised regulations that OSHA is considering is published each April and October in the *Federal Register*. The

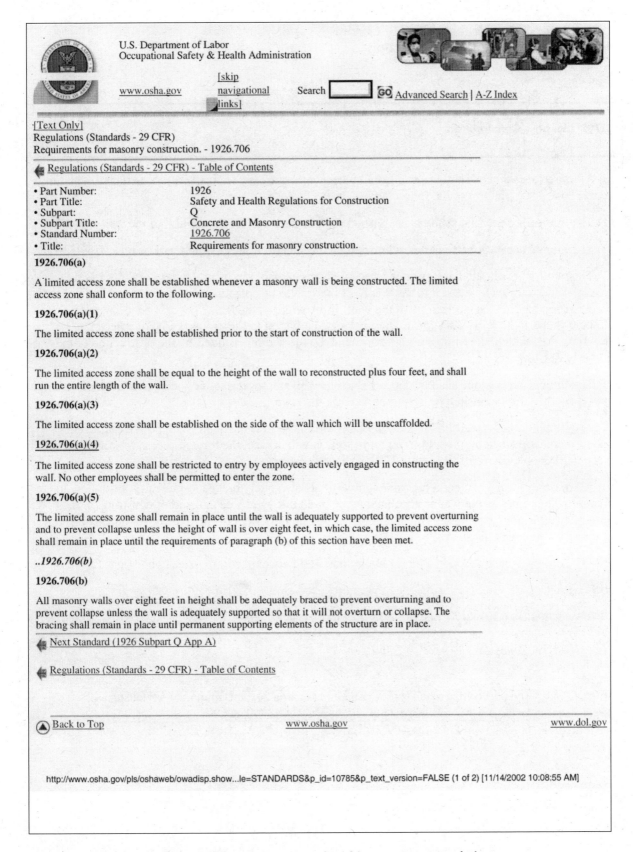

■■ **Figure 1** ■■ **Examples of Federal, California, and Michigan masonry regulations**

Subchapter 4. Construction Safety Orders
Article 29. Erection and Construction

New query

---

§1722. Masonry Construction.

(a) A limited access zone shall be established whenever a masonry wall is being constructed. The limited access zone shall conform to the following:

(1) The limited access zone shall be established prior to the start of construction of the wall.

(2) The limited access zone shall be established on the side of the wall which will be unscaffolded.

(3) The width of the limited access zone (measured perpendicularly from the base of the wall on the unscaffolded side) shall be equal to the height of the wall to be constructed plus four feet. The limited access zone shall run the entire zone of the wall. If the width of the limited access zone can not be attained because the wall being constructed is located adjacent to the property line or adjacent to a structure, the width of the limited access zone shall be the width permitted by the obstruction of the property line or structure.

(4) The limited access zone shall be restricted to entry by employees actively engaged in constructing the wall. No other employee shall be permitted to enter the zone.

(5) The limited access zone shall remain in place until the wall is adequately supported to prevent collapse unless the height of wall is over eight feet, in which case, the limited access zone shall remain in place until the requirements of section 1722(b) have been met.

(b) All masonry walls over eight feet in height shall be adequately braced to prevent overturning and to prevent collapse unless the wall is adequately supported through its design and/or construction method to prevent overturning or collapse. The bracing shall remain in place until permanent supporting elements of the structure are in place.

NOTE: Authority cited: Section 142.3, Labor Code. Reference: Section 142.3, Labor Code.

HISTORY

1. New section filed 10-22-90; operative 11-21-90 (Register 90, No. 48).

Go Back to Article 29 Table of Contents

■ ■ **Figure 1** ■ ■ (cont.)

DEPARTMENT OF CONSUMER AND INDUSTRY SERVICES
DIRECTOR'S OFFICE
CONSTRUCTION SAFETY STANDARDS

Filed with the Secretary of State on **November 15, 1989**

These rules take effect 15 days after filing with the Secretary of State

(By authority conferred on the director of the department of consumer and industry services by sections 19 and 21 of Act No. 154 of the Public Acts of 1974, as amended, and Executive Reorganization Order No. 1996-2, being §§408.1019, 408.1021 and 445.2001 of the Michigan Compiled Laws)

PART 2. MASONRY WALL BRACING
TABLE OF CONTENTS

GENERAL PROVISIONS

R 408.40201. Scope.
   Rule 201. This part pertains to the temporary bracing of unsupported masonry walls during construction which are exposed to wind forces.

R 408.40202. Applicability.
   Rule 202. This part is designed to ensure a safe work environment for all personnel on the construction site through the use of temporary bracing of unsupported masonry walls. The requirements of this part are as follows:
   (a) Identifying masonry walls requiring temporary bracing (R 408.40204(1)).
   (b) Proposing an acceptable temporary bracing system (R 408.40207).
   (c) Vacating the collapse area during winds of 35 mph or more (R 408.40204(9) and R 408.40205).
   (d) Standard sign requirements for collapse areas (R 408.40208), which are all designed to prevent on-site injury. While winds of more than 35 mph may cause collapse of walls braced in accordance with this part, compliance with all of the other provisions of this same part will ensure that no one will be within the collapse area.

R 408.40203. Definitions; C to U.
   Rule 203. (1) "Cavity wall" means a masonry wall with a continuous insulated or uninsulated air space of 2 to 4 1/2 inches between wythes that are connected with rigid metal ties.
   (2) "Collapse area" means that area which is within the height of the wall, plus 4 feet, measured at right angles to the wall on both sides.
   (3) "Composite wall" means a bonded masonry wall with 2 or more wythes of different masonry units.

   (4) "Qualified person" means a person who, by knowledge, training, or experience, has the ability to solve or resolve problems relating to the subject matter or the work.
   (5) "Single wythe hollow masonry" means a masonry wall 1 unit in thickness made up of units with bearing surfaces that are less than 75% solid.
   (6) "Solid masonry unit" means a masonry unit with bearing surfaces that are 75% or more solid.
   (7) "Unsupported masonry wall" means a masonry wall that has not obtained its final lateral stability from design features when required, such as roofs, floors, buttresses, crosswalls, and piers.

R 408.40204. Maximum unsupported height tables.
   Rule 204. (1) The maximum unsupported height of a masonry wall shall not be more than the height shown in tables 2 to 5 of this rule. Unbraced walls exceeding the heights specified in these tables are in imminent danger of collapse.
   (2) The exposure to which a wall is subjected for use in tables 2 to 5 shall be determined from table 1, which reads as follows:

TABLE 1
Exposure Selection

Exposure	Example
A	Center of large cities and very rough hilly terrain.
B	Suburban areas, towns, city outskirts, wooded areas, and rolling terrain.
C	Flat, open country, open flat coastal belts, and grasslands.

   (3) Exposure A shall not be used in Michigan.
   (4) Table 2 reads as follows:

■■ **Figure 1** ■■ (cont.)

TABLE 2
Single Wythe Hollow Masonry

Width of Wall	Minimum Weight psf	Maximum Unsupported Height			
		Exposure B		Exposure C	
		(1)	(2)	(1)	(2)
4 in.	25	6 ft.	(2 ft.)*	6 ft.	(1 ft.)*
6 in.	34	6 ft.	(5 ft.)*	6 ft.	(2.5 ft.)*
8 in.	40	7 ft.	(7 ft.)*	6 ft.	(4 ft.)*
10 in.	48	10 ft.	(10 ft.)*	6 ft.	(6 ft.)*
12 in.	56	14 ft.	(14 ft.)*	9 ft.	(9 ft.)*
16 in.	75	24 ft.	(24 ft.)*	16 ft.	(16 ft.)*

* See subrule (8) of this rule.

(5) Table 3 reads as follows:

TABLE 3
Solid Brick Walls

Width of Wall	Minimum Weight psf	Maximum Unsupported Height			
		Exposure B		Exposure C	
		(1)	(2)	(1)	(2)
4 in.	40	6 ft.	(3 ft.)*	6 ft.	(2 ft.)*
8 in.	80	12 ft.	(12 ft.)*	8 ft.	(8 ft.)*
12 in.	120	20 ft.	(20 ft.)*	19 ft.	(19 ft.)*
16 in.	160	26 ft.	(26 ft.)*	26 ft.	(26 ft.)*

* See subrule (8) of this rule.

(6) Table 4 reads as follows:

TABLE 4
Composite Walls — 4-inch Brick and Hollow Block Units (Various Widths)

Width of Wall Total	Brick	Block	Min. Weight psf	Maximum Unsupported Height			
				Exposure B		Exposure C	
				(1)	(2)	(1)	(2)
8 in.	4 in.	4 in.	65	9 ft.	(9 ft.)*	6 ft.	(5 ft.)*
10 in.	4 in.	6 in.	74	13 ft.	(13 ft.)*	9 ft.	(9 ft.)*
12 in.	4 in.	8 in.	80	16 ft.	(16 ft.)*	11 ft.	(11 ft.)*
14 in	4 in.	10 in.	88	19 ft.	(19 ft.)*	14 ft.	(14 ft.)*
16 in.	4 in.	12 in.	96	26 ft.	(26 ft.)*	17 ft.	(17 ft.)*

* See subrule (8) of this rule.

(7) Table 5 reads as follows:

TABLE 5
Cavity Walls — 4-inch Brick and Hollow Block Units (Various Widths)

Wall Section Brick + Block		Minimum Weight psf	Maximum Unsupported Height			
			Exposure B		Exposure C	
			(1)	(2)	(1)	(2)
4 in.	4 in.	65	6 ft.	(2.5 ft.)*	6 ft.	(1.5 ft.)*
4 in.	6 in.	74	6 ft.	(5 ft.)*	6 ft.	(2.5 ft.)*
4 in.	8 in.	80	8 ft.	(8 ft.)*	6 ft.	(4.5 ft.)*
4 in.	10 in.	88	11 ft.	(11 ft.)*	7 ft.	(7 ft.)*
4 in.	12 in.	96	27 ft.	(27 ft.)*	18 ft.	(18 ft.)*

* See subrule (8) of this rule.

(8) If employees within the collapse area are working from elevations that are lower than the bottom elevator of the wall, the maximum unsupported height of a masonry wall shall be determined from values given in column (2) of tables 2 to 5.

(9) No one shall be permitted within the collapse area of an unbraced or braced wall subjected to winds of more than 35 miles per hour.

R 408.40205. Wind velocity; determination by qualified person.

Rule 205. For the purpose of this part, the wind velocity shall be determined by a qualified person.

R 408.40206. Wall bracing design.

Rule 206. (1) When the height of a masonry wall exceeds maximum unsupported height as shown in tables 2 to 5 of R 408.40204, the masonry wall shall be braced on both sides upon completion. Crosswalls are acceptable instead of bracing an interior wall if the crosswalls are not spaced more than 20 feet apart. If crosswalls are spaced more than 20 feet apart, wall bracing in accordance with the requirements shall be provided.

(2) On masonry projects that require temporary bracing, the wall bracing system shall be determined before a masonry wall exceeds the maximum unsupported height limits specified in tables 2 to 5 of R 408.40204.

(3) The wall bracing system for a masonry wall shall be designed by a qualified person in accordance with acceptable engineering practices or as prescribed in this part and shall be capable of providing stability to the wall for a wind with a velocity of 35 miles per hour.

(4) If pilasters, buttresses, or other reinforcing is part of the wall design, the unsupported height of walls according to tables 2 to 5 of R 408.40204 may be exceeded by complying with accepted engineering practices. Calculations or plans and specifications shall be available at the jobsite.

(5) If scaffolding, because of work operations, remains erected on 1 side of the completed wall, the collapse area shall be identified and marked. No one shall be permitted within the collapse area when the wind velocity is more than 35 miles per hour.

(6) The height of a masonry wall above the intersection of the diagonal support with the vertical plane of the wall shall not be more than the maximum unsupported height as shown in tables 2 thru 5 of R 408.40204.

R 408.40207. Typical wall bracing system.

Rule 207. (1) A typical wall brace may consist of 4 essential parts as follows:

(a) A 16-foot, 2-inch by 10-inch vertical upright.
(b) A 16-foot, 2-inch by 10-inch diagonal strut.
(c) A 2-inch by 4-inch stiffner.
(d) A deadman.

(2) The angle of intersection of the 16-foot, 2-inch by 10-inch diagonal strut and the ground should be between 35 degrees and 45 degrees and the diagonal strut should not intersect the vertical brace below the midpoint of the masonry wall.

(3) When using this typical wall brace, the total wall bracing system shall be designed in accordance with the provisions of this rule and R 408.40206.

(4) Other materials and designs may be used in the construction of a wall bracing system if the design requirements of this rule and R 408.40206 are met.

(5) The following figure is an example of a typical wall brace.

FIGURE 1
Typical Wall Brace for Masonry Wall

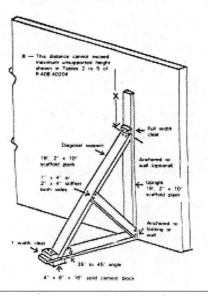

■ ■ **Figure 1** ■ ■ (cont.)

(6) The maximum spacing for typical exterior wall bracing shall not be more than 20 feet. Table 6 specifies typical exterior wall bracing requirements and reads as follows:

TABLE 6
Typical Exterior Wall

8 inch wall .... 18 feet maximum height
12 inch wall .... 22 feet maximum height

R 408.40208. Signing.
Rule 208. (1) Each unsupported masonry wall that is more than 6 feet in height, braced or unbraced, and 50 feet or less in length shall be posted with a danger sign on each side of the wall.
(2) Each unsupported masonry wall that is more than 6 feet in height, braced or unbraced, and more than 50 feet in length, shall be posted with danger signs at each end of the wall and at intervals of not more than 100 feet along each side of the wall.
(3) When scaffolding is in place along an unsupported masonry wall, the posting requirements of subrule (1) or (2) of this rule are only required for the unscaffolded portions of the wall.
(4) The danger sign shall be placed in a conspicuous location either on the wall or anywhere within the collapse area.
(5) The danger signs shall be maintained in place and in a legible condition until the masonry wall is permanently supported.
(6) A danger sign as required by subrule (1) or (2) of this rule shall comply with all of the following requirements:
  (a) Be 10 inches in height by 14 inches wide.
  (b) Have the word "DANGER" in white characters which are, 2 1/6 inches high and which appear within a red oval which is 4 1/8 inches high by 11 7/8 inches long and which is in the top 1/2 of the sign.
  (c) Have the lower 1/2 of the sign state, "This Unsupported Wall is Unstable in Windy Conditions."
(7) An illustration of a danger sign which compiles with the requirements of subrule (6) of this rule is shown in the following figure:

R 408.40209. Inspection.
Rule 209. An unsupported masonry wall, including the wall bracing system, shall be inspected for damage by a qualified person after each windstorm if the wind velocity was more than 35 miles per hour. If any movement of the wall or other physical damage, including damage to the wall bracing system, is found, only those persons repairing the wall or wall bracing system shall be permitted to work within the collapse area until repairs have been made.

R 408.40210. Wall bracing system; responsibility for installation; responsibility for replacing system and danger signs.
Rule 210. The masonry contractor shall be responsible for the initial installation of the wall bracing system. After a wall bracing system and danger signs have been installed in accordance with the provisions of this part, any party, including a subcontractor, general contractor, or owner, who alters or removes the bracing system or danger signs shall be responsible for replacing the bracing system and danger signs in accordance with the provisions of this part.

FIGURE 2

These masonry regulations can be accessed at:
http://www.osha.gov/pls/oshaweb/owadisp.show_document?p_table=STANDARDS&p_id=10785&p_text_version=FALSE, http://www.michigan.gov/cis/0,1607,7-154-11407_15368-39938--,00.html (click on Part 2, Masonry Wall Bracing, which is hyperlinked to the PDF file), and http://www.dir.ca.gov/title8/1722.html.

▪▪ **Figure 1** ▪▪ **(cont.)**

Advisory Committee on Construction Safety and Health (ACCSH) is a permanent committee that advises the Secretary of Labor on potential revisions. Per statute, this committee is composed of government, labor, and employer organizations. In addition, temporary official committees that focus on specific issues may be formed, and all interested parties may present OSHA with documents relevant to the revisions. Formal hearings may occur before OSHA issues the final wording of a new or revised standard and mandates the date that it becomes effective.

Individual employers may petition OSHA to receive permission to deviate from a standard currently in effect or scheduled to be implemented. Such permission, called a *variance*, may be granted temporarily (if, for example, equipment or labor is not available to meet a new standard) or permanently if they are using methods that are demonstrated to be as effective in preventing injury but that are not allowed by the standard. Policies and procedures for variances are covered in 29 CFR 1905.

U.S. Department of Labor
Occupational Safety & Health Administration

www.osha.gov   [skip navigational links]   Search [ ] GO Advanced Search | A-Z Index

[Text Only]
Standard Interpretations
02/23/1999 - Insulation issues: definition; plastic sheaths;and guarding requirements.

Standard Interpretations - Table of Contents

• Standard Number: 1926.416(a)

February 23, 1999

T. Michael Toole, Ph.D., P.E.
Director of Construction Systems
Packer Engineering, Inc.
1950 North Washington Street
P.O. Box 353
Naperville, Illinois 60566-0353

Re: 29 CFR 1926.416(a); whether plastic conduit sheathing is insulation; definition of insulation

Dear Dr. Toole:

This is in response to your letter dated May 1, 1998, to the Occupational Safety and Health Administration (OSHA) requesting clarification of the term "insulation" as used in 29 CFR 1926.416(a).

Definition of insulation

The first part of that standard states:

(1) No employer shall permit an employee to work in such proximity to any part of an electric power circuit that the employee could contact the electric power circuit in the course of work, unless the employee is protected against electric shock by deenergizing the circuit and grounding it **or by guarding it effectively by insulation or other means** [emphasis added].

The term "insulation," as used in this provision, means a material that will protect employees from shock hazards associated with electrical power circuits. Factors such as use and location will determine the proper type of conductor insulation required. Insulation integral to conductors must withstand exposure to atmospheric and other conditions of use without detrimental leakage of current.

Whether plastic sheathing is considered insulation

You also ask about a situation in which a "circuit" has "plastic sheathing over the conductor" and is not in any type of rigid conduit. You ask if "the plastic sheathing over the conductor [would] be considered insulation" under §1926.416(a).

In most 120 volt and 240 volt circuits, current is carried through two of three circuit conductors. One is ungrounded and one is neutral.[1] An equipment grounding conductor is also usually included in the cable. The two circuit conductors are made of solid or stranded conducting material -- wire. Each wire is encased in insulating material. The equipment grounding conductor is normally either bare wire or wrapped in paper. The two insulated conductors, plus the equipment grounding conductor, are collectively encased in either a flexible plastic sheath or a flexible metal sheath.

We assume that when you use the term "conductor" in your question, you are referring to the current carrying conductors (consisting of wire individually encased by insulating material), plus the equipment grounding conductor. We further assume that when you refer to the "plastic sheathing over the conductor," you are referring to the typical plastic sheath that collectively encases the insulated conductors plus the grounding conductor. In that circumstance, the plastic sheathing that holds the conductors plus the equipment grounding conductor is not designed to serve as insulation. While it may or may not have some

■ ■ **Figure 2** ■ ■ **Example letter of interpretation**

insulating properties, its purpose is to (1) conveniently package the three conductors into one cable, (2) provide protection against atmospheric and other conditions of use without detrimental leakage of current and (3) protecting the conductors -- both the wires and the insulation that encases them, as well as the equipment grounding conductor -- from mechanical damage.

Guarding against electric shock

You ask if the term "insulation," as used in 1926.416(a), refers only to the insulating material that encases a wire, or if the term refers to "a separate insulating system worn or used by the workers?"

Where a circuit has not been deenergized, section 1926.416(a) requires that employees be protected from electric shock by "guarding it [the circuit] effectively by insulation." This means that the employer must ensure that insulation already covers the energized parts and will protect the employee. That insulation must be sufficient/appropriate for the working conditions. If it will not protect the employee, then the employer must use insulating material, such as an insulating blanket, to protect against the shock hazard. Where that is not feasible, this provision, in conjunction with §1926.95(a), requires employers to protect employees with appropriate insulating personal protective equipment.[2]

If you require any further assistance, please do not hesitate to contact us again by writing to: Directorate of Construction - OSHA Office of Construction Standards and Compliance Assistance, Room N3621, 200 Constitution Avenue N.W., Washington, D.C. 20210.

Sincerely,

Russell B. Swanson, Director
Directorate of Construction

Footnote (1) In a solidly grounded neutral system, the neutral conductor is also called the grounded conductor. (Back to text)

Footnote (2) Article 100 of the National Electric Code defines guarded as "covered, shielded, fenced, enclosed, or otherwise protected by means of suitable covers, casings, barriers, rails, screens, mats or platforms to remove the likelihood of approach or contact by persons or objects to a point of danger." (Back to text)

Standard Interpretations - Table of Contents

Back to Top www.osha.gov www.dol.gov

Contact Us | Freedom of Information Act | Information Quality | Customer Survey
Privacy and Security Statement | Disclaimers

Occupational Safety & Health Administration
200 Constitution Avenue, NW
Washington, DC 20210

This letter can be accessed at:
http://www.osha.gov/pls/oshaweb/owadisp.show_document?p_table=INTERPRETATIONS&p_id=22698&p_search_type=InterpTextPolicy&p_search_str=Toole&p_text_version=FALSE#ctx1

■ ■ **Figure 2** ■ ■ (cont.)

OSHA -- Occupational Safety & Health Administration
U.S. Department of Labor
Directives CPL 2-1.34 - Inspection policy and procedures for OSHA's steel erection standards for construction

- Record Type: Instruction
- Directive Number: CPL 2-1.34
- Title: Inspection policy and procedures for OSHA's steel erection standards for construction
- Standard Number: 1926
- Information Date: 03/22/2002

OSHA INSTRUCTION

DIRECTIVE NUMBER: CPL 2-1.34	EFFECTIVE DATE: Friday, March 22, 2002
SUBJECT: Inspection policy and procedures for OSHA's steel erection standards for construction.	

ABSTRACT

Purpose:	This instruction describes OSHA's inspection policy and procedures and provides clarification to ensure uniform enforcement by field enforcement personnel of the steel erection standards for construction.
Scope:	OSHA-wide
References:	Construction Safety and Health Standards, Subpart R, 29 CFR 1926.750-761, Subpart M, 1926.502 and §1926.105; Federal Register, Vol. 66, No. 12, January 18, 2001, pages 5196-5280, Final Rule; Safety Standards for Steel Erection; Federal Register, Vol. 66, No. 137, July 17, 2001, pages 37137-37139, Final Rule; Delay of Effective Date; OSHA Instruction CPL 2.103, The Field Inspection Reference Manual (FIRM>>); and Occupational Safety and Health Act of 1970, Section 5(a)(1).
Cancellations:	All interpretations (including letters of interpretation and memoranda) of the previous version of Subpart R issued prior to January 18, 2001.
State Plan Impact:	This instruction describes a Federal Program change for which State adoption is not required.
Action Offices:	National, Regional and Area Offices
Originating Office:	Directorate of Construction
Effective Date:	The effective date for the steel erection standard is January 18, 2002 except that §1926.754(c)(3) will not take effect until July 18, 2006. Certain other provisions are subject to a phase-in period (see Chapter 1, Section X).
Contact:	Mark Hagemann (202) 693-2345 Directorate of Construction N3468, FPB200 Constitution Ave., N.W.Washington, D.C. 20210
Approval:	By and Under the Authority of John L. Henshaw, Assistant Secretary

Executive Summary

This instruction implements the inspection policy and procedures necessary for uniform enforcement of OSHA's new steel erection standard. To achieve this objective, the Agency has included in this instruction a list of anticipated questions and answers along with a Compliance Officer Guide containing inspection tips.

Significant Changes

The new standard addresses the hazards that have been identified as the major causes of injuries and fatalities in the steel erection industry. Concepts addressed by the standard include:

- Site layout and construction sequence
- Site-specific erection plan
- Hoisting and rigging
- Structural steel assembly
- Column anchorage
- Beams and columns
- Open web steel joists
- Systems-engineered metal buildings
- Falling object protection
- Fall protection
- Training

This excerpt on new steel erection can be accessed at:
http://www.oshaslc.gov/pls/oshaweb/owadisp.show_document?p_table=DIRECTIVES&p_id=2730&p_search_str=FIRM&p_search_type=DIRECTTEXTPOLICY&p_status=CURRENT&p_text_version=TRUE#ctx2

■ ■ **Figure 3** ■ ■ **New steel erection standard from FIRM**

29 CFR 1926 Subparts

The safety standards for construction consist of 26 subparts, designated by both a title and a letter of the alphabet (A through Z; see Table 1). (As explained later in the chapter, it is important to note that per 29 CFR 1926.11, Subparts A and B apply only on construction projects that have federal funding or loan guarantees.) Having portions that have been written and rewritten by bureaucrats over many years, the OSHA standards are annoyingly vague in some sections and annoyingly detailed in other sections. Locating the exact paragraphs that address a specific topic is often difficult using only the subpart titles. Fortunately, the bound versions of standards include reasonable indexes and the OSHA Web site includes a search engine.

OSHA Letters of Interpretation

The previous paragraph stated that portions of the OSHA standards are vague. To help employers clarify and better implement the standards, OSHA formally responds to questions sent to the agency by individual employers and trade organizations. The responses are made public on OSHA's Web site (http://www.osha-slc.gov/OshDoc/toc_interps.html). Letters of interpretation on specific topics can be found through OSHA's search engine. (See Figure 2 for a sample letter.)

The Field Inspection Resources Manual (FIRM)

OSHA standards are enforced by OSHA field inspectors who conduct inspections of construction sites within their assigned region on a scheduled or unscheduled basis. Any citations that are issued by OSHA inspectors reflect the applicable paragraphs in 29 CFR 1926 or 1910 as well as guidance provided by directives contained in an internal OSHA document called the Field Inspection Resources Manual (FIRM). [Until the mid-1990s, the FIRM was referred to as the Field Operations Manual (FOM).] The contents of the FIRM can be reviewed by purchasing a compact disk from OSHA—or at no cost through OSHA's Web site (http://www.osha.gov/Firm_osha_toc/Firm_toc_by_sect.html).

One of the key portions of the FIRM articulates OSHA's Multiemployer Policy (MEP). As discussed in the chapter "Multiemployer Work-Site Issues," the MEP is not found in 29 CFR 1926, but it is a very important policy because it allows OSHA to issue citations to companies other than the one that exposed employees to a hazard. Specifically, the MEP allows field inspectors to cite firms that created a hazard, were responsible for correcting a hazard, or were in control of the operations where a hazard was located. (See Figure 3 for an excerpt from FIRM covering the new steel erection standard.)

Other Standards Referenced by the OSHA Text

A number of standards established by private and semipublic organizations are referred to in 29 CFR 1926, most commonly the American National Standards Institute (ANSI). Compliance with the referenced standards is explicitly mandatory in some paragraphs. That is, the OSHA text requires employers to comply with a referenced standard just as if the OSHA text contained the actual text in the referenced standard itself. For example, paragraph 1926.96 requires that safety-toe footwear comply with ANSI standard Z41.1-1967. On the other hand, some paragraphs in the OSHA text reference standards published by other organizations as helpful but nonmandatory guidelines. For example, Appendix A to subpart L "provides non-mandatory guidelines to assist employers in complying with the requirements of subpart L of this part."

It is important to note that there are many standards published by prominent national standards organizations that are not referenced in the federal safety standards. For example, the ANSI A10 series includes many individual standards that are not referenced in 29 CF 1926. While complying with nonreferenced standards may improve a company's safety program, the employer does not have a duty to do so unless the company has entered into a contract that specifically references a standard.

KEY DUTIES OF EMPLOYERS

Regarding safety responsibilities on construction projects, the OSHA text establishes two key tenets. First, the standards for construction nearly exclusively address the responsibilities of employers, not of the entities who are assuming various roles on the project, such as owners, design professionals, general contractors, or subcontractors. Second, the OSHA text clearly requires employers to proactively manage the health of their employees. These tenets are discussed further below.

Employers Have Primary Responsibility

Regarding the first tenet, there is a common mis–perception within the construction industry that the general contractor has primary responsibility for the safety of subcontractors' employees. This misperception likely reflects the wording of the general conditions commonly used in the private sector that are published by the American Institute of Architects (AIA A201, 1997), the Associated General Contractors (AGC 200, 2000), and Engineering Joint Contract Documents Committee (EJCDC 1910-8, 1996) (Toole, 2002B). For example, paragraph 10.1.1 of AIA A201 states: "The Contractor shall be responsible for initiating, maintaining and supervising all safety precautions and programs in connection with the performance of the Contract" (AIA A201, 1997).

The OSHA text, however, places primary safety responsibility for a subcontractor's employee directly on the employer, as illustrated in 1926.20(b)(1): "It shall be the responsibility of the employer to initiate and maintain such programs as may be necessary to comply with this part." Explicitly making construction employers responsible for the safety of their employees follows the requirements set forth in the original OSH Act. The General Duty Clause within the OSH Act (U.S. Code, Title 29, Section 654, Paragraph 5) states:

> Each employer shall furnish to each of his employees employment and a place of employment which are free from recognized hazards that are causing or likely to cause death or serious physical harm to his employees.

The portions of the OSHA text that apply on private-sector projects do not even include the words "general contractor" or "subcontractor," except in the portions relating to Asbestos [1926.1101(d)], Cadmium [1926.1127(m)], and MDA [1926.60(d)]. One generalized section in 29 CFR 1926 that includes these terms, but is frequently misapplied, is 1926.16 ("Rules of Construction"), which states,

> **1926.16(a)** The prime contractor and any subcontractors may make their own arrangements with respect to obligations which might be more appropriately treated on a job site basis rather than individually. Thus, for example, the prime contractor and his subcontractors may wish to make an express agreement that the prime contractor or one of the subcontractors will provide all required first-aid or toilet facilities, thus relieving the subcontractors from the actual, but not any legal, responsibility (or, as the case may be, relieving the other subcontractors from this responsibility). In no case shall the prime contractor be relieved of overall responsibility for compliance with the requirements of this part for all work to be performed under the contract.
> **1926.16(b)** By contracting for full performance of a contract subject to section 107 of the Act, the prime contractor assumes all obligations prescribed as employer obligations under the standards contained in this part, whether or not he subcontracts any part of the work.
> **1926.16(c)** To the extent that a subcontractor of any tier agrees to perform any part of the contract, he also assumes responsibility for complying with the standards in this part with respect to that part. Thus, the prime contractor assumes the entire responsibility under the contract and the subcontractor assumes responsibility with respect to his portion of the work. With respect to subcontracted work, the prime contractor and any subcontractor or subcontractors shall be deemed to have joint responsibility.
> **1926.16(d)** Where joint responsibility exists, both the prime contractor and his subcontractor or subcontractors, regardless of tier, shall be considered subject to the enforcement provisions of the Act.

As stated above, this section is frequently misapplied because individuals quoting it fail to note that it is found in Subpart B. An excerpt from an OSHA interpretation letter dated August 31, 1990, shows that 1926.16 applies only on projects that are funded or guaranteed by the federal government.

Subparts A and B have pertinence only to the application of section 107 of the Contract Work Hours and Safety Standards Act (the Construction Safety Act) (emphasis added). For example, the interpretation of the term "subcontractor" in paragraph (c) of 29 CFR 1926.13 of this chapter is significant in discerning the coverage of the Construction Safety Act and duties thereunder. However, the term "subcontractor" has no significance in the application of the Act, which was enacted under the commerce clause and which establishes duties for "employers" which are not dependent for their application upon any contractual relationship with the Federal Government or upon any form of Federal financial assistance (emphasis added). Therefore, 29 CFR 1926.16, as part of Subpart B (General Interpretation), has pertinence only to the application of section 107 of the Contract Work Hours and Safety Standards Act (The Construction Safety Act) and has no direct significance in the enforcement of the OSH Act. (see http://www.osha.gov/pls/oshaweb/owadisp.show_document?p_table=INTERPRETATIONS&p_id=20068&p_text_version=FALSE)

The previous paragraphs are not intended to imply that the general contractor has no responsibility whatsoever for the health of subcontractors' employees. As discussed in the chapter in this book on OSHA's multiemployer policy, OSHA can cite general contractors as the controlling or correcting employer, even if no general contractor employees were exposed to the hazard.

Proactive Safety Management Required

The second tenet introduced earlier was that employers must *proactively* manage the health and safety of their employees. In other words, employers can neither assume that employees can manage their own safety nor assume that employees will report unsafe situations before an injury occurs. The broadest and most direct statement of this principle is 1926.20(b)(1), which states, "It shall be the responsibility of the employer to initiate and maintain such programs as may be necessary to comply with this part."

The paragraphs that follow this broad mandate address more specific duties that employers must fulfill. One such duty is to inspect employees' work and working conditions on a regular basis to identify potential hazards. Specifically, 1926.20(b)(2) states, "Such programs shall provide for frequent and regular inspections of the job sites, materials, and equipment to be made by competent persons designated by the employers." OSHA does not define the terms "frequent" and "regular" but it does define what a "competent person" is. This definition is discussed later.

The paragraph quoted above referred to the *equipment*, which includes equipment used solely to prevent injuries and equipment used to perform the work that may cause injuries. The former is called personal protective equipment (PPE) and is an employer responsibility. 1926.28(a) states, "The employer is responsible for requiring the wearing of appropriate personal protective equipment in all operations where there is an exposure to hazardous conditions or where this part indicates the need for using such equipment to reduce the hazards to the employees." 1926.95(b) further states, "Where employees provide their own protective equipment, the employer shall be responsible to assure its adequacy, including proper maintenance, and sanitation of such equipment."

Another set of specific duties is related to hazards that result from employees' use of equipment that increases the efficiency or effectiveness of their work. One potential hazard is associated with unsafe equipment. 1926.20(b)(3) states, "The use of any machinery, tool, material, or equipment which is not in compliance with any applicable requirement of this part is prohibited. Such machine, tool, material, or equipment shall either be identified as unsafe by tagging or locking the controls to render them inoperable or shall be physically removed from its place of operation." A second potential hazard is associated with an unsafe operator. 1926.20(b)(4) therefore requires, "The employer shall permit only those employees qualified by training or experience to operate equipment and machinery."

The preceding sentence refers to employee training, which is an employer duty mentioned repeatedly throughout the OSHA text. Employers must train every employee on the recognition and prevention of any hazards they might be exposed to. Specifically, 1926.21(b)(2) requires "The employer shall instruct each employee in the recognition and avoidance of unsafe conditions and the regulations applicable to his work environment to control or eliminate any hazards or other exposure to illness or injury." Several specific paragraphs within Subparts D–Z require that training be documented. Contrary to the perception held by some construction professionals, neither the written standards nor OSHA specify a minimum amount of training (such as a 10-hour safety course) that all individuals working in construction must receive.

An excerpt from the original OSHA act was provided earlier in this chapter and referred to as the General Duty Clause (GDC). This clause is an important component of OSHA's enforcement of safety standards because it is used to cite employers when a hazard exists, yet no specific OSHA paragraph has been violated. To justify a citation under the GDC, OSHA must demonstrate there was a recognizable and correctable hazard that has caused or was likely to cause serious harm or death. The GDC has often been used as a stopgap measure during the period between when a specific hazard is starting to be recognized within the industry and before OSHA promulgates specific safety standards for the hazard.

Administrative Duties

The previous paragraphs discussed the different ways in which employers were required to proactively manage the tasks their employees perform and the conditions under which they work. OSHA standards also require employees to comply with a number of administrative processes that help OSHA and the company's employees better understand the company's safety management. These processes include the company's record keeping, reporting, and docu-

ment posting requirements, which are set forth in 29 CFR 1904 and summarized below.

Employers must post the following information where they normally post notices to employees:

- copies of OSHA citations received within the past three days or until the cited violations are corrected, whichever is longer
- a summary of employees' work-related injuries and illnesses during the previous year
- the Job Safety and Health Protection workplace poster (OSHA 2203), which summarizes employees' health and safety rights and responsibilities.

Employers with over ten employees on the job are required to:

- maintain a log recording all injury and illness records
- make records accessible to employees and OSHA
- report each fatality and each accident that hospitalizes three or more employees.

SPECIFIC QUALIFICATIONS

The previous paragraphs indicate that the federal safety standards require employers to actively manage the occupational health and safety of their employees. The various duties discussed above relate to broad areas of responsibility found in the beginning of subpart C (General Safety & Health Provisions) of 29 CFR 1926. The vast majority of 29 CFR 1926, namely subparts D–Z, consist of very detailed and technical requirements associated with avoiding specific hazards or performing specific tasks. Some of these technical paragraphs explicitly require that individuals with specific qualifications or competencies be involved in ensuring safety management. Specific qualifications repeatedly referenced in the text include competent persons, qualified persons, and engineers.

■■ **Table 4** ■■

29 CFR 1926 Paragraphs that Refer to *Competent Person*

Topic	Paragraphs Referencing a Competent Person
General inspections	1926.20.b.2
Material handling	1926.251(a)
Welding	1926.354(a)
Scaffolding	1926.451.(f), 1926.454
Fall protection	1926.502*, 503(a)
Crane	1926.550*, 552(c)
Excavation	1926.651*, 1926.652*, Subpart P App A, App B
Lift slab operations	1926.705(i)
Structural steel	1926.753, 1926.754(d),
Assembly	1926.755(a), 1926.756(a)
Underground	1926.800*, 803(a)
Construction	1826, 850(a), 1926.852,
Demolition	1926.859(g)
Ladders	1926.1053(b), 1926.1060(a)
Asbestos	1926.1101*
Cadmium	1926.1127

* Denotes there are multiple references within this section.

Competent Person

1926.32(f) defines a *competent person* as, "One who is capable of identifying existing and predictable hazards in the surroundings or working conditions which are unsanitary, hazardous, or dangerous to employees, and who has authorization to take prompt corrective measures to eliminate them." As quoted earlier, 1926.20(b)(2) requires that the work be frequently inspected by competent persons. Table 4 lists other paragraphs that require a competent person be involved.

Qualified Person

1926.32(m) defines a *qualified person* as, "One who, by possession of a recognized degree, certificate, or professional standing, or who by extensive knowledge, training, and experience, has successfully demonstrated his ability to solve or resolve problems relating to the subject matter, the work, or the project." Table 5 lists the paragraphs that reference a qualified person.

■■ **Table 5** ■■

29 CFR 1926 Paragraphs that Refer to *Qualified Person*

Topic	Paragraphs Referencing a Qualified Person
Gases, vapors, fumes, dusts, and mists	1926.55(b)
Electrical, general requirements	1926.403(j)
Scaffolds	1926.451(d), (f), 1926.452*
Fall protection	1926.502*, Subpart M App C, App E
Cranes, derricks, hoists, elevators, and conveyors	1926.552*
Excavations	1926.650*, 1926.651*, 1926.652*, Subpart P, App B*, App F*
Lift slab operations	1926.705*
Steel erection	1926.752(2), 1926.757*, 1926.758(g), 1926.761(a), Subpart R App G
Blasting	1926.900

* Denotes there are multiple references within this section.

■ ■ **Table 6** ■ ■

**29 CFR 1926 Paragraphs that
Refer to *Engineers***

Topic	Paragraphs Referencing an Engineer
Scaffolding	1926.451(d), (f), .452*
Fall protection	Subpart M, App C, App E
Material hoists, personnel hoists, and elevators	1926.552*
Cranes and derricks	1926.550(a),(g)
Excavation	1926.651(i), 1926.652*, Subpart P App B, App D, App F
Cast-in-place concrete	1926.703(b)
Lift slab operations	1926.705(a)
Structural steel assembly	1926.755(b), 1926.756*, 1926.757(a)

* Denotes there are multiple references within this section.

Engineer

Although the OSHA text does not provide a definition of engineer or professional engineer in the Definitions section of subpart C (1926.32), a substantial number of paragraphs allow or mandate the involvement of an engineer. The text does not indicate whether such engineers are to be employees or consultants, or who should retain them. Table 6 lists the paragraphs that reference engineers.

CONCLUSION

Federally mandated construction safety standards are published in 29 CFR 1926 and portions of 29 CFR 1910. Both are frequently revised. These documents reference both mandatory and nonmandatory standards published by other organizations. Safety and operational managers seeking insights into how OSHA standards are drafted and enforced should review the OSHA Field Inspection Resource Manual (FIRM), formal OSHA letters of interpretation, and other sets of information available at www.osha.gov. A list of states with their own OSHA agencies and/or standards that must be at least as rigorous as federal standards has been provided.

In addition, excerpted are portions of the OSHA standards demonstrating that, contrary to common misperceptions, primary responsibility for the safety and health of construction workers rests with employers, not with a project's general contractor. An employer's key safety duties include:

- ensuring the workplace is free of hazards through regular inspections
- ensuring employees are trained in hazard avoidance and the proper use of equipment
- providing personal protective equipment
- using individuals with specialized safety qualifications, such as competent persons, qualified persons, and engineers
- performing administrative duties (e.g., posting mandated documents and submitting annual accident logs).

These excerpts and discussion provided in this chapter should make it clear that the minimum level of safety management mandated by the federal government cannot be attained by merely possessing specified documents or memorizing technical requirements proscribed within the documents. Rather, employers should be sufficiently familiar with the standards to effectively and proactively manage the health and safety of their employees. Effective safety and operational managers integrate the mandated standards into a safety-conscious culture supported by well-functioning organizational systems.

REFERENCES

American Institute of Architects, Inc. "Standard general conditions of the construction contract," AIA A201. Washington, DC: AIA, 1997.

American Society of Civil Engineers, American Consulting Engineering Council, and National Society of Professional Engineers. "Standard general conditions of the construction contract," EJCDC 1910-8, 1996.

Associated General Contractors of America, Inc. "Standard form of agreement and general conditions between owner and contractor," AGC Doc. No. 200. Alexandria, VA: AGC, 1999.

Toole, T. M. "Construction Site Safety Roles," *ASCE Journal of Construction Engineering and Management,* May/June 2002.

_____. "A Comparison Of Site Safety Policies of Construction Industry Trade Groups," *ASCE Practice Periodical in Structural Design and Construction,* May 2002.

U.S. Department of Labor, Occupational Safety and Health Administration. "Occupational safety and health standards for the construction industry," 29 CFR 1926. Washington, DC: U.S. Government Printing Office, 2002.

_____. "OSHA Field Inspection Reference Manual" (CPL 2.103), 29 CFR 1926. Washington, DC: U.S. Government Printing Office, 2001.

http://www.osha-slc.gov/Firm_osha_toc/Firm_toc_by_sect.html (November 14, 2002).

REVIEW EXERCISES

1. What is the primary federally mandated construction safety standard?
2. What is OSHA's publication, FIRM, and why should safety managers be familiar with it?
3. If a safety manager is not sure what a specific paragraph within 29 CFR 1926 means, how can he or she access documents that indicate how OSHA generally interprets this paragraph?
4. Does your state have its own construction safety standards? If so, how do these standards compare to the federal OSHA standards?
5. Identify and discuss the paragraphs in 29 CFR 1926 that indicate that the subcontractors, not the general contractor, have primary responsibility for the safety of their employees.
6. List five key duties that employers have regarding the safety and health of their employees.
7. Summarize the differences in OSHA's definitions and requirements for a competent person, a qualified person, and an engineer.

■■ PART IV ■■

Technical Issues in Construction

■■ **About the Author** ■■

FALL PROTECTION: COMPLIANCE AND SOLUTIONS
by J. Nigel Ellis, Ph.D., P.E., CPE, CSP

J. Nigel Ellis is President of DSC/Ellis Fall Safety Solutions with local service nationwide. His firm is committed to the possibility of reducing fall deaths by assessing unrecognized fall hazards and engineering solutions for their elimination or control, installing the systems under P.E. supervision, and training affected workers. His textbook, *Introduction to Fall Protection*, 3rd edition, is published by ASSE. Nigel is Chair of ANSI Z359.2 and a member of ANSI Z359.0 and Z359.1, A10.32, A14.3, and A1264.1. He is former president of the Lower Delaware Valley chapter of the American Society of Safety Engineers and former chairman of the Fall Protection Group, ISEA. He is a member of the National Safety Council Construction Division and the ASTM F13 Committee.

Nigel studied at the University of Manchester, England, for his B.S., M.S., and Ph.D. in the sciences. He has maintained his CSP (Safety) and CPE (Human Factors) peer certifications and is a Registered Professional Safety Engineer in California and Massachusetts.

CHAPTER 15

Fall Protection: Compliance and Solutions

LEARNING OBJECTIVES

- List the key places where workers are killed at work from falls.
- List the hierarchy of fall protection to set priorities for planning.
- Explain why holes are so deadly.
- Discuss how to watch out for hazards when using fall equipment.
- List the most cited OSHA fall standards.
- Discuss the responsibility of the parties in construction relating to fall hazards.
- Describe the root-cause principles of fall-incident investigation.

INTRODUCTION

This chapter on fall protection includes principles applying to all construction work. However, some construction areas have *particular* needs when it comes to fall protection. Different standards will apply in different fields (e.g., OSHA standards that address fall hazards for shipyard work are the most inclusive; MSHA standards are the most liberal). Nevertheless, all standards are minimum, and the purpose of fall protection is to determine any fall hazard that could reasonably do harm before it can. Adopting OSHA 5(a)(1) for all suspected fall hazards is the smartest way to address known hazards. When they do occur, emergencies in every size corporation should elicit a measured response, and employees should hold an all-level huddle to examine new hazards before proceeding.

The larger the construction project, the easier it is to achieve fall protection through adequate preplanning, having suppliers deliver steel and concrete sections with fall protection already included, and having more funds available to expeditiously correct hazards as they are identified.

DEFINITIONS

Terms commonly associated with the topic of fall protection include the following:

Access: the protected means of reaching the overhead work zone by climbing ladders or stairs, or using aerial lifts or scaffolds.

Fall: a sudden movement quickly downwards from a higher place, usually by accident (Macmillan); impacts resulting from falls have a high probability of producing multiple injuries that can be fatal.

Fall hazards: zones where a sudden drop or structural collapse is reasonably possible as a result of a misstep, tripping, slipping, or some other design flaw or human frailty. These must be addressed by OSHA, ANSI, and other voluntary standards, or otherwise be recognized by safety professionals established in the science of fall protection and statistics of fall fatalities, and/or registered structural engineers with a record of resolution of fall hazards. (*Note:* A chance of a fatal fall greater than one in a million is sufficient to create an urgent need for a solution that provides elimination or control, including addressing foreseeable behavior).

Fall protection: the means for elimination or effective control of fall hazards on a case-by-case basis. That means can be design, procedures, or equipment systems. Falls can occur from the same level, steps, a ladder, or from a height (i.e., over a trigger height of 4–6 ft). Falls can be caused by tripping, stepping back, stumbling, stepping over something, or slipping. Tripping is an interruption of the walking or stepping motion by an obstruction that causes the body to move off its center of gravity. For example, an overlapped plank on a scaffold presents a 1.5- to 2-inch tripping hazard that could send a worker through the guardrails; also a 3/4-inch edge on a plywood cover for a hole could cause very serious body reactions and resulting impacts. The solution is to create a sloped transition or use aluminum picks or a steel plate. Likewise, stepping on an empty pan in a stair section where concrete has not yet been poured creates a tripping hazard for the heel. The solution here is to add 2-inch × 10-inch wood blocks or a filler that fits tightly, or prevent use of the stairway. Shear or Nelson studs offer a perfect example of a defective walking surface [e.g., 5-inch-high studs with a 10-inch separation for 12-inch shoes (when shop-installed) can promote catastrophic tripping and falling during girder erection]. These are illegal in the OSHA regulations and various state DOT requirements; they must be field-installed after deck pans have been secured

Engineered fall protection: installed fall arrest equipment that has been included in a fall protection system and engineered by a registered structural engineer in accordance with manufacturer's instructions with anchor points and design drawings that are as-built and that is maintained and inspected as required by instructions listed on the drawings.

Fall prevention: the reasonable protection of persons from falls from heights by limiting falls to the same level; it deals with floors, walls, guardrails, and other physical barriers. (*Note:* In no way does caution/warning/danger tape qualify as a physical barrier. Such tape is used only for directing people away from a zone on a temporary basis where there is no threat of serious injury and death from falls if breached for any reason.)

Fall arrest: the process of slowing a fall to a stop; it deals with equipment designed to be competently installed for protection from injury in accidental falls as the fall occurs, dynamically stressing the walking/working structure, and before the structure permits fall prevention to be reasonably applied. Safety nets and harness systems supply fall arrest protection, and engineered, properly installed systems are the focus of what makes up an approved system. Harness systems are *active* in that they require personal physical attachment, whereas *passive* nets cover an opening or reach beyond an edge.

Personal fall arrest system: a documented system that has been tested to standards, incorporating a full body harness with features for the application, such as a frontal D-ring for climbing, if necessary, and a back D-ring for fall arrest, used with a connecting device such as a lanyard or self-retracting lanyard/lifeline or rope grab, and attached to an approved fixed anchor point (such as a D-ring or an anchorage connector strap or wire), or a cable (such as a horizontal lifeline), or a rail system by means of a snaphook or other connector. (*Note:* a harness and lanyard available to or issued to workers is not a fall protection system.)

Trigger height: the height exposure at which fall protection is needed. There is a simple way to determine a best-practice trigger height: it is any height at which a fall would produce serious injury or death. OSHA statistics show that only 1 percent of falls up to 6 feet result in death, whereas 9 percent of fall deaths occur at 10 feet or less. Thus 6 feet is a reasonable general construction (29 CFR 1926) fall protection trigger height—4 feet for general industry (29 CFR 1910). (Other trigger heights may apply, e.g., shipyard work [1915], steel erection and scaffolds [1926].)

Fall restraint: a temporary, rigged system that allows working on a walking surface so that the access to a fall-hazard zone cannot be entered. (*Note:* Since this is very difficult to accomplish successfully in construction geometries, fall arrest equipment should be used, perhaps considering a lower anchor point after the consequences are discussed.)

Work positioning: the use of equipment designed for the stress of leaning against or supporting a worker in a rigid chair or full body harness with seat strap so that the hands are free to do work while the worker remains exposed to a fall hazard during transfer or failure of the support system—examples are pole work, tree work, or form work. A rigid seat and/or back support is required for more than 5 minutes of exposure.

Rope access system: equipment with harnesses with frontal D-rings used for access and fall arrest in very short falls and where the capability exists for self-rescue or crew rescue when a rope-access, certified crew member is on the crew (*see* Figure 26).

FALL HAZARDS

Unprotected fall hazards have resulted in a steady rise of fatal falls over the past decade, including an approximately 15 percent rise in 2001 despite the increased availability of harnesses and lanyards.

Generally, the fall-hazard fatality toll comes from incidents involving ladders, roofs, and scaffolds, which account for approximately 80 percent of fall fatalities in the United States. More specifically, these fatalities are the result of inadequate ladder access above scaffold guardrails or roof edges; skylights and skylight/utility/smoke vent openings; or small and large floor/roof openings, with or without coverings that are not secured down and offer no way to ascertain the 6-ft safety range or a proper warning marker of any adequate cover.

The Cost of Falls

The workers' compensation pay-out multiplied by 2 to 20, depending on the organization or operation, accounts for the hidden costs of a serious fall injury (refer to the Business Roundtable's *Report A-3-1981*). Costs can average $400,000 for the work-

ers' compensation portion (based on one major carrier's records in 2001). The proactive and effective safety manager needs to know that management's understanding of these numbers from a productivity viewpoint can change its attitude from "Have we had an injury or OSHA citation?" to "Let's get started on fixing this hazard before we have a serious or fatal injury."

Planning for Fall Protection

A fall protection work plan is a study of potential fall hazards that is prepared before any phase of work begins. It highlights vague safety concepts and permits hazard awareness and analysis that can suggest defensive measures, which should be noted in writing for reference when the actual work is done, thereby integrating safety and operations. While fall protection work plans often are required by law in the western United States, they are gradually being incorporated into work processes nationally.

(*Note:* fall protection work-plan expectations are not fixed; revision is needed when the work method changes or occurs on a changed schedule, or when the fall protection system changes.)

	Passive	Active
Elimination	x	
Prevention		
Aerial lifts/platforms	x	
Parapets	x	
Guardrails	x	
Warning lines	x	
Fall Arrest		
Anchorage		x
Lifelines		x
Lanyard		x
Body support		x
Connecting hardware		x
Nets	x	x
Administrative Techniques		
Fall protection program	x	
Training	x	
Safety monitors	x	
Warnings	x	

■ ■ **Figure 1** ■ ■ **Hierarchy of fall protection**

What Is Continuous Fall Protection?

Continuous fall protection uses one or more of the following means:

Fall Prevention
Warning lines at least 6 ft from edge
Barriers, covers
Guardrails
Perimeter cables
Walls, fences
Platforms, buckets, aerial lifts
Scaffolds, planking

Fall Arrest
Safety nets
Fall arrest equipment system

■ ■ **Figure 2** ■ ■ **Maintaining fall protection throughout the work task**

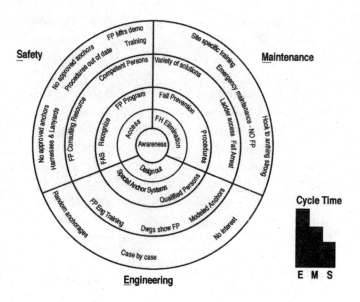

■ ■ **Figure 3** ■ ■ **Fall protection bullseye**

Hazard recognition by forecast and inspection consists of engineering deficiency observation and work-method evaluation. Too low a railing can be found easily. However, predicting worker shortcuts, even if not directly observed, is an important and more difficult task. For example, cutting a hole in the deck could be done from above or below, likewise the replacement of a particular fan; each has different hazards associated with it that can be determined in advance, and protection can be linked to training and possibly a mandatory work method. Where holes are cut late, stairs put in late, fixed ladders put in late and out of proper sequence, a huddle must be held with the general contractor (GC) present to reevaluate the hazards to various trades. This is when a fall protection work plan proves invaluable.

Fall protection can be classified as *passive* or *active* and should be considered using the hierarchy of fall protection (see Figures 1 and 2).

A fall protection bullseye (Figure 3) that encompasses engineering, maintenance, and safety as critical factors should be utilized on construction projects.

Fall protection design is an integral factor to the success of any project; therefore, the safety professional should analyze each fall hazard using the five-step process illustrated in Figure 4. A flowchart

of one fall protection system is also provided in Figure 5.

Sources of Fall Hazards

Currently, fatal and serious falls occur most often around:

- tower construction
- scaffold ladders and platforms
- holes
- skylights
- edges
- roofs
- elevator shafts
- ladder side rails
- decking and plywood
- perimeter cables.

TOWER CONSTRUCTION

Ten to twenty-five percent of construction fall fatalities occur during tower construction for cell phones and paging systems. This represents the highest death toll on construction sites.

Workers often believe they can switch employers at will and sometimes do not respect a "100% fall protection" policy. Foremen and lead climbers often do not feel duty-bound to direct peer climbers in

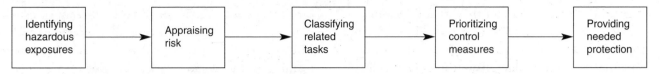

■■ **Figure 4** ■■ Analyzing a fall hazard takes these five steps

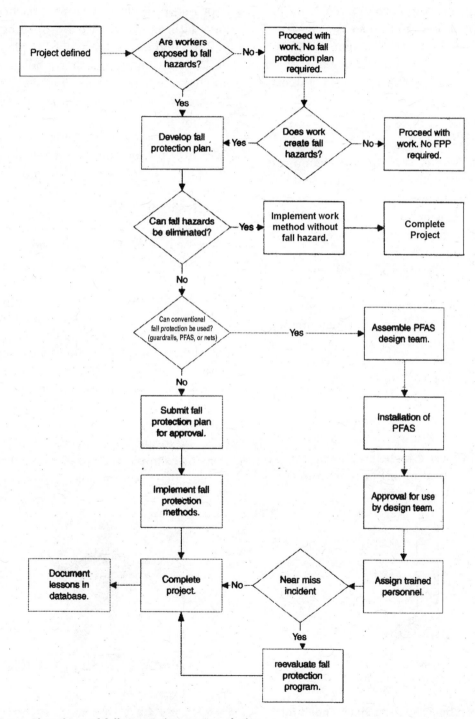

■■ **Figure 5** ■■ **Flowchart of fall protection system design**

how they work and whether they treat their work as an avocation by rappelling down the lifeline, one slip away from a free fall. More frequent and professional training is one answer, and maturing to rope access systems that permit descent with a deadman lever and fall protection is beginning to gain acceptance. Organizations currently requiring 4–8 hours of training or less should specify a classroom curriculum and field experience that will qualify employees for such work and determine the proper level of competence. (OSHA has a CPL to cover site inspections.)

SCAFFOLD LADDERS

When a scaffold worker is required to climb over guardrails, the scaffold ladder must be extended 3 feet above the guardrails or platform level and utilize swing gates. This helps avoid shifting the ladder and destabilizing balance while accessing a lower level. Fall protection anchorages can be made compatible, and scaffold suppliers should have tested and certified this attachment (see Figure 6).

Scaffold legs are stronger than rails or cross braces. Scaffold turnkey suppliers frequently warrant finished scaffolds for fall protection anchorage capacity. Limited access by agreement may be appropriate even if guardrails or decking is not complete, provided structural erection is complete.

HOLES

Holes of any size over 2 inches × 2 inches pose a danger (severe or fatal injuries occur from holes over 12 inches × 12 inches): Humans cannot estimate the danger from holes and floor openings and will approach to within 6 feet of an edge despite training not to do so. Work in an opening should be conducted with an aerial lift platform under the opening (conditioned upon a manufacturer-approved procedure) or a net attached to the bar joists sufficient to reduce any opening to 2 inches or less. If there is no aerial lift platform or net underneath, any approach to a hole cover is uncertain because the cover may not be secured and only upward pressure will determine that fact for nailed or screwed covers.

Nets attached to bar joists can cover holes too large for a plywood board but not large enough for guardrails. Even small holes with covers should have nets added to protect authorized work in the hole. No injury should be reasonably caused in the fall to the net, otherwise personal fall arrest systems must also be used. Crossing to bridge girders at cross braces requires net systems.

Holes disguised with loose boards or equipment over them are probably more hazardous than open holes because loose covers are scrounged and unconsciously lifted by walking up them and stepping into the opening. An engineer must approve a temporary cover with a drawing specifying the cover material, its dimensions, and its securing method (specification of bolts or nails), as well as its capability to sustain twice the load of equipment, materials, and torque from wheels of mobile equipment. (See the example in Figure 7.) Other safety options include guardrails for holes approaching 4 feet × 8

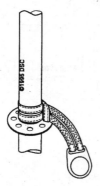

■■ **Figure 6** ■■ **Compatible attachment point for fall protection hardware**

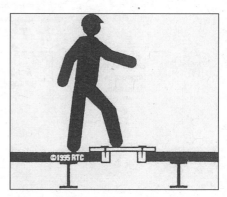

■■ **Figure 7** ■■ **Covers must be secured in a way that can be reasonably observed by workers**

feet or larger, or a personal fall arrest system with a designed anchor system with a permanent marker showing the type of approved attachment for compatibility and safe use. Walking into holes is an often-repeated incident that can produce death and serious injury because holes are not an open and obvious hazard. People inevitably walk, back up, or step into holes because the brain does not appreciate the danger when the victim is preoccupied (such as with work).

SKYLIGHTS

Serious falls can occur from skylights at any stage, from fully open to an unsecured, covered opening to plastic domes or corrugated translucent panels. The problem is the plastic design specification that crazes around the frame screw holes and leads unexpectedly to a sudden failure when a worker sits, stands, steps on, or backs into the dome design. For corrugated panels, weathering can turn galvanized steel panels yellow-brown, and the plastic sheet also turns yellow to brown. Crazing at stressed frame screw holes lies in wait for an unsuspecting worker. The results are usually fatal. No installation of a skylight should be made unless a compatible tested grille is available (see Figure 8).

EDGES

Open sides are best protected by guardrails. These can be either permanent, temporary, and/or removable. Edges seem to act like magnets for people, who often approach closer than 6 feet (i.e., enter a fall-hazard zone to view and communicate by line of sight). Caution tape is never a barrier against a serious or fatal fall-hazard exposure. Tape of any kind is only a guide for temporary personnel access; it should never mean a death sentence for a person who does not see it, walks through it, or ignores it. Tape does not meet OSHA requirements for a guardrail (200-lb strength) or a warning line (500-lb tensile strength) or a CAZ control line (200-lb tensile strength) placed 6 feet from an exposure when constantly patrolled by a monitor for leading edge, precast concrete erection, brick- or block-laying work, or an unprotected roof edge. Fall protection is required for railing installation (see Figure 9).

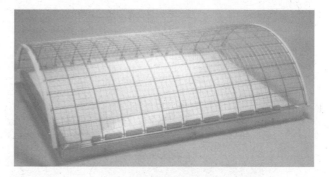

■ ■ **Figure 8** ■ ■ **Plastic skylights must be protected from catastrophic failure (standing, stepping, sitting, etc.) using grilles for structural support**

■ ■ **Figure 9** ■ ■ **High-rise apartment railing installation (web strap attached to rebar embedded in concrete floor pour may be removed after use by cutting)**

ROOFS

Parapets with a minimum height of 30 inches plus a width that must equal 48 inches or more are ideal edge protection for commercial roofs (see Figure 10). Temporary railings and, of course, permanent railings are the next best thing for edge protection. The use of restraint systems for single-point or horizontal travel is possible with a designed and engineered system that reasonably prevents fall hazards with less than 2 feet free fall.

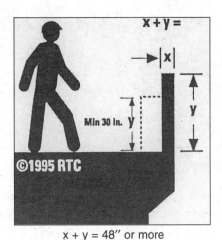

x + y = 48" or more

■ ■ Figure 10 ■ ■ Parapets can provide protection from falling if the sum of height and width exceeds 48 inches and the height is minimally 30 inches (proposed OSHA 1910 Subpart D)

■ ■ Figure 11 ■ ■ Horizontal grab bars can be gripped; the gate provides fall-back protection

ELEVATOR SHAFT HAZARDS

The self-closing elevator door is often left open, or no door may have been installed on any given floor, creating a lethal opening; a curious worker looks in and ends up down the shaft. Even dropping from an elevator car down to floor level a few feet below can be lethal because of the danger of falling back into the "hidden-from-view" open shaft.

Often, the 2 × 4 barricades are removed by a particular trade. The nets that are used early on in shafts are frequently pulled aside for hoisting and then typically left open, creating a potentially catastrophic fall hazard. Discipline must strictly be enforced to ensure that all nets and barricades are returned to their proper positions each time they are moved, removed, or left unsupervised. This means that inspections must be frequent.

LADDER SIDE RAILS

Holding the side rails of ladders instead of rungs has been established as a serious hazard based on the incidence of falls from walk-through ladders while workers descend from an upper level. Grab bars must be horizontal for proper holding power under dynamic fall conditions when the foot slips from the rung forcing the arms and hands to save the day with a reliable grip onto the extension side

rail with a suitable attachment (see Figures 11, 12, and 13). This also applies to job-made ladders with no place to gain a reliable grip when accessing to climb down. Only a horizontal grip is reliable. Fall arrest protection is required over 20 feet (lower the trigger height if special hazards exist); ladder cages are considered ineffective and outdated.

MANAGEMENT OF STEEL DECKING AND PLYWOOD

Workers should not have to walk more than one bay to reach loose decking or a pile. According to the Steel Deck Institute, the first piece of decking placed in its final position next to a bundle should be immediately tacked down to avoid stepping and sliding on the piece.

Cutting holes for subsequent trades has been strongly discouraged by industry member policy. Trade hole-cutting can be done from underneath using a scissor lift, but all final covering of holes, with or without curbs, must be done from the top side. Currently, holes are cut only before immediate use, and small nets attached to bar joists are required to ensure that covers have in fact been secured and that skylight and utility installation crews are protected. All temporary holes (less than

■ ■ **Figure 12** ■ ■ Always hold the rungs of any ladder to anticipate a foot slipping off, producing an intolerable dynamic load on the hand, which must act like a rope-grab device

■ ■ **Figure 13** ■ ■ Hands will come off the ladder side rails in any dynamic fall

42 in × 84 in for a 4-ft × 8-ft board) must be covered immediately with 3/4-inch plywood that has been adequately secured and marked "Danger! HOLE, Do Not Remove" to deter casual removal attempts. A specially marked and attached (red) cone is ideal for this application in place of low-attention spray paint. Plywood is usually found unnailed or not screwed down in fall-through cases.

After finishing with plywood, each subcontractor must be required to return decking to the general contractor's agreed storage areas. If there is no general contractor on the job, the contractor–manager must organize this procedure. After use, marked boards must be turned upside down before stacking to avoid confusion.

PERIMETER CABLES

If perimeter cables might possibly be used for fall arrest anchorages, they must be engineered as continuous horizontal lifelines. Washers cannot be used for intermediates because they become too brittle and break at low force—when stepping on the midcable, for instance.

FALL EQUIPMENT HAZARDS

The most common hazards workers encounter when they use fall equipment include:

- harnesses with leg straps undone
- lanyards not attached to the structure (not anchored)
- lanyards attached to improper anchorages or each other
- incompatible equipment
- damaged snaphooks with unreliable self-closing
- nonengineered anchorages
- use of a harness and lanyard alone
- use of a harness and lanyard to attach to climbing cables, rails, or other devices

Harnesses with leg straps undone: Without proper training, workers often take shortcuts; without close supervisor observation, workers may play a dangerous game, making it look like the worker is attached, when, in reality, the leg straps are not secured. It's a game that has cost many lives.

Lanyards not attached to the structure: Not being anchored is a no-win game that can be detected through close observation by supervisors; it warrants immediate discipline (see Figure 14).

Lanyards attached to improper anchorages, such as angle irons and I-beams that will cut rope and webbings in a dynamic fall (see Figure 15) or, alternatively, to weak vents or conduit, are promoted by a fundamental longstanding failure of management to address anchorage points that meet the

■■ **Figure 14** ■■ **Attachments must be compatible! The illustrated attachment can slide off and is highly unreliable. Anchorage points must be provided, usually consisting of an engineered cable.**

■■ **Figure 15** ■■ **Snaphook gate must close fully and no pressure is permitted on the gate**

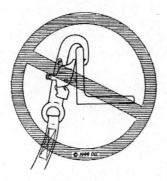

■■ **Figure 16** ■■ **No perching is allowed, as shown here!**

manufacturer's stated requirements for overhead attachment.

Incompatible equipment: Workers do not understand that equipment from different manufacturers has not been designed to be compatible. Pieces are not interchangeable since they have never been tested together (e.g., snaphooks, D-rings, eyebolts, webbing loops, cables). The result of using incompatible equipment can be a sudden disconnection, including some locking snaphooks. (See Figures 14, 15, and 16.)

The hook bowl must always sit at the bottom of an authorized eyebolt attachment. Burst-out can be avoided with D-bolt designs. Burst-out occurs when dynamic force and rope twist is applied, bringing the locked gate into contact with a lower portion of the eyebolt. (Snaphook gate strengths are only 220 lbs nose and 350 lbs side pressure, currently set in ANSI Z359.1. Gate strengths should probably be at least 1000 lbs to exceed the force-level limits of 900 lbf for shock absorbers.) This includes large snaphooks, wire hooks, and carabiners.

Damaged snaphooks that lead to unreliable self-closing: The misuse of snaphooks invariably puts side-load pressure on hook gates, damaging them and making automatic closure sometimes impossible—and perhaps manual closure too. All snaphooks should be pulled from service immediately if they do not close securely and automatically without manual assistance. The consequence of this roll-out or burst-out is sudden catastrophic failure to hold the load. (See Figures 16 and 17.)

If a locking ladder hook is used, the result can be the same because of the low force limits of gates, a result of current U.S. standards. Large snaphooks are intended for leaning hands-free (work positioning) by attachment to ladder rungs backed by additional fall protection. Other makeshift anchorages allow gates to be easily damaged, and hook reliability becomes questionable.

Nonengineered anchorages: Qualified architects and engineers are still not being asked to design anchorages for foreseeable attachment needs of fall protection equipment. Because of the opportunity for errors by workers, it is sometimes best to subcontract with a fall protection engineering firm

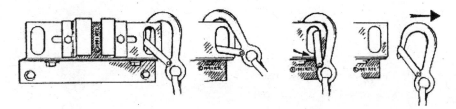

■ ■ **Figure 17** ■ ■ Disconnection sequence for nonlocking ladder hook shown from an elevator rail guide slotted hole

to provide an in-place engineered system or a template for suitable anchorages that a properly trained workforce can utilize.

Use of a harness and lanyard alone: A harness and lanyard in the pick-up truck or tool crib does not constitute a fall protection program. Implementing a *program* consists of an assessment of hazards, engineering an anchorage or railing, installing a proper system, and site-specific fall protection training.

Climbing cables and rails: Using a harness and lanyard to attach to a ladder-climbing cable or rail device is a natural choice if we consider construction training since 1994, which teaches workers to wear harnesses instead of belts. Frequently, both managers and workers fail to understand that the climbing systems are designed for 9-inch attachments and are tested for 18-inch falls, previously from body belts. In field operations, with a harness and a 6-foot lanyard attached between the legs or over the shoulder to the climbing sleeve, a free fall of up to 12 feet is produced and results in a body motion that keeps the fall-arrester sleeve in open mode, resulting in a catastrophic long fall to the ground. ANSI A14.3 was changed in 2002 to require harnesses (not belts) for climbing. OSHA should likewise change the 1926.1053 regulations as soon as possible to avoid the further possibility of loss of life.

HIERARCHY OF FALL PROTECTION

As illustrated in Figure 1, fall protection consists of:

- hazard elimination
- fall prevention
- fall arrest
- monitoring techniques

ELIMINATION

Fall protection safety professionals are committed first to fall hazard elimination, ahead of the need for worker access [e.g., column steel erection, using a removable shackle pin (Figure 18), or installing Nelson studs after the decking and railings have been installed by changing the sequence so that the fabricator merely provides the positioning location]. Once the need for access disappears because of a changed process, there is often no need for fall protection. Consider our example again, if splice plates are strengthened, two columns can be spliced on the ground and erected without placing an erector at risk—and the hazard is eliminated.

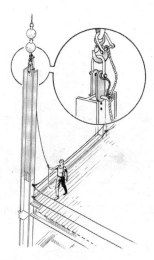

■ ■ **Figure 18** ■ ■ Nonsticking remote release of the crane hook—this method eliminates the hazard of an ironworker climbing the column to release the sling (or a stuck quick release). If the part does jam, the entire column must be returned to the ground.

FALL PREVENTION

Typically fall prevention is necessitated in situations where access is required, such as continuous platforms or stairs (both equipped with guardrails). For example work processes with continuous tripping hazards, such as pipes at knee level—often found in the natural gas industry—require a flat platform over these tripping obstructions, or that the pipes are buried in gravel. Aboveground guardrails (see Figure 19) always must be at or above the center of gravity (approximately 42 inches above foot level), including formwork and rebar operations. Working next to guardrails on a stepladder or other platform requires higher railings or positive fall protection. Restraint systems are designed with fall arrest systems often using lower anchorage heights and rigged so that no free fall is reasonably possible.

FALL ARREST SYSTEM

Fall arrest systems are either personal or communal in their intended fall protection role, that is, catching workers who fall and reducing their injuries primarily by reducing fall distance and adding shock absorption.

■ ■ **Figure 19** ■ ■ Temporary guardrails are especially useful on roofs with no permanent railings

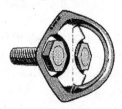

■ ■ **Figure 20** ■ ■ This eyebolt design eliminates perching hazards posed by traditional eyebolts

■ ■ **Figure 21** ■ ■ This anchorage connector is a choker design around a beam. Reliance may seem visually obvious, but is inappropriate if the beam is close to 100% allowable load, requiring careful engineering analysis of the structure.

Personal fall protection systems (PFAS) are for trained individuals who access by climbing structures and ladders, and who walk on unguarded surfaces or platforms, use aerial devices and scaffold structures that present edges, and may encounter equipment failure or collapse potential. These systems must be attached to the structure (that foreseeably will not fail) at an anchorage. (See the acceptable D-bolt design in Figure 20.)

Anchorages are typically D-bolts instead of eyebolts, welded attachments for D-rings, anchorage web connectors, or connectors compatible with the structure and approved for site-specific use (see Figures 21 and 22).

Key elements of a PFAS are a harness with features selected for the application and that has been suspense-tested by the worker, and a connection system that permits travel where the worker will go, limits fall distance, and provides shock-absorption within the geographical space available (see Figure 23). Structural engineer-prepared anchorage templates may fill the need for an anchorage in certain locations. There should be no doubt that the anchorage provided is proper and in the correct location for the necessary movement or work at hand, and that workers are willing to use it unconditionally.

Selection of systems is becoming more than a matter of choosing equipment from manufacturers; total systems selected by a qualified fall protec-

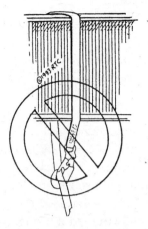

■ ■ **Figure 22** ■ ■ **Wrapping a lanyard is not a safe practice. In addition to sharp cutting edges of beams and angles, the snaphook body itself acts as a sharp edge when dynamically loaded.**

■ ■ **Figure 23** ■ ■ **Temporary horizontal lifeline**

■ ■ **Figure 24** ■ ■ **Work positioning must have fall arrest backup when feasible**

tion engineering provider instead of a manufacturer is required in order to properly assess work safety needs, including compatibility and engineering adequacy. Fall arrest systems assembled in the field must be compatible, meet the OSHA subpart M and Z359.1 standards through testing, and be labeled in accordance with standards.

Unlike elimination or other controls, fall arrest and related systems require workers to attach themselves to the systems. Accordingly, workers must be observed for compliance regularly since the first time they learn about the nature of a fall hazard is when they experience the consequence, which is unacceptable. Shortcuts will occur that destroy the system integrity unless observation is rigorous.

Work-positioning systems are harnesses, lanyards, and lines designed to allow the worker to lean and have free use of the hands while in a fall-hazard-exposure zone. Almost all work-positioning systems must have additional backup fall arrest systems because of movement and wear and tear, (see Figure 24). All nonmanufactured work-positioning systems and fall arrest systems must be tested before use in a recognized application.

Safety nets are used when personal fall arrest systems are impractical and greater passive protection can be applied than individual harness systems. Communal fall protection consists of nets in lieu of platforms (see Figure 25). Their use for stair openings, skylight/utility openings, edge protection, and temporary gaps is very practical.

Key features of nets are the engineered (snaphook or bracket) cable attachments to the structure for easy movement. Pit nets are used in general industry following construction for protection of workers and the public alike. Leaving floor openings and holes uncovered at any time is unacceptable because workers within a few steps of the openings are usually preoccupied with their tasks, and a fall can have severe consequences. Manning a hole is only permitted for a few minutes until the opening can be guarded.

MONITORING TECHNIQUES

Workers foreseeably exposed to fall hazards require vigilant observation. In occasional situations where fall arrest systems are professionally deemed not

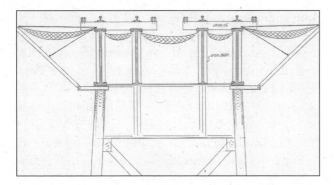

■ ■ **Figure 25** ■ ■ **Net plan for railroad bridge**

practical, a monitoring system may be approved for certain industries, but only if competent personnel are available to closely watch workers' actions in fall-hazard zones without distraction and verbally warn them of imminent stepping or balance hazards. OSHA guidelines for a designated, trained, and visually identifiable monitor include same-level presence, located within earshot, eight or less workers to observe, and no distractions. In almost every case this means the monitor must have no other duties.

THE FALL PROTECTION PROCESS

A capable team is required to undertake the responsibility for providing fall protection. Their duties include:

- assessment
- engineering
- supervised installation
- worker training
- observation of work
- inspection
- training the trainer.

ASSESSMENT

A team of competent persons is required to frequently assess the work areas to document potential fall hazards. Team members typically have a CSP and/or a P.E. safety qualification and are skilled in the field of fall protection.

ENGINEERING

The competent person on site will request that a *qualified person* provide an effective anchor point or design a horizontal lifeline system. The design drawing accounts for the dynamic forces of a fall. This person has at least a B.S. in Engineering and several years of facility engineering experience, or a P.E. (in Structural Engineering), and frequently works with a competent person. A steel erection superintendent often substitutes for a qualified person structural engineer based on his or her experience and ability to read drawings and manuals to identify where welding and anchorage connectors, such as straps, are suitable according to manufacturer's instructions. Where any doubt exists regarding an anchorage, a qualified person skilled in fall protection engineering must be used, including for all cable systems and horizontal lifeline anchorage designs and installations where sag and deceleration must be determined.

SUPERVISED INSTALLATION

The construction crew follows the prints for proper installation, and a registered structural P.E. monitors the work and handles any questions or ongoing modifications as necessary.

TRAINING

Workers are to be trained in company-recognized, site-specific fall hazards; shown the adequate fall protection controls available; and are required to assemble the systems—including any new system. The process takes one day in the beginning with updates and refresher training as needed. In a properly supervised environment, workers agree to hold each other responsible for continued proper use, typically under skilled guidance from qualified fall protection service companies. Management provides the support for proper use of equipment after an engineering survey and remediation has taken place. A harness, when worn properly and part of a competent system, is designed to prevent a person from hitting the ground; workers must have a competent anchor available, one that allows movement even when the worker is attached and anchored. No fall scenario should produce lifeline sag sufficient for a worker to hit the ground or a foreseeable obstruction. Proper harness donning techniques demonstrated, practiced, and repeated frequently, should eliminate any upside-down, back-to-front, and partial attachment mistakes. Training workers with a 1- to 5-hour general awareness program is not adequate where a site-specific training program is required. Training workers does not mean putting a one-time training record into the file for a future inspection.

OBSERVATION OF WORK

Observing workers on site must be regular and unannounced, with the opportunity to reward employees for using proper techniques. Photographs are an irreplaceable source of recording hazards that can be analyzed to help move the program ahead. The classroom retraining of deficient workers and foremen is critical to the process. Learning, application, and commitment are the goals sought.

INSPECTION

Who is qualified to inspect fall protection equipment? The answer has two parts. First, the worker is asked to inspect before each use for apparent defects such as observable cuts, nicks, burns, and holes, or damaged, lost, or nonfunctional parts. The competent person also inspects equipment periodically, considering more detailed issues such as abrasion, distortion, chemical damage, general condition, color of parts, subtle damage, snaphooks that do not close properly, damaged connections, or lack of labels and warnings. People with experience on many types of equipment can be found in independent fall protection service organizations committed to fall protection safety and work analysis.

TRAINING THE TRAINER

Who trains the trainer? Answer: a professional training company with qualified instructors that specializes in fall protection training. It may take several days longer than competent-person training to complete. This type of training is done in two parts: the principles of the technical training that explain the recognition of fall hazards and fall hazard conse-

quences is followed by a session on adult-audience presentation skills with a before and after videotape for the trainer's analysis. The object is to be relevant to the work and address the hazards and the solutions available in a way that is interesting, usable, and testable. Observation of the trainer at work periodically is necessary for proper results.

HISTORY OF FALL PROTECTION

Fall protection is the process of avoiding foreseeable injurious falls. Attention was focused on falls from heights in the 1920s after window cleaners fell from swinging scaffolds close to newspaper reporter locations. Later it was tank and sewer falls and rescues that led to the implementation of harness and extrication gear.

Serious or fatal suspension in a belt was first recognized by the French who instituted regulations in that country, while human testing by Brinkley and others in 1987 led to prohibition of the belt for construction in the United States in 1998.

Personal fall arrest systems became known as such following publication of the ANSI Z359.1 standard in 1992. Full body harnesses became mandatory for construction in the United States in 1998. Shock-absorption devices have been mandated since 1995.

The Purpose of Fall Protection

Fall protection is a backup system that can protect workers from the effects of gravity if they lose their balance where no guardrails are provided, or during possible structural collapse. Protection from collapse can exist when components of the lifeline system are independent of the collapsing structure. The system must include the timely rescue of fallen workers, such as from attendant-monitored confined spaces, whether permit required or nonpermit required.

Standards for Fall Protection

OSHA AND CONSENSUS STANDARDS

Since 1971 OSHA standards have been developed, producing the protective 1926.500, Subpart M (last revision 1994). Nonconstruction standards that also apply to construction work on site include: Longshoring, 1918; Marine Terminals, 1917; Shipyards, 1915; and the General Industry proposals, 1910.23-32 and 1910.128-131 (April 10, 1990), which are due for reassessment in 2003.

ANSI has developed a number of consensus standards: ANSI Z359.1(R1999) is the most complete voluntary standard in general industry; its guidelines are followed by fall equipment manufacturers. ANSI A10.14-1975 for construction gave way to A10.14-1991, which was withdrawn in 2002. The replacement standard, A10.32, has not been issued, following inactivity since 1998 due to a quandary over snaphook gate strength and both the apparent inability of makers to supply 1000-pounds gate strength in some larger connectors and the unwillingness of some utilities to use the locking gate uniformly. Users should demand snaphooks whose parts will exceed the strength/force of fall arrest shock absorption and that are as high strength as possible to help avoid gate damage and subsequent hook failure.

FREQUENTLY CITED OSHA SECTIONS

The following are overwhelmingly the most commonly issued citations among the majority of building and specialty contractors:

- 1926.451/452/453/454—scaffolds, including aerial platforms
- 1926.501/502/503—fall protection
- 1926. 1050/1051/1052/1053/1060—stairs and ladders
- 1926.20/21—general requirements for programs involving frequent inspections and training of exposed workers by a competent person. These are supplemented by Injury and Illness Prevention Programs in the western states, used to identify hazards and propose solutions ahead of the hazard incurrence, and can be required by contract in other states.

Other important standards include: 1926.25—housekeeping to reduce tripping hazards, as well as 1926.200/203—accident prevention signs and tags (and as updated in the ANSI Z535 series standards). A14.1-5 standards are frequently used for construction ladder use in addition to, or as an alternative to, the federal regulations. Walk-through ladders must have horizontal grab bars on the extension rails to avoid hands sliding if the feet slip off.

Trigger Heights

Left to their own experience, workers begin to feel concern for heights in the 12- to 15-foot range. The most extreme discomfort is felt at approximately 30 feet. However, well over half of all fall deaths occur from less than 30 feet so OSHA has imposed minimal standards on trigger heights. (Only 1% of fall deaths occur up to a 6-ft height, but 10% occur at 10 ft.)

Trigger heights are those heights or distances where fall protection or elimination of a fall hazard is required by OSHA and other codes.

- 1/2-inch floors—building codes and ADA regarding tripping hazards with 90-degree lips. Edges should be angled or beveled to reduce risk. Overlapped scaffold planks present a 1.5-inch tripping hazard unless they are treated or replaced with aluminum planks or pans.
- 3/16-inch stair riser difference in height—consistency of temporary and permanent stairs. Stair pans with a 3/16-inch or more lip cause falls; fill completely with 2-inch × 6-inch or 2-inch × 8-inch as a rule, due to heel tripping hazard before being filled with concrete.
- 4-foot general industry edges; 6-foot construction edges, unless falling into machinery or where surfaces might collapse, in which case there is zero distance tolerance.
- 10-foot supported scaffolds—a competent person must prepare a plan following the fall hazard assessment.
- 15-foot steel erectors—bolt-up crews are required to have fall protection.
- 30-foot steel erector connectors (personal fall protection must be available for use between 15–30 ft inside or outside the structure).

In the face of OSHA's lack of complete regulation of all industry workplaces, the OSHA 5(a)(1) requirement for maintaining a hazard-free work site calls for fall protection where any fall hazard exists. Usually a concensus standard will be referenced, such as Z359.1 for fall protection equipment violations. OSHA citations typically reference the alternative to include other possible regulations. For example, construction regulations for painting fall-hazard violations are backed up by general industry regulations.

Same-level falls under OSHA refer to any change in elevation with an opening greater than 2 inches (minimum dimension), with no depth specified. OSHA calls for adequate housekeeping to prevent tripping hazards, such as removing hoses that stretch across fall zones.

CONSTRUCTION METHODS

Types of Construction

Wood, used mostly in residential occupancies, is subject to a CPL inspection protocol interpretation of OSHA regulations, except in Hawaii, or when an owner or general contractor has tighter rules. Framing, decking, trusses, sheeting, and roofing can have sequences that promote proper fall protection, especially for holes and edges, including temporary guardrails and fall and restraint systems.

Steel construction: the fabricator and the steel erector must follow the agreed contract wording with the general contractor for structural steel erection and decking, and for cutting holes; copies of the contract MUST be read and understood by safety supervisors, and any requirements discussed with the steel erector. Anchorage brackets can be installed on beams and raised. Anchorages need to be over shoulder height. Beam-sliding devices underfoot do not achieve a free fall less than 6 feet and pose a very high chance of a collision with the structure beneath when a fall of 11–12 feet occurs and there is no simple method of rescue while a worker is suspended; cooning use may be feasible unless the worker has to stand on the top flange or there is a change in flange width unknown to the worker. Use of anchorage connector straps can assist when a qualified engineer has approved locations and proper methods for attachment. Metal build-

ing manufacturers need to design fall anchorages into their structures and provide netting for holding insulation and falling roofers.

Concrete construction could be forms, piers, columns, floors, and temporary structures, including scaffolds. Guardrails can be installed into sockets after passage of concrete members; these do not need to be at the edge, but can be 6 feet or more back.

When using *suspended scaffolds and bosun chairs* for glass and caulking work, davits need to have instructions for use and must have independent fall protection (one anchor per worker), usually set back in the roof, and an edge protector against lifeline chafe. Where multilevel platforms are used, the scaffold must have backup protection, and workers should be attached to horizontal lines to provide protection from sliding collisions when the scaffold tips at one end. Wire scaffolds used in ship painting are to be addressed as a guidance document in 2003 by OSHA.

Supported scaffolds must be built to manufacturer-tested assemblies, and fall protection should be applied to ladder access, including swing gates on each dismount/access level. *Underhung scaffolds* must be physically prevented from collapse due to tube and coupler connections slipping apart under unexpected concentrated loads. Rope access techniques are being considered for this type of work in Australia (see Figure 26).

■ ■ **Figure 26** ■ ■ **Example of rope access system**

■ ■ **Figure 27** ■ ■   **Harness and lifeline system facing uphill**

■ ■ **Figure 28** ■ ■   **Harness and lifeline system facing downhill**

Aerial platforms need to be checked for proper and complete maintenance history logs before rental to avoid unforeseen possibility of collapse, and fall protection should be applied on all types due to over-reaching hazards. Also, prevention of stepping up to the midrail is required; grating, plates, or other means of deterrence should be used as needed. Access to all buckets on the vehicle must be free of fall hazards. For any aerial electrical hazards in the area, a plastic bucket must be used. All aerial buckets used near power lines need to be insulated for the actual rated voltage; wearing insulated gloves alone will not provide sufficient protection.

Roofing for all kinds of roofs requires special planning to incorporate proper and adequate fall protection that considers stepping into holes or off of edges and falling from gable edges and eaves. Repairs to roofs must be preceded by a recent engi-

neering survey to avoid collapse during access on a roof and subsequent work there. Temporary railings are available and slideguards can provide protection up to an 8 in 12 slope, where positive fall protection becomes mandatory through ridge brackets and harness/lifeline systems (see Figures 27 and 28). Commercial roofs typically require hatches, and the engineer should specify permanent horizontal grab bars on the roof for ascending and descending into the opening.

FALL PROTECTION WORK PLANS

Work plans for fall protection are already required in the Western states. Proper advance planning for hazards should be developed in writing. This distinguishes a submittal to the general contractor more than a generic outline of responsibilities with undefined site-specific circumstances. It is not adequate for a smaller contractor–owner to tell his work crews that he will provide any equipment they need for safety, and that all they have to do is call him on their cell phone; workers typically do what is possible, not what is safe. An assessment is required by a competent person, from which a written work plan is then created.

Subcontractor Monitoring

The general contractor must continually monitor the construction project to sufficiently support any subcontractors' plans for fall protection safety and to ensure that the work plan is more than a "paper" plan. This means that the general contractor needs to do more than simply file the subcontractors' submissions.

Approved Anchorages

It is a nondelegable duty of the general contractor under the 1926.16 Rules of Construction—Multiemployer Work Site Rules—to make sure that only approved anchorages are used on the construction site. Usually, these are also required by the contract between the owner and the general contractor.

Training Requirements

Toolbox meetings explaining the need for fall protection but without specific worker interaction are likely to be ineffective. Training must focus on why workers do not always follow fall protection procedures even when they have been properly applied to properly installed systems. Adequate observation must determine whether any worker needs to be retrained or quickly replaced. Too many times workers will attach fall protection equipment to a structure only when exhortations are received from a foreman. Yet the moment that worker must move, he often detaches and may stay that way. Unless behavioral techniques are introduced by a skilled trainer and then used, with workers accepting their mutual safety responsibility, a worker—or even a foreman—often will not challenge a work procedure or technique that he or she knows is dangerous.

Safety meetings are intended to educate workers in actual site-specific hazards and solutions available from management. Canned programs should be used only for general or refresher training. Hazard identification opportunities in safety toolbox meetings are *not* meant for workers to educate supervisors about potential hazards, but are the realm of skilled safety professionals and structural engineers knowledgeable in the field, who have taken a walk-through prior to such training, and frequently repeat the process as the work-site changes.

Responsibilities of the Competent Person, Qualified Person, Professional Engineer, and Worker

The competent person:

1. Must read all the contract safety provisions and must read the OSHA requirements and interpretations; is responsible for site-specific training and retraining; observes work procedures with a right to stop the work; makes periodic equipment inspections according to manufacturers' instructions.
2. Knowledgeable in how the equipment is being used, is responsible for the inspection of all equipment. For example, tower erectors use 80- to 90-lb harnesses compared to the 5-lb purchased harness because of adding tools and bolts for their remote tasks.
3. Performs visual observation of the work to determine if the training is being followed and what retraining is necessary.
4. Has a knowledge of the safety manual: the standards for the organization need to be adapted to the site-specific conditions.
5. Is responsible for design of equipment and design of interface of equipment concepts with the site-specific work structure.
6. Monitors work in locations where it is infeasible to use positive protection; they must closely observe workers without distraction. OSHA has specific requirements for monitoring in limited types of work.

Qualified persons for the site must read, understand, and apply 1926.501(a)(2) for walking surface reliability, and 1926.850(a) for demolition or renovations. They will typically have mechanical or civil engineering degrees with strong structural engineering skills, be knowledgeable in the practices of fall protection engineering and, particularly, the dynamic force relationships and dynamic line sag. It is possible that the qualified person can be experienced and knowledgeable without having a formal education (such as steel erection superintendents); they can make limited judgments, not including compatibility of manufactured fall equipment parts, cable systems, dynamic load, and dynamic sag. Most fall protection systems require the expertise of a qualified person. For example, horizontal lifeline systems are being devised or purchased from manufacturers and installed frequently by local contractors, often without qualified-person oversight of anchor-point engineering and testing, 1926.502(d)(8). The installation of fall control equipment needs to incorporate structural engineering drawings of the installation and requires field supervision of the contractor's employees. A registered professional engineer in structural engineering, skilled in fall protection engineering and with a demonstrated ability to solve or re-solve engineering fall protection questions in the workplace, is a basic requirement for outside contract engineers retained to supervise horizontal lifeline designs, installations, and usage as a qualified person. If the local on-site contractor is allowed to be his own qualified person, no drawings will be available and no compliant system will be installed relative to dynamic loads and clearances. This could have repercussions if there is an injury, complaint, or question about the validity of the system and no documentation is available. Smaller contractors cannot provide the stamp of a registered professional engineer skilled in fall protection engineering on their work to support the host organization's safety program, but can be supervised by a knowledgeable structural professional engineer.

A worker's responsibility is to be reasonably careful to the limit of his/her safety training knowledge (a worker is not paid specifically to watch his or her step), in accordance with the proper rate of work and the known location of work, and to apply what he/she has learned from site-specific training that has been followed by a bona fide test, a successful grade, and adequate retraining, if necessary, as well as regular safety meetings and personal coaching.

MISCONDUCT

To gauge whether a worker has or has not acted consistent with OSHA regulations (a part of the training program), the misconduct rule is best described from the OSHA *Field Inspection Reference Manual* (*FIRM*):

1. The employer has established work rules designed to prevent the violation.
2. The employer has adequately communicated these rules to employees.
3. The employer has taken steps to discover violations.
4. The employer has enforced the rules when violations have been found and records are kept in employee files.

INSUBORDINATION

An employee is insubordinate if he/she refuses a proper order. An order is proper if it can be proven that it:
  a. is a directive, not simply a request, and the employee understood the instruction.

  b. does not put the employee at more risk than the job typically requires (a mason employee could not justify refusing to go on a metal building construction site just because chimney erection is risky, but might refuse to work if he is directed to go onto the icy roof of the building under construction to lay deck panels without fall protection or training), and is in line with the duties of the employee.
  c. does not violate the law.
  d. does not violate the employee's religious beliefs.
  e. the employee had no good reason to refuse. For example, an employee may refuse to work without proper and adequate protection and where he believes he has inadequate training, provided he is capable of exercising the choice of his own free will and without an environment of employer discrimination.

Preplanning Fall Protection

Fall protection needs should be established prior to bid time and amended throughout the project. Some elements include:

- *Multiemployer work-site* responsibility and communication procedures, with general contractor or contract administrator in control of safety, regardless of contract wording.
- *Subcontractor liability* management, including possible 100% observation for unknown subcontractors.
- *Steel erection* contract assumptions checked concerning fall protection over the 6-foot trigger height, exceeding OSHA minimum requirements—more commonly found in contracts with the general contractor today.
- *Scaffolding erection* completeness, including ladders, platforms, and tagging (colored tags and clear labeling); methods established for fall protection at 10-or-more-feet height exposure, including ladder fall protection.
- *Scaffold designs* meet manufacturer requirements for assembly and use, and workers have fall protection during such assembly and use as provided by a competent person

to supplement manufacturer instructions if necessary.

- *Crane* boom and jib attachments and maintenance procedures, and use of crane-suspended work platforms. Tower crane jumping and gin-pole-use procedures.
- *Electrical safety* access and insulated fall protection anchors when any possible boom extension with any bucket material or personal projection could extend into a live electrical zone.
- *Excavation-pit-edge* setbacks consistent with soil sloping and benching, location of the dirt pile to permit safe access without danger of overturning the concrete delivery truck.
- *Power-line* awareness and choice of insulated systems within proper clearance of a power line, treating all lines as energized until otherwise proven, reducing metal content of harnesses and device attachments used in aerial lifts.
- *Demolition,* including partial renovation, requires a professional engineer's report to assure roof and floors will support workers before any access is made.
- *Concrete:* approved shoring use as an anchorage, rebar structures, and forms relevant to guardrail height. (See Figure 29.)

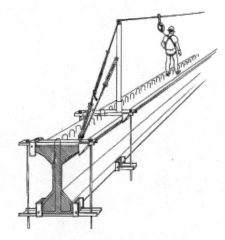

■ ■ **Figure 30** ■ ■ **Concrete or steel girder erection requires engineered lifeline systems that have been tested by the manufacturer. A space of up to 9 feet can allow for angel-wing access but still allow two-lanyard continuous attachment within reasonable reach and move to complete the girder lifeline. Where rebar or Nelson studs are too close for reasonable walking safety, studs *must* be field-welded.**

- *Bridges:* continuous fall protection for girder erection walking and splicing, bridging, decking, edges, and removal of plywood platforms after stripping. Lifeline anchor posts span the girder and are not a C-clamp design that can disconnect under dynamic conditions (see Figure 30).
- *Confined spaces:* proper retrieval systems incorporated into lifeline systems and used after venting is deemed adequate through proper instrumentation.
- *Residential construction:* All states except Hawaii adopted the STD 3-0.1A on residential construction, which provides only for evidence of training for at-risk employees. Hawaii maintained the protections of 1926.502(b)(13) and (k). Owners and general contractors may require hazard control by contract. Roof-decking guardrails and first-floor guardrail systems are available. See Figure 31 for an example of fall protection during truss erection.
- *Tower construction:* The 1926.104/105 regulations apply, although most firms that construct

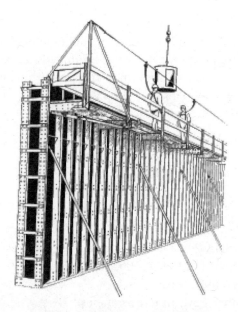

■ ■ **Figure 29** ■ ■ **Platforms with lifelines can be designed not to interfere with the work process**

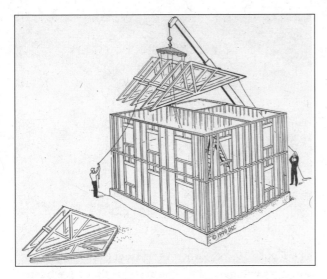

■ ■ **Figure 31** ■ ■ **A spreader bar is often used to spread the load among all truss members. Alternatively, an automatic spacing device can position each truss to reduce exposure.**

telecommunication and TV towers follow OSHA construction regulations and the enforcement guideline CPL 2-1.29. The IEEE standard P-1307 is also used for transmission tower construction, but has exemptions for qualified climbers. The use of rope access techniques with a high skill level is becoming more popular worldwide for access, crew rescue, and shorter falls.

■ *Emergency procedures:* the right and obligation to stop the work for unusual incidents and hold a huddle (an impromptu meeting that includes the general contractor and senior construction officials to determine the best way to resolve a situation when hazards are predictable). For example, after bringing in a pipe section for a heat exchanger under construction, the crane's headache ball becomes lodged in the steel framework—a huddle may be needed.

In summary, contractor bids can be provided with a variety of safety options to present to the owner or general contractor to justify total bid price. Your safety options can protect the owner in most states for minimal cost. It is in the general con-

tractor's interests to provide more safety, and the subcontractor must explain why.

Hazards that are recognized in advance of a phase of work can then be addressed in writing in a work plan. The plan must provide details of the hazards identified and the controls to be applied and presented to the supervising contractor.

PHILOSOPHY OF FALL PROTECTION

Fall protection is simply the result of ideas about keeping people who work in construction from losing their balance so they can avoid the possible injuries from impact or impalement that can result from a fall on the job. There are many options, and workers need to understand clearly what their fall protection choices are for a recognized hazard, including specific anchorage points. This must be for any foreseeable place a worker may reasonably go in carrying out his or her job duties (see Figures 32 and 33). Fall protection planning is *not optional.* It is a sophisticated business, and owners cannot expect workers to recognize and evaluate hazards. Fall arrest systems are not a matter of workers connecting components of fall equipment together to provide convenient reach. Systems that suit a work practice must be designed in advance by a competent person and engineered by a qualified person to avoid capacity overload. At the very least, the application must be tested and documented sufficiently to allow repeat testing or analysis. If we believe the motto of some construction companies, "If we cannot do it safely, then we won't do it," then we need a new paradigm on fall hazards.

INFORMATION SOURCES

Anyone who needs information about fall protection can get it chiefly from three sources:

■ OSHA 29 CFR 1926 standards, especially 1226.20, 21, 451, 500, 760, 1050, and 1060 (go to www.osha.gov)
■ other government agencies
■ incident investigation.

■■ Figure 32 ■■ A dormant fall hazard

■■ Figure 33 ■■ With the presence of a preoccupied worker, the hazard suddenly becomes armed, shortly to become active (irreversible) when the worker steps into the 3-ft × 3-ft unguarded ladder opening

OSHA

Occupational Safety & Health Administration regulations found in 29 CFR 1926 are for employers who are building, erecting, or repairing on a construction project, or who are performing temporary work such as industrial painting. Where no construction regulation is available, the OSHA 1910 General Industry regulations may be effective. If neither apply, then the OSHA Act 5(a)(1) duty to provide a safe workplace applies and may be en-

forced for a recognized hazard that is not controlled or eliminated.

Instruction STDs and CPLs from the OSHA Web site can be helpful, and you should always check for a letter of interpretation or a compliance directive on the regulation being investigated

Other Government Agencies

Familiarity with the jurisdictional areas of the agencies on the following list would be beneficial.

1. *Department of Energy/National Radiation Commission:* OSHA has the right to inspect sites if compliance officers have adequate radiation training.
2. *The Mine Safety & Health Administration:* MSHA has jurisdiction over any property with a mine, including a gravel pit; all workers are trained as miners unless OSHA/MSHA have a local agreement for jurisdiction.
3. *Department of Agriculture:* Safety regulations have been stalled since 1975 for farm employees. OSHA general industry regulations are generally followed, except for construction activity by contractors, in which case the OSHA construction regulations are used.
4. *The Department of Defense:* DOD has jurisdiction over its facilities; its regulations mirror those of OSHA.
5. *U.S. Army Corps of Engineers:* Many federal installation construction contracts are supervised by the Corps. OSHA inspects contractors hired by the Corps of Engineers who otherwise mandatorily use the EM-385-1-1 standard (latest, 2003) by contract. It is very similar to OSHA but has greater detail in some areas of construction.
6. *Department of Transportation/Federal Roads Administration:* Bridge work over 14 feet requires fall protection.
7. *Department of Transportation/U.S. Coast Guard:* Maritime work beyond U.S. territorial waters requires the use of coast guard regulations that are a version of A10.14-1975.

8. ***Department of Transportation/Federal Aviation Administration:*** Most sites employ OSHA requirements.

9. ***The states:*** The following states have their own plans, which must be equal to or may exceed federal OSHA requirements (covering all areas except for longshoring, maritime, and shipbuilding/repair): California, Oregon, Washington, Hawaii. These states have written requirements for Injury and Illness Prevention Plans (IIPP). There are currently 26 states and territories that have state OSHA plans.

Incident Investigation

Progress in hazard recognition requires integrity and thoroughness. Procedures for properly investigating a fall incident incurred by *any person* at the work site, with the objective of understanding the root cause, are as follows. (*Note:* This investigation must include *all* relevant subcontractor employees, owner representatives, independent agents of any kind, and supplier delivery persons, notwithstanding any limited insurance requirements.)

1. After calling for assistance, adequately photograph the entire area without moving anything or adding anything to the scene, including the victim's position.

2. Close off the area with tape and place security guards to prevent entry or exit of personnel who are not documented and listed as present in the work area at the time of the incident.

3. An investigation should be undertaken by police or other independent agency.

4. Statements should be taken from everyone who was a witness or present in the work area.

5. The safety manager should notice and document such things as lighting, noise, visibility, weather, equipment present, temporary conditions, status and progress of work, work method.

6. Organize incident reports completed with factual information.

7. Analyze all data, including programs, procedures, planning, records, training and contract agreements, to conclude what steps may reasonably prevent a recurrence, without placing blame or responsibility.

8. A report is issued to relevant departments and top managers recommending organization changes to reduce the chance of a repeat incident and, except for the facts, without regard to the victim's alleged condition or behavior.

9. Comply with local and federal reporting requirements (e.g., fatalities must be reported within 8 hours to OSHA's federal office or to the state OSHA).

REFERENCES

Construction Safety and Health Association of Ontario. *Construction Health and Safety Manual.* Toronto, Ontario: CSAO (www.csao.org).

Ellis, J. Nigel. *Introduction to Fall Protection*, 3rd ed. Des Plaines, IL: ASSE, 2001 (www.asse.org).

Forrest, Alan. *International Working at Height Handbook.* For rope access technician fall protection: incorporating IRATA levels 1, 2, & 3. Aberdeen Scotland: North Sea Lifting, 2001 (www.nsl-aberdeen.com).

Gagnet, Grace Drennan. *Fall Protection and Scaffold Safety.* Rockville, MD: Government Institutes, 2000 (www.governmentinstitutes.com).

OSHA publication 3146, *Fall Protection.* Washington DC: OSHA, 1998 (www.osha.gov).

Occupational Safety and Health Law, editor-in-chief Randy S. Rabinowitz. Washington, DC: BNA Book, 1999 (www.abanet.org).

Sulowski, Andrew. *Fall-Arrest Systems—Practical Essentials.* Toronto, Ontario: CSA International, 2000 (www.csa.org).

REVIEW EXERCISES

1. What is a fall hazard?
2. What is the process of evaluating a fall hazard?
3. What is the process of training workers effectively?
4. How do you develop a strategy for planning proper anchorage points?
5. What are the critical features of a fall-incident investigation?

6. Why is a harness and lanyard provision not a fall protection program?
7. What is so dangerous about a hole 1-foot square or larger?
8. What fall hazards are expected for a specific roofing job?
9. Why is a 6-foot lanyard not to be used with a fall protection climbing system?
10. What is a fall protection work plan? Are you using one?

■■ **About the Authors** ■■

STEEL ERECTION
by Wayne Rice, CSM, and Brian Clarke, CSP

Wayne Rice, CSM, is Vice-President of Association Services for NEA—The Association of Union Constructors, Arlington, Virginia. Previously, he was Vice-President of Corporate Safety and Health for Modern Continental Construction Companies, Boston, Massachusetts. For over 22 years Wayne was employed by the Bechtel Corporation, serving on several projects in field capacities in the United States and abroad. He was elected vice-president in 1991 and served as the Vice-President of Domestic Safety and Health Services and as Vice-President of International Safety and Health for the Bechtel Group of Companies.

Brian Clarke, CSP, is Corporate Safety Director for Hoffman Construction Co. in Portland, Oregon. He has served two terms as ASSE Region I Vice-President, is a past member of the Board of Directors, and serves on various regional and national committees. He is an OSHA 500 instructor and has presented at several regional and national conferences. Mr. Clarke serves on an Oregon-OSHA external oversight committee, and on OR-OSHA's Multi-Employer Worksite Work Group. Clarke also serves on two national committees for the American Contractors Insurance Group (Ergonomics and Silica in Construction).

CHAPTER 16

Steel Erection

LEARNING OBJECTIVES

- Discuss examples of best practices for steel erection safety, and specifically demonstrate that 100 percent fall protection for all aspects of steel erection is feasible and cost effective.
- Discuss how successful prejob planning for steel erection involves many facets, including adequate protection of erecting and decking crews.
- List the critical documents to use during steel erection.
- Discuss the negotiated rulemaking process.
- Describe how negotiated rulemaking is better for standard promulgation as opposed to standards developed by "others" not familiar with the construction industry and the application of work practices.

INTRODUCTION

The intent of this chapter is to give a brief history and overview of SENRAC (Steel Erection Negotiated Rulemaking Advisory Committee) and to provide possible solutions, guidelines, and examples to assist the safety professional in managing the steel erection processes. This chapter is not a review of the steel erection standards or codes, and it is recommended that additional resources be reviewed for site-specific details such as crane selection and operation, rigging operations, fall protection selection or implementation limits, and structural stability concerns.

DEVELOPMENT OF A STANDARD— SUBPART R

To fully understand 29 CFR 1926, Subpart R and its intent, it is necessary to review the long history of its origin. Many people have been involved with the development of the standard we have today, but two individuals molded and steered the process through the political web in Washington, DC, for more than fifteen years. These individuals were Eric Waterman, NEA (The Association of Union Constructors), and Steve Cooper, Safety Director for the International Association of Bridge, Structural, Ornamental and Reinforcing Iron Workers Union, AFL-CIO (Iron Workers International Union). Working together,

Eric and Steve truly exemplified the type of labor–management cooperation necessary to accomplish the monumental task of bringing consensus to an OSHA standard. They never gave in to the pressures of various special interests and stayed focused. Their mission was to help develop an OSHA standard for steel erection that would benefit the industry and protect the safety and lives of the workers involved. Let's examine the process.

In 1974, OSHA first amended the section of subpart R pertaining to temporary flooring. Subsequently, OSHA received requests for clarification of other provisions of subpart R, including those pertaining to fall protection.

In 1984, OSHA began drafting a proposed rule to revise provisions of subpart R for construction. On several occasions, there were discussions between OSHA and OSHA's Advisory Committee on Construction Safety and Health (ACCSH) regarding a possible revision to subpart R. However, none of the suggested drafts presented were ever adopted or published by OSHA, even with input from ACCSH.

In 1986, OSHA published a proposed rule for subpart M, addressing suggested changes to fall protection for construction. In the proposal the agency announced that the proposed subpart M would apply to all walking surfaces found in construction, except for those in five specific areas. Although none of those specific areas pertains to steel erection, the agency noted that additional requirements for fall protection for connectors and workers on derrick and erection floors used during steel erection would remain in 29 CFR 1926, Subpart R, Steel Erection. Revising fall protection requirements for application in construction while excluding steel erection requirements led to considerable confusion in the industry, and internally in OSHA. Many of the commenters to the proposed subpart M rulemaking noted that they were not sure whether subpart M or subpart R would govern their activities. Because of this uncertainty, the agency (OSHA) decided that it would regulate the fall hazards associated with steel erection in a proposed revision of subpart R.

OSHA hired a consultant to study economic impacts of the proposed "6-foot fall protection rule" in subpart M. The study concluded that implementation of a 6-foot fall protection rule would have no adverse economic impact on the steel erection industry, if adopted. The NEA and the Iron Worker International Union lead a national joint labor–management coalition of the Safety Advisory Committee (SAC) for the ironworking industry to study the 6-foot fall protection rule and its application. This committee, led by Eric Waterman, NEA, and Steve Cooper, Iron Workers International Union, helped to develop comments given to OSHA on a proposed 6-foot fall protection rule and a proposed new subpart M, General Fall Protection Standard, already proposed for revision. The major complaint of the SAC committee was that the 6-foot fall protection rule was a myth, as fall protection users were really falling 9 to 12 feet because of the slack in anchorage, soft stops, lanyard stretch, and other conditions. The myth was that a worker who tied off felt his or her exposure was to a fall of 6 feet.

In January 1988, OSHA announced its intention to regulate the hazards associated with steel erection in a notice published in the *Federal Register*. That notice explained that the rulemaking record developed by OSHA to date indicated a need to gather more information before OSHA could propose a revision to the standard covering all fall protection for employees engaged in steel erection activities. After publication of the 1988 notice, OSHA prepared several draft rule proposals, which were presented to ACCSH for comment. OSHA was also petitioned by the SAC committee to institute mediated rulemaking. In 1989, OSHA denied a request by the NEA and Iron Workers for mediated rulemaking to resolve OSHA's intent to impose a 6-foot fall protection rule for steel erection.

The following year (1990), Congress passed the Negotiated Rulemaking Act, which was supported by the SAC. As a result, the SAC group (NEA and the Iron Workers) submitted the first request for negotiated rulemaking under the new law passed by Congress. At that time it appeared the agency would soon publish a notice of proposed rulemaking in the *Federal Register*. On that basis, the request for negotiated rulemaking by SAC was again denied. However, the SAC committee reiterated its concerns, and again requested negotiated rulemaking. OSHA delayed publication of the notice of proposed rulemaking while it made a more comprehensive study of the concerns raised by the SAC committee and other stakeholders. In 1991, OSHA appointed an additional consultant, Mr. Timothy Cleary, former chairman of OSHA Review Commission, to visit steel erection sites with the SAC committee and other stakeholders to observe their work practices. After a year of visiting sites and reviewing the requirements of steel erection, and meeting with the SAC committee and other stakeholders, Mr. Cleary issued a report supporting the position of negotiated rulemaking for the development of a revision to the existing subpart R. On July 8, 1992, Secretary of Labor Lynn Martin approved the stakeholders' request and Mr. Cleary's recommendation for negotiated rulemaking for the OSHA Steel Erection Standard, which had already been approved by the Negotiated Rulemaking Act of 1990.

Negotiated rulemaking is a process by which a proposed rule is developed through negotiation by members of an industry or stakeholders. A committee is formed by OSHA with representatives of interests that will be significantly affected by the proposed rule. A *key* principle of negotiated rulemaking is that agreement is reached by *consensus.* This process allows interested parties to develop possible approaches to the issue rather than having OSHA ask them to respond to a proposed rule. Each committee member participates in reconciling the interests and concerns of the other members instead of leaving it entirely up to OSHA to resolve conflicting points of view in a standard. OSHA retains an independent consultant to review the fall protection issues raised by the draft revisions to subpart R and to recommend ways to resolve the issues and make suggestions for a course of action.

Based on the consultant's recommendation and continued requests for negotiated rulemaking by various stakeholders (SAC), in December 1992 (six months after the Assistant Secretary of OSHA approval to go forward), OSHA published a *Federal Register* notice of intent to establish a negotiated rulemaking committee for subpart R. After an evaluation of over 225 responses to the notice, it was apparent that an overwhelming majority of the commenters supported this action.

OSHA reluctantly decided to go forward with the negotiated rulemaking process; however, it took almost two years to start solicitations for nominations to the negotiated rulemaking committee. Again, this only happened because of persistent pressure from the NEA and the Iron Workers on ACCSH and OSHA to move forward. During all the activity of standard development via different approaches familiar to OSHA, OSHA was inconsistently issuing citations around the country using the general duty clause for alleged steel erection violations. This confusion arose from OSHA's intent to revise 29 CFR 1926, Subpart M, Fall Protection. In 1994, as a result of the pressure from the NEA and the Iron Workers, Jim Stanley, Deputy Assistant Secretary, issued a letter clarifying subpart R and putting an end to the citing of the general duty clause for steel erection citations. The letter specifically clarified that subpart M does not apply to steel erection activities.

With constant pressure by the NEA and the Iron Workers, OSHA finally solicited nominations for committee members, and on May 18, 1994, it announced the establishment of SENRAC—the Steel Erection Negotiated Rulemaking Advisory Committee—to develop a proposed rule for steel erection. SENRAC, consisting of twenty members, held public meetings to develop the proposed regulatory text. During the course of the meetings, the committee established informal work groups made up of SENRAC members and other interested parties with steel erection expertise or a particular industry interest. The work groups assisted SENRAC by researching technical questions, preparing summaries of literature, and commenting on particular matters before the committee. They also assisted in drafting proposed regulatory language. This gave interested parties who were not selected for membership on the committee an opportunity to contribute to the negotiated rulemaking effort. SENRAC began negotiations in June of 1994. Over an 18-month period, the committee met eleven times. Through negotiations of the full committee meetings, including discussions of recommendations developed by committee work groups, the committee reached consensus on a proposed revision to subpart R.

On May 21, 1996, an industry luncheon was held to celebrate the conclusion of the SENRAC meetings. At that luncheon, OSHA announced the proposed rule would be published for public comment by the end of September 1996. That didn't happen (surprise!), and it took OSHA another two years of dragging its feet to finally publish the final rule in August of 1998. In December of 1998, a public hearing was held in Washington, DC, on the proposed rule. There were no significant issues raised, and it was anticipated the rule would be issued. But, again, OSHA had a new agenda. OSHA called a special SENRAC meeting in December of 1999, one year after the public hearing with no new significant issues, and announced major revisions and changes to the proposed rule.

In January of 2000, OSHA announced that the subpart R final rule would be published by June of that year. Surprise—it didn't happen! Finally, on January 18, 2001, OSHA published the final rule with an effective date of July 17, 2001. However, the new Bush administration imposed a six-month freeze on new regulations in January 2001, further delaying the effective date. Finally, on July 18th, OSHA announced that the effective date for the new standard would be January 18, 2002.

It took general industry (SENRAC) eighteen months to put together a consensus standard for revisions to subpart R, and it took OSHA six years to issue the standard. It is important to note that the entire process of presentng a need to revise the standard, putting together the required revisions, and issuing a standard in final rule form took over fifteen years! Maybe we need to review and fix the process of standard review and revision?

If there is any benefit to this whole protracted process, some significant lessons learned should be documented. Most importantly, it was learned that the records of fatalities kept by both OSHA and the Bureau of Labor Statistics (BLS) are far less than perfect. During the SENRAC process of standard development, 673 OSHA fatality reports for steel erection were reviewed. Amazingly, it was discovered that hundreds of fatalities included in the OSHA database were incorrectly classified by OSHA as steel erection fatalities.

The analysis of these OSHA fatality reports by SAC representatives revealed that the real causes of most of the accidents resulting in fatalities had been included in and were part of the recommendations submitted by SAC to OSHA in the 1980s. Simply put, the major causes of fatalities in steel erection are: improper landing of loads, poor communications between controlling contractors and steel erectors, double connections, poor site conditions, slippery surfaces (paint and coatings), structural collapses, and tripping hazards.

Finally, SENRAC discovered that OSHA's SAC committee had suggested that OSHA needed to revise its fatality report form in order to eliminate confusion as to which craft/trade is involved and to define what specific work activities the deceased had performed and at what skill level. However, this suggestion was given to OSHA in 1995, and the agency is still using the same report form!

The New Subpart R

Since we have explored the long historical record of how the standard was developed, we can now move to the actual standard and its new areas of coverage and focus. As explained in the history, the SAC committee analyzed 673 OSHA fatalities for steel erection to determine if, in fact, each was actually a steel erection fatality. This data was a good source of accident information; however, it was frequently difficult to determine several critical elements of the fatality (e.g., the precise activity being undertaken at the time of the accident, whether the victim was a trained ironworker or another craft person, and the type of structure and whether it was under construction or repair. The importance of reviewing the data is to ascertain the effectiveness of the standard. SENRAC suggested doing an evaluation of the effectiveness of the new subpart R after three years (in 2005) to determine whether additional modifications to the standard may be necessary.

A great deal of thought went into the process of developing the new subpart R. Many dedicated people gave of their time, leading to the development of a standard that is workable for all. It is not perfect, but it is far better than what was written before. The new standard defines the scope of steel erection for the first time and lists all of the activities covered. The scope does exclude transmission towers, communication/broadcast towers, precast concrete, as well as tanks. Following are some of the key provisions of the new subpart R.

SETTING COLUMNS

The setting of columns in preparation to build a structure has contributed to far too many fatalities in steel erection. When the steel erector arrives on a project or site, usually the concrete foundations have been placed and the anchor bolts located for column anchorages. In the past, repairs to anchorages, unknown to the steel erector, were made by others. In some cases only two anchor bolts were designed to secure columns during the steel erection process, causing potential structural failure during the erection process. The number of fatalities studied revealed that a large number of falls occurred during the setting of columns and usually the falls were attributed to column failure (anchor-bolt failure or two-anchor-bolt design).

The new standard requires the controlling contractor (usually the general contractor/construction manager) to notify the steel erector in writing when anchorages (concrete footings, masonry piers, walls, attain 75 percent of their designed strength before steel erection can begin. Additionally, there must also be a minimum of four anchor bolts provided to the steel erector for setting columns (Figure 1).

■ ■ **Figure 1** ■ ■ **A minimum of four anchor bolts must be provided for each column under the revised subpart R requirement**

If there have been any repairs to the anchor bolts, such as straightening or replacing the bolt, the controlling contractor must have the repair approved by the project structural engineer and notify the steel erector in writing prior to the erector beginning work. The controlling contractor/construction manager must also provide a firm, properly graded, drained area, readily accessible to the work, with adequate space for safe storage of materials and safe operation of the erector's equipment.

SITE ERECTION PLANS

Site-specific erection plans can be required under certain conditions:

- Deactivation or making safety latches on hoisting hooks inoperable. *"When a qualified rigger has determined that the hoisting and placing of purlins and single joists can be performed more safely by doing so; or when equivalent protection is provided in a site-specific erection plan"* 1926.753(c)(5), i; ii.
- Setting steel joists 60 feet or more at column spans without all bridging installed. *"Where steel joists at or near columns span more than 60 feet (18.3 m), the joists shall be set in tandem with all bridging installed unless an alternative method of erection, which provides equivalent stability to the steel joist, is designed by a qualified person and is included in the site-specific erection plan"* 1926.757(a)(4).
- Setting of decking bundles on joists without all bridging installed. *"The employer has first determined from a qualified person and documented in a site-specific erection plan that the structure or portion of the structure is capable of supporting the load."* 1926.757(e)(4)(i).

In each of these cases the site erection plan must provide equivalent protection. Nonmandatory "Guidelines for Establishing the Components of a Site-specific Erection Plan" are in Appendix A to subpart R. As a reminder, the OSHA standards are a minimum requirement for any contractor safety program. As a part of doing business, most good steel erectors will have a site-specific erection plan that goes above and beyond the requirements for any contractor safety program.

HOISTING AND RIGGING

Hoisting and rigging is an essential activity in the steel erection process. This section of the standard was to supplement the crane standard 1926.500 by adding specific crane requirements (inspection) for steel erection. It also differentiates between allowing the use of personnel work platforms for steel erection activities and the use of platforms in other activities in construction not governed by subpart R. Iron workers' workstations are typically in high places, far apart, and change fairly rapidly. SENRAC felt that repetitive and short-term activities (i.e, bolting-up) could be performed more safely and reduce fall exposure when done in or from a personnel platform. Additional requirements in 1926.500 for personnel platforms, such as slow lifting speeds, prelift meetings, requirement of anti-two block protection, power down and power up, among others,

■ ■ **Figure 2** ■ ■ **Multiple-lift rigging, so-called "Christmas treeing," shown here is covered in new subpart R**

are still required. The rigging section specifically prohibits the use of the headache ball, hook, or load to transport personnel, except as outlined for the use of personnel work platforms. The change prohibiting anyone from "riding the ball" is significant since this activity is still allowed in some state OSHA plans.

In the section of subpart R that deals with cranes and hoisting equipment, the standard requires the operator (competent person) to make sure safety is part of the rigging and hoisting operations. The operator may shut down the work if, in the opinion of the operator, the specific work activity involving the crane is unsafe. The operator also must perform specific preshift visual inspections of the crane (ANSI B30.5 1994). SENRAC had recommended this inspection be a required written inspection; however, OSHA chose not to require written documentation of the inspection because of the Paperwork Reduction Act. The section also requires a "qualified rigger" to inspect (visually) the rigging to be used. A *qualified rigger* is defined as a "qualified person" who is performing the inspection of rigging equipment. OSHA is relying on the industry to develop specific criteria to further define what a qualified rigger really is. The qualified rigger is also charged with the proper use of rigging equipment. Only in two circumstances under the direction of the "qualified rigger" are employees allowed to work under a suspended load. These exceptions apply to the initial connection work and when hooking or unhooking loads, because overhead exposure is generally unavoidable during these activities. This lends itself to the multiple-lift rigging (Christmas treeing), as shown in Figure 2, which has never been defined by an industry standard before the new subpart R. The SENRAC committee believed multiple-lift rigging, when executed as prescribed in the standard, is a safe and effective method for decreasing total crane swings as well as employee exposure on the steel while connecting.

STRUCTURAL STABILITY

In assembling structural steel, the stability of the structure is paramount to prevent a possible collapse. The standard defines the number of floors to

■■ Figure 3 ■■ Structural steel columns must now be capable of receiving safety cables

be installed with unfinished bolting (no more than 4 floors or 48 feet, whichever is less) and that a permanent floor be installed every 8 stories between the erection floor and the uppermost permanent floor. There also is a requirement for fully planked or decked floor, or nets to be maintained within 2 stories, or 30 feet, whichever is less, directly under the erection work. These safety measures prevent falls of more than 30 feet and the dropping of debris, and provide a staging area in the case of emergency rescue or response.

This is the first standard that gives manufacturers or suppliers of materials design requirements for safety reasons. For example, columns now must have splices to meet a minimum of 300 pounds of eccentric gravity load. This is consistent with the column anchorage requirements for anchor bolts. Perimeter columns must extend 48 inches above the floor to permit installation of perimeter safety cables. Also, the structural steel columns must be capable of receiving safety cables, either in holes prepunched in the shop or attachment-installed prior to erection in the field (Figure 3).

In a review of fatalities in steel erection, SENRAC concluded that failed double connections are a major cause of structural collapse. A double connection is a type of attachment in which two steel members join to opposite sides of a central (carrying) member. An example is a beam, girder, or column web using the same bolts. The erection process is as follows. The first member is bolted to a beam, girder, or column web. Later, a second member is added to the opposite side of the existing connection. This second member is attached through the same holes that are being used to attach the first member. To attach the second member to the structure, the nuts on the first beam's bolts have to be loosened, removed, and the bolts backed most of the way out. The ends of the bolts have to be flush with the surface of the central member so that the second member can be lined up with the existing bolts. Only a fraction of the ends of the bolts are now preventing the first beam from falling. Once the holes in the connection plate of the second member are lined up with the first beam's bolts, the bolts are pushed back through all the holes and the nuts are put back on the bolts and tightened to secure the three pieces of steel together. This maneuver is extremely dangerous. The process often takes place with a worker sitting on the first beam. The risk of collapse is high because of the tenuous grip of the loosened bolts and the possibility that the connector's spud wrench or sleaver bar, used to align the incoming piece, may slip. If at any time the carrying member moves, or is bumped, it could cause the bolts or the wrenches to be dislodged; then you have a structural failure and usually an employee falling.

To address this dangerous connection, the standard requires the connections to be clipped or staggered so that not all of the bolts are removed. There is one common bolt that is not removed during the connection. An additional safety measure is to provide "seats" at these types of connections. A *seat* is either welded or bolted to provide a place for the beam to rest between the flanges of a column until the bolts can be installed. The standard does require one bolt to remain in the connection at all times.

SHEAR CONNECTORS

Initially, the standard allowed only field-installed shear connectors since they are a tripping hazard. Shear connectors, mud lugs, and nelson studs, are installed to bond the concrete floor to the steel structure. Shear connectors (Figure 4) are commonly found on bridge projects and other types of steel structures. The SENRAC committee suggested not allowing shop-installed shear connectors to be installed because they create tripping hazards for the connectors. It makes sense; it is much safer to walk a beam with no obstructions than to walk on a beam with studs installed on the top flange every 6 inches. OSHA has allowed a variance for road/bridge builders to have their shear connectors shop-installed if the employees are tied off. This makes little sense since employees still can fall and possibly become impaled or seriously injured as a result. This is still a major controversy in the steel erection industry.

METAL DECKING

The installation of metal decking, also a very dangerous activity, occurs during steel erection and adds stability to the structure. The standard recognizes the controlled decking zone (restricted work zone), which allows deckers the same exempting as connectors—not to be tied off if working with a fall expo-

■ ■ **Figure 5** ■ ■ **A controlled decking zone permits deckers exemption from tie-off requirements under the new subpart R**

sure at or below 30 feet (Figure 5). The size of a decking job can vary greatly. The standard limits the size of area in the controlled decking zone (CDZ) and the amount of unsecured decking in the CDZ. Specified training must occur for all workers who are limited to the CDZ, including hazard recognition and identification of fall potentials.

JOISTS

The subject of open web steel joists, commonly called bar joists, is the last major revision to subpart R made by the new rule. Items addressed in the new standard are the cause of major structural failures and serious injuries. During erection of joists, the wall or pier (Figure 6) strength must be 75 percent of the designed capacity, as discussed earlier, for column anchorages. The amount of bridging installed is specified in the new standard by joist size and length, to eliminate joists turning when loaded. Requirements for limiting the amount of weight to be placed on the joist and proper securing of the joists are outlined as well. Most importantly, limits are established regarding when the hoist lines can be removed, the number of employees who can access the joists and when, and the size of welds and the amount of bolts required for securing the joists. The amount of erection bridging is specified in the standard as well as horizontal and diagonal bridg-

■ ■ **Figure 4** ■ ■ **Shear connectors (above) continue to be the focus of major controversy**

■ ■ **Figure 6** ■ ■ A number of aspects pertaining to erection of open web steel joists or "bar joists" (shown above) were addressed in the revision of Subpart R

ing. This is also tied to when the hoisting line and employees are allowed on the joists.

TRAINING

Training requirements are specifically required for the connector, deckers working in a CDZ, and employees using multiple-lift rigging. Training frequency, however, is not specified. Each employee involved in the three identified activities must have training. Since the standard was issued, the NEA and International Iron Workers Training Fund has developed eight modules for training on the new OSHA standard, 29 CFR 1926, Subpart R, Steel Erection. Even OSHA has recognized the value of using trainers from the construction industry; they have formed a partnership with the NEA and the International Iron Workers Union, to educate their compliance officers around the country using these eight training modules. Several states have also drawn upon the NEA and its affiliates and the iron workers'

international union to provide this valuable training for their compliance personnel. The movement to utilize a regulated industry to help develop performance standards and provide instruction for compliance officers is a very refreshing approach to safety and policy development.

MANAGING THE STEEL ERECTION PROCESS: GENERAL OVERVIEW OF CONTRACTS

As most seasoned safety professionals will agree, the most successful projects have clearly defined safety expectations identified in the contract. These safety expectations are again reviewed in the prebid, pre-award and premobilization meetings between the erector, suppler, general contractor or construction manager (GC/CM), and often the owner and/or insurer. Samples of agendas and checklists for these meetings are provided in the appendices. No longer should simple contract verbiage, such as "will follow all OSHA and other relevant safety and health codes," be considered adequate. In today's safety conscious and legalistic society, contracts need to be project-specific, supplemented with boilerplate contract language.

GC/CM safety professionals should review the owner/insurer requirements and clearly communicate any and all related findings to involved subcontractors. In recent years, many owners and insurers are taking a more proactive role in the safety of construction projects, especially those projects near operating facilities or operating under an OCIP (owner-controlled insurance plan). GC/CMs will typically include, only by reference, site/owner safety requirements. Often these requirements are not hard-copied in the contracts to the subcontractor, thus owner/insurer safety requirements may not be readily available to subcontractors until after bid submissions.

Progressive GC/CMs will assist bidding subcontractors with a "Special Conditions" or "Minimum Expectations" safety supplement. These supplements highlight safety, health, and environmental requirements that exceed the regulatory minimum standards. Providing these supplements not only ensures a more competitive bid, but can greatly reduce the frustration during premobilization and preerection meetings. This is typically true for subcontractors new to bidding the (safety, health, and environmental) requirements of unfamiliar GC/CMs and/or owners/insurers. These supplements can easily be included in a request for proposal (RFP) and/or distributed as handouts during prebid meetings. Often the subcontractor's field supervisors may not be aware, until mobilization meetings, of site safety requirements. Assurance that the safety supplements were shared multiple times with subcontractors' estimators or operations and/or project managers often defuses initial arguments between supervisors on site.

Appendix A provides examples of line-item health and safety conditions taken from an owner's "Special Conditions Supplement." As can be seen in reviewing this supplement, not communicating these additional requirements can have a devastating effect on the cost of implementing the required safety conditions. The goal is to simply ensure that additional safety requirements are transferred to the subcontractor's estimating team. It should be noted that the examples in Appendix A are approximately 50 percent of the safety "Special Conditions Supplement" from a recently completed project.

Depending on the philosophy of the GC/CM, the level of detail in contract documents and clarifications vary. Some purchasing managers are successful in issuing documents with a minimum of verbiage (e.g., "subcontractor is responsible for structural stability of the structure during erection"), utilizing the "less is better" philosophy. Others may identify that an engineered erection scheme is developed and submitted to the controlling contractor. Still others may require a PE-stamped erection plan verified by third-party engineers. The complexity of contract documents is a company and project-specific decision.

Appendix B identifies examples of site-specific bid document clarifications. These are typically outside the contract's boilerplate attachments. It is highly recommended that local legal council be consulted to review boilerplate, insurance, indemnity, drug/alcohol, safety, and lower-tier contractor attachments. It is advantageous before site-specific

clarifications are submitted as final bid documents, that the preparer of these documents communicates with the site superintendent, engineer, and safety professionals to ensure that all special issues are addressed in the bid document clarifications.

Documentation and Verification of Training

Various regulatory codes identify required training. The intent of this section is not to revisit required training, but to review some of the challenges of training and the verification of employer training for steel erection contractors.

It has been the experience of this author that reliance solely on local apprenticeship training programs and journeymen upgrade training by contractors can be wholly inadequate. In a fatality case involving an ironworker who fell, the employer's on-site supervisor was asked by the chief investigating OSHA compliance officer to provide training documents for those involved. The supervisor's response was, "We rely on the unions' training programs." Next, the compliance officer asked for the employer to identify if/when the training was conducted and the topics covered. The supervisor had no knowledge of content or process of the union training programs. A citation for lack of training was issued and upheld on appeal.

Often, site supervisors are dispatched individuals whose training history and competence level is unknown or lacking. Because of site restraints (i.e., the fall protection class was held last week, or no monies/time identified in the budget for additional training), the needed training is not identified or simply not conducted. There are many challenges for the site supervisor in identifying how much or how little training documentation to request from the steel erector. Judgment calls are made daily on this concern. Some issues to consider when making these decisions should include: knowledge of local apprenticeship and journeyman upgrade training, enforcement and competence of this training, complexity of project, completeness of site-specific orientation for new employees, the comfort level and experience of subcontractor and this specific steel erection supervisor, and simply, the risk level the involved employer and controlling contractor are

■ ■ **Figure 7** ■ ■ **Brewery Blocks Project, Portland, Oregon** (Photo courtesy of Hoffman Construction)

willing to accept. As shown in Figure 7, overhead power lines for a public transportation system, multiple cranes within the radius of each other, adjacent historically recognized buildings, and other related operations can pose additional hazards to construction workers. Site-specific training should be expanded to include these nontraditional safety concerns.

Most construction sites and employers do not have the resources (or may wish not to assume the responsibility) of requesting documentation for *all* OSHA-required training for every craft employee entering a project. Many large projects do require this level of safety excellence and have the safety results that accompany this commitment. However, this represents a small percentage of sites/employers. Safety professionals should have candid dialog with the employer's legal council, risk managers, project superintendents, and project operations managers, regarding what level of training documentation will be required. This dialog should consider site-specific issues as well as companywide expectations. It would be prudent for controlling contractors to ask, at a minimum, if subcontractor employees are adequately trained for the exposures anticipated on site, and whether the employer has documentation of this training at its office.

At a minimum, controlling contractors for all steel erection crafts should at least verify fall protection training (identification, recognition, fall protection system, limitations, capacities, and inspec-

tion). Additionally, verification of training records and competent levels for craft employees involved with rigging and/or crane signaling should be completed by the GC/CM.

Often overlooked is the experience level of the site safety professional (normally hired by the general contractor). The complexity and hazards of the project have a direct relationship on the required experience and education needed in a quality site safety professional. The general contractor would be well advised to ensure this individual has the technical knowledge as well as the ability to effectively communicate with craft workers, project managers, engineers, and regulatory agencies.

Additionally, most general contractors' superintendents, project engineers, and operations managers are assigned to projects from groundbreaking through the completion of final punch-list items. Thus, these individuals may have limited experience or technical knowledge of the latest steel erection concepts and procedures. It may have been years since their last steel erection project. Additional training needs should be evaluated for these individuals. This training should be provided well in advance of project start. The use of a private consultant, well versed in steel erection safety, may be a consideration.

RIGGING

No single document constitutes the entire body of regulations or law pertaining to the safe use of rigging gear and hardware. The following list of statements originates in state or Federal OSHA regulations, American Society for Testing and Materials (ASTM), ANSI, or International Standards Organization (ISO) standards, Federal Procurement Specifications, or any combination of these. By a process of inclusion by reference, many of the various standards listed become part of OSHA regulations. Where there are extreme rigging exposures, it is advisable to have third-party rigging experts, preferably licensed professional engineers (PEs), develop a rigging plan. The canopy erected at the Portland International Airport in Figure 8 presented extreme rigging exposures. Every pick over 300 pounds was an "engineered pick." Third-party PEs identified rigging

■ ■ **Figure 8** ■ ■ **New canopy erected over the operating arrival/departure areas of Portland International Airport** (Photo courtesy of Hoffman Construction)

type/size, pick-points, crane size and location, and travel paths. Hoffman Construction required PEs to include site audits in their project pricing. The netting system around the site was engineered to support a 300-pound point-load impact.

If site safety professionals are unfamiliar with basic rigging principles, they should identify and attend quality rigging training (e.g., Industrial Training International or the Training and Inspection Resources Center Inc., 800-727-6355). Additional information can also be found at any one of the selected Internet sites listed in Appendix E.

The following are basic principles the site safety professional must understand when assigned to a steel erection project.

   ■ Inspection of rigging gear and hardware before each use. This needs to be only a quick visual or hands-on check to make sure that the previous use did not cause damage.
   ■ Ability to identify excessive wear, nicks or gouges deeper than the manufacturers' tolerances, kinks, cracks, breaks, stretch, heat discoloration, bird caging, weld splatter, fraying, or disfigurization from overload.
   ■ Familiarity with slings made of wire rope; having the knowledge to identify excessive breakage of individual wires in specified lengths of the rope.
   ■ Basic knowledge of chains when used in sling applications. Basic knowledge of components

such as hooks or collector rings and the relationship component parts have with the chain itself with respect to grade and load rating.
   ■ Familiarity with proper marking, stampings, labels, or tags that must be present and legible on all rigging gear and hardware at the time of use. This now includes, at a minimum, a designation of lifting capacity, usually expressed as Safe Working Load (SWL) or Working Load Limit (WLL); a diameter measured across the sling or in the bowl of the hook or shackle; and a manufacturer's name or trademark symbol (e.g., CM for Columbus-McKinnon). It is believed that domestic and European manufacturers include this information stamped or forged onto their products. Most Asian import products only show the SWL, a diameter, and the name of the producing country.
   ■ Knowledge of how to verify load weights, or close approximates; the ability to question the person in charge of the lift (often the rigger) and the lifting-equipment operator; and knowledge of the process for identifying rigging components with proper capacities for that load.
   ■ Ability to identify if an attachment to the load or lifting machinery (hardware positioning) is in accordance with industry-accepted rigging practices.
   ■ Authority to ensure that only competent, qualified riggers are allowed to rig loads.

SLING SAFETY

Sling angle is of paramount importance when determining the overall capacity of the sling. Site safety professionals must have a basic understanding of sling angles and the issues involved in determining proper sling selection. The following step-by-step safety analysis is recommended (adapted from the *AGIC Safety Bulletin*, October 2002).

   ■ Determine the weight of the intended load.
   ■ Determine the available head room.

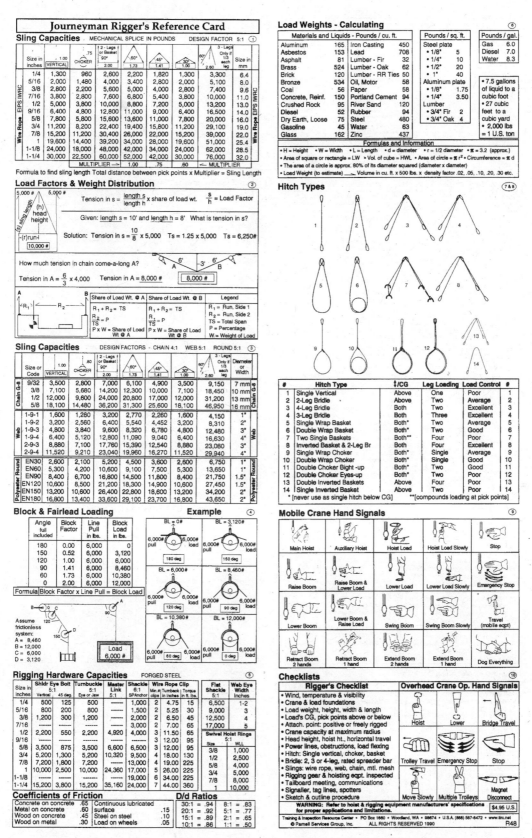

■ ■ ■ **Figure 9** ■ ■ ■ **Journeyman rigger's reference card** (Courtesy Industrial Training International; not for reproduction)

Rigging Gear Inspection Reference Card

Wire Rope Slings — *29 CFR 1910.184 ASME B30.9 Good Practices* ①

1) Do not exceed rated capacity.
2) Min. of 10 dia. between sleeves/splices.
3) Consult mfg. below -60°F, over 400°F.
4) Weld/proof test fittings prior to assembly.
5) Remove if:
   - Evidence of heat damage
   - Broken wires = 10/lay, 5 in 1 strand/lay
   - 1/3 wear of outside wires
   - Kinks • Crushing • Unstranding
   - Birdcaging • Corrosion
   - Distorted rope structure
   - Damaged end attachments
   - Hooks with 15% spread or 10° twist

1) Slings shall be marked showing mfgr, rated load and angle upon which it is based, diameter or size.
2) Do not use knots to form eyes.
3) Use WR clips only if can't prefabricate.
4) Remove if:
   - **Broken wires for multi-part slings:**
     Less than 8-part and Cable Laid = 20/lay, 20/braid/lay, 1 strand/sling
     8-part or more = 40/lay, 40/braid/lay, 1 strand/sling
   - **Core damage • Severe abrasion**

1) Use chaffing gear and avoid load slippage.
2) Don't drag slings. Avoid shock loads.
3) Don't choke on fittings, avoid pinch points.
4) Bent hooks or more than 5% spread.
5) Don't place small sling eye on large hook.

Synthetic Web Slings — *29 CFR 1910.184 ASME B30.9 Good Practices* ②

1) Do not exceed rated capacity.
2) Do not use nylon web or aluminum fittings near acids, nor polyester web around caustics.
3) Repaired slings shall be proof tested to twice rated capacity.
4) Stitching is the only acceptable method to attach end fittings and form eyes.
5) Fittings shall be of minimum breaking strength equal to that of the sling.
6) Do not use at temp. in excess of 180°.
7) Remove if:
   - Acid or caustic burns
   - Melted or charred
   - Snags, punctures, tears or cuts
   - Distortion of fittings
   - Broken or worn stitiches

1) Slings shall be permanently marked with:
   a) Manufacturer name and stock number
   b) Rated load for types of hitch(es) and angle upon which it is based.
   c) Type of synthetic web material
2) Remove if:
   - **Holes, tears, cuts, snags, crushing**
   - **Missing or illegible tag**
   - **Knots in any part of the sling**
   - **Excessive pitting or corrosion, or cracked distorted, or broken fittings**
   - **UV/sunlight damage**
   - **Other visible damage that causes doubt as to the strength of the sling**

1) Use padding between sharp edges and sling.
2) Don't drag slings. Avoid shock loads.
3) Bent hooks or more than 5% spread.
4) Do not shorten or lengthen using knots.
5) Personnel stand clear of suspended load.

Alloy Chain Slings — *29 CFR 1910.184 ASME B30.9* ③

1) Do not exceed rated capacity.
2) Slings shall be permanently marked with size, grade, rated capacity and reach.
3) Slings shall be thoroughly inspected at intervals no greater than once every 12 months. Records of such inspections must be kept on file.
4) Makeshift links or fasteners made from bolts or rods shall not be used.
5) Worn or damaged slings or attachments shall not be used until repaired.
6) Mechanical coupling or low carbon steel repair links shall not be used to repair broken lengths of chain.
7) Remove if:
   - Heated above 1000°
   - Cracked/deformed master links, couplings or other components
   - Hooks are cracked and have been opened more than 15% or twisted more than 10° from plane of unbent hook
   - Reduction in size of links at any point

1) Prior to use, welded components of new slings shall be proof tested to twice rated load.
2) Repaired slings shall be permanently marked with name of repairing agency.
3) Latches on hooks should seat properly, rotate freely, and show no permanent distortion.
4) Slings shall be marked to also show mfgr., and number of legs.
5) Check chain and attachments for wear, nicks, cracks, breaks, gouges, stretch, bends, weld splatter, discoloration from excessive temperature, and throat opening of hooks.

1) Bent hooks or more than 5% spread.

Roundslings — *ASME B30.9* ④

1) Do not exceed rated load (RL).
2) Slings shall be marked with mfgr's I.D., code/stock #, rated loads for types of hitch(es) and angle upon which it is based, core material and cover material (if different than core).
3) Always consult the manufacturer when using a roundsling in chemically active environments, such as acids or caustics.
4) Repairs of load bearing yarns or fittings are not permitted. Repairs to protective covers shall be done by mfgr. or qualified person, then marked by repair agent and proof tested to 2xRL.
5) Fitting surfaces shall be cleanly finished and sharp edges removed.
6) Fittings shall be of minimum breaking strength equal to that of the sling. Previously used fittings shall be free from defects and proof tested 2xRL.
7) Do not use above 194°F, or below -40°F. Some mfgrs. vary.
8) Remove if:
   - **Missing or illegible tag**
   - **Chemical or heat damage**
   - **Holes, tears, cuts, abrasion, snags, or broken/worn stitching that expose core yarns**
   - **Fittings that are stretched, worn, cracked, pitted or distorted**
   - **Knots in any part of the sling**
   - **Other damage that causes doubt as to the sling's strength**
9) Prevent cutting with padding.
10) Do not constrict or bunch roundslings by the load, hook or fittings.

Below-the-Hook Lifting Devices — *ASME/ANSI B30.20* ⑤

1) A nameplate or other permanent marking shall be affixed displaying the following:
   a) Manufacturer's name and address
   b) Serial number
   c) Lifter weight, if over 100 lbs (45kg)
   d) Rated load
2) Design factor shall be a minimum of 3, based on yield strength, for load bearing structural components.
3) All welding shall be in accordance with ANSI/AWS D1.1.
4) Exposed moving parts constituting a hazard under normal operating conditions should be guarded.
5) Electrical components and wiring shall comply with Article 610 of ANSI/NFPA 70.
6) During frequent or periodic inspections, any deficiencies, such as listed below, shall be carefully examined, and determination made as to whether they constitute a hazard:
   a) Structural deformation, cracks, or excessive wear on any part of the lifter
   b) Loose or missing guards, fasteners, covers, stops, or nameplates
   c) All functional operating mechanisms for misadjustments interfering with operation
   d) Loose bolts or fasteners
   e) Cracked or worn gears, pulleys, sheaves, sprockets, bearings, chains and belts
   f) Excessive wear of friction pads, linkages, and other mechanical parts
   g) Excessive wear at hoist hooking points and load support clevises or pins

Hoist System - Mobile Cranes — *29 CFR 1910.180 ASME B30.5* ⑥

1) Prior to initial use, all new and altered cranes shall be inspected.
2) Crane hooks shall be inspected periodically for cracks and wear.
3) Check for hooks bent or twisted more than 10° or increased throat opening exceeding 15%.
4) Sheaves and drums should be inspected periodically for cracks and wear.
5) A thorough inspection of all ropes in use shall be made at least once a month; records must be kept.
6) When inspecting wire rope, check for the following items:
   a) Worn outside wires
   b) Corroded or broken wires at end connections
   c) Severe kinking, crushing, cutting or unstranding

1) Remove if number of broken wires in:
   Running Hoist Ropes = 6/lay or 3 in 1 strand/lay.
   Standing Rope = more than 2/lay in one section beyond end connections, more than 1 at end connections.
2) Remove for 1 valley break.
3) Sheave grooves shall be free from surface defects which could cause rope damage.
4) If a load is supported by more than one part of rope, tension in the parts shall be equalized.
5) Hoist rope shall not be wrapped around load.
6) Hoist rope shall not be kinked.
7) Multiple-part lines shall not be twisted around each other.
8) Slack in the rope shall be removed while the rope is seated on the drum and in sheaves.

1) Bent hooks or more than 5% spread.

Hoist System - Overhead Cranes — *29 CFR 1910.179 ASME B30.2* ⑦

1) Prior to initial use, all new and altered cranes shall be inspected.
2) Check for hooks bent or twisted more than 10° or increased throat opening exceeding 15%.
3) Any deterioration on wire rope resulting in appreciable loss of original strength shall be carefully observed and determination made as to whether further use would constitute a safety hazard.
4) Any unsafe conditions disclosed upon inspection shall be corrected before operation of the crane is resumed.
5) Wire rope should be inspected for reduction of rope diameter due to loss of core support, internal or external corrosion, or wear of outside wires.

1) Remove if number of broken wires = 12/lay or 4 in 1 strand/lay.
2) Remove for 1 valley break.
3) No less than two wraps of rope shall remain on the drum.
4) The rope end shall be anchored by a clamp attached to the drum.
5) Hooks should rotate freely and shall not be overloaded.
6) If repairs of load sustaining members are made by welding, identification of materials shall be made and appropriate welding procedures shall be followed.
7) Dated inspection reports shall be made on critical items such as hoisting machinery, sheaves, hooks, chains, ropes and other lifting devices. Records should be available to appointed personnel.

1) Bent hooks or more than 5% spread.

Shackles — *Good Practices*

1) Do not exceed rated load.
2) Shackles to be marked with size, mfg. and rated load.
3) Remove if:
   - Missing mfg. name, rated load
   - Heat, corrosion, elongation damage
   - Bent, twisted, 10% reduction of dia.
   - Nicks, gouges, thread damage
4) Users trained in shackle selection, inspection, and rigging practices.
5) The screw pin shall be fully engaged.
6) All portions of human body to be kept from between shackle, load, and other rigging during the lift
7) Applied load should be centered.
8) Reduce rated load if side loaded.
9) When used in a choker hitch, the pin shall be in the eye, not on the body.

Turnbuckles — *Good Practices* ⑧

1) Do not exceed rated load.
2) Turnbuckles to be marked with mfgr. and rated load or grade or size.
3) Remove if:
   - Missing or illegible identification
   - Heat, corrosion, elongation damage
   - Bent, twisted, 10% reduction of dia.
   - Nicks, gouges, thread damage
4) Users trained in turnbuckle selection, inspection, and rigging practices.
5) The end fitting threads shall be fully engaged, with extra care in pipe bodies.
6) All portions of human body to be kept from between turnbuckle, load, and other rigging during the lift
7) Applied load should be in line and in tension. No side loading.
8) Shock loading should be avoided.

Rigging Blocks — *Good Practices*

1) Do not exceed rated load.
2) Blocks to be marked with rope size(s), mfgr. and rated load.
3) Users trained in block selection, inspection, and rigging practices.
4) Metal rigging blocks generally used between 32°F and 150°F.
5) Stand clear of suspended loads.
6) Load applied to be in-line with sheave.
7) Block's rated load is based on rated connection: eye, bail, hook, or shackle.
8) Remove if:
   - Missing mfgr. name, rated load
   - Heat, corrosion, elongation damage
   - Bent, twisted, cracked, distorted
   - Broken bearings or sheave wobble
9) • Nicks, gouges, wear 10% of original dimension.

Swivel Hoist Ring — *Good Practices* ⑨

1) Do not exceed rated load.
2) Swivel hoist ring to be marked with mfg., rated load and torque value.
3) Remove if:
   - Missing or illegible identification
   - Pitting, corrosion, weld spatter
   - Elongation, cracked, broken load bearing components
   - Nicks, gouges, thread damage
   - Bent, twisted, 10% reduction of dia., lack of free rotation or pivot.
4) Users training in swivel hoist ring selection, inspection, and rigging practices.
5) Bushing flange shall fully contact load.
6) Applied load in center of bail.
7) Through-hole, washer/nut shall meet or exceed mfg. specification.
8) Threaded hole, engage 1.5 times bolt dia.

Eyebolts — *Good Practices*

1) Do not exceed rated load.
2) Eyebolts to be marked with mfg. and size, grade or rated load.
3) Remove if:
   - Missing or illegible identification
   - Pitting, corrosion, weld spatter
   - Elongation, cracked, broken load bearing components
   - Nicks, gouges, bent, twisted
   - Thread damage, heat damage
   - 10% reduction of original dimension
   - General temperature range is 30° - 275°F.
4) Users trained in eyebolt selection, inspection, and rigging practices.

Wedge Sockets — *Good Practices* ⑩

1) Do not exceed rated load.
2) Socket and wedge to be marked with mfg. and size, applicable model numbers.
3) Remove if:
   - Missing or illegible identification
   - Pitting, corrosion, weld spatter
   - Elongation, cracked, broken load bearing components
   - Nicks, gouges, bent, twisted
   - Thread damage, heat damage
   - Indications of wire rope slippage
   - 10% reduction of original dimension
   - Unauthorized replacement component
4) Users trained in wedge socket selection, inspection, and use.

Warning: This card is provided for quick reference only. OSHA regulations and ASME standards have been paraphrased. Inspectors should be well trained in rigging gear inspection.

Training & Inspection Resource Center • PO Box 1660 • Woodland, WA, 98674 • (888)567-8472 • www.tiro.net $4.95

© Parnell Services Group ALL RIGHTS RESERVED 1996. V9

■ ■ **Figure 10** ■ ■ **Rigging gear inspection reference card** (Courtesy Industrial Training International; not for reproduction)

- Determine the correct hitch based upon the size, shape, and weight of the intended load.
- Determine the correct lifting device necessary to adequately cover all weights applied to the unit during the lift.
- Determine the correct sling length.
- Review your plan—whatever you do, don't guess.

There are numerous *rigging reference* and *rigging gear inspection* reference cards available (see Figures 9 and 10). It is highly recommended that these references be made available to all personnel who are authorized to rig.

CHRISTMAS-TREEING IRON (MULTI-PICKS)

The act of Christmas-treeing iron (or multi-picks) has been used in the construction industry for years. In their rulemaking process, SENRAC has outlined safe procedures for industry to follow. Various state OSHA plans had accepted and provided guidance for Christmas-treeing as a process before SENRAC. Hoffman Construction Company has a procedure for Christmas-treeing that was developed and distributed companywide on June 8, 1994. Since that time, there has not been a Christmas-treeing incident on a Hoffman project. When the lift is planned, craft and site supervisors are trained, and the lift is within the capacity of the crane, Christmas-treeing is a safe, efficient, and productive practice. There are various concerns that arise when the process of Christmas-treeing iron is discussed. This is especially true when those involved have limited exposure to the process. Those concerns are reviewed below.

There are construction safety experts who conclude that the only safe operation allowss *no overhead loads*. Additionally, it is argued that the OSHA regulations (and enforcement of these codes) do/do not allow overhead exposure. In project preplanning, work scheduling, and basic field communication, every effort should be made to ensure no individuals are exposed to overhead loads. There are projects, both large and small, that can achieve this level of safety excellence.

However, picking up single beams one at a time is not always practical, and tandem loads significantly increase efficiency. Some safety benefits are associated with this procedure, including a reduction in the length of time connectors and othes are exposed to the hazards posed by overhead loads because fewer swings are required, a reduction in the time connectors must spend out on the iron because tandem loading allows them to complete their tasks more quickly, and reduced stress on the crane operator because fewer mechanical operations are required.

Rigging

There are publications available (e.g., *Bob's Rigging & Crane Handbook;* see reference section at the end of this chapter) that provide rigging design, manufacture, and assembly criteria that must be met for a multiple pick to be permitted. A multiple lift cannot involve hoisting more than five members during the lift. Limiting the number of members hoisted is essential to safety. The industry and regulators have determined that five members is the maximum number that should be hoisted safely, taking into account the necessity of controlling both the load and the empty rigging. In addition, this limit on the number of members recognizes that a typical bay, consisting of up to five members, could be filled with a single lift. Too many members in a lift may create a string that is too awkward to control or allow too much empty rigging to dangle loose, creating a hazard to employees. In addition, only structural members should be lifted during a multiple lift. Other items, such as bundles of decking, do not lend themselves to the multiple-lift procedure.

The employer must ensure that each multiple-lift rigging assembly is designed and assembled with a maximum capacity for the total assembly and for each individual attachment point. This capacity, certified by the manufacturer or qualified rigger, would be based on the manufacturer's specifications and would have a 5 to 1 safety factor for all components. Since multiple-lift rigging is special rigging used only for the purpose of performing a multiple-lift rigging procedure (MLRP), the rigging should be certified by the qualified rigger who

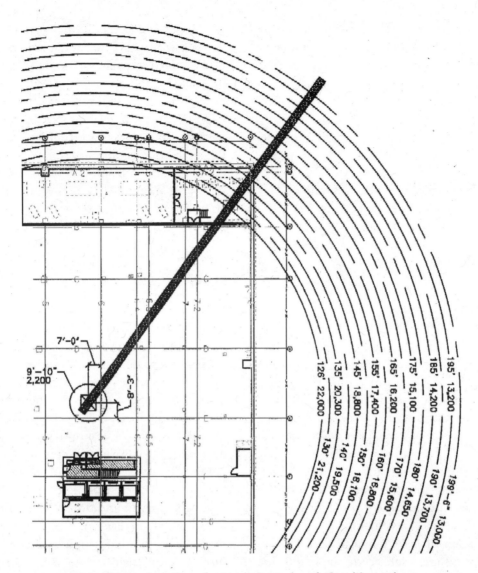

■ ■ Figure 11 ■ ■ Location of a tower crane in relationship to the building footprint, the crane's boom length, and load limits at various radii from the building

assembles it or the manufacturer that provides the entire assembly to ensure that the main line is capable of supporting the whole load and each hook is capable of supporting the individual members.

As each member is attached, accepted practice is that it would be lifted approximately 2 feet off the ground to verify the location of the center of gravity and to allow the choker to be checked for proper connection. Adjustments to choker location would be made during this trial lift procedure. The choker length would then be selected to ensure that the vertical distance between the bottom flange

of the higher beam and the top flange of the next lower beam is never less than 7 feet.

Training

Employees engaged in a multiple-lift operation must be trained in these procedures. Due to the specialized nature of multiple lifts and the knowledge necessary to perform them safely, this training requirement is necessary to ensure that employees are properly trained in all aspects of multiple-lift procedures.

Employers should not *rely completely* on training done at local apprenticeship programs. As noted throughout this chapter, all training should be documented and conducted by competent instructors. Supervisors should require retraining when new crew members are involved, when the process changes, and/or when errors are identified.

The Crane

Several crane manufacturers have recognized that MLRP is becoming an industry practice and have accepted the use of their cranes for this purpose, provided that the crane is utilized in a manner consistent with the safe practices defined in the operator's manual and crane-capacity chart.

When the operator is working in the blind (i.e., where the connectors cannot be seen), reducing the number of swing cycles is particularly important because it minimizes the opportunity for a communication error, which could precipitate an accident.

There are several additional benefits of MLRPs. For example, the increased weight of the load being hoisted using the MLRP results in reduced swing, boom, and hoist speeds, which increases the amount of control the operator has over the lift. Crane operators report that the swing operation has the greatest potential for operator error and loss of load control; therefore, reducing the number of swings enhances safety.

Other basic items such as crane capacity, operator qualifications, communication, and environmental considerations must be included in the preplanning (task plan) when Christmas-treeing iron will be allowed. Figure 11 shows a training aid for identifying crane capacity at different radii. A simple laminated drawing showing the crane location compared to the footprint of the building will ensure that load limits are identified and clearly communicated to field employees.

Nonerection Personnel

The total suspended load time and the frequency of loads passing overhead are reduced for all non-erection personnel on the job when Christmas-treeing is allowed. This is particularly important

■ ■ **Figure 12** ■ ■ **Reducing the number of picks in congested areas, such as this U.S. courthouse project in downtown Portland, Oregon, is safer because the load spends less time over the street**

because these workers normally are occupied with other tasks and often do not pay attention to suspended loads that may be passing overhead. This group of employees includes those working under canopies and partially completed floor systems that cannot see hoisted material passing overhead, but could be injured if a load were dropped. Although controversial, if accepted safe practices are followed, the act of Christmas-treeing, or multiple picks, can be a safe and productive process. As shown in Figure 12, Christmas-treeing reduces exposure to the public when the project is in confined areas.

Connectors

According to the workgroup, a great safety benefit of multiple lifting is that the manipulation of the members at the point of connection limits the movement of the hoist hook, in most cases, to an area less than 10 feet in diameter, and additionally requires that such movement be done at a slow speed and with maximum control. The hazard that connectors consider the most serious, that of a high-speed incoming beam, is thus minimized using the MLRP process.

In addition, when single pieces are hoisted, the emphasis is often on speed. Frequently the lift is hoisted, swung, and boomed at maximum crane speed in an effort to maximize production. Under these circumstances, single-piece hoisting increases the potential for problems in the hoist sequence and in the final placement of each member, and also contributes to operator fatigue.

An MLRP is treated as an engineered lift and accordingly receives the full attention of the entire raising gang. The lifts are made in a more controlled fashion due to the special rigging and physical size of the assembled load.

Tag Lines

It is common industry practice to include the words "tag lines should be used when conditions warrant" in safety programs. Preferably, contractors' policies should require tag lines be used at all times (with the exception of shakeout), unless the qualified rigger identifies individual picks where the tag line presents an increased hazard. Tag lines should be long enough to allow the connector easy access to the tag, yet short enough not to become entangled with objects while being flown.

TEMPERATURES

As building schedules become more compressed and owners are building in environments with extreme temperature fluctuations, safety managers should be conscious of the effects that temperature has on steel expansion. For instance, to minimize the effects of cold temperature, the entire construction site at Eklutna, Alaska, shown in Figure 13, was enclosed and heated by Hoffman Construction. Note the tires that were installed at the tip of the jib. The tires acted as a buffer if the crane jib ever encountered the tent fabric.

Temperature concerns are not limited to extreme hot or cold environments, which would be expected in the far northern or southern parts of the United States; they are also factors in areas that have temperature fluctuation during the day.

On a recent truss project, the tilt-up walls were installed and as-built measurements were supplied to the fabricator. However, the detailer had not allowed for thermal extremes. When the first trusses arrived on site, they were lifted into place late in the afternoon. Only then did the erector identify that the truss had expanded approximately one inch, not allowing the bolting pattern to match.

The engineer of record should be consulted to identify when the erector should initially torque the truss base bolting. This is not typically done until after thermal and loading relief takes place.

DOCUMENTATION

Site-specific Steel Erection Safety Plan

Most regulatory agencies are very limited in what they require in a site-specific steel erection safety plan. Additionally, these requirements are mostly focused on employee protection. Appendix A to subpart R does provide an outline for a steel erection plan, and should be reviewed. However, there are many additional areas of concern that should be included in the site-specific steel erection safety plan. Appendix C of this chapter provides a sample agenda for a steel erection project's premobilization meeting. This sample could also be used as a guide when developing a site-specific steel erection plan. The level of detail in the plan will vary with the details and complexity of the project, as well as the expectations of the GC/CM and/or owner. For example, the complexity of the Experience Music

■ ■ **Figure 13** ■ ■ **The Eklutna Water Project in Eklutna, Alaska, was covered and heated, permitting year-round operations** (Photo courtesy of Hoffman Construction)

■ ■ **Figure 14** ■ ■ **Experience Music Project, Seattle, Washington, under construction, presented the challenge of "ribs" of various size and shape used as structural support**

Project in Seattle, Washington, required months of prejob planning. Three-dimensional software was used throughout this planning process.

The safety professional should also identify if local city ordinances require that a site-specific safety plan be submitted to their engineering department before steel erection begins. This is often the case when steel is erected in or around public buildings or transportation structures. The Seattle Monorail required such a safety plan for building the Experience Music Project (see Figure 14). Service on the monorail was not interrupted throughout the duration of the project.

If working in a state that requires a "state plan," safety professionals should always identify what site-specific plans are needed under that state's jurisdiction. As noted in Appendix C under "Others to Invite," consideration should also be given to inviting a local OSHA consultant to the prejob planning meeting, as long as you determine that this consultant has steel erection experience before the meeting. Inviting the OSHA consultant not only shows that the project wants to work with OSHA to ensure a "safe and healthful project," but including OSHA in the prejob planning process (and inviting them to tour the project during the steel erection

process) can pay large dividends if problems arise later in the project.

INITIAL ERECTION INFORMATION

OSHA's subpart R identifies the controlling contractor's responsibilities for notification to the steel erector. The controlling contractor's supervision should be familiar with these requirements. Contract documents from the steel erector should also identify what information is needed/required by the erector for that specific project. Appendix D of this chapter provides sample "Erection Procedure Notes" from a steel erecting subcontractor's engineer to the site's general contractor. It is appropriate, and should be expected, for the steel erector to ask for clarification.

All notifications should be done through the *document transmittal process* for the project. However this (and other) information is transferred, it must be *in writing*. To paraphrase an attorney's recent conversation, "if it was not in writing, it did not happen."

MEETINGS AND TRAINING

Every safety meeting and training program should be documented. At a minimum, the topic of discussion, date, and attendees should be recorded. It is often said that when questions are posed, whoever has the most paper (documented information) wins. Though safety should not be a win–lose situation, those safety professionals involved in major regulatory investigations or legal proceedings know good documentation (including pictures) is valuable.

SITE AUDITS

It is obvious that the project safety professional and superintendents should document in-compliance and out-of-compliance issues continuously throughout the project. In addition to these minimums, others who visit the site (e.g., insurance loss-control representatives, GC/CM's upper management, corporate safety directors, OSHA consultants, etc.) should be aware of the steel erection safety plan. These individuals and their issues (both in/out of compliance) should be documented on the project's daily logs. The level of documentation is always an area of controversy. It has been the experience of this author that more is always better.

FALL PROTECTION

The ability to complete a project profitably and on schedule is possible with 100 percent fall protection at 6–10 feet. More and more GC/CMs are requiring 100 percent fall protection at 10 or 6 feet. A few progressive steel erection subcontractors are also implementing 100 percent fall protection requirements. These contractors have seen positive results, not only in the saving of lives, but also in the reduction of insurance-related costs. To achieve this higher level of safety, many factors must be accomplished. This includes, but is not limited to, commitment from top management; clear, specific direction in contract language; enforcement of policies equally with all contractors; early communication of these requirements to all bidders to ensure adequate resources are allocated; and trained staff to assist in problem solving and constructability issues.

Additionally, some progressive GC/CMs are *not* allowing safety monitors or controlled decking zones for steel/decking operations. In a recently published fall protection training video demonstrating the correct use of the monitoring system, the safety monitor on the video was wearing his tool bags while the crew was working. While this in and of itself is not an OSHA violation, why would a safety monitor, who should have no other duties, be wearing his tool bags?

Often the contract requirement of 100 percent fall protection at 6 feet is interpreted to mean that personal fall protection is the only allowable option. Figure 15 shows connectors achieving 100 percent tie-off through the use of extendable boom lifts. Installing the concrete slab before steel erection allowed the erector a level, stable surface from which to operate man lifts.

A common oversight with safety professionals is the lifting, receiving, and placement of panels (as shown in Figure 16). To varying degrees, many

on a job site drop their guard once all the structural steel is in place. However, the prejob plan for lifting, receiving, and placing panels should have as much depth and detail as the structural steel erection plan, since most of the same hazards still exist.

As should be identified in the project's prebid/premobilization coordination and planning meetings, there are many alternatives to body harnesses—such as handrails shown in Figure 16—as the means for achieving 100 percent fall protection at 10 feet. Some of these alternatives were identified in the chapter on fall protection. Additional options can be identified by simply asking for assistance from others on and off the project. Often, when asked, the iron workers themselves offer practical approaches to safe steel erection. Most safety professionals do not appreciate, respect, or engage the years of experience a journeyman or foreman iron worker has. A site-specific safety plan developed with the involvement of a lead journeyman and/or foreman will be more accepted than one issued from the corner or safety office. Fall protection manufacturers and vendors are also an excellent resource for options on fall protection systems.

Craft employees often travel from job to job and employer to employer with their own personal protection equipment, without regard to inspection and acceptance by the general contractor or owner representative. It is not unusual to discover fall protection equipment that is damaged and does not comply with the latest codes and standards and sometimes is modified or remanufactured, but not in accordance with the original manufacturer's specifications. Some employers may allow the use of employee-owned fall protection equipment. If this is allowed, site safety professionals should have processes in place for the inspection and written acceptance of employee-owned equipment. The selection and acceptance of employee-owned fall protection equipment is often overlooked and can become of paramount importance following an incident or accident. It only takes a moment to verify this very important aspect during new employee orientation.

When selecting the personal fall protection system(s) for craft use and the overall project, cost

■ ■ **Figure 15** ■ ■ **At the Expo project in Portland, Oregon, Hoffman Construction coordinated with members of the Iron Workers Union to ensure adequate access was possible with 100 percent fall protection**

■ ■ **Figure 16** ■ ■ **Placement of panels at the U.S. Court House, Portland, Oregon (Hoffman Construction General Contractor)**

tary standard and not always enforced, it would be prudent to acquire the manufacturer's certified third-party testing results, thus ensuring that the manufacturer has complied and the employees are protected.

The selection, care, inspection, and training of each component are essential to good safety practice. It is also imperative to determine the compatibility of the different manufactured components or among various manufacturers. Mixing and matching different manufacturers' equipment without regard to each manufacturer's instructions, warnings, and installer/user instructions should not be attempted.

SAFETY IN DESIGN

The inclusion of safety concerns in *constructability* is becoming a more frequently considered concept during the design stages of structures. An article published by Steven Hecker, Associate Professor at the University of Oregon, and John Gambatese, Assistant Professor at Oregon State University, does an excellent job of describing the current state of activity in safety in design for construction. This article reviews some of the barriers and the efforts underway to overcome these barriers, and presents preliminary findings from a case study of safety-in-design efforts on a large industrial construction project. For information on this project and the article, contact Steven Hecker at the Labor Education and Research Center, University of Oregon, Eugene, OR 97403 (541-346-0274,shecker@oregon. uoregon.edu).

should not be the only consideration. The safety professional should consider at least the following: compliance with ANSI A-10-14-91, adequate supply and delivery, ability to list the employer as an "additional insured" on the manufacturer's certificate of insurance, ability of manufacturer or distributor to provide on-site fall protection training, and compliance with ANSI Z-359.1-92. Because this is a volun-

REFERENCES

American National Standards Institute. ANSI Standard A10/13-2001 "Safety Requirements for Steel Erection." Itasca, IL: NSC, 2001.
Bob's Rigging & Crane Handbook. Leawood, KS: Pellow Engineering Services, Inc., 2000.

REVIEW EXERCISES

1. What is SENRAC? What was SENRAC's purpose? What did SENRAC accomplish aside from the rulemaking?
2. What is the purpose of a preerection meeting? Who should be in attendance? What is the goal of a prejob erection meeting?
3. List ten safety concerns for steel erection in cold (below-freezing) environments. What are cost-effective controls for these safety concerns?
4. List ten steel erection safety issues that should be included in the contract between the general contractor and the steel erection subcontractor.
5. List ten sources for information or assistance that are available for steel erection safety.

APPENDIX A

Appendix A provides examples of line-item safety and health conditions taken from an owner's "Special Conditions Supplement." As can be seen in reviewing this supplement, not communicating these additional requirements can have a devastating effect on the cost of implementing the required safety conditions. The goal is simply to ensure that additional safety requirements are transferred to a subcontractor's estimating team. It should be noted that the example below is approximately 50 percent of the safety "Special Conditions Supplement" from a recently completed project.

1. Subcontractor shall complete and submit a Subcontractor Job Hazard Analysis (JHA) prior to commencement of the work on the Project site of any awarded work. This plan will be reviewed for compliance with the structure of the overall site safety programs. This plan may be incorporated into the overall Project Safety Plan. Additional task-specific safety plans shall be required, depending upon the hazard and work environment changes, at the discretion of the Contractor.
2. All task activities require written Task Plans (TP) and/or permits prior to work beginning.
3. All new Subcontractor employees shall attend the (Owner) and Contractor New Employee Orientation (Safety, Badging, Drug and Alcohol Test). Badging will be provided by Owner and will require an additional hour. Total New Employee Orientation activity will take approximately 15 hours over a two-day period to complete.
4. Subcontractors shall provide training or ensure that each employee has an appropriate level of knowledge to meet OSHA and Project site requirements. This includes, but is not limited to, Confined Space, Scaffolding, Mobile Elevated Work Platforms (Scissors Lifts), Electrical Hot Work (EEW), forklifts, Lock-out/Tag-out, Ladders, Rigging and Signaling,

(Owner) Best Known Methods (BKMs), Fall Protection and Hazard Communication.
5. Subcontractor employees shall be required to attend a weekly safety meeting. These meetings will be held on Mondays at start of shift and will alternately be conducted by their Foreman/Safety Group Leader and the Contractor Project Superintendent. An additional 1-hour meeting will be held following the biweekly Foreman/Group Leader Safety Meeting and all Foremen shall attend.
6. Daily Foremen Meetings:
    6.1 Subcontractor Foremen shall begin their shifts with a brief crew meeting. At these meetings, the crew will prepare for their work by stretching in accordance with the "Stretch & Flex" program. Written Task Plans of each crew shall be reviewed at these sessions.
    6.2 After these daily crew meetings, all Foremen will attend a Hall Meeting with their respective Contractor Area Superintendent. The purpose of these Hall Meetings is to review the activities for the current and following day and to ensure the necessary planning and coordination is complete and that each Subcontractor working within the area is aware of other activities.
7. Construction Incident Prevention Program and Site Incident Prevention Program:
    7.1 This project utilizes both a Construction Incident Prevention Program (CIPP) and a Site Incident Prevention Program (SIPP). All CIPP or SIPP activities shall be identified by the Subcontractor on their four-week look-ahead schedule and coordinated with the Contractor Area Superintendent and CIPP/SIPP Manager. In addition, these activities will be reviewed at the Contractor Weekly Work Coordination meeting. Determination as to whether CIPP or SIPP is used is

modified during the construction activity. As such, the correct process will be chosen at the time when needed.

7.2 Subcontractor shall utilize the CIPP for all activity performed on the Project, which potentially could cause harm to utilities, building systems, or the manufacturing capabilities of the owner.

7.3 Subcontractor shall utilize the SIPP for all Project activities that have a direct interface with any Owner utilities. SIPP is required for all work regardless of location of performance, which involves any tie-ins to live manufacturing facility system, or utilities, which could potentially impact the manufacturing facilities.

7.4 Each Wednesday, Owner will provide a one-hour training class to explain the CIPP and SIPP process and expectations.

8. All personnel working at elevations over six (6) feet in height shall have a 100% fall protection system. Only Full Body-harnesses ANSI A10.14 approved shall be used.

9. Subcontractor shall designate a person(s) responsible for EHS activities for this project. Immediately upon award of Subcontract, Subcontractor shall submit to Contractor that person's name, telephone number, cellular telephone number, and pager number. *Subcontractors with 20 or more employees on the Project site shall have a dedicated full-time safety representative at the project site. In addition, for each incremental increase of 50 workers, Subcontractor shall provide an additional dedicated full-time on-site safety representative.*

10. Designated or dedicated, as the case may be, safety representative's qualifications are subject to prior approval by the Contractor's EHS Department. Minimum qualifications for dedicated safety representative shall include OSHA 500 Training or 40 hours of construction safety training, ability to demonstrate skill in training others and a minimum of 3 years of construction safety experience.

11. Safety Team Meetings will consist of daily whiteboard meetings and weekly 2-hour informational/training meetings. Subcontractor designated or dedicated safety representative shall attend.

12. All subcontractor EHS programs and EHS training programs are subject to review by Contractor EHS for adequacy and completeness.

13. Training records shall be submitted to Contractor EHS for all training required by OSHA, or the Project (per type of training and employee name). All training whether Contractor or Subcontractor administered shall be in compliance with the current Site Training Matrix.

APPENDIX B: SAMPLE CONTRACT INCLUSIONS

The examples below only include safety, health, and environmental examples. Other nonsafety examples were excluded from this appendix.

1. Subcontractor will provide, without limitation, all labor, equipment, tools, and miscellaneous installation materials (weld rod, temporary bracing if required, etc.) required to furnish and install all items.

2. Subcontractor acknowledges the requirements of the "AISC Code of Standard Practice" as it relates to steel erection and specifically Section 7.9 "Temporary Support of Structural Steel Frames." Prior to commencing on-site work, Subcontractor shall provide the Contractor a detailed erection plan that is in compliance with these requirements. Subcontractor shall continually monitor conditions on the site and if conditions should change on the project from those anticipated by the original erection plan, Subcontractor shall be responsible for submitting a modified erection plan which complies with Section 7.9 "Temporary Support of Structural Steel Frames." (*Author's note:* this AISC references an earlier edition of this AISC publication.)

3. The project has an SK-400 tower crane and operator on the job as the primary means of hoisting. Subcontractor shall provide hoisting as needed to complete beams and columns outside of the tower crane radius as identified on the bid documents at the northeast building corner. The use of the tower crane would be made primarily, but not exclusively, available to Subcontractor for the eight-hour regular workday during the period required for erection of the structural steel, decking, and joists of this package. The Contractor will utilize the crane during any down time of the steel erector and as needed for minimal support functions of the project. These functions would be discussed and agreement reached with the erector at daily coordination meetings to eliminate

impact to the Subcontractor. The balance of hoisting activities for the project will be performed on weekends or after regular work hours.

4. Subcontractor includes, but is not limited to the following:

   a. The project has implemented a 100% glove policy for all work done on the site. Subcontractors shall supply protective gloves to suit their respective tasks of work on site. Subcontractor's on-site employees will be expected to cooperatively comply with the program.

   b. Subcontractor shall provide "certified" flaggers as needed for Subcontractor's work at the site.

   c. Subcontractor shall erect permanent handrails (for stairs) with the stairs as they are erected.

   d. Subcontractor understands that annual street use moratoriums imposed by the City of _____ are in effect for _____ week(s) in _____ and from approximately Thanksgiving through New Year's Day. Subcontractor will schedule deliveries in a manner to avoid violation of the moratorium restrictions.

   e. Subcontractor shall include furnishing and installation of an engineered perimeter safety rail system to occur from Levels 1 to uppermost roof levels at the perimeter of each floor and all interior openings. The rail system shall consist of two strand levels of cable per OSHA standards. The strand levels shall be made up of a minimum of four strands per floor to enable cables to be removed on one side of the building for access without compromising safety on the entire floor. Where columns are to be used in lieu of posts, Subcontractor has included fabrication of holes through columns at locations to be verified by structural engineer. As curtainwall completes, erector shall collect cables and posts from a central location on each floor. Cables, clamps, turnbuckles, and other components shall be maintained by others commencing once the Subcontractor has completed all work on the floor. Controlled access points will be coordinated on site.

5. Contractor shall provide suitable access in and around the erection area for erector's cranes and trucks. Contractor shall be responsible to pay for street use permit required at the NE corner of the project for Subcontractor's additional crane.

6. Contractor to supply, at no cost to the Subcontractor, adequate sanitary facilities. These facilities shall meet all area codes and regulations.

7. Contractor to maintain a stairway to the uppermost erected level as required by state and local regulations. Subcontractor agrees to keep installation of permanent stairs as close as possible to uppermost erected level for the duration of the project.

8. Traffic barricades, if required, are to be supplied and maintained by others. Any permits and/or permission for street access that affect steel erection procedures are to be obtained and paid for by others. Subcontractor will provide written notification of these requirements in adequate time to obtain the permits and/or permission. Subcontractor will provide trained flaggers and removal of fence for off hours worked if any.

9. Overhead protection, screens for the perimeter of the building, sidewalks, streets, adjacent buildings and the like, if required, are to be furnished, installed, maintained, and removed by others.

10. Overhead wires and aboveground obstructions, which interfere with access and erection, shall be removed by others prior to commencement of erection except as follows. Subcontractor understands that no access from _____ street will be provided. Subcontractor is aware that construction of the concrete core will precede overhead of steel erection below.

11. The inspection agency will be expected to perform their services immediately after Subcontractor completes its tasks, and while ladders and scaffolding are still in position.

12. Clean-up: Subcontractor shall continually and thoroughly clean up and remove to Contractor supplied drop box, at its expense, all waste, debris, surplus equipment and surplus materials resulting from Subcontractor's operations. Recycling of debris will also be utilized by Subcontractor on this project. Such clean-up shall occur on at least a weekly basis and more often if, based on the judgment of Contractor's Superintendent, more frequent clean-up is necessary to prevent risk of injury to employees or crafts people or personnel on the job site. If Subcontractor fails to clean up such waste, debris, surplus materials and surplus equipment, such clean-up and removal will be done by others and costs for this work will be deducted from the Subcontract.

13. Traffic Control and Flagging: Subcontractor shall be responsible for providing all traffic control to comply with all applicable codes and regulations as it pertains to its work.

14. Vehicle Parking: Subcontractor is to make its own arrangements for employee and company parking. No parking is available on site.

15. Subcontractor shall supply perimeter railings at all perimeter conditions and at all openings $3' \times 4'$ and larger. Openings smaller than $3' \times 4'$ shall be covered by Contractor with plywood. Subcontractor shall incorporate Contractor-provided turnbuckles into wire railing system for future tensioning of the system as required. Contractor will remove all Subcontractor railing materials and return to Subcontractor when the fall protection is no longer needed.

16. Subcontractor hereby acknowledges that the schedule and complexity of this project will require a high level of coordination, efficiency and expediency throughout preconstruction and submittal preparation

phase. Subcontractor will coordinate the preparation of all shop drawings and calculations with the skylight system requirements as provided in information and detail requirements to be furnished by Contractor and skylight subcontractor. Subcontractor will work mutually with Contractor and skylight subcontractor to ensure that all drawings, calculations, and submittals are fully coordinated.

17. Subcontractor will assist Contractor in developing a comprehensive work plan and Public Safety Plan for all field-assembled/erected canopy steelwork. This plan shall include without limitation:

    a. A detailed phasing plan showing the planned sequence of the work, complete with milestone dates for planned start and completion of each subphase of the work.

    b. Access and egress routes, temporary traffic control measures and traffic plan configurations for deliveries of equipment and materials at various phases of the project.

    c. Locations of staging areas for materials and equipment in and around the construction work area.

    d. Scheduled dates and times for major deliveries.

    e. Scheduled dates and times for major erection picks for all main and edge trusses.

    f. Schedule and sequence for the erection of all infill work.

    g. Locations, sizes, and configurations of cranes, shoring, load-distributing framework, etc., of all major pieces of equipment when in use, and when not in use.

    h. Locations and configurations of temporary protection as required inside the Terminal, outside the Terminal west wall, over the Terminal core areas, over the lightwells, over the Enplaning Roadway slot adjacent to the Garage, etc.

    i. Detailed sequencing of the pedestrian bridge erection work.

18. Subcontractor shall prepare and furnish a fully-engineered erection plan for this scope of work as required, complete with calculations, bearing the stamp of a professional structural engineer registered in the State of _____. Subcontractor will provide a fully qualified steel erection superintendent to supervise the erection of all work, including all edge and main trusses.

19. Contractor has applied for a preliminary FAA approval for the use of cranes required to perform this work, assuming that the top of hoisting equipment will not exceed elevation 180.0'. In developing the erection plan and selecting hoisting equipment, Subcontractor will be mindful of this requirement and will attempt to select equipment that will not exceed elevation 180.0'. Subcontractor will assist Contractor in obtaining final FAA approval for all cranes and hoisting equipment that Subcontractor will utilize in the performance of this work.

20. Subcontractor acknowledges that due to the project's immediate proximity to the public and Terminal Building occupants, the City of _____ will require Contractor to submit and obtain approval for a detailed Public Safety Plan for approval. Subcontractor will fully cooperate with Contractor in the preparation of this plan and will provide any and all calculations, narrative and drawings as may be required to satisfy City requirements. Subcontractor acknowledges that all erection work will conform to all UBC seismic requirements during construction. Subcontractor will work expeditiously with Contractor to ensure that the Public Safety Plan is complete and approved well in advance of the construction commencement.

21. Subcontractor is aware of the need to maximize the implementation of preassembly and prefabrication opportunities during the fabrication of the steel truss sections prior to completion of finish painting and delivery to the job site. Subcontractor will work mutually with Contractor for attachment of

junction boxes to fabricated truss sections prior to final finish painting in the fabrication shop. Furthermore, Subcontractor will cooperate with Contractor and his subcontractor's to deliver prefabricated truss sections to a staging area in time for the installation of other Canopy components, i.e., conduit and light fixtures, etc., as required prior to final delivery and erection. Subcontractor shall require layout information for junction boxes prior to issue of shop drawings to fabrication shop. Subcontractor shall not be responsible for paint damaged during installation of components "by others."

22. Subcontractor has proposed voluntary value engineering alternates for truss joint welding as currently specified for Owner's consideration. Subcontractor will provide all data, samples and proposed welding procedures as required for the Owner and engineer to make informed decisions relative to selecting the value engineering proposals. Should Owner accept any or all of the value engineering proposals as presented by Subcontractor, the Subcontract price will be reduced by the amount as proposed by Subcontractor and accepted by Owner.

23. Subcontractor acknowledges that all steel erection work will be performed from the Enplaning Roadway. The Deplaning Roadway Level must remain unobstructed and open for public use at all times, except as required to install temporary protective barriers and materials. Subcontractor will be solely responsible for developing the erection plan and ensuring that all loads introduced into the Enplaning Roadway during steel erection are within allowable load limits of the structure. Subcontractor shall prepare all drawings and calculations as required to specifically show supplemental support requirements for the cranes, hoisting equipment, shoring materials and all other loads. Subcontractor will engineer, furnish, and remove all supplemental supports as required. Subcontractor shall perform all hoisting in accordance with Contractor's corporate safety plan, and in accordance with the following requirements:

a. Subcontractor shall provide a certificate from an independent third-party inspection crane agency to certify that the equipment operates safely and is fit for use.

b. Subcontractor shall conduct inspections of all hoisting equipment daily to ensure that equipment is safe and fit for use. Subcontractor shall maintain records of all daily inspections to verify that the inspections occurred.

c. All loose parts and articles shall be hoisted in crates or other containers with engineered rigging and attachments sized and checked for appropriate capacity.

d. All loads over 2000 pounds hoisted to the Canopy level will require the following:
   1) Submittal and approval of a critical lift plan, including rigging diagram and crane radius diagram.
   2) Engineered pick points and center-of-gravity calculations.
   3) New rigging, specifically engineered for this project, inspected daily.
   4) A designated, qualified rigger and crane operator.

APPENDIX C: STEEL ERECTION SAFETY PLAN

(The following steel erection safety plan is intended only as an example.)

HOFFMAN CONSTRUCTION COMPANY

JOB #_____

SUBCONTRACTOR: _____

MEETING DATE AND TIME: _____

WORK LOCATION: _____

ATTENDED BY (NAME AND COMPANY): (attendance required by Site Supt. (HCC); Site Supt. (Steel Erection Sub); Regional Safety Manager (HCC). (*Other invitees to consider: OSHA Consultant, Insurance Loss Control Representative, Owners Representative, Project Manager, QC/Building Inspector.*)

_____   _____

_____   _____

_____   _____

_____   _____

Action items are identified by the name of responsible party as a subparagraph to the bullet item.

Steel erection sequencing and stability plan is attached. Crane load limits are attached.

JOB-SITE ANALYSIS

- Site constraints
- Crane and material accessibility
- Material laydown
- Material handling
- Power lines or underground utilities
- Special permits required?
- Consideration for transportation of large members to site
- Truck access into site [flagging required?]
- Delivery schedule and times
- Craft drug & alcohol/new employee orientation
- Work shifts (special lighting needs)
- Environmental impact issues
- Does the erector understand all of the site constraints?

ERECTOR'S SITE-SPECIFIC SAFETY PLAN

- Is it an engineered fall protection system?
- All changes in plan must be approved/documented.
- It is 100% tie-off rule [6 feet above ground]–how are transition points going to be addressed, double lanyards?
- Beamers or static lines?
- Safety system installation, can shop modifications be made?
- Final perimeter cable rail systems (posts, cable size, turnbuckle's elevated temp, loading zones planned)?
- Can perimeter cable be used as anchor point for fall protection?

- Who will "maintain" perimeter cable guardrails? Consider using turnbuckle to reduce time required to maintain perimeter.
- Provide a rescue plan for an injured worker [still on the iron or deck]. Is there high-angle rescue capacity from local EMS? Who provides manbasket, aerial lift, etc.?
- Task Plans—When/How often are task plans required?
- Is there any required netting areas for material/tools to protect others below.

CRANES AND RIGGING

- Crane size – reach – radius – type?
- Crane size – capacity?
- What is wind speed limit for crane(s)? Where are wind speed indicators?
- Experienced operator [copy of credentials]?
- Has the crane been certified? Inspections daily?
- Crane outriggers firmly on ground with proper pads? (Who supplies pads?) Where are pads needed?
- Is the crane position firm and level?
- If tower crane – was crane re-torqued per manufacturers recommendations?
- Does your site require special lights – flags or permits for crane?
- Has erector preplanned rigging methods?
- Know the weight of members. How is this verified?
- Shake-out hooks are for shaking out only!
- Is the rigging new? If not who inspects?
- Has the rigging been taken out of service properly when it is deemed unusable? How often is it inspected?
- What is used for sharp—or hard—edges to protect rigging?
- Extreme or heavy or difficult picks' critical lifts must be engineered. (Who identifies these picks?)
- Crane and rigging daily inspection is a must.
- Are column jerk pins being used?
- Level ground for shake out—is there adequate dunnage?
- All pieces required to have tag line.
- An FAA permit may be required depending on crane height and location of crane in relationship to Airport.
- Are there potential issues with flight path of local helicopters (TV, police, private, military, emergency medical transport)?
- Christmas trees—each piece must be rigged separately back to the hook. See Christmas tree policy/section.
- Spreader bars been approved for maximum load?
- Controls for working under suspended loads (Christmas treeing)?
- Identify which crane picks need a specific plan showing crane setup location and data to support the pick of the load (major trusses, odd shape steel, etc.).
- Identify where utility locates are needed for outrigger placement.
- Hoffman will be advised of any change in the pretask plans and provided with updated pretask plans.

STEEL ERECTION   [THREE CATEGORIES FOLLOW]

1. BOLT UP

- Has the concrete had enough cure time?
- Provide a smooth level surface for shims. Who supplies the shims?
- Leveling nuts are not made to take the place of shims.
- How far behind are grouting base plates compared to erection? Who makes decision when to grout?

- Two bolts minimum for each connection per OSHA. What will your bolt-up procedure be?
- Locations where engineer requires more than two bolts?
- If a bolt is unmakable, has the appropriate amount of weld been installed to replace it? What is inspector's procedure?
- Best on long connections to install 1 top and 1 bottom in lieu of 2 top.
- Snug—(wrench-tight) bolts are ok?
- What are engineer specifications (number of stories steel erection should be ahead of the decking, when can the slabs be poured)?
- How far ahead of the bolt-up crew is the raising gang?
- MIS—fits are the responsibility of the erector to make safe.
- Assure that bolt buckets and tools are secured to structure.
- How will others below be protected?
- Lejunebolts/tcbolts—who will test lots?
- Holes in deck, recommend block/frame, cut decking later as needed.

2. ERECTION BRACING

- Erection bracing is 100% of the erectors responsibility. Is there a bracing plan? Who is the engineer of record? When can deviations from the erection bracing happen? Who approves deviation? When/Who identified release of temporary bracing/guying?
- Is the building top heavy? Is there too much material stocked in one location?
- How far ahead of the bracing installation is the raising gang?
- Are the cable clamps properly installed?
- How many cable clamps per connection?
- Are drop-forged cable clamps required?
- Identify how certain areas will be left braced before leaving for the day, and is engineering completed for bracing requirements?

3. WELDING AND DECKING

- Fall protection for decking operation?
- Fall protection training for crew and inspectors?
- Will warning line systems be used? Who will monitor compliance?
- A decked floor shall be maintained within two stories or 30 feet, which ever is less, below and directly under that portion of each tier of beams on which any work is being performed—OSHA requirement, reference subpart "R".
- Tack weld leading deck.
- Nelson studs are a fire hazard when being installed.
- Ensure supplier for decking knows erector's deck sequence and bundling requirements.
- No cantilevered deck bundles.
- Clean up loose deck scraps (secure from winds).
- Two-inch-minimum bearing on deck ends on support members.
- More ironworkers are injured while installing decking than on any other operations.
- Check for adequate wind tacking.
- Deck span [verified by supplier]?
- Check for open holes in the decking. How will they be protected? Who will provide protection? Do not make penetrations until whatever is going to go in the floor is ready to be installed. Some penetrations will be made by other contractors, not the erector.
- Ventilation procedure.
- Fire prevention and hot work?
- Indoor air-quality issues?

OTHER ISSUES?

- How will public/other craft be protected from welding (spark and light)?
- Scissors lift training?
- Fall protection training?
- Special inspector accesses? [He provides his own safety gear.]
- Identify if a 6-foot warning line will be used and establish where and how?
- Identify any noise hazards.
- Protection of existing property.
- Welders gas or electric (review requirement especially for nelson stud gun, ventilation for gas welders).
- Will netting be used (personnel or debris)? Who installs/inspects? Protection for netting (welding slag, sharp objects, etc.)?
- Welding units secured how?
- Will respiratory protection be needed for any operation (fit test, medical examination, training, selection)?
- Weather issues (snow/ice, thermo expansion of steel)?
- Lightning potential?

APPENDIX D: STEEL ERECTION PROCEDURE

Sample erection procedure notes from steel erection subcontractor to general contractor. The samples below include three recent transmittals from three different projects. Not all notes were included from any one project.

- Subcontractor understands that steel will not be lifted or set when the wind speed exceeds 25 MPH. Other work, such as the installation of temporary cable braces, bolts, welding and other activities, will continue. In addition, the seismic load design criteria, as directed by the Owner, is based on UBC Seismic Zone 2B.

- The erection of structural steel included in this procedure is based upon paragraph 7.10.3 of the AISC Code of Standard Practice for Steel Buildings and Bridges, adopted effective March 7, 2000. In part the section reads: "The Erector shall determine, furnish, and install all temporary supports, such as temporary guys, beams, false work, cribbing, or other elements required for the erection operation. These temporary supports shall be sufficient to secure the bare Structural Steel framing or any portion thereof against loads that are likely to be encountered during erection, including those due to wind and those that result from erection operations. The Erector need not consider loads during erection that result from the performance of work by, or the acts of, others, except as specifically identified by the Owner's Designated Representatives for Design and Construction, nor those that are unpredictable, such as hurricane, tornado, earthquake, explosion or collision."

- This erection plan addresses erection of the structural steel, metal decking and associated connections only. This plan does not address shoring or bracing for any cast-in-place concrete members (i.e., topping slabs, concrete beams, concrete walls, concrete columns, etc.), or any other structural members or loads other than the structural steel and metal decking specifically addressed in this erection plan.

- The location and configuration of all temporary cable bracing are shown in Figures 1 through 4. The temporary bracing and column shim packs have been designed to resist the loads associated solely with the structural steel and the metal decking (i.e., the bracing has not been designed to resist the lateral loads that can potentially be developed in the structure during and after the placement of the concrete slabs.)

- Any modifications to this erection plan must be approved by (*Steel Erection Subcontractor*), and verified in writing by (*Engineers*). Periodic structural observation will be conducted by (*Engineers*) and copies of the site visit reports will be forwarded to (*Steel Erection Subcontractor*). (*Engineers*) anticipates performing site visits early in the steel erection sequence, periodically during the steel erection sequence and during the early stages of the Node Truss erection sequence.

- (*Steel Erection Subcontractor*) is responsible for any required Fall Protection Work Plan and/or a Site-Specific Safety Plan associated with their work. Work will be performed per site safety parameters and per (*Steel Erection Subcontractor*) Safety procedures. These plans and procedures are not engineered by or reviewed by (*Engineers*).

- The General Contractor, _____, is responsible for providing stable, cast-in-place concrete members (walls, columns, beams, etc.) to which (*Steel Erection Subcontractor*) can attach the structural steel. The stability of cast-in-place concrete columns and walls has not been reviewed by (*Engineers*). The structural steel, during erection, does not stabilize concrete columns or walls.

- All structural members shall be stable prior to releasing the crane from the member. Steel beams shall have a minimum of two erection bolts (size and grade to match final connection bolts) per connection before being released from the crane. An erection bolt shall be placed in both the top and bottom holes of each framing connection, for a total of 2 erection bolts per connection. Open web steel joists shall be secured back to stable structure with bridging and joist-bearing connections shall have a least one erection bolt in place prior to releasing the crane. All erection bolts shall be "snug tight" such that the two pieces of steel being connected are firmly in contact.

- Where connections are added by (*Steel Erection Subcontractor*) solely for the purpose of erection, all bolts used shall be 7/8-inch diameter A325 bolts, unless otherwise noted. All erection bolts shall be spaced as far apart as practical for maximum "inherent moment capacity" and shall be snug tight such that the two pieces of steel being connected are firmly in contact.

- Each column will be set on shim packs and leveling nuts as shown in Figure 14. All anchor bolt nuts and leveling nuts shall be secured to a "snug tight" position before releasing the crane. The base plates shall be grouted by the General Contractor, after (*Steel Erection Subcontractor*) has indicated the base plates are ready to be grouted.

- Columns are considered stable when they are adequately braced in both orthogonal directions. Adequate bracing is achieved with beams and girders tying the column back to a stable structure. The W14 x 730 jumbo columns shall be considered stable when all bolts connecting the W36 beam to the column are installed (i.e., welds in progress). Moment capacity at the base of a column due to shim packs and anchor bolts (or leveling nuts) shall only be used for "short-term" stabilization (within the time frame of a single work shift). At the end of any working day (either a normal work shift or if work is stopped for any reason), columns shall not be left free standing. Columns must be tied back to the existing and stable structure with girders or beams, unless otherwise noted in this erection plan.

- All temporary cable bracing shall be installed, when specified, by the steps outlined below and in accordance with Figures 5 through 7. The cables shall be pretensioned using a turnbuckle to remove any sag in the cable. Temporary cable braces, where required, will be a minimum of 1/2″-diameter IWRC with 5/8″-diameter shackles and 3/4″ turnbuckles, and shall be installed and removed according to this erection plan. Additional cables used solely for plumbing the structural steel may be installed and removed as required by the steel erector.

- The number and order of erecting columns in a zone prior to hanging beams and girders is at the discretion of (*Steel Erection Subcontractor*), unless noted otherwise in this erection plan. Please note that all columns must be stable at

the end of the working day (See Erection Note #8) and the steel erector should not stand more columns on any particular day than can be stabilized by the end of the working day.

- When a bay of structural steel framing is complete (every steel member in that bay is stable), the steel erector may land bundles of metal decking (for that bay only) and the decking may be laid out and tack welded to the steel framing at the steel erector's discretion. If additional bundles of metal decking are to be temporarily staged in that bay, the steel erector must ensure that all bolted connections within that bay are fully stuffed and all fillet welds, that are part of moment connections, are in their "final" state (full penetration welds may be in progress).

- Upon completion of the erection of the structural steel and the metal decking, the General Contractor is responsible for providing any additional shoring or bracing that may be required for placement of additional loads, such as cast-in-place concrete, on this level or any level above.

- Shear tabs that are to be attached to embedded plates within certain cast-in-place concrete members (i.e., columns, beams, and walls) shall be welded to their "final" state prior to attaching any structural steel to the shear tab. Welding of these shear tabs shall be coordinated between (*Steel Erection Subcontractor*) and the General Contractor.

- When erecting beams that are part of a gravity moment connection, install all bolts that are part of the connection. Where noted in this erection plan, the moment connection must be in its "final" state prior to erecting any beams that frame into the cantilever. A moment connection must be in its "final" state prior to erecting the deck above the connection.

- Decking can be installed within the sequence at (*Steel Erection Subcontractor*)'s discretion provided the steel is plumb and all connections below the deck are in their final state.

- The erection of the miscellaneous structural steel members, such as stairs and handrails, have not been accounted for in this procedure and is left up to the "means and methods" of (*Steel Erection Subcontractor*). If (*Steel Erection Subcontractor*) desires engineering advice for any of these miscellaneous items, (*Engineers*) would provide this additional service upon request.

- Prior to beginning structural steel erection in any zone, the following items must be complete, where applicable:
  - All cast-in-place concrete members within that zone must be stable and able to support and stabilize the steel during steel erection.
  - All bolted connections and fillet welds that are part of moment connections, within the zone(s) immediately below, must be in their "final" state. Complete and partial penetration welded connections shall be in progress.

- Within each Step there may be points at which the zone, or portions of the zone, becomes stable after the completion of a sub-step. This juncture is indicated with "STABLE" after the sub-step. At this point the steel that has been erected up to that time may be considered stable and the work shift may end, or the remainder of the zone may be placed prior to the end of the work shift.

- The installation of the temporary cable braces may be completed after the Step has been completed. However, the temporary cable braces must be installed either prior to the end of the working day, or before the erecting crew has finished the next Step in the erection plan.

- Prior to erection of open web steel roof joists, all node roof truss diagonal and chord member connections must be in their "final" state.

- For the purpose of this erection plan, connections shall be considered to be in their "final" state after all bolting and/or welding, as specified by the construction documents, has been completed, but prior to receiving special inspection, if required.

Area Turnover to General Contractor:

- Areas of the building will be turned over for access of other trades after the connections are in their "final" state and the metal decking has been set in place and permanently attached to the steel framing. Areas are to be inspected by the General Contractor and the Owner's testing agency and accepted in the same sequence as completion. For safety and efficiency, the need to remobilize to a completed area must be minimized. *Author's Note: many general contractors require pictures being taken and/or check-off lists completed and signed before accepting a floor being turned over.*

- It shall be the responsibility of the General Contractor to monitor current and forecasted wind speeds.

- (*Engineers*) has not reviewed the picking means and procedures intended to place steel members into final positions. Also, (*Engineers*) has not reviewed any of the means or procedures used to move the steel to the job site or from location to location on the job site.

- Figure _____ depicts minimum and typical required temporary diagonal erection cable bracing to be used as called for in this erection plan.

- Prior to beginning structural steel erection in any zone, all members within that zone must be stable and must be able to support and stabilize the steel during steel erection.

- Stabilize column base plates with 2-inch × 2-inch steel shim stacks using four (4) shim stacks per base plate. Locate a shim stack near the edge of each side of the base plate to create a balanced and symmetrical pattern per Figure _____.

Below is a section from a sequence plan:

12) Release tower crane from T12 and rehook to T12 near grid CB/2.
13) Release conventional crane from T12.
14) While holding onto T12 with tower crane, erect T13 with conventional crane. All cranes may be released. STABLE. Temporary lateral brace at north end of T8 may be removed. Conventional crane may leave work area.
15) Erect T15 with tower crane. CONNECTION OF T9 TO T8 MUST BE IN "FINAL" STATE PRIOR TO ERECTING T15. STABLE.
16) Erect T14 with tower crane. STABLE.
17) Erect T11 with tower crane. STABLE.
18) Erect T6 with tower crane. STABLE.
19) Erect T7 with tower crane. STABLE.
20) Erect T16 with tower crane.

21) Erect T17 with tower crane. DO NOT RELEASE TOWER CRANE FROM T17 UNTIL NEXT SUB-STEP IS COMPLETE.

22) Complete all welds for connection of T16 to T8. STABLE.

23) Release tower crane from T17.

24) Erect T18 with tower crane. STABLE.

25) See Erection Notes #8 & #22. Erect remaining open web steel joist and other structural steel pieces in zone in any order.

APPENDIX E: WEB RESOURCES FOR STEEL ERECTION SAFETY

Steel Joist Institute	www.steeljoist.com
U.S. Census Bureau	www.census.gov/epcd/ec97/def/23591.HTM
OSHA – Steel Erection	www.osha.gov/steelerection/index.html
Operation Safe Site – Subpart R	www.opsafesite.com/steel
4-Safety – Steel Erection	www.4-safety.com/steelerection.htm
Winter Simulation Conference – Structural Steel Erection Process	www.informs-cs.org/wsc99papers/136.PDF
Singapore Structural Steel Society	www.ssss.org.sg/
Industrial Machinery Company	www.imac.ca/technofocus/steel%20types%20discussed.htm
Association of Iron & Steel Engineers	www.aise.org/
Steel News	www.steelnews.org/
American Iron & Steel Institute	www.steel.org/
American Society Testing Materials	www.asme.org/education/
Light Gauge Steel Engineering Assoc.	www.lgsea.com/
Training and Inspection Research Center – Specifically related to Cranes, Rigging and Heavy Equip.	www.tirc.net
Wire Rope & Rigging Consultants	www.wrrc.com
AGC Web site	www.agc.org/safety-info/index.asp
American Institute of Steel Construction	www.aisc.org
Modern Steel Construction	www.aisc.org/MSCTemplate.cfm

■ ■ **About the Author** ■ ■

SCAFFOLDING SAFETY
by John Palmer, CSP

John Palmer, CSP, serves as Director of the Scaffold Training Institute, conducting scaffold safety training seminars and providing scaffold safety consulting services. His background includes a broad range of experience in the scaffolding industry, including general manager of scaffold rental and erection companies, safety trainer, consultant, and designer of patented scaffold components. He has a Bachelor of Science degree in Industrial Safety and Hygiene and is a Board Certified Safety Professional.

Scaffolding Safety

INTRODUCTION

A scaffold is defined as any temporary elevated (supported or suspended) work platform and its supporting structure (including points of anchorage) that is used for supporting employees or materials, or both. Note that there are three main points to the definition: it is elevated, it is temporary, and it supports either personnel or materials, or both.

Scaffolds are divided into two main categories: those supported from underneath or those suspended from above. OSHA promulgates specific rules for 25 different types of scaffolds in 29 CFR 1926.452. In addition, aerial lifts are a type of scaffold.

This chapter is divided into three main sections: identification of the various types of scaffolds, safety guidelines for supported scaffolds, and safety guidelines for suspended scaffolds. The first section below identifies some of the various types of scaffolds.

The fabricated frame (tubular welded frame) is the most popular supported scaffold in use in commercial construction. It is called frame scaffolding because it consists of prefabricated frame panels attached longitudinally by cross braces or other means (see Figure 1).

A tube and coupler scaffold consists of individual tubes, two inches in diameter, of various lengths attached by couplers. It is primarily used in industrial plants, refineries, and offshore work sites (see Figure 2).

System-type scaffolds, such as those shown in Figures 3 and 4, are used in commercial and industrial settings. These are similar to tube and coupler in that they are job-built from individual two-inch tubing of various lengths. However, system scaffolds have a proprietary, built-in locking device that is used to attach the horizontal components to the verticals. Therefore, couplers are not required to attach the horizontal components to the verticals.

■ ■ **Figure 1** ■ ■ **Scaffold supported from underneath**

■ ■ **Figure 3** ■ ■ **A system scaffold**

■ ■ **Figure 2** ■ ■ **Tube and coupler scaffold**

Suspended scaffolds come in various types as shown in Figures 5–13 (drawings courtesy of OSHA). Two-point suspended scaffolds are commonly used on the exterior of tall buildings while single-point suspended scaffolds are most often used in industrial settings.

Drawings of other types of scaffolds covered in OSHA 1926.452 are shown in Figures 14–27. Some of these are seldom used today but are still referenced in the regulation. Two types of aerial platforms are illustrated in Figure 28.

GENERAL SAFETY GUIDELINES FOR SUPPORTED SCAFFOLDS

This section will address some of the general safety guidelines to consider when planning a scaffold job. The first consideration is training personnel. All personnel who might work on a scaffold must have user training that covers such topics as fall protection, loading, electrical safety, material handling, falling-object protection and safe work practices [1926.454(a)]. All personnel involved in inspecting, erecting, or modifying scaffolds must be trained in design criteria, assembly procedures, loading, scaffold hazards, OSHA regulations, and manufacturer's recommended assembly instructions applicable to the type of scaffold being used [1926.454(b)]. A *competent person* must supervise erection and perform inspections every workshift [1926.451(f)(7) and 1926.451(f)(3)].

■ ■ **Figure 4** ■ ■ **Example of another type of system scaffold**

■ ■ **Figure 5** ■ ■ **Two-point suspended scaffold**

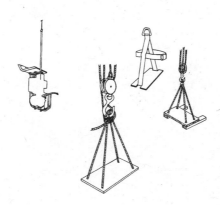

■ ■ **Figure 6** ■ ■ **Boatswain's chair scaffold**

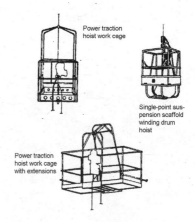

■ ■ **Figure 7** ■ ■ **Single-point suspended scaffolds**

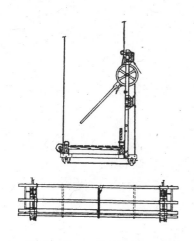

■ ■ **Figure 8** ■ ■ **A stone setter's adjustable multipoint suspended scaffold**

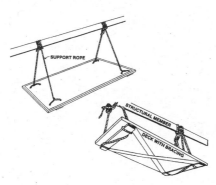

■ ■ **Figure 9** ■ ■ **Floating scaffolds**

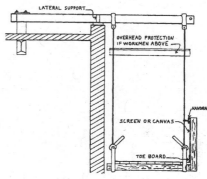

■ ■ **Figure 10** ■ ■ **A mason's adjustable multipoint scaffold**

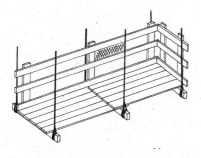

■ ■ **Figure 11** ■ ■ **Interior hung scaffold**

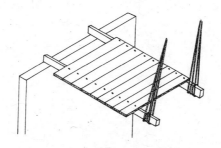

■ ■ **Figure 12** ■ ■ **Needle point scaffold**

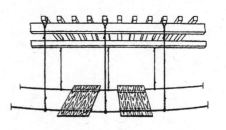

■ ■ **Figure 13** ■ ■ **A catenary scaffold**

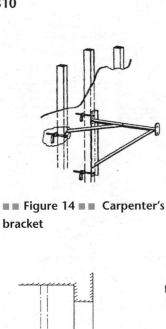

■■ Figure 14 ■■ Carpenter's bracket

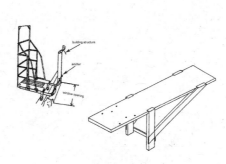

■■ Figure 15 ■■ A window jack

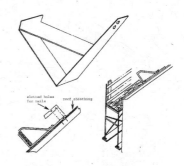

■■ Figure 16 ■■ Roof brackets

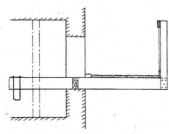

■■ Figure 17 ■■ An outrigger scaffold

■■ Figure 18 ■■ A pump jack

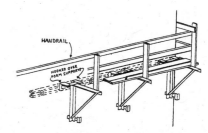

■■ Figure 19 ■■ A form scaffold

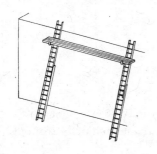

■■ Figure 20 ■■ A ladder jack

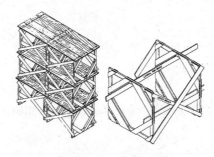

■■ Figure 21 ■■ A bricklayer's square

■■ Figure 22 ■■ Trestle ladder

■■ Figure 23 ■■ Mobile frame scaffold

■■ Figure 24 ■■ Plasterer's or decorator's scaffold

■■ Figure 25 ■■ Wood pole scaffold

■■ **Figure 26** ■■ **Horse scaffold**   ■■ **Figure 27** ■■ **Chicken scaffold**   ■■ **Figure 28** ■■ **Aerial platforms**

WARNING: SERIOUS INJURY OR DEATH CAN RESULT FROM IMPROPER ERECTION OR USE OF SCAFFOLDING EQUIPMENT. ERECTORS AND USERS MUST BE TRAINED IN, AND MUST FOLLOW SAFE PRACTICES. PROCEDURES, AND SPECIFIC SAFETY RULES.

A *qualified person* should design the scaffold job [1926.451(a)(6)]. Since unique conditions exist on each job site, consider the following items:

- proximity of electric lines 1926.45(f)(6), process piping, or overhead obstructions
- adequate access to the job site
- weather conditions and wind/weather protection
- openings, pits, and ground conditions
- adequate foundations of sufficient strength to support scaffolds from a sound, stable surface that assures support of the intended loads
- interference with other jobs or workers
- environmental hazards
- proper bracing that is rigid in all directions
- safe and easy means of access and egress to the platform
- fall protection for all workers using the scaffold
- adequate decking materials and overhead protection, where required
- falling-object protection of people passing or working near or underneath the scaffold
- planning for the loading (weight) on the scaffold.

Loading on the scaffold is a major item to consider when planning a scaffold job. Historically, the scaffold structure loading calculations have been based on one of three anticipated load ratings. *Light duty* is the term for up to 25 pounds per square foot. *Medium duty* is the term for up to 50 pounds per square foot. *Heavy duty* is the term for up to 75 pounds per square foot. The user should know how much weight they will place on the platform with workers, tools, and materials, and plan for the corresponding rating. The following example illustrates a 7-foot long by 5-foot wide platform with 25 pounds per square foot loading.

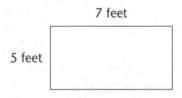

7 feet

5 feet

7 ft x 5 ft = 35 sq ft
35 sq ft x 25 lbs/sq ft = 875 lbs

In this 7-foot by 5-foot example (which happens to be the most popular frame scaffold platform size), up to 875 pounds could be placed on the platform. Since a worker and tools are rated at 250 pounds by industry standard, three workers could be using this platform, and that would still leave 125 pounds of additional capacity without exceeding the 25 pounds per square foot (PSF) usage.

Most crafts would not need to exceed this loading during normal usage. Consequently, 25 PSF or "light" duty is adequate for most crafts. The term light duty is somewhat misleading; standard duty would probably be a more fitting term for 25 PSF usage.

The next rating is called medium duty, or 50 PSF. In our example of a 7-foot by 5-foot platform, up to 1750 pounds could be placed on a medium-duty platform. Crafts that may place heavy stacks of materials on the platform will need this capacity. For example, a brick mason will place pallets of bricks, or possibly wheelbarrows full of mortar, on the platform. A brickmason should plan for 50 PSF usage.

The highest rating is called heavy duty, or 75 PSF. A heavy-duty 7-foot by 5-foot platform could support 2625 pounds. It is extremely rare that this much capacity is required (stonesetters may be the exception).

The scaffold user should communicate to the scaffold designer and erector which category of loading they will be imposing on the scaffolds, including the number of levels to be loaded simultaneously and any special point loads or cantilevered loads that might be imposed. Scaffold frame legs vary in the amount of weight they can carry. Scaffold users should obtain leg-loading information from the manufacturer of the product they are using. Some scaffolds will have load limits printed on a label directly on the scaffold. Many manufacturers provide simple tables giving a maximum number of light-, medium-, or heavy-duty platforms to be used simultaneously, depending upon scaffold height and configuration. For example, OSHA gives the following guidelines in the appendix for tube and clamp scaffolding:

Light Duty:
With one L. D. level in use at a time only, a maximum of 16 other levels with planks on them (additional planked levels).
With two L. D. levels in use simultaneously, a maximum of 11 other planked levels ready to use.
With three L. D. levels in use simultaneously, a maximum of 6 other planked levels ready to use.
With the maximum four L. D. levels in use simultaneously, only 1 additional planked level.

Medium Duty:
With one M. D. level in use at a time only, a maximum of 11 other planked levels ready to use.
With the maximum two M. D. levels in use simultaneously, one additional planked level ready to use.

Heavy Duty:
With the maximum one H. D. level in use at a time, 6 additional planked.

What this means is that there is a limit to how many levels can be loaded with the weight of workers or materials simultaneously. Notice that as the number of loaded levels increases, there is a decrease in the number of other levels that can be planked out (additional planked levels). The standard allows a scaffold to be erected up to 125 feet high before a design by a registered engineer is required. However, that does not mean as long as a scaffold is under 125 feet high that all of the levels can be planked out and loaded! Similar guidelines are available for frame scaffolds from the manufac-

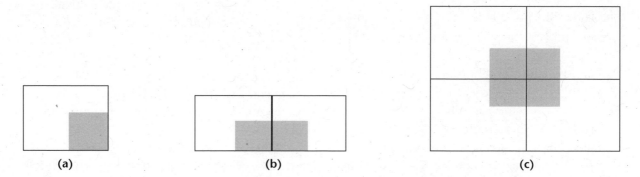

■■ **Figure 29** ■■ **(a) Four-leg scaffold contributory area, (b) Six-leg scaffold contributory area, (c) Center-leg scaffold contributory area**

turers. For example, one major frame manufacturer recommends placing no more than 100 pounds of live load per leg, with two additional planked levels allowed. This would equate to four working levels at 25 PSF, two at 50 PSF, or one at 75 PSF.

A leg-load calculation may also be done. This process involves determining the anticipated weight that can be imposed on the scaffold leg in that design. This should include the weight of scaffolding (referred to as dead load), and the weight from workers and materials (live load). The anticipated load is then compared to the manufacturer's published rated load for that frame leg in that configuration. The anticipated load must be less than the rated load. An explanation of how to calculate the leg load is beyond the scope of this chapter, but is available from scaffold training programs. Several factors come into play, such as the configuration of the scaffold. On rows of scaffolding the inner legs will be supporting the load from more than one platform, as shown in Figure 29.

The area which the leg is supporting is referred to as the contributory area. As the diagrams indicate, the contributory area for a center leg and the resultant weight that it carries is much greater than for the corner leg. Therefore, the weight imposed on a scaffold leg must be evaluated by determining the share of the weight of all the components that attach onto that leg that is being calculated (on a leg-by-leg basis). For example, on a tube and clamp scaffold, the leg (post) itself, the horizontals and diagonals that attach onto that post, the associated clamps, that leg's share of the weight of planks, toeboards, ladders, and other accessory equipment must all be included in the calculation of the weight for that specific leg. That leg's expected share of

live load (workers and materials) must also be included. This will vary depending upon the anticipated loading (25 PSF, 50 PSF, or 75 PSF) and the contributory area.

Other factors could include special loads, such as wind loads (particularly on enclosed scaffolds), or point loads if the scaffold is used as a staging area for heavy equipment.

If the limits given by the manufacturer's guidelines will be exceeded, a design must be done by a person qualified in leg-load calculation. The design might include changes such as putting in more posts so that each post carries less weight, "double legging" the posts, or using a higher capacity material such as heavy-duty shoring frames.

In addition to leg loading, the spans on planks are affected by the loading, as given in the generic table from OSHA for scaffold-grade planking (see Table 1).

Manufactured planks often have a label stating the maximum capacity. For natural wood planks, a maximum 250-pound point load would be a good conservative limit.

On tube and clamp scaffolds the bearer span is affected by loading. A maximum unsupported bearer span of 4 feet is recommended.

On system scaffolds, special truss-shaped bearers are recommended on longer spans.

The leg loading will also affect the size of the footing (mudsill) required, as discussed later.

In summary, planning for the proper loading is critical and requires communication between the scaffold user and the scaffold designer and erector. Areas of concern are overloading the posts (legs) by building the scaffold too high or placing weight on too many platforms simultaneously, overloading

■■ **Table 1** ■■

Effects of Loading on Scaffold Planks

Maximum intended nominal load (lb/ft^2)	Maximum permissible span using *full* thickness undressed lumber (ft)	Maximum permissible span using *nominal* thickness lumber (ft)
25	10	8
50	8	6
75	6	

a platform by putting more pounds per square foot on it than originally planned for, overloading a plank by point loading, or exceeding the recommended span on a plank or bearer. For specific loading information for the type of scaffold being used, consult the manufacturer.

PLANNING FOR FALL PROTECTION

Taking fall protection into consideration is extremely important when planning, erecting, and using a scaffold. By definition, a scaffold platform is elevated. Consequently, both scaffold users and erectors are exposed to fall hazards.

For most supported scaffolds, the guardrails provide the fall protection system for the user. If the guardrails are installed, a personal fall arrest is not required. If the guardrails are not in place, then a personal fall arrest system must be employed. Guardrail installation and requirements are covered later in this section. For boatswain's chair, catenary, float, needle beam, or ladder jack scaffold, a personal fall arrest system is required (instead of guardrails). For single-point or two-point suspended scaffolds, both guardrails and a personal fall arrest system are required. On suspended scaffolds, a vertical lifeline in combination with a rope grab is used. The anchorage point for the vertical lifeline system must be separate from the anchorage used for the scaffold support lines.

For scaffold erectors and dismantlers, the issue of fall protection becomes quite complex. The regulation states that fall protection for scaffold erectors is required if the competent person determines that it is "feasible and does not create a greater hazard" to employ it. The use of a personal fall arrest system (PFAS) is covered in detail in another chapter. The basic elements of a PFAS include a proper anchorage point that will hold 5000 pounds (or two times the impact force), proper body wear such as a full body harness, and an attachment device to connect the harness to the anchorage point, such as a shock-absorbing lanyard, self-retracting lifeline, vertical lifeline, or horizontal lifeline system. The difficult part of employing a PFAS during scaffold erection is finding a proper anchorage

point. Most scaffold manufacturers will not endorse the use of their product as a PFAS anchorage point (consult the manufacturer). The ability of a scaffold component to function as an anchorage point depends upon many factors, such as the type of scaffold, the materials used, the configuration of the scaffold, and required safety factors. In many cases, no anchorage point other than the scaffold itself will be available. Each scaffold must be evaluated separately to determine if it is feasible to implement a fall protection system during erection. In some cases, fall protection for erectors will be feasible, in other cases it will not. If it is determined to be feasible, then it is mandatory.

The photo in Figure 30 shows a full body harness and shock-absorbing lanyard being attached to the post of a system-type scaffold. As stated earlier, most manufacturers will not endorse their product as an anchorage point. However, if the scaffold is to be used as an anchorage, the post is the strongest component, compared to a guardrail or cross brace.

Figure 31 shows a variety of lanyards, snaphooks, and self-retracting lifelines. Lanyards are available with a variety of features such as built-in shock absorbers, large-opening snap hooks, "Y" configuration for 100 percent tie off, webbing, rope or steel line.

Figure 32 shows a rope grab. This is used in conjunction with a lifeline that is attached to an anchorage above the platform. For example, on suspended scaffolds, the lifeline may be anchored

■ ■ **Figure 30** ■ ■ **Attaching a lanyard to a scaffold post**

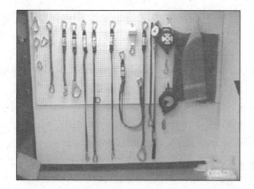

■ ■ **Figure 31** ■ ■ **Examples of lanyards, snaphooks, and self-retracting lanyards** (Photo courtesy of Spider, a Division of Safe Works, LLC)

■ ■ **Figure 32** ■ ■ **A rope grab** (Photo courtesy of Spider, a Division of Safe Works, LLC)

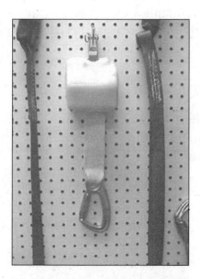

■ ■ **Figure 33** ■ ■ **Self-retracting lifelines** (Photo courtesy of Spider, a Division of Safe Works, LLC)

(a)

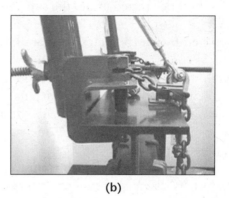

(b)

■ ■ **Figure 34** ■ ■ **(a) and (b) A horizontal lifeline system** (Photo courtesy of Spider, a Division of Safe Works, LLC)

on the roof a few hundred feet above. The rope grab allows travel up and down the lifeline, but will dig in and catch if a fall occurs. Lifeline elongation and pendulum effects must be considered.

Self-retracting lifelines like those shown in Figure 33 work like the seat belt in a car. The line will play out slowly. But if the line is pulled quickly, as would occur in a fall, the clutch engages and the line stops. The one on the left is lightweight and designed to be carried by the users. The anchorage must be directly above the employee or pendulum effects may occur.

Figures 34a and 34b show the posts to a horizontal lifeline system. The posts are attached to an I-beam and a horizontal lifeline is run between the posts.

In summary, planning for fall protection is one of the most important aspects of a scaffold job. The fall protection system might be guardrails, PFAS, or both. The type used will depend upon the type of scaffold. Providing fall protection for scaffold erectors may be particularly challenging due to the difficulty of locating an anchorage point that meets the OSHA requirements. The competent person must determine if it is "feasible and does not create a greater hazard" to provide fall protection for the scaffold erectors. If it is feasible, then it is mandatory.

Other Safety Guidelines

Scaffolds must be erected under the supervision of a competent person, and performed by a trained crew selected by the competent person. *Competent person* means one who is capable of identifying existing and predictable hazards in surroundings or working conditions that are unsanitary, hazardous, or dangerous to employees, and who has the authorization to take prompt, corrective measures that will eliminate them.

Personal protective equipment must be worn (i.e., hard hats, safety glasses, gloves, personal fall arrest equipment, and other clothing or devices as required). Tethering of tools, such as levels used above ground, is a good safety practice to reduce falling-object hazards.

Inspecting Equipment

Scaffolds are often used on construction sites, in industrial settings, and in other harsh environments. As a result, the equipment may be damaged during use as a result of erection, dismantling, handling, transporting, or environmental conditions.

All scaffolding equipment must be carefully inspected before use to ensure that it is serviceable and in good condition. Damaged or deteriorated equipment must be removed from service.

- Do not use scaffold equipment or accessories that are obviously damaged.
- Do not use rusty or corroded scaffold equipment. The strength of such equipment is unknown. If any areas show pitting, flaking, powdering, or excessive rust, discard the equipment. Certain atmospheric conditions (such as might be present in industrial plants) may corrode steel after a brief exposure. This corrosion may appear to be rust, only brighter in color.
- Check for bent components, in particular where the tube is kinked, flattened, or crushed.
- Check for cracks around welds, joints, or around the circumference.
- Look inside the tube and inspect for rust.
- Check moving parts, such as gravity locks, for freedom of movement.
- Check for brackets with deformed attachment hooks.
- Check the holes in cross braces for splitting out.
- Check manufactured planking for missing hooks, locks, or rivets; bent siderails; or a damaged walking surface. If the surface is plywood, check for rotten areas.
- Check castors for damaged brakes, axles, or stems.
- Look for any painted areas that appear blistered, cracked, or crazed, which may indicate prior damage.
- When in doubt about the condition of scaffold equipment, either discard the component, or consult your scaffolding supplier. Do not take chances with potentially defective equipment. Remember that scaffolding components are relatively inexpensive commodities to replace, especially compared to the cost of an accident.

THE FOUNDATION

Mud sills must be sized to distribute scaffolding loads to the ground or support structure.

The use of 2-inch × 10-inch pads, between 12 and 18 inches long, is common for scaffolds four levels or less in height. An example is shown in Figure 35. Another common type of sill is composed of planks running down the length of the scaffold and across under the posts. Do not reuse sill planks as scaffold planks. The following may be used to determine the size pads needed. First, calculate the weight imposed by the scaffold leg on the sill (leg load). Then divide that number by the square footage of the sill to determine the PSF imposed on the soil. Plan for a large enough sill to reduce the PSF imposed on the soil to match the soil capacity. The soil capacity should be determined by a competent person using guidelines from 29 CFR 1926 (Subpart P). As a conservative guideline based on a maximum 3000-lb leg load, a scaffold grade 2 inches × 10 inches that is 18 inches long is adequate for Type A soil, 18-inch-square pads should be constructed for Type B soil, and 3-foot × 3-foot pads should be used on type C soil (assumed for construction sites).

Don't use unstable objects (loose bricks, etc.) as a sill.

Base plates with screw jacks (Figure 36) should always be used so that you will be able to make leveling adjustments later.

1. Start at high ground; try to keep screw-jack extension to a minimum. While some jacks may extend to 18 inches, the capacity decreases as extension increases.
2. Make sure the jack handles firmly contact the frame leg. Settling or uneven leg loading may cause a leg to "raise up" off the handle slightly, as shown in Figure 37. In this case, recheck the level on the other legs as well. Check manufacturer's recommendations.

Plumb, level, and square the scaffold at the base.

1. Level the frame across from side to side and the legs in both directions.
2. Ensure the corner-to-corner measurement is identical. This guarantees 90-degree corner angles. In other words, the scaffold is "square" (actually a rectangle).

TYING THE SCAFFOLD TO A STRUCTURE

The scaffold must be secured to prevent it from tipping. This may be accomplished by tying the scaffolding to an adjacent structure, using guy wires, or increasing the base width. The narrower the scaffold, the more likely it will tip over. It is less stable in the narrow direction. Consequently, the vertical location of the first tie-in is a factor of the minimum base width (the narrower direction). As the scaffold

■ ■ **Figure 35** ■ ■ **An example of a 2 x 10 pad commonly used for scaffold sills**

■ ■ **Figure 36** ■ ■ **A base plate with screw jack**

■ ■ **Figure 37** ■ ■ **Scaffold leg that has lifted off of the handle due to settling or an uneven leg load**

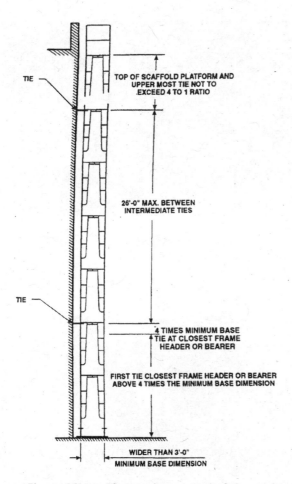

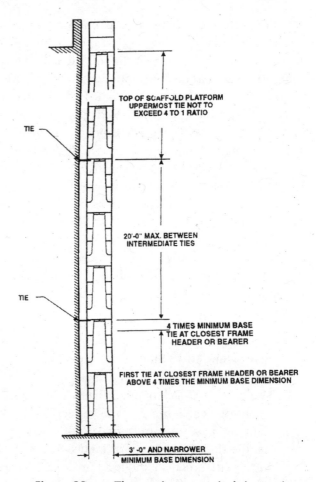

■■ **Figure 38** ■■ The maximum vertical tie spacing for scaffolds with bases wider than 3 feet

■■ **Figure 39** ■■ The maximum vertical tie spacing for scaffolds with bases of 3 feet or narrower

rises, the first tie must be no higher than four times the minimum base width. For example, if the scaffold is 5 feet wide and 21 feet long, the first tie should be at 20 feet (4 × 5). (*Note:* California requires stability bracing when the scaffold height exceeds three times the minimum base width.)

1. After securing the first tie-in at an elevation of four times the minimum base, two separate requirements are given for the next interval up.
   a. If the scaffold is wider than 3 feet, subsequent tie-ins must be secured to the building or structure at intervals no greater than 26 feet vertically, as shown in Figure 38.
   b. If the scaffold is less than 3 feet wide, the subsequent tie-ins must be at no greater

than 20-inch maximum vertical intervals (see Figure 39).

2. Ties should be placed as close as possible to the top of the scaffold, but no further down from the top than four times the minimum base dimension.
   a. These ties must be placed at both ends and every 30 feet, horizontally. For example, if the scaffold were 120 feet long, the ties would be at 30, 60, and 90 feet, and at both ends.
   b. Ties should be installed during the erection process and should not be removed until the scaffold is dismantled to that height.
   c. Anchoring, guying, tying off, or bracing of scaffolds must be affixed to structurally sound components. It is crucial

that the tie is properly attached and able to carry the tension and compression loads. If guying is necessary, have an engineer design the guy system.

d. Overturning forces are imposed by side brackets, hoist arms, cantilevered plat-forms, and machinery in use (e.g., hydro-blasting). Additional ties may be required. Consult an engineer for a design on enclosed scaffolds.

e. Circular scaffolds that are erected com-pletely around or inside a structure may be prevented from tipping by installing "stand-off" braces.

Beam clamps, such as the one shown in Figure 40, can be used to attach a scaffold tube to a beam. Wire could also be used to attach a scaffold to a structure, as shown in Figure 41.

Figures 42–46 show details of various butt/tie arrangements. Note that these may be adapted for

■■ Figure 40 ■■ Beam clamp

■■ Figure 41 ■■ A scaffold that is secured to a building using #9 gauge wire attached to an anchor bolt in combination with tube and clamp butter tube

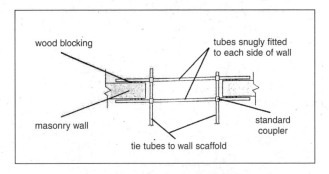

■■ Figure 42 ■■ Butt/tie detail, Type A-1

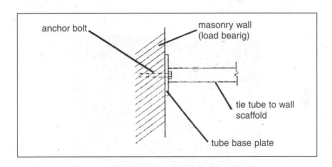

■■ Figure 43 ■■ Butt/tie detail, Type A-2

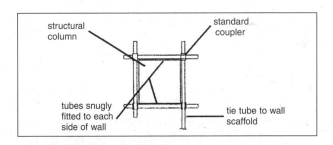

■■ Figure 44 ■■ Butt/tie detail, Type A-3

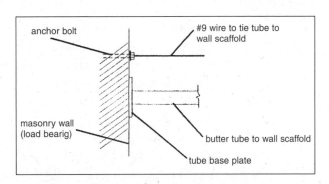

■■ Figure 45 ■■ Butt/tie detail, Type A-4

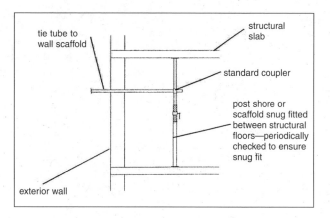

■■ **Figure 46** ■■ **Butt/tie detail, Type A-5**

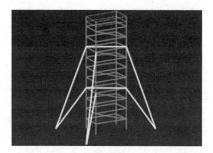

■■ **Figure 47** ■■ **Guy cables can provide lateral stability**

■■ **Figure 48** ■■ **Scaffold with access ladder**

industrial plant applications. Based on specific projects, the butt/tie arrangements in Figures 42–46 may be combined or used alone, such that the entire exterior wall scaffold is tied to the existing structure at intervals not to exceed 30 feet horizontally and 26 feet vertically, or as recommended by the scaffold manufacturer and/or the engineering design for the scaffold.

Wind Force

Wind can generate tremendous forces on scaffolds, especially on a scaffold enclosed by tarps, poly wrap, or other covering. For example, on an enclosed scaffold with a horizontal length of 100 feet and a vertical height of 100 feet, there is 10,000 square feet of surface area. A wind of 50 mph will exert a force of 12.3 PSF on the scaffold. This means the total force exerted on the scaffold would be equal to 123,000 lbs. Wind forces, such as described in the example, can easily push the scaffold against,

or pull it away from a structure. Scaffold failures have occurred when enclosed scaffolds were exposed to strong winds. Enclosed scaffolds will require additional horizontal diagonal bracing and additional tie-ins to the building or structure. Consult an engineer for a design on enclosed scaffolds.

When there is no structure available for anchor ties, guy cables may be used to provide lateral stability, as shown in Figure 47. (*Caution:* overtightening the cables can preload the legs and cause failure.)

1. Guying is not a substitute for vertical or horizontal bracing.
2. Guying requires careful analysis and design, which should be performed by a competent engineer. Unanticipated forces can easily overstress scaffold components.
3. Use the proper size wire rope or cable for guying.
4. Three clips should be used to make each cable connection. When tying to a structure,

make sure the structure is adequate to withstand the loads.

5. Provide guys to ensure stability as the scaffold is erected. Remove all slack from the cables, but do not overtighten them.

PLATFORM ACCESS

A platform should not be installed on a scaffold without a safe and easy means of access. An access ladder, such as the one shown in Figure 48, or equivalent safe access, must be provided for any platform more than 2 feet above a lower level.

Access may be provided by ladders specifically made for scaffold access by scaffold manufacturers, portable ladders, stair towers, stairway-type ladders, ramps, walkways, integral prefabricated rungs in a frame, or direct access from another structure.

1. Cross braces shall not be used as a means of access.
2. Follow the manufacturer's recommendations for ladders that are specifically manufactured for scaffold access and egress.
3. Attachable ladders made specifically for scaffolding must meet the following:
   a. be positioned so as not to tip the scaffold
   b. have the lowest rung within 24 inches of the lower level
   c. if higher than 35 feet, have a rest platform at 35 feet maximum
   d. have a horizontal rung length of at least 11 1/2 inches.
   e. have a maximum vertical rung spacing of 16 3/4 inches. (*Note:* Most manufacturers set the vertical spacing at 12 inches, to conform to fixed ladders.)
   f. Most manufacturer's recommend that the ladder extend 3 feet above the platform level. Ladders vary in strength of material and standard length. Consult the manufacturer for maximum ladder extension above the top of tube and coupler or system scaffold posts.
4. Stair towers, when used, shall meet the criteria in 1926.451(e)(4).

5. Ramps, when used, shall meet the criteria in 1926.451(e)(5).
6. Stairway-type ladders must meet the criteria in 1926.451(e)(3).
7. Most frames should not be climbed. However, some frames do have integral prefabricated scaffold access rungs and must meet the following criteria:
   a. must be specifically designed by the manufacturer as a ladder rung (*Note:* most horizontal members found on frames are intended for structural bracing or shelving platforms, not as ladder rungs.)
   b. have a rung length of at least 8 inches
   c. be uniformly spaced
   d. not be used as a work platform
   e. have a rest platform available at the maximum 35-foot intervals (Scaffold Training Institute (STI) recommends 20 feet).
   f. have maximum vertical spacing of 16 3/4 inches. (*Note:* This rules out climbing the end horizontals of most mason-style frames.)

The scaffold shown in Figure 49 has two means of access. A stair tower is on the left and a hook-on-style ladder is on the right. Though not required, the author believes that two means of access should be considered for large scaffolds. The photos in Figures 50 and 51 show the staircase.

■ ■ **Figure 49** ■ ■ **Scaffold with two means of access**

■■ Figure 50 ■■ Exterior view of staircase

■■ Figure 51 ■■ Interior view of staircase

■■ Figure 52 ■■ Platform access through a ladder swing gate

■■ Figure 53 ■■ Access to platform through a swing gate is shown

The next two photos show access through a swing gate through the ladder's side-rail extensions (Figure 52) and through an independent swing gate (Figure 53). The horizontal members on this frame cannot be used as a ladder.

The picture in Figure 54 shows a ladder built into a frame-type scaffold. It is used in combination with a trap door built into the plank.

The photos in Figures 55 and 56 show standard attachable scaffold ladders on tube and clamp and system scaffolds. The ladder bracket provides approximately 7 inches of standoff for foot clearance. Ladders vary in strength of material and standard length. The bracket attaches onto the post with a built-on clamp. Consult the manufacturer for maximum ladder extension above the top of the ladder bracket.

Selecting Planks

Planks may be solid sawn wood, manufactured wood, or manufactured steel or aluminum planks. If solid sawn is used, it must be of a specific grade as determined by a certified grading association or agency. The most popular solid sawn is southern pine. The

■■ Figure 54 ■■ Ladder on frame-type scaffold

■ ■ **Figure 55** ■ ■ **Ladder attached to system-type scaffold**

■ ■ **Figure 56** ■ ■ **Ladder affixed with clamps to a tube and coupler scaffold**

■ ■ **Figure 57** ■ ■ **Aluminum hook-on planks**

■ ■ **Figure 58** ■ ■ **Manufactured wood plank**

minimum grade is Dense Industrial 65 (determined by the Southern Pine Inspection Bureau).

Steel hook-on-type planks are available for system scaffolds from many manufacturers. These are made in various lengths to fit the runner length of the system. Most are approximately 9–10 inches in width, similar to a standard wood plank.

Aluminum hook-on planks are also very popular. These are often used with fabricated frame-type scaffolding. The plank may be all aluminum as in Figure 57, or it may have an aluminum frame with plywood deck. Some have other types of composite decks. The most common lengths are 7 feet and 10 feet. Most of these are the width of two regular wood planks, approximately 19 inches. Therefore, three aluminum planks fill out a 5-foot-wide scaffold, as opposed to six wood planks.

Manufactured wood planks made of laminated veneer lumber (LVL) are also available. Figure 58 shows a manufactured wood plank.

Longer aluminum planks and modular truss planks are also available. Single planks may be over 30 feet in length. Modular truss planks may be connected up to 80 feet. The drawing and photo in Figure 59 show the key components of one modular truss plank. Consult the manufacturer for specific

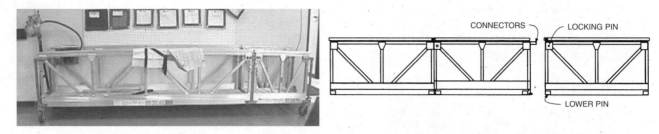

■ ■ **Figure 59** ■ ■ **Modular aluminum truss plank** (Photo courtesy of Spider, a Division of Safe Works, LLC)

Grade stamp courtesy of Southern
Pine Inspection Bureau

Grade stamp courtesy of West Coast
Lumber Inspection Bureau

■ ■ **Figure 60** ■ ■ **Two scaffolding-grade lumber stamps**

assembly, use, and loading instructions for these longer planks.

Figure 60 shows examples of two grade stamps that appear on lumber that can be used for scaffold planking.

The following guidelines should be followed for scaffold planking. Section (b) applies to natural wood. Section (c) applies to man-made decks and laminated planks.

WOOD SCAFFOLD PLANK

From 29 CFR 1926 Subpart L, Appendix A:

(b) Solid sawn wood used as scaffold planks shall be selected for such use following the grading rules established by a *recognized lumber grading association* or by an *independent lumber grading inspection agency*. Such planks shall be identified by the *grade stamp* of such association or agency. The association or agency and the grading rules under which the wood is graded shall be certified by the Board of Review, American Lumber Standard

Committee, as set forth in the American Softwood Lumber Standard of the U.S. Department of Commerce.

(c) Fabricated planks and platforms may be used in lieu of solid sawn wood planks. Maximum spans for such units shall be as recommended by the manufacturer based on the maximum intended load being calculated as follows:

	Intended load
Light-duty	• 25 pounds per square foot applied uniformly over the entire span area.
Medium-duty	• 50 pounds per square foot applied uniformly over the entire span area.
Heavy-duty	• 75 pounds per square foot applied uniformly over the entire span area.
One-person	• 250 pounds placed at the center of the span (total 250 pounds).
Two-person	• 250 pounds placed 18 inches to the left and right of the center of the span (total 500 pounds).
Three-person	• 250 pounds placed at the center of the span and 250 pounds placed 18 inches to the left and right of the center of the span (total 750 pounds).

■■ Figure 61 ■■ Work platform

PLATFORM PLANKING

Work platforms (such as the one shown in Figure 61) must be tightly planked over the full width of the scaffold. Edges must be close together (maximum 1-inch gap). If the last plank next to the posts will not fit completely, OSHA does allow a gap of up to 9 1/2 inches on the side next to the posts [1926.451(b)(ii)]. If a pipe or line extends vertically up through the deck, 3/4-inch plywood may be used to cover the gap, up to 18 inches. The plywood should extend all the way across the adjacent planks.

1. If an overhead hazard above the platform exists, overhead protection must be provided.
2. All planking should be scaffold grade or equivalent manufactured planking.
3. All platforms must be at least 18 inches (two boards) wide.
4. The author recommends that planks and/or platforms should be fastened to the scaffold as necessary to prevent uplift or displacement due to wind or other job conditions. A common way to accomplish this is by wiring the toeboard down and nailing the toeboards to the planks. The toeboards are cut to fit, with the long toeboard running across the ends of the planks and abutting the posts. On overlapped planks, a common practice is to "toenail" the planks together so that a tripping hazard is not created. However, OSHA does not require the planks to be secured, as long as the following provisions are met.

   a. Planks shall extend at least 6 inches beyond their support unless cleated or secured.
   b. Planks that are shorter than 10 feet may extend no more than 12 inches unless the platform is guardrailed to prevent access to the cantilevered area. Planks that are longer than 10 feet may extend no more than 18 inches unless the platform is guardrailed to prevent access to the cantilevered area.
   c. Planking on runs of scaffold must overlap a minimum of 12 inches.
5. Any damaged or weakened planks must be immediately removed and replaced.
6. Any spills or slippery conditions on the planking must be eliminated as soon as possible after they occur.
7. Where the scaffold changes points of direction, such as at a corner, the planks that would lay across the bearer at other than a right angle should be laid first. The planks that lay at a right angle are then laid on top. The result is that the ends of the top planks form a straight line rather than being sawtoothed, reducing the tripping hazard. This also ensures that the bottom planks overlap the bearer (see Figure 62).
8. Scaffold platforms and planks should not be painted to obscure the top or bottom since this might hide a defect.
9. Scaffold plank spans should be in accordance with the following: When using *nominal thickness* planking, the maximum span is 8 feet

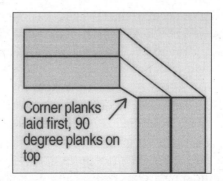

■■ Figure 62 ■■ Correct sequence for laying corner scaffold planks at other than a right angle

for normal 25 PSF loading, or 6 feet for 50 PSF loading. When using *full thickness* (*rough cut*) lumber, the maximum span is 10 feet for 25 PSF loading, 8 feet for 50 PSF loading, and 6 feet for 75 PSF loading.

a. As a general guideline, STI recommends loads on an individual plank should not exceed one worker or 250 pounds per board. (Some planks may be rated to carry 500 pounds.) The total platform load is limited by the uniform loading (e.g., 25 PSF). Consult the manufacturer for allowable loads on manufactured planks such as 7-foot aluminum boards. (Most are rated for 75 PSF.)

b. Loads on the planks should be evenly distributed when possible.

c. Try to stand on two boards whenever possible.

d. Do not overload individual planks or use them on too long a span. Do not overload the platform as a whole.

e. Plank the platform all the way across so the planks are secure when loaded.

Falling-object Protection

Platforms must provide the following safeguards against falling objects:

1. Toeboards must be installed on the open sides when the platform is more than 10 feet off the ground. Toeboards must be at least 3 1/2 inches in height with a maximum 1/4-inch gap to the platform. Toeboards must be secured to withstand 50 pounds applied in a downward or horizontal direction. (Toeboards on a platform are shown in Figure 63.)

2. If materials are piled higher than the toeboard, paneling or screening must be installed from the platform to the handrail.

3. Barricading the area below and not permitting employees to enter the area is given in the standard as an alternative to toeboards. This option should be used only in select areas.

4. Canopies, debris nets, or catch platforms may be alternatives in some areas.

Railings

Guardrails and midrails must be installed on all open sides (open sides are defined as more than 14 inches from a solid-faced structure) of scaffold platforms more than 10 feet high (CalOSHA specifies 7 1/2 feet).

1. Guardrails should be strong enough to withstand 200 pounds applied in a downward or horizontal direction.

2. Guardrails manufactured or placed in service prior to January 1, 2000, can be between 36 and 45 inches tall. After January 1, 2000, the minimum height is 38 to 45 inches. (CalOSHA requires gaurdrails at 42 to 45 inches).

3. The midrail is installed halfway between the platform and the guardrail.

4. Wire rope may be used as a guardrail but must not deflect more than the allowable heights.

5. Guardrails should be smooth-surfaced to prevent lacerations. Steel or plastic banding should not be used for a guardrail.

6. On frame scaffold intermediate levels where cross braces are present, STI recommends that both guardrail and midrail be installed in addition to the cross brace, as shown in

■ ■ **Figure 63** ■ ■ **Toeboards installed on a work platform**

■ ■ Figure 64 ■ ■ **Intermediate level of a scaffold with both guardrails and midrail installed in addition to cross braces, as recommended by STI**

Figure 64. However, OSHA does allow the cross brace to be substituted for either the guardrail or midrail, depending on the height of the cross-brace center point and other provisions [see 1926.451(g)(xv)].

The poles, legs, or uprights of scaffolds must be plumb and rigidly braced to prevent swaying or displacement.

Putlogs must not be used for material storage; they are designed for personnel use only.

1. Putlogs must overhang support points by a minimum of 6 inches.
2. Putlogs used for side or end brackets need special bracing.
3. Putlog spans greater than 10 feet require knee-bracing and lateral support.
4. Putlogs are made in lengths up to 22 feet. But that does not mean a putlog is designed to be used on standard 7-foot or 10-foot spans. The spacing between putlogs will vary depending on the load imposed (e.g., 25 PSF). In many cases the putlogs must be spaced closer together than the frame spacing, such as 2 feet on center.

(*Note:* Consult your scaffolding company for advice, special requirements, and engineering support when using putlogs.)

OSHA regulations state that components made by different manufacturers shall not be intermixed unless the components fit together without force and a competent person determines that the resulting structure still maintains the 4:1 safety factor [29 CFR 1926.451(b)(10)].

It should be noted that some manufacturers recommend against intermingling components.

Do not intermix components of dissimilar metals (such as steel clamps on aluminum tubing) unless a competent person has determined that a galvanic-action problem will not occur. (*Note:* Since scaffolds are erected for a short duration, galvanic action should not be a problem. Of more concern may be the possibility of a steel clamp crushing an aluminum tube.)

Improperly Erected Scaffolds

Figures 65 through 69 show scaffolds that were not erected properly. The most common hazards include poor foundation, lack of plumbness, lack of full planking, lack of fall protection (guardrails), lack of falling-object protection, lack of proper ladder access, and failure to secure the scaffold to the building. *Never use a scaffold with any of the defects shown in these photographs!*

Figure 65 is a picture of a scaffold that was poorly constructed: the foundation is not good; it does not have adequate planking, no guardrails or toeboards have been installed, and there is no ladder access.

■ ■ Figure 65 ■ ■ **A poorly constructed scaffold**

■ ■ Figure 66 ■ ■ A
scaffold with incom-
plete planking and
no guardrails or
toeboards

■ ■ Figure 67 ■ ■ Scaffold tower is not tied
to the building along with many other
apparent violations

■ ■ Figure 68 ■ ■ Close-up of the scaffold
in Figure 67—shows missing planking and
no ties

You can see from the photo in Figure 66 that
the planking is not complete, and it is obviously not
scaffold grade. This is also another example of a
scaffold without guardrails and toeboards in place.

Figure 67 shows a free-standing tower built on
the back side of the same house. It is four sections
high (approximately 26 feet), but it is not tied in to
the house. It is also not fully planked, has no guardrails or toeboards, and there are no screw jacks or
mudsills at the foundation.

Figure 68 is the end view of the same stand-
alone tower, again showing that it is not tied in to
the building at any location and that it is not fully
planked.

Figure 69 shows a close-up of the foundation
of the scaffold. Notice that loose cribbage is piled
up in an attempt to level the scaffold as opposed
to the proper mudsill, base plate, and screw jack
combination.

SUSPENDED SCAFFOLDS

Suspended scaffolds are supported by cables from
above. There are many types of suspended scaffolds,
but the most popular types are powered platforms
suspended by wire rope. Examples include work
cages suspended by one cable (single-point sus-

■ ■ Figure 69 ■ ■ Scaffold leg placed
directly on loose cribbage

pended scaffold), longer platforms supported by two cables (two-point suspended scaffold), and special platforms supported by multiple cables (multiple-point suspended scaffold). Two-point suspended scaffolds are often used to do maintenance on the outside of an office building, such as glazing or waterproofing. Inside industrial plants single-point suspended scaffolds are used inside boilers, stacks, and other vessels for sandblasting, painting, welding, and other repair work. On offshore oil rigs single-point suspended scaffolds are used under the rig platforms for maintenance.

The photos in Figures 70 and 71 show single-point suspended scaffolds. The scaffold in Figure 70 is electrically powered while the one in Figure 71 is air-assisted.

Figure 72 shows a two-point suspended scaffold. With modular platforms, as shown in Figure 73, the scaffold can be extended and erected to fit the building.

Figure 74 shows a single-point suspended scaffold hanging from I-beams under an offshore platform with its associated fall protection lifeline. The other rigging shown is used to transfer the support-line anchorage point from one I-beam to another. Consult the manufacturer about specific rigging equipment and instructions.

■ ■ **Figure 72** ■ ■ **Two-point suspended scaffold (Photo courtesy of Spider, a Division of Safe Works, LLC)**

■ ■ **Figure 73** ■ ■ **Modular scaffold providing full access to interior corner surfaces of a building (Photo courtesy of Spider, a Division of Safe Works, LLC)**

■ ■ **Figure 70** ■ ■ **An electrically operated single-point scaffold (Photo courtesy of Spider, a Division of Safe Works, LLC)**

■ ■ **Figure 71** ■ ■ **An air-powered single-point scaffold (Photo courtesy of Spider, a Division of Safe Works, LLC)**

■ ■ **Figure 74** ■ ■ **A single-point suspended scaffold (Photo courtesy of Spider, a Division of Safe Works, LLC)**

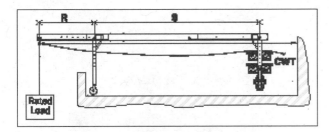

■ ■ **Figure 75** ■ ■ Illustration of the reach and span of a scaffold hoist (Photo courtesy of Spider, a Division of Safe Works, LLC)

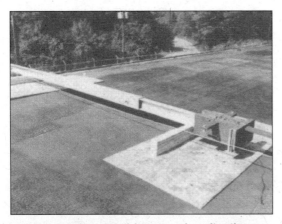

■ ■ **Figure 76** ■ ■ Cribbing used to distribute the load of a counterweight on a roof surface (Photo courtesy of Spider, a Division of Safe Works, LLC)

Basic Safety Guidelines

The rigging device and the adjoining structure should be capable of supporting four times the maximum rated load of the hoist motor. Hoists typically range in capacity from 1000 to 1500 pounds. Check the label on the hoist motor for the rating of any specific hoist unit. The cable must support six times the load.

The following formula shows that the amount of counterweight equals

$$\frac{4 \times \text{Rated hoist capacity} \times \text{Reach in feet}}{\text{Span in feet}}$$

Reach and span are illustrated in Figure 75.

Confirm that the roof deck is rated sufficiently to support the counterweight that will be used. It may be necessary to spread out the counterweight load on the roof by cribbing with planks as shown in Figure 76.

Tiebacks with strength equivalent to the hoisting ropes should be installed without slack at right angles to the building and firmly secured to a sound portion of the structure (Figure 77).

If the tieback cannot be installed at right angles to the structure face, "two tiebacks, without slack, shall be attached to each rope-supporting device to prevent movement in any direction." (see Figure 78).

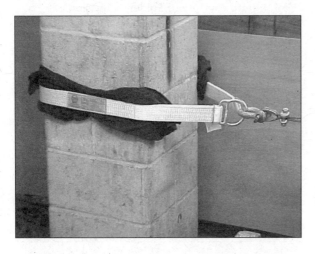

■ ■ **Figure 77** ■ ■ A tieback secured to a structural building component (Photo courtesy of Spider, a Division of Safe Works, LLC)

■ ■ **Figure 78** ■ ■ Secondary tieback attached to the back of the beam with directional anchoring (Photo courtesy of Spider, a Division of Safe Works, LLC)

Counterweights must be properly attached to ensure they cannot accidentally slip off (Figure 79).

Wire Rope Inspection and Service

The wire rope support line must have a 6:1 design/ safety factor: "Each suspension rope, including connecting hardware, used on adjustable suspension scaffolds shall be capable of supporting, without failure, at least six times the maximum intended load applied or transmitted to that rope. . . ." [1926.451(a)(4)]

Figure 80 shows the typical components of wire rope. Inspecting the suspension wire rope is very important. Wire rope is a consumable item—it wears out with normal use. The rate at which a wire rope weakens depends on where and how often it is used, how it is cared for, and the condition of the hoist it is used with. Regular inspection is needed to determine, to the greatest extent possible, whether the wire rope has enough useful life left to support a scaffold with the desired safety factor until the next inspection.

The typical wire rope used for many hoists is 5/16 of an inch, 6 × 19, right regular lay, Seale construction, made of improved plow steel with a rated breaking strength of 4.26 tons (see Figure 81). Never use a wire rope on any personnel hoist beyond

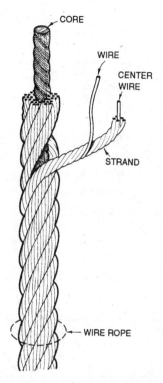

■ ■ **Figure 80** ■ ■ **The three basic components of a typical wire rope (Drawing courtesy of the Wire Rope Technical Board)**

■ ■ **Figure 79** ■ ■ **A shackle being attached to a counterweight unit. A stop (left) bolted to the rod keeps the counterweights from sliding off. (Photo courtesy of Spider, a Division of Safe Works, LLC)**

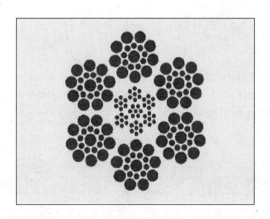

■ ■ **Figure 81** ■ ■ **Typical 6 x 19 Seale wire rope construction (Drawing courtesy of the Wire Rope Technical Board)**

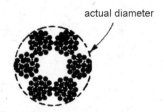

■ ■ Figure 82 ■ ■ The actual diameter of wire rope is shown in this cross-sectional illustration (Drawing courtesy of the Wire Rope Technical Board)

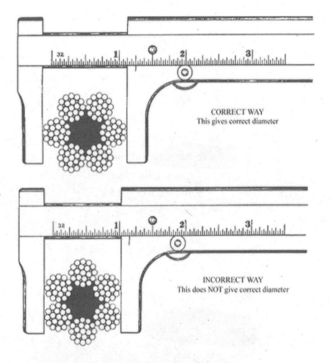

■ ■ Figure 83 ■ ■ How to correctly measure the diameter of wire rope (Drawing courtesy of the Wire Rope Technical Board)

■ ■ Figure 84 ■ ■ Fatigue fractures in a wire rope subjected to heavy loads. Here the usual crown breaks are accompanied by breaks in the valleys between the strands, caused by strand nicking—a result of the heavy loads (Drawing courtesy of the Wire Rope Technical Board)

the point where it can no longer support six times the intended load. The larger outer wires provide greater resistance to abrasion and crushing.

WHEN TO REPLACE THE WIRE ROPE

A wire rope can become unusable due to abrasion, corrosion, scrubbing, flattening, peening, kinking, exposure to excessive heat, and broken wires. More detail on wire rope is available in publications by the Wire Rope Technical Board (e.g., *Wire Rope Users Manual*, 3rd edition).

Replace the wire rope when it shows any signs of damage beyond the manufacturer's listed tolerances. Any wire rope that cannot safely support six times the hoist capacity should immediately be replaced.

The most common cause of premature wear is abrasion. Abrasion, corrosion, scrubbing, flattening, or peening that causes a loss of more than 1/3 of the diameter of the outside individual wires necessitates replacement of the rope. Figure 82 illustrates that the actual diameter of wire rope is proscribed by a circle.

Normal wear on the rope from the hoist sheaves will eventually reduce the rope diameter. When there is a reduction from the original diameter of 0.0156 inches, the rope should be replaced. An improperly maintained hoist motor can prematurely wear out the rope. Figure 83 shows both the correct and incorrect ways of measuring the diameter of wire rope.

Ropes with valley breaks (broken wire on the inside) should not be used for suspension. This is an indication that inside the rope there are wires that are fatigued or broken. The development of this condition might be reduced if a suitable cut-off practice were employed.

Wire ropes with six randomly distributed broken wires in one rope lay, or three broken wires in one strand within one rope lay (see Figure 84), must be removed from service and discarded.

Do not kink or bend the wire rope over a sharp edge. Allow wire rope to rotate so that it does not unravel or kink when tensioned. Align stirrups with rigging points to avoid excessive sideways wear on the rope or guides.

Additional Safety Guidelines for Suspended Scaffolds

- When you are welding from suspended scaffolds, be sure the platform is grounded to the structure.
- Insulate the wire rope above and below the platform to protect it from damage by a welding torch or electrode.
- Insulate the wire rope at the suspension point and be sure it does not contact the structure along its entire length. Coil any excess and hang it in the air to keep it from contacting the structure.
- Do not abuse, misuse, or use suspended scaffold equipment for purposes or in any ways for which it was not intended.
- Do not use a suspended scaffold unless you are wearing a properly attached fall arrest system.
- Erected suspended scaffolds should be continuously inspected by the user to ensure that they are maintained in a safe condition.
- Report any unsafe condition to a supervisor.
- Care must be taken when operating and storing equipment during windy conditions.
- Powered platforms must never be operated near live power lines unless proper precautions are taken. Consult the power service company for advice.
- Do not work on scaffolds if you feel dizzy, unsteady, or are impaired in any way by drugs or any other substances.
- Never take chances! If in doubt regarding safety or use of suspended scaffolds, consult your instructor/trainer/supervisor.
- Ensure the structure is checked for strength and security, and is capable of supporting loads.
- Ensure the scaffold is properly assembled under rigging points.
- Ensure that all mechanical fastenings are checked.
- Ensure the electrical source is identified and checked for proper connections.
- Ensure that tag lines are used to lower the cable drum hook end of the wire rope and the female end of the electric cord (for taller buildings) to prevent damage to the building. Use kellum grips to secure power cords.
- Ensure the outrigger systems are assembled with adequate counterweights installed to provide 4:1 safety margin.
- Ensure the hoist wire ropes are secured to rigging equipment with the proper screw-pin or bolt-type shackle.
- Ensure the screw-pin shackle is equipped with safety tie/bolt-type shackle with cotter key.
- Ensure that the tiebacks are looped through a rigging thimble eye (loop secured with a minimum of three fist grip clips). The tieback must be installed without slack and perpendicular to the building (or use two tiebacks). The connection of tiebacks to the building should be made with a loop and a minimum of three fist grip clips.
- Ensure that cornice hooks/roof irons have tieback secured to the opposite end from rigging end, as designed.
- Ensure that an electrical connection is made. Ground check the winch operations.
- Ensure that when using outriggers, or rolling outriggers, the tieback cable is first installed through the thimble eye and then through the eye or shackle on the outrigger before being secured to the building.
- Ensure that tag lines are connected to a stage if necessary to "float" the stage into position.
- Ensure that personal fall arrest includes safety lines that are properly secured, and that body harnesses, lanyards, and rope grabs are available.
- Ensure that saw horses, safety lines, and warning signs are positioned around the scaffold.
- Ensure that wire mesh screen is installed on the scaffold between the toeboard and top hand rail. If required, install wire mesh between the toeboard and the hand rail.
- Ensure that the scaffold is raised to a suspend position. Retorque wire rope connections and check all rigging.

INSPECTION

All scaffolds, whether supported or suspended, must be inspected every workshift by a competent person. OSHA describes the expectations for this "competent inspector" in the preamble to the regulation. The individual must have a thorough knowledge of the OSHA regulations, a knowledge of the structural integrity of the scaffold, and an ability to evaluate incidents, such as the effects of a dropped load on the scaffold. This type of training goes beyond user-only training. While each user should do a basic inspection of the scaffold before use, the official workshift inspection must be done by someone trained at the competent-person level. This inspection applies to all scaffolds. By definition, there is no such thing as a "permanent scaffold" or a temporary elevated work platform that is not a scaffold.

Training requirements for scaffold users are found in 1926.454(a). Training requirements for erectors and inspectors are found in 1926.454(b).

OTHER SCAFFOLDS

This chapter has focused primarily on fabricated frame scaffolds, tube and coupler (clamp) scaffolds, system-type scaffolds, and single-point and two-point suspended scaffolds.

These are the most widely used types of scaffolds. However, as illustrated in the first part of this chapter, there are over twenty-five different types of scaffolds listed in the OSHA regulation. Space does not allow detailed coverage of all of them.

The general requirements for scaffolds are covered in 1926.451. Additional requirements, separated by specific type of scaffold, are covered in 1926.452. Readers are encouraged to review 1926.452 for specific rules for these other types of scaffolds (e.g., ladder jack, pump jack, etc.). Also, the nonmandatory appendix at the end of the regulation contains recommendations by specific type of scaffold.

SUMMARY

This chapter has attempted to provide an overview of the basic safety issues related to scaffolds. The subject of scaffolding is so broad and encompasses so many different types of equipment that it cannot be examined in detail in one short chapter. The OSHA regulations and specific manufacturers' instructions should always be obtained and followed for the particular type of scaffold being used. More thorough training materials are available from a variety of sources.

GLOSSARY

Adjustable suspension scaffold: a suspension scaffold equipped with a hoist(s) that can be operated by employees on the scaffold.

Bearer (putlog): a horizontal transverse scaffold member (which may be supported by ledgers or runners) upon which the scaffold platform rests and that joins scaffold uprights, posts, poles, and similar members.

Boatswain's chair: a single-point adjustable suspension scaffold consisting of a seat or sling that is designed to support one employee in a sitting position.

Body belt (safety belt): a strap with the means both for securing it about the waist and for attaching it to a lanyard, lifeline, or deceleration device.

Body harness: a design of straps that may be secured about the employee in a manner that distributes the fall arrest forces over at least the thighs, pelvis, waist, chest, and shoulders, with a means for attaching it to other components of a personal fall arrest system.

Brace: a rigid connection that holds one scaffold member in a fixed position with respect to another member, or to a building or structure.

Bricklayer's square scaffold: a supported scaffold composed of framed squares that support a platform.

Carpenter's bracket scaffold: a supported scaffold consisting of a platform supported by brackets attached to building or structural walls.

Catenary scaffold: a suspension scaffold consisting of a platform supported by two essentially horizontal and parallel ropes attached to structural members of a building or other structure. Additional support may be provided by vertical pickups.

Chimney hoist: a multipoint adjustable suspension scaffold used to provide access to work inside chimneys. (See *Multipoint adjustable suspension scaffold.*)

Cleat: a structural block used at the end of a platform to prevent the platform from slipping off its supports. Cleats are also used to provide footing on sloped surfaces such as crawling boards.

Competent person: one who is capable of identifying existing and predictable hazards in surroundings or working conditions that are unsanitary, hazardous, or dangerous to employees, and who has authorization to take prompt corrective measures to eliminate them.

Continuous run scaffold (run scaffold): a two-point or multipoint adjustable suspension scaffold constructed using a series of interconnected braced scaffold members or supporting structures erected to form a continuous scaffold.

Coupler: a device for locking together the tubes of a tube and coupler scaffold.

Crawling board (chicken ladder): a supported scaffold consisting of a plank with cleats spaced and secured to provide footing for use on sloped surfaces such as roofs.

Deceleration device: any mechanism, such as a rope grab, rip-stitch lanyard, specially woven lanyard, tearing or deforming lanyard, or automatic self-retracting lifeline lanyard, which dissipates a substantial amount of energy during a fall arrest, or limits the energy imposed on an employee during fall arrest.

Double pole (independent pole) scaffold: a supported scaffold consisting of a platform(s) resting on cross beams (bearers) supported by ledgers and a double row of uprights independent of support (except ties, guys, braces) from any structure.

Equivalent: alternative designs, materials, or methods to protect against a hazard, which the employer can demonstrate will provide an equal or greater degree of safety for employees than the methods, materials, or designs specified in the standard.

Exposed power lines: electrical power lines that are accessible to employees and that are not shielded from contact. Such lines do not include extension cords or power-tool cords.

Eye or Eye splice: a loop with or without a thimble at the end of a wire rope.

Fabricated decking and planking: manufactured platforms made of wood (including laminated wood, and solid sawn wood planks), metal, or other types of materials.

Fabricated frame scaffold (tubular welded frame scaffold): a scaffold consisting of a platform(s) supported on fabricated end frames with integral posts, horizontal bearers, and intermediate members.

Failure: load refusal, breakage, or separation of component parts. Load refusal is the point where the ultimate strength is exceeded.

Float (ship) scaffold: a suspension scaffold consisting of a braced platform resting on two parallel bearers and hung from overhead supports by ropes of fixed length.

Form scaffold: a supported scaffold consisting of a platform supported by brackets attached to formwork.

Guardrail system: a vertical barrier, consisting of, but not limited to, toprails, midrails, and posts, erected to prevent employees from falling off a scaffold platform or walkway to lower levels.

Hoist: a manual or power-operated mechanical device to raise or lower a suspended scaffold.

Horse scaffold: a supported scaffold consisting of a platform supported by construction horses (saw horses). Horse scaffolds constructed of metal are sometimes known as trestle scaffolds.

Independent pole scaffold: (see *Double pole scaffold*).

Interior hung scaffold: a suspension scaffold consisting of a platform suspended from the ceiling or roof structure by fixed-length supports.

Ladder jack scaffold: a supported scaffold consisting of a platform resting on brackets attached to ladders.

Ladder stand: a mobile, fixed-size, self-supporting ladder consisting of a wide, flat tread ladder in the form of stairs.

Landing: a platform at the end of a flight of stairs.

Large area scaffold: a pole scaffold, tube and coupler scaffold, systems scaffold, or fabricated frame scaffold erected over a substantial work area. For example, a scaffold erected over the entire floor area of a room.

Lean-to scaffold: a supported scaffold that is kept erect by tilting it toward, and resting it against, a building or structure.

Lifeline: a component consisting of a flexible line that connects to an anchorage at one end to hang vertically (vertical lifeline), or that connects to anchorages at both ends to stretch horizontally (horizontal lifeline), and which serves as a means for connecting other components of a personal fall arrest system to the anchorage.

Lower levels: areas below the level where an employee is located and to which an employee can fall. They include, but are not limited to, ground levels, floors, roofs, ramps, runways, excavations, pits, tanks, materials, water, and equipment.

Mason's adjustable supported scaffold: (see *Self-contained adjustable scaffold*).

Mason's multipoint adjustable suspension scaffold: a continuous run suspension scaffold that is designed and used for masonry operations.

Maximum intended load: the total load of all persons, equipment, tools, materials, transmitted loads, and other loads reasonably anticipated to be applied to a scaffold or scaffold component at any one time.

Mobile scaffold: means a powered or unpowered, portable, caster or wheel-mounted supported scaffold.

Multilevel suspended scaffold: a two-point or multipoint adjustable suspension scaffold with a series of platforms at various levels resting on common stirrups.

Multipoint adjustable suspension scaffold: a suspension scaffold consisting of a platform(s) suspended by more than two ropes from overhead supports and equipped with a means to raise and lower the platform to desired work levels. Such scaffolds include chimney hoists.

Needle beam scaffold: a platform suspended from needle beams.

Open sides and ends: the edges of a platform that are more than 14 inches (36 cm) away horizontally from a sturdy, continuous vertical surface (such as a building wall) or a sturdy, continuous horizontal surface (such as a floor), or a point of access. *Exception:* For plastering and lathing operations, the horizontal threshold distance is 18 inches (46 cm).

Outrigger: the structural member of a supported scaffold used to increase the base width of a scaffold in order to provide support for and increased stability of the scaffold.

Outrigger beam (thrustout): the structural member of a suspension scaffold or outrigger scaffold that provides support for the scaffold by extending the scaffold point of attachment to a point out and away from the structure or building.

Outrigger scaffold: a supported scaffold consisting of a platform resting on outrigger beams (thrustouts) projecting beyond the wall or face of the building or structure, the inboard ends of which are secured inside the building or structure.

Overhand bricklaying: the process of laying bricks and masonry units such that the surface of the wall to be jointed is on the opposite side of the wall from the mason, requiring the mason to lean over the wall to complete the work. It includes mason tending and electrical installation incorporated into the brick wall during the overhand bricklaying process.

Personal fall arrest system: a system used to arrest an employee's fall. It consists of an anchorage, connectors, a body belt or body harness, and may include a lanyard, deceleration device, lifeline, or combinations of these.

Platform: a work surface elevated above lower levels. Platforms can be constructed using individual wood planks, fabricated planks, fabricated decks, and fabricated platforms.

Pole scaffold: [see definitions for *Single-pole scaffold* and *Double (independent) pole scaffold*].

Power-operated hoist: a hoist that is powered by other than human energy.

Pump jack scaffold: a supported scaffold consisting of a platform supported by vertical poles and movable support brackets.

Putlog scaffold: a type of scaffold erected on base plates and tied into brickwork by means of putlog tubes which have a flattened end, or have been fitted with a blade. This feature allows the end of the tube to be inserted within the brickwork, through spaces left by bricklayers for this purpose, or to rest upon the brickwork of a structure. The holes are later filled when the scaffold is dismantled.

Qualified person: someone who, by possession of a recognized degree, certificate, or professional standing, or who by extensive knowledge, training, and experience, has successfully demonstrated his or her ability to solve or resolve problems related to the subject matter, the work, or the project.

Rated load: the manufacturer's specified maximum load to be lifted by a hoist or to be applied to a scaffold or scaffold component.

Repair bracket scaffold: a supported scaffold consisting of a platform supported by brackets that are secured in place around the circumference or perimeter of a chimney, stack, tank, or other supporting structure by one or more wire ropes placed around the supporting structure.

Roof bracket scaffold: a rooftop supported scaffold consisting of a platform resting on angular-shaped supports.

Runner (ledger or ribbon): a lengthwise horizontal spacing or bracing member which may support the bearers.

Scaffold: any temporary elevated platform (supported or suspended) and its supporting structure (including points of anchorage) that is used for supporting employees or materials, or both.

Self-contained adjustable scaffold: a combination supported and suspension scaffold consisting of an adjustable platform(s) mounted on an independent supporting frame(s) not a part of the object being worked on, and which is equipped with a means to permit the raising and lowering of the platform(s). Such systems include rolling roof rigs, rolling outrigger systems, and some mason's adjustable supported scaffolds.

Shore scaffold: a supported scaffold which is placed against a building or structure and held in place with props.

Single-point adjustable suspension scaffold: a suspension scaffold consisting of a platform suspended by one rope from an overhead support and equipped with means to permit the movement of the platform to desired work levels.

Single-pole scaffold: a supported scaffold consisting of a platform(s) resting on bearers, the outside ends of which are supported on runners secured to a single row of posts or uprights, and the inner ends of which are supported on or in a structure or building wall.

Stair tower (scaffold stairway/tower): a tower comprised of scaffold components that contains internal stairway units and rest platforms. These towers are used to provide access to scaffold platforms and other elevated points such as floors and roofs.

Stall load: the load at which the prime mover of a power-operated hoist stalls or the power to the prime mover is automatically disconnected.

Step, platform, or trestle ladder scaffold: a platform resting directly on the rungs of step ladders or trestle ladders.

Stilts: a pair of poles or similar supports, with raised footrests, used to permit walking above the ground or working surface.

Stone setter's multipoint adjustable suspension scaffold: a continuous run suspension scaffold that is designed and used for stone setter operations.

Supported scaffold: one or more platforms supported by outrigger beams, brackets, poles, legs, uprights, posts, frames, or similar rigid support.

Suspension scaffold: one or more platforms suspended by ropes or other nonrigid means from an overhead structure(s).

System scaffold: a scaffold consisting of posts with fixed connection points that accept runners, bearers, and diagonals that can be interconnected at predetermined levels.

Tank builder's scaffold: a supported scaffold consisting of a platform resting on brackets that are either directly attached to a cylindrical tank or attached to devices that are attached to such a tank.

Top plate bracket scaffold: a scaffold supported by brackets that hook over or are attached to the top of a wall. This type of scaffold is similar to carpenter's bracket scaffolds and form scaffolds; it is used in residential construction for setting trusses.

Tube and coupler scaffold: a supported or suspended scaffold consisting of a platform(s) supported by tubing, erected with coupling devices connecting uprights, braces, bearers, and runners.

Tubular welded frame scaffold: (see *Fabricated frame scaffold*).

Two-point suspension scaffold (swing stage): a suspension scaffold consisting of a platform supported by hangers (stirrups) suspended by two ropes from overhead supports and equipped with the means to permit the raising and lowering of the platform to desired work levels.

Unstable objects: items whose strength, configuration, or lack of stability may allow them to become dislocated and shift and therefore may not properly support the loads imposed on them. Unstable objects do not constitute a safe base support for scaffolds, platforms, or employees. Examples include, but are not limited to, barrels, boxes, loose brick, and concrete blocks.

Vertical pickup: a rope used to support the horizontal rope in catenary scaffolds.

Walkway: a portion of a scaffold platform used only for access and not as a work level.

Window jack scaffold: a platform resting on a bracket or jack that projects through a window opening.

REFERENCES

U.S. Department of Labor. *29 CFR 1926.* Washington, DC: U.S. Government Printing Office.

_____. *29 CFR 1910.* Washington, DC: U.S. Government Printing Office.

Scaffold Training Institute. *Competent Person Manual.* Houston, TX: STI, 1994.

Spider, Division of Safeworks, LLC. *Competent Person Manual.* Tukwila, WA: Safeworks, LLC, 2001.

Credits

Pictures with no credit assigned were reprinted with permission of the Scaffold Training Institute. Scaffolds shown in pictures taken at the Scaffold Training Institute were manufactured by various companies, including Betco Scaffold, Patent Construction System/SGB, Aluma Scaffold, Baker Scaffold, and Theil Manufacturing.

REVIEW EXERCISES

1. What safety factor must be built into all supported scaffolds?
2. Scaffolds must be constructed and loaded in accordance with their design. Who is responsible for designing a scaffold?
3. Under whose supervision and direction can scaffolds be erected, moved, dismantled, or altered?
4. How frequently should scaffolds and their components be inspected for visible defects by a competent person?
5. Which scaffold personnel must undergo training?

■■ **About the Author** ■■

CRANE SAFETY ON CONSTRUCTION JOB SITES
by David V. MacCollum, P.E., CSP

David MacCollum, P.E., CSP, is a licensed Professional Engineer in
two states and a Certified Safety Professional. He has had a professional
career of over fifty years in safety, construction, and related high-hazard
industries. He served as one of the original members of the U.S. Department of Labor's Construction Safety Advisory Committee established
under the Construction Safety Act, the forerunner to OSHA. His experience includes safety engineering for the U.S. Army Corps of Engineers
and other worldwide commands of the U.S. Army. In the 1950s he developed the design of rollover protective canopies for a variety of mobile
construction equipment, which was some ten years prior to the SAE
standard for rollover protective structures (ROPS). In private practice,
he served as a consultant to a Swiss tunnel-support manufacturer and
became familiar with construction practices worldwide. For over thirty
years he has served as an expert on construction safety matters. He is
concluding his career as one of the principal founders of the Hazard
Information Foundation, Inc. (HIFI), which is a not-for-profit public
service group. HIFI's principal function is to develop public awareness
of available engineering that can save lives. David MacCollum was
recognized as an ASSE Fellow in 1999.

CHAPTER **18**

Crane Safety on Construction Job Sites

LEARNING OBJECTIVES

- Discuss how most crane accidents are the result of management's failure to address crane safety during:
  - initial project planning
  - procurement of equipment
  - preparation of the work site to ensure safe crane use.
- Discuss how crane hazards are unsafe physical conditions that are best overcome during the design stage by management's choice of the safest methods and equipment available.
- Identify the six most frequent causes of crane injury and death.

INTRODUCTION

Examination of several thousand serious injuries and deaths involving the use of cranes on construction sites reveals one tragic fact—*most of them were the result of omissions of hazard prevention measures that occurred long before the work began at the job site.* This is not intended to place blame, but to give a wake-up call. Executive management must fulfill its obligation to play an active leadership role in all construction planning. Training operating personnel has done wonders to further control of construction hazards, but now is the time to redirect priorities to include crane safety training for both construction management and designers. This training should present detailed information to them on the humanitarian and moral aspects of hazard prevention and its profitability. Also, when management has failed to address safety, it loses its public credibility when disaster strikes. The current business climate seems to favor short-term economics, which results in many hazards being accepted as the *custom or practice* in construction, with any resulting injury or death considered a *cost of doing business.* It often seems that nothing is done until these costs become so excessive that the hazards have to be eliminated to keep the business from failing.

The American work ethic includes performing tasks in a safe manner. It is widely held that employees are personally responsible for their own safety, and usually this is a valid assumption. However, many workplace hazards are beyond a worker's perception, ability to avoid, or authority to eliminate. It is easy to recognize that driving a car under the

341

influence of alcohol or drugs is unsafe behavior that causes many roadway injuries and deaths. Unsafe construction-worker behavior is a pattern that is also easily recognized. Management should view any dangerous use of heavy equipment as a red warning flag, and any such conduct should be considered unacceptable for continued employment.

Rather than attempt to cover all aspects of crane safety, such as understanding load charts, licensing of crane operators, planning, training, and accident investigations, this chapter provides an overview for the reader of the dramatic changes that are taking place in crane operations. The majority of cranes are leased, not owned, and operation varies from skilled professionals to incompetent operators. Many excellent references are readily available that focus on operator performance and safe rigging practices, and are too numerous to list in this chapter. To save lives, management's attitude toward safety and its responsibility for it must be reshaped: It must realize that by loosening the purse strings to provide safe cranes and ensure their safe use, it has opened the door for more profitability.

CRANE DESIGN AND SAFE USE

This chapter focuses on the *urgency* of improving executive management's safety leadership in the design of cranes and their use (either directly or indirectly). To improve safety performance, management needs to know the root cause of workplace injuries so it has a clear understanding of its safety role. Unfortunately, instead of eliminating hazardous conditions through safer design or methods, their first attempt is often to *kill the messenger* by ignoring the advice of the safety professional. Winston Churchill once made a comment that you can trust Americans to always do the right thing—after they have tried everything else. This truism is often apparent in the process of construction safety when the adoption of safer equipment and methods is delayed. Progressive management avoids these traps and immediately focuses its attention on utilizing available technology to control hazardous conditions and circumstances. This is far better than allowing

hazards to exist for years, maiming and killing until minimal consensus safety standards are upgraded. This chapter gives a broad overview of some of the more repetitive crane hazards that have caused crippling injuries and deaths and emphasizes use of current hazard prevention technology that will make the construction site safer.

Two major routes are available to achieve safety in crane operation on construction sites:

- management involvement
- engineering improvements.

Management's Role

Certain duties cannot be delegated to someone else. One such duty is ensuring the safe use of cranes brought onto construction sites. *Construction management* owns this duty, and no one else can be responsible for it. Managers who oversee the design, manufacture, sale, and leasing of cranes also have the same duty when they provide cranes for use on construction sites because all cranes must be safe for their intended use. Management is often tempted to adopt a quick fix as a *cure-all*; however, this can easily backfire, making management accountable when crane disasters result in crippling injuries, fatalities, and costly property damage. These quick fixes may also lead project supervisors to the false belief that workers can be trained to cope with dangerous circumstances or unsafe equipment. Instead of falling for quick fixes, management must be vigilant in seeing that proper equipment is provided and safe methods used. Blaming the victim never eradicates the hazard. In plain English, the responsibility for crane safety is often delegated downward to the lowest operating level where hazard prevention options are very limited. The assumption that the worker is responsible for his own safety is invalid when management provides an unsafe crane or orders unsafe work methods. For example, many crane-injury investigations indicate that the lifting operation undertaken with the crane was error-provocative and the injured worker had no control over the inherent hazards.

Before any crane is brought onto the job site, it must pass the muster. Each individual in management's chain of command must exercise his or her authority to see that the proper type of crane is selected, that it is equipped with feasible safety devices (operator aids), that the job-site location is safe, and that a qualified operator is employed. Next, *before lifting operations begin,* management must require the development of a safe lift plan and must see that it is reviewed with the work crew.

Management that cooks the books and engages in fraudulent activities is not always above allowing unsafe crane use. Good construction safety planning has no tolerance for any party who is involved in the use of a crane and allows hazardous conditions to exist. Construction safety planning is a methodology in which each step must be analyzed to identify each hazard that may be encountered, starting with the buying or leasing of a crane adequate for its proposed use. From there, each aspect of all proposed lifts must be reviewed, the competency of each crew member must be determined, the details of each movement of the crane must be considered, and the appropriate preventive measures and/or devices must be provided. Prudent and diligent crane management recognizes the cost benefits of investing in a properly equipped crane and the presence of a competent operator and crew. Of utmost importance is providing ongoing top-management oversight. Crane operations always have high visibility. The public loves to watch cranes in action—they are so fascinating to watch. Plenty of eyes are usually around a construction site, and they never miss a thing when someone is injured or killed. Crane safety affects the worker, the public, productivity, and business credibility. Management oversight is a bargain compared to the costly oversight of the courts when liability lawsuits are filed after an injury or death.

A construction safety manager in today's world plays a key role in educating management's decision-makers about their duty to actively oversee and control all aspects of crane use from the moment a crane is purchased or leased. Top management must see that funds are made available so the proper type of crane and necessary safety accessories are provided. Management's ignorance about any life-threatening use of cranes on a project does not always absolve them from strict liability for injury or death.

Toolbox safety meetings and employee training are a farce if management fails to provide crane safety hardware and to coordinate crane safety support functions. We should not train workers to use unsafe equipment. In effect, without management's involvement, crane hazards can become *a recipe for disaster.*

Engineering Advances

Engineering improvements are playing a vital role in the rapid development of sophisticated crane design. Cranes and aircraft have a lot in common when their development is compared. As jet-powered aircraft have become more complex, and as both commercial and private plane ownership has accelerated, there is more and more focus on providing design aids for pilots. Right now, cranes are in a phase of development as dramatic as the change aircraft underwent several decades ago. The incorporation of electronics to make a crane an integrated system of mechanical, hydraulic, and electrical components that depend on computer software has created a new priority for the safe design of cranes. No longer can crane designers consider available technology as "frivolous bells and whistles that give a false sense of security to the crane operator." Just as World War I flying aces and the stunt barnstormers of the 1920s have been replaced by highly trained and licensed pilots for particular aircraft, so must today's crane operators be held to the same level of performance through training and licensing. Located in Fairfax, Virginia, the National Commission for Certification of Crane Operators is considered to be the largest and most authoritative organization that can certify crane operators in the United States [2750 Prosperity Avenue, Suite 620-CCO; Fairfax, VA 22031-4312 (703)560-2391].

No longer can operators judge the lifting capacity of a crane by the "seat of their pants," as they did in the past, assuming a crane was overloaded when it started to become light and raised off its outriggers. On some of the newer cranes, the extra-heavy counterweights increase the tipping load far beyond the designed structural strength of the boom, and use of a load-moment indicator (LMI) is a must to limit its lifting capacity within its rated capacity, which is based upon the structural strength of the boom. Cranes today vary as much as aircraft when we take into consideration helicopters and fixed-wing, jet-powered, and propeller models. Crane design now meets the many needs of construction with telescoping and articulating boom cranes, knuckle boom cranes, tower cranes, and straddle cranes, as well as excavators, back hoes, and many other devices used for mobile lifting, all of which come in a variety of carrier bodies that range from crawler treads to rough-terrain and conventional wheeled transport. Cranes now have a number of attachments—draglines, clamshells, buckets, vibratory pile drivers, aerial lifts, and pumpcrete machines with articulated booms. (It is a well-established safety rule that cranes should lift only freely suspended loads. A crane using a vibratory pile driver to pull piling does not have a freely suspended load, so a load-moment device is necessary to prevent overloading.)

Because cranes on construction sites are used for many purposes, the involvement of the crane manufacturer, the dealer, the rental firm, the construction manager, and the contractor using the crane is paramount in ensuring that all cranes are safe for their intended use (or sometimes their foreseeable misuse).

Sometimes a crane user is unaware that specific hazards may require using a special accessory to control the hazard and may not even be aware that such an accessory is available. An example of this is an anti-two-block device that prevents the headache ball from being pulled into the boom tip with sufficient force to cause the hoist line to part and allow the headache ball and any attached load to fall, endangering those below. Since 2002, OSHA requires an anti-two-block device only on cranes lifting a personnel basket. To control this particular hazard effectively, suppliers of cranes for the workplace need to install anti-two-block devices voluntarily, and construction management must insist that the accessory be installed and used. The crane operator must also be informed so he or she will

understand why it is needed, its use, and the importance of its always being in good working order. In today's construction world with its reliance upon cranes to perform a multitude of tasks in a variety of changing conditions, the following conceptual rule should apply: "Any hazard that has the potential for serious injury or death is always unreasonable and always unacceptable if reasonable hazard prevention devices are available or design modification can be made to eliminate or minimize the hazard."

MOST FREQUENT CAUSES OF CRANE INJURIES OR DEATHS

Six crane hazards are perpetually the primary source of crippling injury and death. They are:

- power-line contact
- upset from overload
- pinch point of the revolving cab and carrier body
- outrigger failure
- upset during travel
- two-blocking.

Power-line Contact

In the 1960s, OSHA instituted a standard that requires equipment operators stay clear of power lines by 10 feet or more. Yet no decline in crane/power-line contacts has taken place during the last 40 years. *Thin air* does not prevent power-line contact. The 10-foot-and-greater-distance clearance rules have proven ineffective.

Thin air is nothing; you can put your hand through it, move through it, drive through it, and move equipment through it. *OSHA's thin-air clearance requirement from power lines is a death trap!* Requirement 1926.550(a)(15) was written with good intentions, but has failed to prevent crippling injuries and deaths from unintentional equipment/power-line contacts.

The current mind set needs changing from emphasis on what employees should do when working near power lines with equipment that can reach and contact power lines to emphasis on what management can do to reduce the hazard before a crane and employees arrive on the job site. Human-factor psychologists have preached for years that people engaged in tasks near overhead power lines cannot be depended upon to make accurate visual judgments about safe clearances when using moving boomed equipment around overhead power lines. Crane booms, noninsulated aerial lifts, pumpcrete machines, and other mobile equipment can be easily raised or rotated into overhead power lines. Even though countless horrifying power-line contacts have been recorded, operators' manuals and warning labels still recite that the 10-foot thin-air clearance must be obeyed.

No one ever questions whether the 10-foot thin-air rule really works. Riggers are watching the load they are guiding, and crane operators are watching the signaler. How are they to know how close the boom has momentarily moved to the power line? Those who say, "Look up and live," don't know that this isn't the answer to this life-taking hazard and is a "kiss of death." Operators' manuals should contain specific instructions for marking a 15-foot danger zone on the ground with engineer's tape or barricades on each side of the power poles. These ground markings will allow the operator and riggers who are guiding the load to see when the load is approaching the danger zone. This visible, 30-foot, barricaded-on-the-ground danger zone eliminates the need to depend upon thin air. (See Figures 1, 2, and 3.)

These figures show how a crane should be located so it cannot be rotated into a power line. When the danger zone is identified on the ground with a barricade or engineer's tape, the crane operator and rigging crew know that a load cannot be lifted or lowered into this hazardous area. Cranes with jibs should not be used in a location close to power lines when the main hook is being used, since the jib may extend into the danger zone beyond where the load is being maneuvered outside the danger zone. When the danger zone is identified by barricades or engineer's tape, it is much easier for workers and crane operators to stay 10 feet away from power lines; without the markings, it is almost impossible for operators to accurately judge the clearance of the hoist line or boom tip from a power line.

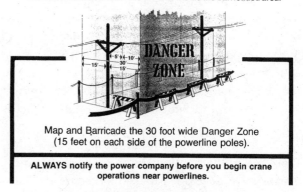

■■ **Figure 1** ■■ **Mapping the danger zone** (Source: David V. MacCollum, *Crane Hazards and Their Prevention*, Figure 4–3A)

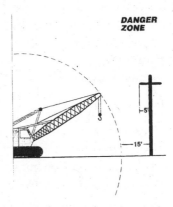

■■ **Figure 2** ■■ **Safe crane location (side view)** (Source: David V. MacCollum, *Crane Hazards and Their Prevention*, Figure 4–3B)

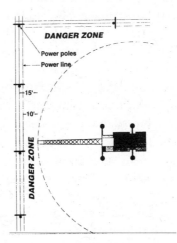

■■ **Figure 3** ■■ **Plan view of mapping the danger zone** (Source: David V. MacCollum, *Crane Hazards and Their Prevention*, Figure 4–3C)

Employers and construction managers deserve praise when they call the electric utility several days before they need to use boomed equipment close to power lines and ask for its help. Utilities can turn off the power, provide temporary insulation, or relocate the power lines to avoid this hazard. (The use of temporary insulation will not relieve the crane user from maintaining a 10-foot clearance, but will prevent a ground fault if unintentional contact is made.)

If the utility company cannot be of assistance, there is another option. In the 1960s, proximity alarms were developed that would sense the electrostatic and/or magnetic fields that surround a power line. The *magnetic field* fluctuates with current flow and was found to be unreliable because the strength of the signal varied. However, using the *electrostatic field* for detection of power lines was a good choice because its voltage did not vary and provided a constant signal. A 7200-volt power line always creates the same electrostatic field strength, and the signal does not vary at any given distance. A proximity alarm that detects the electrostatic field warns all users of boomed equipment that the crane boom is approaching a power line. Cranes equipped with electrostatic proximity alarms allow the crane operator to stop boom movement before it is rotated into a power line. The lift procedure can then be revised. An electrostatic proximity alarm can be wired to intercede and automatically stop all boom movement by installing interlocks. (SigAlarm is presently the commercially available proximity alarm that detects the electrostatic field.)

A number of years ago an employee of a St. Louis highway contractor was instantly electrocuted when unloading steel curb forms from a flatbed truck equipped with a telescoping hydraulic boom. A new concrete highway was being poured and steel curb forms were to provide a continuous curb. This was a stop-and-go operation every few feet as the forms were unloaded. The deceased and the crane operator did not see the single 7200-volt line that crossed the roadway under construction. After this accident, the contractor remained dissatisfied with the 10-foot thin-air rule and bought an electrostatic proximity alarm, tried it out, and bought

fifteen more, one for each crane and one for a pumpcrete machine.

When the U.S. Immigration and Naturalization Service decided it did not want to rely upon thin air when developing border surveillance vehicles, it redesigned a sports utility vehicle (SUV) to include a collapsible pneumatic mast that could be raised 30 feet in the air. These vehicles were to be used for night vision surveillance of illegal border crossings. This agency installed the electrostatic proximity alarm so the mast could not be raised if, by chance, the operator parked in the dark under a power line. Many TV news-gathering vans are now equipped with the same technology after fatal and crippling injury experiences began to occur when masts were inadvertently raised into power lines while covering news events.

Recently, the design of a concrete floodwater holding pond called for a portion of this pond to be located under power lines. A worker who was guiding the hose of a pumpcrete machine while building this pond was electrocuted when thin-air protection failed. If design would have called for relocation of the power lines away from the pond, or if a proximity alarm would have been installed on the pumpcrete boom, this man's life would have been spared.

A manufacturer of a boomed conveyor, which was mounted on a flatbed truck and used for unloading bundles of shingles onto house rooftops, decided that the conveyor boom could be easily and unintentionally slued into power lines because the conveyor boom could be raised, lowered, and rotated 360 degrees like a crane. The manufacturer did not want to rely on thin air and redesigned the boom and conveyor belt with nonconductive materials, eliminating the foreseeable electrocution hazard arising from unintentional power-line contact.

An astute owner of a Berkeley, California, fleet of pumpcrete machines, who did not like thin-air safety, ran his own test to evaluate the electrostatic proximity alarm, and then installed one of them on each of his machines. A construction company in San Antonio, Texas, has equipped some 200 of its cranes with the electrostatic proximity alarm. And the New Orleans District, U.S. Army Corps of Engineers also requires the use of this device.

Insulated links, when installed on crane load lines above or below the headache ball have also proven to be a worthwhile aid in preventing power-line electrocutions and have proven to be invaluable when thin air has been pierced. Eight out of ten crane/power-line contacts occur when the hoist line or boom tip strikes a power line. When this happens, the worker guiding the load is either killed or injured. The insulated link is much better than thin air. In one case, a worker was instantly killed when a choker suspended from a hook brushed over a power line while being lifted over it. A momentary flow of current was enough to cause fatal fibrillation of the heart of the deceased, who was working on a nearby pneumatic drill rig. The drill rig was powered by a compressed air hose protected by a braided wire cover resting on the crane's metal outrigger pad. This congested workplace was a natural setup for a fatal fault current to reach the unsuspecting drill-rig operator. The insulated link would have prevented the wire-covered air hose from becoming energized.

A Canadian approach to solving the 10-foot thin-air problem is to use a range-limiting device that programs movement of the crane boom to avoid power lines. (This device is marketed as "Boom Buoy" by Wyle.)

Operating personnel need all the protection they can get. Proximity alarms, insulated links, and range-limiting devices give operating personnel the backup assistance they need to avoid contact with power lines while working.

Every construction manager wants to avoid crane/power-line contacts. A primary priority is to have the power lines removed, relocated, or deenergized *before* the crane, pumpcrete machine, aerial lift, or other boomed equipment is scheduled to arrive at the construction site. Use of safety devices should not be ridiculed because they can provide reliable safety redundancy for crane safety planning. Many crane safety devices or improvements don't get used for years because of unreasonable performance requirements. A device or design improvement that can prevent serious injury or death in seven out of ten occurrences should not be discarded because three occurrences might still occur.

The old saying, "Half a loaf is better than none," rings true if we are serious about saving as many lives as possible.

Upset from Overload

To avoid upset from overload, a crane needs to be equipped with the practical electronics provided by a load-moment indicator. LMIs are just as valuable to the crane operator as navigational electronics are to aircraft pilots who are qualified by FAA licensing. Computer Assisted Design (CAD) can plan the lift, let you know the particular crane with the right profile that needs to be used, and provide prelift visual orientation for the entire crew. Lifts with multiple cranes can also be planned without guesswork, faulty estimates, or calculations. LMIs and CAD systems are making today's crane operation safer.

Pinch Point of the Revolving Cab and Carrier Body

When the revolving cab of a crane traps an unsuspecting person between it and the carrier body, a life-taking pinch point has been created. The crane operator has no way of viewing what is going on to the right rear of the cab. This blind zone has trapped victims again and again and caused debilitating injuries and death. Infrared sensors can alert the operator to the presence of people in this danger zone, and closed circuit TV allows the operator to see that someone is in the danger zone before he begins to slue the crane. These two devices, when used together, can keep someone from being severely maimed or killed. The hazard of blind spots when backing a crane up or rotating the crane cab can also be overcome by installing Near Object Detection Systems, which are ultra-high-frequency radar. This technology was developed in the early 1980s and needs to be widely adopted. Again, electronics is making it safer to use cranes on the construction site. It is foolish to ignore technology that has been available since the early 1980s. Many well-developed electronic security devices are now mass-produced and are reliable and inexpensive. These should be adapted for use on cranes to prevent people from being trapped in a pinch point or blind zone.

Soil Failure Under Outriggers

Crane upset is often caused by inadequate outrigger support on soil that has insufficient load-bearing ability. Upset due to soil failure occurs when either the ground is too soft or the outrigger pads are not large enough or structurally strong enough to support the crane and the load. Naturally, soil conditions vary. Soft mud offers little support; wet, compacted sand can only support 2000 pounds per square foot; and dry, hard clay may be good up to 4000 pounds per square foot. Subsurface moisture may cause soil that looks dry and solid on the surface to be very unstable, especially if a large crane is being used. Solid blocking should always be used unless a soil compaction test has been performed and a qualified, licensed professional engineer approves the lift. An outrigger that sinks one inch or more in soft mud can create a noticeably greater radius, which reduces the lifting capacity. This hazard can cause a critical overload when using a crane with a long boom. When setting up a tower crane on a high pedestal, be sure the footings are designed by a licensed professional engineer who is qualified in evaluating soil samples to determine their strength. When designing tower crane footings, designers must remember that soils differ. Some can support loads of only a few hundred pounds per square foot and other types of soil can support loads up to several thousand pounds per square foot. This also applies to outriggers on large cranes since soil strengths are not readily apparent to some crane operators or users.

Upset During Travel

Cranes used for pick-and-carry and cranes mounted on flatbed trucks share a common hazard due to their high center of gravity, which can cause upset. Rough-terrain cranes involved in pick-and-carry can easily overturn when traveling along a slope, as the suspended load creates a strong side pull. A crane mounted on a flatbed truck can easily upset on sharp highway curves. Mobile, rough-terrain cranes should have cabs that are crush-resistant to supply operator protection in the event the crane should overturn. The experience gained in the development of rollover protective structures (ROPS) for tractors and other equipment needs to be applied to developing ROPS for rough-terrain cranes and should also be applied to tracked excavators used in construction or converted into log loaders where they are known to be used on slopes. The adoption of life-saving design improvements, such as ROPS, has been unreasonably slow. Loggers in the 1940s used some homemade falling object protective structures (FOPS) to protect themselves from falling objects and soon found that the most sturdy of these FOPS would also usually prevent complete rollover as the tractor would come to rest on the FOPS. They found, in a few instances of complete rollover on steep slopes, that these heavier-designed FOPS did not fail and provided a life-saving compartment for the operator. Today, mobile, rough-terrain cranes and excavators that have been redesigned as log loaders need crush-resistant cabs to protect operators. British Columbia, Canada, has developed standards for crush-resistant cabs for excavators used as log loaders, and the requirement has saved many lives. (For more information, contact Olaf Knezevic, P.E.; Engineering Section, Prevention Division; Workers' Compensation Board of British Columbia; P.O. Box 5350 Stn. Terminal; Vancouver, BC, V6B 5L5; 604-276-3212.)

Two-Blocking

Two-block devices have become an industry success story since manufacturers have made them standard on new equipment and ANSI B30.5 began requiring them. Twenty years ago, in the hope of protecting those who had not invested in the device from the liability claims of widows and orphans who had lost loved ones when a hoist line two-blocked, it was common to hear speculative testimony in court about this device's lack of reliability, the inconvenience it caused the operator, and the false sense of safety it provided. Such testimony never reflected the reality that operators are not perfect and cannot always watch the load and the headache ball at the same time. New safety devices, appliances, and design features should never have to go through this tortuous legal process to gain industry acceptance, but should be investigated, field-tested through use under operating conditions, and adopted as quickly as possible.

MANUFACTURING CONCERNS

One short chapter in a book cannot discuss all of the many crane hazards (refer to *Crane Hazards and Their Prevention*), but it can give some examples of what needs to be done to make cranes safer on the construction site. Crane manufacturers need to be aware of customer or user modifications that create hazards and should not incorporate them into a design to advance sales.

A classic example of this is the hazard of lattice-work boom disassembly. These booms collapse when the lower connecting pins on unsupported booms are removed, often crushing the unsuspecting worker who is removing them. Early latticework boom sections were connected by bolts positioned parallel to the length of the boom. The sections would bind on the long bolts, but not collapse when the nuts were removed. But after World War II, crane users found disassembling nut-and-bolt connections too slow and replaced them with removable pins to speed up the process. Unfortunately, this shortcut was adopted by crane manufacturers. Telescoping crane booms have become a popular substitute for the labor-intensive latticework booms; nevertheless, thousands of these latticework booms are still in use and have been the cause of approximately 100 crippling injuries and deaths. (See Figures 4, 5, and 6.)

Seven known design alternates can be incorporated to minimize this hazard, which so quickly maims and kills:

- Step pins that can only be inserted from the inside, facing out, allowing them to be driven out while standing beside the boom, not under it.
- Welded lugs that permit the pin to be inserted from the inside, facing out, allowing them to be driven out while standing beside the boom, not under it.
- A double-ended pin that can be inserted and driven out while standing beside the boom, not under it.
- A load-sensitive pin with a keyway on both sides that creates friction and prevents the pin from being driven out while the boom is unsupported.

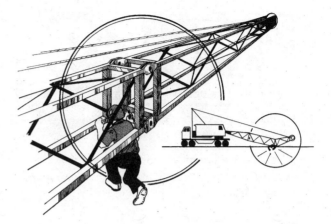

■ ■ **Figure 4** ■ ■ **Unsafe boom assembly** (Source: David V. MacCollum, *Crane Hazards and Their Prevention*)

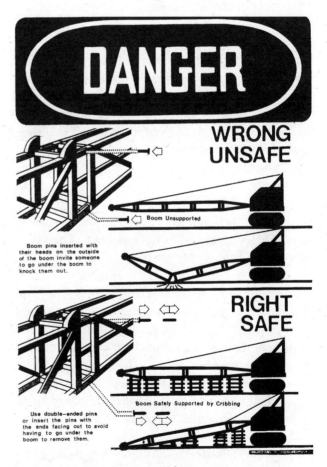

■ ■ **Figure 5** ■ ■ **Boom assembly label for crane cab** (Source: David V. MacCollum, *Crane Hazards and Their Prevention*)

■■ **Figure 6** ■■ **Boom assembly label for crane boom** (Source: David V. MacCollum, *Crane Hazards and Their Prevention*, Figure 14–3)

- A boom supported by cribbing.
- A boom supported on a flatbed fifth-wheel trailer. (Requires a design that when the flat boom is level is the same height as the flatbed.)
- A boom connection that cannot be disconnected unless the boom is flat and fully supported for its entire length.

The safest alternate is to design the boom sections so that they cannot be disassembled when the boom is supported only by the pennant lines at the tip of the boom. Because the deaths or crippling injuries have not been widely publicized, these seven known, alternate design and operating improvements haven't been given much priority.

Industry's perception of this hazard seems to be stuck in a black hole with not much being done to initiate a program to eliminate this hazardous design defect that invites human error. The key is that design philosophy should not rely on perfect worker performance. "To err is human; to forgive, *design.*" Considering the severity of the injuries and the deaths caused by disassembling booms and their connecting pins, booms should be designed to prevent disassembly if the boom is not supported for its entire length. Crane designers and users need to consider that entry-level workers in construction and other industries are usually untrained, temporary help. In addition to prevention of human suffering, serious crippling injuries or deaths create an economic burden on all of society that can be avoided through safe design and use of accessories.

Crane safety on construction sites must revert back to management and its responsibility to see that safe methods are used and available technol-

ogy is applied. Every crane should be designed or equipped to prevent inherent hazards that can cause injury or death. With today's technology, no one has an excuse for not applying appropriate engineering technology to cranes and other types of construction equipment.

Safety engineering professionals get a bad rap every time the *palace guards*, who think they are protecting top management's image, give a press release stating that a fatality was a "freak accident" with no known way to prevent it. Often the CEO is unaware of any hazardous work conditions because months earlier these same palace guards had filtered out warnings that a life-threatening hazard existed and certain measures were needed. The spokesperson then blatantly implies that the accident was the result of an unsafe act and the victim's fault. The greatest danger that a CEO faces is when the palace guards control his or her sources of hazard information. The construction safety manager should be the principal advisor to the CEO on all matters concerning safety. When construction equipment has inherent hazards, the CEO needs to know this so he or she can notify the manufacturer about the need to make the equipment safer. Safety engineers need to set the record straight—that an unsafe act cannot exist unless there is a hazard. Engineers are granted licenses in each of the fifty states to protect the public. Going along to get along leads nowhere. The public consistently supports safety in our justice system when they know the facts about a hazard and how easily design improvements or safety accessories would have saved someone's life.

For those who desire engineering information concerning design improvements and devices that overcome inherent hazards on cranes, the Hazard Information Foundation (HIFI), Inc., can be of assistance. HIFI is a new not-for-profit foundation dedicated to providing a flow of information concerning the successful application of engineering improvements for prevention of hazardous conditions and circumstances encountered in the use of machines, equipment, and facilities. HIFI is a data resource center with an advisory council of engineers and scientists who collectively have knowledge and expertise in an almost unlimited number of diverse areas of hazard prevention by design. (HIFI's Web page is hazardinfo.com.)

REFERENCES

MacCollum, David V. *Construction Safety Planning*. New
York: Van Nostrand Reinhold, 1995. Available
through ASSE.

_____. *Crane Hazards and their Prevention*. Des Plaines,
IL: ASSE, 1993.

_____, "Critical Hazard Analysis of Crane Design,"
paper presented at the Fourth International
System Safety Conference, July 1979.

_____, "Hunting Down Crane Hazards," *Lift
Equipment*, February–March 1992, p. 18.

Ritchie, Dave, "Jobsite Safety," *CraneWorks*, April
2003, p. 12.

REVIEW EXERCISES

1. What are the three modes of a hazard?
2. What methods can management use to prevent:
   a. power-line contact?
   b. a crane upset from overload?
   c. crushing injuries from the pinch point between the revolving crane cab and its carrier frame?
   d. a crane upset while in the travel mode?
   e. two-blocking?
3. Give some examples of dangerous crane design.

■■ **About the Author** ■■

ELECTRICAL SAFETY
by Kraig Knutson, Ph.D., CPC

Kraig Knutson, Ph.D., CPC, is an assistant professor in the Del E. Webb School of Construction at Arizona State University. He holds a Bachelor's and Master's degree in Construction and a Ph.D. in Industrial Engineering from Arizona State University. He is a member of ASSE, ASCE, AACE, and ASHRAE.

Electrical Safety

INTRODUCTION

Electricity has long been recognized as a serious workplace hazard. Workers may be exposed to such dangers as electric shock, electrocution, and burns. According to the Bureau of Labor Statistics (Table A-9), the construction industry claimed 155 deaths from electrocution in 2001. What makes these statistics tragic is that these fatalities could have been easily avoided.

To handle electricity safely, it is necessary to understand the basics of electricity—how it acts, how it can be directed, what hazards it presents, and how these hazards can be controlled. This chapter addresses those basics and also references OSHA's electrical standards to determine how to minimize the potential hazards. The principles contained in these OSHA documents will help further an understanding of electrical safety on construction projects.

In this chapter, you will be directed to OSHA's Web site to study these four publications:

■ *OSHA 3075: Controlling Electrical Hazards*, which explains the common injuries caused by electrical hazards and the safe practices to prevent electric shock-related injuries and deaths in the workplace. It also provides an overview of basic electrical safety for individuals with limited training or familiarity with electrical hazards.
■ *OSHA 3007: Ground-Fault Protection On Construction Sites*, which provides an overview of the use of GFCIs and the Assured Equipment Grounding Conductor Program.

■ *OSHA 29 CFR 1926, Subpart K, Electrical*: Subpart K addresses the electrical safety requirements that are necessary for the practical safeguarding of employees involved in construction work; it is divided into four major divisions and applicable definitions, as follows:

- **1926.400(a) Installation safety requirements.** Installation safety requirements are contained in 1926.402 through 1926.408. Included in this category are electric equipment and installations used to provide electric power and light on job sites.

- **1926.400(b) Safety-related work practices.** Safety-related work practices are contained in 1926.416 and 1926.417. In addition to covering the hazards arising from the use of electricity at job sites, these regulations also cover the hazards arising from the accidental contact, direct or indirect, by employees with all energized lines, above or below ground, passing through or near the job site.

- **1926.400(c) Safety-related maintenance and environmental considerations.** Safety-related maintenance and environmental considerations are contained in 1926.431 and 1926.432.

- **1926.400(d) Safety requirements for special equipment.** Safety requirements for special equipment are contained in 1926.441.

- **1926.400(e) Definitions.** Definitions applicable to this subpart are contained in 1926.449.

■ *OSHA 3120: Control of Hazardous Energy (Lockout/Tagout)*, which presents OSHA's general requirements for controlling hazardous energy during service or maintenance of machines or equipment.

All four of these publications are available on OSHA's Web site (http://www.osha.gov) and may be viewed online, downloaded, or printed free of charge.

BASIC ELECTRICAL PRINCIPLES

Electron Theory

Everything, whether solid, liquid, or gas, is made up of atoms. Each atom contains some number of electrons, protons, and neutrons. The nucleus is the central region of each atom and contains the positively charged protons and neutrons as shown in Figure 1. Negatively charged electrons orbit the nucleus. Each atom prefers to have the same number of electrons as protons to maintain a balance. A particle with more protons than electrons will have a positive charge, and when electrons outnumber protons the particle will have a negative charge. Each charged particle that is out of balance exerts some electrical pressure as it tries to get back into balance. Opposite charges attract each other and like charges repel each other. This is the basis for the electron theory and the flow of electricity.

Conductors

Some materials are willing to let a few electrons move from atom to atom. Materials that let electrons move through them are called *conductors*. Any substance that offers little resistance to the flow of electricity is called a *good conductor*. Although some are better

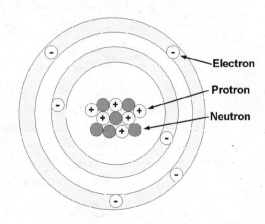

■ ■ **Figure 1** ■ ■ **Structure of a carbon atom**

than others, most metals are good conductors of electricity. Silver, gold, and platinum are very good conductors but are expensive, so they are not often used. Copper and aluminum are reasonably good conductors and are relatively inexpensive. Thus most electrical wiring is copper; the high-voltage transmission lines that we see on the poles along the street are usually aluminum.

Insulators

Other substances keep their electrons under very tight control. Materials that do not let electrons move through them are called *insulators*. Glass is an extremely good insulator and is an example of a type of material that keeps its electrons tightly controlled. Glass is made of silicon molecules, organized very tightly into a crystalline structure. Many plastics are good insulators too. Plastics are cheap, flexible, and durable. That is why most electrical wiring is covered with a layer of plastic, and a wire that is covered is called an *insulated* wire.

The Relationship Between Voltage, Current, Resistance, and Power

VOLTAGE

Electrons don't move from atom to atom without a reason. When electrons are flowing, there is an electrical force, or pressure, pushing them along. We refer to this force as *voltage* in honor of Alessandro Volta, and we measure the amount of electrical pressure in *units of volts*. For example, a single battery from a flashlight creates an electrical force of 1.5 volts and standard household electrical outlets are 120 volts. Volts is often abbreviated as "V" or "E."

CURRENT

Current is the flow rate of the electrons in a circuit. Voltage is the pressure pushing the electrons in a circuit and current is the flow rate of electrons through the circuit. Current is measured in units of *amperes*, which is often shortened to *amps* and abbreviated as "A" or "I."

RESISTANCE

The resistance a conductor offers to the flow of current is determined by the type of material, its length and diameter. The longer a wire is or the smaller its diameter, the more resistance, or friction, must be overcome. Resistance is measured in units of ohms, often represented by "R" or the Greek letter "Ω."

OHM'S LAW

Ohm's law describes the relationship between voltage, current, and resistance, measured in ohms. Ohm's law is stated as follows:

$$E = IR$$

where

E = potential difference in volts
I = current flow in amps
R = resistance to current flow in ohms.

ELECTRIC POWER

The amount of power delivered by an electric circuit depends on the current flow, measured in amps, and the voltage. Power is measured in watts and can be determined from the following formula:

$$W = EI \text{ or } W = I^2R$$

where

W = power in watts.

For example, consider standard light bulbs found in a typical home—most are rated either 60, 75, or 100 watts. Knowing that the voltage in your home is 120 volts, and using the equation above, you could then calculate the current (in amps) that will flow when the light bulb is turned on. Assume you have a 100-watt light bulb, then the current will be:

$$I = W/E$$

$$I = 100 \text{ watts} \div 120 \text{ volts} = 0.833 \text{ amps.}$$

When we know the current in amps, we can then calculate the resistance:

$$R = E/I$$

$$R = 120 \text{ V} \div 0.833 \text{ A} = 144 \text{ ohms.}$$

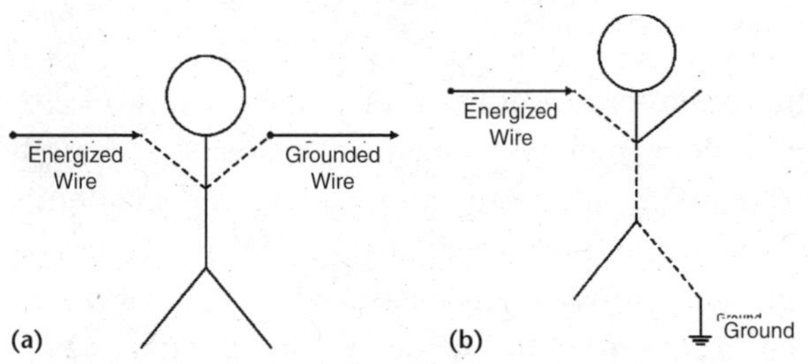

■ ■ **Figure 2** ■ ■ Path of electrical current through the body: (a) completing the circuit, (b) to ground

EFFECTS OF ELECTRICAL CURRENT ON THE HUMAN BODY

Electric shock occurs when the body becomes part of an electrical circuit. This can occur in two ways: when an individual contacts two different wires of an electric circuit and completes the circuit [Figure 2(a)] or when the individual is in contact with the ground and comes in contact with an energized wire or conductor [Figure 2(b)]. The dotted line in each figure indicates the path of the electrical current through the body.

When an electrical current passes through the body, the result can range from a slight tingling sensation to immediate cardiac arrest. The severity of damage to the human body depends on the following factors (OSHA 3075, 2002):

■ The amount of current flowing through the body—as the amount of current increases so do the effects on the human body as shown in Table 1.

■ The current's path through the body—the worst path through the body is a path that is near or through the heart. This path will increase the possibility of ventricular fibrillation and cardiac arrest.

■ The length of time the body remains in the circuit—the longer the exposure, the greater the risk of serious injury.

Table 1 shows the body's reaction to a shock of 1-second duration at various current levels. Remember that 1 amp = 1000 mA (milliamp). Notice that a difference of less than 100 mA (1/10 A) exists between a current that is disturbing, but not painful, and one that can kill! Currents greater than 100 mA can cause ventricular fibrillation and may cause death in a few minutes unless a defibrillator is used. Not being able to "let go" due to muscle contractions increases the duration of shock. A current of 100 mA for 3 seconds is equivalent to 900 mA for 0.03 seconds, in effect causing ventricular fibrillation.

The effect of electricity on the human body is unpredictable and can take several different forms, such as electrocution (death due to electrical shock), electric shock, burns, and falls.

■ *Asphyxia* The electrical shock may cause the body to stop breathing. This condition is defined as asphyxia. The electrical current passing through the body may temporarily paralyze or, if the current is high enough, destroy either the nerves or the area of the brain that controls breathing.

■ ■ **Table 1** ■ ■

Effect of a 1-second Shock on the Human Body

Current Amplitude	Reaction in Body
1 mA	Faint tingle
5 mA	Not painful but disturbing
6–25 mA	Painful shock, muscular reaction
9–30 mA	Frozen to conductor
50–100 mA	Extreme pain, respiratory arrest (asphyxia), contractions
100–200 mA	Ventricular fibrillation
>200 mA	Cardiac arrest, severe burns

■ *Burns—contact and flash* Contact burns are a result of electricity passing through the body. The burns are usually found at the points where the current entered and left the body. Flash burns are the result of flash or an electric arc and the heat generated may burn exposed parts of the body. Burns are the most common nonfatal injury from an electric shock.

■ *Fibrillation—also referred to as "ventricular fibrillation"* When an electrical current passes through the body it may disrupt the rhythm of the heart. If this occurs, the heart stops pumping blood and the pulse disappears (Kurtz and Shoemaker, 1986).

■ *Falls* Electrical shock can cause indirect, or secondary, injuries to the body. If a worker who is on a ladder, scaffolding, or other elevated position receives an electric shock, then that person can lose control of their muscles and fall, resulting in serious injury or death.

Rescue

Because a victim may receive an electrical shock in many different locations—underground in a manhole or trench, on the ground, elevated on a ladder or scaffolding, or on the structure itself—it is almost impossible to describe how to rescue the shock victim. However, there are some guidelines that should be followed.

FREEING THE VICTIM

The first thing to remember is NOT to become a victim yourself. Before touching the shock victim, make sure the victim is clear of the electrical hazard. In most cases the victim will have been thrown clear of the hazard by the force of the shock or because of a muscle spasm from the shock. If not, then find a way to disconnect the electrical power source before attempting to touch the victim.

RESUSCITATION

Once the victim has been freed from the source of electrical power, check to see if the victim is breathing. If the victim is not breathing, then artificial

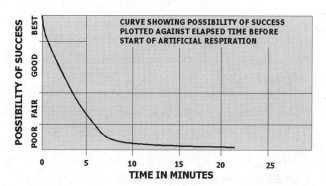

CPR Success

■ ■ Figure 3 ■ ■ Possibility of CPR success decreases rapidly with each passing minute

respiration should be started immediately. To be successful, artificial respiration must be started as soon as possible on a victim who is not breathing. Figure 3 shows that the possibility of CPR success decreases rapidly with each passing minute. The human brain can survive for only a few minutes without oxygenated blood; therefore, every second counts (Kurtz and Shoemaker, 1986). The greater the delay in starting artificial respiration, the less chance there will be of a successful recovery. Training in artificial respiration is best received from an entity such as the American Red Cross, American Heart Association, National Safety Council, local contractor association, or a local hospital. In preparation for an emergency, CPR practice should be kept current.

After the victim is cleared from the source of electrical power and artificial respiration has begun, then either call for help or, if there is another person at the scene, send that person to either call 911 or another emergency number to summon help.

ELECTRICAL HAZARDS ON A CONSTRUCTION JOB SITE

The most common electrical hazards found on a construction job site are related to power tools, extension cords, and temporary lighting. Electrical accidents are usually the result of a combination of three factors: (1) unsafe equipment and/or installation, (2) workplaces made unsafe by the environment, and (3) unsafe work practices. There are various ways of protecting people from the hazards caused by electricity. These include: insulation, guarding, grounding, electrical protective devices, and safe work practices. Let's discuss the safe use of electric power tools first.

(Overhead and underground utilities are also major electrical hazards on construction sites and they are discussed in detail in Chapters 20 and 21.)

Handheld Electric Power Tools

Handheld electric power tools pose a potential danger because they make continuous contact with the hand. Currents as small as 10 mA can paralyze or "freeze" muscles, so when the user is shocked they cannot release the tool (see Table 1). In fact, when the user is shocked, the tool is held even more tightly and this results in a longer exposure to the electric shock, causing additional damage. To protect the user from shock, burns, or electrocution, electric power tools must either:

- have a three-wire cord with a ground and be plugged into a grounded outlet
- be double insulated
  [1910.304(f)(5)(v)(C)(3)].

Grounding Path

The path to ground from circuits, equipment, and enclosures must be permanent and continuous. The three-wire cord cap shown in Figure 4 has two conducting prongs and a grounding prong.

Double Insulation

Double-insulated electrical tools have all exposed metalwork separated from the energized conductors by two layers of insulation so that the exterior metal cannot become energized. There is no earth connection, and the operator's safety depends on the integrity of the two layers of insulation. Double-insulated equipment must be marked to indicate that the equipment uses an approved system of double insulation. The double-insulation marking is shown in Figure 5.

Problems with Extension Cords

With the wide use of portable tools on construction sites, the use of extension cords (Figure 6) often becomes necessary. Hazards are created when cords, cord connectors, receptacles, and cord- and plug-connected equipment are improperly used and maintained.

An extension cord may be damaged by activities on the job, by door or window edges, by staples or fastenings, by abrasion from adjacent materials, or simply by aging. If the electrical conductors become exposed, there is a danger of shocks or burns. A frequent hazard on a construction site is an extension cord with improperly connected terminals. (http://www.osha-slc.gov/doc/outreachtraining/htmlfiles/gfcicon.html)

■■ **Figure 4** ■■ **Cord Cap With A Grounding Prong** (Photo courtesy of Bryant Electric)

■■ **Figure 5** ■■ **Double-insulation mark**

■■ **Figure 6** ■■ **Three-wire grounded electrical extension cord** (Photo courtesy of McGill Electrical Product Group)

Most people never test extension cords prior to use, resulting in the possibility of brand-new, defective units being put into service. Other problems resulting from extension-cord use include overloading, cutting off the grounding prong, and incorrect maintenance and repair. The following safety criteria should be used for training and inspection purposes:

- Before new or repaired extension cords are put into use, they must be tested. In addition, always purchase UL-listed extension cords (this will assure a high standard of quality and safety). If extension cords are repaired by company workers, be sure they test these units prior to returning them to service. The tests involve using (1) a three-prong receptacle circuit tester, (2) a tension tester, and (3) an ohmmeter.
- A visual and electrical inspection should be made of the extension cord each time it is used. Cords with cracked or worn insulation or damaged ends should be removed from service immediately. Cords with the grounding prong missing or cut off must also be removed from service.
- If extension cords are fabricated, only qualified electricians should do the work. The use of receptacles fabricated from junction boxes should be prohibited. Only UL-listed attachment plugs, cords, and receptacle ends should be used when fabricating extension cords.
- Use of two-conductor extension cords should be prohibited. Only three-wire grounding extension cords should be used.
- When cords must cross passageways, whether vehicular and/or personnel, the cords should be protected and identified with appropriate warnings.
- Extension cords must be of continuous length without splices. Personnel should be instructed to remove from service any cords that have exposed wires, cut insulation, or loose connections at the plug or receptacle ends.
- Never overload an extension cord electrically. If it is warm or hot to the touch, have it

and/or the tool being used checked to determine the problem. The wire size of the extension cord may be too small for the amperage of the tool being used. (http://www.conney.com/safetytopics/safetytopic5c.html)
- Extension cords must not be run through standing water, wet floors, or other wet areas.

Employer's Responsibility to Protect Employees from Electrical Shock

Ground-fault circuit interrupters can be used successfully to reduce electrical hazards on construction sites. The use of GFCI protection is required on all construction projects unless a properly approved Assured Equipment Grounding Conductor Program is in place. It is the employer's responsibility to provide either: (a) ground-fault circuit interrupters on construction sites for receptacle outlets in use and not part of the permanent wiring of the building or structure, or (b) a scheduled and recorded Assured Equipment Grounding Conductor Program on construction sites, covering all cord sets, receptacles that are not part of the permanent wiring of the building or structure, and equipment connected by cord and plug that is available for use or is being used by employees. The following sections describe the use of GFCIs and OSHA's Assured Equipment Grounding Conductor Program.

GROUND-FAULT CIRCUIT INTERRUPTERS

The employer is required to provide approved GFCIs for all 120-volt, single-phase, 15- and 20-ampere receptacle outlets on construction sites that are not a part of the permanent wiring of the building or structure and that are in use by employees. Receptacles on the ends of extension cords are not part of the permanent wiring and, therefore, must be protected by GFCIs whether or not the extension cord is plugged into permanent wiring.

The ground-fault circuit interrupter is a fast-acting circuit breaker that senses small imbalances in the circuit caused by current leakage to ground and, in a fraction of a second, shuts off the electricity

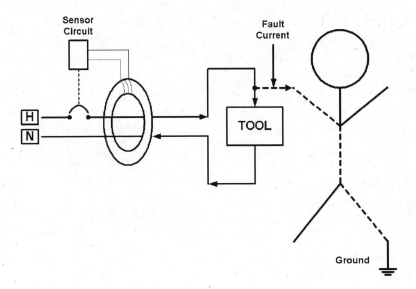

■ ■ **Figure 7** ■ ■ **How a GFCI works**

■ ■ **Figure 8** ■ ■ **A typical GFCI electrical outlet**

■ ■ **Figure 9** ■ ■ **A GFCI tester** (Photo courtesy of Ideal Industries)

to the cord or tool that is connected to it. The sensor circuit of the GFCI continually matches the amount of current going to an electrical device against the amount of current returning from the device along the electrical path, as shown in Figure 7. Whenever the amount "going" on the hot wire (H) differs from the amount "returning" on the neutral wire (N) by approximately 5 mA, the GFCI interrupts the electric power within as little as 1/40 of a second.

However, the GFCI will not protect the employee from line-to-line contact hazards (such as a person holding two hot wires or a hot and a neutral wire in each hand). It does provide protection against the most common form of electrical shock hazard—the ground fault. A typical GFCI electrical outlet is shown in Figure 8 where you can see both the test and reset buttons on the face of the GFCI. A GFCI outlet tester is shown in Figure 9. The GFCI tester has three lights which indicate whether the outlet is properly wired, and a button which, when pushed, sends a small current to ground (around 5 mA) to test the operation of the GFCI.

Assured Equipment Grounding Conductor Program

The Assured Equipment Grounding Conductor Program covers all cord sets, receptacles that are not a part of the permanent wiring of the building or structure, and equipment connected by cord and plug that are available for use or being used by employees. The requirements the program must meet are stated in 29 CFR 1926.404(b)(1)(iii), but employers may provide additional tests or procedures. OSHA requires that a written description of the employer's Assured Equipment Grounding Conductor Program, including the specific procedures adopted, be kept at the job site. This program should outline the employer's specific procedures for the required equipment inspections, tests, and testing schedule.

The required tests must be recorded, and the record maintained until replaced by a more current record. The written program description and the recorded tests must be made available at the job site to OSHA and to any affected employee upon request. The employer is required to designate one or more competent persons to implement the program.

Electrical equipment noted in the Assured Equipment Grounding Conductor Program must be visually inspected for damage or defects before each day's use. Any damaged or defective equipment must not be used by the employee until repaired.

Two tests are required by OSHA. One is a continuity test to ensure that the equipment grounding conductor is electrically continuous. It must be performed on all cord sets, receptacles that are not part of the permanent wiring of the building or structure, and on cord- and plug-connected equipment that is required to be grounded. This test may be performed using a simple continuity tester, such as a lamp and battery, a bell and battery, an ohmmeter, or a receptacle tester like the one shown in Figure 10.

The other test must be performed on receptacles and plugs to ensure that the equipment grounding conductor is connected to its proper terminal. This test can be performed with the same equipment used in the first test.

These tests are required before first use, after any repairs, after any damage is suspected to have occurred, and at 3-month intervals. Cord sets and receptacles, which are essentially fixed and not ex-

posed to damage, must be tested at regular intervals (not to exceed 6 months). Any equipment that fails to pass the required tests shall not be made available or used by employees.

Included in the Appendix at the end of this chapter is the complete text of OSHA §1926.404, Wiring Design and Protection from the Construction Safety and Health Regulations, Part 1926, Subpart K. Section 1926.404 describes the requirements for the utilization of both GFCIs and the Assured Equipment Grounding Conductor Program.

TEMPORARY LIGHTING

During the construction process it is usually necessary to provide temporary lighting for the workforce. The installation of the temporary lighting must be in accordance with OSHA and the National Electric Code (NEC) requirements. The greatest source of electric hazards associated with temporary lighting is failure to protect the lamps from damage. If the glass of the lamp is broken, the energized filament will be exposed. If a person makes accidental contact with the energized filament they will receive a shock.

Several manufacturers offer a variety of temporary job-site lighting systems, from reusable, low-cost

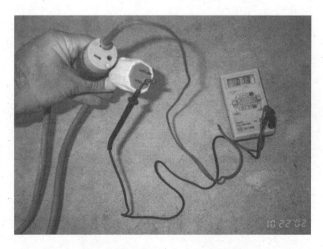

■ ■ **Figure 10** ■ ■ **Testing an electrical extension cord for continuity**

cable versions (Figure 11) to an assortment of open conductor styles (Figure 12). They are available in various lengths with 5 to 10 light assemblies each.

Another key item to remember when dealing with temporary lighting is that the lights should not be suspended by their electric cord. A separate support cable or messenger wire must be used when it is not possible to support each individual light socket from the structure. When selecting temporary lighting for your construction project make sure that the lighting system meets these criteria:

- meets current NEC requirements
- is UL-listed
- meets OSHA requirements
- is convenient and economical
- has messenger wire available to support light strings where desirable
- has cages (as shown in Figure 13) that are attached to weather-resistant sockets.

CLUES THAT ELECTRICAL HAZARDS MAY EXIST ON YOUR CONSTRUCTION PROJECT

There are usually warning signs or clues that electrical hazards exist on your construction project. Look for these warning signs:

- tripped circuit breakers or blown fuses
- warm tools, wires, cords, connectors, or junction boxes
- GFCI that shuts off a circuit
- worn or frayed insulation on a tool or extension cord.

TRAINING

Train employees working with electrical equipment in safe work practices, including:

- deenergizing electric equipment before inspecting or making repairs
- using only electric tools that are in good repair
- using good judgment when working near energized lines and maintaining the recommended clearances

■ ■ **Figure 11** ■ ■ **Prepackaged temporary cord set suspended from ceiling** (Photo courtesy of McGill Electrical Product Group)

■ ■ **Figure 12** ■ ■ **Prepackaged temporary open wire set shown in the box** (Photo courtesy of McGill Electrical Product Group)

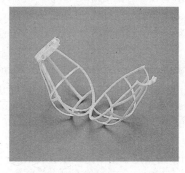

■ ■ **Figure 13** ■ ■ **Plastic light cage** (Photo courtesy of McGill Electrical Product Group)

■ using appropriate protective equipment when operating electrical equipment.

OSHA has delineated work-practice requirements relating to electrical safety in 29 CFR 1910.331–.335.

DEENERGIZING ELECTRICAL EQUIPMENT

Before any inspections or repairs are made on electric tools, extension cords, or temporary lighting systems, the electrical current must be turned off, locked off, and then tagged so that others will know who is working on the equipment. This must be done to prevent accidental or unexpected start-up or turning on of electrical equipment, which can cause severe injury or death to those inspecting or repairing the equipment. OSHA refers to this procedure as Lockout/Tagout (LOTO). It is described in detail in OSHA standard 1910.147 and in OSHA Publication 3120.

If the electricity to the equipment can be turned off at a safety switch, then a tag, a safety lockout hasp, and a padlock (Figure 14) can be applied directly to the safety switch. The safety lockout hasp shown in Figure 14 has six positions, or openings, for padlocks and will allow up to six workers to lock out a single energy source. If needed, additional lockout hasps can be added for additional workers.

If the equipment is connected with a cord that can be unplugged, then a cover (like the one shown in Figure 15) that can be locked must be placed over the cord's plug to ensure that others will not plug in the cord while you are working on or inspecting the equipment. Figure 16 shows the 1-2-3 installation instructions for the LOTO device for cord-connected equipment.

Guarding of Live Parts

On the construction site, all live parts of electrical equipment operating at 50 volts or more must be

■■ **Figure 14** ■■ **Four-lock station kit with tags, safety lockout hasps, and padlocks** (Photo courtesy of Ideal Industries)

■■ **Figure 15** ■■ **LOTO device for cord-connected equipment** (Photo courtesy of McGill Electrical Product Group)

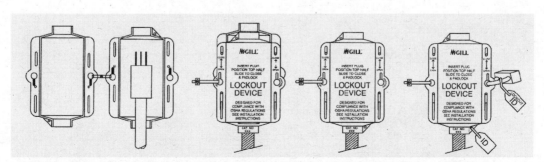

■■ **Figure 16** ■■ **1-2-3 installation instructions for LOTO device for cord-connected equipment** (Photo courtesy of McGill Electrical Product Group)

guarded against any accidental contact. Guarding can be accomplished by:

■ using approved cabinets or enclosures
■ selecting a location or using permanent partitions that ensure live parts are accessible only to qualified persons
■ elevating the parts 8 feet or more above the floor or working surface.

For additional information on this topic, refer to OSHA 1910.303(g)(2)(i) and 1910.303(g)(2)(iii). OSHA requires that the entrances to these guarded locations are marked with conspicuous warning signs (like those shown in Figure 17). OSHA also requires that any electrical equipment exposed to physical damage must be protected from possible damage (1910.303(g)(2)(ii)).

■ ■ **Figure 17** ■ ■ **Danger—High Voltage warning signs** (Photo courtesy of Ideal Industries)

REFERENCES

OSHA 3075: Controlling Electrical Hazards (2002). PDF file format available at http://www.osha-slc.gov/Publications/osha3075.pdf.

OSHA 3007: Ground Fault Protection On Construction Sites (2002). (http://www.osha-slc.gov/Publications/osha3007.pdf).

OSHA 29 CFR 1926, Subpart K, Electrical (2002). (http://www.osha-slc.gov/pls/oshaweb/owastand.display_standard_group?p_toc_level=1&p_part_number=1926&p_text_version=FALSE).

OSHA 3120: Control of Hazardous Energy (Lockout/Tagout)

Table A-9. Fatal occupational injuries by event or exposure by major private industry division, 2001. (http://www.bls.gov/iif/oshwc/cfoi/cftb0153.pdf).

Basic Electrical Principles. (http://www.reprise.com/host/electricity/atoms.asp).

Kurtz, E. and T. Shoemaker. *The Lineman's and Cableman's Handbook.* (New York: McGraw Hill, 1986).

Problems with Extension Cords (http://www.conney.com/safetytopics/safetytopic5c.html).

Ground-Fault Protection on Construction Sites. Describes why and how to create proper ground-fault protection on construction sites. (http://www.osha-slc.gov/doc/outreach training/htmlfiles/gfcicon.html).

Electrical Standards for Construction. Describes the National Electrical Code's provisions included in OSHA's electrical standard. (http://www.osha-slc.gov/doc/outreachtraining/html files/elecstd.html).

REVIEW EXERCISES

1. A substance that offers little resistance to the flow of electricity is called a _____. Give three examples of this type of material.

2. Materials that do not let electrons move through them are called _____. Give three examples of this type of material.

3. Explain the relationship between voltage, current, resistance, and power.

4. Given a 60-watt light bulb, and assuming the voltage is 120 volts, calculate the current that will flow when the light bulb is turned on. Determine the resistance of the light bulb.

5. What are the effects of electrical current on the human body?

6. If you discover someone who has been the victim of an electrical shock, what is the first thing you should do?

7. Electrical accidents are usually the result of a combination of three factors. What are they?

8. How can you protect people from the hazards caused by electricity?

9. A visual and electrical inspection should be made of the extension cord each time it is used. What should you look for during this inspection?

10. Draw a diagram showing how a GFCI works.

11. What does an Assured Equipment Grounding Conductor Program cover?

12. What is the greatest source of electric hazard associated with temporary lighting? How do you protect against this hazard?

13. List the warning signs that electrical hazards may exist on your construction project.

14. Describe the steps of the lockout/tagout procedure.

APPENDIX: CONSTRUCTION SAFETY AND HEALTH REGULATIONS, PART 1926, SUBPART K (PARTIAL)

§1926.404 Wiring design and protection.

(b) Branch circuits—(1) Ground-fault protection—
(i) General.

The employer shall use either ground-fault circuit interrupters as specified in paragraph (b)(l)(ii) of this section or an assured equipment grounding conductor program as specified in paragraph (b)(l)(iii) of this section to protect employees on construction sites. These requirements are in addition to any other requirements for equipment grounding conductors.

(ii) Ground-fault circuit interrupters. All 120-volt, single-phase, 15- and 20-ampere receptacle outlets on construction sites, which are not a part of the permanent wiring of the building or structure and which are in use by employees, shall have approved ground-fault circuit interrupters for personnel protection. Receptacles on a two-wire, single-phase portable or vehicle-mounted generator rated not more than 5 kW, where the circuit conductors of the generator are insulated from the generator frame and all other grounded surfaces, need not be protected with ground-fault circuit interrupters.

(iii) Assured equipment grounding conductor program. The employer shall establish and implement an assured equipment grounding conductor program on construction sites covering cord sets, receptacles which are not a part of the building or structure, and equipment connected by cord and plug which are available for use or used by employees. This program shall comply with the following minimum requirements:

    (A) A written description of the program, including the specific procedures adopted by the employer, shall be available at the job site for inspection and copying by the Assistant Secretary and any affected employee.

    (B) The employer shall designate one or more competent persons [as defined in §1926.32(f)] to implement the program.

    (C) Each cord set, attachment cap, plug and receptacle of cord sets, and any equipment connected by cord and plug, except cord sets and receptacles which are fixed and not exposed to damage, shall be visually inspected before each day's use for external defects, such as deformed or missing pins or insulation damage, and for indications of possible internal damage. Equipment found damaged or defective shall not be used until repaired.

    (D) The following tests shall be performed on all cord sets, receptacles which are not a part of the permanent wiring of the building or structure, and cord- and plug-connected equipment required to be grounded:

        (1) All equipment grounding conductors shall be tested for continuity and shall be electrically continuous.

        (2) Each receptacle and attachment cap or plug shall be tested for correct attachment of the equipment grounding conductor. The equipment grounding conductor shall be connected to its proper terminal.

    (E) All required tests shall be performed:

        (1) Before first use;

        (2) Before equipment is returned to service following any repairs;

        (3) Before equipment is used after any incident which can be reasonably suspected to have caused damage (for example, when a cord set is run over); and

        (4) At intervals not to exceed 3 months, except that cord sets and receptacles which are fixed and not exposed to damage shall be tested at intervals not exceeding 6 months.

    (F) The employer shall not make available or permit the use by employees of any equipment which has not met the requirements of this paragraph (b)(l)(iii) of this section.

(G) Tests performed as required in this paragraph shall be recorded. This test record shall identify each receptacle, cord set, and cord- and plug-connected equipment that passed the test and shall indicate the last date it was tested or the interval for which it was tested. This record shall be kept by means of logs, color coding, or other effective means and shall be maintained until replaced by a more current record. The record shall be made available on the job site for inspection by the Assistant Secretary and any affected employee.

■■ **About the Author** ■■

EXCAVATION, TRENCHING, AND SHORING
by Michael W. Hayslip, Esq., P.E., CSP

Michael W. Hayslip, Esq., P.E., CSP, is president and owner of NESTI, a national safety training institute and litigation support firm in Dayton, Ohio. Previously, he served as Safety and Risk Director of a large national concrete placement firm.

Mike is a licensed Professional Civil Engineer, attorney, and Certified Safety Professional. The majority of Mike's construction experience is hands-on in the field as a laborer, pile driver, carpenter, surveyor, or project manager. Besides working for a concrete specialty subcontractor and a heavy/marine general contractor, Mike has worked for a design/build contractor constructing office buildings and light manufacturing projects.

Mike is a professional member of ASSE, serving as past president of the Kitty Hawk chapter. He has actively served on the Society's National Governmental Affairs and Standards Development committees. ASSE's Construction Division, Region VII, All-Ohio Safety Council, and Kitty Hawk chapter have recognized Mike as their Safety Professional of the Year.

Mike is also an active member of the American Society of Civil Engineers, National Federation of Independent Businesses, American Bar Association (Construction Law Forum), Dispute Review Board Foundation, and the World Safety Organization. He sits on the Associated Builders and Contractors National Safety Committee, the ANSI A-10 committee, the National Construction Dispute Resolution Committee of the American Arbitration Association, and several Boards of Directors. He has been an adjunct faculty member at three local colleges.

Mike welcomes and invites your comments, questions, or requests for support at Hayslip@aol.com.

CHAPTER 20

Excavation, Trenching, and Shoring

LEARNING OBJECTIVES

- Discuss real-world job hazards associated with excavation work.
- Discuss the importance of preplanning and continuous inspection in excavation work.
- Explain soil mechanics theory.
- List acceptable means for establishing soil stability (namely sloping, shoring, and shielding).
- Discuss relevant federal OSHA standards found in 29 CFR 1926.
- Identify and explain suggested trench rescue techniques.

INTRODUCTION

Soil stability is the foundation of excavation safety. Simply stated, excavation safety relies on the ability of craftspeople to identify and properly employ the means and methods for establishing and maintaining soil stability, thereby avoiding the "Running Rats." "Running Rats" is the author's descriptive term for the rocks, debris, or clods of dirt that "run" down the face of a sloped excavation. This is similar to OSHA's use of the term "small spalls" or "raveling" in Appendix A and B from Subpart P of 29 CFR 1926. These "Running Excavation Rats" are sometimes the last hazard indicator before a potentially serious failure occurs. Heed their warning.

The reason soil stability is so important is that *all cuts made in the earth will eventually try to heal themselves.* This may occur in a matter of minutes, several hours, a few days, even months or years; but it will occur. Unstable soil will fail, leading to a collapse that may result in undesired property loss, other increased costs, or worse. This chapter discusses in some detail the topic of soil stability along with other relevant subject matter in the area of excavation safety.

This material has been prepared for the American Society of Safety Engineers with three underlying themes in mind: (1) to offer real-world insights into spotting excavation hazards derived from hands-on experiences commonly not found in everyday textbooks; (2) to acquaint the reader with a working knowledge of the federal OSHA construction safety standards found in Subpart P of 29 CFR 1926; and (3) to convey other relevant and useful topical information to time-pressured safety professionals and students of safety in a manner that is clear, concise, and easy to understand.

The chapter's organization and subject matter spans a wide spectrum of issues in the field of excavation work. A bullet format is used for rescue techniques and OSHA guidelines. Thoughtful write-ups address significant issues such as basic soil mechanics theory, protective systems, and soil classification methods. Numerous sketches and photographs are offered to further develop and illustrate significant concepts of hazard awareness.

EXCAVATION WORK

To start off on the correct footing, we should clearly define the subject matter of this chapter, which is twofold: First, to assist in timely identification of hazardous conditions or unsafe practices in excavation work, and second, to offer reasonable suggestive solutions that can be acted upon by those involved with excavation work who are ready, willing, and authorized to use them.

What Is Excavation Work?

Excavation work is basically the entire man-made process of moving earth (a heavy nonhomogeneous material, with dramatically variable engineering properties that can be controlled but not sculpted to a permanent form) for the purpose of making improvements to the land. Excavation work is part of everything from digging for the placement of concrete footings and walls, to the ultimate backfilling against them (Figure 1). It also includes cutting and filling sections of embankments for a roadway, bridge, or dam. Excavation work is generally involved with the installation of underground utilities and tanks along with some types of soil stabilization methods. The hazards of excavating, like the different types of excavation work, vary greatly.

Excavation safety issues may include, but are not limited to, crushing and suffocation, being struck by falling objects or overhead loads, vehicle and equipment mishaps, exposures to falls, electrocution, drowning, and respiratory hazards. These dan-

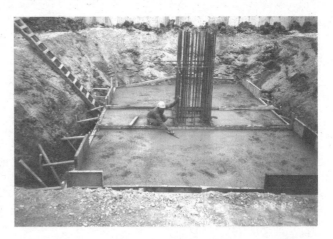

■ ■ **Figure 1** ■ ■ **Bracing of footings may involve some excavation**

gers can directly affect craftspeople, owners, construction managers, governmental officials, and the general public in and around excavation work.

CASES, COSTS, AND TRAGEDIES

Injuries to people and property occur too often when safe work practices or procedures are not followed. This certainly presents a moral and ethical dilemma. Likewise, there are always critical *costs* associated with these calamities.

The following section catalogs miscellaneous direct and indirect costs and a few of the many unfortunate tales from the trade. Risk managers commonly refer to these costs in terms of *losses*. Losses are generally controlled in one of four ways, through (1) avoidance, (2) abatement or mitigation, (3) shifting or transferring of the risk to a third party, and (4) acceptance of the risk and potential loss. Consider how you might handle each of these direct and indirect losses.

DIRECT COSTS

Direct costs are often referred to as *hard* costs. The following is a sampling of these hard (or direct) costs:

- Compensation to injured employee
  - direct wages
  - work transition

- wage loss (difference between regular job and accepted light-duty position)
- temporary total
- permanent total
- permanent partial
- facial disfigurement
- difference due to a change in occupation

- Medical
  - physician care
  - nursing care
  - medicine
  - hospital room and procedures
  - appliances

- Death benefits
- Property damage
- Rehabilitation of employee
- Handicap reimbursement

INDIRECT COSTS

Indirect costs are often referred to as *soft* costs. The following is a sampling of these soft (or indirect) costs:

- Training
- Recruitments
- Rehiring
- Orientation
- Retraining
- Insurance (General Liability/Builder's Risk/Umbrella)
- Reputation
- Lost production
- Lower morale
- Poor quality craftsmanship
- Additional drug testing and physicals
- Loss of good will
- Reputation
- Lost opportunities
- Mandated overtime
- Tools
- Supplies
- Equipment (purchase and rental)
- Fines, citations, and penalties
- First aid
- Rescue
- Increased security
- Redrafting of policies and procedures

- Survivor's remorse
- Creating new policies and procedures
- Administrative
- Incident investigation
- Abatement means and measures
- Bonding and surety costs
- Workers' compensation premiums
- Employee turnover and absenteeism
- Liquidated damages
- Schedule delays
- Employee's potential loss of livelihood
- Death
- Criminal prosecution
- Third-party suits
- Intentional tort claim
- VSSR claim (in Ohio)
- Negligence claim
- Friends and family respective loss of consortium
- Consultant fees
- Legal fees and cost of litigation, arbitration, mediation, or negotiation

A final word on costs: It is not uncommon for soft costs to exceed hard costs by a factor of five (5 x). When money is paid for costs out of bottom-line net profits (typically ranging from 2% to 8% of sales), a single incident with $1650 in direct costs can equate to an additional $125,000–$500,000 in sales required to simply *break even*.

How many of your single-incident direct costs (especially in excavation work) are held to only $1650? Then ask yourself, "Do unsafe contractors with lower bids actually represent a savings when the day is done?"

Tales from the trade need not be limited to the financial costs of an occupational workplace mishap in the arena of excavation work.

Consider an individual's discomfort when confronted by a small mass of loose soil that weighs 3000 pounds/cubic yard.

Consider the double-digit number of individuals who die each year with their mouths open as they suffocate from the weight of dirt wedged up against their chests.

Consider individuals who have been crushed in a trench by falling, dropped, or rolling overhead loads.

Consider watching a mother pick up and drive away her child's (or husband's) car from a parking lot several days after his death or hospitalization.

Consider the infant who won't have both of her parents to play with and learn from.

Consider a spouse who now needs to get a first or even second job to make ends meet.

Consider a recovering craftsperson who is now unable to ever work in his or her selected trade.

Consider the fatal blood clots that are released once the pressure of dirt is removed from a partially buried victim.

Consider dying alone in darkness while buried in a trench surrounded by mud, dirt, water, and muffled sounds.

HAZARD IDENTIFICATION

Working in and around an excavation site can be very hazardous. But it doesn't have to be that way if the threats can be anticipated or spotted early and controlled.

The first step in avoiding or abating a hazard is to identify it. To that end, this section of the chapter details numerous hazards common to excavation work. They are categorized in two simple ways: (1) hazards from being directly exposed to unstable soil and (2) those hazards incidental to the work being performed.

Direct Exposure to Hazards

So what is the big deal if a bit of dirt falls on a coworker? Well, consider that a single cubic yard of dirt weighs between 3000 and 4000 pounds. The exact weight depends on its composition—how much water (moisture content) and air (percent voids) are present, plus the density of the solids present (called specific gravity, which indicates how many times heavier the solid is than water).

Regardless, this amounts to about the weight of a small pick-up truck! Statistics indicate that significant numbers of victims who die from an excavation collapse will suffocate or be crushed by the soil around them. Nearly 100 people die each year from excavation incidents, impacting thousands of friends, family, and co-workers.

Although the section on soil mechanics will go into more detail describing the composition and engineering properties of soil, a few brief words on what soil is and how it behaves are in order.

Soil is a very complex material, composed of varying amounts of solids, water, and air voids. With respect to solids, soil can be composed of various-sized rock particles such as clay, silt, sand, gravel, cobbles, and boulders. Each range of different-sized soil particles behaves independently and as a unit with a unique set of engineering properties. Engineering properties of soil vary based on particle size, shape, texture, and composition of the parent material. But that is not all.

Often, knowing the manner in which the soil was deposited (i.e., wind blown, erosion, uplift), or if the soil has been previously disturbed, becomes important to an engineer designing a lateral earth retaining structure, as does the impact of environmental conditions (like freeze/thaw cycles and vibration). Add to this that some degree of water is found between the voids of these particles, not to mention possible surcharge loading, and quickly the combination of permutations greatly increases. It has been said that understanding geotechnical issues is more of an art than a science. (Terzaghi, K.)

Finally, soil may contain materials other than different types of rock. It may contain organic material such as peat, roots, and other plant matter as well as man-made materials that have been discarded or intentionally placed beneath the ground.

The engineering properties of soil commonly vary greatly both in the horizontal and vertical directions. Modes of failure are often as unique as the individual job site. For example, an excavation may fail from the top (from a tension crack), in the middle (from bulging), or at the bottom (from weak pockets) (see Figure 2). Rotational failures are especially dangerous because they occur in waves—first with a primary failure and then with more deadly secondary failures (Figure 3).

To summarize, soil's stability can vary along the length of an excavation as well as at different elevations. Its engineering properties will vary with the material itself, the manner in which it was deposited, and factors that are external to the soil itself.

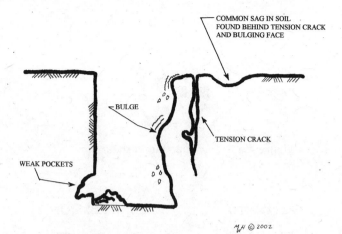

■■ **Figure 2** ■■ **Three soil conditions that could lead to failure at the top, middle, or bottom of an excavation**

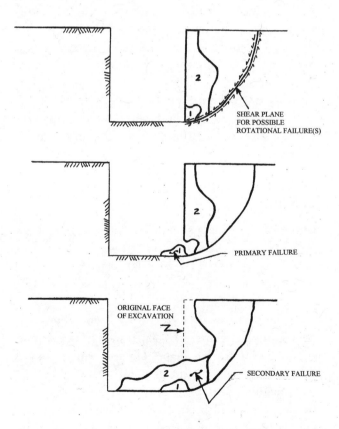

■■ **Figure 3** ■■ **A rotational failure along a shear plane can trigger a primary failure, potentially setting up a more catastrophic secondary failure**

External factors are those engineering properties beyond particle size, shape, texture, or chemical composition that influence a soil's stability.

Some of these external factors are:

■ water from any source (precipitation, a high water table, flooding, surface runoff, or aquifers)
■ vibrations from construction equipment, vehicular traffic, pile driving, blasting, or miscellaneous machines
■ previously disturbed soil
■ surcharge loads from spoils, equipment, stockpiled materials, or neighboring structures
■ external-point loads from equipment and materials
■ seepage pressures or quick conditions
■ degree of soil compaction
■ gravity
■ tension cracks (especially when holding water)
■ location of utilities or existing structures
■ layering
■ depth
■ time
■ fissures
■ weather (including freezing and thawing cycles)
■ slip surfaces and sloped bedding planes.

A question often asked when teaching this topic is: "Which of these external factors presents the greatest hazard?"

Of the factors listed immediately above, many professionals (including this author) believe that the presence of water is the single greatest concern.

But a word of caution is in order. The presence of water *by itself* may not be the sole issue. What makes the art of understanding soil so challenging is being able to identify the interaction of multiple causative factors. For instance, soil failure and collapse may be due to several external elements acting together. A competent person may encounter a combination of vibration sources that are linked with previously disturbed and saturated soil. (The topic of water will be discussed in much greater detail later in this chapter).

Incidental Exposures

This section discusses hazards related to excavation work. Readers are encouraged to review their own job-specific hazards to develop a more thorough list. Incidental exposures include:

- fall hazards (i.e., walking too close to an unprotected edge or walking across a plank that spans a trench without appropriate fall protection)
- public or private vehicle operation (such as being run over, getting caught in an equipment's swing radius, or the equipment itself rolling into the excavation)
- being struck by items that fall from a higher level (such as unsecured loads, pieces of equipment, tools, supplies, backfill material, and excavated spoils)
- drowning that occurs with a sudden rise of water elevation or if the dewatering system fails
- excessive noise from equipment or machinery
- fire resulting from poor housekeeping practices or other means
- improper lifting techniques or procedures
- electrocution (resulting from equipment contacting a power source or from faulty, frayed, or worn electrical cords)
- bad atmospheres and other respiratory hazards (including particulates from silica-based products, equipment exhaust, or landfill remnants).

BASIC SOIL MECHANICS THEORY

The stability of an excavation depends primarily on the soil's ability to resist shearing forces. When the capacity of a soil's shear strength is exceeded it fails. That's it, plain and simple.

Shear is an engineering term used to describe forces that seem to "slide" past each other along a given surface called a failure plane (Figure 4). This is in contrast to forces acting under tension or compression (Figure 5).

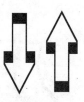

■ ■ **Figure 4** ■ ■ Forces acting in shear

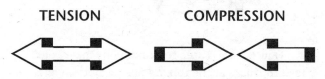

■ ■ **Figure 5** ■ ■ Forces acting in tension and compression

Note: the term *unconfined compressive strength* found in the OSHA standard is used for classifying cohesive soils to indicate shear strength. The value of a soil's unconfined compressive strength is two times the value of its shear. Both terms are measured in units of tons per square foot.

There are two predominate factors contributing to a soil's shear strength: cohesion and internal friction.

Cohesion

Cohesion is a soil's stickiness. For example, clays are cohesive while silts, sands, gravel, cobbles, and boulders are not. In particular, clay particles act differently upon each other than sand or silt particles do.

Clay particles directly attract each other. The cohesion, or stickiness, of clay results from its small size and relatively flat shape. The following example is provided to flesh out this point.

Imagine a clay particle the size and shape of a shelled sunflower seed. Now, relative to that particle of clay, a medium-sized sand particle would be well over 100 ft in diameter! Due to clay's small size and relatively flat shape, it develops particle-to-particle attractions from static electrical charges that collect on the faces of these microscopic minerals.

Sands and silts are generally rounded particles that do not exhibit cohesion in the truest sense of the word. Instead, they exhibit what is called *apparent cohesion* (also called the sandcastle effect). An intermediate substance between the particles, such as water, temporarily binds them together through surface tension. However, when there is too little or too much water (e.g., from a high water table), the apparent cohesion (false stickiness) fades.

Friction

Internal friction is a geotechnical engineer's way of indicating the tendency of individual particles to resist movement past each other due to their shape and texture. Rounded and smooth particles are more apt to slide past each other than angular or rough-faced particles. Therefore, rounded and smooth particles are said to have less internal friction.

Imagine rolling smooth and rounded particles like bank-run material in your hand compared to crushed aggregate. Bank run, like marbles, tend to be rounded and smooth; they move about more easily, thus, they are said to have less internal friction, or a lower resistance to shear, than crushed gravel.

Water

Excessive water is very dangerous! Excessive water reduces soil's shear strength in several ways. First, it may act as a lubricant between particles to decrease the soil's internal friction. Second, as a dipolar molecule (H_2O), water will travel upward against gravity through surface tension and capillary action to disrupt the apparent cohesion (sandcastle effect) in sand and silt. Third, water may disrupt the particle-to-particle attraction (cohesion) between clay particles. Fourth, water may cause a soil to expand. This may be a direct reaction to water "overfilling" existing air voids ("floating" particles apart), or fifth, water freezing and expanding inside the cavities of a soil mass (Figure 6).

Water can also alter a soil's weight, density, and lateral earth pressure by entering the air voids of

the soil. This may render shores and trench boxes inadequate.

Water can generate increased pore pressure, yielding a *quick* condition. Water causes erosion and potentially may undermine the support gained from the toe (base) of a slope.

Water that fills tension cracks will exert downward and lateral pressure that forces the cracks to become deeper and wider. (*Note:* Tension cracks run parallel to the edge of an excavation and may be difficult to locate, especially if stored materials, brush, or pavement inhibit a clear inspection.)

Finally, water may even chemically alter some mineral constituents.

However, the presence of water is not always bad. A couple of good things about water warrant discussion. First, adequate amounts of moisture may often promote stability by serving to develop an *apparent* cohesion in some cohesionless (granular) soils. Second, when excavations in cohesive soils (claylike soils) experience standing water, that water asserts a supporting hydraulic static pressure against the face of the walls (Figure 7).

In summary, a competent person can draw one significant conclusion from the facts above. Given that standing water in the bottom of a trench exerts supporting pressure to help hold it open, will tend to add weight and weaken the surrounding soil. *As a trench is dewatered and vibrating equipment is started*

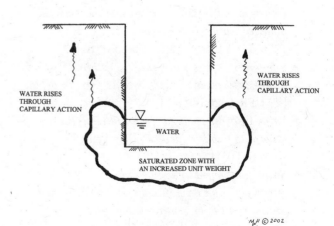

■ ■ **Figure 6** ■ ■ **What happens to the soil when water accumulates in the bottom of a trench**

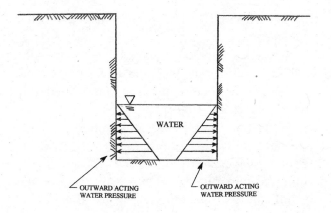

■■ Figure 7 ■■ Standing water in an excavation in cohesive soils provides a supporting hydraulic pressure

up, individuals may be expected to enter an area at the very time the trench may be at its most hazardous!

OSHA'S FEDERAL EXCAVATION STANDARDS

On the Federal level, OSHA's construction excavation standard 29 CFR 1926.650–652 (Subpart P) contains some of the requirements excerpted below. [*Note:* some states (called "State-Plan" states) have chosen to enforce more stringent excavation safety standards than those listed here.]

- *Underground installations must be located, protected, supported or removed [29 CFR 1926.651(a) & 29 CFR 1926.651(b)].*
- *Ladders, ramps or stairs shall be within 25 feet of an employee when a trench exceeds 4 feet in depth [29 CFR 1926.651(c)].*
- *Safety vests are required when individuals are exposed to public vehicular traffic [29 CFR 1926.651(d)].*
- *Spoils, equipment and other materials shall be restrained or kept a minimum of 2 ft. from the edge of a cut to prevent traveling into an excavation [29 CFR 1926.651(j)(2)].*
- *Excavations of 5 ft. in depth or with a potential for cave-in, shall be protected [29 CFR 1926.652(a)].*

- *Support systems shall be immediately backfilled and removed from the bottom up [29 CFR 1926.652(e)].*
- *A professional engineer shall design protective systems greater than 20 ft. in depth (Appendix F).*

The competent person [specifically defined in 29 CFR 1926.650(b) and 29 CFR 1926.32(f) means an individual that can spot bad stuff and then has authority to do something about it] is mandated to perform the following work activities:

- *Each day before individuals may begin work in an excavation, a competent person shall inspect the excavation, protective systems and surrounding areas. Inspection is also required after a hazardous occurrence such as a rainstorm or damage to a structural member of a shore or trench box [29 CFR 1926.651(k)].*
- *A competent person shall design structural ramps [29 CFR 1926.651(c)(1)(i)].*
- *A competent person shall test for hazardous atmospheres [29 CFR 1926.651(g)].*
- *A competent person shall monitor water removal [29 CRR 1926.651(h)(2)].*
- *A competent person shall have the assigned duty to remove individuals exposed to possible cave-ins or other hazardous conditions until necessary precautions have been taken to ensure their safety [29 CFR 1926.651(k)(2)].*
- *A competent person shall conduct both visual and manual tests/analysis (Appendix A).*
- *A competent person shall classify soil then reclassify soil as necessary (Appendix A).*
- *A competent person shall evaluate actual slope stability under impacting surcharge loads and shall cause adjustment to be made as warranted (Appendix B).*

The OSHA standards represent the minimum level of acceptable practice. Businesses may always choose to be more restrictive.

SOIL CLASSIFICATION

The description of "a method of classifying soil and rock deposits based on site and environmental conditions, and the structure and composition of the earth deposits" is from Appendix A of 29 CFR 1926 Subpart P. This classification system is a conservative field *guesstimate* that provides a soil stability rating. A competent person is directed to classify (or reclassify) soil. And although the following description of this classification procedure may seem overly complex, in actuality a competent person is often able to classify soil by using common-sense training and a thumb.

Soil stability is evaluated because of its direct impact on protective system selection and usage. The classification system found in the OSHA standard lists four classes. They are ranked here in a hierarchy from most stable to least stable using tons per square foot (tsf):

- Stable rock
- Type A, (unconfined compressive strength greater than 1.5 tsf)
- Type B, (unconfined compressive strength between 1.5 tsf and 0.5 tsf)
- Type C, (unconfined compressive strength less than 0.5 tsf).

Soil classification includes performing at least one visual and one manual analysis/test on the soil. Many of these tests are highlighted below. The reader is urged to perform a visual and manual analysis/test at different elevations and in more than one location along an excavation, because not all soil (even in the same region) is guaranteed to have the same degree of stability.

Visual and Manual Tests

Visual tests are intended to provide initial qualitative information (overall character, nature, or condition) of a given soil mass. Manual tests offer the competent person more detailed information into the quantitative properties (specifically how much of a given element is present). A competent person is required to perform at least one visual and manual analysis or test as recommended in Appendix A.

Visual tests are performed over the entire range of the excavation work, including the adjacent, removed, or existing material. Some of these tests include:

- Estimating the composition and range of particle sizes.
- Observing the soil as it is excavated; if it stays in clumps that might indicate the existence of cohesive material.
- Determining the presence, direction, and slope of layered systems: The properties of the soil mass is governed by the weaker soil mass.
- Looking for weak points or signs of potential failure such as tension cracks, fissures, running rats, seeping, or open sources of water.
- Looking for the presence of previously disturbed soil, especially where existing utilities or underground structures are concerned.
- Looking for the presence of unacceptable sources of vibration in and around the area that may affect the stability of the excavation.

Manual tests are performed on individual samples of soil. However, it is important to note that not all soil is created equal. Soil may be composed of varying combinations of cohesive and/or granular materials. Some tests may only be effective on certain types of soil samples.

The basic types of manual tests include:

- A plasticity test (relating shear strength or unconfined compressive strength to the ability to "mold" a given sample).
- A dry strength test (relating a dry sample's resistance to crumbling under hand pressure to indicate the degree of fissures present)
- A thumb penetration test (used to estimate unconfined compressive strength in tsf per ASTM D2488). Estimates of unconfined compressive strength can be directly used to classify soil under the OSHA designation of type A, B, or C.
- Likewise mechanical devices such as pocket penetrometers or shearvanes provide readings that indicate a sample's unconfined compressive strength.
- A competent person is permitted to make available "other recognized methods of soil classification and testing such as those adopted

by the American Society for Testing Materials, or the U.S. Department of Agriculture textural classification system." These systems include the use of sieve analysis, mason jars, standard penetration tests, and hydraulic testing machines not described here.

- The "bite test" is *not approved* as an acceptable means of soil classification. (Bite test: After biting into a soil sample, if one must brush his teeth to remove the material, it indicates the presence of cohesive material in the soil. Again, this is not recommended.)

Note: It would be wise for a competent person to perform all, not just one, of the visual tests listed above. It would also be wise for a competent person to perform several manual tests along the length of an excavation both in the vertical and horizontal direction.

Detailing Categories and Conditions

The classification of soil as listed above is based on two sets of variables. For purposes of this chapter they will be called "categories" and "conditions." It is a combination of evaluating categories *and* conditions that finally leads to the classification of soil and ultimately the selection of a protective system.

CATEGORIES

OSHA references five general categories of soil in Appendix A of Subpart P:

- rock
- granular soil
- cohesive soil
- loam
- cemented soil.

Each of these general categories can be broken out further.

Rock is a natural solid mineral. Yet not all rock is stable. Therefore, the OSHA standard does not use origin-based terminology (sedimentary, igneous, or metamorphic) to distinguish between rock formations. Rather, OSHA simply considers rock to be either stable or unstable. The difference between stable and unstable is that stable rock will not cave in when excavated. Unstable rock can cave in and requires a protective system. For classification purposes, dry unstable rock with fissures, fractures, or faults could be a Type B soil. However, moist unstable rock (submerged or seeping), and rock with layered systems sloped greater than 4H:1V, would be considered a Type C soil. For example, igneous rock such as granite could potentially be a more stable rock, while sedimentary rock such as sandstone or shale is generally a more unstable form of rock.

Granular cohesionless soils are broken out into subcategories of gravel, sand, and silt. Note that angular gravel (gravel shaped like crushed rock) and silt are granular soils that are initially considered to be Type B soils. Other gravels and sand are considered to be Type C soils.

Cohesive soils are clay soils, or soils with a high clay content. (*Note:* There are several different parent materials that can give birth to a clay deposit, each with differing engineering properties.) If a soil has enough clay in it, it will *act* like a clay soil, and therefore be considered cohesive. Cohesive soils that have enough moisture content can be evaluated based on their unconfined compressive strength. Cohesive soils could potentially be either Types A, B, or C depending upon their strength, the presence of water, and the effect of site conditions on the deposit.

A *loam* soil is simply a combination of granular (sand and silt) and cohesive (clay) soils. Recall that if a soil has enough clay in it the soil will act cohesively, so saturated loam "cohesive" soils can be tested for their unconfined compressive strength.

Loams could potentially be Type A, B, or C. Perhaps the one thing that makes loams simple to classify is that only "loamy sand" is listed as a Type C soil. All other loams mentioned in Appendix A are either Type A or Type B. The differences in the A and B categories will be due to either the type of loam in the deposit or the fact that, absent "other factors," there are five site-condition limitations on a Type A soil.

Type A soils cannot be previously disturbed, subject to vibrations, fissured, layered at more than 4H:1V, or exposed to excessive water.

Many times these site-condition limitations help simplify the classification process because where there could be a question of whether a soil is Type A or Type B, any of those common site conditions will at least cause the soil to be placed in the less stable Type B classification.

Cemented soil is generally defined as a caliche or a hardpan. These soils are hardened by nature and could potentially qualify for Type A soil classification. Caliche soil is a cemented soil that is created over time by water with a soluble cementing agent (often the calcium carbonate from limestone) that leaches through a given soil matrix. This soil can be extremely dangerous in the presence of water when rehydrated and weakened by a rainstorm or rising water table.

Hardpan is a soil hardened by the force of natural causes. For example, soil below the surface and subject to tons of the compressive weight of the soil above could be hardened into a solid matrix and then, due to geological activity, could be brought to the earth's surface. This type of hardpan soil is often so hard that water cannot penetrate it. Each of these types of cemented soil can fit within the strict guidelines of a Type A classification. However, rarely will one ever encounter enormous areas of Type A soil in the continental United States.

CONDITIONS

Site and environmental conditions play an extremely important role in determining soil stability. There are five conditions that appear in the definitions of Types A, B, and C soil:

- water content (including dry, moist, seeping, saturated, and submerged)
- layered (at a 4H to 1V ratio)
- vibration
- previously disturbed
- fissures.

A quick look at these definitions shows that any one of five site conditions may keep any soil from being classified as Type A. In other words, if soil is originally stable enough to be classified as Type A, and is exposed to any of these site conditions, the soil is going to be classified at least as a Type B and possibly a Type C soil.

Excessive water in the trench makes any soil Type C. Of course, the effect of water may be greater on granular soils than on cohesive soils, but excessive water can make either one Type C. (See the earlier discussion under "Basic Soil Mechanics Theory.")

A 4H to 1V sloped layer means that, if the excavation is in layered soils and the layers dip one foot for every four feet across the excavation, the slope is steep enough to deserve the classification of a Type C. For example, if a trench excavation is four feet wide, and the layers appear to drop at least one foot over that four-foot span, the entire trench would be classified as Type C soil.

A simplified example of soil classification could be conducted as follows:

1. There are two site conditions that cause any soil to be classified as a Type C. These are excessive water and 4H:1V sloped layers. If either of these is present, the soil is Type C.
2. Absent the two site conditions listed above, examine the soil to see if it is better than Type C soil. Type C soils include gravel (except for angular gravel), sand, loamy sand, and weak cohesive soils (cohesive soils measured with 0.5 tons/square foot of unconfined compressive strength). If the soil does not fit these Type C soils, go to step three.
3. At this point the soil and the site conditions place the soil into either the A or B categories. To easily eliminate Type A soils, look for one or more of the following limiting factors. If the soils are previously disturbed, subject to vibrations, or if the soils are fissured, they cannot be classified as Type A; therefore, the soil is a Type B soil.
4. In the absence of any of those (common) site conditions, look at the soil to see if it fits into one of the two Type A soil categories. For the soil to be Type A, it must either be a strong cohesive soil (1.5 tsf of unconfined compressive strength or greater), or it must be a cemented soil such as a caliche or hardpan. At this point, soils that do not meet these strict Type A criteria are classified as Type B soils.

Lateral Earth Pressure

One way to distinguish between classifications of soil is by the amount of force required to maintain its stability. Again, soil tries to heal itself when its natural support of surrounding soil has been removed. The force exerted by the existing soil is termed *lateral earth pressure*. It is important to understand this concept because knowing how much force (or push) it will take to keep soil in place indicates the needed strength of the structural members of a protective system. The following are lateral earth pressures and their corresponding soil classification:

- Type A soil requires at least 25 psf per foot of depth to hold it in place (in equilibrium).
- Type B soil requires at least 45 psf per foot of depth to hold it in place. This means that at the bottom of a 10-foot-deep excavation the active lateral earth pressure of soil moving against a shield could potentially be 450 psf.
- Type C soil requires a minimum of either 60, 80, 120, or 150 psf per foot of depth to hold it in place. The C-60 soil is accepted in the industry as a "better" Type C soil and is more commonly used for Type C designs. The Type C-60 soil is not flowing nor is it submerged. Further, as a rule of thumb, C-60 soil can be cut with near-vertical sidewalls that will stand open long enough for a shoring or shield system to be installed.

    It is very important to match your designed protective system with the actual Type C soil present. In other words, do not use a Type C-60 design if you have Type C-80 or weaker soil. Shores and shields based on a Type C-80 design are commonly very bulky, or stout, and generally are not in-stock rental items.

Note: Engineers may use the psf ratings of 25, 45, 60, 80, 120, and 150 to determine soil pressures when calculating loads on protective systems. The competent person might simply use the designations when referring to protective system specifications and limitations to determine how to safely use the system based on soil classification. The competent person must consult with manufac-

■ ■ **Figure 8A** ■ ■ **Waler systems provide shoring support with greater working room between cylinders than vertical shores (in addition, Waler systems can be used in C-80 soils)** (Photo courtesy of Speed Shore Inc.)

■ ■ **Figure 8B** ■ ■ **Slide rail is a versatile module shoring system that has proven useful in the most challenging excavations—particularly under poor soil conditions and at sites with adjacent structures or utilities** (Photo courtesy of Speed Shore Inc.)

turer's tabulated data to determine recommendations, specifications, and limitations of manufactured protective systems, such as shoring or shielding products.

Even after selection of a protective system, inspections are to be conducted by a competent person at least daily (before work has begun) and after a hazardous occurrence (such as a rainstorm or structural breakdown).

PROTECTIVE SYSTEMS

In review, a protective system is a means to (1) maintain soil's natural stability, (2) provide additional stability to avoid collapse, and/or (3) create a secure environment in case of an excavation failure. When an excavation is 5 feet or greater in depth (or indicates a reasonable potential for cave-in) a protective system must be used, provided the excavation is not entirely made in stable rock. Protective systems used in excavations with a depth of 20 feet or greater must have the approval of a licensed professional engineer.

Protective systems are commonly referred to in the context of the three S's:

- sloping (including benching)
- shoring
- shielding.

Sloping means to remove soil while still maintaining its stability. Forces are kept in equilibrium, which prevents soil from caving in under its own weight. Sloping and benching (or stair-stepping) are common terms for cutting the face of the excavation back to an acceptable slope angle. Table 1 expresses the maximum slopes for each soil type.

Shoring excavated walls prevents soil movement. Shores may run vertically. They may run horizontally as in a waler system (Figure 8A). Shores may be aluminum/hydraulic, steel/air, timber, or a hybrid (Figure 8B).

In Appendix B, OSHA offers tabular information that can be used for installing timber shoring (either oak or Douglas fir) when performed in accordance with 29 CFR 1926.652(c)(1). A notable limitation exists. These tables do not apply when adjacent loads exceed the weight of a 2-foot soil surcharge or in the presence of equipment that weighs more than 20,000 pounds (e.g., a half-full concrete ready-mix truck or a trailer loaded with reinforcing steel).

In Appendix C, OSHA likewise offers information for selection and installation of aluminum hydraulic shoring for trenches. A trench is defined as an excavation that is deeper than it is wide and less than 15 feet in width.

Shielding provides a safe work area within an excavation, even if the soil fails. Trench shields (or trench boxes) do not necessarily prevent soil movement, but they are designed by manufacturers to be strong enough to handle a given amount of lateral earth pressure from a failure (Figures 9A–C). Shield design is not addressed in the OSHA standard. It is important to know and trust your trench shield supplier.

■ ■ Table 1 ■ ■
Maximum Allowable Slopes

Soil or Rock Type	Maximum Allowable Slopes (H:V) for Excavations Less than 20 Feet Deep
Stable rock	vertical (90°)
Type A	0.75:1 (53°)*
Type B	1:1 (45°)
Type C	1.5:1 (34°)*

*measured from horizontal

■ ■ **Figure 9A** ■ ■ **Significantly lighter than steel trench shields, aluminum trench shields are designed for ease of handling with a rubber-tired backhoe or light excavator**

■■ Figure 9B ■■ Manufacturer's (steel frame) modifications to this shield provide for a more rugged aluminum shield that maintains the lightweight feature while providing a safe working environment

■■ Figure 9C ■■ Some shields, like this manhole shield, are designed to meet a particular need but are often found to be useful in other applications

Several concerns with respect to trench shields include:

1. Ensuring the shield(s) is/are rated for the type of soil and depth of trench (especially if staked).
2. Keeping individuals clear of overhead loads.
3. Proper access into and out of the shield.
4. Not being in a trench box when it is being moved vertically.
5. Manufacturer's tabulated data is present on the job site and matches the shield being used.
6. Maintaining at most a minimal distance between the exterior face of the shield and the soil.
7. Open cuts are backfilled immediately after work is completed.
8. Spoils remain at least 2 feet from the edge of the cut.
9. Backfill material being placed in an occupied trench box.
10. Unguarded equipment permitted near the edge of an excavation.

11. Ensuring properly rated slings that are used to smoothly lower the box into place are not worn, frayed, knotted, or kinked.
12. Excavation buckets do not "tap" the horizontal struts of the shield to secure the shield into the soil, rather the side shields are used.
13. Any structural modifications are approved by a licensed professional civil engineer.
14. Cross braces are not bent or deformed.
15. Soil sloping upward and away from the top of a shield must be held 18 inches lower than the top edge of the shield.
16. Not permitting co-workers to leave the safety of a shield when in the bottom of an excavation.
17. Excavation equipment stays clear of overhead power lines.
18. Being aware of and taking precautions against open-ended boxes in close proximity to exposed excavation head walls (use end plates).

19. Ensuring that stacked boxes are adequately secured together and designed for that purpose.

There are some protective systems not found in the OSHA standard—such as driven steel sheet piling. Therefore, steel sheet piling falls into a special category known as an "Engineered Design System" or a "site-specific" engineered system. This is also the case for H-beam, or soldier pile systems (see Figure 10), screw jacks, tie backs, rock bolting and Krings systems.

The competent person must always consider the site conditions in and around an excavation when making protective system selection decisions. One important site condition that should always be considered is the effect of a surcharge load on the stability of the trench.

Usually a surcharge load, such as an adjacent spoil pile or machinery, does not figure into the soil classification, but it does greatly impact protective system decisions. Loading in the adjacent area creates a greater downward pressure that subsequently could generate greater lateral earth pressure. That is why engineers may choose to limit the amount of loading with protective systems.

Generally speaking, tabulated data might limit the height of the spoil pile in the adjacent area of the trench to no more than 3 feet, or it might limit the weight of machinery operating within the adjacent area to 10 tons or less. These kinds of limitations ensure that the protective system is not overloaded with lateral earth pressure.

Finally, tabulated data such as tables, charts, and write-ups of selection criteria for all protective systems must be maintained on the job site. If one chooses not to use a protective system as detailed and designed by OSHA in 29 CFR 1926, then copies of other relevant tabulated data must be maintained on the job site, presumably including the identity of the approving professional engineer authorized in that state.

TRENCH RESCUE

Consider this: an 8-foot-deep, nearly vertical trench in previously disturbed silty clay has been open for less than an hour, with no shoring or shielding where water begins to collect in the bottom, and vibration from operating equipment causes dirt to roll back down one side of the trench face. Would you enter or expect others to enter? Of course not, because you know the danger—how terrible it must be to slowly suffocate in a collapsed trench—to be all alone, feeling the earth close in against your body with each outgoing breath, and to know, for a brief moment, that you may never again see a loved one's smile. You are wise to heed the warnings of those "Running Excavation Rats"!

However, if an excavation fails, co-workers may be trapped. What steps should be taken before beginning rescue or recovery efforts when this occurs? The following is suggested as one approach to avert just such a disaster. It was prepared strictly as a guide and should not be used in place of an adequate emergency action plan.

Initial Stage of Collapse—Assessment

- Remove exposed individuals from the excavation.
- Call 911 and provide directions to the site. Someone should be sent to meet the rescue vehicle(s) since many new developments are on unmarked streets. This is valuable time that can be saved.

■ ■ **Figure 10** ■ ■ **H-beam or soldier pile system**

- Account for everyone.
- Appoint someone (i.e., a competent person) to gather and retain critical facts for rescue personnel (e.g., type of soil, number and location of victims, length of time trapped, excavation width and depth, available resources).
- Determine whether the failure damaged a utility; if so, take appropriate action.
- *Shut down all equipment* and reroute traffic to eliminate harmful vibration.
- When looking into a collapsed trench, do so from the short end rather than along the length of the trench, because the short end is usually more stable; however, note that looking into a trench from the short end does not guarantee a safe vantage point.
- Begin/continue to remove seeping groundwater.
- Do not remove hand tools, personal protective gear, or other items that may be used to locate a victim.
- Rescue attempts are typically performed by hand and take a considerable amount of time. Order lights, these may be needed if the rescue continues into nighttime hours.

Intermediate Stage of Collapse—Preentry Phase (after calling 911)

- Establish a perimeter (see zones described below).
- Assign an observer to watch for secondary failures; often a rescuer becomes another victim.
- Ensure that the site is accessible for rescue vehicles (wide enough without other obstacles such as soft roads, buildings, pipes, spoils, or equipment).
- Cooperate with the rescue team.
- Call the home office to speak to an officer of the company and inform him/her of the situation. Locate the crisis management or emergency action plan and use it.
- Designate an experienced person to speak to the media and family members.

- Span tension cracks with planks or a sheet of plywood as necessary.
- Do not enter a failed excavation without adequate protection.
- Test for hazardous atmospheres as necessary.
- Do not park equipment too close to the edge.
- Consider lowering a digging tool tied to a rope to the victim so he or she may dig out without help from another person.
- Do not pull a partially buried victim out by a rope, belt, sling, or choker. The human body is not built to resist tension very well. Static loads, skin friction, and vacuums all serve to put a victim into tension when pulled. Even if one was able to move a trapped person, this might only serve to loosen surrounding soil and create another collapse.

When an excavation collapses, decisions often need to be made quickly, considering the facts available at the time. It can be valuable to preplan for an emergency by staging a mock excavation collapse. At that time, try to introduce the use of zones.

Zones help secure a site so those performing the rescue can work without distractions. Establish several zones from the incident site (perhaps using caution tape or other markings), permitting access within each zone only to a designated group. For example:

- Zone 1 (Hot Zone) is closest to the failure and is accessible to rescue and first-aid professionals only.
- Zone 2 (Warm Zone) is a rest station for rescue workers as well as a staging area for materials, supplies, and equipment.
- Zone 3 (Cool Zone) is for immediate family members, owner representatives, and interested third parties.
- Zone 4 (Beyond the Cool Zone and farthest from the incident) is intended for media, general public, and others.

To ensure that this system works, appoint several individuals to prevent unauthorized parties from entering a given zone.

PREPLANNING

As stated earlier, soil is a complex material. However, there is much more to be concerned with in excavation work besides the stability of the soil. The following are some of these "other" hazards, with suggestions for handling them and measures that can help you prevent them.

The best way to combat a hazard is never to confront it. Anticipate potential hazards and preplan for them. Get a commitment of resources from management. Develop the awareness to identify potential hazards through training, locate the proper resources, and act affirmatively.

One primary duty of the competent person is to prevent hazards by pinpointing them, then avoiding, eliminating, or controlling them. Below is a list of measures a competent person should be prepared to undertake:

- Preplan the work at bid time even before you mobilize. This includes maintenance of equipment and any attachments. (For example: Are excavation buckets without cracks and do they have sound welds? Are the teeth secured?) Has a professional civil engineer been consulted if necessary? Have your people been trained for this type of work?
- See that equipment is actually kept away from overhead power lines. Training and company policies do not always keep people out of harm's way in the absence of adequate supervision and support of management.
- Secure necessary street, building, and environmental permits.
- If homeowners or businesses are affected, notify them in advance.
- Double check that all the proper equipment, machines, tools, PPE, and other supplies are on site. Then confirm these goods are without defects.
- Notify the required utility locators of underground installations to mark them, and do not forget to contact the property owner for the location of any "private" utilities.

- Be able to identify and act based on the color-coded significance of utility markings on the ground (typically Red = Electrical, Yellow = Gas/Oil, Orange = Communications, Blue = Water, Green = Sewers, White = Proposed Excavation).
- Have co-workers manually dig any time you are in close proximity to utilities, paying special attention to local or state building codes.
- Remove, support, or control utilities. When digging, look for concrete, gravel, flagging, seeping soil, or a dramatic change in the color of surrounding soil that may indicate buried utilities. Know that if you find concrete (especially dyed red) there may be utilities below. Sometimes seeping water indicates utilities because the flow of water often follows the gravel surrounding utilities.
- Maintain storage of shoring and shielding equipment consistent with manufacturer's recommendations to avoid damage to the equipment and to safeguard employees and passersby.
- Confirm stairs, ramps, or ladders are present and secured, within 25 feet of co-workers for trenches greater than 4 feet in depth.
- Monitor water removal and ensure potential free surface water will be diverted from re-entering the trench. (This may mean using diversion ditches and dikes; or using longer, or better located, section(s) of dewatering pump hose; or grading the surrounding area away from the excavation).
- "Crumb" or scale back loose soil from the face of excavated surfaces.
- Classify the soil according to the OSHA standards, then use a protective system (shoring, shielding, sloping, or benching) based upon the strength of the soil as classified by the competent person. OSHA recognizes soil types designated as A, B, C, and rock rather than clay, silt, or sand. (*Note:* Trench-box manufacturers commonly use a C-60 designation in their design.) This is a subclassification of the Type C soil.

- A professional engineer must design protective systems for any excavation greater than 20 feet in depth.
- Look for variations in soil color that might indicate differing soil properties.
- Locate and control weak pockets of material.
- Reclassify soil when conditions dictate.
- Use barricades, signs, flagmen, or warning lights, as necessary, especially when near public access.
- Check to make sure materials (especially pipe) cannot roll into the trench.
- When reasonable, use gravel boxes, and do not permit loaders to directly dump gravel into a trench where individuals are in close proximity without providing a means to identify the exact location of those individuals in the trench.
- Make sure trenches are spanned properly and fall protection is adequate?
- Never permit co-workers to work alone in a trench.
- Remove standing, running, or seeping water whenever possible.
- Limit your exposure. Do not enter an excavation unless absolutely necessary.
- Prevent exposure to overhead loads (including dirt stuck to excavation equipment, unsecured material, inadequate chokers and slings, and detaching buckets or other attachments).
- Reduce sources of vibration where possible (especially when vibration can be felt through the soles of your shoes).
- Reduce surcharge loads by removing spoil material.
- Be aware of the weather's impact on an excavation. Stay updated on local weather forecasts for the short and long term of the project.
- Remove, control, or shield overhead transmission lines.
- Do not undercut existing structures such as buildings, sidewalks, utility poles, and walls without adequate support (i.e., bracing, shoring, or underpinning).

- Support exposed utilities. (To avoid settlement, it may be best to support from above rather than from underneath.)
- Review operator procedures and competence to determine that safe work practices are understood and followed. (For example: Is an operator trained for that type of equipment? Does the operator avoid moving the bucket toward the trench from the blind side of the rig?)
- Confirm that specified equipment is actually used.
- Have a reasonable means of access and egress (i.e., ladders, ramps, or stairways) within 25 feet of each employee.
- Be sure operators have ample means of communication with co-workers, including a single point of reference for responsible signaling.
- Review flagman operations and protective clothing.
- When loading dirt or moving materials, do not move equipment over the head of any worker.
- Ask, "Are excavation cuts made vertically?" Sometimes, if a backhoe or trackhoe sets up on unleveled ground, the sidewall cuts may not be vertical (called scalloped sides), which increases the likelihood that one side may fail because it seems to lean into the trench.
- Follow relevant federal or state OSHA guidelines. (It's the law.)
- Inspect frequently and regularly watch for signs of failure or weakness (e.g., tension cracks, bulging, the groaning or creaking of structural members, and, of course, running rats).
- Train employees how to spot signs of failure and empower them to act. Understand that not all workers read or speak English well, if at all.
- Observe the work practices of co-workers (for instance, signs of confusion, ringing in the ears, or inability to hold on to tools may

be an indication of oxygen deficiency or a bad atmosphere).

■ Make sure outriggers on equipment are used and supported where necessary.

■ Confirm that you have the proper equipment as ordered. (You can do this by looking at serial numbers and comparing to POs or shipping tickets.)

■ Confirm that there are no defects in shoring and shielding materials.

■ Be aware of the potential for hazardous atmospheres. Test for them when it is reasonable to do so and have the proper personal protective equipment (PPE) if the hazard cannot be eliminated administratively or through engineering means.

■ Employ stop logs so equipment cannot roll into an excavation. Often the bucket of a loader keeps the operator from seeing the edge of the cut.

■ Confirm personal protective equipment is worn (hard hats to prevent being hit directly on the head by falling objects or hearing protection to prevent hazards from noise).

■ Inspect chokers, slings, and chains for excessive wear, discoloration, knots, and kinks. Avoid shock loading of any rigging system. Do not lift with the teeth of the bucket.

■ Inspect, inspect, inspect, and watch for signs of failure or weakness: tension cracks, bulging, slip planes—*and don't forget the running rats.* Subtle changes in site conditions can immediately impact a soil's stability.

■ Remove defective materials.

■ Locate who can supply a vacuum truck to remove soil if needed.

■ Anticipate the unlikely event of a collapse and know what to do, even run a mock drill, and establish job responsibilities.

CONCLUSION

Excavation failures hurt, maim, and kill workers. Understanding the basics of how and why exca-

vation failures occur and what to look for will help prevent such catastrophes. Be aware that the stability of soil is impacted by numerous factors such as water, vibration, surcharge, freeze/thaw cycles, and the type and depth of soil.

Given varying degrees of soil strengths (often in both the horizontal and vertical direction) and with no universal mode of failure to draw upon, plus many factors that reduce the strength of soil—not to mention numerous other unrelated hazards—the task of being safe while performing excavation work may seem overwhelming. But it doesn't have to be.

Every day excavation hazards are identified and avoided, eliminated, or controlled. With some knowledge, training, and experience you and your co-workers can work safely and profitably in this field.

Use resources already at your disposal and seek out what you do not know from third parties. There are numerous soil safety training experts and qualified suppliers of earth retention equipment in this country. Training, common sense, and preparedness go a long way toward providing a safe excavation site. Always verify that the craftsperson, manager, supervisor, competent person, or safety professional has received training to identify, and then is authorized to abate, control, or eliminate hazards.

Preplan for hazards, get the needed resources, empower affirmative action, be prepared for emergencies, and always, always, watch for the running rats.

REFERENCES

Abramson, L., T. Lee, S. Sharma, and G. Boyce. *Slope Stability and Stabilization Methods.* New York: Wiley, 1996.

ACI Committee 436, "Suggested Design Procedures for Combined Footings and Mats," *JACI,* October 1996 (now Committee 336).

Barden, L. "Primary and Secondary Consolidation of Clay and Peat," *Geotechnique,* vol. 18, no.1, March 1968.

Barnes G. E. "A Simplified Version of the Bishop-Morganstern Slope Stability Charts," *Canadian Geotechnical Journal*, 28, No. 4, 1991.

Bjerrum, L. and O. Eide. "Stability of Strutted Excavations in Clay," *Geotechnique*, vol. 6, no. 1, March 1956.

Bowles, J. E. *Physical and Geotechnical Properties of Soils.* New York: McGraw-Hill Book Company, 1979.

_____. *Engineering Properties of Soils and Their Measurement*, 2d ed. New York: McGraw-Hill Book Company, 1978.

Carson, A. B. *General Excavating Methods.* New York: McGraw-Hill Book Company, 1961.

Casagrande, A. and N. Carrillo. "Shear Failure of Anisotropic Materials," *Boston Society of Civil Engineers: Contributions to Soil Mechanics.* Boston: SCE, 1944.

De Mello, V. F. "The Standard Penetration Test." 4th Panamerican Conference on SMFE, vol. 1 (pp. 1–86 with 353 references). San Juan, PR: ASCE, 1971.

Den Hertog, J. P. *Advanced Strength of Materials.* New York: McGraw-Hill Book Company, 1952.

Harr, M. E. *Groundwater and Seepage.* New York: McGraw-Hill Book Company, 1962.

Horne, R. A. *Marine Chemistry: The Structure of Water and the Chemistry of the Hydrosphere.* New York: Wiley, 1969.

Janbu, N. "Slope Stability Computations," *Embankment Dam Engineering*, Casagrande Volume. New York: Wiley, 1973.

Jumikis, A. R. *Soil Mechanics.* New York: Van Nostrand-Reinhold, 1962.

Lambe, T. W. "Braced Excavations," *4th PSC.* ASCE, 1970.

Lee, I. K. and A. Casagrande. "Classification and Identification of Soils," *Transactions.* ASCE, 1948.

Mitchell, J. K. *Fundamentals of Soil Behavior*, 2nd ed. New York: Wiley, 1993.

Morganstern, N. R. "Slopes and Excavations." Proceedings, Ninth International Conference on Soil Mechanics and Foundation Engineering, Tokyo, 1977.

OSHA/DOL Excavation standard from *29 CFR 1926* (Subpart P), Washington, DC: OSHA, 2003.

Richart, E. E., Jr., et al. *Vibrations of Soils and Foundations.* Englewood Cliffs, NJ: Prentice-Hall, Inc., 1970.

Schmertmann, J. H. "The Measurement of In-Situ Shear Strength," *7th PSC*, vol. 2. ASCE, 1975.

Seely, F. B. and J. O. Smith. *Advanced Mechanics of Materials.* New York: John Wiley & Sons, Inc., 1952.

Spencer, E. "Circular and Logarithmic Spiral Slip Surfaces," *Journal of Soil Mechanics and Foundations Division.* ASCE: 95, No. SM1, January 1969.

Taylor, D. W. *Fundamentals of Soil Mechanics.* New York: John Wiley & Sons, Inc., 1948.

_____. "Stability of Earth Slopes," *Journal of the Boston Society of Civil Engineers*, July 1937.

Terzaghi, K. *Theoretical Soil Mechanics.* New York: John Wiley & Sons, Inc., 1943.

_____. *From Theory to Practice in Soil Mechanics.* New York: John Wiley & Sons, Inc., 1960.

Terzaghi, K. and R. B. Peck. *Soil Mechanics in Engineering Practice*, 2d ed. New York: John Wiley & Sons, Inc., 1967.

The Civil Engineering Handbook, Section III, "Geotechnical Engineering." W. F. Chen (ed.). Boca Raton, Fla.: CRC Press, 1995.

Tsytovich, N. A. *The Mechanics of Frozen Ground.* New York: McGraw-Hill Book Company, 1975.

Varnes, D. J. "Slope Movement and Type and Processes," *Landslide Analysis and Control*, Special Report 176. Washington, DC: Transportation Research Board, National Research Council, 1978.

Wu, T. H. "Relative Density and Shear Strength of Sands," *Journal of Soil Mechanics and Foundations Division.* ASCE: 83, No. SM1, January 1957.

REVIEW EXERCISES

1. When shall a competent person inspect an excavation? (two-part answer) Identify at least twenty items he/she might visually and manually inspect for.

2. What does it mean when a soil is said to be cohesive? Compare and contrast the terms "cohesion" and "apparent cohesion." (In your answer, please use soil constituents such as clay, silt, sand, gravel, cobbles,

boulders, water, peat, and air voids along with OSHA soil classifications (A, B, C, and rock).

3. Explain the significance of the term "C-60" with respect to the manufacturer's design of shores and shields (trench boxes).

4. What are "running rats" with respect to an excavation and why are they a key indicator of potential failure?

5. How deep may one dig a trench without the aid of a professional civil engineer?

6. Is soil homogeneous? If not, explain the significance of this fact.

7. Name eight things to do when confronted with a trench rescue situation.

8. List the pros and cons when finding the presence of water in a trench. In your answer, describe how water might increase soil stability as well as how it may reduce the stability of the surrounding soil.

9. What subpart letter designates federal OSHA's 29 CFR 1926 standard covering excavation safety?

10. How much might a cubic yard of soil reasonably weigh?

■ ■ **About the Author** ■ ■

POWER-LINE SAFETY
by Samuel J. Gualardo, CSP

Samuel J. Gualardo, CSP, is currently an OSHA consultant and adjunct professor working in the Safety Sciences Department at his alma mater, Indiana University of Pennsylvania. In that capacity, he teaches graduate and undergraduate students in their safety sciences program and provides on-site consultation services to businesses under OSHA's Federal Consultation Program. He has held management positions with Fortune 500 companies including, Director, Corporate Safety and Health, Niagara Mohawk Power; Manager, Safety and Health, Metropolitan Edison; and Manager, Health and Safety, Hershey Foods USA. Sam also serves as President of National Safety Consultants, consults independently and with Dupont Safety Resources, and is a member of the American Society of Safety Engineers' national education faculty. He has lectured extensively throughout his career both nationally and internationally.

Mr. Gualardo earned a B.S. in Safety Management from Indiana University of Pennsylvania and an M.S. in Labor Relations from St. Francis University. He is also a Certified Safety Professional. He is a past national president of the American Society of Safety Engineers.

Power-Line Safety

INTRODUCTION

The occupational safety and health practices discussed throughout this chapter apply to the construction of electric transmission and distribution lines and equipment. As used in this chapter, the term *construction* includes the erection of new electric transmission and distribution lines and equipment as well as the improvement, alteration, and conversion of existing electric transmission and distribution lines and equipment. The information in this chapter reviews minimum requirements for safety and health referenced primarily from OSHA's General Industry and Construction Standards that deal with overhead and underground power-line construction. Note that the most conservative requirements were referenced where conflicts between construction and general industry requirements existed. Also, other specific regulations, national consensus standards, and safe work procedures may need to be referenced to provide further guidance in some areas such as fall protection, confined space entry, excavation, trenching and shoring, National Electric Safety Code, OSHA, the National Fire Protection Association, and others. Although these safe work practices may be similar, some differences do exist. Also note that *maintenance work* performed by qualified workers on electric transmission and distribution lines and equipment is not specifically discussed in this chapter.

GENERAL SAFE WORK PRACTICES AND REQUIREMENTS

Training

Employees should be trained in and familiar with the safety-related work practices, safety procedures, and other safety requirements in this chapter that pertain to their respective job assignments. Employees should also be trained in and familiar with any other safety practices, including applicable emergency procedures (such as pole-top and enclosed space rescue), that are not specifically addressed by this chapter, but that are related to their work and are necessary for their safety.

Qualified employees should also be trained and competent in the skills and techniques necessary to distinguish exposed live parts from other parts of electric equipment; the skills and techniques necessary to determine the nominal voltage of exposed live parts; the minimum approach distances specified in this section corresponding to the voltages to which the qualified employee will be exposed; and the proper use of the special precautionary techniques, personal protective equipment, insulating and shielding materials, and insulated tools for working on or near exposed energized parts of electric equipment. (*Note:* A person must have this training in order to be considered a qualified person.)

The employer should determine, through regular supervision and inspections conducted on at least an annual basis, that each employee is complying with the safety-related work practices required by this chapter. An employee should receive additional training (or retraining) under any of the following conditions: if the supervision and annual inspections indicate that the employee is not complying with safety-related work practices; if new technology, new types of equipment, or changes in procedures necessitate the use of safety-related work practices that are different from those which the employee would normally use; or if he or she must employ safety-related work practices that are not normally used during his or her regular job duties (i.e., tasks that are performed less often than once a year and necessitate some retraining before the performance of the work practices involved).

The training required may be either additional classes or on-the-job exercises. The training will establish employee proficiency in the particular work practices.

Job Briefing

The employer should ensure that the employee in charge conducts a job briefing with the employees involved before they start each job. The briefing should cover at least the following subjects: hazards associated with the job, work procedures involved, special precautions, energy-source controls, minimum clearance distances, and personal-protective-equipment requirements. If the work or operations to be performed during the work day or shift are repetitive and similar, at least one job briefing should be conducted before the start of the first job of each day or shift. Additional job briefings should be held if significant changes, which might affect the safety of the employees, occur during the course of the work. A brief discussion is satisfactory if the work involved is routine and if the employee, by virtue of training and experience, can reasonably be expected to recognize and avoid the hazards involved in the job. A more extensive discussion should be conducted if the work is complicated or particularly hazardous, or if the employee cannot be expected to recognize and avoid the hazards involved in the job. An employee working alone does not need to conduct a job briefing. However, the employer must ensure that the tasks to be performed are planned as if a briefing were required.

Existing Conditions

Existing conditions related to the safety of the work to be performed should be determined before work on or near electric lines or equipment is started. Such conditions include, but are not limited to, the nominal voltages of lines and equipment; the maximum switching transient voltages; the presence of hazardous, induced voltages; the presence and condition of both protective grounds and equipment-grounding conductors; the condition of poles on the site; environmental conditions relative to safety; and the locations of circuits and equipment, including power and communication lines and fire protective signaling circuits. Be aware that all electric equipment and lines should be considered energized until determined to be deenergized by tests or other appropriate methods or means.

Clearances from Energized Lines and Equipment

No employee is permitted to approach or take any conductive object without an approved insulating handle any closer to exposed energized parts than shown in the following table unless (a) the employee is insulated or guarded from the energized part (gloves or gloves with sleeves rated for the voltage involved are considered to be insulation of the employee from the energized part); (b) the energized part is insulated or guarded from the employee and any other conductive object at a different potential; or (c) the employee is isolated, insulated, or guarded from any other conductive object(s), as during live-line, bare-hand work. The minimum working distance and minimum clear hot stick distances listed here should never be violated. The minimum clear hot stick distance is that for the use of live-line tools held by line personnel when performing live-line work.

Conductor support tools—such as link sticks, strain carriers, and insulator cradles—may be used, provided that the clear insulation is at least as long as the insulator string or the minimum distance specified in Table 1 for the operating voltage.

Medical Services and First Aid

The employer should provide medical services and first aid. When employees are performing work on, or associated with, exposed lines or equipment energized at 50 volts or more, persons trained in first aid, including cardiopulmonary resuscitation (CPR), should be available. For field work involving two or more employees at a work location, at least two trained persons should be available. Each site must have a medivac plan and Gobal Positioning System as well as hospital locations posted and filed with the appropriate emergency responder agencies.

For fixed work locations, such as substations, the number of first-aid providers available should be sufficient to ensure that each employee exposed to electric shock can be reached within 4 minutes by a trained person. However, where the existing number of employees is insufficient to meet this requirement (at a *remote* substation, for example), all employees at the work location should be trained.

First-aid supplies should be placed in weather-proof containers if the supplies could be exposed to the weather. Each first-aid kit should be maintained, readily available for use, and inspected fre-

■ ■ Table 1 ■ ■

Alternating Current – Minimum Distances

Voltage Range (phase to phase) (kilovolts)	Minimum Working and Clear Hot Stick Distance
2.1 to 15	2 ft. 0 in.
15.1 to 35	2 ft. 4 in.
35.1 to 46	2 ft. 6 in.
46.1 to 72.5	3 ft. 0 in.
72.6 to 121	3 ft. 4 in.
138 to 145	3 ft. 6 in.
161 to 169	3 ft. 8 in.
230 to 242	5 ft. 0 in.
345 to 362	(1) 7 ft. 0 in.
500 to 552	(1) 11 ft. 0 in.
700 to 765	(1) 15 ft. 0 in.

Footnote (1): For 345–362 kv., 500–552 kv., and 700–765 kv., minimum clear hot stick distance may be reduced provided that such distances are not less than the shortest distance between the energized part and the grounded surface.

quently enough to ensure that expended items are replaced, but checked at least once each year.

Night Work

When working at night, spotlights or portable lights that provide emergency lighting should be available as needed to perform the work safely.

Work Near and Over Water

When crews are engaged in work over or near water where any danger of drowning exists, the OSHA standards for working over water should be followed.

Protective Tools and Equipment

Rubber protective equipment should meet the provisions of the American National Standards Institute (ANSI J6 series), as listed in Table 2.

Rubber protective equipment should be visually inspected prior to each use. In addition, an "air" test should be performed for rubber gloves prior to use. Protective equipment of any material other than rubber should provide equal or better electrical and mechanical protection. Protective hats must meet the provisions of ANSI Z89.2-1971 (Industrial Protective Helmets for Electrical Workers,

■ ■ Table 2 ■ ■

Standards in the ANSI J6 Series

Rubber Item	Standard
Rubber insulating gloves	J6.6-1971
Rubber matting for use around electric apparatus	J6.7-1935 (R1971)
Rubber insulating blankets	J6.4-1971
Rubber insulating hoods	J6.2-1950 (R1971)
Rubber insulating line hose	J6.1-1950 (R1971)
Rubber insulating sleeves	J6.5-1971

Class E); they should be worn at the job site by employees who are exposed to the hazards of falling objects, electric shock, or burns.

LADDERS

Ladders and platforms should be secured to prevent them from becoming accidentally dislodged. All ladders and platforms should be used *only* in applications for which they were designed, and may not be loaded in excess of the working loads for which they were designed. In the configurations in which they are used, ladders and platforms should be capable of supporting, without failure, at least 2.5 times the maximum intended load.

Portable metal or conductive ladders should not be used near energized lines or equipment, except as may be necessary in specialized work, such as in high-voltage substations where nonconductive ladders might present a greater hazard than conductive ladders. Conductive or metal ladders should be prominently marked as conductive, and all necessary precautions should be taken when they are used in specialized work. Hook or other types of ladders used in structures should be positively secured to prevent the ladder from being accidentally displaced.

LIVE-LINE TOOLS

Live-line tool rods, tubes, and poles should be designed and constructed to withstand the following minimum tests: 100,000 volts per foot (3281 volts per centimeter) of length for 5 minutes if the tool is made of fiberglass-reinforced plastic (FRP); or 75,000 volts per foot (2461 volts per centimeter) of length for 3 minutes if the tool is made of wood; or other tests that the employer can demonstrate are equivalent.

Each live-line tool should be wiped clean and visually inspected for defects before use each day. If any defect or contamination that could adversely affect the insulating qualities or mechanical integrity of the live-line tool is present after wiping, the tool should be removed from service and examined before being returned to service.

Live-line tools used for primary employee protection should be removed from service every two years, or whenever examination, cleaning, repair, or testing is required. Each tool should be thoroughly examined for defects. If a defect or contamination that could adversely affect the insulating qualities or mechanical integrity of the live-line tool is found, the tool must be repaired and refinished or permanently removed from service. If no such defect or contamination is found, the tool should be cleaned and waxed.

Each tool should be tested under the following conditions: after the tool has been repaired or refinished; after examination, if repair or refinishing is not performed, unless the tool is made of FRP rod or foam-filled FRP tube and the employer can demonstrate that the tool has no defects that could cause it to fail in use.

The test method used should be designed to verify the tool's integrity along its entire working length and, if the tool is made of fiberglass-reinforced plastic, its integrity under wet conditions. The voltage applied during the tests should be as follows: 75,000 volts per foot (2461 volts per centimeter) of length for 1 minute if the tool is made of fiberglass; or 50,000 volts per foot (1640 volts per centimeter) of length for 1 minute if the tool is made of wood; or other tests that the employer can demonstrate are equivalent.

Note: Guidelines for the examination, cleaning, repairing, and in-service testing of live-line tools are contained in the Institute of Electrical and Electronics Engineers *Guide for In-Service Maintenance and Electrical Testing of Live-Line Tools* (IEEE Std. 978-1984). DO NOT USE lead-based paint; use only proper cleaners.

HYDRAULIC AND PNEUMATIC TOOLS

Safe operating pressures for hydraulic and pneumatic tools, hoses, valves, pipes, filters, and fittings should never be exceeded. (*Note:* If any hazardous defects are present, no operating pressure would be safe, and the hydraulic or pneumatic equipment involved should not be used.) In the absence of defects, the maximum rated operating pressure is the maximum safe pressure.

A hydraulic or pneumatic tool used where it may contact exposed live parts should be designed and maintained for such use.

The hydraulic system supplying a hydraulic tool used where it may contact exposed live parts should provide protection against loss of insulating value for the voltage involved due to the formation of a partial vacuum in the hydraulic line. All hydraulic fluids used for the insulated sections of derrick trucks, aerial lifts, and hydraulic tools that are used on or around energized lines and equipment should be of the insulating type.

Note: Hydraulic lines without check valves, having a separation of more than 35 feet between the oil reservoir and the upper end of the hydraulic system, promote the formation of a partial vacuum. A pneumatic tool used on energized electric lines or equipment, or used where it may contact exposed live parts, should provide protection against the accumulation of moisture in the air supply. Always test dielectric fluid like a HGT-STEL—and keep it free of oil or foreign matter.

Pressure should always be released before connections are broken, unless quick-acting, self-closing connectors are used. Hoses should never be kinked.

Employees should be made aware not to use any part of their bodies to locate or attempt to stop a hydraulic leak.

HANDTOOLS

Measuring tapes or measuring ropes that are metal or contain conductive strands should not be used when working on or near energized parts.

All *hydraulic* tools that are used on or around energized lines or equipment should use nonconducting hoses with adequate strength for the normal operating pressures.

All *pneumatic* tools that are used on or around energized lines or equipment should have nonconducting hoses with adequate strength for the normal operating pressures, and have an accumulator on the compressor to collect moisture.

Cord and plug equipment should be equipped with a cord containing an equipment-grounding conductor connected to the tool frame and to a means for grounding the other end (however, this option may not be used where the introduction of the ground into the work environment increases the hazard to an employee), or it should be double insulated, or it should be connected to the power supply through an isolating transformer with an ungrounded secondary.

Portable and vehicle-mounted generators used to supply cord- and plug-connected equipment must meet the following requirement: The generator may only supply equipment located on the generator or the vehicle and cord- and plug-connected equipment through receptacles mounted on the generator or the vehicle. The noncurrent-carrying metal parts of equipment and the equipment-grounding conductor terminals of the receptacles should be bonded to the generator frame. In the case of vehicle-mounted generators, the frame of the generator should be bonded to the vehicle frame. Any neutral conductor should be bonded to the generator frame.

Mechanical Equipment

Visual inspections should be made of the equipment to determine that it is in good condition each day the equipment is to be used. Tests should be made at the beginning of each shift during which the equipment is to be used to determine that the brakes and operating systems are in proper working condition. No employee should use any motor-vehicle equipment having an obstructed view to the rear unless the vehicle has a reverse signal alarm audible above the surrounding noise level, or the vehicle is backed up only when an observer signals that it is safe to do so. The vehicle or generator should be attached to the station ground grid or a grounding rod driven at least 6-feet deep and a minimum of 15 feet away from all structures.

AERIAL LIFTS

When working near energized lines or equipment, aerial-lift trucks should be grounded or barricaded and considered energized equipment, or the aerial-lift truck should be insulated for the work being performed. Equipment or material should not be passed between a pole or structure and an aerial lift while an employee working from the basket is within reaching distance of energized conductors or equipment that are not covered with insulating protective equipment.

DERRICK TRUCKS, CRANES, AND OTHER LIFTING EQUIPMENT

With the exception of equipment certified for work on the proper voltage, mechanical equipment must not be operated closer to any energized line or equipment than the clearances discussed earlier unless an insulated barrier is installed between the energized part and the mechanical equipment, or the mechanical equipment is grounded, or the mechanical equipment is insulated, or the mechanical equipment is considered to be energized.

Use of vehicles, gin poles, cranes, and other equipment in restricted or hazardous areas should be controlled at all times by designated employees. All mobile cranes and derricks should be effectively grounded when being moved or operated in close proximity to energized lines or equipment, or the equipment will be considered energized. Fenders are not required for lowboys used for transporting large electrical equipment, transformers, or breakers.

The critical safety components of mechanical elevating and rotating equipment should receive a thorough visual inspection before use on each shift. (*Note:* Critical safety components of mechanical elevating and rotating equipment are components whose failure would result in a free fall or free rotation of the boom.)

The operator of an electric line truck may not leave his or her position at the controls while a load is suspended, unless the employer can demonstrate that no employee (including the operator) might be endangered.

Rubber-tired, self-propelled scrapers, rubber-tired front-end loaders, rubber-tired dozers, wheel-type agricultural and industrial tractors, crawler-type tractors, crawler-type loaders, and motor graders, with or without attachments, should have rollover protective structures.

Vehicular equipment, if provided with outriggers, must be operated with the outriggers extended and firmly set as necessary for the stability of the specific configuration of the equipment. Outriggers may not be extended or retracted outside of clear view of the operator unless all employees are outside the range of possible equipment motion. If the work area or the terrain precludes the use of outriggers, the equipment can be operated only within its maximum load ratings for the particular configuration of the equipment without outriggers.

Mechanical equipment used to lift or move lines or other material should be used within its maximum load rating and other design limitations for the conditions under which the work is being performed.

Mechanical equipment should be operated so that the minimum approach distances discussed earlier are maintained from exposed energized lines and equipment. However, the insulated portion of an aerial lift operated by a qualified employee in the lift is exempt from this requirement.

A designated employee other than the equipment operator should observe the approach distance to exposed lines and equipment and give timely warnings before the minimum approach distance is reached, unless the employer can demonstrate that the operator can accurately determine that the minimum approach distance is being maintained.

If, during operation of the mechanical equipment, the equipment could become energized, the operation must also comply with at least one of the following: the energized lines exposed to contact should be covered with insulating protective material that will withstand the type of contact that might be made during the operation, the equipment shall be insulated for the voltage involved, or the equipment shall be positioned so that its uninsulated portions cannot approach the lines or equipment any closer than the minimum approach distances specified earlier.

Each employee should be protected from hazards that might arise from equipment contact with the energized lines. The measures used should ensure that employees will not be exposed to hazardous differences in potential. Unless the employer can demonstrate that the methods in use protect each employee from the hazards that might arise if the equipment contacts the energized line, the measures used should include all of the following techniques: using the best available ground to minimize the time the lines remain energized, bonding equipment together to minimize potential differences, providing ground mats to extend areas of equipotential, and employing insulating protective equipment or barricades that will guard against any remaining hazardous potential differences.

Material Handling

In areas not restricted to qualified persons only, materials or equipment may not be stored closer to energized lines or exposed energized parts of equipment than the following distances plus an amount providing for the maximum sag and side swing of all conductors and providing for the height and movement of material-handling equipment: for lines and equipment energized at 50 kV or less, the distance is 10 feet; for lines and equipment energized at more than 50 kV, the distance is 10 feet plus 4 inches for every 10 kV over 50 kV. In areas restricted to qualified employees, material may not be stored within the working space surrounding energized lines or equipment.

DEENERGIZING LINES AND EQUIPMENT

A designated employee should make a request of the system operator to have the particular section of line or equipment deenergized. The designated employee becomes the employee in charge and is responsible for the clearance.

All switches, disconnectors, jumpers, tapes, and other means through which known sources of electric energy may be supplied to the particular lines and equipment to be deenergized should be opened, rendering them inoperable (unless their design does not permit this), and tagged to indicate that employees are at work. Any disconnecting means that are accessible to persons outside the employer's control (for example, the general public) should be rendered inoperable while they are open for the purpose of protecting employees.

Automatically and remotely controlled switches that could cause the opened disconnecting means to close should also be tagged at the point of control. The automatic or remote-control feature should be rendered inoperable, unless its design does not permit that.

Tags should prohibit operation of the disconnecting means and indicate that employees are at work.

After the lines and equipment are deenergized and tagged, and the employee in charge of the work has been given a clearance by the system operator, the lines and equipment to be worked on should be tested to ensure that they are deenergized. Next, protective grounds should be installed prior to commencing work, and should be visible from the employees' work locations.

If two or more independent crews will be working on the same lines or equipment, then each crew should independently comply with the above requirements.

To transfer the clearance, the employee in charge (or, if the employee in charge is forced to leave the work site due to illness or other emergency, the employee's supervisor) should inform the system operator; employees in the crew should be informed of the transfer; and the new employee in charge will be responsible for the clearance.

To release a clearance, the employee in charge should notify employees under his or her direction that the clearance is to be released; determine that all employees in the crew are clear of the lines and equipment; determine that all protective grounds installed by the crew have been removed; report this information to the system operator, and release the clearance. The person releasing a clearance should be the same person that requested the clearance, unless responsibility has been properly transferred.

Tags may not be removed unless the associated clearance has been released. Only after all protective grounds have been removed, after all crews working on the lines or equipment have released their clearances, after all employees are clear of the lines and equipment, and after all protective tags have been removed from a given point of disconnection, may action be initiated to reenergize the lines or equipment at that point of disconnection.

GROUNDING FOR EMPLOYEE PROTECTION

All conductors and equipment shall be treated as energized until tested or otherwise determined to be deenergized or until tested and grounded. New lines or equipment may be considered deenergized and worked as such where the lines or equipment are grounded, or the hazard of induced voltages is not present, and where adequate clearances or other means are implemented to prevent contact with energized lines or equipment and the new lines or equipment. Bare-wire communication conductors on power poles or structures should be treated as energized lines unless protected by insulating materials.

Before any ground is installed, all lines and equipment should be tested and found absent of nominal voltage, unless a previously installed ground is present. Protective grounds should then be placed at such locations and arranged in such a manner as to prevent each employee from being exposed to hazardous differences in electrical potential.

When attaching grounds, the ground end should be attached first and the other end should be attached and removed by means of insulated tools or other suitable devices. When removing grounds, the grounding device should first be removed from the line or equipment using insulating tools or other suitable devices.

When grounding electrodes are utilized, such electrodes should have a resistance to ground low enough to remove the danger of harm to personnel or permit prompt operation of protective devices. Grounding to tower should be made with a tower clamp capable of conducting the anticipated fault current. A ground lead, to be attached to either a tower ground or driven ground, should be capable of conducting the anticipated fault current and have a minimum conductance of No. 4/0 AWG copper.

(*Note:* Guidelines for protective grounding equipment are contained in American Society for Testing and Materials Standard Specifications for Temporary Grounding Systems to be Used on De-Energized Electric Power Lines and Equipment, ASTM F855-1990.)

Protective grounds should have an impedance low enough to cause immediate operation of protective devices in case of accidental energizing of the lines or equipment.

Grounds may be removed temporarily during tests. During the test procedure, the employer must ensure that each employee uses insulating equipment and is isolated from any hazards involved, and the employer should institute any additional measures that may be necessary to protect each exposed employee in case the previously grounded lines and equipment become energized.

OVERHEAD AND ELEVATED WORK

Prior to climbing poles, ladders, scaffolds, or other elevated structures, an inspection should be made to determine that the structures are capable of sustaining the additional, or unbalanced, stresses to which they will be subjected. Where poles or structures may be unsafe for climbing, they should not be climbed until made safe by guying, bracing, or other adequate means. Before installing or removing wire or cable, strains to which poles and structures will be subjected should be considered, and any necessary action taken to prevent failure of supporting structures.

When setting, moving, or removing poles using cranes, derricks, gin poles, A-frames, or other mechanized equipment near energized lines or equipment, precautions should be taken to avoid contact with energized lines or equipment, except in bare-hand, live-line work, or where barriers or protective devices are used.

Equipment and machinery operating adjacent to energized lines or equipment should be treated as energized unless an insulated barrier is installed between the energized part and the mechanical equipment, the mechanical equipment is grounded, the mechanical equipment is insulated, or the mechanical equipment is considered energized.

Unless using suitable protective equipment for the voltage involved, employees standing on the ground should avoid contacting equipment or any machinery working adjacent to energized lines or equipment. Lifting equipment should be bonded to an effective ground or it should be considered energized and barricaded when utilized near energized equipment or lines.

Pole holes should not be left unattended or unguarded in areas where employees are currently working. Tag lines should be of a nonconductive type when used near energized lines.

Metal Tower Construction

When working in unstable material the excavation for pad- or pile-type footings in excess of 5 feet deep shall be either sloped to the angle of repose or shored if entry is required. Ladders should be provided for access to pad- or pile-type footing excavations in excess of 4 feet. When working in unstable material, provisions should be made for cleaning out auger-type footings without requiring an employee to enter the footing unless shoring is used to protect the employee. All bolts must be tightly installed prior to cutting the load!

A designated employee should be used in directing mobile equipment adjacent to footing excavations. No one should be permitted to remain in the footing while equipment is being spotted for placement. Where necessary to assure the stability of mobile equipment, the location of use for such equipment should be graded and leveled.

Tower assembly should be carried out with a minimum exposure of employees to falling objects when working at two or more levels on a tower. The employer should ensure that no employee is under a tower or structure while work is in progress, except where the employer can demonstrate that such a working position is necessary to assist employees working above.

Tag lines should be used as necessary to maintain sections or parts of sections in position and to reduce the possibility of tipping. Members and sections being assembled should be adequately supported. Tag lines or other similar devices should be used to maintain control of tower sections being raised or positioned, unless the employer can demonstrate that the use of such devices would create a greater hazard.

No one will be permitted under a tower which is in the process of erection or assembly, except as required to guide and secure the section being set. When erecting towers using hoisting equipment adjacent to energized transmission lines, the lines should be deenergized when practical. If the lines are not deenergized, extraordinary caution should be exercised to maintain the minimum clearance distances.

Erection cranes should be set on firm, level foundations, and when the cranes are so equipped, outriggers should be used. Tag lines should be utilized to maintain control of tower sections being raised and positioned, except where the use of such lines would create a greater hazard. The load-line should not be detached from a tower section until the section is adequately secured. Except during emergency restoration procedures, erection should be discontinued in the event of high wind or other adverse weather conditions that would make the work hazardous. Equipment and rigging should be regularly inspected and always maintained in safe operating condition.

Adequate traffic control should be provided when crossing highways and railways with equipment. A designated employee should be utilized to determine that required clearance is maintained in moving equipment under or near energized lines.

Except during emergency restoration procedures, work should be discontinued when adverse weather conditions would make it hazardous in spite of the work practices required by this section.

Note: Thunderstorms in the immediate vicinity, high winds, snow storms, and ice storms are examples of adverse weather conditions that are presumed to make this work too hazardous to perform, except under emergency conditions.

Installing and Removing Overhead Lines

The employer should utilize the tension-stringing method, barriers, or other equivalent measures to minimize the possibility that conductors and cables being installed or removed will contact energized power lines or equipment. For back string, use lay-up methods.

The protective measures for mechanical equipment should also be provided for conductors, cables, and pulling and tensioning equipment when the conductor or cable is being installed or removed close enough to energized conductors that any of the following failures could energize the pulling or tensioning equipment or the wire or cable being installed or removed: failure of the pulling or tensioning equipment, failure of the wire or cable being pulled, or failure of the previously installed lines or equipment.

If the conductors being installed or removed cross over energized conductors in excess of 600 volts, and if the design of the circuit-interrupting devices protecting the lines so permits, the automatic reclosing feature of these devices should be made inoperative.

Before lines are installed parallel to existing energized lines, the employer should make a determination of the approximate voltage to be induced in the new lines, or work should proceed on the assumption that the induced voltage is hazardous. Unless the employer can demonstrate that the lines being installed are not subject to the induction of a hazardous voltage, or unless the lines are treated as energized, the following requirements also apply.

Each bare conductor shall be grounded in increments so that no point along the conductor is more than 2 miles from a ground. The ground shall be left in place until the conductor installation is completed between dead ends. The grounds shall be removed as the last phase of aerial cleanup.

If employees are working on bare conductors, grounds should also be installed at each location where these employees are working, and grounds should be installed at all open dead-end or catch-off points, or the next adjacent structure.

If two bare conductors are to be spliced, the conductors should be bonded and grounded before being spliced.

Reel-handling equipment, including pulling and tensioning devices, should be in safe operating condition and be leveled and aligned. Load ratings of stringing lines, pulling lines, conductor grips, load-bearing hardware and accessories, rigging, and hoists

may not be exceeded. Pulling lines and accessories should be repaired or replaced when defective. Conductor grips may not be used on wire rope unless the grip is specifically designed for this application.

Reliable communications through two-way radios or other equivalent means should be maintained between the reel tender and the pulling rig operator. The pulling rig may only be operated when it is safe to do so.

While the conductor or pulling line is being pulled (in motion) with a power-driven device, employees are not permitted directly under overhead operations or on the cross arm, except as necessary to guide the stringing sock or board over or through the stringing sheave. Stop the pull if it hangs up—DO NOT readjust the stringing sock when in motion.

Installation, Removal, or Repair of Lines Energized at More than 600 Volts

Installation, removal, or repair of deenergized lines if an employee is exposed to contact with other parts energized at more than 600 volts includes: installation, removal, or repair of equipment, such as transformers, capacitors, and regulators, if an employee is exposed to contact with parts energized at more than 600 volts; work involving the use of mechanical equipment, other than insulated aerial lifts, near parts energized at more than 600 volts; and other work that exposes an employee to electrical hazards.

MINIMUM APPROACH DISTANCES

The employer should ensure that no employee approaches or takes any conductive object closer to exposed energized parts than specified earlier, unless the employee is insulated from the energized part (insulating gloves or insulating gloves with sleeves are considered insulation only with regard to the energized part the employee is performing work on); unless the energized part is insulated from the employee and from any other conductive object at a different potential; or unless the employee is insulated from any other exposed conductive object, as during live-line, bare-hand work.

Parts of electric circuits that meet these two provisions are not considered to be "exposed" unless a guard is removed or an employee enters the space intended to provide isolation from the live parts.

TYPE OF INSULATION

If the employee is to be insulated from energized parts by the use of insulating gloves, insulating sleeves should also be used. However, insulating sleeves do not need to be used under the following conditions: if exposed energized parts on which work is not being performed are insulated from the employee and if such insulation is placed from a position not exposing the employee's upper arm to contact with other energized parts.

WORKING POSITION

The employer should ensure that each employee, to the extent that other safety-related conditions at the work site permit, works in a position from which a slip or shock will not bring the employee's body into contact with exposed, uninsulated parts that are energized at a potential different from the employee.

MAKING CONNECTIONS

The employer should ensure that connections are made as follows: in connecting deenergized equipment or lines to an energized circuit by means of a conducting wire or device, an employee shall first attach the wire to the deenergized part; when disconnecting equipment or lines from an energized circuit by means of a conducting wire or device, an employee shall remove the source end first; and when lines or equipment are connected to or disconnected from energized circuits, loose conductors shall be kept away from exposed energized parts. Employees should use a fuse barrel or load pick-up tool when conducting this operation.

APPAREL

When work is performed within reaching distance of exposed energized parts of equipment, the employer should ensure that each employee removes or renders nonconductive all exposed conductive articles, such as key or watch chains, rings, or wrist watches or bands, unless such articles do not in-crease the hazards associated with contact with the energized parts.

The employer should train each employee in the hazards involved with exposure to the dangers of flames or electric arcs, and should ensure that each employee who is exposed to the hazards of flames or electric arcs wear clothing that meets the standard for flame resistance.

Note: Clothing made from the following types of fabrics, either alone or in blends, is prohibited, unless the employer can demonstrate that the fabric has been treated to withstand the conditions that may be encountered or that the clothing is worn in such a manner as to eliminate the hazard involved: acetate, nylon, polyester, rayon.

FUSE HANDLING

When fuses must be installed or removed with one or both terminals energized at more than 300 volts, or with exposed parts energized at more than 50 volts, the employer should ensure that tools or gloves rated for the voltage are used. When expulsion-type fuses are installed with one or both terminals energized at more than 300 volts, each employee should be required to wear eye protection with ANSI UV flash and impact ratings, use a tool rated for the voltage, and remain clear of the exhaust path of the fuse barrel.

COVERED (NONINSULATED) CONDUCTORS

The requirements of this section which pertain to the hazards of exposed live parts also apply when work is performed in the proximity of covered (noninsulated) wires.

NONCURRENT-CARRYING METAL PARTS

Noncurrent-carrying metal parts of equipment or devices such as transformer cases and circuit-breaker housings should be treated as energized at the highest voltage to which they are exposed unless, before work is performed, the employer inspects the installation and determines that these parts are grounded.

OPENING CIRCUITS UNDER LOAD

Devices used to open circuits under load conditions should be designed to interrupt the current involved. Also, an approved-load break tool should be specified—DO NOT use a fuse!

LIVE-LINE, BARE-HAND WORK

In addition to any other applicable standards contained elsewhere, all live-line, bare-hand work should be performed in accordance with the following requirements. Employees should be instructed and trained in the live-line, bare-hand technique and the safety requirements pertinent to it before being permitted to use the technique on energized circuits. Before using the live-line, bare-hand technique on energized high-voltage conductors or parts, a check should be made of the voltage rating of the circuit on which the work is to be performed, the clearances to ground of lines and other energized parts on which work is to be performed, and the voltage limitations of the aerial-lift equipment intended to be used. Only equipment designed, tested, and intended for live-line, bare-hand work should be used. All work should be personally supervised by a person trained and qualified to perform live-line, bare-hand work. The automatic reclosing feature of circuit-interrupting devices should be made inoperative where practical before working on any energized line or equipment. Work should not be performed during the progress of an electrical storm in the immediate vicinity. A conductive bucket liner or other suitable conductive device should be provided for bonding the insulated aerial device to the energized line or equipment. The employee should be connected to the bucket liner by use of conductive shoes, leg clips, or other suitable means. Where necessary, adequate electrostatic shielding for the voltage being worked, or conductive clothing, should be provided. Only tools and equipment intended for live-line, bare-hand work should be used, and such tools and equipment should be kept clean and dry.

Before the boom is elevated, the outriggers on the aerial truck should be extended and adjusted to stabilize the truck, and the body of the truck should be bonded to an effective ground, or barricaded and considered as energized equipment. Before moving the aerial lift into the work position, all controls (ground level and bucket) should be checked and tested to determine that they are in proper working condition.

Arm current tests shall be made before starting work each day, each time during the day when higher voltage is going to be worked, and when changed conditions indicate a need for additional tests. Aerial buckets used for bare-hand, live-line work should also be subjected to an arm current test. This test consists of placing the bucket in contact with an energized source equal to the voltage to be worked upon for a minimum time of three minutes. The leakage current should not exceed 1 microampere per kilovolt of nominal line-to-line voltage. Work operations shall be suspended immediately upon any indication of a malfunction in the equipment. All aerial lifts to be used for live-line, bare-hand work should have dual controls (lower and upper). The upper controls should be within easy reach of the employee in the basket. If a two-basket-type lift is used, access to the controls should be within easy reach from either basket. The lower set of controls should be located near the base of the boom to permit over-ride operation of equipment at any time. Ground-level lift control should not be operated unless permission has been obtained from the employee in the lift, except in emergency situations.

Before the employee contacts the energized part to be worked on, the conductive bucket liner should be bonded to the energized conductor by means of a positive connection that should remain attached to the energized conductor until the work on the energized circuit is completed.

The minimum clearance distances for live-line, bare-hand work are specified in Table 3. These distances should be maintained from all grounded objects and from lines and equipment at a different potential than that to which the insulated aerial device is bonded, unless such grounded objects or other lines and equipment are covered by insulated guards. These distances should be maintained when approaching, leaving, and when bonded to the energized circuit.

▪■ Table 3 ■▪

Alternating Current – Minimum Clearance Distances for Live-Line, Bare-Hand Work

Voltage Range	Distance for Maximum Voltage	
phase to phase (kilovolts)	phase to ground	phase to phase
	(feet and inches)	
2.1 to 15	2 ft. 0 in.	2 ft. 0 in.
15.1 to 35	2 ft. 4 in.	2 ft. 4 in.
35.1 to 46	2 ft. 6 in.	2 ft. 6 in.
46.1 to 72.5	3 ft. 0 in.	3 ft. 0 in.
72.6 to 121	3 ft. 4 in.	4 ft. 6 in.
138 to 145	3 ft. 6 in.	5 ft. 0 in.
161 to 169	3 ft. 8 in.	5 ft. 6 in.
230 to 242	5 ft. 0 in.	8 ft. 4 in.
345 to 362	(1) 7 ft. 0 in.	(1) 13 ft. 4 in.
500 to 552	(1) 11 ft. 0 in.	(1) 20 ft. 0 in.
700 to 765	(1) 15 ft. 0 in.	(1) 31 ft. 0 in.

Footnote (1): For 345–362 kv., 500–552 kv., and 700–765 kv., the minimum clearance distance may be reduced provided the distances are not made less than the shortest distance between the energized part and the grounded surface.

When approaching, leaving, or bonding to an energized circuit, the minimum distances should be maintained between all parts of the insulated boom assembly and any grounded parts (including the lower arm or portions of the truck).

When positioning the bucket alongside an energized bushing or insulator string, the minimum line-to-ground clearances must be maintained between all parts of the bucket and the grounded end of the bushing or insulator string. The use of handlines between buckets, booms, and the ground is prohibited. No conductive materials over 36 inches long should be placed in the bucket, except for appropriate length jumpers, armor rods, and tools. Nonconductive-type handlines may be used from line to ground when not supported from the bucket. The bucket and upper insulated boom should not be overstressed by attempting either to lift or support weights in excess of the manufacturer's rating.

A minimum clearance table should be printed on a plate of durable nonconductive material, and mounted in the bucket or its vicinity so it is visible to the operator of the boom.

It is recommended that insulated measuring sticks be used to verify clearance distances.

UNDERGROUND WORK

Guarding and Ventilating Street Openings

Access to underground lines or equipment is generally achieved through street openings of some sort.

Appropriate warning signs should be promptly placed when covers of enclosed spaces, handholes, or vaults are removed. What constitutes an appropriate warning sign is dependent upon the nature and location of the hazards involved. Before an employee enters a street opening, such as a manhole or an unvented vault, it should be promptly protected with a barrier, temporary cover, or other suitable guard.

Enclosed Spaces

This section covers enclosed spaces that may be entered by employees. It does not apply to vented vaults if a determination is made that the ventilation system is operating to protect employees before they enter the space. It does apply to routine entry into enclosed spaces in lieu of the permit-space entry requirements. If after the precautions specified are taken, the hazards remaining in the enclosed space endanger the life of an entrant or could interfere with escape from the space, then entry into the enclosed space should meet formal permit-space entry requirements not discussed in this chapter.

The employer should ensure the use of safe work practices for entry into and work in enclosed spaces, and for the rescue of employees from such spaces. Employees who enter enclosed spaces or who serve as attendants should be trained in the hazards of enclosed space entry, entry procedures, and enclosed space rescue procedures. Employers must provide equipment to ensure the prompt and safe rescue of employees from the enclosed space.

Before any entrance cover to an enclosed space is removed, the employer should determine whether it is safe to do so by checking for the presence of any atmospheric pressure or temperature differences and by evaluating whether there might be a hazardous atmosphere in the space. Any conditions making it unsafe to remove the cover should be eliminated before the cover is taken up.

Note that this evaluation may take the form of a check of the conditions expected to be in the enclosed space. For example, the cover could be checked to see if it is hot and, if it is fastened in place, whether it could be loosened gradually to release any residual pressure. A determination must also be made of whether conditions at the site could cause a hazardous atmosphere (an oxygen-deficient or flammable atmosphere) to develop within the space.

When covers are removed from enclosed spaces, the opening should be promptly guarded by a railing, temporary cover, or other barrier intended to prevent an accidental fall through the opening and to protect employees working in the space from objects that could enter the space.

Employees may not enter any enclosed space while it contains a hazardous atmosphere, unless the entry conforms to the generic permit-required confined spaces requirements.

While work is being performed in the enclosed space, a person with first-aid training should be immediately available outside the enclosed space to render emergency assistance if there is reason to believe that a hazard may exist in the space or if a hazard exists because of traffic patterns in the area of the opening used for entry. That person is not precluded from performing other duties outside the enclosed space if these duties do not distract the attendant from monitoring employees within the space.

Test instruments used to monitor atmospheres in enclosed spaces should be kept in calibration, with a minimum accuracy of + or − 10 percent. Before an employee enters an enclosed space, the internal atmosphere should be tested for oxygen deficiency with a direct-reading meter or similar instrument capable of collection and immediate analysis of data samples without the need for off-site evaluation. If continuous forced-air ventilation is provided, testing is not required as long as the procedures used ensure that employees are not exposed to the hazards posed by oxygen deficiency.

Before an employee enters an enclosed space, the internal atmosphere should also be tested for flammable gases and vapors with a direct-reading meter or similar instrument capable of collection

and immediate analysis of data samples without the need for off-site evaluation. This test should be performed after the oxygen testing and ventilation to demonstrate that there is sufficient oxygen to ensure the accuracy of the test for flammability.

If flammable gases or vapors are detected, or if an oxygen deficiency is found, forced-air ventilation should be used to maintain oxygen at a safe level and to prevent a hazardous concentration of flammable gases and vapors from accumulating. A continuous monitoring program to ensure that no increase in flammable gas or vapor concentration occurs may be followed in lieu of ventilation if flammable gases or vapors are detected at safe levels. If continuous forced-air ventilation is used, it should begin before entry is made and be maintained long enough to ensure that a safe atmosphere exists before employees are allowed to enter the work area. The forced-air ventilation should be directed to ventilate the immediate area where employees are present within the enclosed space and continue until all employees leave the enclosed space. The air supply for the continuous forced-air ventilation should be from a clean source and may not increase the hazards in the enclosed space.

If open flames are used in enclosed spaces, a test for flammable gases and vapors should be made immediately before the open-flame device is used and at least once per hour while the device is used in the space. Testing should be conducted more frequently if conditions present in the enclosed space indicate that once per hour is insufficient to detect hazardous accumulations of flammable gases or vapors.

LOCATING UNDERGROUND LINES AND EQUIPMENT

During excavation or trenching, in order to prevent employee exposure to the hazards created by damage to dangerous underground facilities, efforts should be made to determine the location of such facilities and work should be conducted in a manner designed to avoid such damage.

When underground facilities are exposed (electric, gas, water, telephone, etc.) they should be protected as necessary to avoid damage. Where multiple cables exist in an excavation, cables other than the one being worked on should be protected as necessary. When multiple cables exist in an excavation, the cable to be worked on should be identified by electrical means, unless its identity is obvious by reason of distinctive appearance.

Before cutting into a cable or opening a splice, the cable should be identified and verified to be the proper cable. When working on buried cable or on cable in manholes, metallic sheath continuity should be maintained by bonding across the opening or by equivalent means. Lockout/tagout procedures must be in place. Marking stand, proper grounding, hot line, and tools must be in place whenever applicable.

CONSTRUCTION IN ENERGIZED SUBSTATIONS

When construction work is performed in an energized substation, authorization should be obtained from the designated authorized person before work is started. When work is to be done in an energized substation, the determination of what facilities are energized and what protective equipment and precautions are necessary for the safety of personnel is needed prior to work.

Extraordinary caution should be exercised in the handling of bus bars, tower steel, all materials and equipment in the vicinity of energized facilities.

DEENERGIZED EQUIPMENT OR LINES

When it is necessary to deenergize equipment or lines for protection of employees, the requirements previously discussed must be complied with.

BARRICADES AND BARRIERS

Barricades or barriers should be installed to prevent accidental contact with energized lines or equipment. Where appropriate, signs indicating the hazard should be posted near the barricade or barrier.

CONTROL PANELS

Work on or adjacent to energized control panels should be performed by qualified and designated employees. Precaution must be taken to prevent accidental operation of relays or other protective devices due to jarring, vibration, or improper wiring.

SUBSTATION FENCES

When a substation fence must be expanded or removed for construction purposes, a temporary fence affording similar protection when the site is unattended, should be provided. Adequate interconnection with the ground should be maintained between the temporary fence and the permanent fence. All gates to all unattended substations should be locked, except when work is in progress.

FOOTING EXCAVATION

Excavation for auger and pad- and piling-type footings for structures and towers require the same precautions as metal tower construction. No employee should be permitted to enter an unsupported auger-type excavation in unstable material for any purpose. Necessary clean-out in such cases should be accomplished without entry. A competent person must make the determination if soil is stable.

FALL PROTECTION

Safety straps, lanyards, lifelines, and body harnesses should be inspected before use each day to determine that the equipment is in safe working condition. Defective equipment may not be used. Lifelines should be protected against being cut or abraded.

Fall arrest equipment, work-positioning equipment, or travel-restricting equipment should be used by employees working at elevated locations more than 6 feet above the ground on poles, towers, or similar structures, if other fall protection has not been provided.

The following requirements apply to personal fall arrest systems. When stopping or arresting a fall, personal fall arrest systems shall limit the maximum arresting force on an employee to 800 pounds. Personal fall arrest systems should be rigged such that an employee can neither free fall more than 6 feet nor contact any lower level. If vertical lifelines or droplines are used, not more than one employee may be attached to any one lifeline. Snaphooks may not be connected to loops made in webbing-type lanyards. Snaphooks may not be connected to each other.

Harnesses, Safety Straps, and Lanyards

Hardware for line personnel harnesses, safety straps, and lanyards should be drop-forged or of pressed steel and have a corrosive-resistant finish tested to the American Society for Testing and Materials Standard, B117-64 (50-hour test). Surfaces should be smooth and free of sharp edges. All buckles should be able to withstand a 2000-pound tensile test with a maximum permanent deformation no greater than $1/64$ of an inch. D rings should withstand a 5000-pound tensile test without failure. Cracking or breaking of a D ring constitutes its failure. Snaphooks should withstand a 5000-pound tensile test without failure. Failure of a snaphook is determined by any distortion sufficient to release the keeper. All fabric used for safety straps should withstand an A.C. dielectric test of not less than 25,000 volts per foot, "dry," for 3 minutes without visible deterioration. All fabric and leather used should be tested for leakage current and should not exceed 1 milliampere when a potention of 3000 volts is applied to the electrodes positioned 12 inches apart. Direct current tests may be permitted in lieu of alternating current tests. The cushion part of the body belt should contain no exposed rivets on the inside, be at least 3 inches in width, be at least $5/32$ of an inch thick if made of leather, and have pocket tabs that extend at least $1 1/2$ inches down and 3 inches in back of the circle of each D ring to allow riveting on of plier or tool pockets. On shifting D belts, this measurement for pocket tabs should be taken when the D-ring section is centered.

A maximum of four tool loops should be situated on the body belt so that 4 inches of the body belt in the center of the back, measuring from D ring to D ring, are free of tool loops and any other attachments. Suitable copper, steel, or equivalent liners should be used around the bar of D rings to prevent any wear between these members and the leather or fabric enclosing them. All stitching should be done with a minimum 42-pound-weight nylon or equivalent thread and should be lock-stitched. Stitching parallel to an edge should not be less than $3/16$ of an inch from the edge of the narrowest member caught by the thread. The use of cross-stitching on the leather is prohibited. The snaphook keeper should have a spring tension that will not allow the keeper to begin to open with a weight of $2 1/2$ pounds or less, but should begin to open with a weight of 4 pounds (when the weight is supported on the keeper against the end of the nose).

Testing of line personnel safety straps, body belts, and lanyards should be in accordance with the following procedure. Attach one end of the safety strap or lanyard to a rigid support, the other end should be attached to a 250-pound canvas bag of sand. Allow the 250-pound canvas bag of sand to free fall 4 feet for the safety-strap test and 6 feet for the lanyard test, in each case stopping the fall of the 250-pound bag: failure of the strap or lanyard will be indicated by any breakage, or slippage sufficient to permit the bag to fall free of the strap or lanyard. The entire *body-belt assembly* should be tested using one D ring. A safety strap or lanyard that is capable of passing the *impact loading test* should be attached. The body belt should then be secured to the 250-pound bag of sand at a point that simulates the waist of a man and allowed to drop. Failure of the body belt will be indicated by any breakage or slippage sufficient to permit the bag to fall free of the body belt.

DEFINITIONS

Alive or live (energized): The term means electrically connected to a source of potential difference, or electrically charged so as to have a potential significantly different from that of the earth in the vicinity. The term "live" is sometimes used in place of the term "current carrying," where the intent is clear, to avoid repetition of the longer term.

Automatic circuit recloser: A self-controlled device for automatically interrupting and reclosing an alternating current circuit with a predetermined sequence of opening and reclosing followed by resetting, holding closed, or locking out.

Barrier: Any physical obstruction intended to prevent contact with energized lines or equipment.

Barricade: A physical obstruction such as tapes, screens, or cones intended to warn and limit access to a hazardous area.

Bond: An electrical connection from one conductive element to another for the purpose of minimizing potential differences or providing suitable conductivity for fault current, or for mitigation of leakage current and electrolytic action.

Bushing: An insulating structure, including a through conductor, or providing a passageway for such a conductor, with provision for mounting on a barrier—conducting or otherwise—for the purpose of insulating the conductor from the barrier and conducting current from one side of the barrier to the other.

Cable: A conductor with insulation, or a stranded conductor with or without insulation and other coverings (single-conductor cable), or a combination of conductors insulated from one another (multiple-conductor cable).

Cable sheath: A protective covering applied to cables.

Circuit: A conductor or system of conductors through which an electric current is intended to flow.

Communication lines: The conductors and their supporting or containing structures that are used for public or private signal or communication service, and which operate at potentials not exceeding 400 volts to ground or 750 volts between any two points of the circuit, and the transmitted power of which does not exceed 150 watts. When operating at less than 150 volts, no limit is placed on the capacity of the system. (*Note:* Telephone, telegraph, railroad signal, data, clock, fire, police alarm, community television antenna, and other systems conforming with the above are included. Lines used for signaling purposes, but not included under the above definition, are considered to be supply lines of the same voltage and should be run as such.)

Conductor: A material, usually in the form of a wire, cable, or bus bar, suitable for carrying an electric current.

Conductor shielding: The term means an envelope that encloses the conductor of a cable and provides an equipotential surface in contact with the cable insulation.

Current-carrying part: A conducting part intended to be connected in an electric circuit to a source of voltage. Noncurrent-carrying parts are those not intended to be so connected.

Dead (deenergized): Free from any electrical connection to a source of potential difference, and from electrical charges—not having a potential difference from that of earth. [*Note.* The term is used only with reference to current-carrying parts which are sometimes alive (energized).]

Designated employee: A qualified person delegated to perform specific duties under the existing conditions.

Effectively grounded: Intentionally connected to earth through a ground connection or connections of sufficiently low impedance and having sufficient current-carrying capacity to prevent buildup of voltages that may result in an undue hazard to connected equipment or persons.

Enclosed Space: A subsurface enclosure that personnel may enter and which is used for the purpose of installing, operating, and maintaining equipment and/or cable.

Electric line trucks: A truck used to transport men, tools, and material, and to serve as a traveling workshop for electric power-line construction and maintenance work. It is sometimes equipped with a boom and auxiliary equipment for setting poles, digging holes, and elevating material or personnel.

Enclosed: Surrounded by a case, cage, or fence that will protect the contained equipment and also prevent accidental contact of a person with live parts.

Equipment: This is a general term that includes fittings, devices, appliances, fixtures, apparatus, and the like used as part of, or in connection with, an electrical power transmission and distribution system, or a communication system.

Exposed: Not isolated or guarded.

Electric supply lines: Those conductors used to transmit electric energy and their necessary supporting or containing structures. Signal lines of more than 400 volts to ground are always supply lines within the meaning of the rules, and those of less than 400 volts to ground may be considered as supply lines if so run and operated throughout.

Guarded: Protected by personnel, covered, fenced, or enclosed by means of suitable casings, barrier rails, screens, mats, platforms, or other suitable devices in accordance with standard barricading techniques designed to prevent dangerous approach or contact by persons or objects. (*Note:* Wires that are insulated, but are not otherwise protected, are not considered guarded.)

Ground (reference): A conductive body, usually earth, to which an electric potential is referenced.

Ground (as a noun): A conductive connection, whether intentional or accidental, by which an electric circuit or equipment is connected to reference ground.

Ground (as a verb): The connecting or establishment of a connection, whether by intention or accident, of an electric circuit or equipment to reference ground.

Grounding electrode (ground electrode): A conductor, embedded in the earth, used for maintaining ground potential on conductors connected to it and for dissipating current conducted to it into the earth.

Grounding electrode resistance: The resistance of the grounding electrode to earth.

Grounding electrode conductor (grounding conductor): A conductor used to connect equipment or the grounded circuit of a wiring system to a grounding electrode.

Grounded conductor: A system or a circuit conductor that is intentionally grounded.

Grounded system: A system of conductors in which at least one conductor or point (usually the middle wire or neutral point of transformer or generator windings) is intentionally grounded, either solidly or through a current-limiting device (not a current-interrupting device).

Hotline tools and ropes: Those tools and ropes that are especially designed for work on energized high-voltage lines and equipment. Insulated aerial equipment especially designed for work on energized high-voltage lines and equipment should be considered hotline.

Insulated: Separated from other conducting surfaces by a dielectric substance (including air space), offering a high resistance to the passage of current. (*Note:* When any object is said to be insulated, it is understood to be insulated in a suitable manner for the conditions to which it is subjected; otherwise, it is uninsulated. Insulating conductor covers is one means of making the conductor insulated.

Insulation (as applied to cable): That which is relied upon to insulate the conductor from other conductors, or conducting parts, or from ground.

Insulation shielding: An envelope that encloses the insulation of a cable and at the same time provides an equipotential surface in contact with cable insulation.

Isolated: An object that is not readily accessible to persons unless special means of access are used.

Pulling tension: The longitudinal force exerted on a cable during installation.

Qualified person: A person who by reason of experience or training is familiar with the operation to be performed and the hazards involved.

Switch: A device for opening and closing, or changing the connection, of a circuit. For our purposes, a switch is understood to be manually operable unless otherwise stated.

Tag: A system or method of identifying circuits, systems, or equipment for the purpose of alerting persons that the circuit, system, or equipment is being worked on.

Unstable material: Earth material, other than that of a running consistency, that because of its nature or the influence of related conditions cannot be depended upon to remain in place without extra support, such as would be furnished by a system of shoring.

Vault: An enclosure above or below ground that personnel may enter that is used for the purpose of installing, operating, and/or maintaining equipment and/or cable.

Voltage: The effective (rms) potential difference between any two conductors or between a conductor and ground. Voltages are expressed in nominal values. The nominal voltage of a system or circuit is the value assigned to a system or circuit of a given voltage class for the purpose of convenient designation. The operating voltage of the system may vary above or below this value.

Voltage of an effectively grounded circuit: The voltage between any conductor and ground unless otherwise indicated.

Voltage of a circuit not effectively grounded: The term means the voltage between any two conductors. If one circuit is directly connected to and supplied from another circuit of higher voltage (as in the case of an autotransformer), both are considered of the higher voltage, unless the circuit of lower voltage is effectively grounded, in which case its voltage is not determined by the circuit of higher voltage. Direct connection implies electric connection as distinguished from connection merely through electromagnetic or electrostatic induction.

REVIEW EXERCISES

1. What must be considered prior to working near overhead or underground power lines?
2. What is the minimum working distance allowed by qualified workers when working near 13.2 KV energized lines and equipment.
3. Once lines and equipment are deenergized, what must be done to assure total employee protection?
4. What must be considered prior to working in confined or enclosed spaces?
5. What are the most significant hazards to workers required to construct overhead and underground power lines?

■ ■ **About the Author** ■ ■

MANAGEMENT OF ENVIRONMENTAL IMPACTS
by Jane Beaudry Guerette, CHST, CSP

Jane Beaudry Guerette, CHST, CSP, has worked in both construction safety and project management. Currently, she is the Safety and Environmental Director for Skanska USA Building–New England Division. While Director, Skanska was recognized as the only construction contractor in the United States to be ISO 14001 certified. Skanska is one of the EPA's charter members in the National Environmental Performance Track Program and is a charter member to the state of Massachusetts' Environmental Stewardship Program. In May of 2002 Jane was appointed to chair a safety and health task force for Skanska worldwide. Jane received a B.A. in Environmental Design and Planning from S.U.N.Y. at Buffalo, NY, and a Professional Certificate in Industrial Hygiene from the University of New Haven, Connecticut.

Management of Environmental Impacts

- Describe the significant pitfalls or negative outcomes that may result from an unorganized, unplanned approach to managing a project's environmental aspects.
- List typical environmental impacts that arise from construction activities.
- Explain where in the construction contract procurement process (from the owner to the GC/CM and from the GC/CM to the subcontractor) the identification and management of environmental impacts can be influenced.
- Discuss the major categories of regulatory environmental requirements pertaining to most construction projects.
- Describe the advantages of an Environmental Management System (EMS) approach to addressing environmental hazards in the construction industry.

INTRODUCTION

It is an unfortunate but safe assumption that the management or consideration of environmental issues is not the highest priority for most construction proj-

ect teams. The exceptions to this rule occur when there are obvious, major environmental impacts, such as the abatement of asbestos or lead, the removal of an underground storage tank (UST), soil or water conditions that fall under OSHA's HAZWOPER requirements, or a site that has been designated as a SUPERFUND site. When this is the case, contractors typically depend entirely upon information provided to them by others (i.e., the owner, his consultants and/or engineers) to determine their project's environmental responsibilities or impacts. In practice, however, these sources may not provide the totality of a project's environmental impacts or responsibilities. Often local requirements from a city or conservation commission are not contained or even referenced in project plans, yet the contractor is expected to be aware and in compliance with noise, dust, waste disposal, and street-cleaning requirements and/or limitations. In some cases, owners or developers may attempt to minimize construction costs by underemphasizing or underreporting to the builder the extent of a project's environmental condition. Without a process, system, or methodology in place to internally evaluate a project's environmental aspects, a team may spend much of its time and resources reacting to unexpected requirements or events.

411

In the many cases where owners and developers are conducting themselves responsibly and do provide accurate and timely information about a project's environmental aspects, it is not uncommon to find that the information does not flow down to where it is most needed. If the information *does* find its way to the project trailer, it almost never finds its way to the worker in the field. The workers in the field are the first line of defense against adverse environmental impacts. These are the individuals responsible for implementing, inspecting, and/or maintaining the environmental controls that stand between your company and the front page of the news, or from the scrutiny of some regulatory agency.

When project teams incorporate an owner's, environmental consultant's, or engineer's report and requirements into the contract documents, the assumption is that the requirements will be implemented as intended and be effective. Unless there is a third party on site, such as a Licensed Site Professional (LSP), whose sole responsibility is to monitor a project's environmental requirements, the "if it ain't broke don't fix it" mentality of the contractor's field supervisor may take over. The problem with this latter attitude or approach is that no one takes a leadership role. If no one is monitoring the effectiveness of the controls, the likelihood of a problem arising increases. This contractor will also be forced into a reactive position (to save money, reputation, and/or minimize schedule disruptions).

It has already been acknowledged that contractors have a fair amount of guidance for projects that involve known hazardous materials or contaminated soils or water. But for a "normal" construction project, who is the responsible entity to identify and evaluate environmental risks for the average contractor? Is the method for doing so consistent from project team to project team? When does this evaluation happen:

- During the estimating/purchasing phase? If so, what training and expertise do the majority of estimators or project managers have in researching and evaluating environmental impacts?
- After subcontractor documents are awarded?

- Before premobilization?
- Or, does it happen on a need-to-know basis, as in the case of a federal, state, or local representative showing up on site and shutting down the job amid allegations of a regulatory violation?

There is also the issue of environmental impacts of a project which are *not* regulated, like disposal of nonregulated waste and energy conservation, that can have a significant impact on a project's profit margin. How and when are those being evaluated and/or incorporated into a project's plan?

The most common approach to managing environmental issues is that one person or department of an organization is given the responsibility to "oversee" a project's or company's environmental and safety issues. In most cases this person, or the people in the department, cannot be everywhere at once. When the project team is "on its own," do any of the team members know what all of the environmental aspects are for that project and what the necessary controls are day to day to minimize the likelihood of a problem? In the absence of the company's environmental expert, would the people on this team have, or know how to find, the necessary documentation or evidence for a regulator or inspector to defend against an alleged impropriety?

Obviously, there are many pitfalls in addressing a project's environmental impacts in a haphazard, inconsistent, or uninformed manner. The results of such an approach are similar to those of a company that handles safety in the same manner. The project or company has low morale, has financial problems because of high insurance premiums and too many fines, suffers with inefficiencies, and has a bad reputation with the regulatory agencies and in the community, which includes potential or would-be clients, all of which result in the loss of business.

THE ROLE CONSTRUCTORS CAN PLAY IN POSITIVELY IMPACTING THE ENVIRONMENT

Since the general contractor (GC) and construction manager (CM) are the primary conduit for information about a project to the subcontractors, this is where the *management* of environmental issues should begin. As a rule, GCs and CMs do not assume a subcontractor will know what materials to purchase, how to install them, how their schedule needs to be incorporated into the project schedule, or what the project-specific safety requirements are. Only with very specific, tightly defined requirements will a project's desired outcomes—built safely, on time, under budget, and with high owner satisfaction—be achieved. Ensuring a project's successful management of environmental challenges begins with the contract documents as well. But to *maximize* control and management of environmental impacts, first one must recognize what those impacts are.

Environmental Issues with Legal Implications

The first step is to identify a project's legal requirements. They may fall into two categories, those that are regulated—federal (EPA), state (DEP or DEM), or local requirements (i.e., city ordinances or conservation commission orders of conditions), or those that are driven by a project's specific characteristics.

For the former, one need go only as far as the Internet. Most if not all state and federal requirements can be found on Web sites that are fairly easy to navigate. General searches for topics such as storm water, sediment control, or dust control will bring the inquirer to specific statues or regulations that are applicable.

A slightly more expensive approach would entail the services of an environmental consultant who can do the research and identify all of the federal, state, and local requirements that apply to the environmental impacts related to that company's overall activities. Of course the fee would be more if the company works in multiple states. In addition, the consultant could be asked to translate the regulations into "plain English." A mere list of statutes and regulations will not further the layperson's understanding of what is required.

It is also advisable for a contractor to develop a relationship with an environmental consulting company that can make itself available to provide guidance in case of an unexpected emergency. It is common practice for contractors to rely on the consultative services of the emergency spill/release clean-up company that is going to do the work, but it is important to remember that emergency clean-up services typically come in the form of time and materials contracts. These companies have a potential conflict of interest and may clean up more than is actually needed, which means more labor and higher costs associated with disposal. A consultant is a third party who will not take advantage of a panic-stricken client who may feel compelled to do whatever the clean-up company's "expert" says.

For the identification of *project-specific* environmental requirements, the owner or client in most cases will provide a baseline assessment of a project, such as an AHERA asbestos audit or a Phase I or Phase II environmental assessment. It may be for an interior renovation project or a new construction project which entails the disturbance of soil for utilities, foundations, or removal of subsurface objects like underground storage tanks (USTs). In these cases, the typical contaminants are asbestos, lead and other heavy metals, and petroleum products or volatile organic compounds (VOCs). Spreading contaminated soils can exacerbate preexisting contamination. Contractors should take special care not to cross-contaminate soils to previously uncontaminated areas of a site. In cases where this is not managed or controlled properly, the contractor may be held liable for cleaning up and would be considered "the operator of an uncontrolled hazardous waste site." (See the *Construction Guidebook for Managing Environmental Exposures*.) Sweeping construction-site entrances frequently reduces the likelihood of sedimentation impacts to wetlands areas. Also, properly installed curbing can prevent runoff from flowing *into* a construction site, which might have serious consequences for sites with adjoining wetlands (Figure 1).

In cases where hazardous materials are pervasive, the client/owner will usually have a third-party consultant such as an LSP provide a site assessment that identifies quantities and locations of contaminants, specifications, and sometimes oversight, to properly manage and dispose of regulated materials. If the owner has not made this information available for whatever reason, the contractor should consider not proceeding with the project until the

■ ■ **Figure 1** ■ ■ **Curbing can prevent storm-water runoff from entering a job site where it might overwhelm project-specific design systems**

issue is addressed. Failure to do so may result in the contractor performing work it is not licensed or trained to do. It is true that the owner will always be the "generator" and "cradle-to-grave owner" of any hazardous materials originating from and leaving the property, but that fact will not protect the contractor from legal obligations to handle, report, and/or dispose of these materials in accordance with the applicable regulations. Such an example would be the deleading of bridges. The EPA has stated that the bridge owner and the contractor removing the lead will be considered *co-generators* of the waste. In the absence of an owner-provided environmental survey, the contractor always has the option of negotiating with the client to have site assessments conducted and then including the cost of the assessment and any resulting remediation in the proposed cost of the work.

Often the project-specific requirements coming out of the site assessment will be formatted and provided in accordance with federal and/or state requirements. Local requirements may or may not be included or referenced in these documents, so it is incumbent upon the GC/CM to research and be familiar with these requirements so that they too can be incorporated into contract documents. One can determine local requirements by contacting the building department for the city where the work is going to take place. Depending on the size of the city or municipality, there may be a separate conservation commission or other office of environmental affairs.

It is important to note that even when the owner employs an environmental consultant, the GC/CM still must play an active role, primarily because they—not the third-party consultant or the engineer—hold the contract(s) with the subcontractor(s). The GC/CM is responsible for ensuring that the subcontractor is in compliance with the specifications and full scope of the work, as well as some of the safety and waste-disposal requirements. Even in the case where the third party takes full responsibility for all of the aforementioned, it is still prudent for the GC/CM to require copies and testimonials from the third party that the work was done in accordance with all applicable regulations so that if a legal claim is made some time in the future the GC/CM will have the documentation to mount a defense. Examples of records that the GC/CM should require the subcontractor to submit prior to release of fee retainage are:

- employee training records
- environmental sampling results (for soil, air, and/or water if it was discharged)
- personal sampling results that were done for exposure assessments
- waste disposal manifests showing that regulated materials removed from the site made it to a licensed disposal facility.

If this work and these records are in the control of a consultant, the contractor should consider including in his agreement with the owner a requirement to be indemnified for errors or omissions made by the consultant, as well as a provision that requires the consultant to release this information upon the contractor's request.

It is important to understand that an owner or client has a cradle-to-grave responsibility for hazardous materials originating from his or her property. An owner on the hot seat with the media or the EPA is unlikely to provide repeat business to a GC/CM who allowed materials with the client's name on them to be disposed of illegally.

Environmental Liability and Risk Assessments

The following discussion of specific environmental regulations is excerpted from The Associated General Contractors (AGC) of America *Construction Guidebook for Managing Environmental Exposures*, prepared for the AGC by ESC Risk Control, Inc.

Contractors face a variety of environmental exposures, including many they are unaware of. A Pollution Incident, according to the insurance industry definition, is "the discharge, dispersal, release, seepage, migration, or escape of smoke, vapors, soot, fumes, acids, alkalis, toxic chemicals, liquids or gases, waste materials, or other irritants, contaminants, or pollutants into or upon the land, the atmosphere, or any watercourse or body of water. . . ." These exposures occur when a party claims damages against a company for an incident that causes bodily harm, property damage, or cleanup costs. Environmental incidents need not be hazardous or dangerous in nature. Claims can be filed from: clients claiming negligence, regulatory agencies demanding the cleanup of a release, the public for presenting a risk, or other contractors for placing their employees at risk. The exposures faced by a contractor are not limited to the following:

RCRA
The Resource Conservation and Recovery Act (RCRA) of 1976 identified wastes considered to be hazardous and established acceptable disposal procedures for those wastes. RCRA contains extensive requirements for the design, construction, maintenance, and closure of different disposal facilities, such as landfills. RCRA also established a "cradle-to-grave" responsibility for hazardous waste generators. The cradle-to-grave responsibility states that generators are responsible for hazardous wastes and any environmental damages caused over the life of the wastes. In order to track the course of hazardous wastes, a manifest system was developed. The manifest system documents the transfer of hazardous waste from generator to transporter to the treatment, storage, or disposal facility (TSDF). This enables regulatory agencies to determine the amount of hazardous waste generated, its location, the final treatment or disposal method, and enables responsible parties to be identified.

CERCLA and SARA
The Comprehensive Environmental Response, Compensation and Liability Act (CERCLA) of 1980 and its amendments, the Superfund Amendment and Reauthorization Act (SARA) of 1986, are collectively known as the Superfund Laws. The laws were established in order to handle emergencies at closed (inactive or abandoned) hazardous waste sites, remediate the sites, and deal with environmental problems associated with the sites. A major component of Superfund legislation is devoted to the identification of Potentially Responsible Parties (PRP) and the power the EPA has to make those PRPs financially responsible for the remedial actions. The following parties were identified as PRPs, according to the EPA and the federal court systems:

- Generators
- Owners/Operators (past and present) of hazardous waste sites
- Transporters
- Parties Arranging for Disposal – This PRP is rarely liable for remediation costs, but, nevertheless, can be identified as such. These parties could be consultants or contractors who either contract or recommend a waste hauler or a TSDF for hazardous wastes. It is important to note a contractor does not have to be negligent to be found liable.

Clean Water Act
The Clean Water Act (CWA) of 1977 and its related amendments are significant tools for the EPA to regulate discharges into the nation's waters. The CWA regulates point source discharges including effluent discharges as well as storm water discharges.

Effluent Discharges – The CWA requires the testing of effluent discharges for toxicity, acidity, and other parameters. A national pollutant discharge elimination system (NPDES) permit is required for effluent discharges. If a contractor intends to discharge any wastewater into a waterway either at a project site or at an owned location, the CWA must be referenced and the NPDES permit received, if required.

<u>Storm Water Discharges</u> – Storm water discharges from industrial facilities, mines, hazardous waste TSD facilities, and construction sites disturbing less than 5 acres are covered under the NPDES system. Permits may be in one of three forms: an individual permit for an individual discharger, a group permit for groups of like industries or dischargers, or a general permit issued to dischargers which do not require individual permits. Requirements for an NPDES storm water permit include the development of a storm water pollution prevention plan (SWPP), the monitoring and sampling of storm water discharges, and the submission of discharge monitoring reports to regulatory bodies.

If a contractor has reason to believe that they fall under CWA regulations, assistance should be sought from regulatory agencies or consulting firms. The program can be very cumbersome to those unfamiliar with the regulation, and can be easily misunderstood.

NPDES General Permit

On February 17, 1998 the NPDES General Permit for storm water discharge from construction sites became effective. The permit authorizes discharge of pollutants in storm water runoff from construction sites disturbing greater than 5 acres of land, in accordance with permit terms and conditions. Phase II regulations, issued in the December 8, 1999 Federal Register and effective in February 7, 2000, expanded the scope to include construction sites less than 5 acres in size.

The primary "pollutant" to our Nation's surface water from construction sites is sediments. This is because erosion, sediment transport, and delivery are the primary pathway for introducing key pollutants, such as nutrients (particularly phosphorus), metals, and organic compounds into aquatic systems.

- Waste materials resulting from demolition or renovation activities may include asbestos from insulation, surfacing materials, ceiling and floor tiles, and roof shingles.
- Lead may be present in paint on bridges or tanks or may be in the soil surrounding these structures.
- PCBs are located in transformers that may have leaked in the past.
- Concrete washout needs to be properly managed to avoid impacting surface waters of municipal storm sewer systems at the construction site. (Figures 2 and 3). In addition to waste materials, materials used in construction can impact runoff if improperly stored.
- Tracking sediments off site should be minimized at construction site entrances by either stabilizing the entrance or by providing a truck wash area (Figure 4).

To determine whether a construction operator needs to obtain coverage under the General Permit for storm water discharges, information regarding NPDES permitting authorities can be obtained at the following web site: www.epa.gov/owm/sw/contacts/index.htm. A Storm Water Pollution Prevention Plan (SWPP) must then be developed and implemented, a Notice of Intent submitted, and a Notice of Termination once the operations are complete.

■■ **Figure 2** ■■ **Sedimentation and snow fencing protect an environmental receptor— the catch basin—and protect vegetation from construction activities**

■■ **Figure 3** ■■ **Protecting environmental receptors such as catch basins ultimately benefits wetlands and aquatic life at storm water outfalls**

■ ■ Figure 4 ■ ■ Many cities and municipalities require the use of construction-site entrance/exit wheel washes, gravel aprons, and/or street sweeping to keep dust, dirt, and mud off of city streets

Clean Air Act

The Clean Air Act (CAA) was first developed in 1970 and amended in 1990. The Act requires companies to take actions to prevent or minimize the consequences of an accidental release of pollutants into the air. The Act was also designed to reduce air pollution by 75% over the next 20 years. Contractors using mobile concrete or asphalt plants or other emissions sources may fall under the CAA.

Spill Reporting and Management

Spill reporting requirements are set forth by the following laws: CERCLA, the Emergency Planning and Community Right to Know Act (EPCRA), SARA, and the CWA. Environmental protection laws require that spills of hazardous materials be reported to the government if they are above a certain quantity. If a spilled material could cause harm to people or the environment, the government must be contacted to organize the response and mitigate the damage. Not reporting a spill or delaying in reporting is a federal offense punishable by both civil and criminal penalties. Spill reporting procedures should be established so that spills are quickly documented and remediated. Procedures should be in place to ensure that subcontractors also report any spills to an emergency coordinator.

Answering the following questions provides a simplified demonstration of how to identify if a spill must be reported:

- **Was the material released?** This is defined as spilling, leaking, pumping, pouring, emitting, emptying, discharging, injecting, escaping, leaching, or disposing into the environment (eg, the ground, water, or air); this excludes spills into a containment area.
- **Was the material released into the environment?** Only materials released to the environment should be reported.
- **Is the spilled material an EPA-classified hazardous substance, extremely hazardous substance, or is the material a petroleum product?** The EPA has established two lists of materials that must be reported when spilled: the list of hazardous substances (40 CFR Part 302, Table 302.4) and the list of extremely hazardous substances (40 CFR Part 355, Appendices A and B).
- **Is the spilled material in excess of its reportable quantity (RQ)?** Each material on the two lists is assigned a threshold amount that when released within a 24 hour period constitutes a RQ. If the release meets or exceeds the RQ, it must be reported. For petroleum product spills, a reportable quantity is considered to

be any amount that causes a visible sheen on water. This includes any nearby creeks or ponds, but would not include a rain puddle on a parking lot. Although a material may not be specifically listed on one of the two lists, it may still need to be reported if it contains a listed material or if the material falls in one of the lists general categories. Hazardous wastes and their reportable quantities are also found on the lists.

Almost all levels of government have some level of reporting requirements, from the federal government to the municipal government. The following are the major response centers that must be contacted:

- **The National Response Center (NRC):** All spills that constitute a release of a listed material's reportable quantity must be reported to the NRC.
- **The Local Emergency Planning Committee (LEPC):** The LEPC should be contacted if the spill has potential to harm persons off site. The LEPC is the local level organization established to prepare for and respond to spills. The LEPC may be a city, county, or regional organization and will work in concert with the State Emergency Response Commission (SERC).
- **The State Emergency Response Commission:** The SERC should be contacted if the spill has the potential to harm persons off site.

Unregulated Environmental Impacts

It is not uncommon to find that many contractors do not provide any environmental management for impacts beyond what is required by the contract documents. The misconception is that to do anything more will surely make them uncompetitive, will hamper their ability to get the job done on time, or will translate to higher subcontractor costs. The truth is that by thinking outside the box and thoroughly evaluating *all* of a project's impacts, including those that are unregulated, contractors are likely to find opportunities to *save* money. Two examples are resource conservation and waste disposal.

Resource conservation can take on several forms, including the preservation of natural vegetation as shown in the photo in Figure 5. For projects that have significant site work, where there is a lot of earth moving or ledge removal, for example, by looking ahead to final grade requirements, one may determine that materials that were originally planned to be removed from the site can be stockpiled and used later for grading, temporary or permanent roads, underlayment, or use in balancing the site when the time comes. For demolition projects that have concrete, brick, or block, many of these materials can be crushed on site and also reused as described above before new building construction begins. This approach saves natural resources (dirt, loam, or gravel), and saves the cost of buying new materials. By reducing or eliminating the need to truck new materials to the project, there is also a

positive impact to the environment by saving fuel (another natural resource) and reducing emissions from exhaust. A related example for conserving resources is to determine the extent to which project materials can be purchased locally, again with an eye toward reducing project-related air-quality detriments. Locally purchased materials again may be less expensive because transportation costs are minimized.

■ ■ **Figure 5** ■ ■ **An example of an unregulated environmental aspect that must be managed is one driven by an owner's needs or requirements**

The last two examples of resource conservation relate to project utility requirements. In the cases where electrical needs and water consumption are not provided (paid for) by the owner, there is an opportunity to reduce a GC/CM's overhead costs. By planning ahead and incorporating energy conservation into the electrical contractor's scope of work, significant dollars can be saved. Putting building lights on timers (in most cases the entire project does not need to be lit during off hours), using energy-efficient lamps or ballasts, and installing motion sensors in the project offices can have a positive impact on the environment and increase a project's profit margin. Water conservation can be considered similarly. By requiring subcontractors that have significant water requirements to provide new hoses and institute a hose maintenance and inspection protocol daily, one can literally save money from washing down the drain!

Waste disposal is needed for every construction project. But waste disposal is having more and more of an impact on a project's bottom line because landfill space around the country (and world) cannot meet the demand. Landfill closings and increasing restrictions on disposal of construction and demolition (C&D) debris are resulting in upwardly spiraling costs for waste removal because materials must be hauled across the country to facilities that meet the demand. In many cases contractors are already facing problems because disposal of unseparated C&D debris is being restricted by the states or municipalities. By using waste haulers that have markets for sorted materials or ground-up fines, the contractor can contribute to the longevity of local landfills and keep his costs down by not having to pay for materials to leave the region. By incorporating waste reduction strategies in all subcontract documents, the GC/CM has a better chance of minimizing costs associated with trash removal. Such strategies may include requesting packaging minimization from suppliers and mandatory participation for subcontractors in source separation of materials that have postconsumer use such as pallets, cardboard, plastic, metal, sheetrock, asphalt, and brick or concrete. Many cities have building resource centers or other facilities that support programs like Habitat

for Humanity, where salvaged or left-over construction materials in good condition can be donated rather than thrown away. The bottom line is that minimizing what gets put in the dumpster minimizes disposal costs. Recent experiences of a Boston-based division of an international construction company that is certified to the ISO 14001 standard are illustrative. Once this company began focusing on waste reduction and recycling of C&D, savings were immediately apparent. They were able to redirect and/or recycle approximately 35,000, 32,000, and 20,000 tons of mixed C&D waste in the years 2000, 2001, and 2002, respectively, and corporately saved millions of dollars in each respective year.

There is one other category of unregulated project-related environmental impacts that may be driven by the owner or client. More and more owners are looking at how to maximize their investment by considering sustainable designs and demanding efficiencies in building operating systems. Looking beyond traditional *short-term* savings efforts like "value engineering (VE)" to lower up-front construction costs, owners are becoming more sophisticated. They are recognizing that lowering long-term operating costs with efficient systems and those that minimize waste (i.e., using gray water for building cooling or irrigation) results in much larger savings than those typically resulting from VE efforts, lower occupant turnover, and higher worker morale. The latter is arguably the most significant owner return on investment because of reductions in retraining requirements and sick days and increases in productivity. If a developer is the project proponent, then the upside of accepting a higher initial investment translates to less frequent tenant turnover. The contractor who is educated about "green building," meets an owner's needs during preconstruction, teams with the architect, and incorporates these issues in the estimating/purchasing phase is still an exception—which means this company also has a competitive advantage.

Table 1 presents a sample of environmental aspects that can assist a project team in systematically evaluating its activities to determine which environmental aspects are considered significant and should be incorporated into its project management plan.

■ ■ Table 1 ■ ■

**Project X – Terminal Consolidation
Aspects Identification Table**

Activities, Operations, Services	Environmental Aspects	Potential Environmental Impacts	Significant Yes/No If no, why?*
DIVISION 1 GENERAL REQUIREMENTS			
Temporary facilities	Spills and leaks into ground/water from temporary toilets, excavation for temporary utilities	Contaminated soil, water, dust	Yes
Office supplies, office facilities	Waste disposal, energy usage	Landfill depletion, resource depletion	No – A
Final cleaning	Hazardous materials	Contamination from improper disposal	Yes
Temporary utilities	Water/energy usage	Resource depletion	No – A
DIVISION 2 SITE WORK			
Earthwork	Noise, dust, runoff to sewers/waterways, contaminated soil, mud on roadways, soil erosion	Contaminated soil/air, permit violations	Yes
Stockpiled soil	Dust, runoff	Contaminated water/air	Yes
Heavy equipment use	Spills, leaks into ground or water from fueling, air emissions, fuel consumption	Contaminated soil, water, air emission can contribute to smog, depletion of petroleum supplies	Yes
Demolition	Noise, dust, lead, asbestos, or silica exposure, solid and hazardous waste disposal, storage tanks, process piping, scrap materials (metals, ceiling tiles, carpet, concrete, batteries), contaminated ductwork or hoods	Noise/dust regulated, depletion of landfills, emissions air/water	Yes

* Reasons why significance criteria do not apply:
A. Not a Law and/or required by client
B. Solid waste stream cannot be profitably recycled or reused
C. Has not received two or more complaints by client, regulatory agent or other entity within 30 days
D. Cannot be profitably conserved
E. Publicity or public opinion will not be adversely affected

A Solution: Industry Collaboration

One organization in the United States, the U.S. Green Building Council (USGBC), has developed a "green" building program called LEED (Leadership in Energy and Environmental Design). What follows is the organization's description of its goals.

> The Council is the nation's foremost coalition of leaders from across the building industry working to promote buildings that are environmentally responsible, profitable, and healthy places to live and work.

The LEED Green Building Rating System is a voluntary, consensus based national standard for developing high performance, sustainable buildings. It was created to:
- Define "green building" by establishing a common standard of measurement
- Promote integrated, whole building design practices
- Recognize environmental leadership in the building industry
- Stimulate green competition
- Raise consumer awareness of green building benefits, and
- Transform the building market

LEED provides a complete framework for assessing building performance and meeting sustainability goals. Based on well-founded scientific standards, LEED emphasizes state of the art strategies for sustainable site development, water savings, energy efficiency, materials selection and indoor environmental quality. LEED recognizes achievements and promotes expertise in green building through a comprehensive system offering project certification, professional accreditation, training, and practical resources.

AWARENESS IS THE KEY

The approach beginning to unfold in this chapter for effectively managing environmental impacts is unlikely to be implemented successfully if only one person on the project team or in the organization is working to get it done. It is infeasible, as well as unrealistic, to propose that one person is going to:

- research a project's environmental impacts
- communicate with the owner and his third-party representatives
- research a project's legal requirements
- develop the project's logistics plan
- estimate the project's budget
- develop the project's management plan
- identify each subcontractor's scope of work
- conduct premobilization meetings

- inspect daily all of the operational controls that have been identified or developed to mitigate environmental impacts.

Since this is clearly a team effort, the entire project team needs to be educated on the following:

1. What environmental impacts have been identified.
2. What their drivers are—is it a legal requirement, does the owner require it, or does it make good business sense?
3. How the impacts are going to be managed, including delegation to subcontractors.
4. What each member of the team is going to be responsible for.
5. What the emergency procedures will be in the event of an unexpected spill, release, or excursion.

Once the GC or CM's project team is aware of these elements, then it is time to make sure the subcontractors and their employees are also informed. The first step is to include what will be required of them in their contract documents. Federal, state, or local requirements or conditions imposed upon the owner need to be passed down to each appropriate subcontractor. Soil remediation, erosion control, dust control, noise control, street cleaning (Figure 6), spill management, utility conservation, and/or

■ ■ **Figure 6** ■ ■ Sweeping construction-site entrances frequently reduces the likelihood of sedimentation impacts to wetlands areas

waste-management requirements are all examples of measures that may be defined by a regulatory agency, an owner's consultant, and/or the GC/CM to mitigate a project's environmental impacts. This is a good place to incorporate environmental control checklists into the project's requirements. Many state and conservation commission erosion control or storm water pollution prevention plans require regular inspections of control measures and attach a required checklist as an appendix to the plan. GCs or CMs may also choose to develop their own checklists, either for their own internal use or to be incorporated into subcontracts. Examples might include erosion, water, or noise control checklists, or a table that delineates what the reporting requirements are for spills or releases of typical construction materials. See Tables 2 and 3 for specific examples.

■■ Table 2 ■■
Erosion Control Checklist*

Project Name: _____

City and State: _____

Project Number: _____

Is the silt fence in good condition? (No holes or tears that can allow sediment or materials to pass through unfiltered)	Yes _____	No _____
Are hay bails and stakes in good condition?	Yes _____	No _____
Are all inlets, waterways, and catch basins protected?	Yes _____	No _____
Are other siltation devices/methods (i.e., sedimentation basins, flow reducers, etc.) in proper working order?	Yes _____	No _____

How often is the erosion control monitored?

List all nonconformances or corrective actions implemented.

General remarks concerning the condition of erosion control.

Inspection performed by: _____ Inspection date: _____

Inspector's employer: _____

*Sample erosion control checklist developed by a construction manager to proactively manage siltation control devices.

▪▪ **Table 3** ▪▪

Water Conservation Checklist*

Project name: _____ Inspection date: _____

Inspection made by: _____

Inspector's employer: _____

What are the current uses of water on the project? _____

_____

_____

_____

_____

How is water distributed to the areas where it is needed? (Check all that apply)

❑ Fire hydrant
❑ Fire hose
❑ Hose bib
❑ Water hose
❑ Other _____

If a hydrant has been used, was a permit required? Yes _____ No _____
Is there a backflow preventor in use? Yes _____ No _____

Who is responsible for the calibration of the flow meters? _____

Who is responsible to make the hydrant dis/connection? _____

Check which items were leaking/defective, list corrective action, party responsible for correction, and date corrected:

Defective Equipment	Recommended Corrective Action	Who Corrected Defect	Date Corrected
Fire hydrant connection			
Fire hose tear and/or connection(s)			
Hose bib			
Water hose tear and/or connection(s)			
Other			

General remarks concerning water usage and/or conservation.

*Sample water conservation checklist developed by a construction manager to proactively manage and conserve natural resources and to minimize the general conditions budget.

Once a subcontractor is contractually bound to the management of an environmental control measure, the last step is to ensure each worker in the field understands his/her role in its implementation and why it is necessary. This can be accomplished using existing worker awareness methods such as project orientations and toolbox talks. Orientations are already conducted on many job sites to communicate information to workers about project-specific safety and health provisions and a project's logistical requirements, such as parking and working hours. Toolbox talks are practically a defacto OSHA requirement and they are the typical method that employers use to meet OSHA's hazard communication and safety training requirements. This is a perfect opportunity to pass along project-specific environmental control information so that workers—the ones most likely to observe, correct conditions, and/or notify a GC's superintendent—can do so in a timely manner. For example, most workers are oblivious to the need for sediment control unless they are told it is important. When they are properly trained, if they knock over hay bales or observe a hole in siltation baskets, they will know why it is important to correct the problem, and how to do it. There is a much higher likelihood these controls will be effective if there are 50 people on the site aware of their necessity than if it is the responsibility of just one person. By handling worker awareness via the toolbox talks, it is also possible to focus information where it will be most effective. It is not necessary for the workers of the civil contractor, for example, to be aware of or notified about a painting contractor's requirement to ensure paints or solvents are not washed into drains or disposed of in the general site trash dumpsters. By focusing information where it is needed, a contractor will minimize the likelihood of information overload for workers trying to be cognizant of safety, constructability concerns, and now a project's environmental control requirements.

AN ALTERNATIVE TO THE TRADITIONAL

The U.S. Environmental Protection Agency (EPA) has recently embraced and recognized a new ap-

proach to environmental management. Realizing that Environmental Management Systems (EMS) provide organizations of all types with a structured approach for managing environmental and regulatory responsibilities to improve overall environmental performance, including those areas not subject to regulation, for the past several years the EPA has been involved with a wide range of activities designed to facilitate EMS adoption. Its Web site states the following in "About EMSs."

> An Environmental Management System (EMS) is a continual cycle of planning, implementing, reviewing, and improving the processes and actions that an organization undertakes to meet its business and environmental goals. Most EMSs are built on the "Plan-Do-Check-Act" model (Figure 7). This model leads to continual improvement based upon:
> - Planning, including identifying environmental aspects and establishing goals [plan];
> - Implementing, including training and operational controls [do];
> - Checking, including monitoring and corrective action [check]; and
> - Reviewing, including progress reviews and acting to make needed changes to the EMS [act].

In the late 1990s, the EPA—a historic "command and control" agency—kicked off its National Environmental Performance Track Program, to recog-

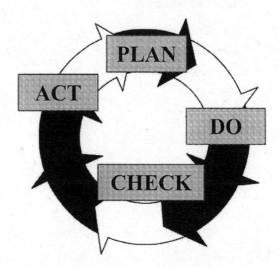

■■ **Figure 7** ■■ **The Environmental Protection Agency's "Plan-Do-Check-Act" model**

nize and reward top environmental performers. In addition, its statement of principles in the "EMS Position Statement," indicates that the EPA "will lead through by example, by implementing EMSs at appropriate EPA facilities." (See the EPA Web site designations in the reference section of this chapter.)

Although the application of Environmental Management Systems has historically been limited to industries other than construction, there is no reason for such limitation. The EPA's "EMS Position Statement" points out that the development of an EMS for use in construction results in the same potential benefits realized by other industries:

- improvements in overall environmental performance and compliance
- a framework for using pollution-prevention practices to meet EMS objectives
- increased efficiency and potential cost savings when managing environmental obligations
- predictability and consistency in managing environmental obligations and customer requirements
- fewer accidents and reduced liability
- improved public image
- competitive/trade advantages
- increased employee involvement and morale
- enhanced image with regulators, and improved compliance
- reduction in insurance rates
- more favorable credit terms.

"Section 1: Why Your Organization Should Have an EMS," on the Environmental Protection Agency's Environmental Management Systems Web page goes on to state:

> An effective EMS makes sense, whether your organization is in the public or private sector. By helping to identify the causes of environmental problems and then eliminate them, an EMS can help a company save money. Think of it this way:
> - Is it better to *make a product (or provide a service) right the first time,* or to fix it later?
> - Is it cheaper to *prevent a spill in the first place* or to clean it up afterwards?
> - Is it more cost effective to *prevent pollution* or to manage it after it has been generated?

The initial response to such a radical change in approach might be one of trepidation because it is new, it sounds complicated, and to some it may sound like a very separate, onerous program. For many field supervisors, the initial perception might be that this new program is to be implemented while the goal of building the project appears to be getting lost in standard operating procedures (SOPs) such as loss prevention programs, job hazard analyses, and crisis management programs. The reality is this approach improves the project's overall performance if implemented as intended. As the name suggests, it is a *management system*, and the intent is that it gets incorporated, functions, and is tracked using existing management tools, such as:

- preplanning processes (estimating, purchasing, logistics plan development)
- progress and coordination meetings
- daily reports
- site inspections.

The International Standards Organization defines what a management system is and how it impacts an organization:

> A *management system* refers to what the organization does to manage its processes, or activities. In a very small organization, there is probably no "system" as such, just "our way of doing things," and "our way" is probably not written down, but all in the manager's or owner's head. The larger the organization, the more people involved, the more likelihood that there are some written procedures, instructions, forms, or records. These help ensure that everyone is not just "doing his or her thing," but that there is a minimum of order in the way the organization goes about its business. So that time, money, and other resources are utilized efficiently.
> To be really efficient and effective, the organization can manage its way of doing things by systemizing it. This ensures that nothing important is left out and that everyone is clear about who is responsible for doing what, when, how, why, and where.

Using a project's safety program as an analogy, if the EMS is a stand-alone program with one or two people dedicated to it's implementation and oversight, it is unlikely to have the desired outcome(s). As with safety, the program can only be successful if everyone on the team knows what the requirements and expectations are, what their role is, participates in the identification of potential problems and solutions before they arise, and participates in identifying the root cause of problems and their correction.

Key Elements of an EMS

While there is more than one environmental management system model, they are similar in many respects. One well-known example is the ISO 14001 series of voluntary standards and guidelines. It is outlined on the Environmental Management Systems Home Page (see the Reference section):

> ISO stands for the International Standards Organization, located in Geneva, Switzerland. ISO is a non-governmental organization established in 1947. The organization mainly functions to develop voluntary technical standards that aim at making the development, manufacture and supply of goods and services more efficient, safe and clean.
>
> ISO 14000 refers to a family of voluntary standards and guidance documents to help organizations address environmental issues. Included in the family are standards for:
> - Environmental Management Systems (ISO 14001),
> - environmental and EMS auditing (ISO 14011 and 14012),
> - environmental labeling (ISO 14020)
> - performance evaluation(ISO 14031), and
> - life-cycle assessment (ISO 14040–43).

While other management systems have many of the attributes listed below, the most significant difference between ISO 14001 and other environmental management systems is that, to be certified to the ISO standard, a company must subject itself to third-party external audits to verify conformance.

ISO 14001 is the specification standard within ISO 14000. Its seventeen elements, listed on the Environmental Management Systems Home Page, include:

1. Environmental Policy – develop a statement of the organization's commitment to the environment
2. Environmental Aspects and Impacts – identify environmental attributes of products, activities, and services, and their effects on the environment
3. Legal and Other Requirements – identify and ensure access to relevant laws and regulations
4. Objectives and Targets – set environmental goals for the organization
5. Environmental Management Program – plan actions to achieve objectives and targets
6. Structure and Responsibility – establish roles and responsibilities within the organization
7. Training, Awareness, and Competence – ensure that employees are aware and capable of their responsibilities
8. Communication – develop processes for internal and external communication on environmental management issues
9. EMS Documentation – ensure effective management of procedures and other documents
10. Document Control – ensure effective management of procedures and other documents
11. Operational Control – identify, plan, and manage the organization's operations and activities in line with the policy, objectives, and targets
12. Emergency Preparedness and Response – develop procedures for preventing and responding to potential emergencies
13. Monitoring and Measurement – monitor key activities and track performance
14. Nonconformance and Corrective and Preventative Action – identify and correct problems and prevent recurrences
15. Records – keep adequate records of EMS performance
16. EMS Audit – periodically verify that the EMS is effective and achieving objectives and targets
17. Management Review – review the EMS

The Web site also addresses the integration of existing environmental activities into the EMS:

> An EMS is flexible and does not require organizations to necessarily "retool" existing activities. An EMS establishes a management framework by which an organization's impacts on the environment can be systematically identified and reduced. For example, many organizations, including counties and municipalities, have active and effective pollution prevention activities underway. These could be incorporated into the overall EMS.

CONCLUSION

Although construction projects and company's do not enjoy the advantages of fixed manufacturing facilities which have a stable workforce, consistent and known inputs and outputs, and measurable environmental detriments that can be targeted for reduction, this does not mean that managing their environmental impacts has to be a Herculean task. The willingness to realize the shortcomings and negative or unsatisfactory results from a haphazard approach to managing environmental issues is a significant first step in improving its performance. Once a company adopts an approach that includes a systematic and thorough evaluation of its activities and how they impact the environment, and then develops and implements a system to mitigate or eliminate the negative effects, the outcome will certainly include improved environmental management. Beyond that, however, it will also enjoy better operating efficiencies, better employee morale, positive perception in the regulatory community and the communities in which it operates, and thus with its clients as well.

REFERENCES

Associated General Contractors of America (AGC). *Construction Guidebook for Managing Environmental Exposures*. ECS Risk Control, Inc., 2000.

Environmental Management Systems Home Page, 8/6/02, "What are some potential benefits of an EMS," "What are the 17 elements of the ISO standard?" and "Can existing environmental management activities be integrated into the EMS?" http://www.p2pays.org/iso/faqs/htm.

International Organization for Standardization Web page, "The Basics: Generic Management System Standards." http://www.iso.org/iso/en/iso9000-14000/tour/generic.html.

_____, 7/25/02. "ISO 14000 Information, General Information." http://www.p2pays.org/iso/main/isoinfo.htm.

U.S. Environmental Protection Agency Environmental Management Systems Web page, 12/4/02, "EMS Position Statement." http://www.epa.gov/ems/policy/position.htm.

_____, 12/4/02, "About EMSs." http://www.epa.gov/ems/info/index.htm.

_____, 5/20/02, "EMS Position Statement" and "Section 1: Why Your Organization Should Have an EMS." http://www.epa.gov/ems/reasons/sme1.htm.

U.S. Green Building Council LEED Web page, "Leadership in Energy & Environmental Design." http://www.usgbc.org/LEED/LEED_main.asp.

Watkins, R. V. "ISO 14000 The Coming of Worldwide Environmental Management System Standard – Part III," *EHS Management News*, vol. 1, #1, 1996.

REVIEW EXERCISES

1. Name 4–6 negative outcomes or pitfalls relating to unidentified and/or mismanaged environmental impacts arising from construction activities.
2. Name 4–6 sources and categories of legislated environmental requirements.
3. Name and describe 3–5 unregulated environmental impacts that can significantly impact a project's financial outcome.
4. Describe 5 positive results or outcomes related to approaching the management of a project's environmental impacts with the application of an EMS.
5. Identify several areas within the current construction project management framework or model that can influence a project's success in identifying and managing any environmental hazards.
6. How does the management of environmental hazards parallel the management of a project's safety hazards?

■■ **About the Author** ■■

DEMOLITION
by David B. Korman, CSP

David B. Korman, CSP, is a professional member of ASSE and a Certified Safety Professional. Actively involved with the Construction Practice Specialty for the last six years, he is currently the Assistant Administrator of the ASSE Construction Practice Specialty. Mr. Korman is employed as the Safety and Environmental Manager with Skanska USA Building Inc., a national construction management firm headquartered in Parsippany, New Jersey. In this position Mr. Korman provides safety, health, and environmental support to projects located throughout the United States and Puerto Rico. While in this position Mr. Korman was instrumental in the Skanska companies becoming the first construction group in the United States to have their environmental management system certified to the ISO 14001 Standard, and recognized by OSHA by achieving the VPP STAR designation at one of their projects. Mr. Korman has a Bachelors Degree from the University of Maryland and has been involved in safety, health, and environmental work for sixteen years.

Demolition

CASE STUDIES

Case #1

OSHA cited a Syracuse, New York, contractor for work associated with demolishing Three Rivers Stadium in Pittsburgh, Pennsylvania. The contractor received six willful health violations with a proposed penalty of $324,000 for overexposing workers to lead. Employees were totally unprotected and at risk to the dangerous effects of lead. The contractor had no plan to protect the employees.

Willful Health Violations Received

- Exposing employees to lead above the permissible exposure limit
- Lack of engineering or work-practice controls
- No compliance plan
- Employer ignored complaints of lead exposure symptoms
- Improper use of objective data to exposure assessment of torch-cutting operation
- No representative air monitoring
- No employee notification within five days
- No respirators provided
- No respiratory protection plan
- No medical exams for respirator use
- No fit test or respirator training
- No protective clothing, hand wash, shower facilities
- No blood-test preassessment
- No employee notification of blood-test results
- No lead training; no signs posted

(Source: *OSHA Region III News Release: III-01-06-29—62-PGH*)

Case #2

A front-end loader was being used to remove a piece of pipe and conduit near two 13-foot-high non-load-bearing cement block walls. The front-end loader, equipped with a "clam" attachment was pulling on pipe and conduit, which ran through the upper part of the block walls. Pulling on the piping caused the walls to collapse, crushing and killing the laborer who was working 3 feet from the base. (Source: *OSHA Fatal Facts Accident Summary Report No. 7*)

Case #3

Employees were working on a scaffold near the top of a 250-foot smoke stack when a section of concrete being removed fell onto the scaffold, knocking the employee off. The employee was not tied off with a harness and a lanyard, and he fell to the ground below. The demolition procedures utilized should have been developed to ensure that the concrete sections fell inside the stack instead of outside. (Source: *OSHA Fatal Facts Accident Summary Report No. 64*)

These cases are examples of the potential hazards that are associated with demolition activities.

INTRODUCTION

What is demolition? The *American Heritage Dictionary of the English Language* defines *demolition* as "The act of overthrowing, pulling down, or destroying a pile or structure by violence; utter overthrow. Demolition is the systematic removal and destruction of structures and/or buildings."

Demolition makes up one of the largest sectors of specialty construction work and can also pose some of the greatest construction hazards. According to the National Association of Demolition Contractors (NADC), the demolition industry's total case incident rate (TCIR) is about the same as that in the nonresidential commercial construction industry sector (2000 statistics showed 8.1 accidents per 100 full-time workers). However, these Bureau of Labor Statistics do not track the demolition industry separate from other specialty construction sectors. The difficulty with tracking the demolition industry is that, of the 1000–1200 demolition companies oper-

ating in the United States, a majority of them are also involved in other business areas. Some examples are waste hauling, recycling, and hazardous material remediation. The major hazards associated with demolition work are in line with the four leading causes of fatalities for construction: falls, electrocutions, being struck, or getting caught between objects. These will be discussed in greater detail later in this chapter. A major focus has been on the trend indicating an increased number of fatalities from being struck, particularly those accidents directly caused from being struck by moving equipment such as skid steer equipment, dump trucks, and other moving vehicles.

Many contractors have moved away from traditional demolition work to other types of construction in large part because of the tremendous increase in workers' compensation coverage for the demolition industry. The rate of injuries, as discussed earlier, is comparable to those of other construction industry sectors. However, workers' compensation manual rates are significantly higher than for other types of construction specialty contracting work, indicating that the insurance industry recognizes that demolition work is high risk with a great potential for catastrophic losses.

Demolition operations involve one of the most hazardous types of operations in the construction industry. They cover a wide spectrum of work from minor selective operations involving the simple removal of sheet-rock wall partitions to the use of explosives to implode structures for total collapse.

Hazards of Demolition Work

The common hazards associated with demolition work are:

- falls
- being struck by an object
- material handling
- hazardous materials
- noise and vibration
- electrical shock
- fires and explosions
- moving equipment.

FALLS

A risk assessment to focus on the prevention of falls should be undertaken prior to the commencement of demolition operations. Fall protection needs to be provided for all employees who may be exposed to fall hazards of 6 feet or more. Specific attention should be placed on the hierarchy of controls, in particular the use of engineering controls aimed at eliminating this hazard by providing a safe work platform.

Examples of fall hazards that may be seen at demolition project sites include the following:

- falling through a structurally unstable roof and/or floor
- falling through penetrations or open holes/shafts
- falling from leading edges (stairways, landings, fixed platforms, scaffolding, and roofs)
- falling out of elevated work platforms (scissor and boom lifts)
- falling while accessing an elevated area from a ladder or stairs
- collapse of flooring
- tripping over debris and materials.

The hazard of being struck by moving equipment and/or debris also poses a significant hazard during demolition operations. Some of the hazards include the possibility of being struck by falling demolition debris, the uncontrolled collapse of a building structure, contact with moving equipment (cranes, skid steer equipment, and loaders), the failure of structural members from overloading the floors with debris during demolition operations.

MATERIAL HANDLING

Any manual material-handling task performed on demolition sites should be assessed and controlled. Mechanical means of material handling should be planned into the scope of the demolition operations where feasible in order to reduce those hazards associated with manual lifting. Typical types of hazards associated with material handling include:

- moving equipment
- manual handling of debris

- lifting material
- loading operations.

HAZARDOUS MATERIALS

All hazardous materials should be identified during the engineering survey or, if discovered during the course of the demolition operations, they should be handled in accordance with the particular legal requirements. The building may contain hazardous materials as a result of its previous use. Any residues of materials contained in storage tanks/vessels, storage areas, or pipelines, and in ductwork, need to be identified and evaluated before commencement of operations in these areas. The safe working procedures shall then be detailed, including the use of controls to eliminate the hazards and to properly handle and dispose of the hazardous materials.

Hazardous materials create an exposure danger for demolition workers from being inhaled or ingested, or by being absorbed through the skin. Air monitoring may need to be performed during the operations to determine the appropriate protective clothing requirements and/or work practices to be used. The most common types of hazardous materials encountered during demolition operations include the following:

- *Lead:* The most dangerous substance when it is in the air as a fume or dust, such as during the burning or cutting of structural steel with lead-paint-coated surfaces. Exposures can also result from the cutting of tanks covered with lead-based paint and from general demolition of walls and other structural members with lead-paint-coated surfaces.
- *Asbestos:* Where possible, all asbestos should be removed prior to the start of demolition operations. Asbestos may be found in sprayed coatings, thermal and acoustic insulation materials, fire-resistant partitions, asbestos cement sheets, flooring, and in mastic materials under flooring and other surfaces.
- *PCBs:* A toxic substance typically found in electrical components such as transformers, capacitors, ballast, and heating equipment.

■ *Silica:* A material that naturally occurs in concrete and masonry materials. Any destructive demolition of these materials will create the potential for exposure to silica dust through inhalation of its particles. Proper controls need to be established to eliminate these hazards.

NOISE AND VIBRATION

The source of noise and vibration on demolition sites can be caused by the use of heavy equipment and tools, falling debris, and explosives. Compressors, pneumatic handheld tools, loaders, excavators, and other equipment can create noise levels of more than 85 dB and may at times create a significant hazard from impact noise in excess of 120 dB. When the threat of vibration may cause a hazard of premature building collapse or damage to adjacent structures, then work plans must be developed to eliminate those hazards. Seismic monitoring may be required in adjacent areas to measure the effect vibrations will have on existing structures.

ELECTRICAL SHOCK

All wiring—except where temporary installations are required—should be disconnected and made safe prior to the commencement of demolition work. Temporary electrical installations must comply with all applicable codes. Where there is a potential for exposure to live wires, the areas shall be conspicuously tagged and signs posted warning of the hazards. A detailed plan of these areas should be developed.

Equipment and tools should be used with caution to ensure no part of them comes into contact with overhead or underground wires or cables. When working near any live services, appropriate authorities should be notified prior to work. All temporary electrical services for power tools should be protected through the use of a ground-fault circuit interrupter (GFCI).

FIRE AND EXPLOSIONS

Where existing buildings have contained flammable and combustible materials, precautions should be maintained to avoid the hazards of fires and explosions. The potential for the accumulation of flammable vapors or substances should be identified and assessed and then made safe prior to operations.

The burning of structures may not be used as a method of demolition. Performing hot-work operations (burning, welding, and other spark-producing operations) will present the potential for fires. All flammable materials should be identified and the risk of fires assessed prior to these operations being undertaken. The release of explosive vapors from damage to pipes during demolition processes is yet another source of fires and explosions. All services, including gas, need to be identified during the engineering survey process prior to the start of work. If active utility services are identified during demolition operations then all operations must immediately cease until further investigation determines the extent of the services. A work plan needs to be developed in order to proceed with the demolition operations in a safe manner. The prompt removal of the accumulation of demolition debris from the structure will help to prevent the potential for hazards associated with fires. A charged fire hose, portable fire extinguishers, and access to fire hydrants should be maintained at all times during the course of the demolition project.

MACHINERY AND EQUIPMENT

The hazards associated with the use of machinery and equipment used on demolition projects are numerous. Only qualified, trained equipment operators should be allowed to operate machinery and equipment. Examples of hazards associated with machinery and equipment on demolition sites include the following:

■ electrocution—coming in contact with power lines
■ equipment failure—crane exceeding its safe working capacity
■ dropping material—incorrect rigging of loads
■ being struck by moving equipment
■ noise and vibration—loaders, pneumatic hammers
■ flying particles and debris

- dust and other airborne hazards—including carbon monoxide from operating equipment in poorly ventilated areas
- objects falling onto operators—from the premature collapse of structures
- structural collapse of floors—equipment operations on floors which are not rated for the imposed loads exerted by the equipment and materials
- welding and cutting hazards—structural failure of steel work, fire, and explosions
- falls—from elevated work platforms, ladders, and scaffolding.

PREPARATORY OPERATIONS

Before the start of any demolition project, steps should be taken to safeguard the safety and health of each and every worker at the job site. The overall demolition plan should include all elements to successfully complete the project, including the methods used to bring down the structure or to remove the various components of the building structures. The sequence of demolition work to be followed, the equipment necessary to perform the project, and the measures to be used to perform the work safely should also be included in the plan. Since the plan provides for the safety of all employees involved in the operation, it should also include the personal protective equipment to be used (respirators, eye and face protection, hearing and head protection). The abatement of and the protection from hazards such as glass, shafts, skylights, floor/wall openings, equipment, and hazardous materials need to be addressed, as well as protective measures to be taken to ensure that the public is protected, including the use of fences, signs, barricades, and debris netting. Specific instructions should be included on how to respond to emergencies involving personal injuries, fires, premature collapse, and timely retrieval of personnel from a structurally compromised building.

Engineering Survey

Planning for a demolition job is as important as performing the work. A competent person with the necessary experience and/or training in demolition operations should therefore plan all demolition work.

The American National Standards Institute, in its ANSI A10.6-1983, *Safety Requirements for Demolition Operations* states: "No employee shall be permitted in any area that can be adversely affected when demolition operations are being performed. Only those employees necessary for the performance of the operations shall be in those areas."

Prior to the start of every demolition project a competent person must conduct an engineering survey of the structure. (See Figure 1 for a sample of an engineering survey form.) The purpose of this survey is to determine the condition of the framing, floors, and walls so that measures can be taken to prevent the premature collapse of any portion of the structure. Any adjacent structures as well as improvements and alterations to the existing structures should similarly be checked. The contractors involved with the demolition activities must maintain a written copy of this survey. Photographs and a videotape of the condition of existing structures is also recommended.

The engineering survey allows the contractor to evaluate the project in its entirety. The contractor needs to plan for the wrecking of the structure, the equipment to do the work, manpower requirements, and the protection of the public. The safety of all workers at the project site should be a prime consideration. During the preparation of the engineering survey, the contractor should plan for potential hazards such as fires, cave-in, or premature collapse.

If the structure to be demolished has been damaged by fire, flood, explosion, or some other causes, appropriate measures, including bracing and shoring of walls and floors, should be taken to protect workers and any adjacent structures. Another critical aspect to the survey is including a thorough investigation for the presence of any hazardous chemicals, gases, explosives, flammable materials, or similar dangerous substances to determine whether they have been used or stored on the site. If the nature of the substances cannot be easily determined, samples should be taken and analyzed by a qualified person prior to demolition.

ENGINEERING SURVEY

JOB NAME_____   JOB LOCATION _____

JOB CONTACT_____   PHONE NUMBER _____

NAME OF STRUCTURE _____ DATE BUILT _____

STRUCTURE DESCRIPTION_____

STRUCTURE DIMENSIONS   LENGTH _____   WIDTH _____   HEIGHT _____

MATERIALS   FOUNDATION _____   WALLS _____

FLOORS _____   ROOF _____

METHOD OF DEMOLITION _____

UNDERGROUND UTILITY CONFIRMATION NO. _____

POTENTIAL HAZARDS (IE. COLLAPSE, STRUCTURAL FAILURE) _____

_____

SAFETY EXPOSURES	YES	NO	LOCATION/DESCRIPTION
OXYGEN LINES			
NATURAL GAS LINES			
ELECTRICAL LINES			
WATER LINES			
OTHER UTILITY LINES			
FIRE HAZARDS			
ADJACENT WALKWAYS			
ADJACENT ROADWAYS			
ADJACENT BUILDINGS			
COMBUSTIBLES			
WATER TOWERS			

■ ■ **Figure 1** ■ ■ **Suggested engineering survey** [From the *Demolition Safety Manual* published in 2000 by the National Association of Demolition Contractors (NADC), reprinted with permission.]

SAFETY EXPOSURES	YES	NO	LOCATION/DESCRIPTION
SMOKE STACKS			
ELEVATORS			
PARTY WALLS			
BASEMENTS			
PITS OR TRENCHES			
BULKHEADS			
MANHOLES TO PRESERVE			
MSDS PROVIDED			
FALL HAZARDS			
LEAD EXPOSURES			
CHEMICALS EXPOSURE			
PROCESS HAZARDS			
LIVE UTILITIES			
UTILITIES TO SHUT DOWN			
LIGHTING			
TELEMETERING LINES			
FIBER OPTIC CABLES			
TRENCHING/EXCAVATING			
SHORING REQUIRED			
PRE-EXISTING STRUCTURAL DAMAGE			

OTHER SAFETY ISSUES

■■ **Figure 1** ■■ (cont.)

ENVIRONMENTAL ISSUES	YES	NO	LOCATION/DESCRIPTION
ASBESTOS			
USTs			
BACKFILL MATERIAL			
STORMWATER RUNOFF			
WASTE OIL			
SOLVENTS			
CONTAMINATED SOIL			
UNIDENTIFIED DRUMS			
SPECIAL WASTE			
HAZARDOUS WASTE			
LEAD DISPOSAL			
PCB TRANSFORMERS			
PCB BALLASTS			
GALBESTOS			
CONTAMINATED WATER			
HEAVY GREASES			
CONTAMINATED BRICKS			
PROCESS WASTES			

OTHER ENVIRONMENTAL ISSUES

_____

_____

_____

_____

COMPLETED BY   DATE

■ ■ **Figure 1** ■ ■ (cont.)

During the planning of the demolition job, all safety and personal protective equipment needs should be determined. At a minimum, this cataloging should include the required type of respirators, fall protection systems, warning signs, and safety nets, any specialized personal protective equipment requirements, adequate-sized equipment to maintain a safe distance from work tasks, and any other protective devices outlined in the engineering survey.

Utility Location

One of the most important elements in prejob planning is the location of utility services. All electric, gas, water, steam, sewer and other service lines should be shut off and capped, or otherwise controlled, before demolition begins. Many demolition projects involve older buildings that do not have as-built drawings, records, or other information that helps to determine the exact locations of utilities. Additional testing with magnetometers and other instruments may assist in the identification. The utility companies should be notified in advance of the demolition project, and its services should be used in coordinating and making the project safe. If utilities such as water and electric are needed during the demolition activities, then those utilities should be relocated as necessary, or protected, to ensure the safety of the operations. The location of all overhead power lines should be identified, especially since they are extremely hazardous in the presence of equipment operations. All workers need to be informed about the location of any existing utility services.

A plan should be developed for marking what is to be removed and what is to stay on selective demolition projects. One example of a marking scheme is to use the color RED for items that need to stay (STOP) and GREEN for those that should be removed (GO).

Medical Services and First Aid

Prior to commencement of any demolition activities, plans should be established to ensure that prompt medical services are available in case of a serious injury. The nearest hospital, clinic, or physician should be identified as part of the engineering survey and preproject plan. The plan should also include the provisions for a communications system in order to contact any necessary ambulance services, and the phone numbers for the hospital and physicians must be made available at the site.

In the absence of a readily accessible hospital or clinic in terms of proximity or time from the project site, then there should be access to personnel who are certified in first aid and cardio pulmonary resuscitation (CPR) by a recognized entity, such as the American Red Cross or the Bureau of Mines.

The job site should be equipped with first-aid supplies, as determined by an occupational physician. The kit should contain approved supplies in a weatherproof cabinet with individual sealed packages for each type of item. The kit should also make available rubber gloves to prevent the potential transfer of infectious diseases. Provisions should be made to provide for quick drenching or flushing of the eyes. The contents of the first-aid kit should be checked at least weekly to ensure consumed items are replaced. In many areas the local municipalities require that a fire detail be hired to assist in fire protection during the demolition operations. This local fire department can also assist in responding to other emergencies that may occur on site.

Fire Prevention and Protection

A fire plan should be established prior to beginning any demolition project. It should outline the assignments of key personnel in the event of a fire and provide an evacuation plan for workers on site. Key elements in the prevention of fires include:

- All potential sources of ignition should be evaluated and the necessary corrective measures taken.
- Electrical wiring and equipment for providing light and heat or power should be installed by qualified electricians and inspected at frequent, regular intervals.
- All internal combustion engine equipment should be located so that the exhaust discharges well away from combustible materials and away from workers.
- All equipment should be shut down prior to refueling. All fuel and flammable liquids should be properly stored away from sources of ignition and heat.
- Sufficient fire-fighting equipment should be located near flammable storage areas.
- Only approved containers should be used to store flammable or combustible liquids.
- Smoking should be eliminated in any areas where flammable or combustible materials are stored.

■ A permitting system should be used to control all hot-work activities; it should specify procedures to use when performing any such work.

Sufficient identification of all emergency exits should be done prior to start of work and reviewed during orientations. High-visibility arrows painted on the floors can serve as an effective method to indicate the emergency routes.

SPECIAL STRUCTURES DEMOLITION

When preparing to perform demolition work on special structures such as stacks, a silo, a cantilevered structure, or a cooling tower, the first step must be a detailed inspection of the structure by an experienced competent person. If possible, engineering/architectural drawings should be reviewed. Attention should be paid to the condition of the stack and chimney. Workers should be trained to recognize the signs of any structural defects such as weak or acid-laden mortar joints; exposed, twisted, or rotted rebar; and any cracks or openings. The interior brickwork in some sections of large industrial chimneys can be extremely weak. If the stack has been banded with steel bands, these straps should be removed only as work on the stack progresses from the top down. Sectioning of the chimney should be considered.

SAFE WORK PRACTICES

When hand demolition is required, it should be performed by working from a work platform.

■ Experienced personnel should install a self-supporting tubular scaffold, suspended platform, or knee-braced scaffolding around the entire chimney.

■ Close attention should be paid to the design, support, and tie-in (braces) of the scaffold.

■ A competent person needs to be present at all times while scaffold erection is being performed.

■ Access to the top of the scaffold should be provided by means of properly protected walkways.

■ All work platforms should be decked solid and the area from the work platform to the wall bridged with a minimum of 2-inch-thick lumber.

■ Platforms should be protected with the use of a standard guardrail system, including a top rail (42 inches above the work surface) and a midrail. The guardrail should have protective canvas or mesh to prevent debris from falling from the work platform. Debris netting should be installed directly below the work platform to prevent debris from falling to the surfaces below.

Demolition of Prestressed Concrete Structures

The different types of construction used in a variety of buildings built during the last few decades will create a variety of problems when the time comes for them to be demolished. Prestressed concrete structures fall into this category. The most important aspect of demolishing a prestressed structure takes place during the engineering survey. During the survey, a qualified person should determine whether the structure to be demolished contains any prestressed members.

The demolition contractor is responsible for alerting all workers on site to the presence of prestressed concrete members within the building structure. The workers should also be trained in the safe work practices that must be followed to safely perform the demolition work. Also, included in the training should be a total review of the hazards that might result if workers deviate from prescribed procedures.

The type of prestressed concrete contained in the building should be determined before demolition is attempted.

When conducting a demolition project on progressively prestressed structures, it is essential to obtain the services of a professional engineer, and to demolish the structure in strict accordance with the engineer's method of demolition. The stored energy in this type of structure is tremendous and can precipitate a catastrophic event. In some cases, the inherent properties of the stressed sections may delay failure for some time, but the presence of these large prestressing forces may cause sudden and complete collapse with little warning. Professional engineers should also be advised before attempting to expose the tendons or anchorages of structures in which two or more members have been stressed together. Temporary supports may have to be provided so the tendons and the anchorage can be exposed.

Safe Blasting Procedures

Demolition using explosives should be conducted in accordance with 29 CFR 1926, Subpart U. The transportation, handling, storage, and use of explosives are subject to numerous federal, state, and local regulations. All regulations pertaining to your area of operations must be consulted. The manufacturer's recommended procedures should be followed through each stage of the blasting operation.

Prior to the blasting of any structure, a complete written survey needs to be completed by a qualified person of all adjacent structures and underground utilities. When the possibility exists of excessive vibrations due to the blasting operations, seismic or vibration tests should be conducted to determine the proper safety limits to prevent damage to nearby buildings, utilities, or other property. The preparation required to demolish a building with explosives may require the removal of structural members or other building components. A structural engineer must direct this work, or a qualified competent person may be required to direct the removal of these structural elements. Extreme caution should be taken during preparatory work to prevent the premature collapse of the structure. Blasting and imploding a building is a very specialized operation and should only be performed by a qualified company that specializes in this work. In all cases, you can only implode steel or concrete structures that are six stories or taller in order to control the implosion and still be economically feasible.

FIRE PRECAUTIONS

All efforts should be made to ensure that no fires or sparks are allowed anywhere near the presence of explosives. Smoking, matches, firearms, open-flame lamps and other fire-, flame-, or heat-producing devices must be prohibited near explosive magazines or in areas where explosives are being handled, transported, or used. A safe distance of a minimum of 100 feet should be maintained around all explosives. Evacuation procedures should be developed before any blasting operations. If electrical detonators are in use, then procedures should be implemented to ensure all radio frequency (RF) sources are restricted from or near the demolition site.

A key to the success of the explosives operations is the selection of personnel. A competent blaster must be used for all explosive demolition operations. A blaster must be qualified on the basis of experience, knowledge, and training in the field of explosives. In addition, the blaster should have a thorough understanding of all applicable federal, state, and local requirements for blasting and the use of explosives. Training courses are available from the manufacturers of explosives, and blasting safety manuals are available from the Institute of Makers of Explosives (IME) and other organizations. All blasters should be capable of providing evidence of their training, experience, or licensing in the use of the type of explosives and operations required for a project.

USE OF EXPLOSIVES

Protection of the public is an essential part of all blasting operations. Blasting operations should be conducted between sunup and sundown, whenever possible. Proper signs should be posted to alert the public to the hazard presented by the blasting, but care should be given not to overtly publicize the blast (which will result in drawing more pedestrians to the blast site). Protective measures such as the use of blasting mats, berms, or other similar structures should be used where there is a danger of rocks or other debris being thrown in the air. Every blasting project is unique and requires its own protective measures. The safety measures should be the responsibility of the blasting contractor. Radio, television, and radar transmitters create fields of electrical energy that can, under certain circumstances, detonate electrical blasting caps. Precautions must be taken to prevent the premature discharge of blasting caps from current induced by radar, radio transmitters, lightning, adjacent power lines, dust storms, or other sources of energy or static electricity. Some precautions include:

- Ensuring that transmitters on the job site less than 100 feet away from electric blasting caps are deenergized and locked.
- Prominently displaying warning signs against the use of mobile radio transmitters on all roads within 1000 feet of the blasting area.
- Maintaining minimum recommended distances between the nearest transmitter and electric blasting caps.
- Suspending all operations during the approach or progress of any electrical storm.
- Inspecting all connections, as well as the blast site, prior to firing to ensure that everyone is clear of the site before giving the order to fire. Standard signals that indicate a blast is about to be fired, and an all-clear signal afterward, are used. All personnel in the area should be trained in the use of these signals and in seeing that they are strictly enforced.

PROCEDURES AFTER BLASTING

Inspection
Immediately after the blast has been fired, the firing line should be disconnected and short-circuited. Power switches should be locked open in the off position. Sufficient time should be allowed for the settling of dust, smoke, and fumes to leave the blast areas before returning. Inspection of the entire area should be performed by the blaster to ensure that all charges have exploded before workers are allowed to return.

Disposal
Explosives, blasting agents, and blasting supplies that are obviously deteriorated or damaged should not be used; they should be properly disposed of. Explosives distributors will usually take back old stock, or the local fire officials or representatives from the U.S. Bureau of Mines or Alcohol, Tobacco or Firearms (ATF) may also assist in their disposal.

Under no circumstances should any explosives be abandoned or buried.

Protective Structures

A variety of structures are used to protect workers, as well as the public, from the hazards created at a demolition site from falling debris. Ladders, scaffolds, and powered manlifts, when used properly, provide a safe means for access. Guardrails and roof guards enable workers to work safely on sloped roofs. Sidewalk canopies, such as the one shown in Figure 2, temporary walkways, dumping sand 6-feet-thick around the site perimeter, catch platforms, and fences protect the public from the hazards of demolition activities. Shoring is a means of retaining structures that are unstable or may become that way during the course of the demolition operations.

Signs and Lighting

The use of both signs and lighting can assist in the identification of predictable hazards.

To clearly mark every hazard on a job site with a sign would be impossible. The key is to make sure

■ ■ **Figure 2** ■ ■ **Falling debris protection provided by a sidewalk canopy**

that both the site workers and the public can be made aware of unexpected hazards and precautions to take to prevent injuries through the use of signs. Signs should use the standardized ANSI color and definitions for "DANGER" and "CAUTION." Danger signs should be used where an immediate hazard exists. Caution signs should be used to warn against potential hazards or to discourage unsafe practices. Additional signs should include "EXIT" signs to identify egress points and doors that lead to safety.

Sufficient lighting is vital to a safe working environment, and natural light may be sufficient for some work areas. However, light levels during early morning, as well as insufficient light due to bad weather, adjacent structures, and dust can easily result in unsafe working conditions. Additional lighting should be provided if there is any uncertainty that adequate lighting cannot be maintained during all work operations. It is essential that lighting be adequate in all walkways and paths of egress. OSHA requirements for providing adequate illumination in all work areas, offices, and ramps mandate the site be in accordance with the requirements indicated in 1926.65(m) Table D-65-1, SHORING.

Shoring

Maintaining the structural integrity of the building under demolition and any nearby structures is critical during the demolition process. The stresses created during the removal of walls and supporting members and the loads imposed by equipment and falling or stored debris must be taken into consideration. In the event that the structural integrity of the building does not provide a sufficient factor of safety for imposed loads, shoring should be implemented. Shoring systems can provide either vertical or horizontal support. The equipment used for shoring should be inspected prior to each use. If made of steel it should be inspected for rust, bends, and defective welds; wood shoring should be checked for splitting, warping, and cracking. The stability of the shoring, as well as the structure it is retaining or supporting, should be inspected on a regular basis during the course of demolition.

Vertical shoring may be required when walls or columns are removed, or when the weight of mech-

anical equipment or debris exceeds the safe loading capacity of the floor. Attention should be paid to the impact loading of falling debris, which is considerably greater than the load imposed by the debris at rest. Common types of vertical shoring use screw jacks, threaded collars, or wedges to secure the shoring in position. Individual posts must not be stacked or tiered to reach ceiling heights. A competent person should determine the placement of shoring on a solid footing.

Shoring should be placed on base plates, which are positioned on the footing, and the shorehead should be centered on the beam above the base plate to distribute the load over a wider area. The shoring must be plumbed before tightening and plumb-checked frequently to ensure safe conditions.

Lateral shoring may be required when a building has a shared wall with another building. Foundation walls, which serve as retaining walls to support the earth or adjoining structures, should not be demolished until proper bracing is in place. Free-standing wall sections more than one story high must have lateral bracing, unless the wall was originally designed and constructed to be self-supporting and stand higher than one story without the need for lateral support. Walls must be left in a stable position at the end of each shift. Shoring may be required after the removal of interior floors due to the potential for shifting of lateral stresses. Two common types of lateral shoring are generally used: raking shores and horizontal shores. *Raking shores* have a solid base, which rests on the ground opposite the wall to be braced, and shoring members fan out to reach the various levels. *Horizontal shores* extend from a wall on a nearby structure, which is opposite the wall being braced. The advantage of horizontal shores is that they provide for a clear path on the ground around the shored wall. Horizontal shores must be anchored to a load-bearing member of that structure. For raking shores, a foundation should be excavated or laid with attention to soil compaction and stability.

Protecting the Public

Before starting any demolition work, every sidewalk or public thoroughfare adjacent to or near enough to be affected by the demolition work should be closed, relocated, or protected. Thoroughfares used by the public should be kept clean and unobstructed at all times. *Whenever possible, pedestrian and vehicular traffic should be prohibited from any area that is closer to the perimeter of the structure being demolished than 1/4 the height of the structure.* When pedestrian traffic is required to use an area closer to the perimeter than 1/4 the height of the structure being demolished, a substantial protective canopy must be constructed. The canopy must cover the length of the route adjacent, adjoining, contiguous, or abutting the structure to be demolished. The canopy should be wide enough to accommodate pedestrian traffic without causing congestion. The pedestrian canopy must be lighted to ensure adequate illumination for safe travel. The roof must have a loading capacity of a minimum of 150 pounds per square foot plus the added weight of any additional materials stored on top. If pedestrian canopy gates enclosing the site are installed, they should be kept closed at all times when not actually being used for operations.

On some sites a temporary walkway is used in lieu of a canopy. This can only be used in those areas that are a greater distance from the perimeter of the structure than the 1/4 the height of the structure criteria. The temporary walkway should consist, at a minimum, of a fence on the inner side and a continuous concrete or equivalent traffic barrier and barricade (Jersey Barrier) at the street side. The walking surface must be kept free of slip, trip, and fall hazards. When in partially occupied buildings, a safe area for the public must be maintained. This may include the use of separate entrances that are protected by a pedestrian canopy. If the adjacent occupied building is lower than the building being demolished; its roof must be protected with barriers, debris catchers, and/or other substantial and effective coverings. All doors and gates should swing toward the site and not toward the line of pedestrian or vehicle traffic. When the 1/4 height of the structure is not possible to obtain during demolition operations, then a catch platform or scaffold should be erected in addition to the pedestrian canopy. Platforms should be no more than 40 feet below the demolition activities. The catch platforms must be built to withstand a live load of at least 150

pounds per square foot. Debris must not be stored or allowed to accumulate on catch platforms. They are designed solely to retain falling debris. Any accumulation will affect the capacity of the platform; therefore, all platforms should be maintained in a clean condition.

DEBRIS REMOVAL

Debris removal operations conducted inside the walls of a structure are usually accomplished through floor openings. When available, existing openings such as elevator shafts and mechanical shafts should be used for this purpose. If floor openings must be cut for debris removal, then a qualified person should be consulted for this work. The floor openings must not take up more than 25 percent of the total floor space on each level unless the lateral supports are left intact. Support beams must be left intact wherever possible, but when floors are weakened or otherwise made unsafe they must be shored to carry the load from the demolition operations. Precast or posttensioned concrete must never be cut unless a professional engineer is consulted. All entrances to each floor level must be marked with "WARNING" signs, which indicate the presence of a hazardous condition. All floor openings must be barricaded by a substantial guardrail system and toe boards located at least 6 feet from the opening. All floor openings that are not in use as material drops should be covered with materials capable of withstanding two times the maximum intended load. Covers must be properly secured to prevent movement and should be marked according to OSHA requirements [29 CFR 1926.502(i)].

Trash chutes used for debris removal should be designed and constructed to eliminate failure due to the impact from the materials or debris being loaded. To prevent this, the chute should be designed to change direction every 120 feet, and baffles should be installed as needed. The chute should be installed so that it is adequately secured to its supporting members. Areas where debris drops outside of the building structure must be adequately protected to prevent pedestrians from entering them. When material is dumped into a chute by means of a wheelbarrow or mechanical device, a securely attached bumper should be provided at each opening. All chute openings should be kept closed when not in use. At the discharge end of the chute there should be a gate. This gate should be closed when the debris removal chute is not in use, or when the container or dump truck is being changed. A competent person should be assigned to control the traffic and the flow of personnel in and around the discharge area from the chute.

REFERENCES

The American Heritage Dictionary of the English Language, 4th ed. Boston: Houghton Mifflin Co., 2000.

American National Standards Institute. *Safety Requirements for Demolition, ANSI A10.6 1983.* New York: ANSI, 1983.

MacCollum, D. *Construction Safety Planning.* New York: Van Nostrand Reinhold, 1995.

National Association of Demolition Contractors, *Demolition Safety Manual.* Doylestown, PA: NADC, 1989.

The St. Paul Technical Guides. *Demolition Planning and Engineering Surveys.* St. Paul, MN: St. Paul Companies, Inc., October 2002.

_____. *Demolition: Manual vs. Explosives.* St. Paul, MN: St. Paul Companies, Inc., August 2002.

U.S. Army Corps of Engineers. *Safety and Health Requirements Manual* (EM 385-1-1). Washington, DC: Government Printing Office, September 3, 1996.

U.S. Department of Labor, Occupational Safety and Health Administration. *Safety and Health Standards for Construction, (29 CFR 1926 Subpart T, Demolition; Subpart U, Blasting and the Use of Explosives).* Washington DC: U.S. Government Printing Office.

REVIEW EXERCISES

1. Discuss the importance of preparatory operations to the overall safe execution of any demolition project. Give specific examples of items that should be included in all preparatory operations for a demolition project.
2. What are the some of the effective means for eliminating potential dangers in demolition projects? Give a specific example of the means used before, during and after the demolition activities.
3. Discuss the role of *engineer* in development of the demolition plan. Give two examples of key issues that should be addressed by an engineer.
4. Discuss the different types of health hazards associated with demolition operations. List some of the controls that can be used to reduce the exposures to these health hazards.

■■ **About the Author** ■■

CONCRETE
by Robert Eckhardt, P.E., CSP

Robert Eckhardt, P.E., CSP, received his Bachelor of Science and Master of Science degrees from Angelo State University. An author of numerous publications, he currently serves as monthly EHS Contributing Editor to *Concrete Products* magazine, and has also authored the book *Safety and Health in the Concrete Products Industry.* He is credited with establishing a universal tie-down procedure for concrete pipe transportation that was tested by the CVSA and later adopted by the DOT. Mr. Eckhardt has been honored with two President's Awards from the Gulf Coast chapter of the ASSE and he also received the Chapter and Region Safety Professional of the Year Awards. He has served on numerous committees with the Texas Safety Association as well as having served as a member of its board of directors.

Concrete

INTRODUCTION

Since concrete is the second most widely used building substance on earth, it is no wonder that almost every construction project includes concrete work. Grade beams, foundation piers, slabs, wall panels, masonry, and roads are only a few of the uses of concrete familiar to construction. Many construction projects utilize cast-in-place structures, although there are numerous other structures that are not commonly cast on the job site, such as wall panels, bridge beams, concrete pipe, culverts, utility vaults, and specialty structures cast at concrete products plants. Once these structures arrive at the job site, however, they become part of the construction process during installation.

Construction projects that mix concrete on site are usually the very small or the very large construc-

tion projects. Small projects still mix concrete on site with a mixer and bagged cement, while some very large construction projects, using many thousands of yards of concrete, will mix the concrete on site using portable batch plants. By far the most common method of delivering concrete to the construction site is via the ready-mix truck. Ready-mix plants are located throughout the United States and are designed for delivering only a few yards up to very large quantities of concrete to construction projects any time of the day, and usually on very short notice.

For these reasons, this chapter primarily addresses the hazards associated with working with ready-mix concrete in the construction environment. Safety in concrete batch plants is better addressed in general industry publications.

GENERAL SAFETY CONSIDERATIONS

Accidents associated with concrete construction include caustic burns caused by skin contact with wet concrete; cuts from handling rebar and tie wire; back strains typically caused from handling, installing, and tying rebar; and trips and falls. Trips and falls occur under many circumstances, including tripping

▪▪ Table 1 ▪▪

Summary of OSHA Requirements

Rule	Section Number
Placing loads on concrete structures: No load can be placed on any concrete structure unless "a person who is qualified in structural design" identifies the structure as being capable of carrying the load. **Exposed rebar and wire ends:** • Protruding rebar, which could be fallen upon, must be protected to prevent impalement. • Wire mesh must be uncoiled with the exposed end down, or securing the end to prevent cuts and punctures. **Post-tensioning:** Do not stand behind the jack when post-tensioning. **Riding buckets:** Do not ride concrete buckets. **Suspended loads:** Do not work or walk under a suspended load.	701 General
Bulk container entry: Bulk containers must have a mechanical or pneumatic means to start the material. Permit-required confined space entry procedures must be observed when entering a bulk container. **Loading skips 1 cubic yd or larger:** Must have mechanical skip clearing device and guardrails on each side of the skip. **Powered concrete trowels:** Must be equipped with a hand safety release switch. **Concrete buggies:** Shall not extend beyond the wheels on either side. **Pumping pipes:** Must be designed for 100% overload. **Concrete buckets:** Must be equipped with positive safety latches. **Bull float handles:** Must be nonconductive if overhead electrical wires are present. **Masonry saws:** Must be equipped with a saw guard.	702 Equipment and Tools
Drawing and plans: Must be at the job site for jack work, formwork, reshoring, and scaffolding. **Shoring equipment inspection:** Required before each use, during concrete placement, and immediately after concrete placement. Damaged equipment must not be used. Additional shoring shall be added for any equipment damaged or weakened after pouring. **Tiered single-post shores:** • designed by a qualified designer and inspected by an engineer qualified in structural design • vertically aligned • spliced to prevent misalignment • adequately braced in two mutually perpendicular directions at the splice level. Each tier must also be diagonally braced in the same two directions. **Shoring equipment:** Must be properly sized and used correctly. **Reinforcing steel columns:** Must be adequately braced to prevent tipping or collapse.	703 Cast in Place
Panel lifting inserts: Must be designed with a 2x safety factor. **Other inserts:** Must be designed with a 4x safety factor. **Lifting hardware:** Must be designed with a 5x safety factor. **Suspended panels:** Do not walk under suspended panels.	704 Precast Concrete
Lift-slab operations: Must be designed by a PE. **Lift-slab plans:** Must include the method of erection and lateral stability during construction. **Jacks and lifting units:** Must be designed to stop lifting when exceeding lift capacity. **Leveling:** • Jacking operations must be coordinated within $1/2$ inch to level. • Manual leveling must be conducted by a competent person. • Automatic leveling must be able to stop the operation at $1/2$ inch out of level. **Jacks/lifting units:** "The maximum number of annually controlled jacks/lifting units on one slab shall be limited to a number that will permit the operator to maintain the slab level within specified tolerances. . . ." **Temporary connection:** When making temporary connections to support slabs, wedges shall be secured with tack welding by a certified welder or equivalent method.	705 Lift-slab Construction
Limited access zone for wall construction: • zone established prior to start of construction • shall be the height of the wall plus 4 feet • must be on the opposite wall side of the scaffold • entry limited to masonry workers. **Limited access zone on walls over 8 feet:** • must be adequately braced • bracing must remain in place until permanent supports are in place.	706 Masonry

while walking on tied rebar mesh, loose construction materials in the work area, and water and form oils that make walking surfaces slippery. Once crews are accustomed to working with concrete and become familiar with the associated hazards, they can work safely, achieving excellent safety records.

Typical regulatory safety topics that involve concrete workers include the following:

- hazard communication
- hearing protection
- respiratory protection
- other personal protective equipment
- confined space entry
- excavation safety
- hand- and power-tool safety
- mobile equipment safety
- fall protection program, ladder and scaffold use
- lockout/tagout if working with mixers or batch plants
- safe operating procedures for equipment used, whether working with a conveyor truck, pump truck, ready-mix truck, prestressing operation, or other specialized equipment
- concrete and masonry construction.

OSHA regulations applicable to concrete and masonry are contained in 29 CFR 1926, Subpart Q, in sections 700 through 705, and are summarized in Table 1.

In addition to regulations specified in OSHA's 29 CFR 1926, Subpart Q, many parts of other OSHA regulations apply to concrete crews, in addition to standards and guidelines issued by the Prestress Concrete Institute (PCI), the American National Standards Institute (ANSI), American Society of Concrete Constructors (ASCC), and the Associated General Contractors (AGC). The knowledgeable safety professional will know that only a fraction of safe work performance is presented by the OSHA federal regulations alone. Table 2 provides an example of training requirements for typical concrete crews.

Personal Protective Equipment (PPE)

CUT PROTECTION

When rebar is cut with shears, the cut end typically leaves a razor-sharp ridge. Similarly, when rebar is cut with a metal saw, a sharp edge is frequently left on one side of the rebar. This tendency from mechanical cutting is the source of many, usually superficial, cuts.

Many cuts are reported every year and there are a variety of affordable, comfortable, cut-proof gloves that are available from all major safety suppliers. They are inexpensive and very comfortable and workers often prefer cut-proof gloves to leather gloves. However, the manual dexterity required to tie rebar or use hammers and nails for form construction sometimes precludes the use of gloves for these jobs. The use of rotating equipment such as drills or saws also precludes the use of gloves. Cut-proof gloves are porous and do not protect the hands from wet concrete.

■■ Table 2 ■■

Training Requirements

Subject	Source Recommendation or OSHA Requirement
Back injury prevention	ASCC
Compressed air	1926.803, PCI
Concrete forming	PCI
Concrete placement	PCI
Fall protection	1926.503
Fire protection	1926.150
General safety and health provisions	1926.20
Hazard communication (construction)	1926.59
Hazards of concrete	OSHA Special Emphasis Program, AGC
Ladders	1926.1053
Hearing conservation	1926.101
Medical services and first aid	1926.50
Overhead lines	1926.955
Power-operated hand tools	1926.302, PCI, ASCC
Personal protective equipment, general for eyes and face, head, feet, and hands	PCI, ASCC, ANSI Z87.1, Z89.1, Z42.1
Prestress equipment and care	PCI
Pretensioning	PCI
Respiratory protection	1926.103, ANSI Z88.2
Scaffolding personnel	1926.451, ANSI A10.8
Silicosis, causes and prevention	OSHA Special Emphasis Program on Silicosis
Slings	OSHA 3072, ANSI B30.9, B30.10, B30.20
Vapors, dusts, and mists	1926.55
Welding (arc) and cutting operators	1926.351, .352
Welding (gas) and cutting operators	1926.350, .352

EYE PROTECTION

The requirement for the use of safety glasses is meant to address the risk of concrete splatter entering the eyes. Safety glasses should be equipped with side shields to protect the eyes from the caustic, corrosive nature of wet concrete. The application of a face shield should be considered for grinding or chipping applications in addition to the use of standard safety glasses. A face shield and hard hat are required when pumping concrete.

HEARING PROTECTION

Before requiring the use of hearing protection, a noise problem should first be approached by reducing the noise at the source, or isolating the noise from the worker. Some of the following simple noise-reduction steps can be exercised to reduce exposure.

1. Position noisy equipment, such as ready-mix, conveyor, and pump trucks with the front of the truck pointed in an unoccupied direction. Much of the noise from this equipment comes from the front of the truck through the radiator.
2. Replace burned-out mufflers on mobile equipment.
3. Install all manufacturers' recommended noise-suppression equipment on powered hand tools.
4. Use remote wire-controlled tampers when possible rather than the gasoline-powered jumping-jack tamper.

Hearing protectors are typically required for workers operating or working near vibrators, wood and metal saws, and tampers, or working in the immediate vicinity of the engine of a ready-mix, conveyor, or pump truck.

FOOT PROTECTION

Concrete producers should consider the requirement for steel-toed shoes with a nonslip sole and a steel shank in the sole of the shoe to protect against nail and wire punctures. Steel shanks are standard equipment in most brands of safety footwear. Toes can be protected from hazards like dropping tools such as sledgehammers, boards, rebar, and other equipment on them.

During a concrete pour, workers should wear rubber boots that are equipped with a steel toe. Care should be taken to inform all workers regarding the hazard of getting a small amount of concrete down the rubber boot and continuing to work with the concrete inside the boot. The abrasive nature of rubbing the concrete against the foot during walking can cause a severe caustic burn. To prevent such an event, many workers tape the tops of their boots to their pants if the pour is deeper than about 6 inches.

When using the jumping-style tampers, a metatarsal guard should also be used in addition to the steel-toed shoe to protect against an accidental foot injury due to a misguided compression.

KNEE PROTECTION

A terrific invention that became popular in the 1970s was the curved, reinforced kneepad for employee use in jobs that require kneeling. The pads distribute the weight evenly on the knee, which keeps the knees from becoming sore. When kneeling, they also protect the knees from exposure to cement dust, concrete, and sharp objects.

Severe caustic burns to the knees can occur when novice concrete workers get their knees wet with concrete and continue to work in the kneeling position. Kneeling causes the caustic, soaked pants to work against the skin, creating a severe burn.

HEAD PROTECTION

The use of hard hats should be required when employees are working in areas where an object could be dropped from above and when pumping operations are being conducted. Headgear also protects the worker from electrical shock and burns to the head in some circumstances.

HAND PROTECTION

The most common, preventable hand injuries are caustic burns, minor cuts, and abrasions. Leather gloves or cut-proof gloves are frequently selected for protection against minor cuts by wire, nails, and rebar, and abrasions from handling cured concrete

▪ ▪ **Table 3** ▪ ▪
Respirator Requirements

Operation	Respirator Required	Comments
Pneumatic unloading to storage silo	No	Spilled cement in the area can be a source of fugitive windblown dust and care should be taken to prevent spillage.
Feed hopper loading	No	Feed hoppers must have negative air pressure systems exhausting through a filter system. Working near feed hoppers without this system may require the use of respirators.
Mechanical mixing	No	Mixers must seal tightly and must be equipped with a filter system for the displaced air created when the materials are dropped into the mixer. Otherwise respirators are probably needed.
Manual mixing	Yes	Bagged cement or concrete mix contains respirable silica and Portland cement dust.
Pouring and finishing	No	
Abrasive blasting	Yes	Abrasive blasting should be conducted with nonsilica or some low-silica abrasives. Respirators must be hooded with supplied air.
Sawing concrete dry	Yes	Dry-sawing should not be conducted.
Sawing concrete wet	No	
Sawing wood forms	Sometimes	Proper direct ventilation usually eliminates the need for respirators
Grinding dry	Yes	Vacuum systems on hand grinders should be used.
Grinding wet	No	

products such as block. There are abrasive-resistant coatings provided on some fabric gloves that resist abrasion far better than leather. Nonpermeable gloves are typically recommended for new concrete workers since they provide protection against wet, caustic concrete or other materials. Most skilled concrete workers have learned how to keep their hands free from contact with the wet material and prefer to use leather or breathable cut-proof gloves when pouring concrete.

RESPIRATORY PROTECTION

Concrete contains large amounts of silica that can become airborne when concrete is crushed to dust. Cement also contains high silica concentrations. Silica, in a respirable form, can cause a debilitating and permanent lung disease known as silicosis (in addition to other ailments and cancers) if it is breathed over a period of time. It is important to be aware of dusty situations when working with concrete or cement and take precautions to eliminate the dust problem or, as a last resort, utilize adequate respiratory protection if the dust problem cannot be eliminated. Table 3 presents a listing of general processes and requirements for respiratory protection.

Concrete spilled on the job site should be expeditiously cleaned up since bits of concrete that are in areas where they can be repeatedly crushed and pulverized into very fine particles by mobile equipment can create a significant dust problem. Many times spilled concrete is simply mixed or crushed into the foundation soil, although this can present problems with compaction, depending on how the material is broken and the soil selected for fill. Spilled concrete and dried concrete-mixer washout material are actually considered construction waste. They should be disposed of in a dumpster designated for concrete so it can be forwarded to a local recycling plant or used as landfill, depending on local environmental regulations.

Welding fumes from cutting mild steel, such as wire mesh, cage wire, and rebar do not present a significant respiratory problem if local ventilation is used. This may include an area box fan for outdoor applications or a fume extractor for indoor applications.

Wet concrete generates no need for the use of a respirator. However, since concrete and cement both contain silica, dust generated from cement, or sawing or grinding solidified concrete, should be monitored by an industrial hygienist or similar professional who can measure and document the presence or lack of these dusts in accordance with regulatory limits. Typical operations will not create overexposures to respirable dusts (particle size <10 microns), although some may have overexposures to specified limits for nuisance dusts (particle size >10 microns). Personnel who use handheld grinders typically exceed the limits for both respirable and nuisance dusts, especially when a pneumatic exhauster is absent. Workers who saw concrete are also exposed to dust concentrations that often grossly exceed the regulatory limits, although these conditions can be controlled through wet-sawing. If either respirable or nuisance dust exposures exceed the regulatory limits, respirators must be carefully chosen for each application. Paper dust masks do not provide adequate protection against respirable silica dust. In addition, a silicosis program and/or a written respiratory protection program must be implemented in conjunction with the requirement for respiratory protection.

Sawing wood to construct forms can also generate significant finely divided wood dust, which is now recognized as a potential health hazard. This hazard can typically be controlled with a ventilation fan, such as a box fan, or, if controls cannot be used, then the worker should be provided with a NIOSH-approved respirator for wood dust.

Supervisors of projects requiring abrasive blasting of concrete structures should take particular care with respiratory protection. Although sand works very well as a blasting agent, nonsilica blasting agents are recommended. In addition, the person operating the blasting wand must wear a hood with supplied air when blasting concrete. This practice requires a well-prepared respiratory protection program that addresses training, exposure control to the operator and others working in the area, equipment, and other factors.

BODY PROTECTION

Cement or concrete dust against damp, perspiring skin can cause an irritating dermatitis. In fact, some people are allergic to cement, creating a problem for which there is no known solution—except to find other employment or jobs that do not expose the worker to wet concrete, concrete dust, or cement dust. Most construction crews wisely require employees to wear long pants and shirts that cover the shoulder. Although long-sleeved shirts are preferable for protection against possible concrete splatter and sunburn, the construction supervisor might be best advised to evaluate the requirement for long shirts against local customs since the actual risk of significant exposure is relatively minor, except with sensitized individuals. Sunburns to the tops of exposed shoulders in summer months can be prevented by the requirement of shirts that cover the shoulder.

Fall Protection

Fall hazards are common to concrete workers, not only from slips and trips addressed earlier, but also due to exposures from constructing or removing forms. Fall prevention calls for the use of ladders as opposed to climbing on the forms. The forms should not be used as a work platform since utilizing ladders or scaffolds is by far the safer practice.

Due to the tendency of rebar to develop a very sharp edge when cut with either shears or a steel saw, rebar caps should be placed on the ends of any rebar that protrudes upward. This protects a worker from impalement should the worker fall onto the rebar. Wire mesh should also be unrolled so that the arc of the mesh points the sharp, cut end into the ground.

Confined Space, Trench, and Excavation Considerations

All concrete crews eventually come into contact with confined spaces and excavations. This chapter will not elaborate on the confined space entry procedures or excavation safety procedures, except to point out that all concrete crewmembers must be trained in the dangers and procedures of entering confined spaces and excavations. A responsible, competent person in each category must be involved with the jobs for which each concrete crew is working.

Toxicity Issues

Procedures that include cutting or welding coated rebar need extra precautions for ventilation and even respiratory protection, if necessary, to protect against the smoke from the coatings. The coatings typically include a variety of chlorinated organics and other carcinogenic compounds. Additionally, operations that must weld stainless steel components into the forms or rebar need to take similar precautions since welding stainless steel can emit chromium welding fumes that are also carcinogenic and toxic to the liver.

A variety of chemicals used in concrete should also be reviewed for safe work practices and appropriate personal protective equipment. These chemicals are outlined in Table 4.

Due to the toxicity of these materials, the material data sheet should be consulted before use, in order to minimize the health hazards and protect workers with the proper personal protective equipment. In addition, storage requirements for many of these compounds are specific in nature.

Ergonomic Considerations

Existing technology for universal tool improvements offer significant improvements for the concrete in-dustry. These include the use of handheld tool isolation vibration dampers, such as padded handles on vibrators, padded gloves, or improved tool handle size and design, and better procedures for reducing concrete build-up on portable batch plant mixers and equipment since the removal of this material requires varying degrees of lower-torso and upper-extremity physical effort.

The fabrication of cages involving the tying of rebar can be vastly improved by elevating the work platform to approximately knee level so that the fabricator works at waist level and does not stoop or squat as frequently. Constructing and tying cages on the ground is grueling on the feet, knees, and back. The first improvement in wire-tying was the looped tie wire, which was connected with a manual tie wire-twister. Battery-operated, automated, wire-tying tools are quick and keep the worker from a continual wrist-twisting action with pliers or the manual, swiveling, wire-twister. A simple carpenter's apron for the looped wire, or placing the box of looped wire on a waist-level drum or platform, saves repetitive stooping to pick up the ties.

■ ■ **Table 4** ■ ■

Chemicals Commonly Used in Concrete and Their Toxic Properties

Product(s)	Typical Ingredients	Hazards
Curing compounds	Either resin based or chlorinated rubber based. Includes mineral spirits, organic solvents (xylene and/or toluene).	Vapor inhalation hazard to internal organs. Skin absorption hazard to internal organs. Skin rashes and burns.
Bond breakers	Mineral spirits, organic solvents (xylene and/or toluene). Some contain VM&P naphtha.	
Retarders	Solvent vehicle such as toluene.	
Sealants, waterproofing agents, and adhesives	Silicone, asphalt-based liquids and epoxies. Sometimes Polyurethane and polysulphide compounds.	Wide variety of health hazards. Irritants to eyes, nose, throat, and skin. Vapor inhalation hazard to central nervous system and internal organs. Asphalt-based carcinogenic hazards.
Silicone coatings and bonding agents	Solvent base used singularly, or mixtures of acetone, hexane, xylene, toluene, etc.	
Sealers	Contains methylene biphenyl diisocyanate or diisocyanate. Lead in the paste.	*See material safety data sheet for specifics.*
Epoxy compounds in coatings, grouts, binders, sealants, bonding agents, patching materials, and adhesives	Phenylallyl, butyl glycidyl ether, styrene, methylene dianiline, and epoxy.	

Source: Associated General Contractors of California. *Health Hazards Associated with the Use of Concrete and Mortar, A Manual for the Construction Industry.* AGC of California, March 1985.

Masons are at real risk of musculoskeletal injury from numerous repetitive motions involving significant weight. These include the repetitive action of mixing mortar, carrying mortar, applying mortar, and placing block. The key element to reducing fatigue is the layout of the job site so that the movement of block and mortar is limited in carry distance. This includes placing mortar on a mortarboard positioned at waist level. Placing mortar on a board at foot level simply creates unnecessary repetitive stooping and bending. Lightweight concrete block and a cast, lifting handle on the center divider of the block have been great ergonomic improvements for masons.

Shovels are indispensable tools in the concrete business, but they are a back-breaker since they operate as a lever with the back serving as the fulcrum. The best solution to shoveling is to eliminate the reasons for which shovels are used. This includes the careful prejob planning of concrete placement during slab pours and care in preventing concrete spillage.

Using a shovel should be a last resort for cleaning up spills. The use of the small Bobcat front-end loader has been a terrific improvement in eliminating the manual shoveling of concrete spills. Short-handled scoop shovels should be avoided whenever possible since they require working in a stooped position, although when their use is unavoidable, they should always be equipped with a D-ring handle to assist the wrist since the wrist is forced into a supporting action for the back. Shovels with broken handles should be discarded.

Bull floats constructed with aluminum handles for ease of handling are ideal, except that the OSHA standard specifically prohibits the handle from being constructed of a conductive material. Therefore, a nonconductive handle such as a lightweight, hollow fiber material is ideal. A rubber cover over the end of the long handle is also recommended since the handle tends to protrude into others' workplaces during use and could injure a worker. Replacing or repairing a float handle with a wood handle creates an unnecessary physical burden to the concrete finisher's shoulders and arms and therefore should not be considered as a replacement for a broken handle unless the float handle is relatively short.

■ ■ Table 5 ■ ■

Flammable Components of Chemicals Used with Concrete

Material	Flammable Components
Curing compounds	Either resin based or chlorinated rubber based and includes mineral spirits and/or organic solvents such as xylene and/or toluene.
Bond breakers	Mineral spirits, organic solvents (xylene and/or toluene). Some contain VM&P naphtha.
Retarders	Contains a solvent vehicle such as toluene.
Sealants, waterproofing agents, adhesives, nonwater-based glues	Silicone, asphalt-based liquids and epoxies. Sometimes Polyurethane and polysulphide compounds.
Silicone coatings and bonding agents	Solvent base used singularly, or mixtures of acetone, hexane, xylene, toluene, etc.
Cleaners	Various, including xylene, alcohols, or acetone.
Lubricants	Oils and greases used on slip tie rods, spray lubricants such as WD-40.
Fuels	Diesel and gasoline used in compressors and small engines.

Source: Associated General Contractors of California. *Health Hazards Associated with the Use of Concrete and Mortar, A Manual for the Construction Industry.* AGC of California, March 1985.

Fire Prevention Considerations

Fires associated with concrete are not common, although many materials associated with concrete are flammable. Fires that do occur with flammable liquid components are typically associated with the mishandling of the material while pouring. Grounding the pouring container and the receiver container can usually prevent these fires. Table 5 is a listing of various materials associated with concrete and their flammable components.

GENERAL FORMWORK

Formwork must be capable of supporting without failure all vertical and lateral loads that might be applied to it. Formwork must be designed, fabricated, erected, supported, braced, and maintained in conformance with Sections 6 and 7 of the American National Standard for Construction and Demolition Operations—Concrete and Masonry Work (ANSI A10.9, 1983).

Oiling Forms and Equipment

A variety of oils are used in the oiling process, including supplier-grade release agents, or form oils, specifically used for nonstick purposes. Most of these supplier form oils are petroleum-based products. The application of the oil is normally conducted with hand-pump garden sprayers. It is important that the MSDS form for the particular oil be carefully evaluated since most of the oils are combustible and will expand the storage drums if left in the sun or other hot environment. These drums are required to have a pressure-relief bung, and they must be stored out of the sun. Some agents also include xylene, naphtha, isopropyl alcohol, or other solvents that pose other health problems and therefore require special personal protective equipment, including dual-cartridge respirators with organic filters and full body protection, plus the requirement for an emergency shower. Simply avoiding the use of form oils with additives such as these will eliminate the additional hazards. Depending on the oil used, the spray operator is generally required to use safety glasses with side shields or goggles and long latex or neoprene gloves.

Drums should not be handled manually, but should be handled by "mount and tilt" drum racks equipped with drip pans, mobile lifting equipment, or other ergonomic handling equipment. The drums should be raised onto elevated racks by forklift or backhoe with the proper rigging. All racks should be equipped with ground wires to the drum and receiving container to prevent the accumulation of a static charge during dispensing.

Vertical Slip Forms

The following excerpt is taken from the OSHA bulletin 3106:

> The steel rods or pipes on which jacks climb or by which the forms are lifted must be specifically designed for that purpose and adequately braced where not encased in concrete. Forms must be designed to prevent excess distortion of the structure during the jacking operation. Jacks and vertical supports must be positioned in such a manner that the loads do not exceed the rated capacity of the jacks.
>
> The jacks or other lifting devices must be provided with mechanical stops or other automatic holding devices to support the slip forms whenever failure of the power supply or lifting mechanisms occurs.
>
> The form structure must be maintained within all design tolerances specified for plumbness during the jacking operation. The predetermined safe rate of lift must not be exceeded. All vertical slip forms must be provided with scaffolds or work platforms where employees are required to work or pass.

Removal of Formwork

Formwork removal for vertical surfaces must only be conducted in accordance with the plans and specifications that stipulate their removal, or after the concrete has been properly tested using ASTM testing methods and has proven to be strong enough to remove the forms. Nails should be removed and the boards either stacked for reuse or discarded.

MIXING AND BATCHING

Second only to ready-mix delivery operations, the best safety invention for large projects in the concrete construction industry has been the automated mixer, which has undoubtedly prevented many back injuries and falls, as well as silicosis and dermatitis cases. This invention has also saved production time and manpower costs and improved quality, although a new safety problem was introduced to the workplace. Since mixers are heavy, powered equipment, they can cause horrific injuries and fatalities when a batch plant operator gets caught in the moving blades or gates.

Three basic considerations for batching safety include: (1) engineering out hazards by providing the available safety equipment for mixers; (2) providing proper maintenance of that equipment; (3) the application of administrative procedures, which includes proper maintenance and well-defined lockout procedures (control of hazardous energy), followed by adequate training and supervision. If hazards are engineered out and adequate maintenance, training, and supervision are provided, then the risk of a life-threatening accident is remote.

The basic causes of mixer accidents usually include, either ignorance regarding the operation of the equipment, or the assumption of too much risk on the part of either the company or the operator.

1. Proper mixer safety equipment is not provided or not maintained.
2. Adequate lockout/tagout procedures are not established and enforced. (Mixers require lockouts/tagouts for the control of electric, mechanical, pneumatic or hydraulic and gravity energy sources. Not just an electrical motor switch lock!)
3. Adequate training and supervision are not provided.
4. Personal protective equipment is not provided.

Other hazards associated with mixers include muscle strains that are frequently caused by the ergonomic design of mixer accessibility for maintenance and cleaning. Ideal mixer design includes side and top units that hinge open allowing access to the interior mixer parts while the mixer cleaner stands up, outside the mixer. The floor of the mixer should be positioned at a person's waist level as he or she stands on the deck, which would help prevent repetitive stooping and bending injuries.

All safety equipment recommended by the manufacturer should be provided with the system. Depending on the type of batch plant, the safety equipment in Figure 1 is usually recommended.

PRESTRESS OPERATIONS

Prestress operations are not addressed in the construction standard since these operations are usually conducted at prestress plants. However, safe operating procedures for every portion of the stressing operations are explicit due to the danger of extreme amounts of stored energy in the stressing procedure. The following information is provided since prestressing concrete is gaining popularity in

1. Electric drive motors must have a mixer motor lockout switch mounted next to, or on, the mixer, and the switch clearly labeled.
2. For operations designed with the batch controls being located away from the mixer, a mixer service control box should be installed at the mixer for accessible lockout/tagout of the electrical power by maintenance and cleaning personnel.
3. Red mushroom emergency stop buttons should be installed on either side of the mixer and clearly labeled.
4. Manual disconnect hose couplings must be provided at waist level for all pneumatically controlled doors and gates.
5. An easy and quick disconnect must be provided for all mechanical couplings on the mixer pneumatic or hydraulic arms that operate the gates. A lock must be installed in the ram coupling, which will prevent the ram from being reconnected while a person is inside the mixer.
6. Weigh hopper feed and discharge gates must be mechanically locked closed and the controls locked/tagged out.
7. A double-acting solenoid must be provided on the gate and controls.
8. An exterior concrete batch-sampling tray should be provided on the mixer.
9. The mixer doors should be interlocked, that is, equipped with an automatic power cut-off switch that shuts down the electrical power to the motors whenever the doors are opened.
10. In the United States, and more commonly in Europe, a time-delay mechanism on the electrical switch is installed with an audible alarm near the mixer to alert personnel in the vicinity that the mixer is about to be started.
11. In addition to the alarm, a revolving light is also placed on the mixer to indicate when the mixer is in operation. However, equipment suppliers in the United States reasonably argue that the noise and vibration of the mixer itself is ample notice.
12. When the material enters the hopper, the displaced dust-laden air must be safety handled so that it will not enter the breathing air. This precaution will include installing seals on the door edges to the mixer with sealing latches to provide an airtight compartment. The displaced air can be handled with either a simple bellows dust bag (which may not contain microscopic respirable silica) or an exhaust fan to collect the dust and transport it to a baghouse for particulate removal and external exhaust. To prevent dust exposures to the batch plant operator, both the mixer and the weigh hopper must have a dust control system and the system must be adequately maintained.

■ ■ **Figure 1** ■ ■ **Batch plant safety equipment**

residential construction and in some commercial applications. This chapter cannot contain the lengthy procedural and safety requirements to conduct prestress operations safely. The author therefore refers the reader to the Prestress Concrete Institute, to provide safety information regarding stressing operations. A brief summary of stressing operations follows to impress upon the prospective prestress worker the intricacies of safe operation.

Chucks and Components

Chucks and chuck components should be cleaned, inspected, and lubricated as recommended by the manufacturer, and on a schedule that is recommended by the manufacturer. Worn, bent, cracked, or broken components must be replaced before returning them to service. Chucks with excessive hammer dents should also be replaced.

Strand

The strand is subject to precision strength capacities; therefore, it must be handled carefully so that it is never dropped or banged. It should be stored off the ground and kept dry. Strand used for stressing should be inspected as it is being placed to detect faults that might cause strand failure when under tension. Defects include excessive corrosion, nicks, broken wires, kinks, and crushing. Strand should never be subjected to welding, cutting, or even sparks or slag drip since these heat processes can compromise the structural integrity of the strand. Strand storage should be away from welding operations because welding sparks, drips, or slag can permanently destroy the structural integrity of the steel. These flaws in the wire are not always visible.

Cutting the strand itself should be conducted with a fast-cutting abrasive wheel. Strand vices used to anchor strand at the pulling head should be placed at least 12 diameters of the strand away from a burned end (usually 6 inches or more). All strand vices should be checked to assure that they are gripping securely and are in line with the intended pull.

Jack

The jack should be carefully aligned so that it is parallel to the strand before and during tensioning. Stressing should not require that any personnel are in the area of the jack. No one should stand behind the jack or in front of a chuck because a strand failure will cause it to retract along the line of pressure and will recoil the jack and chuck away from the pulling head at a dangerous velocity.

When removing the jack, the crew should observe the chuck to assure it is holding the strand securely. Strand elongation must be accurately measured. Variances in elongation or gauge reading from that calculated by the prestress engineer should be reported immediately to the prestress engineer so a solution can be derived to prevent overstressing. When the stressing jack is removed from the strand, the crew should watch the chuck to assure it is holding the strand.

Releasing strand tension can be conducted through dejacking or detensioning, depending on the application. Dejacking is a dangerous operation that should be avoided if possible and conducted with utmost care. Detensioning is similarly a dangerous operation, although precautions can be taken prior to stressing to minimize the need for detensioning. Preplanning and the utilization of special fixtures can greatly reduce the exposures to injury.

PRECAST CONCRETE

All precast units require handling, which differentiates them from being cast in place. The two critical elements for safe handling include gaining the assurance that the concrete has adequate strength before handling, and that the handling procedures are sound. The amount of hardness necessary for a precast unit to be lifted depends on the design engineer's specifications. All precast units (including tilt-up units), should be strength-tested before handling to assure they meet the engineer's minimum strength requirements for lifting. Lift-slab construction regulations require they be designed by a registered professional engineer and

Step	✔
1. Verify the crane operator's credentials and the qualifications of the rigger, lift supervisor, and crew.	
2. Confirm that the lift plan and risk assessment is complete.	
3. Verify that the placement structure and the panels have adequate strength. Typically, the quality assurance engineer can verify the integrity of the structure before lifting.	
4. Barricade the lift, swing, and potential fall area, and provide a safety spotter to keep onlookers and passing pedestrians away from the lifting zone.	
5. Verify correct placement of locating dowels, leveling shims, and the immediate availability of temporary bracing. To avoid the problem of erecting and removing temporary bracing, many crews conduct the permanent weld to the structure before the crane lets go of the panel.	
6. Verify that the temporary base restraint is provided for any precast unit to prevent a sliding failure (kick-out) at the base or support of the unit.	
7. If strongbacks (stiffbacks) are used, verify that the rigger understands the correct installation of the strongbacks per the drawings or specifications for the unit.	
8. Inspect the inserts and verify that they are clean and that all inserts were cast in the correct location. Typically, the quality assurance engineer can verify the correct location of the lifting inserts.	
9. Inspect the crane, crane prelift checklist, overhead lines, adequacy of mudsills, counterweight swing, other crews working in the area, pick points, and placement site. The lift plan must have a specified wind limit for safe lifting.	

■ ■ **Figure 2** ■ ■ **Preparation checklist for tilt-up panels**

are separated in the OSHA standard from precast concrete. Figure 2 provides a preparation checklist for tilt-up panels.

Once these procedures have been completed, the safety rules in Figure 3 should be observed *during* the lift:

READY-MIX CONCRETE PLACEMENT

At the construction site a person should be assigned to direct the driver of the ready-mix truck while the truck is backing up and during other close-quarter operations. Wearing an orange vest should easily identify the driver signal person. The National Ready Mixed Concrete Association has established universal hand signs for easy communications between the driver and the signal person. Duties of the signal person include the following:

1. To assure that the area behind the truck is safe for backing operations.
2. To check for overhanging lines (also the driver's responsibility).
3. To allow plenty of room from the edge of an excavation to prevent the weight of the truck from collapsing the excavation wall.
4. To assure the truck is never moved with the chute in the extended position.

In accordance with the rules provided by the American Concrete Pumping Association, the

Step	✔
1. The lift supervisor must be in charge of the lift.	
2. Barricade the fall zone of the panel where it will be positioned for final placement, in the event temporary bracing should fail.	
3. Keep all persons away from the lifting zone except essential personnel. It is necessary that essential lifting personnel always face the panel and be alert to any signs of structural failure such as cracking.	
4. No one should ever be allowed on the drop side of the panel being lifted.	
5. When taglines are used, personnel should position themselves away from underneath the load. No one should allow any part of their body to be under the unit.	
6. Assure that temporary bracing is correctly positioned in accordance with the structural engineer's specifications.	
8. Inspect the inserts and verify that they are clean and that all inserts were cast in the correct location. Typically, the quality assurance engineer can verify the correct location of the lifting inserts.	
7. Panels should be welded into place as soon as practical.	

■ ■ **Figure 3** ■ ■ **Lift procedure checklist for tilt-up panels**

1. Do not kink the line.
2. Do not open clamps under pressure when the hose is plugged.
3. Hose frays, breaks, or any sign of hose damage should be reported to the operator.
4. Hoses must not be dragged by the clamps.
5. The boom must never be used for pulling the concrete-placing hose or line.
6. Gaskets must always be used with clamps.
7. Plugged hoses typically build pressure behind the plug. The operator must always notify the hose handler if a plug occurs so personnel working in the immediate vicinity of the hose can move away from the hose until it clears. The ends of plugged hoses must always be pointed in a safe direction. Never look into the end of a plugged hose.
8. Never sit, straddle, lean on, ride, or stand on a concrete hose. The hose must never be rested on a person's shoulder.
9. Always start at the farthest point when installing placing line.
10. Make sure the lead hose person has good communications with the pump operator. All air-cleaning procedures should first be discussed with the pump operator.
11. The hose should be directed by a tag line so that a plug and sudden release of pressure will not cause the hose to whip and strike the hose handler.

■■ **Figure 4** ■■ **Concrete pumping: hose-handling procedures**

hose-handling safety rules in Figure 4 should be observed.

Pumping Accidents

One of the most frequent fatal incidents in the concrete products industry occurs to concrete pumping operators through electrocution. According to a Canadian research project conducted by Joseph-Jean Paques that studied 31 fatalities from electrocution of pumpers, the following preventive measures can be taken to reduce the hazard exposure.

1. Cut off the power and ground conductors or remove the mobile equipment or conductors in order to eliminate the possibility of contact.
2. Simple limit switches can limit rotation and elevation movement of the boom's first section. In addition, limiting boom movement requires the operator to set and adjust the system, which implies the operator has previously identified power-line presence and location.
3. A line detector can be used to alert the operator of a power line within 20 meters. Appropriate safety precautions can then be taken.
4. Operators and ground helpers should be trained to detect power lines and use appropriate safety measures. These protective measures include the following:
   • Use a fully insulated remote control panel operated via radio signals. The operator can then stand away from the vehicle and operate the boom and outriggers.
   • Place appropriate warning signs on the vehicle to warn personnel in the area not to lean against the truck.
   • Operators and ground helpers should be trained to seek power shutdown before rescue in order to reduce additional potential for injury once contact has occurred.

Although pumping equipment has extraordinary versatility in placing concrete, the use of a telescoping conveyor truck eliminates the inherent hazards associated with hose pressures.

CONCLUSION

Concrete workers are subject to a large number of training requirements, which means significant training is needed just to meet OSHA compliance. In addition to training, concrete and concrete components also possess toxicity hazards that require special considerations for PPE and training in concrete handling, pouring, and finishing. Concrete used in construction also involves the use of a variety of hand tools, power tools, and heavy equipment. For these reasons, concrete workers require significant training and supervision to ensure a safe construction site.

REFERENCES

American Concrete Pumping Association. "Hose Handling Safety Rules." Flyer, undated.

American Conference of Governmental Industrial Hygienists. *Industrial Ventilation*, 16th ed. Lansing, Michigan: ACGIH, 1980.

American General Contractors Association. *Health Hazards Associated with the Use of Concrete and Mortar*. AGC of California, March 1985.

American National Standards Institute. *Construction and Demolition Operations—Concrete and Masonry Work*, A10.9. New York: ANSI, 1983.

_____. *Eye and Face Protection*, Z87.1. New York: ANSI, 1989.

_____. *Fall Arrest Systems and General Fall Protection Requirements*, A10.14. New York: ANSI, 1992.

_____. *Men's Safety-Toe Footwear*, Z41.1. New York: ANSI, 1967.

_____. *Safety Requirements for Industrial Head Protection*, Z89.1. New York: ANSI, 1969.

_____. *Standard Practice for Respiratory Protection*, Z88.2. New York: ANSI, 1969.

American Society for Concrete Construction *Hazardous Materials*, Safety Bulletin. Northbrook, Illinois: ASCC, undated, p. 2.

_____. *Safety Manual for Concrete Constructors*. Northbrook, Illinois: ASCC, undated.

_____. *Tool Box Meetings*, Safety Manual. Northbrook, Illinois: ASCC, undated.

Eckhardt, Robert F. "Mixer Safety," *Concrete Products* magazine, November 1997, pp. 31–35.

_____. *Safety and Health in the Concrete Products Industry*, (numerous references). San Angelo, TX: ProSafe Environmental Publishing, March 2000.

Leszcynski, John. Personal communication with the author regarding mixer equipment safety, September 30, 1997.

Occupational Safety and Health Administration. *Concrete and Masonry Construction*, 29 CFR 1926.700-.706. Washington, D.C.: U.S. Government Printing Office, July 1, 2000.

_____. *Fall Protection*, 29 CFR 1926 and 29 CFR 1910. Washington, D.C.: U.S. Government Printing Office, July 1, 2000.

_____. *Hazard Communication Program*, 29 CFR 1926.59 and 1910.1200. Washington, D.C.: U.S. Government Printing Office, July 1, 2000.

_____. *Respiratory Protection*, 29 CFR 1910.134. Washington, D.C.: U.S. Government Printing Office, July 1, 2000.

_____. *The Control of Hazardous Energy (lockout/tagout)*, 29 CFR 1910.147. Washington, D.C.: U.S. Government Printing Office, July 1, 2000.

_____. *Concrete and Masonry Construction*, OSHA Bulletin no. 3106. Washington, D.C.: U.S. Government Printing Office, 1998.

_____. *Fall Protection in Construction*, OSHA Bulletin no. 3146. Washington, D.C.: U.S. Government Printing Office, 1995.

_____. *Training Requirements in OSHA Standards and Training Guidelines*. OSHA Bulletin no. 2254. Washington, D.C.: U.S. Government Printing Office, 1992.

(Available from area OSHA offices or via telephone at 202.219-8615 or the United States Government Printing Office, Washington DC.)

Paques, Joseph-Jean. "Protecting Concrete Pump Operators Against Electrocution," *Professional Safety*, journal of the American Society of Safety Engineers, May 1995, pp. 41–43.

Prestressed Concrete Institute. "Safety and Health Provisions," *PCI Safety and Loss Prevention Manual*. Chicago, Illinois: PCI, 1972.

_____. *Safety and Prevention Manual*, 4th ed., Chapter IV, "Pretensioning." Chicago, Illinois: PCI, 1976.

REVIEW EXERCISES

1. Name at least five health hazards associated with working with concrete.
2. What types of accidents are most common among concrete workers?
3. Name five or more major regulatory standards that apply to concrete crews.
4. What kinds of PPE should concrete workers be provided with?
5. What are the major hazards associated with mixers?

■ ■ **About the Authors** ■ ■

CONFINED SPACES
by Bradley D. Giles, P.E., CSP, and William Piispanen, CIH, CEA, CSP

Bradley D. Giles, P.E., CSP, is the Corporate Vice President of Environmental, Safety and Health (ES&H) for Washington Group International. He has more than 26 years of ES&H management experience in the engineering/construction and mining industries, and has worked in all of Washington's six business units. Mr. Giles has received several professional honors during his career and was named 1994 "Safety Professional of the Year" by the Construction Division of the American Society of Safety Engineers for his work as Program Manager for the Washington Group on the $3.1 billion Denver International Airport construction project. Bradley Giles graduated from Southern Illinois University, where he also earned a master's degree in industrial safety. He is a Certified Safety Professional and a Registered Professional Engineer.

William Piispanen, CIH, CEA, CSP, is the Corporate Industrial Hygienist for Washington Group International at the corporate office in Boise, Idaho. Formerly, Mr. Piispanen served as the ES&H Manager for the Boise office of Morrison Knudsen (MK) Corp. (now Washington Group International).

Prior to his employment with Washington Group International (MK), he was a Principal Research Scientist with Battelle Memorial Institute. Mr. Piispanen is a Diplomat of the American Board of Industrial Hygiene, designated as Certified Safety Professional by the Board of Certified Safety Professionals, and is recognized as a Certified Ergonomics Associate by the Board of Certification in Professional Ergonomics.

William Piispanen holds a B.S. degree from the University of Toledo and an M.S. degree from Washington State University. He is the author or coauthor of 26 publications in the fields of environmental sciences, aerosol science, and industrial safety.

Confined Spaces

INTRODUCTION

Confined space entry is potentially one of the most hazardous activities that can occur at any work site. A January 1986, NIOSH Alert document (DHHS Publication No. 86-110) cited the greatest risk of death during work in confined spaces was posed by would-be rescuers. As a result of this study, it was recommended that worker training for confined space activities include specific information on (1) recognition of hazards; (2) testing, evaluation, and monitoring of the hazards; and (3) preestablished rescue procedures.

On January 14, 1993, OSHA first issued a confined space standard for general industry under 29

CFR 1910.146. On December 1, 1998, OSHA issued a final rule, amending this standard to include a number of changes to air monitoring and the criteria for rescue services. In the construction standards, confined spaces are addressed in 29 CFR 1926.21(6) and assigned a pre-rule status as a standard under 29 CFR 1926.36, "Confined Spaces in Construction: Preventing Suffocation/Explosions in Confined Spaces" (OSHA Unified Agenda, 2002, 1218-AB47-1615).

The risks to employees are great unless proper work practices, procedures, controls, and training are required and strictly enforced. Since 1979 NIOSH has recognized the need for a safety standard that would address the hazards of confined space work.

An effective confined space program has two basic goals. The first is to control confined space hazards by using good engineering and work-practice controls. The second goal is to verify the safety of each confined space by instituting an aggressive program of atmospheric testing, monitoring, and inspection. Guidelines for the permit confined space entry, alternate confined space entry, and non-permit confined space entry are defined by OSHA in the 29 CFR 1910.146 standard as well as in other procedure documents.

APPLICABLE SAFETY STANDARDS AND REGULATIONS

Confined space work is covered by OSHA under 29 CFR 1910.146, "Permit-required Confined Spaces" (January 14, 1993) and an amendment to 29 CFR 1910.146 (December 1, 1998), and in a section on confined space work in construction under 29 CFR 1926.21(6). Department of Defense procedures for confined space work are found in Section 06.I of the U.S. Army Corps of Engineers (USACE) *Safety and Health Requirements Manual* (EM 385-1-1). For Department of Energy sites, confined space requirements are included in Chapter 4 of the *Department of Energy OSH Technical Reference*. In addition, guidance can be found in American National Standards Institute (ANSI), "Safety Requirements for Confined Spaces," ANSI Z117.1-2003.

DEFINITIONS

The following definitions are used in the OSHA standard and in other procedures describing confined spaces. In some cases the definitions represent OSHA regulatory requirements and in others the definitions provide guidelines for performance.

Alternate entry confined space: A confined space, initially classified as a permit-required confined space, presenting only an atmospheric hazard that can be completely controlled through ventilation.

Asphyxiation: Suffocation from the lack of oxygen. Chemical asphyxiation is produced by a chemical that binds with the blood's hemoglobin in place of oxygen. Simple asphyxiation is the result of an inert gas displacing oxygen in the lungs.

Attendant: A person stationed outside one or more permit spaces. An attendant monitors the location and condition of authorized entrants and performs all other assigned duties listed in the permit-space program.

Authorized entrant: An employee designated by the employer to enter a permit space. The duties and training for an authorized entrant are specified in the permit-space program.

Blanking or blinding: The absolute closure of a pipe, line, or duct by the fastening of a solid plate (such as a spectacle blind or a skillet blind) that completely covers the bore and that is capable of withstanding the maximum pressure of the pipe, line, or duct with no leakage beyond the plate.

Confined spaces: Spaces large enough and so configured that an employee can bodily enter and perform assigned work, has limited or restricted means for entry and exit, and is not designed for continuous employee occupancy. Confined spaces include, but are not limited to, storage tanks, vessels, manholes, pits, bins, boilers, digesters, ventilation ducts, utility vaults, tunnels, pipelines, trenches, and vats; open-top spaces more than 4 feet deep, such as pits, tubs, and excavations; or any space with limited ventilation or suspect atmosphere.

Double block and bleed: The closure of a line, duct, or pipe by closing and locking, or tagging, two in-line valves and by opening and locking, or tagging, a drain or vent valve in the line between the two closed valves.

Emergency: Any failure of hazard control or monitoring equipment, or other event(s) inside or outside of a confined space, that could endanger entrants within the space.

Engulfment: The surrounding and effective capture of a person by a fluid (i.e., liquid or finely divided particulate) substance that can be aspirated and cause death by filling or plugging the respiratory system, or that can exert enough force on the body to cause death by strangulation, constriction, or crushing.

Entry: The action by which a person passes through an opening into a permit-required confined space. Entry includes ensuing work activities in that space and is considered to have occurred as soon as any part of the entrant's body breaks the plane of an opening into the space.

Entry permit: The written or printed document that is provided by the employer to allow and control entry into a permit space.

Entry supervisor: An entry supervisor is an employee, foreman, or crew chief that authorizes and/or supervises confined space entry operations. After the initial entry authorization, the duties of an entry supervisor may be passed from one individual to another during the course of an entry operation.

Entry supervisors can serve as attendants, or as authorized entrants, as long as they are properly trained.

Flammable atmosphere: Any atmosphere that contains a concentration of flammable or combustible material in excess of 10 percent of the lower explosive limit (LEL), or lower flammable limit (LFL).

Hazardous atmosphere: An atmosphere that can expose employees to the risk of death, incapacitation, injury, impairment of ability to self-rescue (i.e., escape unaided from a permit space), or development of an acute illness from one or more of the following causes:

1. Flammable gas, vapor, or mist in excess of 10 percent of its LFL.
2. Airborne combustible dust at a concentration that meets or exceeds its LFL, or obscures vision at 5 feet.
3. Atmospheric oxygen concentration below 19.5 percent or above 23.5 percent.
4. Atmospheric concentration of any substance for which a threshold limit value (TLV) or a permissible exposure limit (PEL) is exceeded.
5. Any other atmospheric condition that is immediately dangerous to life or health. The material safety data sheet (MSDS) can provide guidance in establishing acceptable atmospheric conditions when a TLV or PEL is not given.

Hot-work permit: The employer's written authorization to perform operations (e.g., riveting, welding, cutting, burning, and heating) capable of providing a source of ignition.

Immediately dangerous to life or health (IDLH): Any condition that poses an immediate or delayed threat to life, could cause irreversible or adverse health effects, or might interfere with an individual's ability to escape unaided from a confined space.

Inerting: The displacement of the atmosphere in a permit space by a noncombustible gas (such as nitrogen) to such an extent that the resulting atmosphere is noncombustible. This procedure produces an IDLH oxygen-deficient atmosphere.

Isolation: Process by which a permit-required space is removed from service and completely protected against the release of energy and material into that space by such means as: blanking or blinding; misaligning or removing sections of lines, pipes, or ducts; a double block-and-bleed system; lockout or tagout of all sources of energy; or blocking or disconnecting all mechanical linkages.

Line breaking: The intentional opening of a pipe, line, or duct that is or has been carrying flammable, corrosive, or toxic material, an inert gas, or any fluid at a volume, pressure, or temperature capable of causing injury.

Lockout/tagout: A method for keeping equipment from being energized or actuated while workers are performing maintenance or repairs. A disconnect switch, circuit breaker, valve, or other energy-isolating mechanism is placed in the safe, or off, position; then it is locked out using a lock or locks controlled by those who are working on the equipment. A tag with a warning of "Do Not Operate" (or similar wording), and signed and dated by the authorized employees, is attached to the disconnection or isolation means. Tagout should always be used in conjunction with lockout. Often a device is placed over the energy-isolating mechanism to hold it in the safe position while a lock is attached so that the equipment can't be energized.

Nonpermit-required confined space: A confined space that does not contain or have the potential to contain any hazard or hazardous atmosphere capable of causing death or serious physical harm.

If it can be demonstrated that the only hazard posed by the permit space is an actual or potentially hazardous atmosphere and that continuous forced-air ventilation can safely maintain the permit space, then the space can be considered nonpermit required pending satisfactory air-monitoring results.

Oxygen-deficient atmosphere: Any atmosphere having 19.5 percent or less available oxygen content. Such a space should not be entered without wearing an approved self-contained breathing apparatus (SCBA), or an approved supplied-air, full-face respirator.

Oxygen-enriched atmosphere: Any atmosphere with 23.5 percent or more available oxygen content.

Oxygen-enriched atmospheres will cause flammable materials to burn violently when ignited. If oxygen is being used in a confined space, it may be necessary to check for excess oxygen and ventilate the work area with a clean source of air.

Rescue service: The personnel designated to rescue employees from permit spaces. This may be a contracted service, a host-client-provided service, a public emergency service (fire rescue service), or trained employees. Specific requirements apply to qualifications of rescue personnel and equipment.

Retrieval system: Any equipment such as retrieval lines, harnesses, wristlets (if appropriate), lifting devices, and anchors used for nonentry rescue of persons from permit spaces. Note that a retrieval line differs from a lifeline, which is a type of fall arrest system.

Self-rescue: The act of escaping unaided from a hazardous atmosphere or IDLH situation in a permit-required space.

Serious safety hazard: Any nonatmospheric hazard that may expose entrants to the risk of death, incapacitation, or impaired ability for self-rescue. Examples of serious safety hazards include:

1. Energized and exposed electrical systems.
2. Fall hazards.
3. Extreme temperatures.
4. Personal protective equipment that may cause a physical or health hazard or may compromise an individual's ability to react appropriately to a hazard.
5. Unguarded mechanical systems.
6. Performing piping connections on hazardous systems such as steam, inert gas, and hazardous materials.
7. Space configurations with inward-converging walls that slope down to a smaller cross section, or any physical conditions that could cause entrapment of the occupant of the space.

Testing: The process by which the hazards that may be faced by entrants of a permit space are identified and evaluated. Testing includes specifying the tests to be performed in the permit space.

Toxic atmosphere: An atmosphere inflicting poisonous physical effects that may be immediate, delayed, or a combination of both. Substances such as poisonous liquids, vapors, gases, mists, dusts, fumes, and biological agents in the air should be considered hazardous in a confined space. Hydrogen sulfide and carbon monoxide are the most common toxic agents that can be found in a confined space.

PERMIT-REQUIRED CONFINED SPACES

A confined space containing hazards that may expose workers to risk of injury or death may be classified as a permit-required confined space under OSHA's definition (see 29 CFR 1910.146). The permit system is designed to identify hazards and implement controls to reduce these risks.

Any confined space that has one or more of the following characteristics should be considered as a possible permit-required confined space:

1. Contains or potentially could contain a hazardous atmosphere.
2. Contains a material that potentially could engulf an entrant.
3. Contains any other recognized serious safety or health hazard.
4. Offers a potential for toxic or oxygen-deficient atmosphere.
5. Presents the potential for engulfment, such as hoppers and silos used for sand and gravel.
6. Has an internal configuration that could result in an entrant being trapped or asphyxiated by inwardly converging walls or by a floor that slopes downward and tapers to a smaller cross section.
7. A work space with unfavorable natural ventilation may be considered a confined space in certain conditions.

By following the procedures for confined space entry using the OSHA permit system, these hazards can be identified, and procedures to control or reduce the risks can be implemented using a task-based approach.

Entry Procedure

The first step is to determine whether there are any confined spaces that will require entry. The decision logic shown in Figure 1 may be used to assess the potential confined space. The existence of a potential confined space does not pose as significant a hazard if personnel are not required to enter the space. In the case of a potential confined space work area, it may be necessary only to identify the work space as a "confined space" through posted warning signs and barricades or locks. When it becomes necessary to enter an identified confined space to perform a task, it must be classified as permit required or nonpermit required. The decision logic of Figures 1 and 2 should be used to assist in this determination. The critical input to the decision logic is the identification of the poten-

tial workplace hazards at the time that the task will be performed. Because conditions can change over time, it is also important that the hazards be monitored to assure controls are adequate and that conditions have not changed.

HAZARD CHARACTERISTICS

A thorough inspection of a confined space must be performed to verify acceptable entry conditions. This frequently includes a physical inspection, not only of the space to be entered, but any adjacent or

TYPE OF CONFINED SPACE DECISION FLOW CHART

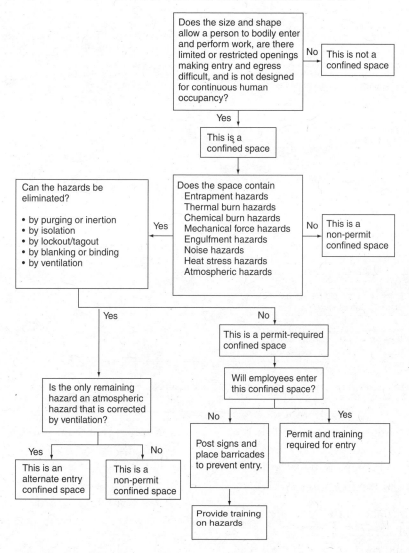

■ ■ **Figure 1** ■ ■ **Decision flow chart**

PERMIT-REQUIRED CONFINED SPACE DECISION FLOW CHART

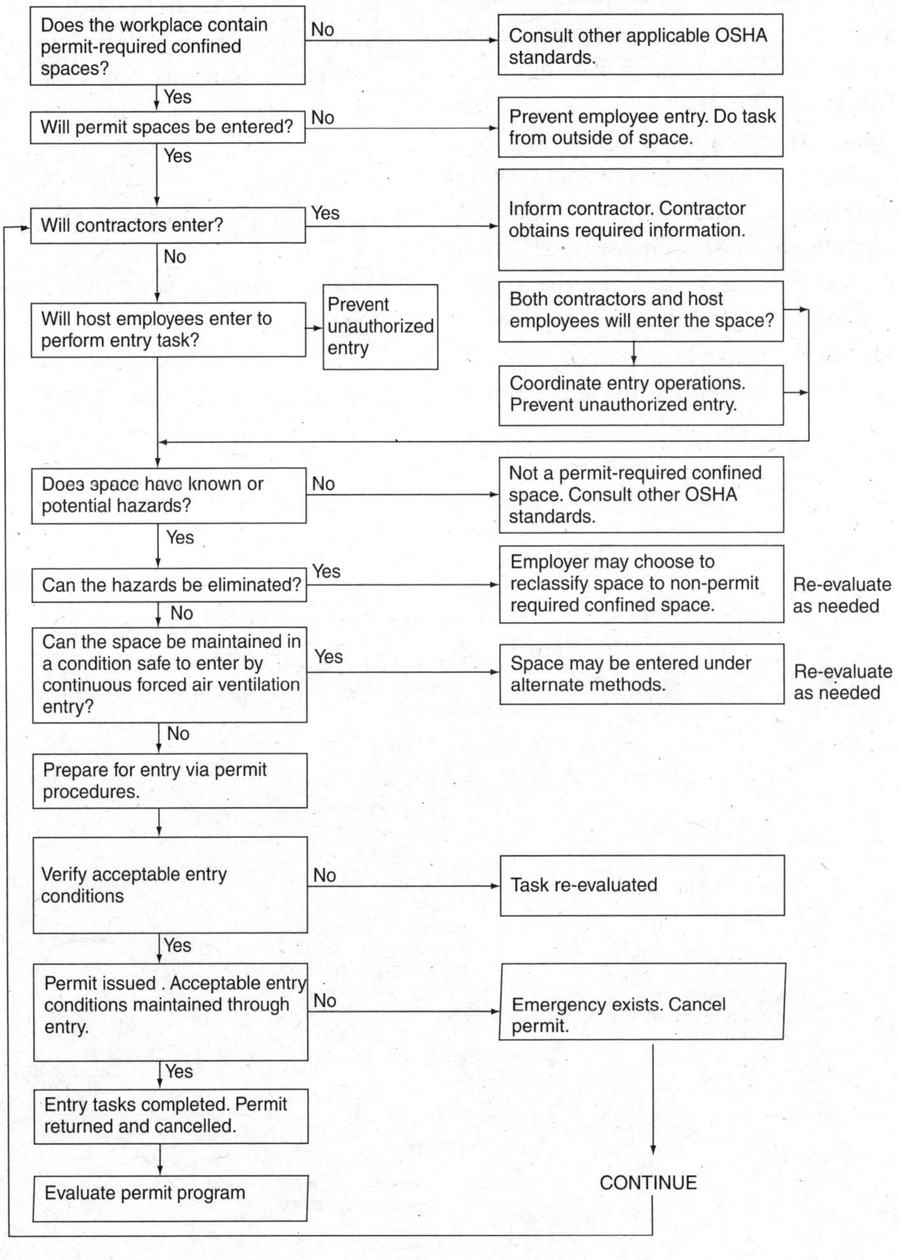

■ ■ Figure 2 ■ ■ Decision flow chart where a permit is required

connected spaces that might pose a threat to the entrants. The hazards of greatest concern are those considered immediately dangerous to life or health that pose one or more of the following threats:

- An immediate or delayed threat to life (see the definition of IDLH).
- A threat that would cause irreversible adverse health effects.

- A threat that would interfere with an individual's ability to escape unaided from a permit space.

The major IDLH hazards workers can encounter when they enter confined spaces include:

- atmospheric hazards
- thermal or chemical hazards

- mechanical force or electrical hazards
- engulfment in liquids or finely divided solid particles.

Atmospheric Hazards

A hazardous atmosphere in a confined space can expose workers to the risk of death, injury, acute illness, incapacitation, or impairment of their ability to escape unaided from the permit space (self-rescue). One or more of the following can cause a hazardous atmosphere:

- A flammable gas, vapor, or mist in excess of 10 percent of its lower flammable limit (LFL).
- An airborne combustible dust at a concentration that meets or exceeds its LFL (approximated as a condition in which dust obscures vision at a distance of 5 feet [1.52 m] or less).
- An atmospheric oxygen concentration below 19.5 percent (oxygen deficient) or above 23.5 percent (oxygen enriched).
- An atmospheric concentration of any substance that could result in employee exposure to toxic air contaminants in excess of permissible exposure limits (PEL) or threshold limit values (TLV).

OXYGEN-DEFICIENT ATMOSPHERES

Oxygen deficiency (less than 19.5% O_2) within a confined space can be caused when oxygen is:

- Absorbed by other substances, such as activated charcoal.
- Consumed by chemical reactions such as rusting and burning, or biological processes such as bacterial decomposition.
- Displaced by another gas (e.g., when a confined space is intentionally inerted by a nitrogen blanket or other nonreactive atmosphere that contains no oxygen).

Breathing oxygen-deficient air causes poor judgment, loss of coordination, fatigue, vomiting, unconsciousness, and, ultimately, death. Asphyxiation from insufficient oxygen frequently occurs when victims, unaware of the problem, reach the point where they cannot save themselves or call for help. (See Table 1.)

OXYGEN-ENRICHED ATMOSPHERES

The atmosphere in a confined space can also have *too much* oxygen. An oxygen-enriched atmosphere is not an asphyxiation hazard; however, an oxygen concentration greater than 23.5 percent can create a serious fire hazard—oxygen-enriched air can cause combustible materials to burn violently.

▪▪ Table 1 ▪▪
Oxygen Respiratory Action Levels
for Confined Space Entry

% Oxygen	Condition	Action
>23.5%	Oxygen enriched	(1) check for source (2) control source (3) dilute if needed (4) manage fire hazards
20.9 %	Normal atmosphere	(1) manage other hazards
<19.5 %	Oxygen deficient	(1) ventilate area and recheck (2) enter only when >19.5
16%	High Hazard-impaired judgment increased fatigue likely	(1) Remove entrants unless supplied air is being used
12.5%	IDLH Atmosphere	Emergency evacuation unless supplied air and escape bottles or SCBAs are being used
6%	Extremely Dangerous	Only supplied air and escape bottles or SCBAs permitted

TOXIC ATMOSPHERES

Toxic atmospheres in confined spaces can cause serious health problems, even death. Poisonous physical effects may be immediate, delayed, or a combination of both.

Toxic contaminants can be gases, vapors, fumes, or airborne dusts. The most common toxic gases encountered in confined spaces are carbon monoxide and hydrogen sulfide. Other sources of toxic atmospheres in confined spaces include fuel vapors, protective tank coatings, inerting media, fumigants, and residues from previous tank contents.

Engulfment Hazards

Engulfment in a confined space occurs when the victim is immersed in liquid or trapped and enveloped by finely divided dry bulk materials, such as grain or sawdust.

Engulfment hazards include asphyxiation from aspirating (inhaling) the engulfing material, which causes death by filling or plugging the respiratory system. Another asphyxiating effect of engulfment is compression of the torso by the weight of the engulfing material, preventing the victim's lungs from moving.

Noise Hazards

Work performed inside a confined space can be deafening. Scaling, chipping, grinding, hammering, riveting, power scrubbing, the use of power and pneumatic tools, and air-line leaks create hazardous noise levels. When the work is done inside a vessel, tank, or other space with nonabsorbing surfaces, the noise increases when it bounces off the walls.

Even ventilation adds sound and noises from outside the space, which can sound louder inside a confined space.

Heat/Cold Stress Hazards

A tank or vessel can become a health hazard to the entrant if the ambient temperature of the tank or vessel exceeds acceptable work limits. The sun on a metal tank, lack of air circulation, or performing hot work, can contribute to an entrant being overcome by heat stress. Tank or vessel work in cold weather can present increased risks because work-

ers may need insulated gloves and clothing that may limit movement and impede escape. If preentry hazard identification indicates heat stress can become a problem, it is a safe practice to plan periodic temperature testing. In cold conditions, warm-up regimens should be considered and all emergency retrieval equipment should be sized to fit over any insulating clothing.

Electrical Hazards

The effects of electrical energy frequently contribute to confined space accidents. It is difficult in confined spaces to avoid contact with electrical components. For sources of electrical energy that already exist in the work area, an effective lockout/tagout program is necessary. For electrical hazards that are introduced into the workplace, such as drop lights and power cords, the use of ground-fault circuit interrupters (GFCIs) or low-voltage lighting are usually standard practice. If flammable atmospheres are present, the lighting and electrical power tools will need to be classified for use in the hazard class of the work space according to NFPA 497A or NFPA 497B.

Flammable Hazards

Flammable and explosive atmospheres contain gases, vapors, or airborne dusts at concentrations great enough to burn rapidly upon contact with ignition sources such as heat, open flames, or electrical sparks. The LFL is the lower limit at which a flammable substance will ignite into sustained combustion.

Changes in oxygen concentrations must also be monitored. While it is not flammable itself, oxygen is necessary for all combustion to take place. Materials that are normally nonflammable, such as clothing, can burst into flames at the smallest spark in a confined space containing a high volume of oxygen.

Mechanical Hazards

The effects of mechanical energy are also a frequent contributor to confined space accidents. Mechanical energy usually involves stored potential energy in the form of compressed air or gas lines, mechanically assisted doors or closures, and mechanical lifting or conveying devices. Unguarded machinery such as fans, flywheels, or conveyors may constitute a mechanical hazard. An effective lockout/tagout program can prevent nearly all mechanical hazards.

HAZARD IDENTIFICATION

Following the evaluation of the workplace and the identification of all confined spaces, the nature and severity of the hazards present in each space must be determined before employees can enter the space. Some or all of the spaces may present the same hazards, but hazardous characteristics of each space must be determined individually. Some of the equipment that may be required to identify the hazards could include the following:

> Explosimeter or combustible gas instrument (readout as percent LEL)
> Oxygen monitor (readout as percent oxygen)
> Hydrogen sulfide monitor (readout as ppm H_2S)
> Carbon monoxide monitor (readout as ppm CO)
> Organic vapor analyzer—OVA (readout as ppm organic equivalent)
> Ambient temperature (readout in °F or °C)
> Additional monitoring instruments for specific chemical contaminants.

Figure 3 shows an example of external monitoring of a confined space using a handheld, direct-reading instrument. In this photo the attendant is in communication with the entrants using a portable radio communication system.

Testing of the confined space should consider the following approaches:

- *Evaluation Testing:* The atmosphere of a confined space should be analyzed using equipment of sufficient sensitivity and specificity to identify and evaluate any existing or potentially hazardous atmospheres so that appropriate permit entry procedures can be developed and acceptable entry conditions stipulated for that space. Evaluation and interpretation of this data and development of the entry procedure should be performed

and reviewed by an experienced safety professional with knowledge of the chemical hazards and physical characteristics of potential contaminants. The authorized entrant, or a designated representative, must be provided the opportunity to observe the testing and to review the results.

- *Verification Testing:* Any residual concentrations of contaminants identified in a permit space should be verified to be within the range of acceptable entry conditions. Results of testing (i.e., actual concentration) should be recorded on the permit.

- *Testing Stratified Atmospheres:* When monitoring for entries involving a descent into stratified atmospheres, the atmospheric envelope should be tested at a minimum distance of approximately every 4 feet (1.22 meters) in the direction of travel and to each side. If a sampling probe is used, the entrant's rate of progress should be slowed to accommodate the sampling speed and detector response. Since gases have different densities and may stratify in layers, it is necessary to test spaces before entering and at least at 4-foot vertical intervals in the direction of travel from side to side.

■ ■ **Figure 3** ■ ■ **Direct-reading gas monitoring being used to monitor a confined space** (Photo courtesy of Industrial Scientific Corporation)

Hazardous atmospheres may be controlled by continuous forced-air ventilation from a clean air source. In this case, testing should be redone periodically to ensure that ventilation is sufficient if conditions change, or if the source of clean air is potentially contaminated by another source of air hazards, such as welding fumes or exhaust products.

Additional tools or instruments may include flashlights, camera, tape measure, piping, and instrument drawings, process information, utility locations, and a list of potential chemical hazards that may be present in the workplace. Critical to assessing the hazard is ensuring that the person doing the assessment is not placed in a risk situation. Therefore, the collection of data is often by remote means, such as lowering probes into the work space and collecting the information at a remote location. It is important that the work space be well characterized and that factors such as gas density and hidden pockets or sources of hazards be included in the evaluation to the extent feasible without increased risk. It is also important that the instruments used for collecting the data are accurate and representative of the conditions.

Limitations of Testing

Most instruments used for testing confined space atmospheres and physical conditions are designed to be field-portable and easy to use. In rare cases, it may be necessary to collect samples that are sent for laboratory analysis, but most confined space work necessitates obtaining quick and accurate information in order to assess a hazard.

While the instruments available are generally accurate and reliable, there are some limitations that should be considered:

■ *Oxygen meters:* These instruments usually rely on an electrochemical cell to measure the oxygen concentration in the air. These cells can become contaminated by acidic or alkaline gases or may deteriorate over time due to exposure to ambient air. Also, the instruments are sensitive to ambient air pressure so an instrument calibrated at a high altitude will over-read oxygen if used at a lower altitude. It is important to follow the manufacturer's instructions for maintaining and replacing the cells as well as the calibration procedures.

■ *Combustible gas instruments (CGI):* These instruments contain a catalytic combustion filament and provide a method for measuring the relative change in resistance from a wheatstone bridge. The instrument measures the differential between the unburned gas and the burned gas. Since the instrument requires that a combustion process occurs, it is critical that sufficient oxygen is present. This is why, in confined space testing, oxygen levels are always measured first and then the LEL. Insufficient oxygen will affect the LEL reading. Another factor in these instruments is calibration. Like the oxygen meters, altitude and air pressure are variables in the gas concentration. A unique element of CGIs is that the choice of calibrant will affect the read-out of the measured gas. Ideally, the calibrant gas should have a similar heat of combustion as the gas being measured. If this is not possible, the instrument manufacturers supply correction factors that are applied to the readings. CGIs also are subject to poisoning from organic metals and silicates, or silanes; if this occurs, the catalyst material is poisoned, or coated.

■ *Carbon monoxide monitors:* These monitors may include either electrochemical cells, solid-state sensors, or infrared sensors. Like the other instruments, calibration and maintenance are critical. It is also important to check the instrument specifications for cross-interfering gases, that is gases that will produce either a false-positive response or may mask the actual response. If a cross-interfering gas is suspected in the confined space, and it is necessary to measure carbon monoxide, it may be necessary to select an alternative method. In some electrochemical cells, unsaturated hydrocarbons (like acetylene) will produce false readings. In some cases very high humidity may affect readings.

■ *Hydrogen sulfide monitors:* These monitors also use electrochemical cells, but other measurement methods can be employed. Electrochemical cells for hydrogen sulfide are not subject to as many cross-interfering gases as carbon monoxide. The problem with hydrogen sulfide is that it is an extremely toxic gas with a low permissible exposure level and ceiling limit. Therefore it is necessary to assure that the results are accurate in order to assure that an overexposure does not occur. Hydrogen sulfide is also slightly heavier than air so it will tend to accumulate in lower levels and may dissolve in water to some degree.

HAZARD CONTROL

When all existing and potential hazards have been identified, the appropriate measures must be developed and implemented to protect confined space entrants. If it is necessary to enter the permit space to eliminate hazards, the permit-required confined space program should be followed. If any of the control measures to correct the hazard fail, or prove less than totally effective, entry should be postponed and not resumed until all hazards are adequately controlled. Some of the control measures that may be necessary as part of the planned entry may include the following items:

1. Real-time, handheld, monitors for O_2, LEL, UEL, H_2S, CO, and any other potential toxic air contaminants that might be present. These monitors should have preset audible and visible alarms.
2. SCBA equipment, or air packs, or supplied-air full-face respirator and 5-minute escape approved for IDLH situations.
3. Adequate access/egress equipment.
4. Rescue and emergency equipment, such as body harnesses, backboards, and 5-minute-escape air systems, when applicable.
5. Appropriate fixed-location, real-time, or other chemical gas alarm monitors as indicated by the contents of the confined space.
6. Work space and emergency back-up lighting if needed. This lighting may require

explosion-proof design for the type of hazard present.

Control Measures

To prevent sudden changes in workplace conditions, continuous observation of the work area is necessary, and periodic atmospheric tests should be conducted.

Examples of other hazard control measures include:

- Cleaning, purging, or inerting hazardous atmospheres in the space.
- Isolation of the permit space from hazardous energy and materials.
- Lockout/tagout programs to prevent unexpected operation of equipment inside the space.
- Blanking or blinding of pipes that carry materials into the space to prevent accidental entry of hazardous materials.
- Mechanical ventilation of the space to control toxic or flammable gases and vapors, or to ensure adequate oxygen supply.
- Providing personal protective equipment, such as hard hats, safety shoes, respirators, eye and face protection.
- Communication equipment to allow attendants to maintain contact with entrants when they are not in direct view of each other.
- Acquiring hot-work permits and using additional fire-control procedures.

Responding to Unacceptable Atmospheres

Persons performing initial testing or periodic monitoring should not rely on a single air sample. If the atmosphere in any part of a confined space should become unsafe during work, evacuate the space immediately if the alarm on a testing device goes off, or if tests show the atmosphere has become hazardous. Stop all hot work at once if there is any possibility of a flammable atmosphere. Evacuate the space immediately if toxic concentrations rise above preset action levels. If physical hazards develop, such as a water-line break, utility penetration, or engulfment or entrapment hazards, it will be necessary to evacuate workers in the confined space by a safe evacuation route.

After evacuation, identify and control the source of the hazard. Then retest the atmosphere before reentering the space. If the air is unsafe, control the hazard before reentering. One important method for controlling hazardous atmospheres is appropriate ventilation, which can be accomplished using large air-moving systems. For physical or utility hazards, it will be necessary to control the source of the hazard through either lockout/tagout or diversion and removal. The introduction of any new hazard, or the identification of inadequate control, will require a new hazard analysis.

Preventing Unauthorized Entry

Unauthorized personnel who enter a confined space expose themselves to life-threatening dangers and threaten the safety of authorized entrants.

Actions to prevent unauthorized entry include training all employees and providing necessary information to all visitors. Other options are posting warning signs, erecting barriers, and installing covers with locks at entry points.

A system for documenting the implementation of these measures and ensuring that they remain in place is necessary. A survey conducted during routine safety inspections is a good means of checking that all areas are covered and that signs or barriers have not been removed or damaged.

Ventilation

No hazardous atmosphere should be allowed to develop within a space at any time when an employee is inside that space. To ensure this condition is met, continuous forced-air ventilation is often used to control this work-space hazard. The guidelines listed here should be followed when setting up a forced ventilation system for a confined space task:

- No employee may enter the space until continuous forced-air ventilation has controlled any hazardous atmosphere and air monitoring has documented that the atmospheric conditions are acceptable.
- Blowers, vents, or ducts should be placed and directed in a manner that will ventilate

the immediate areas where workers will be present. Forced air is preferable to air suction unless there is a problem with high dust generation. Forced-air ventilation must continue until all employees have left the space.

- The air supply for the forced-air ventilation system must originate from a clean source and may not increase the atmospheric hazards in the space. It is important that internal combustion engines or chemical tank venting not be in close proximity to the forced-air intake location. The air intake should be at least 6–8 feet beyond the exhaust of the ventilation in order to prevent reentrainment of the exhausted air.
- The air supply should be designed to provide twenty air changes per hour and the outlet of the air supply line should be placed to allow maximum mixing and displacement of the confined space air. Commercially available air movers will usually specify the maximum air-delivery tube length, but it is important to prevent bends or obstructions to the air-delivery system.
- The atmosphere within the space must be periodically tested to ensure that continuous forced-air ventilation is preventing the accumulation of atmospheric hazards. Allow 15 minutes of ventilation prior to initial testing of the vented atmosphere.
- Nonpermit confined spaces must be reevaluated whenever changes in the use or configuration of the space (including ventilation) might increase hazards to entrants.

NONPERMIT CONFINED SPACES

Every confined space identified during the initial evaluation should be considered a permit-required confined space until further investigation reveals the nature and extent of its specific hazards. In the event that the hazard assessment determines that all hazards are controlled and there is minimal risk of worker injury during the task, the work space may be classified as a nonpermit confined space. Planned tasks may then be conducted in the workplace with minimal monitoring and entry procedures, provided conditions do not change.

ALTERNATE-ENTRY CONFINED SPACES

The confined space standard specifies the conditions to be met before an employer may use OSHA's alternate procedures for entering a permit space with a potentially hazardous atmosphere. If it is determined after initial entry that all hazards within the space, including actual or potential atmospheric hazards, can be completely eliminated rather than merely controlled, the employer may reclassify the permit space to a nonpermit confined space.

If, however, testing and monitoring results show a controllable hazardous atmosphere as the only hazard, the space must remain a permit space, but may be entered using the alternate procedures. To justify the use of alternate entry procedures, monitoring and inspection data must demonstrate both of the following:

1. An actual or potentially hazardous atmosphere is the only hazard posed by the permit-required confined space.
2. Continuous forced-air ventilation alone is sufficient to maintain the permit space safe for entry.

The criteria and test results used to demonstrate that the space is safe to enter by either a nonpermit or an alternate procedure must be documented and available for review by authorized entrants.

To ensure that the entrant will have sufficient time to escape the space should ventilation fail for any reason (loss of power, for instance), the atmosphere within the space can never be allowed to approach a hazardous level.

CONTROL OF PERMIT-REQUIRED CONFINED SPACES

When a work space demonstrates hazard characteristics of a confined space, and the hazards require control measures in addition to mechanical ventilation to reduce the risk of injury, the work space requires that all entry and work be managed by a task-specific permit system. This permit system is basically a job hazard analysis that includes specific requirements for air testing, hazard monitoring, worker training, access control, and rescue services. The logic diagram in Figure 2 may be used to assist in assessing the hazards of a permit-required confined space.

The permit should include information on location of work, description of work, employee(s) allowed to enter, entry date and time, work practice and proper controls checklist, hazardous work, hazards expected, safety precautions, atmospheric testing, proper authorizations, and expiration time and date. The permit system also requires specific functional duties for employees involved in the task as follows:

ENTRY SUPERVISOR: Any individual empowered by the employer to authorize or directly supervise entry operations in a permit space.

The entry supervisor makes sure conditions are safe.

- Before entry, the supervisor verifies that the permit is filled out completely and all safety steps listed on it are taken, then signs the form.
- The entry supervisor makes sure that any unauthorized people are removed.
- Every entry supervisor is responsible for canceling the entry authorization and terminating entry whenever acceptable entry conditions are not present. If conditions become unsafe, the permit is canceled and everyone is ordered out of the space.
- The entry supervisor directly in charge of entry operations at the time the work authorized by the permit is completed must terminate the entry and cancel the entry permit. This includes taking necessary measures for concluding the entry operation and closing off the permit space.
- The entry supervisor on each shift must determine, at appropriate intervals dictated by the hazards and operations performed within the space, that entry operations remain consistent with the terms of the entry permit and that acceptable entry conditions are maintained.

■ Whenever responsibility for a permit-space entry operation is transferred, the outgoing entry supervisor determines that entry operations are still consistent with the terms of the permit and that acceptable entry conditions are present before turning operations over to the incoming entry supervisor.

Each entry supervisor should be aware of all potential hazards during entry and work. If acceptable entry conditions are present at a permit space where entry is planned, the entry supervisor then authorizes the entry and oversees entry operations. The entry supervisor must determine that entry and work operations remain consistent with entry-permit terms and that acceptable entry conditions are maintained.

ATTENDANT: This person is specifically trained and assigned to oversee and monitor the employees performing the work in the confined space. The attendant stays at his or her post to observe conditions and support the entrants.

The attendant must:

■ Know the hazards of the permit space and the signs of exposure.
■ Keep a current count and be able to identify all entrants.
■ Stay in continuous contact with the entrants.
■ Be sure only authorized people enter the space or the area surrounding the space.
■ Order all workers out of the space in any of these situations:
  • a condition not allowed by the entry permit
  • any signs of exposure in any entrant
  • something outside the permit space that could cause danger inside.
■ Never leave the observation post for any reason and remain outside the permit space.
■ Knows behavioral effects of exposure.
■ Performs no conflicting duties.

In case of emergency, confined space rescue should be undertaken only if the attendant has received rescue training, has the proper emergency equipment, and ensures that there is a replacement attendant.

The attendant continuously maintains an accurate count of authorized entrants in the permit space and guarantees that the means used to identify authorized entrants can accurately identify those in the permit space. This requires that the attendant keep track of entrants as they enter and exit the space. The attendant must know the exact count at all times so that no one is accidentally left in a confined space. During emergencies, an accurate count also ensures that no useless searches are made to find entrants no longer in the permit space.

The attendant must remain outside the permit space during entry operations until he or she is relieved by another attendant. Keeping unauthorized persons out of the space, being alert for hazards, and providing information to rescue services are three duties requiring the attendant to remain posted until actually replaced by another attendant.

A well-trained attendant always monitors the permit space itself—as well as the immediate areas around the space—to detect potential hazards. Knowing that all attendants have adequate training, frees up an entrant's attention for work and ensures the entrant's confidence that hazards will be detected.

The attendant orders entrants to evacuate the permit space immediately whenever the attendant:

■ Detects a prohibited condition.
■ Detects the behavioral effects of hazard exposure in an entrant.
■ Detects a situation outside the permit space that could endanger entrants in the space.
■ Cannot effectively and safely perform all the duties required.

The attendant must summon rescue and other emergency services as soon as he or she has any concern for an entrant who may need assistance escaping from permit-space hazards.

The attendant takes the following actions when unauthorized persons approach or enter a permit space while entry is underway:

■ Warn unauthorized persons to stay away from the permit space.
■ Advise all unauthorized persons to exit immediately if they have entered the space.

■ Inform the authorized entrants and the entry supervisor if unauthorized persons have entered the permit space.

The attendant cannot perform any other duties that might interfere with his or her primary duty to monitor and protect authorized entrants.

ENTRANT: A worker who goes into the confined space and is dually responsible for controlling hazards of confined space entry. Authorized entrants must maintain contact with their attendant in order to improve their chances of safe exit. Such systems as two-way radios, television, or other continuous electronic monitoring equipment, used in combination with alarms and voice contact, are considered to be effective methods of communication between attendant and entrant.

Authorized entrants must be trained to alert the attendant whenever:

■ The entrant recognizes any warning sign or symptom of exposure to a dangerous situation.
■ The entrant detects a prohibited condition. A prohibited condition is any condition in a permit space that is not allowed by the permit during the period when entry is authorized.

Authorized entrants must exit the permit space as quickly as possible when:

■ An order to evacuate is given by the attendant or the entry supervisor.
■ The entrant recognizes any warning sign or symptom of exposure to a dangerous situation.
■ The entrant detects a prohibited condition.

Any change in work conditions will result in cancellation of the permit. The entry supervisor will cancel the permit and reevaluate the task and condition as necessary. Permits are also canceled at the conclusion of the task. Canceled permits must be retained as records for one year, and the permits should be reviewed annually as part of an overall confined space program.

Permit Cancellation

Upon completion of the entry covered by the permit, and after all entrants have exited the permit space, the entry supervisor should cancel the permit. While the permit must be canceled upon completion of the work, it may also be canceled any time situations develop that are in violation of the permit.

The cancellation of the permit after all entrants have exited requires that each authorized entrant is accounted for, so that no one is inadvertently left in the space.

LIMITATIONS OF THE PERMIT SYSTEM

The permit-required confined space entry procedure was designed to reduce the risk of serious injury or death for workers required to work in hazardous work spaces. The procedure is effective if, once a hazardous work space is identified, the following procedures are followed:

1. The confined space entrant and entry supervisor consider all potential physical and health hazards, such as flammable, combustible, reactive, corrosive, caustic, toxic, and radioactive materials; oxygen content; electrical or mechanical equipment; and fire suppression systems (CO_2 and halon).
2. No one enters a confined space until all hazards have been identified, the confined space has been classified, and all necessary precautions have been taken for a safe entry.
3. All workers engaged in a confined space operation are knowledgeable of potential hazards they may encounter and have been trained in their specific duties regarding the confined space operation.
4. All permit entry spaces in the workplace have been posted with a sign reading, "Danger, Permit-Required Confined Space, Do Not Enter"—or similar language.
5. Effective measures have been taken to prevent employees from entering permit spaces without proper training.

TOOLS AND EQUIPMENT RESTRICTIONS AND PRECAUTIONS

- Only explosion-proof lighting or equipment should be used in confined or enclosed spaces unless atmospheric tests have proven the space to be nonexplosive.
- When welding or cutting is to be performed in any confined space, the gas cylinders and welding machines should be stored outside the space. Local exhaust ventilation, adequate eye and skin protection, and fire protection are required.
- In order to eliminate the possibility of gas escaping through leaks or improperly closed valves during welding or cutting, torch valves and cylinder valves should be closed whenever the torch is not to be used for a substantial period of time, such as during the lunch break. The torch valves should be shut off outside the confined space. Torches and leads must be removed from the confined space at the end of the shift and whenever not in use, where it is practical to do so.
- When arc welding is to be suspended for any substantial period of time, such as for a lunch break or overnight, all electrodes should be removed from the holders, and the holders located so that accidental contact cannot occur. The welding machine should be disconnected from the power source.
- Use air-driven tools whenever possible. Use grounded or double-insulated power tools along with ground-fault circuit interrupters.
- The nozzles of air hoses, inert gas hoses, and steamline hoses, when used in the cleaning or ventilation of tanks and vessels that contain explosive concentrations of flammable gases or vapors, should be bonded to the tank or vessel shell. Bonding devices should not be attached or detached in the presence of hazardous concentrations of flammable gases or vapors.
- Ladders, scaffolds, working platforms, and rigging must be completely secured. Ladders or portable stairs used for access and egress must be secured so that they are not removed during the work activities.

If hazardous materials (paints, thinners, etc.) are to be used in a confined space, the safety and health information from the MSDS should be incorporated in the confined space entry procedure. If new materials are introduced to the confined space during a task, the permit must be reviewed for any additional hazards and protective requirements.

TRAINING

Training for permit confined space work is critical and required under the OSHA regulations.

The following are general training topics that should be covered when planning for work in confined spaces:

- permit-required space hazards
- hazards of the job site, location, or entry operation
- proper use and limitations of PPE and other safety equipment
- the permit system and other procedural requirements
- response to emergencies
- duties and responsibilities of each member of the permit-required confined space team
- how workers can recognize air-contaminant overexposure symptoms in themselves and co-workers
- methods for alerting the attendant.

RESCUE

Rescue of injured or incapacitated workers in a confined space represents one of the most dangerous aspects of confined space work. OSHA refers to the term "rescue service" as the designated responder to a confined space accident or emergency. The rescue service can be either an on-site service or an off-site service, such as a fire department or an emergency medical team, as long as certain criteria are met:

- The responding team must have the proper tools, equipment, and PPE necessary to make the rescue from the confined space.
- Each member of the rescue service must be trained in the use of the rescue equipment and PPE along with the requirements of confined space operations.
- The rescue service must practice making confined space rescues at least once every 12 months, using simulated accidents in representative confined spaces.
- Rescue personnel must be trained in first-aid and CPR procedures, and at least one member of the rescue team must be currently certified in first aid and CPR.

The choice to use on-site or off-site services is left to the contractor performing the task, but the expectation is that the service will be readily available and able to locate the work area quickly. The OSHA rule preamble states that the goal of a rescue operation is to remove an incapacitated entrant from a confined space within 4 minutes. Unless the service is immediately available at the work location, this is not practical. Each task needs to be assessed for the degree of hazard present and what the acceptable adequate response time would be to minimize the risk. In many cases, standby rescue equipment can be staged in the work area to facilitate rescue operations.

Rescue and appropriate first-aid equipment must be accessible. A rescue test should be performed to increase awareness of the demands of rescue operations and to ensure that rescue equipment will fit through the confined space entryway and that communications flow as anticipated.

A retrieval system (i.e., tripod/winch system) is frequently used as a means of retrieving injured victims from a confined space. Figure 4 shows a tripod retrieval system in an outdoor application. In some cases the installation of the retrieval system requires some modification to the work area. Support brackets may need to be welded to structural beams, support for the tripod legs may be required for uneven or soft surfaces, and for pipe runs in sublevels, it may be necessary to remove floor grating to allow access to the work areas.

To facilitate nonentry rescue, retrieval systems or methods must be used whenever an authorized entrant enters a permit space. This requirement applies in all cases, except when retrieval equipment would increase the overall risk of entry or would not contribute to the rescue of the entrant. The ANSI Z117.1 standard addresses this particular problem, and the above approach was adopted by the OSHA standard.

Retrieval systems, therefore, must meet the following requirements:

- Each entrant must use a chest or full-body harness with a retrieval line attached at the center of the entrant's back near shoulder level or above the entrant's head.
- The other end of each entrant's retrieval line must be attached to a mechanical device or fixed point outside the permit space so that nonentry rescue can begin as soon as the attendant becomes aware that rescue is necessary.
- Wristlets may not be used unless it can be demonstrated that the use of a chest or full-body harness is not feasible, or is unsafe,

■ ■ **Figure 4** ■ ■ **A tripod retrieval system is provided to assist in rescue operations.** (Photo courtesy of Industrial Scientific Corporation)

and that the use of wristlets is the most effective alternative.

On-site rescue services are often available on a work site and can be used as long as all OSHA requirements for rescue are met. Otherwise, arrangements for contract or municipal rescue services must be secured for the permit-space rescue. In order to assure that the rescue service is adequately prepared for the anticipated hazards, information on location of the work, chemical hazard information (MSDSs), and special precautions or restrictions of the planned work need to be provided to the rescue-service incident commander. If off-site services are part of the rescue plan, it is important that notification methods be established for summoning the service, and that information on locating and accessing the work area are adequately established. Many times a project will assume that the local fire department will be the source of rescue without ever verifying that the service is available or qualified to perform the rescue.

CONCLUSION

In confined space work there are a number of responsibilities required of the host employer, the contractor performing the work, and the employees responsible for the task. Table 2 lists some of the responsibilities of the facility's host employer or owner and the corresponding contractor or subcontractor performing the work.

UNUSUAL CHALLENGES

In many cases of permit-required confined space entry, the approach is straightforward; air testing determines the level of hazard, ventilation is used to control the contamination, lockout/tagout programs control energy sources, and entry and rescue programs are implemented and managed. It is commonly found in process industries where scheduled maintenance operations would necessitate this type of program. In construction work or unscheduled maintenance or repairs, there can be unexpected challenges to both hazard assessment and program design.

Some unusual conditions might include the following:

- Trenches and excavations may be considered a confined space. In an OSHA interpretation letter (National Utilities Contractors Assoc., June 15, 1992) an excavation was considered a confined space if it exhibited conditions of poor ventilation and limited egress and access. On construction sites it is also important to look for external sources of contaminants that could enter an excavation and make the work area a hazardous site. In an actual situation, a propane filling facility near a utility road cut was releasing propane that subsequently accumulated in a 12-foot-deep trench. Only through vertical interval testing of the trench prior to a pipe-joining task was the hazard detected. Air movers were placed in the trench and the filling facility was asked to suspend venting during the confined space work.
- Pipe assemblies and run areas were identified as a particularly unrecognized and hazardous work space for construction and plant renovation activities (USDOL, *Anatomy of Confined Spaces in Construction*, May 1996). This is because the pipe runs themselves create obstacles to emergency evacuation and impede communication and observation of work activities. In a major process upgrade of a metal cleaning and coating operation, it was necessary to work under plating tanks and piping systems during ongoing operations. There were significant hazards of overhead spills and potential toxic atmospheres from pipe-break releases. Workers were supplied with full chemical protective clothing and air respirators. In order to provide for retrieval devices, floor gratings were removed over the work area to provide access and direct observation by the attendant.
- Vaults and enclosures are frequently found on job sites. Often there is a potential hazard of asphyxiation or toxic gas exposure

▨▨ **Table 2** ▨▨

Host Employer/Owner and Contractor Confined Space Responsibilities

Host Employer/Owner Responsibilities	Contractor Responsibilities
Develop an inventory of all confined spaces on the property. Document inventory list.	Develop a written confined space entry program. Include all elements of OSHA 1910.146.
Evaluate hazards for each of the identified confined spaces. Document identified hazards and appropriate control measures.	Obtain and review the owner's permit-required confined space entry program and precautions for entering the on-site permit-required confined spaces prior to any planned entry.
Post signage on all permit-required confined spaces. Secure against unauthorized entry.	Inform the host employer or owner of the entry program that the contractor will be following.
Develop a written confined space entry program for the site. Include all elements of OSHA 1910.146.	Provide equipment necessary for safe entry into the permit-required confined space.
Coordinate entry operations when the host employer and/or one or more outside contractors are working simultaneously or together in a permit-required confined space so that employees from different employers do not endanger one another.	Ensure provisions are in place for emergency rescue in a timely manner.
Apprise the contractor of the elements, including the hazards identified and the host employer's experience with the space, that make the space in question a permit space.	Perform entry into permit-required confined space in accordance with applicable policies and regulations.
Apprise the contractor of any precautions or procedures that the host employer has implemented for the protection of employees in or near permit spaces where contractor personnel will be working.	Inform the host employer or owner of confined space hazards that are confronted or created in or near the space.
Ensure rescue services have timely access to all permit-required confined spaces.	Coordinate entry operations with the host employer when both host employer personnel and contractor personnel will be working in or near permit spaces to prevent unsafe or prohibited confined space conditions.
Debrief the contractor at the conclusion of the entry operations regarding the permit space program followed and regarding any hazards confronted or created in permit spaces during entry operations.	Inform the host employer of the permit space program that the contractor will follow and of any hazards confronted or created in permit spaces, either through a debriefing or during the entry operation.
Ensure contractors' permit-required confined entry procedures comply with OSHA 1910.146.	

Table courtesy of Kevin Stroup, CSP

when secondary sources of gas accumulate in the vault. Sometimes this can be from a source brought into the vault, such as welding gas cylinders. If the cylinders leak, the vault will accumulate the gas. In an investigation of a welder killed in an explosion in a vault during a water treatment plant construction, it was determined that the source of the explosion was a leaking regulator on an acetylene cylinder that was left in the vault overnight. In another situation, a vault that was protected with a fire-suppressant system, required a lockout/tagout (LO/TO) procedure in order to conduct hot work in the vault. In this case the LO/TO system was designed so that plant personnel were informed of the work and would not attempt to reestablish the fire-control system.

REFERENCES

American National Standards Institute. ANSI Standard Z117.1-1989 and Z117.1-2002. New York, NY: ANSI, 1989 and 2002.

National Fire Protection Association, "Recommended Practice forClassification of Class I Hazardous (Classified) Locations for Electrical Installations in Chemical Process Areas" (NFPA 497A). Boston, MA: NFPA, 1992.

_____. "Recommended Practice for Classification of Class II Hazardous (Classified) Locations for Electrical Installations in Chemical Process Areas" (NFPA 497B). Boston, MA: NFPA, 1992.

National Institute for Occupational Safety and Health. NIOSH Alert, "Request for Assistance in Preventing Occupational Fatalities in Confined Spaces," DHHS Publication No. 86-110, DSDTT, Cincinnati, Ohio: NIOSH, 1986.

U.S. Army Corps of Engineers, *Safety and Health Requirements Manual*, Manual No. EM 385-1-1. Washington, DC: Department of Army, 1996.

U.S. Department of Energy, *Department of Energy (DOE) OSH Technical Reference*, Chapter 4, "Confined Space Entry," DOE G 440.1. Washingon, DC: DOE, July 1997.

Occupational Safety and Health Administration (OSHA). 29 CFR 1910.146. "Permit-Required Confined Space Entry for General Industry." Final Rule, 1993. (Preamble to Final Rule is found in FR January 14, 1993.) Washington, DC: U.S. Department of Labor, 1993.

_____. 29 CFR 1926.21. "Safety Training and Education." Washington, DC: U.S. Department of Labor.

_____. "Confined Spaces in Construction (Part 1926): Preventing Suffocation/Explosions in Confined Spaces." Prerule unified agenda entry 1218-AB47-2625. Washington, DC: U.S. Department of Labor.

_____. "Permit Required Confined Spaces" Final Rule amendment to 29 CFR 1910.146. FR 63: 66018-66036. Washington, DC: U.S. Department of Labor, December 1, 1998.

REVIEW EXERCISES

The following three questions may be used to assess comprehension of the material presented.

1. In confined space entry, the primary hazards are chemical hazards, stored energy hazards, physical hazards, and entrapment or restraint of movement. List an example of each hazard as it might be encountered in a confined space entry and specify a practical method of controlling the hazard.

2. For gaseous chemical hazards, the three most common concerns are oxygen deficiency, flammable atmospheres, and toxic atmospheres. While monitoring with a typical multigas meter, what order should be followed in measuring these components? What effect would the relative density of the hazardous gaseous component have on the monitoring strategy?

3. The OSHA regulation for confined space entry lists specific task requirements that are assigned to participants in a permit-required confined space entry. List the four specific tasks that are required under the OSHA regulation and at least one specific duty assigned exclusively to each task.

■■ **About the Authors** ■■

INDUSTRIAL HYGIENE ISSUES IN CONSTRUCTION
by Adele L. Abrams, Esq., CMSP; Jeffery C. Camplin, CPEA, CSP;
and Richard K. Hofman, CIH, CPEA, CSP

Adele L. Abrams, Esq., CMSP, is president of the Law Office of Adele L. Abrams, P.C., Beltsville, MD. Ms. Abrams is a Certified Mine Safety Professional and MSHA-approved trainer, a professional member of the ASSE's Construction and Mining practice specialty groups (2000 "SPY" Award recipient, Mining practice specialty), an active member of the International Society of Mine Safety Professionals, the National Safety Council (past chairman, Cement, Quarry & Mineral Aggregates section), the Energy and Mineral Law Foundation, Women in Mining, the Holmes Safety Association, and the Washington Metropolitan Area Construction Safety Association. She earned a Juris Doctor degree from George Washington University National Law Center in Washington, DC, and a B.S. from the University of Maryland, College Park. She contributed the section on asbestos, hexavalent chromium, and diesel particulate matter.

Jeffery C. Camplin, CPEA, CSP, is President, Camplin Environmental Services, Inc., Rosemont, IL, holding a B.S. in Occupational Safety and Health from Northern Illinois University. His firm provides environmental and safety consulting, specializing in asbestos, lead, indoor air quality, and mold issues. He has been a USEPA accredited instructor on lead and asbestos since 1988. An ASSE Professional Member, Mr. Camplin received the 2003 ASSE President's Award for work with the USEPA on their lead paint initiative. He is currently serving a second term as Assistant Administrator of ASSE's Environmental Practice Specialty, and is a member of the Editorial Review Board for *Professional Safety*. He also serves on ASSE's Mold Taskforce, which is currently developing an ANSI standard for mold remediation. He contributed the section on lead-based paint.

Richard K. Hofman, CIH, CPEA, CSP, has over 25 years experience in environmental, safety, and industrial hygiene. He has a B.S., Environmental Science Degree from Wilkes University. He is the Vice President of Hofman Safety and Industrial Hygiene Consulting, Inc. (Hofman), located in Lebanon, Pennsylvania. He has managed and operated a hazardous waste landfill and its construction, including contractor oversight; he was the Manager of Safety and Industrial Hygiene for a specialty chemical company; and has serviced clients across the country for a large EHS consulting company and for Hofman. He is an Affiliate at Penn State University and Wilkes University. As a consultant, he has developed safety and industrial hygiene programs for clients and has trained on those programs. He has been active in the ASSE on the chapter, regional, and national levels and, as of this writing, is the administrator for the ASSE Industrial Hygiene Practice Specialty.

CHAPTER 26

Industrial Hygiene Issues in Construction

LEARNING OBJECTIVES

- Discuss the need to consider worker exposure scenarios in construction activities.
- Discuss the four concepts of anticipation, recognition, evaluation, and control.
- Explain how to perform a basic industrial hygiene assessment.
- List some of the industrial hygiene issues related to construction activities.
- Discuss the impact the following hazards may have at a construction site: silica, asbestos, hexavalent chrome, diesel particulate matter, and lead.

INTRODUCTION

Industrial hygiene issues in the construction industry are many. One of the biggest concerns about industrial hygiene issues is that the health impact on the worker is usually not detected for many years. These delayed effects do not get the attention they deserve. For example, if you are standing on an aluminum straight ladder putting a light fixture on a building, what is the hazard? Many would say the worker could fall off the ladder, or be electrocuted.

These two hazards can definitely have an immediate impact on the worker. Did you anticipate, however, that the siding on the house may contain asbestos, or that the paint on the house contained lead, and that while scraping, drilling, or running the wires that asbestos, lead, or even bird droppings in the attic could lead to health problems in the future?

INDUSTRIAL HYGIENE
—Isn't That the People Who Clean Teeth?

No! Industrial hygiene is the science and art of anticipation, recognition, evaluation, and control of health hazards. A construction worker is potentially exposed to hundreds of chemicals, physical hazards such as noise and radiation, and biological hazards. The industrial hygienist's job is to protect the worker and the company by ensuring that the workplace is healthy, that proper engineering controls are implemented, and/or personal protective equipment is selected to protect the worker.

In this chapter we will discuss the basics of what industrial hygiene is, review some of the risks, and discuss what is meant by the anticipation, recognition, evaluation, and control of hazards. A sample

Industrial Hygiene Audit will be reviewed and air-monitoring protocols discussed, followed by more detailed information on industrial hygiene hazards. The chapter will finish with a detailed review of silica and asbestos.

Definitions

Industrial hygiene: science and art of anticipating, recognizing, evaluating, and controlling environmental factors or stresses from the workplace that can cause sickness, impaired health, or significant discomfort among workers and the community.

Industrial hygienist: the occupational health professional who can, through training and education, anticipate, recognize, evaluate, and control environmental health factors and stressors.

The chapter makes repeated references to activities done by an industrial hygienist. This is not to imply that only personnel with a job title of Industrial Hygienist can conduct these activities; it is the level of training and experience that qualifies an individual, as noted in the definition above. Certain tasks, such as sampling, could be conducted by personnel with basic air-sampling training, preferably overseen, at some level, by an industrial hygienist. However, a person conducting detailed exposure assessments, establishing sampling protocols, overseeing monitoring programs, and the like, should be an industrial hygienist (i.e., person trained to that level). Some large construction firms employ corporate-level hygienists, while many small firms use a part-time consultant to help oversee their programs. It is becoming more common for major corporations that employ contractors to expect those contractors to have industrial hygiene programs already in place.

There are a number of chemical, physical, and biological hazards that face the construction industry today. In a recent conversation with a construction safety professional, we identified a number of specific issues in the construction trades that could impact the health of construction workers as well as other workers who may be near the area where construction work is being performed. Many hazards are well known, and there are specific regulations and/or guidelines that can be followed to protect workers. Unfortunately, many hazards are *not* well defined and the health and safety professional must do his or her homework to determine if a hazard exists and what that hazard is. Table 1, "Common Industrial Hygiene Issues," lists some of the chemical, physical, and biological hazards that can be present on the construction site.

■■ **Table 1** ■■

Common Industrial Hygiene Issues

Chemical	Physical	Biological
Chemicals listed in 29 CFR 1926.55	Ionizing radiation (1926.53) Nonionizing radiation (1926.54)	Bloodborne pathogens (bacteria, viruses)
Asbestos (1926.1101)	Ergonomics	Poison plants (ivy, sumac)
Lead (1926.62)	Noise (1926.52)	Bird droppings
Welding fumes	Vibration	Animal bites (snakes, spiders)
Silica	Heat stress	Mice feces, urine
Solvents	Cold stress	Animal feces, urine, carcasses
Corrosives: acids/bases		Mold
ACGIH TLVs		Bacteria, such as Legionella
NIOSH RELs		
Other chemicals 1926.1102–1926.1152, and 1926.60		

Anticipation

If you work with chemicals, earth products (sand, brick, stone), use mechanical equipment, use hand tools, carry heavy loads, reach into awkward positions, strike an object with a tool, use abrasive blast agents, grind, or weld, you may have an industrial hygiene issue. It is possible that because of the dose (exposure time and concentration) of the hazard, there may not be a risk. However, extend the work time or the concentration, or increase the toxicity of the chemical, physical, or biological hazard, and a worker may be at risk of illness, disease, or death.

There are thousands of environmental factors, or stressors, (chemical, physical, biological) that can cause harm. *Anticipation* is the ability to recognize the potential for exposure to a hazard. An industrial hygienist tries to anticipate the risk of exposure to a hazard prior to the work being performed by understanding the work and gathering information on the tasks to be performed. Control measures can then be utilized to reduce the risk to workers.

Recognition

An industrial hygienist observes the workplace where the hazards are anticipated. Based on observed work practices, controls in place, job tasks, chemicals being used, physical and biological hazards present, the industrial hygienist determines if a potential exposure exists.

For example, while abrasive blasting paint off of a tank, lead-based paint dust was blowing out of the containment area. In this instance, the control was not working properly and the risk to workers and the community was increased.

In another situation, a worker needed to move bundles of fiberglass roofing shingles to a second-floor roof. The worker faces a fall hazard and an ergonomic hazard posed by the weight and handling of the shingles, but is not exposed to fiberglass because the fibers are coated in a tar base and are not friable, nor can they become airborne at this time. The shingles should be mechanically moved to the roof to reduce any fall and/or muscle-strain hazard.

Recognition of the hazard requires knowledge and understanding of the hazard and how it affects the body. When looking at chemical exposure hazards we look at the "route of entry" into the body. Can the chemical be inhaled, ingested, or cause injury through skin contact. Once the chemical enters the body, what will it do—break down or metabolize into a harmless chemical, become more toxic, or accumulate in a body organ? When working with chemicals, a material safety data sheet (MSDS) should be available that will describe how the chemical can affect the body.

A painter using paint with an organic solvent base turns off the ventilation and closes the door to keep the bugs off the surface being painted. The room no longer has ventilation. What are the hazards? There is a potential fire hazard and a health hazard from high chemical concentrations, which could result in central nervous system depression or other health-related hazards. In this last example, you recognized that a hazard was present and informed the worker to modify the work environment. But even with the ventilation on, is that adequate protection? To determine the extent of the hazard to the worker, you need to evaluate the environment.

Evaluation

Evaluation is the act of determining the *extent* of an exposure to a chemical, physical, or biological hazard. This is usually accomplished by monitoring the worker in the work environment. The room the painter was in could be tested using handheld, real-time instruments that will give an indication to the industrial hygienist instantly or within a few minutes. The results can be compared to regulatory or manufacturer-recommended limits to provide an indication of whether the control was adequate. While these real-time, or grab, samples are helpful in evaluating the potential for overexposures, they are generally not used to determine compliance with regulatory limits because they are not longer-term personal samples. Unless the initial grab samples show low levels of air contaminants and little potential for higher levels, further sampling involving longer-term personal monitoring devices would be needed. The results would then be compared to regulatory as well as recommended exposure limits.

In the painter example, methyl ethyl keytone (MEK) was present at 400 parts of MEK per million parts of air (ppm). The regulatory OSHA permissible exposure limit (PEL), the NIOSH recommended exposure limit (REL), and the recommended threshold limit value (TLV) of the American Conference of Governmental Industrial Hygienists (ACGIH), all have a time-weighted, average (TWA) exposure limit of 200 parts per million (ppm). In addition, both ACGIH and NIOSH recommended a short-term exposure limit (STEL) of 300 ppm. A STEL is a time-weighted exposure limit that should never be exceeded during the day because longer exposure presents an adverse health effect of the chemical. A worker's exposure can exceed the TLV and go as high as the STEL for no more than a 15-minute period of time, four times per day. There must be a break of at least 60 minutes between 15-minute exposures (ACGIH). In the painter example, MEK has a STEL because of central nervous system and irritation effects. A worker is not to be exposed to MEK above an average of 300 ppm for any 15-minute period of time. A worker can be exposed to the 200 ppm TLV up to the 300 ppm STEL for no more than 15 minutes at a time, four times a day. In this example the exposure was found to be over acceptable limits; controls would need to be evaluated and implemented to reduce MEK to an acceptable level.

Evaluation of hazards can be performed by use of direct reading instruments and by collecting samples for later laboratory analysis. Direct reading instruments include colorimetric indicator tubes, gas meters, sound-level survey meters, radiation meters, dust monitors, photoionization detectors for organics, wet bulb globe temperature (WBGT) for heat stress, and more. These instruments are handy but may not provide actual employee exposure results unless they can be placed next to a worker or in a worker's breathing zone. Sample collection for laboratory analysis is used to evaluate a worker's time-weighted average and to determine compliance. For chemicals, it is usually more accurate than direct-read instrument sampling, and many more chemicals can be seen. However, it takes longer to obtain results, since it takes time to send samples to an accredited laboratory and for the lab to run the analysis. Samples are collected on sample media such as filters and sorbent tubes. Pumps are sometimes used to pull air through the sample media and must be calibrated before as well as after sampling. Passive dosimetry badges may also be used at times. Since they work by air diffusion into the sample media, a pump is not required. The body's exposure to some chemicals can also be determined through the collection and analysis of urine, blood, or breath samples. A list of biological exposure indicators can be found in ACGIH's TLV and BEI booklet.

Based on a comparison of exposure results to exposure limits, the industrial hygienist is able to recommend additional controls as needed.

An important factor to consider when evaluating jobs is the normal variation in exposures. Just as every rain shower produces varying amounts of rain, each day a job is done the exposures will vary. This is especially true for construction activities. One good sample result does not mean there are no overexposures. Likewise, one bad sample result does not mean major changes need to be made. The good sample could have resulted from an unusually light day while the bad one could have been caused by one abnormal event. Good documentation of task activities and environmental conditions during the sample period is important. Of similar importance is the establishment of a sound sampling strategy, including what job activities should be sampled and how many samples are needed to be statistically valid.

Controls

Controls are used to prevent exposure of an environmental factor or stressor to workers and the community. They should be used to prevent exposure. Table 2 shows the preferred hierarchy of controls.

CHEMICAL SUBSTITUTION

The best control option, which should be used whenever feasible, is *chemical substitution*, which *eliminates* a potential hazard. In one case, a painting contractor was hired to strip paint from a new half-mile-long enclosed conveyor because of a manufacturer problem with poor paint adhesion. The chemical stripper they planned on using was methylene chloride. Following a review of the chemical hazards and regulatory requirements for use of this stripper, the contractor found a safer alternative. As a result, the job was completed safely and with a side benefit of better environmental quality.

ENGINEERING CONTROLS

It would be ideal to have engineering controls for every job to protect workers from hazards. However, short-term jobs and the expense of the controls may be prohibitive and personal protective equipment may at times be an acceptable alternative. Another issue with engineering controls is that they must work properly. For example, if a ventilation control is recommended for collecting dust from a diamond-bit saw used for cutting concrete, the velocity of the air at the pick-up point needs to be fast enough to capture the dust, then fast enough in the duct to transport the dust so it does not plug the duct, and dust must be collected in an efficient collecting device so that workers near the collection device are not exposed to dust blowing out of it. A portable welding hood that is located close to the work being performed can be very effective. However, if the hood is located too close, weld quality will suffer, and if it is several feet away from the welding operation, the amount of fume captured will be minimal. If used improperly, workers will only see it as a worthless nuisance and soon discontinue its use. It is important that workers be properly trained on control devices and that their use and proper functioning be periodically checked. The workers are often the best experts on the jobs they do and should be involved in developing engineering controls. This will help ensure getting it right the first time and ending up with something the workers will use.

ADMINISTRATIVE CONTROLS

Administrative controls are used by management to reduce exposures to a safe level. Since these controls are dependent upon people consistently following procedures, they are generally not as reliable as the first two types. Some examples of these types of controls include:

- Workers may be asked to switch positions every 15 minutes with a co-worker to use other muscles and to prevent injury of any one muscle group.
- A worker performing a hard manual task in extreme heat may be instructed to work 15 minutes and break for 45 minutes in a cool environment. (ACGIH heat-stress guidelines)
- A worker dumping bags of silica-containing refractory mix is instructed not to squeeze down the bags as they are put into a waste container. This is to minimize creating a dust cloud in his breathing zone each time a bag is squeezed.

The goal is to limit exposure to the hazard through changes in how a job is done. Periodic job audits should be done to ensure the correct procedures continue to be followed. If not stressed, exposures may again occur and respiratory protection would then be required.

■■ **Table 2** ■■

Hierarchy of Controls—Preferred to Least Preferred

Control	Example
Eliminate the hazard	Chemical substitution with lower-hazard chemical or process
Engineering	Ventilation, robotics, wet methods
Administrative	Work-schedule modification in hazard zone, extended breaks, work practices
PPE	Respiratory protection

PROTECTIVE EQUIPMENT

Personal protective equipment (PPE) is the control of last choice because the hazard is still present and you need to rely on the worker to use the PPE properly. It is up to the company to determine the proper PPE to be worn through a hazard determination process. Of specific concern to the industrial hygienist is the proper selection and use of respiratory protection. If respirators are to be used (even optional usage), specific respiratory regulatory requirements must be followed. (1910.134)

Industrial Hygiene Audit

After having the chance to review the process of anticipation, recognition, evaluation, and control, how can you use this information in the field? One of the most effective tools for an industrial hygienist is the Industrial Hygiene Audit. The audit can be designed as a preliminary walk-through of a work area or job site, or it can be designed to be more detailed in nature, based on known or identified hazards.

Preparing for the audit is accomplished in the anticipation phase by gathering information on potential hazards, jobs to be performed, monitoring results from previous similar jobs, control measures in place, MSDSs, and toxicology information from both regulatory agencies and science-based resources. Industrial-hygiene-related programs can also be reviewed to identify if a program exists and if acceptable work practices are included. Of course, if a needed program does not exist, one may have to be developed. Once the background information is obtained, an audit checklist can be compiled to address the issues of concern for the hazards identified. A number of industrial hygiene audit checklists can be obtained through the various resources, including the OSHA Web site (OSHA.gov), where the inspection criteria for some chemicals can be found in the field operations manual and the enforcement guidance documents. Your insurance carrier will usually have checklists that can be used and topical checklists can be found on the Internet. Figure 1 is a generic checklist that can be modified for your needs.

Once the audit checklist is complete, conclusions should be drawn from your findings. If all controls are working and no potential for exposure exists, then all is well. When will a follow-up audit be conducted to ensure that controls remain in place or if new hazards are introduced? If a hazard is identified in the audit, was it controlled properly or is monitoring required to evaluate the extent of the exposure? Make recommendations and follow up on those recommendations.

A final note: Whenever industrial hygiene monitoring is performed, workers must be informed of their personal monitoring results, or results that were collected that represent their exposure.

AIR MONITORING

Industrial hygienist—isn't that the person who hangs a pump? Isn't that where you take a pump and a filter and find out what is in the air? Unfortunately there are safety practitioners who think that is all there is to exposure monitoring—grab a pump and sample. Once you've determined that air monitoring is the way to go, you should develop a sampling strategy that helps to identify the objective(s) for taking samples. There are many ways you can sample for some chemicals and no method of sampling for others. With the development of thousands of new chemicals each year, there may not always be a sampling protocol available, and there will probably not be an OSHA PEL or ACGIH TLV to compare your results to.

Of the many questions you should be asking, here is a possible list:

- What are you going to be using the data for? Information? Developing controls? PPE determination?
- What will you be sampling for?
- How is it transported in the environment?
- Is there an analytical method?
- Under what conditions should the samples be collected? Worst case? Routine?
- Who should be monitored and how many samples should be taken?
- Can you properly perform the collection technique once the parameters are established? Equipment available? Collection methods? Accuracy? Reproducibility?

Name of Facility: _____   Date: _____

Name of Contact Person: _____   Phone No.: _____

Individuals Conducting Survey:

_____

Describe the job location, type of work to be performed, number of employees, shifts, hours of each shift.

_____

What chemicals are being used as raw materials? Include naturally occurring minerals and wood products as well as other materials that may result in exposure, such as steel beams that will be cut or welded. Do you have material safety data sheets for these chemicals and materials? Is there a PEL, TLV, REL, or other exposure parameter for the chemical?

_____

Based on the list of chemicals and the activities, do you expect that a worker may have an exposure to a chemical (inhalation, ingestion, skin absorption)? List the chemical(s) of concern per task.

_____

For chemicals listed, what are the control methods that you will use to remove or reduce exposures to the chemical(s)? List types of controls to be used, including substitution, engineering controls, administrative controls, and personal protective equipment.

_____

Do you need to conduct monitoring (noise, chemical, radiation, vibration, etc.) to determine if an exposure hazard exists (for example, welding fumes) or the level of PPE to be worn?

_____

Are chemical containers labeled with the name of the chemical and hazard?

_____

Is there a respiratory protection program? Have workers who are required to wear respirators been medically qualified, trained, fit-tested, and provided with the proper respirator for the hazard? Is air monitoring conducted?

_____

List any physical hazards that may be present, including noise, vibration, radiation (ionizing, nonionizing), ergonomic, and heat or cold stress.

_____

Provide details on noise levels and duration; ergonomic details, including the frequency and weights being pushed, pulled, carried, lifted, and the types of moves expected.

_____

List any biological hazards that may be present, including plants, insects, animals, feces, mold, human pathogens, and potential sources of high bacterial growth.

_____

Is a medical monitoring program needed to determine baseline exposure to chemicals and current health status (i.e., blood lead level, audiograms)?

_____

■■ **Figure 1** ■■ **Industrial Hygiene Audit checklist sample**

- What are the limitations of the monitoring device or direct-read instrument?
- Do you expect there to be a lot of day-to-day variation in the task and the environmental factors?

Before you collect the sample, have you chosen a laboratory?

There are a number of laboratories that will claim they can run your industrial hygiene samples. Some can do it better than others. You need to determine what is important for you. The working relationship you have with your laboratory can be a significant component to getting the sampling strategy correct. For example: Have you sampled the right chemical on the right sample media at the correct air-sample rate for the correct duration? Has the sample been properly handled after being collected, such as sealing it or keeping it cold if required? Based on the information that you provided to the laboratory, they will analyze for the correct substance(s) and report the information back to you in a timely manner in the format that is useful to you.

Some Things to Look for When Selecting a Laboratory

You can see that laboratory selection is an important part of the process of sampling. If all you are doing is shopping for price, you may lose on quality assurance or support. If you have a large volume of samples, you can usually work out a better pricing schedule.

- Ask for copies of the laboratories certification/accreditations. Verify that these are applicable to the substance(s) you want analyzed.
- Does the laboratory have a quality assurance plan that includes standard operating procedures for the equipment they use? Ask how it would be used for samples you submit to the lab.
- Does the lab provide sampling media?
- Is there support for you when you need to review the sampling strategy?

- Is the turn-around time acceptable? Are there charges for faster turn around?
- Do you understand the lab report format?
- How are sample blanks handled by the lab?
- Can they analyze for the chemicals you want?

Hang the Pump?

It might be time to hang the pump. But first be sure to check the calibration of the pump and other sampling devices, as applicable. This should also be checked following sampling to ensure the equipment continued to work properly. Don't forget the field observations. Do not hang the pump and run. Remember that *industrial hygiene* is the anticipation, recognition, evaluation, and control of workplace environmental factors. If at the end of the sampling period a worker was above OSHA's PEL, your observations may have helped to identify that the worker was exposed to the hazard during an activity that lasted 15 minutes, not 8 hours. Document any unusual events that occurred and any task activities that could contribute to a high exposure during the day. For example, was the job shut down for 2 hours due to rain, or was a welding fab job moved into a tight enclosure just this one time to avoid the rain? One event may greatly reduce the exposure, while the other greatly increases it. Neither may be truly representative of the actual exposure. While several additional samples would normally be collected from different days, and would hopefully be low, the one *high* sample may still indicate some action is needed, since it may represent 33 percent of the samples if a total of three days' sampling was completed. Good documentation would prove this was an isolated extreme case and indicate that the solution for the high welding-fume exposure may simply be not to weld in an enclosed area again—especially when other air-sample results collected in an open area indicate low exposure levels. The cost of unneeded engineering controls would be saved in this instance and emphasis could be placed on other issues that might have a much greater safety or health impact on the worker. Of course the opposite may also be true, and a task with high exposures could have gone unidentified, possibly impacting the long-term health of a worker.

INDUSTRIAL HYGIENE HAZARDS

Earlier in this chapter on industrial hygiene a number of chemical, physical, and biological exposure items were identified. The hazards at your job site may vary, and may not even be caused by your operation, but by another contractor. If your people are potentially exposed to a hazard, they may be at risk. Following is a brief description of several of the hazards discussed. More detailed information can be found on the Internet through many resources, including the cdc.NIOSH.gov Web site.

Noise

In high noise areas, sound can be a physical hazard that affects many workers. Excessive noise levels (higher than 85 decibels) over an extended period of time, or high-impact noise, can lead to permanent hearing loss. As a rule of thumb, if you are arm's length from another person and you have to raise your voice to talk to him/her, then your work environment is at or above 85 decibels. Protection against levels exceeding 90 decibels of noise is required in the OSHA Construction Standard. Because some workers over time could lose some of their hearing from noise exposures between 85 and 90 decibels averaged over an 8-hour day, the use of hearing protection is recommended at or above 85 decibels. In the construction industry, feasible administrative and engineering controls must be put in place at 90 decibels, and a hearing conservation program (recommended at 85 decibels) developed, which should include a written program, annual audiograms, annual training, noise monitoring, and the proper selection and use of hearing protection devices (see OSHA 29 CFR 1926.52).

Chemical Hazards

Chemicals are present in many forms, such as gases, vapors, fumes, mists, and dusts that are airborne. These respiratory hazards, plus direct skin contact or ingestion, can lead to exposure. When evaluating specific chemicals, obtain information from the MSDSs, manufacturer, and regulatory and science-based resources. The OSHA standard 1926.55, plus a number of specific chemical standards, are found in these regulations. It is the chemical user's responsibility to ensure that employees are not overexposed to the listed chemicals.

Solvents

Solvents are chemicals that dissolve into other chemicals. Water is the universal solvent. Solvent exposures that are generally a concern are the organic solvents that tend to become airborne through evaporation of liquid vapors or from spraying operations where a mist can be inhaled. Solvents can affect various organs such as kidney, liver, spleen, blood, and central nervous system, depending on the composition of the specific solvent. Many common solvents are listed in the OSHA rules and their PELs, NIOSH RELs, and ACGIH TLVs are published. Ventilation controls, substitution to water-based products, and respiratory protection are common methods for reducing the potential for excessive solvent inhalation exposure. Air-sampling methods for many listed chemicals are available through the NIOSH Web site at cdc.niosh.gov.

Acids and Bases

These are corrosive chemicals that can cause burns and blistering to the skin on contact. A dilute mixture of corrosives can cause irritation to the skin and, eventually, with extended contact, cause blistering. Protect the eyes from splashes of corrosives; they can cause blindness. As a hazard recognition issue, any operation that uses, heats, or sprays acids or bases can cause skin irritation and burns. Sulfuric acid has been identified as an "A2 Suspect Human Carcinogen" by the ACGIH in the TLV and BEI guidebook. Some of these corrosive materials will have a delayed burning effect and may also be toxic when absorbed. Be sure to review the MSDS for each chemical since not all corrosives are the same.

Heat Stress

Heat usually becomes a physical hazard when work is performed in a hot environment. There are a number of factors that affect a person's ability to control body heat and not get overloaded. Factors include: air temperature, humidity, air movement, and radiant heat exchange [these factors can be measured using WBGT (Wet Bulb Globe Temperature)], clothing worn, health status, acclimation to hot conditions, and a person's metabolic load. Metabolic load is based on how physically hard a person is working A worker putting a roof on at 1:00 pm in the summer has a higher heat load than a person standing in the shade reading a blueprint. Key factors in this case include the radiant load increase from the sun, temperature increase from the hot roof, and metabolic load increase for the roofer compared to the worker in the shade.

Workers who are overheated may be more susceptible to accidents, and if the body's core temperature reaches critical degrees, life-threatening heat stroke is likely. There are many guidance documents available through OSHA.gov on protecting workers from heat stress. Recommended work–rest regimens are available through the ACGIH TLV and BEI guidebook. One of the key steps in preventing heat-stress illnesses is to make sure workers drink plenty of fluids. Cool water is usually the best, but if a lot of heavy sweating is expected, an electrolyte replacement drink may be desired. Be sure to avoid carbonated beverages as well as those containing caffeine. Workers should also wear loose-fitting clothes and take breaks in a cooler area. It should also be noted that someone recovering from the flu or other illness may be at a much higher risk for heat stress. When they return to work, they may not be able to handle the heat as well as before and may need to take it a bit easier for a few days until they become reacclimated.

Cold Stress

A cold environment will cause the body to lose heat and can lead to hypothermia, a life-threatening drop in deep-body core temperature to below 96.8°F (ACGIH TLVs and BEIs). Lack of activity, poor insulation, damp clothing, exposed skin, and immersion in cold water are primary factors in cold stress leading to hypothermia. Also, exposed skin can lead to frostbite. To fight off cold stress, wear layered clothing, cover the skin, take periodic warming breaks, and remember that an increased metabolic rate increases body heat.

Biological Hazards

Dangerous biological elements cover a wide variety of agents/organisms. The news has many articles on anthrax and monkey pox, but these are not as common as other hazards. Viruses, such as human immunodeficiency virus (HIV), hepatitis, and tuberculosis (TB) are several of the many pathogens carried in the human body. These pathogens can be transmitted to co-workers if there is an injury resulting in the transfer of body fluids, and in some instances (TB, for example) when coughed aerosols of the infected individual are inhaled.

Insect bites may also lead to infection and illness, including West Nile virus from mosquitoes and Lyme disease or Rocky Mountain spotted fever from ticks. Other insects and animals may have a toxic bite or sting that can cause an adverse reaction in the body. Clearing brush can expose workers to insect bites as well as contact with plants such as poison ivy and poison sumac.

Mold has received a lot of attention, and some types of mold may cause allergic reactions in some at-risk individuals under certain conditions. It may often be encountered during the demolition, or renovation, of water-damaged buildings.

Animal feces from livestock, mice, and birds have also been known to cause illnesses and infections. Construction demolition work performed where there is extensive bird feces (histoplasmosis) requires an understanding of the potential hazards and proper PPE, determined through a hazard evaluation. If a neighboring building has a cooling tower, Legionella (legionnaire's disease) could be present. Work involving contact with biologically active sludges, such as cleaning out sumps, cooling tower basins, or pits, may present a serious health risk. Any skin cut in the presence of biological material can result in a severe infection. If material is ingested, such as by touching the mouth with a contaminated glove (gloves should always be worn for this kind of work), a severe gastrointestinal illness may result. Even the water-based cutting fluids that are used in some metal-working equipment may present a biological hazard if it is not changed periodically and treated with a bactericide. Simple sampling methods are available for evaluating many of these potential biological hazards.

Radiation

Radiation is used in the construction trades to evaluate the thickness of materials; to examine welds on steel and other items; to move particles around at high velocity to seal, weld, or mold plastics; and to activate light-sensitive materials. Different forms of radiation can cause different levels of harm to the worker. Ionizing radiation is electrically charged or neutral particles from radioactive source material, or may be generated, for example, by using an X-ray tube. If you are providing service to a nuclear energy plant, you are required to go through training on ionizing radiation, which is capable of causing cell damage, including cancer, and includes alpha particles; beta particles; and gamma, X-ray, and neutron radiation. Ionizing radiation may be present at many industrial sites in the form of fixed-level gauges and portable measuring equipment. Radiography contractors often bring relatively large radioactive sources onto a construction site when critical welds need to be X-rayed. Using X rays in construction requires specialized training [OSHA 1926.53 and Nuclear Regulatory Commission (NRC) regulations]. Other site construction workers must be made aware of these radiation sources and trained on proper precautions for working around them. For example, during a radiography shot, a high radiation area may exist, which will be blocked off with barrier rope and signs. No unauthorized personnel are allowed to enter this restricted area. While there is an exposure limit for radiation, the guiding principle to be followed is called ALARA—as low as reasonably achievable. If all reasonable steps to ensure exposures are *not* ALARA, then a regulatory violation (most likely from state or federal NRC) may be issued, even though an overexposure did not occur. For construction work in some parts of the world, background radiation levels may be high in natural soils.

Nonionizing Radiation

Because they do not produce ions, electromagnetic energy, which includes ultraviolet (UV) light, lasers, infrared light, microwaves, and radio frequency, are classified (OSHA 1926.54) as nonionizing radiation. Overexposure to UV light can cause sunburn; lasers can cause eye damage; and microwaves cause heating that can literally cook a person's flesh. The key to radiation safety is to recognize the hazard, keep your distance, and reduce your contact time. Spend 15 minutes in the sun in Chicago in the fall and you may feel a little warmth on your skin; fly to Puerto Rico and spend 15 minutes in the sun and you could be seriously burned—you are closer to the sun. Ultraviolet light is also generated in welding operations and can cause sunburn on the skin and on the lens of the eye. This is also referred to as welder's flash. Precautions are required to protect welders and nearby workers from arc flash.

CRYSTALLINE SILICA

For many of us silica is dirt, or the sand on the beach. The most abundant mineral at the earth's surface, crystalline silica has been assigned to an Occupational Safety and Health Administration Special Emphasis Program. According to OSHA, over 2 million workers are exposed to this dust, which claims 200 to 300 lives a year in the United States. In this text, the term *silica* refers to *crystalline silica* unless otherwise indicated.

What exactly is silica? And what's the big deal?

Silica refers to the chemical compound silicon dioxide (SiO_2), which occurs in a crystalline or noncrystalline (amorphous) form. Crystalline silica (see Figure 2) may be found in more than one form. The polymorphic forms of crystalline silica are alpha quartz, beta quartz, tridymite, cristobalite, keatite, coesite, stishovite, and moganite (NIOSH).

When workers inhale crystalline silica, the lung tissue reacts by developing fibrotic nodules and scarring around the trapped silica particles. This fibrotic condition of the lung is called silicosis. If the nodules grow too large or too numerous, breathing becomes difficult. While silicosis may not present symptoms in its early stages, long-term overexposure may eventually result in death. Silicosis victims are also at higher risk of developing active tuberculosis. Although many cases of silicosis result from years of exposure to a low dose of silica, there is also a disease form known as acute silicosis. Acute silicosis is a result of inhaling high concentrations

■ ■ **Figure 2** ■ ■ **Detailed microscopic view of silica particles on a filter** (Photo courtesy of NIOSH. Scanning electron micrograph by William Jones, Ph.D.)

of silica, resulting in death within one to several years (NIOSH). After reviewing the International Agency for Research on Cancer (IARC) findings and other studies, the National Institute for Occupational Safety and Health has presented information to OSHA that recommends silica be considered a potential occupational carcinogen (NIOSH).

There are other health-related symptoms from exposure to silica that can be found in the literature. It is recognized by the various interest groups and agencies that there is still a lot of work to be done in evaluating the health effects of silica. A number of research programs are evaluating how the body reacts to silica particles. For example, is freshly fractured (cut, ground, crushed, processed) silica more of a hazard than silica particles that have been processed days earlier? Why do smokers exhibit a higher risk for developing silicosis than nonsmokers?

On May 2, 1996, a memorandum from Assistant Secretary for OSHA, Mr. Joseph A. Dear, to the OSHA regional administrators, outlined a Special Emphasis Program (SEP) for Silica. The strategy targets at-risk operations in both construction and manufacturing. These facilities in high-risk SIC codes have their names added to a list. If your name is selected, OSHA will evaluate your program, and identify the inspection as part of the SEP process.

If you have a program in place and have protected workers, the compliance officer will not spend a lot of time at your facility. If you do not have a program, OSHA will spend more time with you.

On the regulatory side, OSHA uses the following formula to determine the permissible exposure limit for dust containing silica quartz. In this equation, the percentage of silica quartz determines the PEL. OSHA's PEL for respirable dust not containing silica is 5 mg/m^3, or 10 ÷ 2. Respirable dust is dust that is less than 10 microns (< 10 μ) in size. A micron equals 1/1000 mm or 1/25,000 inch. The effect of this PEL equation is to lower the allowable respirable dust concentration as the percentage of silica in that dust increases.

$$PEL = \frac{10 \text{ mg/m}^3}{\% \text{ quartz} + 2}$$

Sample Calculations:
If a sample has 20% silica quartz,

$$PEL = \frac{10 \text{ mg/m}^3}{20\% \text{ quartz} + 2} = 0.45 \text{ mg/m}^3$$

If a sample has 35% silica quartz,

$$PEL = \frac{10 \text{ mg/m}^3}{35\% \text{ quartz} + 2} = 0.27 \text{ mg/m}^3$$

As the examples indicate, the higher the concentration of quartz in the sample, the lower the PEL.

The current NIOSH recommended exposure limit is 0.05 mg/m^3 for the silica fraction in respirable dust as a time-weighted average for up to a 10-hour day for a 40-hour work week (NIOSH). The American Conference of Governmental Industrial Hygienists has established the threshold limit value of 0.05 mg/m^3 for the silica fraction in respirable dust for an 8-hour day for a 40-hour work week. ACGIH has also listed crystalline silica as a suspect human carcinogen.

With silica all around us, is everyone at risk? As a safety professional providing industrial hygiene services at a construction project, you would need to look at the whole picture. Let's look at anticipation, recognition, evaluation, and control as means to protect workers.

Anticipation

To be able to anticipate a health risk, such as that presented by silica, you need to be aware of the potential for exposure. The information presented earlier is just the start of your search for knowledge about the risks of silica. To take this a step further, if you work in an *at-risk* industry, such as construction, you can anticipate that some of the activities that are performed can put workers at risk of exposure. If you work with dirt, you may be at risk. If you work with earth-building materials such as brick, stone, tile, sand, and so on, you may be at risk. For example, a concrete floor at a new construction site is not a significant risk until you learn that the plumbing contractor needs to cut a trough in the concrete to run some pipes. At this time your antennae should pick up and you should be saying to yourself, "there might be a silica exposure problem with this operation."

Recognition

Recognition of a silica exposure hazard is once again based on some knowledge of the hazard. If you have silica in the workplace, the recognition of exposure is based on the activities performed. The hazard arises when silica dust becomes airborne and available to the respiratory tract. Following is a list of jobs and related activities that usually generate dust in and around construction sites:

- chipping, hammering, and drilling of rock
- crushing, loading, hauling, and dumping of rock
- manufacture of products containing silica such as cement, block, brick, paving stone, glass, pottery, and ceramics
- abrasive blasting using silica sand as the abrasive
- abrasive blasting of concrete or masonry (regardless of abrasive used)
- sawing, hammering, drilling, grinding, and chipping of concrete or masonry
- demolition of concrete and masonry structures
- dry sweeping or pressurized air blowing of concrete, rock, or sand dust

- mixing refractory for spray-on or cast applications, or removal of it.

Recognizing at-risk activities can take place as part of a work-site audit. If you walk through the work site and see dust coming from an operation where silica may be present, there is a potential for silica exposure. Remember that the dust you see is not the microscopic ($< 10\ \mu$) size particle that can get deep into the lungs. One of the ways to recognize the presence of silica is to take a bulk sample of the media (i.e., concrete) and send it to a laboratory for analysis to determine the percent silica. However, the percent in the bulk sample is not a reliable indicator of the percent silica in a respirable dust air sample. Another way to recognize a dust-generation hazard is to shadow a worker using a handheld respirable dust monitor. By holding the monitor at or near the worker's breathing zone while a task is performed, you can determine if this is an at-risk activity that may need further evaluation. Real-time, dust-monitoring equipment can be found through environmental equipment rental companies and through companies that manufacture air-monitoring equipment. Based upon field observations, if you see visible dust around a worker for more than a brief period of time, and the material is known to contain silica, a more detailed evaluation is warranted.

During a site evaluation for silica exposure, a sand mixer had exposures to silica over the PEL. Although he was protected with a respirator, the goal was to identify the sources of exposure. By using a handheld monitor, a number of tasks were identified that collectively led to the worker's full-shift overexposure. The specific high-risk tasks could then be individually controlled to reduce overall exposure levels. When a second worker performed the same tasks, using different methods or techniques, there was an increase or decrease in exposure based on how the work was performed. This helped to identify personal work practices that resulted in some workers having a higher exposure than others. An evaluation of work-practice techniques should be included in the recognition of silica exposure hazards. Additional tasks that were identified as being "at risk" included: how a bag of sand additives was opened and poured, how the bag was disposed of (placed directly into a trash container, or was first squeezed and folded), how employees had worked around conveyor systems, how housekeeping was performed, how quality-control samples were collected, and how much dirt was allowed to accumulate on a person's clothing.

Evaluation

Do you have workers exposed above the OSHA PEL, ACGIH TLV, or NIOSH REL? To evaluate the workers' exposure, a sampling strategy must be developed. The handheld monitor discussed earlier has helped to identify workers that may be at risk of overexposure to respirable silica. You do not know if the dust sampled actually contained silica until you sample the air and have the filter analyzed by a competent laboratory. The sampling strategy should include: what job tasks are to be sampled, how many samples should be collected and over what time frame, what sampling and analytical methods will be used, and the certification status of the laboratory that will run the analyses. Once you receive the results from the lab, you need to calculate the OSHA PEL, and/or compare the results to the ACGIH TLV, or NIOSH REL, or another guideline that may apply to your situation.

Remember that you need to collect a respirable dust sample. The dust is therefore collected using a cyclone or equivalent device that will collect particles in the respirable fraction range. The most common method to sample and analyze silica is NIOSH Method 7500, which uses XRD (X-ray diffraction) to analyze the sample; this method has been fully evaluated. Method 7601 uses VIS (Visual Absorption Spectrophotometry) to analyze the sample and Method 7602 uses IR (Infrared Absorption Spectrophotometry) to analyze the sample. Methods 7601 and 7602 have not been fully evaluated (NIOSH NMAM). In all of these methods, the sample collection technique is similar. The air-flow collection rate will vary slightly based on the make and model of the particle-size selection device (e.g., cyclone) being used. You must use the sample flow rate as specified by the manufacturer.

The published methods outline specific problems that may occur during sampling. If the cyclone

■ ■ **Figure 3** ■ ■ **Worker wearing a cyclone sampler**

is tipped into a horizontal position, or even upside down, there is a possibility that some large, nonrespirable particles will get onto the filter, resulting in higher exposure numbers. The correct location for placing the collection apparatus is in the worker's breathing zone, as you can observe in Figure 3.

Sampling should take place for the entire work shift. The ACGIH TLV and OSHA PEL are based on up to an 8-hour day and 40-hour work week. The NIOSH REL is based on a 10-hour day and 40-hour work week. From an industrial hygiene sampling viewpoint, it may be appropriate to collect several samples during the day. For example, if a worker would perform an operation in the first 2 hours of the shift, a second operation in the next 4 hours of the shift, and a third operation in the last 2 hours of the shift, three samples would be collected. The total sampling time would still be 8 hours. If any of the three operations generate more dust, you could then spend time correcting the problem area or task. A calculation for adding exposures through the total shift is found in the OSHA SEP. Additional sampling on other days and involving other workers should also be considered to ensure the results are representative of the typical exposures. This is part of the sample strategy and the decision would be based upon expected variation in the dust levels, job variations, worker variations, work-practice variations, and past dust exposure levels.

There have been a number of errors made determining compliance with the OSHA PEL. The most common error is to compare the concentration of silica quartz from a sample to the calculated OSHA PEL. You must compare the concentration of respirable dust to the calculated OSHA PEL on a per-sample basis. Yes, you may have ten different OSHA PELs for ten dust samples. Typically, the laboratory will give you the following information:

- total weight of dust on the filter
- concentration of dust on the filter in mg/m^3
- concentration of silica in mg/m^3
- percent silica on the filter.

You may then calculate the OSHA PEL based on the formula previously given.

EXAMPLE OF A SILICA EXPOSURE CALCULATION FOR OSHA COMPLIANCE

If a sample has 0.25 mg/m^3 silica quartz, and 1.50 mg/m^3 respirable dust (or 17% silica), then:

$$\text{PEL} = \frac{10 \text{ mg/m}^3}{17\% \text{ quartz} + 2} = 0.53 \text{ mg/m}^3$$

Wrong: Compare 0.53 mg/m^3 PEL to 0.25 mg/m^3 silica.
Correct: Compare 0.53 mg/m^3 PEL to 1.50 mg/m^3 respirable dust.

In this example you could have improperly determined that you were 50% of the PEL when in actuality you were almost 300% of the PEL.

If you want to compare your sample result to the ACGIH TLV or the NIOSH REL, you would compare the concentration of silica quartz (not respirable dust) directly to the TLV of 0.05 mg/m^3. In the example provided, the silica concentration at 0.25 mg/m^3 is 500% of the TLV and REL.

More than one day of sampling may be needed to adequately characterize the potential for exposure to workers. Variables may include products or materials used, workload, equipment, air movement, humidity, and individual work practices. Sometimes, even under the same conditions, individuals may perform their jobs differently. Do you remember the Charles Shultz character "Pig Pen" on the Charlie Brown comic strip? I have monitored many workers who resemble that character.

Figure 4 shows a construction worker cutting earthen tiles while using a dust-collection device and a respirator. The exposure to dust clearly was anticipated and recognized. I do not know if the worker or equivalent work were evaluated for dust or silica concentration. (*Note:* this may be why the ventilation system and respirator are being used). A further evaluation would show that the ventilation system was not very effective in controlling the dust. Did you ask yourself if the respiratory protection was adequate? What can be done to improve the ventilation system? Is the shop vacuum equipped with a high-efficiency particulate air (HEPA) filter?

Controls

Through observation and industrial hygiene monitoring, you have defined the problem areas and tasks. Now what can you do to correct the dust problems? Engineering controls are always the first priority. In production and construction operations the most common controls are: capture ventilation, shrouding, and wet methods. These methods, if used appropriately, can reduce exposure. If not designed properly, and/or not used according to

■ ■ **Figure 4** ■ ■ **Photograph of a construction worker cutting through earthen tile** (Photo by Richard K. Hofman)

their design, ventilation systems will not work. Ventilation controls should be periodically tested for air flow and capture velocity. Ventilation systems should be periodically opened and inspected for material build-up from settled dust, and for wear and tear. Sand moving at high velocity inside ducts tends to wear out the joints and the impellers of the blowers. Shrouding is a cover at the source of the dust-generating device that helps minimize dust creation while improving ventilation dust control.

Ventilation systems (dust collectors) are used to collect dust at the source, or in a room generally. Source collection is preferred for keeping dust away from the worker's breathing zone. Dust collectors come in a number of shapes and sizes, and can be selected based on your needs. Contact a ventilation/pollution control company in your yellow pages or through the Internet. Some dust-collection issues to be aware of are:

1. You need enough velocity (transport) to carry the dust in the duct without it settling out.
2. Although you are trying to capture particles, the particle size that is harmful is < 10 μm and acts like a gas when it is airborne. It can stay afloat for hours, even days.
3. Larger dust particles can also inhibit your body's natural defenses from working properly and should be lower than the permissible total dust (particulate) concentration. The OSHA PEL = 15 mg/m^3, the ACGIH TLV = 10 mg/m^3, and the NIOSH REL = 10 mg/m^3. OSHA also has a formula to calculate the PEL for total dust containing silica (29 CFR 1910.1000, Table Z-3); however, the respirable fraction, as discussed above, is the recognized hazard.
4. Try to reduce the number of turns in the duct and try not to use slide gates, which get damaged by the sometimes abrasive dust.
5. Keep the pick-up points as close to the source of the dust as possible.
6. Large dust-collection systems tend to vent outside. You may need to check air-pollution control permits and requirements. Dust-collection systems that return air into the workplace (i.e., recycled air) must pass

through a collection device such as a high-efficiency (99.99% efficient at 0.3 μm) filter. The question is, however, can you tell if the filter is working? Remember the size particle that you are trying to protect yourself from is < 10 μm in diameter. ANSI/AIHA Z9.7-1998 discusses the recirculation of air from industrial processes. Specifically, the guidance document indicates that recirculated air should not exceed 10 percent of the regulated exposure concentration.

A fine water mist can be applied over the operating work area. In several operations evaluated, even where the misting system was quite close to the workers, the workers did not get wet. The mist appears to have caused the microscopic dust particles to agglomerate, or be attracted to each other, so that they dropped out of the air. When applied near the dust-generating area, dust levels dropped dramatically. It took several tests with the sprayer head placement, sprayer size, and water and air pressure to get the system fine-tuned. The end result was that respirators were no longer needed.

Keeping the workplace clean is easier said than done. If you are considering sweeping—think twice. The dust generated during sweeping can easily exceed 10 mg/m^3 of respirable particles. These particles, being microscopic, stay airborne for a long time. Several options to reduce the dust are:

1. Use a vacuum with a high-efficiency filter. Avoid sweeping. Canister vacuums with regular filters can increase exposure to respirable particles because the microscopic particles can blow back into the air through the vacuum filter.
2. If you must sweep, use a sweeping compound. The sweeping compound mixed with the dust will keep small particles from staying airborne. A test that I conducted with a person sweeping with and without a sweeping compound showed that without a sweeping compound the respirable dust concentration rose to 10 mg/m^3 during the sweeping and after one minute was still above 1 mg/m^3 in the air even though the sweeping had

stopped. The dust results when sweeping compound was used showed that the dust cloud was 10 mg/m³ during sweeping and the dust concentration dropped off rapidly to < 1 mg/m³ when the sweeping stopped.

3. If you wet a surface, dust is greatly reduced. When it dries you may still have a problem.

4. Using water to wet the cutting surface of a rock saw may lower the dust and therefore the exposure; however, the dry material below the surface being cut may not get wet and can still lead to dust exposure.

5. Minimize air movement. Do not use fans or air-supply ductwork in dusty areas.

6. Do not use air to blow off surfaces or clothing; it creates a lot of airborne particles.

7. Keep your clothing clean or use laundered or disposable clothing to keep the dust at the work site, not at home.

8. Wash up before leaving the work site.

Personal protective equipment is the last resort in terms of controls and protection for workers. If a respirator is used, you must follow OSHA 29 CFR 1910.134. The construction standard 29 CFR 1926.103 refers directly to 29 CFR 1910.134. You must select the respirator based on the hazard, and therefore you should know the concentration of silica in the air. If the shift exposure is going to be at or above the PEL, you are required to provide respiratory protection until you can implement permanent controls. In addition, if there are feasible temporary controls that could help lower the exposures, these should be implemented in the interim, along with respirators, until the problem can be permanently resolved. The respiratory protection program requires that employees have medical approval when required to use a respirator. Workers are to be trained, and fit-tested annually for tight-fitting respirators. All workers must be clean shaven in the respirator seal area. An alternative to tight-fitting respirators may be the use of loose-fitting positive pressure hoods. These same issues would be audit findings when evaluating a work site. Also, respirators need to be cleaned daily, stored properly, and maintained properly. Workers should be checked frequently for correct use of respirators.

Silica is all around us. You can reduce the hazard to workers if you properly evaluate the workplace and engineer out the hazard.

ASBESTOS

Asbestos is the name for a group of naturally occurring silicate minerals that can be separated into fibers. The fibers are strong, durable, and resistant to heat and fire. There are several types of asbestos fibers, three of which are used for commercial applications: (1) chrysotile, or white asbestos, comes mainly from Canada, and has been very widely used in the United States—it is white-gray in color and found in serpentine rock; (2) amosite, or brown asbestos, comes from southern Africa; (3) crocidolite, or blue asbestos, comes from southern Africa and Australia. Amosite and crocidolite are called *amphiboles*. This term refers to the nature of their geologic formation. Other asbestos fibers that have not been used commercially are tremolite, actinolite, and anthophyllite, although they are sometimes contaminants in asbestos-containing products. In addition to asbestos mines, asbestos is found as a contaminant mineral in the host rock in nonasbestos mining operations.

Asbestos has long been recognized as a human carcinogen. (Refer to the U.S. Department of Health and Human Services, Public Health Service, the National Toxicology Program report of December 2002, "10th Report on Carcinogens," at http://ehp.niehs.nih.gov/roc/tenth/profiles/s016asbe.pdf.) OSHA estimates that 1.3 million employees in construction and general industry have significant asbestos exposure on the job—those workers involved in construction, renovation, and demolition have the most risk of exposure. Other high-risk jobs include manufacture of asbestos products (such as building materials and insulation) and performing automotive brake and clutch repair. Of particular utility is a list of suspected asbestos-containing materials prepared by the Environmental Protection Agency (EPA), which can be found in Table 3 of this chapter and at http://www.epa.gov/Region06/6pd/asbestos/asbmatl.htm. Construction employers whose projects involve materials of

▪ ▪ Table 3 ▪ ▪

EPA Sample List of Suspect Asbestos-containing Materials

Cement Pipes	Elevator Brake Shoes
Cement Wallboard	HVAC Duct Insulation
Cement Siding	Boiler Insulation
Asphalt Floor Tile	Breaching Insulation
Vinyl Floor Tile	Ductwork Flexible Fabric Connections
Vinyl Sheet Flooring	Cooling Towers
Flooring Backing	Pipe Insulation (corrugated air-cell, block, etc.)
Construction Mastics (floor tile, carpet, ceiling tile, etc.)	Heating and Electrical Ducts
Acoustical Plaster	Electrical Panel Partitions
Decorative Plaster	Electrical Cloth
Textured Paints/Coatings	Electric Wiring Insulation
Ceiling Tiles and Lay-in Panels	Chalkboards
Spray-applied Insulation	Roofing Shingles
Blown-in Insulation	Roofing Felt
Fireproofing Materials	Base Flashing
Taping Compounds (thermal)	Thermal Paper Products
Packing Materials (for wall/floor penetrations)	Fire Doors
High Temperature Gaskets	Caulking/Putties
Laboratory Hoods/Table Tops	Adhesives
Laboratory Gloves	Wallboard
Fire Blankets	Joint Compounds
Fire Curtains	Vinyl Wall Coverings
Elevator Equipment Panels	Spackling Compounds

the types listed should anticipate potential asbestos exposure and carefully review their obligations under applicable OSHA, EPA, and/or state standards.

Asbestos, or fibrous dust, is created and released into the ambient air by the breaking, crushing, grinding, drilling, or general abrasive handling of a solid material that has fibrous components. Chrysotile is the type of asbestos most commonly found in commercial products. Amosite and crocidolite are generally considered to be the most toxic. Fibrous dust particles do not readily settle out of the air, but can remain suspended for long periods of time. As a result, accumulations of fibrous dust can continue to present an inhalation hazard when they are stirred up by vehicular traffic, by persons walking through them, or by the wind. Overexposure to asbestos can result in development of asbestosis (a type of pneumoconiosis that results from the inhalation of asbestos fibers); cancers of the lung, larynx, and gastrointestinal tract; and mesothelioma, another cancer associated with asbestos exposure.

The asbestos regulations of OSHA and the EPA are somewhat intertwined, and both should be carefully reviewed by safety and health professionals who are developing an asbestos compliance program. OSHA last modified its asbestos standard in 1994 and lowered the permissible exposure limit and the excursion limit to reflect increased asbestos-related disease risk to asbestos-exposed workers. The OSHA asbestos standards include medical surveillance, exposure monitoring requirements, and employee training (29 CFR §1910.1001 and §1926.1101).

Section 1926.1101 covers construction work, including alteration, repair, renovation, and demolition of structures containing asbestos. Employee exposure to asbestos must not exceed 0.1 fibers per cubic centimeter (f/cc) of air, averaged over an 8-hour work shift. Short-term exposure must also be limited to not more than 1 f/cc, averaged over 30 minutes. The administrative control of rotating employees to achieve compliance with either PEL is prohibited. In construction work, unless the employer is able to demonstrate that employee exposures will be below the PELs (a "negative exposure assessment"), it is generally required to conduct daily monitoring for workers in Class I and II regulated areas. For workers in other operations where exposures are expected to exceed one of the PELs, the employer must conduct periodic monitoring. In no case can periodic monitoring exceed intervals greater than 6 months for employees exposed above a PEL. (See the *OSHA Fact Sheet*, "Asbestos," U.S. Department of Labor, 2002.)

The OSHA standard for the construction industry classifies the hazards of asbestos work activities and prescribes particular requirements for each classification. (There are equivalent regulations in states with OSHA-approved state plans.)

- *Class I* is the most potentially hazardous class of asbestos jobs and involves the removal of thermal system insulation and sprayed-on or toweled-on surfacing asbestos-containing materials or presumed asbestos-containing materials.
- *Class II* includes the removal of other types of asbestos-containing materials that are not thermal system insulation, such as resilient flooring and roofing materials containing asbestos.
- *Class III* focuses on repair and maintenance operations where asbestos-containing or presumed asbestos-containing materials are disturbed.
- *Class IV* pertains to custodial activities where employees clean up asbestos-containing waste and debris.

Construction employers and contractors must create controlled zones known as regulated areas that are designed to protect employees where certain work with asbestos is performed. Access must be limited to regulated areas by authorized persons who are wearing appropriate respiratory protection. Activities such as eating, smoking, drinking, chewing tobacco or gum, and applying cosmetics are prohibited in these controlled zones, and the employer must display warning signs at each regulated area. In construction, workers must perform Class I, II, and III asbestos work (and all other operations where asbestos concentrations might exceed a PEL) within highly regulated areas (*OSHA Fact Sheet*, "Asbestos").

OSHA favors engineering controls and work practices to control worker exposures to or below the applicable PEL. If, after using all feasible controls, levels still exceed the PEL, the employer must supplement these methods with respiratory protection. Generally, the level of exposure dictates which type of respirator must be used (see, for example, 29 CFR §1910.134). Employees who must wear respirators are required to be trained, have medical approval, and undergo fit testing. In addition, workers exposed to levels above the PEL must also use protective clothing (coveralls, gloves, foot covers, head covers, face shields, goggles, etc.). (OSHA is still considering rulemaking that would require employers to pay for personal protective equipment that workers must use on the job, pursuant to 29 CFR §1926.95 and §1910.132.) The work site must also have decontamination areas and established hygiene practices for exposed workers. OSHA's construction standard also requires training for workers exposed above the PEL, and for other employees, depending upon their specific work classification, as shown above.

Construction workers who are exposed to asbestos for 30 or more days per year, and who work in Class I, Class II, or Class III activities, must receive medical examinations. The employer is required to keep accurate records of:

- all air sampling or other measurements taken to monitor employee exposure (maintain for 30 years)
- medical records, including physician's written opinions (maintain for duration of worker's employment, plus 30 years)
- training records (maintain for 1 year past the last date of employment).

The Department of Labor's Inspector General has recommended that future asbestos regulations address take-home contamination from asbestos and utilize Transmission Electron Microscopy to analyze fiber samples that may contain asbestos, rather than the Phase Contrast Microscopy method currently specified in the OSHA standard.

HEXAVALENT CHROMIUM

OSHA estimates that approximately one million workers are exposed to hexavalent chromium on a regular basis in all industries, and that as many as 34 percent of those workers could contract lung cancer if exposed for 8 hours a day for 45 years at OSHA's current exposure limit (1995 OSHA study).

The major uses of hexavalent chromium are as a structural and anticorrosive element in the production of stainless steel, ferrochromium, iron, and steel, and in electroplating, welding, and painting. The chemical is present in Portland cement, as well as in other cement-based construction materials, and is coming under increasing scrutiny by the federal Advisory Committee on Construction Safety and Health. Occupational exposure to Chromium 6 occurs primarily by inhalation, but can also occur to a lesser extent through dermal and oral routes, including through exposure to construction materials containing Chromium 6 and by having chromium dust contaminate hands, clothing, beards, food, and other things. The major illnesses associated with occupational exposure to hexavalent chromium are lung cancer and dermatoses. Exposure to Chromium 6 also is known to cause bronchial asthma, nasal septum perforations, eye irritation and/or perma-

nent eye damage, and skin ulcers. Information on the occupational health effects of Chromium 6 can be reviewed in: *Documentation of the Threshold Limit Values and Biological Exposure Indices*, Sixth Edition, Volume 1 (American Conference of Governmental Industrial Hygienists) and in the National Toxicology Program's "10th Report on Carcinogens" (U.S. Department of Health and Human Services, December 2002).

In 1971, OSHA adopted and made applicable to general industry a national consensus standard, ANSI Z37.7-1971, which sets a permissible exposure limit for hexavalent chromium compounds at 100 $\mu g/m^3$ (0.1 mg/m^3) as a ceiling concentration. This remains OSHA's PEL for general industry, codified at 29 CFR §1910.1000, Table Z. The substance is measured as Chromium 6 and reported as chromic anhydride (CrO_3). The amount of chromium (VI) in the anhydride compound equates to a PEL of 52 $\mu g/m^3$. This ceiling limit applies to all forms of hexavalent chromium, including chromic acid and chromates, lead chromate, and zinc chromate. The current PEL for hexavalent chromium (Chromium VI) in the construction industry is 100 $\mu g/m^3$ as a TWA PEL, which also equates to a PEL of 52 $\mu g/m^3$, codified at 29 CFR §1926.55. The EPA has set a limit of 100 μg of Chromium VI per liter of drinking water (100 $\mu g/L$). In addition, two construction welding standards, 29 CFR §§1926.57 and 1926.353, have provisions dealing with ventilation and protection while working on materials containing chromium. OSHA, NIOSH, and ACGIH exposure limits for hexavalent chromium are presented in Table 4.

■ ■ Table 4 ■ ■
Current Exposure Limits for Hexavalent Chromium

8-hour TWA	Chromic Acid and Chromates	Calcium Chromate	Chromium Trioxide	Lead Chromate	Strontium Chromate	Zinc Chromate
OSHA PEL (ceiling)	0.1 mg/m³					
NIOSH REL		0.001 mg/m³	0.001 mg/m³	0.001 mg/m³	0.001 mg/m³	0.001 mg/m³
ACGIH TLV		0.001 mg/m³ A2 carcinogen	0.05 mg/m³ A1 carcinogen	0.012 mg/m³ A2 carcinogen	0.0005 mg/m³ A2 carcinogen	0.01 mg/m³ A1 carcinogen

Source: U.S. Department of Labor, OSHA, April 2003 (http://www.osha.gov/SLTC/hexavalentchromium/pel.html)

OSHA has examined the biological mechanisms of hexavalent chromium toxicity in conjunction with NIOSH, and it is working to refine the sampling and analytical method. In June 1988, OSHA revised and validated the ID-215 analytical method to evaluate airborne occupational exposures of hexavalent chromium. The method includes pulling air through a PVC (polyvinyl chloride) filter analyzed by ion chromatography with an ultraviolet detector. The method reportedly has a qualitative detection limit of 0.001 $\mu g/m^3$ for a 960-liter air sample. The quantitative detection limit is 0.003 $\mu g/m^3$ for the same size sample.

On August 22, 2002, OSHA published a "Request for Information" concerning occupational exposure to hexavalent chrome. OSHA's RFI included topics such as health effects, risk assessment, analytical methods, investigations into occupational exposures, control measures, and technical/economic feasibility, use of PPE and respirators, employee training, medical screening programs, and environmental and small business impacts. The standard is now on a fast track for rulemaking because of a court-imposed deadline of January 2006.

DIESEL PARTICULATE MATTER

Diesel-powered equipment is quite common at construction sites in the United States. Examples of such equipment include vehicles such as haul trucks, front-end loaders, hydraulic shovels, load-haul-dump units, face drills, and explosives trucks. Diesel engines are also used in support equipment, which might include generators, air compressors, crane trucks, ditch diggers, forklifts, graders, locomotives, lube units, personnel carriers, hydraulic power units, scalers, bull dozers, pumps (fixed, mobile, and portable), elevating work platforms, tractors, utility trucks, water spray units, and welders.

The emissions from diesel engines are a complex mixture of compounds, containing gaseous and particulate fractions. The specific composition of the diesel exhaust varies according to the type of engines being used, the usage pattern, and other factors, including type of fuel, load cycle, engine maintenance, tuning, and exhaust treatment. This complexity is compounded by the multitude of environmental settings where diesel-powered equipment is operated.

The U.S. Department of Labor has reported basic facts about diesel emissions that are of general applicability. The gaseous constituents of diesel exhaust include oxides of carbon, nitrogen, and sulfur; alkanes and alkenes (e.g., butadiene); aldehydes (e.g., formaldehyde); monocyclic aromatics (e.g., benzene, toluene); and polycyclic aromatic hydrocarbons (e.g., phenanthrene, fluoranthene). The oxides of nitrogen (NOX) also can precipitate into particulate matter. The particulate components of the diesel exhaust gas include the so-called diesel soot and solid aerosols, such as ash particulates, metallic abrasion particles, sulfates and silicates. The vast majority of these particulates are in the invisible submicron range of 100 nm. Diesel particulate matter (DPM) has a solid core, mainly consisting of elemental carbon, and a very surface-rich morphology. This surface absorbs many other toxic substances that are transported with the particulates, and can penetrate deep into the lungs (66 Fed. Reg. 5715–5716, Jan. 19, 2001).

In 1988, NIOSH issued a *Current Intelligence Bulletin* recommending that whole diesel exhaust be regarded as a potential carcinogen and controlled to the lowest feasible exposure level. In its bulletin, NIOSH concluded that, although the excess risk of cancer in diesel-exhaust exposed workers has not been quantitatively estimated, it is logical to assume that reductions in exposure to diesel exhaust in the workplace would reduce the excess risk. NIOSH stated that "given what we currently know, there is an urgent need for efforts to be made to reduce occupational exposures to DEP [DPM]."

Diesel exhaust has long been known to contain carcinogenic compounds (e.g., benzene in the gaseous fraction and benzopyrene and nitropyrene in the DPM fraction), and a great deal of research has been conducted to determine if occupational exposure to diesel exhaust actually results in an increased risk of cancer. Evidence that exposure to DPM increases the risk of developing cancer comes from three kinds of studies: human studies, genotoxicity studies, and animal studies.

OSHA has not yet regulated worker exposure to diesel particulate matter. However, on January 19, 2001, the Mine Safety and Health Administration (MSHA) released a final rule governing diesel particulate exposure at underground metal/nonmetal mines (66 Fed. Reg. 5706 et. seq., Jan. 19, 2001). *The MSHA health standard is designed to reduce the risks to workers of serious health hazards that are associated with exposure to high concentrations of DPM.* MSHA has concluded that underground workers exposed to current levels of DPM are at excess risk of incurring the following three kinds of material impairment: (1) sensory irritations and respiratory symptoms (including allergenic responses); (2) premature death from cardiovascular, cardiopulmonary, or respiratory causes; and (3) lung cancer (66 Fed. Reg. 5823, Jan. 19, 2001).

According to MSHA's risk assessment, DPM, rather than the gaseous fraction of diesel exhaust, is assumed to be the agent associated with any excess prevalence of lung cancer observed in the epidemiologic studies (66 Fed. Reg. 5774, Jan. 19, 2001). After a comprehensive evaluation of the available scientific evidence, the World Health Organization's International Agency for Research on Cancer concluded in 1996 that: "Carbon black is possibly carcinogenic to humans (Group 2B)." (See also, 66 Fed. Reg. 5820, Jan. 19, 2001.)

DPM is a very small particle in diesel exhaust. Underground employees (e.g., those in mining, construction, and tunneling) who work in areas where diesel-powered equipment is used may be exposed to far higher concentrations of this fine particulate than any other group of workers. The best available evidence indicates that such high exposures put these miners at excess risk of a variety of adverse health effects, including lung cancer. Higher-than-average exposures have also been recorded in individuals who had long-term employment driving diesel-powered equipment, such as over-the-road truck drivers.

MSHA's 2001 diesel standard for underground metal/nonmetal mines establishes an average 8-hour equivalent, full-shift, airborne concentration limit (CL) of 400 $\mu g/m^3$ of total carbon, effective July 20, 2002 through January 19, 2006. The CL drops to 160 $\mu g/m^3$ of total carbon, effective January 20, 2006. MSHA has recently reopened the rule and proposes using elemental carbon as a surrogate for DPM, rather than total carbon because of sampling and analytical difficulties experienced since the interim rule took effect.

MSHA inspectors determine compliance by analyzing a single sample drawn from active working areas. The compliance officer samples for DPM using a respirable dust sampler equipped with a submicrometer impactor, and analyzes the sample using the NIOSH 5040 method or "any methods of collection and analysis subsequently determined by NIOSH to provide equal or improved accuracy" for the measurement of DPM [30 CFR §57.5061(b)]. The MSHA rule also provides that diesel fuel used to power equipment in underground areas may not have a sulfur content greater than 0.05 percent.

Because of the similarities in heavy equipment and working conditions between some underground mines and construction equipment and tunneling operations, construction companies should ensure that workers are not overexposed to DPM in order to avoid injurious health effects that may be associated with long-term exposure to DPM above 160 $\mu g/m^3$.

LEAD PAINT REGULATIONS AND BEST PRACTICES

Since the early 1970s, the federal government has taken several steps to reduce worker and building-occupant health risks due to exposure to lead-based paint (LBP). Workers encounter LBP in a variety of occupational situations, which require compliance with EPA, OSHA regulations, and possibly housing and urban development (HUD) guidelines. HUD and EPA have developed regulations and guidance for abating LBP hazards and conducting safe work activities when disturbing LBP in various types of residential properties, ranging from large public housing apartments to private single-family homes. According to OSHA, inhalation of airborne lead is generally the most frequent source of occupational lead exposures to workers. In 1991 and 1992, NIOSH released reports documenting lead poisoning in workers engaged in abrasive blasting, maintenance, repainting, and demolition of bridges (NIOSH 1,2). These studies were able to identify worker exposures to lead that were nearly 600 times current OSHA exposure levels. In 1990, NIOSH set a national goal to eliminate worker exposures resulting in blood lead concentrations greater than 25 micrograms per deciliter (25 μg/dl) of whole blood. OSHA then began to develop a comprehensive standard regulating lead exposures in construction.

OSHA has had regulations pertaining to occupational lead exposure since 1971 for construction and general industries. In October 1992, Congress passed Section 1031 of Title X of the Housing and Community Development Act of 1992, requiring OSHA to issue an interim, final lead standard for the construction industry. This interim rule was published on May 4, 1993, adding the new section 1926.62, which will remain in effect until OSHA issues a final standard. OSHA's lead in construction standard applies to all construction work where an employee may be occupationally exposed to lead. This standard is not specific to lead-based paint; it also includes metallic lead, all inorganic lead compounds, and organic lead soaps. All construction, alteration, and repair activities are covered by the standard.

The OSHA lead standard establishes maximum limits of exposure to lead for all workers covered, including a PEL and action level (AL). The PEL requires worker exposure to lead at airborne concentrations no more than 50 μg/m^3, averaged over an 8-hour period. The standard requires employers utilize engineering, work practices, and administrative controls, when feasible, to reduce and maintain employee lead exposure at or below the PEL.

If a worker is exposed over the PEL, the employer must provide respiratory protection, protective work clothing and equipment, change areas, hand-washing or shower facilities, biological monitoring, and training. An action level is the level at which an employer must begin to take certain actions or compliance activities, such as medical surveillance and training of employees. Until an employer performs an exposure assessment and documents worker exposures below the PEL, the employer must treat employees performing certain operations as if they were exposed above the PEL,

triggering compliance activities. Work tasks involving LBP that employers must treat as exceeding the PEL (unless exposure monitoring proves otherwise) are:

- manual demolition of structures (e.g., dry wall), manual scraping, manual sanding, and use of a heat gun where lead-containing coatings or paints are present
- abrasive blasting, including clean-up activities and enclosure movement and removal
- power-tool cleaning
- spray painting with lead-containing paint
- rivet busting or welding, cutting, or burning on any structure where lead-containing coatings or paint are present.

If the initial determination through air monitoring is that the employee exposure is below the action level, further exposure determination or OSHA compliance activities need not be repeated unless there is a change in processes or controls. This is an important level to achieve for an employer who has work practices that typically do *not* generate significant levels of airborne lead. If employee exposure is found to be between the AL and PEL, or above the PEL, then additional exposure monitoring and compliance activities are required.

Jobs involving exposures over the PEL, or where no initial determination is conducted, must establish and implement a written compliance program to reduce employee exposures to or below the PEL. The written program must be revised every 6 months and include the following components:

- a description of each activity in which lead is emitted (e.g., equipment used, materials involved, controls in place, crew size, employee job responsibilities, operating procedures, and maintenance practices)
- specific plans to achieve compliance and engineering plans and studies where engineering controls are required
- information on the technology considered to meet the PEL
- air-monitoring data that document the source of lead emissions

- a detailed schedule for implementing the program, including copies of documentation (e.g., purchase orders for equipment and construction contracts)
- a work-practice program, including regulations for the use of protective work clothing and equipment, housekeeping, and hygiene facility guidelines
- an administrative control schedule for job rotation, if used
- a description of arrangements made among contractors on multicontractor sites to inform affected employees of potential exposure to lead and their responsibility to comply with this standard
- any other relevant information.

The OSHA Web site has interactive computer software that can assist an employer with compliance, including writing a job-specific compliance program.

HUD Guidelines and Targeted Housing

The Lead-Based Paint Poisoning Prevention Act (LBPPPA) of 1971 was adopted to reduce levels of lead in paint in federally financed and subsidized housing. It was amended in 1973 to require HUD to reduce or eliminate the hazards of LBP poisoning in federally financed and subsidized housing. In 1989, HUD developed comprehensive technical guidelines on testing, abatement, clean up, and disposal of LBP in public and Indian housing. An official guideline was published in 1990, followed by extensive revisions in 1995.

The 600+ pages of "Guidelines for the Evaluation and Control of Lead-Based Paint Hazards in Housing" provide detailed, comprehensive, technical information on how to identify and address LBP hazards safely and efficiently. The guide lists three types of LBP hazard control: interim controls, abatement of LBP hazards, and total LBP abatement. Interim controls are designed to address hazards quickly, inexpensively, and on a temporary basis, while abatement is intended to produce permanent solutions. The goal of the document is to help property own-

ers, private contractors, and government agencies sharply reduce children's exposure to lead without unnecessarily increasing the cost of housing.

HUD developed this document to compliment regulations, other directives, and other guidelines to be issued by HUD, EPA, OSHA, and CDC. The guidelines were developed to provide more comprehensive and complete guidance than most regulations do on how activities related to LBP should be carried out and why certain measures are recommended. The guidelines are *not* enforceable by law unless a federal, state, or local statute or regulation requires adherence to certain parts of them. Employers and building owners need to realize that the guides are designed for housing where small children are present. The HUD guidelines are not designed for commercial or industrial LBP disturbance activities. Employers working in single-family and multifamily LBP activities can benefit from the mountain of guidance provided by HUD.

EPA on Training, Clearance Testing, Safe Work Practices, and Waste Disposal

The overall purpose or policy of the EPA lead office is to formulate and execute programs that will promote the reduction of human exposure to lead hazards. The Residential Lead-Based Paint Hazard Reduction Act (Title X) developed a comprehensive federal strategy for reducing lead paint exposure. Title X also provided the authority for the following regulations by amending the Toxic Substances Control Act (TSCA) to include Title IV (Lead Exposure Reduction). Issues that have been delegated to the EPA include:

- National Lead Laboratory Accreditation Program [405(b)]: Establishes protocols, criteria, and minimum performance standards for laboratory analysis of lead in paint, dust, and soil.
- Hazard Standards for Lead in Paint, Dust, and Soil (403): Establishes standards for lead-based paint hazards and lead dust clean-up levels in most pre-1978 housing and child-occupied facilities. *Clearance levels*

specified here currently only apply to HUD/ state/local-targeted LBP projects and are a guide for other residential LBP projects.

- Training & Certification Program for Lead-Based Paint Activities (402/404): Ensures that individuals conducting lead-based paint abatement, risk assessment, or inspection are properly trained and certified, that training programs are accredited, and that these activities are conducted according to reliable, effective, and safe work-practice standards.
- Pre-Renovation Education Rule [406(b)]: Ensures that owners and occupants of most pre-1978 housing are provided information concerning potential hazards of lead-based paint exposure before certain renovations are begun on that housing.
- Disclosure Rule (1018): Requires disclosure of known lead-based paint and/or lead-based paint hazards by persons selling or leasing housing constructed before the phase-out of residential lead-based paint use in 1978.
- Lead-Based Paint Debris Disposal: Regulatory Status of Waste Generated by Contractors and Residents from Lead-Based Paint Activities Conducted in Households.

Proposed Revisions for Training, Accreditation, and Certification

Section 402(c)(3) of TSCA directs EPA to revise regulations codified at 40 CFR 745, Subpart L, to ensure that individuals engaged in renovation and remodeling activities that create lead-based paint hazards are properly trained; that training programs are accredited; and that contractors engaged in such activities are certified. As of this writing, only those contractors, supervisors, and workers engaged in federally funded HUD projects, or those working on federal public or Indian housing, are required to be certified by accredited training providers. Most states also require licensing for workers, supervisors, and contractors who perform HUD work as well. *This proposed revision would require training and certification for LBP work in most other residential, public, and commercial properties, including bridges and steel structures.*

The EPA has developed standards, guidance documents, and training programs for employers performing renovation, remodeling, painting, and other activities in an attempt to minimize creation and dispersal of lead-containing dust and protect residents, especially children, from possible exposure. The EPA mandates safe work practices on HUD and other targeted housing and recommends them elsewhere. There is currently an initiative to expand safe work practices to other nontargeted housing and LBP activities.

The Lead-based Paint Debris Disposal Proposed Rule

The EPA regulates LBP waste/debris as hazardous or nonhazardous depending upon where and how the LBP is disturbed. This rule allowed for easier disposal of lead-based paint debris generated in residences or public and commercial buildings. The rule was developed out of a recognition that the Resource Conservation and Recovery Act (RCRA) subtitle C regulations for such disposal as hazardous waste were burdensome, time consuming and costly—especially for average homeowners who are considering whether or not to have their homes abated. The proposed standards will allow disposal of lead-based paint debris in specified alternative, nonhazardous landfills [i.e., construction and demolition (C&D) landfills], without requiring a hazardous waste determination. To accomplish this, the proposed rule shifted the regulations for management and disposal of lead-based paint debris from RCRA to a program under the Toxic Substances Control Act (TSCA). The proposed TSCA standards do not apply to the lead-based-paint debris or soil generated by homeowners or contractors engaging in renovation activities in homes. Rather, this debris is covered under the household hazardous waste exclusion in RCRA subtitle C [40 CFR 261.4(b)(1)].

Until the EPA finalizes a disposal rule, employers and homeowners can refer to The Cotsworth Solid Waste Memo/July 2000. This memorandum clarifies the regulatory status of waste generated as a result of lead-based paint activities (including abatement, renovation, and remodeling) in homes and other residences. Since 1980, EPA has excluded household waste from all types of RCRA hazardous wastes under 40 CFR 261.4(b)(1). The household exclusion applies to waste generated by either residents or contractors conducting lead-based paint activities in residences. As a result of this clarification, contractors may dispose of hazardous lead-based paint wastes from residential lead-paint abatements as household garbage subject to applicable state regulations. Note, however, that some lead-based paint waste from residential LBP activities may still be subject to more stringent hazardous waste requirements in certain states, localities, territories, and tribal areas. *LBP waste and debris generated from nonresidential projects must be tested to determine if RCRA hazardous waste regulations apply. The only exclusion is for LBP adhered to building components during demolition, which can be handled as C&D waste.*

Best Practices for Lead-based Paint Activities

The regulatory review above is a brief description of possible regulatory requirements or guidelines that may apply to a maintenance, construction, renovation, or demolition project that involves LBP. The question then becomes, "How do we work with LBP in compliance with regulations while protecting workers and the public?" This can be best illustrated with case studies. The two projects selected for discussion identify two extremes of LBP disturbance: removing a lead painted window in a home by a small renovation/rehab contractor versus the removal of lead paint from a structural steel bridge over a river using professional lead abatement contractors.

REPLACING A WINDOW THAT HAS BEEN PAINTED WITH LBP

The removal of windows painted with LBP involves HUD and EPA guidelines for safe work practices, clearance testing, and waste disposal. The compliance components of OSHA's lead in construction standard apply unless air monitoring is performed and worker exposures to airborne lead are not above the permissible exposure level. Fortunately, there are several resources available to assist employers

with compliance with these regulations and for establishing best practices.

NEVER FORGET—Lead dust is your enemy on LBP residential projects! The "safe work practices" teach simple principles: (1) avoid activities that produce high levels of lead dust or fumes by prohibiting certain activities and avoiding disturbing any dust and spreading lead-based paint to the greatest extent possible; (2) protect occupants and workers; and (3) perform proper cleaning, and verify with clearance testing.

Prior to work starting:
- Determine if the work will take place from the interior of the residence or from the exterior. Exterior work may appear to be more difficult (especially for windows above grade); however, there are great advantages to avoiding interior work. Interior work involves relocating occupants and furnishings. Interior work also requires clearance testing while most exterior projects do not.
- OSHA requires a lead compliance program with components as outlined previously. HUD projects also require a work-site compliance plan with similar components to OSHA, except that HUD additionally addresses occupant notification and clearance testing. If respirators are used, a written respiratory protection plan is required by OSHA. OSHA also requires a hazard communications plan on site, including material safety data sheets (MSDS) for chemicals used on the job site.
- EPA and HUD would require training in safe work practices if the window is being replaced for renovation purposes. An employer could easily become an approved trainer to instruct his or her own employees. Certified contractors, trained in abatement by an accredited training provider, would be required by EPA for targeted housing (federally funded or child-occupied) if the intent of removing the window was to eliminate a lead hazard (note that some states also require lead licenses). If respirators are used, OSHA requires employees to have documented respirator training. Finally, hazard communication training

on chemicals used on the job site is also an OSHA requirement.

- Personal protective equipment, specific for lead, would be required, including washable or disposable coveralls, N100 respirators, gloves, and wash facilities, unless the employer conducts exposure monitoring, documenting compliance with the PEL. HUD and EPA recommend a disposable painter's hat, coveralls (either disposable or stored in a bag on site), and a N100 disposable respirator. Gloves are optional; however, frequent hand washing is recommended (premoistened wipes are also suggested). If a respirator is issued, then OSHA requires the employer have a written respiratory protection program, including medical surveillance, fit testing, and training for employees. Chemical use (such as paint removal products) may require eye/face/hand/body and respiratory protection above and beyond protective equipment discussed above.
- Medical surveillance, including biological monitoring (blood tests to measure lead levels) is required by OSHA if no employee lead-exposure monitoring has been performed. If exposure monitoring indicates employee exposure to airborne lead is over the AL or PEL, medical surveillance is also required. Finally, as mentioned above, the use of respirators triggers the medical surveillance requirements of OSHA's Respiratory Protection Standard.

Work-site preparation and setup for interior work:

- Restrict access to the work area. OSHA requires establishing a regulated area where the PEL is expected to be exceeded. Eating, drinking, and smoking are prohibited in the restricted area. This requirement is performance based. Some jobs require a sign or barrier tape, while other jobs may require fences or lockable barriers.
- Remove drapes, curtains, furniture, and rugs within 5 feet of the work area. Large furniture that cannot be moved should be covered

with protective sheeting. A protective drop cloth should be securely placed immediately under the work area, extending at least 5 feet out (lead dust typically falls within 5 feet of a work area).

- Avoid spreading lead dust! Bring in necessary supplies prior to the start of the job to avoid having to step off of the protective sheeting. Protect shoes with disposable shoe covers and remove the covers upon leaving the work area. Other options are to wipe the bottom of shoes with a damp towel, clean off shoes with a tack pad (large sticky pad that removes dust from bottom of shoes), or remove shoes when leaving the work area.

Clean up and waste disposal:

- Use special cleaning techniques. Pick up large paint chips with damp paper towel and/or mist, then push into dust pan. Wet wipe and HEPA vacuum all surfaces. Lead cleaners and heavy scrubbing will be necessary multiple times to remove the lead dust. Using dry sweeping or regular shop vacuums are prohibited cleaning methods.
- All debris, protective sheeting, and other disposables should be bagged prior to leaving the work area.
- Consider painting the surfaces prior to performing clearance testing.
- The waste including the removed window is classified by EPA as a nonhazardous residential waste. Refer to local agencies for disposal requirements in your area.

Clearance testing:

- HUD and targeted housing projects require clearance testing on interior surfaces and exterior surfaces such as porches. Dust wipes are used to determine if surfaces are safe for reoccupancy.
- Clearance levels established by the EPA are extremely low. Surfaces must be clean enough for a toddler to lick, literally! Clearance on a floor surface is 40 μg/ft^2, a window sill is 250 μg/ft^2, and window troughs are 400 μg/ft^2. Clearance tests have to be

conducted by certified inspectors or risk assessors and analyzed by EPA accredited laboratories.

- Clearance testing documents cleaning that is not just construction-site clean or even reoccupancy clean. Clearance is nearly clean room clean. (Example: Take 100 rooms, 10 feet by 10 feet. Evenly spread a half-packet of sugar (filled with lead dust) over all 100 rooms. The floors of all 100 rooms would fail clearance testing! Remember, dust is your enemy on residential projects.

This "routine" project can become very tedious, time consuming, and expensive if OSHA exposure monitoring does not indicate low lead exposure to workers. Most guidance information provided by HUD and EPA only mentions OSHA in passing. Staying under OSHA's action level will make window replacements involving lead paint safe, efficient, and cost effective for all.

ABRASIVE BLASTING OF LBP ON A STRUCTURAL STEEL BRIDGE

Abrasive blasting on structural steel is a process that usually involves compressed air to propel the blasting media (coal slag, silica sand, steel shot, or steel grit) to the surface being cleaned. Abrasive blast cleaning is used because it is a fairly quick and cost-effective way of removing paint, rust, and mill scale from a steel structure. Abrasive blasting can result in worker exposures over 100 times higher than OSHA's permissible limit. There are also potential issues with environmental contamination of air, soil, and water. The potential for high employee exposures to lead along with significant pollution concerns makes abrasive blasting a heavily regulated activity. Associations have developed several guidance documents to assist employers with regulatory compliance and best work practices for abrasive blasting activities on steel structures. Refer to the resources listing for more information.

Prior to work starting:
- Determine if the work will take place over ground, water, roads, or railways. Consider weather conditions and loads expected to

act on the work site. Environmental testing may be necessary to establish background levels prior to starting, and clearance levels upon conclusion of the project.

- OSHA requires a lead compliance program with components as outlined previously. The EPA and many local regulatory agencies require containment plans to focus on pollution control as well. Respirators are mandatory, so a written respiratory protection plan is required by OSHA. OSHA also requires a hazard communications plan on site, including material safety data sheets (MSDS) for chemicals used on the job site. The use of hearing protection also requires a written program.

- EPA recommends the use of certified contractors trained in abatement by an accredited training provider. OSHA requires job-specific training that addresses the hazards at the job site (note that some states also require lead licenses). OSHA requires respiratory protection, which necessitates the need for employees to have documented respirator training. Also, hazard communication training on chemicals used on the job site is also an OSHA requirement. Finally, job-site hazards may require many more OSHA compliance issues such as fall protection, scaffolding, hearing protection, and silica exposure.

- Personal protective equipment, specific for lead, would be required, including washable or disposable full-body coveralls, gloves, shoe covers, and hats. Supplied-air-type CE abrasive-blasting hood respirators are required for those in contained work areas, and air-purifying P100 respirators are required for those outside of the blasting area. Wash facilities, including showers, are required (when feasible). Abrasive blasting involves high noise levels, so hearing protection and a written hearing conservation program is usually required. Finally, if silica is used as the blasting agent, additional precautions should be taken.

- Medical surveillance, including biological monitoring (blood lead-level tests) is required by OSHA. As mentioned above, the use of respirators triggers the medical surveillance requirements of OSHA's respiratory protection standard. Finally, silica use and excessive noise will require additional medical surveillance, as specified by OSHA.

Work-site preparation and setup:

- A partial or full containment is usually required by OSHA and EPA for worker protection and pollution control. Most enclosures require ventilation and filtration to reduce lead exposures.
- Restrict access to the work area. OSHA requires establishing a regulated area where the PEL is expected to be exceeded. Eating, drinking, and smoking are prohibited in the restricted area. This requirement is performance based. Some jobs require a sign or barrier tape, while other jobs may require fences or lockable barriers.
- A system has to be constructed to move waste from the blasting area to the storage area through large vacuums, funnels, or conveyors.

Clean up and waste disposal:

- OSHA requires that all surfaces be kept as free as practical of accumulations of lead with clean-up methods that minimize the likelihood of lead becoming airborne.
- Shoveling, dry or wet sweeping, and brushing may only be used when the use of HEPA-filtered vacuuming is not effective.
- The use of compressed air during these activities is prohibited, except under rare conditions.
- Waste should be segregated into hazardous and special classifications. Wastes typically considered hazardous include paint chips, dust, and contaminated blasting agents. Those materials that typically are not considered hazardous include paint still adhered to structural components, materials used in constructing containments, and disposable suits.
- Clearance testing is not required on this project by the EPA.

Abrasive blasting of LBP has a great potential for excessive employee exposures to airborne lead. There is an equal concern for environmental pollution to soil, water, and the air. The project's proximity to water, wetlands, hospitals, schools, daycare, private property, and food sources must be incorporated into a sound containment plan. Waste handling and disposal is also a concern. Finally, elevated job sites near roads, railways, and other hazardous areas pose more complicated compliance plans. Failure to properly plan and implement a sound project can result in significant human health injuries and environmental pollution.

LBP CONCLUSION

Disturbance of lead-based paint can trigger several federal, state, and local regulations. The examples cited in this chapter show the range of work practices—from minimal to extreme—depending upon the type and location of LBP work being performed. Best work practices should incorporate applicable regulations and guidelines along with specific work practices developed by associations, unions, and other groups.

REFERENCES

American Conference of Governmental Industrial Hygienists. *2002 Threshold Limit Values for Chemical Substances and Physical Agents & Biological Exposure Indices.* Cincinnati, Ohio: ACGIH, 2002.

Anderson, Kim, et al. *Fundamentals of Industrial Toxicology,* 2nd ed. Ann Arbor, MI: Ann Arbor Science Publishers, 1982.

ANSI/AIHA. *Z9.7-1998 American National Standard for the Recirculation of Air from Industrial Process Exhaust Systems.* New York, NY: ANSI, 1998.

Department of Environmental Health, Office of Continuing Education. *Lead Abatement Training for Workers: Volume I.* Cincinnati: University of Cincinnati, 1997.

Moeller, Dade, W. *Environmental Health.* Cambridge, MA: Harvard University Press, 1992.

National Institute of Occupational Safety and Health. *NIOSH Alert—Request for Assistance in Preventing Lead Poisoning in Construction Workers,* Cincinnati: U.S. Department of Labor, April 1992.

_____, Department of Health and Human Services, Centers for Disease Control and Prevention. *Publication No. 2002-129, Health Effects of Occupational Exposure to Respirable Silica.* Cincinnati: NIOSH DHHS, updated December 19, 2002.

Occupational Safety and Health Administration. *29 CFR Part 1926.* Washington, DC: US Government Printing Office, 2002.

_____. Special Emphasis Program (SEP) for Silicosis. Memorandum from Joseph A. Dear, May 2, 1996.

REVIEW EXERCISES

1. List the four-step process an industrial hygienist uses when working on a construction project, and discuss how each step reduces workers' health risks.

2. List the industrial hygiene risks at a job you are working on, and discuss how you would determine if workers were over the exposure limits.

3. When should an industrial hygienist be involved in a construction project, and what role should he or she play?

4. Discuss dust-management options, and how they can be used to reduce exposure.

▪▪ PART V ▪▪

Other Considerations

▨▨ About the Author ▨▨

WORK ZONE SAFETY/TRAFFIC CONTROL
by George Wolff, CSP

George Wolff, CSP, has 20 plus years of professional safety training and loss prevention consulting experience. He has held positions in public and private industry, including a municipality, manufacturing companies, a fully integrated petroleum company, and an insurance carrier with construction clients representing nearly half of its business. Currently, George is a Principal and Safety and Health Manager of Rule Engineering, LLC, an environmental engineering, safety and health consulting firm located in Denver, Colorado.

While he was pursuing his Master of Industrial Safety (MIS) degree from the University of Minnesota, Duluth, George was awarded a fellowship grant from the National Institute of Occupational Safety and Health.

George has served on various committees of the Minnesota Safety Council (Northern section), as well as the Colorado Safety Association (CSA), and chaired diverse projects such as the Model City Program for Safety Belt Use in Duluth, Minnesota, prior to seat-belt law enactment. He was also chairman of the CSA Annual Safety and Health Conference in 1992.

As a Professional Member of ASSE and a Board Certified Safety Professional in comprehensive practice (CSP # 9994), George has served as ASSE Colorado chapter President 1991–92, Area Director in Region II 1997–98, and Colorado delegate to the annual National ASSE member meeting and Professional Development Conference 1990–98.

CHAPTER **27**

Work Zone Safety Traffic Control

LEARNING OBJECTIVES

- List key elements of roadway work zone hazards on both sides of the barrier.
- Discuss sources of regulations and other documents that impact roadway construction companies and roadway construction workers.
- Discuss traffic control principles as stated in the Federal Highway Administration's *Manual on Uniform Traffic Control Devices.*
- Describe potential injury exposures and mitigation techniques that can be used in roadway construction to protect motorists and workers.

ROADWAY WORK ZONES

As drivers, when we see roadwork signs we react to the prospect of navigating a work zone. Our reactions could range from very negative (approaching anger) to viewing the work zone with professional and academic interest. We might simply accept the work zone with a positive attitude toward the temporary inconvenience for a roadway improvement.

The slogan on the license plates of Minnesotans proclaims the state the "Land of 10,000 Lakes," but residents know it as the land of two seasons—winter

and roadwork. Currently, motorists driving on our interstates and highways will, on average, encounter a work zone every 40–50 miles. Based on our experience and use of roadways, we assume that states and other political subdivisions repair potholes and other hazards, plus keep bridges structurally sound and wide enough for our personal vehicles as well as the commercial vehicles with whom we share them. We may sometimes lose patience with the process of repair, improvement, modification, or new construction. Urban expansion causes changes in roadway design from low-volume rural roads to high-volume urban arterial routes. Reconstruction of buildings as well as construction of tunnels and new buildings affect all of us as motorists, workers, or pedestrians. As safety professionals, some of us recognize that the activities of roadway construction put many people with whom we interact at risk because of the location of the work in relation to traffic. Unfortunately, the inherent risks of working in and around roadway traffic or heavy equipment may become routine for the construction worker, just as backup alarms from equipment working on construction sites become background or white noise and may no longer provide the intended warning to workers on foot.

Workers typically are performing tasks in close proximity to vehicles traveling at highway speeds. Their personal protective equipment may include a hard hat, substantial work boots, safety glasses, retro-reflective clothing, and gloves. Imagine the injury potential created by the energy generated by even a small vehicle constructed of 3000 pounds of steel, plastic, and glass with blunt or sharp body shapes traveling at 65 mph and coming into contact with the flesh and blood of workers protected by a hard hat, mesh vest, and leather boots. The worker side of the collision equation does not hold up well. The results of such a collision are often disastrous; seldom are there minor injuries when vehicles and workers collide. The number of incidents, fatalities, and injuries related to work zones is on the rise. According to data compiled by the BLS and quoted in the American Road & Transportation Builders Association (ARTBA) trainer's manual, "roadway construction is one of the most hazardous occupations in the nation. Accident and fatality rates for heavy and highway contractors top the list of construction rates, while construction rates generally, exceed the rates for other U.S. industries." Training, education, and altering work practices are all required to change these trends.

Who Gets Injured and Killed?
What Injuries Can Be Expected?

The following cases are just a few of the many road construction motor vehicle incidents that occur every year in the United States:

Sterling, Colorado, spring of 1998: A traffic control supervisor (TCS) and flag person were killed in a work zone on I-76 by a truck with faulty brakes that could not stop for the work zone.

Thornton, Colorado, March 2002: A worker for a traffic control contractor was modifying a work zone during an evening rush hour. While retrieving a traffic control device that had been displaced by wind, the worker was struck and killed by a motorist who was passing stopped traffic.

Pueblo, Colorado, September 2002: A traffic control supervisor for a paving contractor was struck and killed by a motorist on I-25 just after midnight.

Drums as well as a portable, changeable message sign were used in this temporary traffic control zone.

Denver, Colorado, November 2002: A striping company working nights on metro area reconstruction of the I-25 corridor had five truck-mounted-attenuator (TMA) shadow vehicles equipped with flashing lights struck in a 6-week period. Employees in the trucks with TMAs were not injured seriously. The striping trucks and employees were not hit.

Michigan, August 2002: A Michigan motorist who drove into a work zone, killing a highway worker on August 9, 2002, may face a maximum penalty of 15 years in prison and a $7500 fine. The motorist killed a 26-year-old safety engineer on her first day on the job and critically injured her co-worker. The workers were near a traffic-signal trailer that the motorist struck. At the time of the accident, the injured worker was installing a lane-merge system.

Even though motorists represent the majority of fatalities in the temporary traffic control zone (TTCZ), this chapter focuses on protection of workers, with the caveat that appropriately designed and implemented TTCZs will protect motorists as well. The *Manual of Uniform Traffic Control Devices (MUTCD)* clearly states that its purpose and design is to protect motorists and workers. The typical applications are designed to provide efficient traffic flow, minimizing disruption while providing protection for motorists from road construction hazards. Traffic control devices are designed to minimize damage to the motorist if his or her vehicle is struck. Even the concrete, temporary traffic barriers used in a TTCZ should be installed to separate vehicles from hazards and workers from vehicles. They must be equipped with energy-absorbing crash-attenuation treatment for barrier ends facing traffic, or installed with sufficient flare angles that redirect the vehicle when struck.

In general, construction is a hazardous industry. It involves heavy materials, large trucks and equipment, awkward work positions, inhospitable work conditions, and nonstationary work sites. According to a NIOSH study, "Workers represent 18% of all fatalities relating to roadway work zones. Of those roadway workers who are killed by vehicles, about half are killed by vehicles that breach the barricades and strike workers, while the other half are killed by construction trucks and equipment operating within the work zone. Workers on foot (WOF) represent approximately 57% of the worker fatalities in work zones; another 30+% of the fatalities involve equipment operators (ARTBA trainer's manual).

Types of Hazards and Worker Exposures

Outside the barriers, hazards are: workers being struck or caught between equipment, worker injury from overuse and poor body positions (i.e., while placing devices on the roadway), and environmental exposures to heat, cold, and sun. Workers and vehicles are in constant conflict. Increasingly, political subdivisions (states, counties, municipalities) are under pressure from the motoring public to minimize the impact to daytime travel; subsequently, contracts often require work to be performed at night. In addition to the obvious issues of nighttime work shifts, there are special issues of working on roadways at night. Lighting for work and illumination of workstations may be helpful to the workers on site, but unless controlled appropriately may interfere with drivers' sight while they're adjusting from dark to light to dark again. If the light plants are not properly adjusted, the glare may blind drivers momentarily.

Every night drivers operate vehicles at speeds exceeding the distance of their headlight illumination. As our population matures, the factor of aging becomes an issue. Unfortunately, because it occurs so gradually, affected individuals are seldom aware of it. For each 13 years after age 20, the average adult requires approximately twice the light to see adequately. At age 33 we need 2X the light to see what a 20-year-old sees; at age 46 we need 4X the light; at age 59 we need 8X the light (CCA *Instructor's Manual for Traffic Control Supervisor Certification*). Many drivers have not made the adjustment to their driving techniques to accommodate the changes in physical abilities that age brings.

Another hazard that is unique to night roadway work is caused by two very different elements. One element creating the hazard is the requirement that the flag-person workstation be illuminated for visibility during night operations. The time when bars or nightclubs are closing and when workers' alertness level may be low because they're in the middle of the shift becomes a time of heightened risk. The second element is the impaired driver who, like a moth, will focus on and drive toward the light. This combination of the flag person's low-level of alertness and impaired drivers heading for the light is a recipe for disaster. There is a large volume of anecdotal data expressed by safety directors, workers of roadway striping companies, flagging companies, and public entities that relates to specific incidents where drivers have hit truck-mounted attenuators or other devices equipped with flashing lights. The drivers typically have not applied the brakes and will often emphatically state they didn't see the unit or the flashing lights.

Workers placing the traffic control devices on the roadway are exposed to falls, sprains, and strains from bending, reaching, and turning (the average traffic cone weighs 8–12 pounds) during the installation and removal of the devices. Highway applications require dozens of devices to guide the motorist though the work zone and to separate them from the worker and the work.

Additionally, some hazards may be innocuous but a threat to the health of the worker nonetheless. Noise, silica, and lead all pose hazards that, depending on the project, may affect a worker not necessarily directly involved in the work creating the hazard. For example, bridgework that involves the sandblasting of lead-based paint, or mandatory roadway-marking removal throughout a project by sandblasting, grinding, and other methods, creates health hazards that roadway workers may be exposed to even though they are not directly involved in the work activity.

Inside the barriers, a dynamic internal traffic control plan (ITCP) is recommended to prevent a worker equipment conflict during the various stages of all construction projects. As the project changes, the ITCP must address parking areas for workers; public road crossings for worker access to the site, and width and structural integrity of haul roads. Requirements that address worker access for areas to cross the haul roads to the construction area and break areas are also part of the ITCP. This internal traffic control plan should be adopted and enforced to prevent injuries and fatalities. Each stage of the construction project may affect the plan design; consequently, the plan must be developed to accommodate changing requirements. The traffic control specialist or practitioner must be flexible in approach and seek input from workers, on-site safety representatives, construction managers, and other reliable sources in order to gain a grasp of the schedule and be able to adjust the plan at the appropriate times during the project.

The same hazards that expose workers on the traffic side of the barrier are also a threat to the worker on the construction side of the barrier. Noise, silica, and lead can endanger the health of workers, unless they are properly trained and equipped with appropriate PPE. Often construction workers do not recognize these hazards and may be exposed to contaminant levels that exceed the OSHA PELs. Each of these physical and chemical hazards may present injury potential that, depending on the project, expose the worker who may not be directly involved in the specific work that creates the hazard. A worker in the general work area may be ill-prepared and unprotected from a hazard created by others. For instance, unless exposures are controlled, sandblasting lead-based paint on a bridge can pose a health hazard to an equipment operator moving dirt nearby. Falls on the same level or to lower levels as well as strains or sprains while lifting or working in awkward positions are also dangers facing roadway workers. Workers on foot around fast- or slow-moving construction trucks or other equipment with limited visibility are constantly exposed to the obvious injuries from being struck or caught between the equipment.

OSHA construction standards require training for all employees engaged in construction activities. All employees must be able to recognize hazards and employ techniques to prevent injuries from those hazard exposures. In addition, there are *specific* OSHA construction standards that cover such things as noise, hazard communication, scaffolds, silica, and lead, for which contractors must establish policies and procedures to achieve compliance.

Roadway construction also requires contractors to consider the standards found in the *Manual of Uniform Traffic Control Devices* (*MUTCD*), published by the Federal Highway Administration (FHA).

This document provides the minimum standard for construction, maintenance, and incidents on public roadways. The first *MUTCD* was published in 1935, with revisions starting in 1939. New editions were published in 1941, 1948, 1961, and 1971. The 1971 edition of the *MUTCD* was adopted by reference in OSHA standards 1926.201, 202, and 203 (Signs, Signals and Barricades). The millennium edition, published in December 2000, was adopted by final rule with an effective date of December 11, 2002 (OSHA, 29 CFR 1926). According to an OSHA press release announcing the final rule, "the millennium edition" *or* the 1993 edition may be used for enforcement purposes. (This approach of accepting both editions creates ambiguity and conflict in OSHA enforcement actions due to the differences in some basic requirements of the two editions). One significant change in the new edition is a shift from designs primarily for motorist safety to the current dual focus of safety for both the motorist and workers.

Organization of the millennium edition of the *MUTCD* includes 10 distinct parts. Practitioners in the field use the minimum standards illustrated in the various printed versions that include Part 1, General information; Part 5, Rural Roads; and Part 6, Temporary Traffic Control Zones. All 10 parts of the *MUTCD* are available on the Internet at http://mutcd.fhwa.dot.gov//. The format from the internet version can be printed and contains approximately 1000+ pages. This is meant to be a dynamic document and presently offers several hundred proposed changes to the current millennium edition that range from editorial to typographical in nature.

There is movement within various organizations to improve the training of traffic control practitioners. Because of the number of deaths and injuries in work zones, this movement has gained national attention through groups such as the American Society of Safety Engineers, the Associated General Contractors of America, the National Safety Council, and the American Road & Transportation Builders Association. This training includes standardizing the application of traffic control devices on the roadways, properly applying the principles that provide motorists with advance warning and positive guidance through work zones to separate and protect workers on the site. In addition, several states require certification programs that ensure workers meet their DOT specifications for roadway construction projects.

Learning how to apply the principles set forth in the *MUTCD* requires:

(a) Recognition of the alpha–numeric system of identifying pages, sections, and figures.

(b) Use of information titled *standard* that can be recognized by bold print, and the use of the verb *shall* to indicate a mandatory condition or situation. The information titled *guidance* uses the verb *should* for recommended but not mandatory conditions. The information titled *option* and using the verb *may* in the *MUTCD* indicates an optional condition that is left to the discretion of the practitioner.

(c) Applying the information included in the various charts, graphs, and typical applications that are included in Part 6 of the *MUTCD*. (Note that primary distances are in metric notation whereas the distance numbers in parentheses use the American approach to distance measurement).

(d) Using basic math skills to determine length of tapers, length of the tangent that runs parallel to the activity area (longitudinal buffer and work zone), and the minimum number of devices required for the entire temporary traffic control work zone.

(e) The command of certain basic definitions: *upstream traffic*—If you are in the work zone, this is the traffic traveling toward you (i.e., from the first advance warning sign); *downstream traffic*—If you are in the work zone, this is the traffic traveling away from you from the activity area to the termination or downstream taper, as shown in this diagram.

⇨ Upstream traffic ⇨ / Work zone \ ⇨ Downstream traffic ⇨

What Is a Temporary Traffic Control Work Zone (TTCZ)?

For this chapter, and as a basis of understanding the term *temporary traffic control work zone*, the "work zone" is made up of four basic elements (illustrated in Figure 1):

1. *Advance warning*, often consisting of three signs that advise, warn, and instruct the motorist.

2. The *transition*, or *taper*, that moves traffic from one lane to another.
3. The *activity area* that consists of the optional longitudinal buffer (safety space) and the work zone.
4. The *termination*, or *downstream taper*, provides the motorist a clear path back to the lane from which he or she was diverted.

All dimensions and charts in the *MUTCD* are based on roadway speed and traffic volume. A strong argument can be made for the rationale of increased distances providing an additional margin of safety when warning-sign distances increase; longer transitions are utilized and longer distances for buffer

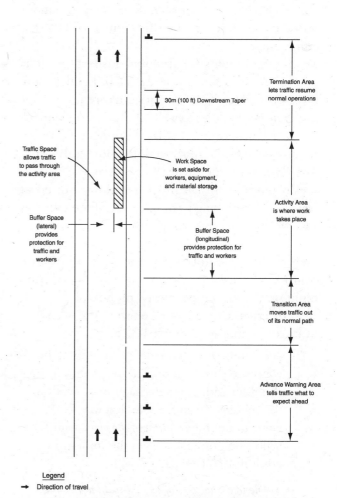

■ ■ **Figure 1** ■ ■ **Component parts of a temporary traffic control zone** (From the Federal Highway Administration's *Manual of Uniform Traffic Control Devices*, Part 6, Figure 6C-1)

zones and for the flagger station location are also designed into the system.

A *traffic control plan* (TCP) is a detailed plan of what traffic controls are to be included throughout the work zone and what changes will be necessary during the dynamic process of construction. There may be special night-shift considerations and changes for weekends based on local activities or changes in volume. Typically, there is a responsible person identified on the plan with emergency contact numbers should something occur in the work zone that requires repair, adjustment, or replacement of devices. This plan includes: drawings of the site (usually not to scale), the number and type of traffic control devices to be used, and lists or illustrations of any special personnel requirements such as flag-person stations or the use of uniformed traffic control (UTC—law enforcement).

A requirement for frequent inspections (day and night) is stated in the *MUTCD*. Political subdivisions such as states and cities may require other inspection schedules as well. On Colorado Department of Transportation (CDOT) projects there is a requirement for daily inspections of the traffic control work zone for each calendar day the devices are in place, with a night inspection required at least weekly for long-term projects. Some cities or specific projects may require different inspection frequencies (e.g., Boulder, Colorado, has a requirement to perform documented inspections of work zones and device placement every two hours during night operations on certain roadways).

The criteria for designing effective traffic control work zones and preparing a TCP are:

1. Determine the *location of the work*, (i.e., beyond the shoulder, on the shoulder, on the shoulder with lane encroachment, work in the median, or in the traveled way.
2. Determine the *duration of the work*, using the following guidelines from the millennium edition *MUTCD* (page 6G-2):

   > *The five categories of work duration and their time at a location shall be:*
   > *A. Long-term stationary is work that occupies a location more than 3 days.*
   > *B. Intermediate-term stationary is work that occupies a location more than one daylight period up to 3 days, or nighttime work lasting more than 1 hour.*
   > *C. Short-term stationary is daytime work that occupies a location for more than 1 hour, but less than 12 hours.*
   > *D. Short duration is work that occupies a location up to 1 hour.*
   > *E. Mobile is work that moves intermittently or continuously.*

   (Excerpted from the electronic version of *MUTCD*, referencing Colorado revised statutes.)

3. Determine the *type of roadway* (i.e., freeway, rural road, urban high speed, or urban low speed).
4. Determine the *appropriate typical application(s)* that are to be used as a basis for design of the traffic control plan.
5. Determine the *appropriate traffic control devices* that would be most effective in warning the motorist and protecting workers. TCDs include signs, traffic tubes, cones, vertical panels, drums, barricades (types I, II, and III), and temporary traffic barriers. As we move up through the hierarchy of devices, we gain more respect from the motorist. While people will avoid a collision with a drum, they may try to pick off the last cone in the tangent. Both NIOSH and OSHA promote the increased use of temporary traffic barriers to protect workers.

THE ADVANCE WARNING AREA

In this portion of the TTCZ, an appropriate warning of the hazard is provided to the motorist. Usually three distinct signs that warn, advise, and then instruct are used to convey the information to the motorist. The distances separating the signs in the advance warning area are dependent on the speed and volume of traffic of the roadway.

Table 1 identifies the distances between the advance warning signs used for roadway construction to identify and warn motorists about roadway-related hazards. These distances (A, B, C) are detailed under the typical applications listed in the *MUTCD*.

■ ■ Table 1 ■ ■

Suggested Spacing of Advance Warning Signs

Road Type	Distance Between Signs**		
	A	B	C
Urban (low speed)*	30 (100)	30 (100)	30 (100)
Urban (high speed)*	100 (350)	100 (350)	100 (350)
Rural	150 (500)	150 (500)	150 (500)
Expressway/Freeway	300 (1000)	450 (1500)	800 (2640)

* Speed category to be determined by highway agency

** Distances are shown in meters (feet). The column headings A, B, and C are the dimensions shown in Figures 6H-1 through 6H-46. The A dimension is the distance from the transition or point of restriction to the first sign. The B dimension is the distance between the first and second signs. The C dimension is the distance between the second and third signs. (The third sign is the first one in a three-sign series encountered by a driver approaching a temporary traffic control zone.)

Source: Federal Highway Association's *Manual of Uniform Traffic Control Devices*, Part 6, Table 6C-1.

The distances noted in the above chart start at the upstream side of the transition, or lane restriction, and are measured upstream from that point. The A distance is the distance from the start of the transition or constriction of the lane upstream to the specific instruction (e.g., the merge/transition symbol sign—the last sign seen by the motorist before the transition starts; this is where they must move from one lane to another). In a typical set-up the B distance is from the last sign in a series of three signs to the second sign the motorist sees and that provides them with a specific instruction (e.g., RIGHT LANE CLOSED). The C distance is measured upstream from the second sign in a series of three signs to the first sign the motorist sees and that provides the appropriate advance warning (e.g., ROAD WORK AHEAD). If the practitioner determines more information should be provided to the motorist and additional signs are to be used in the advance warning area, the B distance is duplicated for the additional sign-spacing requirements.

The point must be made again that the *MUTCD* provides minimum requirements for traffic control and by increasing distances, increasing the number of signs or devices (within reason to avoid confusing the motorist), the practitioner can improve safety on the work site for motorists and workers. Moving through the advance warning, the motorist first sees the warning, travels through the C distance to the advisory sign, and continues through the B distance to get to the instruction sign. The motorist then travels through the A distance to get to the point of transition, or the start of the taper.

THE TRANSITION AREA

The primary purpose of the transition area is to move traffic from one part of the road to another in a positive manner. Most often traffic is moved from one lane to another. The transition area requires tapers of various types. Table 2 identifies the different types of tapers that are used in traffic control. The basis for determining the distance covered by merging, shifting, and shoulder tapers is *L*, or the *length* of the taper. The merging taper is designed to move traffic from one lane to another and requires at least one *L* for the taper length. Once the *L* is determined, a multiplier is used to determine the length for a shifting taper (0.5) or for a shoulder taper (0.33). The other tapers identified in the chart are for the downstream, or termination, taper (100 feet) or the one-lane, two-way taper (100 feet). Again, please note that the primary numbers used in the formulas are in metric notation while the numbers in parentheses are the American measurement notation that we use for illustrations. Speed is the other factor that must be considered before completing the calculations for the *L* of the

■ ■ **Table 2** ■ ■

Taper-length Criteria for
Temporary Traffic Control Zones

Type of Taper	Taper Length (L)*
Merging	at least L
Shifting	at least $0.5\,L$
Shoulder	at least $0.33\,L$
One-lane, two-way	30 m (100 ft) maximum
Downstream	30 m (100 ft) per lane

* Formulas for L are as follows:

For speed limits of 60 km/h (40 mph) or less

$$L = \frac{WS^2}{155} \qquad \left[L = \frac{WS^2}{60} \right]$$

For speed limits of 70 km/h (45 mph) or greater

$$L = \frac{WS}{1.6} \qquad (L = WS)$$

where L = taper length in meters (feet)
    W = width of offset in meters (feet)
    S = posted speed limit, or off-peak, 85th percentile
         speed before work starts, or the anticipated
         operating speed in km/h (mph)

Source: Federal Highway Administration's *Manual of Uniform Traffic Control Devices*, Part 6, Table 6C-2.

taper for the TCP. The calculation uses the width of the lane (or offset—how much of the lane the project will impact) as one factor and speed as the other factor. The high-speed (45 mph or greater) calculation is straightforward $L = WS$, where: W = width of the lane, and S = the posted speed (some plans use the 85th percentile of peak or measured speed). While training traffic control supervisors, we normally use a 12-foot lane to practice and compare the different lengths of tapers based on the speed of the roadway. The formula for low-speed applications is $L = WS^2 \div 60$. With some basic math skills, traffic control practitioners can master the low-speed (40 mph or less) formula in a short time.

High Speed	Low Speed
$12 \times 45 = 540$	$12 \times 25 \times 25 \div 60 = 125$
$12 \times 50 = 600$	$12 \times 30 \times 30 \div 60 = 180$
$12 \times 60 = 720$	$12 \times 35 \times 35 \div 60 = 245$
$12 \times 70 = 840$	$12 \times 40 \times 40 \div 60 = 320$

The taper distances are measured along a straight line on the roadway (i.e., the white line), not the hypotenuse or line of the taper on which the devices are placed.

Once the length of the advance warning and transition, or taper, is determined based on the type of project and speed, the length of other elements must be determined. Typically, the longitudinal buffer zone is an optional portion of the temporary traffic work zone. To determine the distance of the buffer zone, the chart for the flagger's distance from the work zone in the current *MUTCD* is used. We use the same distances from the chart for both the buffer zone and the flagger-station location from the current edition of the *MUTCD*.

When developing a TCP, the total distance of the temporary work zone must be considered to determine the number of devices to place on the roadway. All the elements include the advance warning (identified in the advance warning chart), the length of the taper (by calculation), and the buffer zone (from the flagger workstation chart), and all are based on speed and roadway type. The length of the work zone will be given or determined by the project. It could vary from hundreds of feet to miles in length. The termination taper or downstream taper is 100 feet long to allow traffic back in their lane as soon as possible.

To calculate the number of devices used in the TCP, two calculations are used. In the taper or transition area the number of devices is determined by the length or L divided by the speed limit expressed in feet rather than miles per hour (i.e., 55 mph becomes 55 feet between the devices). The practitioner employs a generally understood principle that one device is to be used to establish a starting point from which to measure the distance between the devices. To illustrate, for a 12-foot lane closure on a roadway that has a 55 mph speed limit, the length of the taper is 660 feet. The 660-foot L is divided by 55 ft (mph converted to feet) to reach a product of 12, but you must add 1 device to start the line for a total of 13 devices as a minimum for this taper. The distance for a tangent line from the taper, or line parallel to the direction of traffic flow, is the total of the activity area comprised of the optional buffer zone and the work zone. In keeping with the above illustration, on a 55-mph roadway the buffer zone distance is 335 (see chart titled the "flagger station from work area") and the work zone would be, hypothetically, 500 feet. Thus the *total dis-*

tance of the activity area is 835 feet. This 835-foot distance for this portion of the traffic control zone is then divided by *two times the speed limit*, or 110 feet, between the devices. The line of devices for the tangent starts at the last device in the taper. In this illustration the tangent line of devices is made up of 8 units (835 ÷ 110 = 7.6—rounded up to 8). The downstream taper is 100 feet and is defined by 5 devices (the minimum for any taper). Approximately 500 feet downstream of the termination taper is the optional END WORK sign. Colorado's CDOT allows a sign that typically is a thank you from the contractor (e.g., "THANK YOU KORN CONSTRUCTION") to indicate the end of a work zone.

Traffic control plans detailing the minimum standards provide concise, positive guidance through work zones. An adequate warning of hazards ahead is provided to motorists by the advance warning signs. Positive guidance for motorists through the work zone is achieved with proper use of traffic control devices and also provides separation of the motorist from hazards and workers from motorists. Positive protection is achieved with temporary traffic barriers. The successful warning, guidance, and separation of motorists and workers thus achieves the stated purpose of the millennium edition of the *MUTCD*.

The above illustration reflects the minimum identified in the millennium edition of the *MUTCD*, but most practitioners agree that the number of devices calculated is too few. Most of the people in this profession use the dashed lines (often called skips) as a measurement tool. In Colorado, the skip measurement is a 10-foot painted line with a 30-foot space for a total of 40 feet from the start of one skip to the start of the next skip. This 40-foot distance (or skip) is *not* a standard measurement used throughout the United States. Typically, the practitioners on the roadway utilize a skip, no matter what the length, as a field measurement.

HAZARDS OF INSTALLING, MODIFYING, AND REMOVING WORK ZONES

Installing traffic control devices is the most hazardous part of the work zone process and puts people who have the responsibility for installation and re-

moval at extreme risk. Once a person (flag person) or device (cone, drum, barricade) is placed on a roadway or in the traveled way, a hazard is created and must be preceded by advance warning. While installing, modifying, or removing devices, some areas of the country use a high-level warning device to alert the driver that something is happening on the roadway ahead. A roadway worker is most at risk from a driver not expecting anything in the roadway. The driver might be concentrating on what has to be accomplished at work or could be engaged in conversation with passengers or, as is the case more often today, is distracted by using their hands-free cell phone conversation.

To properly set up a work zone, the signs for advance warning should be positioned first; working downstream, the remainder (or other elements) are placed in order from upstream to downstream. The transition for a right-lane closure is created by placing the first device either on the shoulder, (if it is included in the design) or on the white line on the far right of the roadway. Determine the distance along the white line to the next device based on the speed limit of the roadway and place each subsequent device one foot further into the lane until the lane is closed. The tangent is a line of devices that parallels the roadway configuration—straight or curved—moving downstream. The termination taper is the final portion of the traffic control work zone in which devices are used close to the activity area. The termination taper is 100 feet long, developed with a minimum of 5 devices. The sign END WORK is optional, but if the design calls for this option, these signs are installed last. The next step for the practitioner will be to drive through the work zone to assess the effectiveness of the setup. Modifications may have to be made if the plan is not accomplishing the goal of safety for the motorist and safety for the workers in the work zone. When traffic control work zones must be modified for nighttime or weekend use, the specific information or warning signs must be disabled, taken down, turned on their side, or removed. RIGHT LANE CLOSED, FLAGGER AHEAD, and the merge symbol are examples of signs that may mislead or at the very least diminish the credibility of the setup to the motorist. In these situations the ROAD WORK

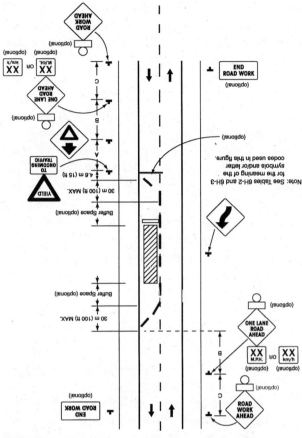

Figure Notes

Option:

1. This temporary traffic control zone application may be used as an alternate temporary traffic control plan to the lane closure with flaggers (Figure 6H-10), when the following conditions exist:

a. Motor vehicle traffic volume is such that sufficient gaps exist for motor vehicle traffic that must yield.

b. Drivers from both directions are able to see approaching motor vehicle traffic through and beyond the work site.

[Standard]

2. When flaggers are used, the flagger symbol sign shall be used in place of the YIELD AHEAD sign.

Option:

3. The Type B flashing warning lights may be placed on the ROAD WORK AHEAD and the ONE LANE ROAD AHEAD signs whenever a night lane closure is necessary.

■■ **Figure 2** ■■ **Lane closure on low-volume, two-lane road** (From the Federal Highway Association's *Manual of Uniform Traffic Control Devices*, Part 6, Figure 6H-11)

AHEAD sign should remain in place to inform the motorist that there could be a hazard ahead that may affect their ability to transverse the roadway without incident.

There are some specific situations that require modification of the above approach. One-lane, two-way traffic is a situation where a two-lane roadway is turned into a temporary one-lane road and must handle two-way traffic (see Figure 2 and the typical application that follows this explanation). Assume an east–west two-lane roadway with the eastbound lane temporarily closed for repairs. Opposing traffic must be controlled first to prevent head-on collisions. The eastbound traffic is to be controlled by flaggers or signs or signals until the travel lane delineated by traffic control devices can be set up. The transition chart shows the one-lane, two-way taper as a short transition (typically 100 feet) to allow the efficient flow of traffic. A goal of the *MUTCD* standards is to minimize the impact on traffic. To achieve this goal in the one-lane, two-way traffic situation, the length of the one-lane portion of the work zone should be as short as possible. The termination, or downstream taper, is also 100 feet to allow traffic to quickly move through the work zone. Advance warning signs such as ROAD WORK AHEAD, ONE LANE ROAD, and FLAGGER AHEAD are used if appropriate. The westbound lane is addressed first (opposing traffic); the eastbound traffic will soon be traveling on the roadway or lane that is designated for westbound traffic. The advance warning signs must be installed to indicate a one-lane road ahead with appropriate warnings of what type of system is to be employed to control the traffic (flag person, yield signs, or signals). The traffic control devices are placed first in the one-lane, two-way taper, then along the center line or tangent to the downstream taper for the eastbound lane. The termination taper is then completed with a minimum of five devices. Once the devices are in place, alternating eastbound or westbound traffic can be stopped (by flaggers, signs, or signals) while allowing the opposite direction of traffic to proceed in the open lane. This traffic control plan is only used in situations where traffic speed and volumes are low and can be effectively controlled. There are many variations of this theme, but placing the traffic control for opposing traffic first is the best practice to prevent incidents while installing this type of system. Knowing the duration of the project, the schedule for both the number of hours per day as well as the number of days for the project, will enable the practitioner to decide which typical application to use. The choices identified in the *MUTCD* for the one-lane, two-way traffic scenario are the flag persons (TA 10), yield signs (TA 11), or motion detector or timer-controlled traffic signals (TA 12).

Using a Flag Person

One of the reasons OSHA has given for adopting the millennium edition of the *MUTCD* is its recommendation for flaggers' clothing. A recently published ANSI standard identifies three categories of clothing for roadway work—types I, II, and III. The rationale for using one type of clothing versus another is based on speed and traffic volume of the roadway on which the work is performed. Starting with type I and moving through type III, the clothing has more area covered by retro-reflective material. The visibility afforded by day operations versus night operations and flat light or limited-sight situations, such as fog, rain, or snow, also affects the clothing choice. Advances in retro-reflective clothing have increased visibility. Larger areas of high-visibility materials are now required on vests or jackets worn for high-speed roadway work. High-speed operations performed at night require retro-reflectorized materials on hard hats, jackets, and pants.

The *MUTCD* recommends that flaggers should be used only when other methods are not adequate to protect the motorist or workers. Specific types of applications where the use of flaggers is appropriate are: one-lane, two-way traffic; slowing traffic through limited-sight situations; or stopping traffic where a haul road crosses traveled lanes or where equipment must be moved into or out of a work zone. A flag person is the representative of the contractor and the project. It's in everyone's best interest to properly train the flaggers in appropriate contact with the public, proper clothing, and standard flagging techniques. Many state departments of transportation require certification or a minimum amount of training for flaggers to fulfill certain contracts. Flaggers must be stationed with two means of escape and highlighted at night. Nighttime flaggers must have retro-reflective clothing and a means of alerting and communicating clearly with the driver, such as light wands (flashlights with orange tubes attached). Communications are required whenever flag persons are used for equipment entering or exiting lanes of traffic adjacent to the construction zone. When the project requires one-lane, two-way traffic, such as a lift of asphalt applied to one lane, flag persons will stop one direction of traffic and release the opposite direction. Radio contact or hand signals by flag persons are acceptable means of communication. Another technique used in these situations is to give the last driver in the stopped lane of traffic a token of some type to be given to the flagger at the other end of the project. According to experienced flag persons, the token given should be something of limited value—a rock, for instance—because it may not be returned. An additional technique used in this situation, if the project covers a great distance, is to use a pilot car to lead traffic back and forth through the work zone. The pilot car must be signed appropriately (PILOT CAR, FOLLOW ME) and appropriate advance warning should indicate that a pilot car is being used on this project. Unfortunately, flaggers are often ignored by drivers and at times must be aggressive in getting the drivers' attention. Mobile operations depending on speed and length of the project may also use flaggers to protect workers. Typically, in mobile operations without flaggers a truck-mounted attenuator with flashing caution arrows is used to protect workers and get the drivers' attention. Because of their job and the location of their station, preventing flagger injuries is a challenging prospect. To minimize exposure to injury, several basic training techniques must be reinforced: always be alert, wear proper clothing, always face the traffic, and stand in the proper location with two means of escape.

Driver Awareness

This chapter has frequently mentioned the variable levels of ability and attention of the driving public. Throughout the country there has been an attempt to create a consistent and positive approach to the use of work zone devices and changeable message boards, but it appears that as a nation we have been remiss in continuing to require formal driver training, stringent testing, and enforcement. Drivers demonstrate great variability in experience, physical ability, and attitude (many of us have had experience with road rage and aggressive drivers). Impaired individuals (those using drugs—prescription or illegal—and alcohol) are all too often operating vehicles on our roadways day and night. Types of vehicles also change our approach to the driving task. The driver of a three-quarter-ton truck will take mountain curves a bit differently than the driver of a sports car or motorcycle. The warning of reduced speed for curves may or may not be heeded depending on the type of vehicle being driven.

The frame of reference for drivers can be very different as well. Some of us have had driver training in high school as a mandatory course, others have had no formal training, others learned to drive on a ranch or farm where operating equipment was permitted once the pedals could be reached, and still others learned to drive in countries with very different driving techniques. The latter may be a tourist without a clue how to get through town and can easily lose his or her way if a detour is necessary. With the tourist and the weakest driver among us in mind, the work zone must be set up with geometrics that allow a driver with limited experience who may be pulling a 30-foot camper with a 20-foot truck to get through the project without incident. The TCP must be designed with the worst driver in mind to enable all drivers on the road to efficiently navigate the construction work zone. When the TCP is properly developed, the choice of direction should be clear to every driver, and a clear choice of travel without ambiguity promotes efficient traffic control work zones.

LIABILITY AND LITIGATION

In today's litigious society the issue of liability often enters into conversations with traffic control supervisors and other practitioners who set up traffic control work zones. The question often asked is "What if something goes wrong and there is an incident in the work zone we designed?" This question is closely followed by the concern that the practitioner could potentially have a plaintiff's attorney include them in deposition for an incident that happened years ago. Documentation is one of the loss mitigation techniques that can be achieved without great difficulty. Most companies doing this type of work have standard forms for the inspections, which should become part of the owner's project files; subcontractors and experienced practitioners should retain a copy for their personal files as well. Many of the questions of liability exposure may be positively addressed by using the *MUTCD* guidelines as a minimum for design, and then documenting any decision to go beyond the standards. By increasing the number of devices, extending the length of tapers, adding additional signs in the advance warning area, or adding electronic devices such as message boards and flashing arrow signs, you will exceed the minimums.

There are other considerations for the practitioner developing and implementing a traffic control plan. Liability suits against local governments and contractors responsible for the design, maintenance, and operation of highways and street systems have increased dramatically. Judgments against deep-pockets political subdivisions (even with sovereign immunity) and contractors with bonds, insurance policies, or other assets that could be attached can reach astronomical amounts with a sympathetic jury. These judgments can be awarded for failing to provide: advance warning, adequately lighted areas, or properly maintained retro-reflective traffic control devices.

Consider the elements of tort liability and how they may relate to your construction projects, including the traffic control work zone. A tort is negligent or wrongful conduct that causes a bodily injury or property damage incident for which financial judgments may be awarded in a civil suit. Liability is not an absolute. To determine liability you must consider at least four elements.

Element 1: Duty to the public. The law generally holds that any person or agency working in or near the traveled way has an obligation to use a standard of care that a "reasonable and prudent" person would use under similar circumstances. Juries however may have difficulty deciding how reasonable or prudent a contractor, represented by an insurance company attorney, acted in the work zone when faced with a widow or a quadriplegic teenager.

Frequently, attorneys use well-accepted and published standards to establish "ordinary care" to develop their case. The *MUTCD* standard for roadway construction and maintenance work has been adopted by most states with some additions (state requirements cannot be less stringent than the federal standards); OSHA has adopted the millennium edition by reference. The *MUTCD* is easily accessible through organizations such as the Associated General Contractors of America or the American Traffic Safety Services Association and on the Web.

Element 2: A breach of the duty must be proven to establish a case. To develop a case for a breach of duty there must have been a hazard. The agency must have knowledge of the hazard, or should have known about the hazard. Complaints by citizens or deficiencies noted by inspectors must be addressed to establish fulfillment of responsibility and obligations to provide a roadway reasonably safe for the traveling public.

Second, the defendant must have failed to remedy the hazard or warn the motorist. An agency has an obligation to address a known hazard immediately. Even if a total remedy is not feasible, the agency has a duty to provide adequate warnings of the hazard. Response time is critical.

The plaintiff's attorney may try to show that the defendant violated some section of the *MUTCD* or other accepted standard. Violations of the printed standards, which are detailed throughout the manual in bold type, are usually easy to prove. A mitigation technique for the defendant is to document the care taken to effectively sign or protect the workers from motorists or protect the motorist from the hazard. Documentation of the traffic control used on a construction project is critical to covering or protecting your organization's assets. The *MUTCD* requires inspections on a frequent basis, day and night. Typically, the traffic control practitioner uses a daily diary to document actions that fulfill contractual obligations and to identify changes in the project, changes in work zone traffic control,

and any improvement or replacement of devices. The *MUTCD* has established mandatory standards that are printed in bold type and use "shall" as the operative verb. The guidance statements are not printed in bold, but are recommended with "should" as the operative word. An option statement allows the practitioner, based on experience and judgment, to decide how much to add to the minimum traffic control plan, thereby adding to the safety of the work zone. Documentation of work zone traffic control techniques that exceed the minimum requirements will serve the defendant well in case of a lawsuit.

Element 3: Proximate cause. Just because an incident occurs at an inadequately or improperly marked work zone does not necessarily impute an agency's negligence. To be successful, a plaintiff must show that the hazard did, in fact, contribute to the incident. This statement is not absolute. A drunk driver or a vehicle defect may not negate recovery if the work zone hazard plays a "substantive factor" in causing the incident. The work zone must be properly protected to provide the motorist positive guidance through the construction project with separation from hazards that also provides protection for workers.

Element 4: Damage. Another element that must be satisfied to establish liability is the damage sustained by the plaintiff. This could be physical, emotional, or mental damage; pain and suffering a loss of income or income potential; even loss of comfort from the injured spouse. In all cases, measurable damage to either persons or property must be established.

TRAINING

Skill training for workers is required in the construction field to successfully complete quality projects. OSHA standards in construction also require training for all employees to be able to recognize hazards and have the knowledge to protect themselves from injury when exposed to those hazards. This chapter has detailed various exposures for workers both on the traffic side and the construction side of traffic barriers. The chapter does *not* attempt to develop an exhaustive coverage of every possible exposure to workers and associated solutions. Each project must be considered for its unique as well as common hazards and their potential to injure workers. Quality of work completed, Cost control, Safety of workers and the public, and Production schedules (QCSP) must all be considered to have a successful project. When all QCSP elements are considered as a total system approach to the work (not one in favor of another), typically, the construction project is completed on or ahead of schedule, on or below budget, with fewer items on punch lists, and with few incidents and injuries to workers and the public. Managers making expectations clear, maintaining constant vigilance, and coaching workers to improve behavior outcomes will promote a safe workplace for all employees and the public. Controlling known hazards to workers and developing systems and a work environment where the employees will communicate hazards they see on the site, or correct them without being instructed to do so, creating peer pressure to work safely, serves as a road map to success.

Many states have training and certification programs that comply with the specific state requirements to work on DOT projects. There are various national, state, and local organizations that provide training and are authorized by individual states to certify people for designing and implementing traffic control plans.

REFERENCES

American Road & Transportation Builders Association (ARTBA). *Trainer's Manual for OSHA 10-hour Training for the Road Construction Industry.* Washington, DC: ARTBA, 2001.

Colorado Contractors Association (CCA) *Instructor's Manual for Traffic Control Supervisor Certification.* Denver, CO: CCA, 2003.

Federal Highway Administration (FHA). *Manual of Uniform Traffic Control Devices* (millennium ed.). Washington, DC: FHA, 2000. (Also available as an electronic version with referenced Colorado revised statutes at www.trafficgraphics.com.)

"Hazards on both sides of the barrier," a paper presented by George Wolff, CSP, at Safety Fest in the West symposium sponsored by OSHA, Region V, Colorado AGC and Colorado Contractors Association, April 2002.

Humphreys, Jack B. "Take the Liability Out of Roadway Construction," *The American City & County,* December 1976.

OSHA, *29 CFR 1926.* Standards for the construction industry. Washington DC: OSHA, 2003.

REVIEW EXERCISES

1. What are the four basic elements of a temporary traffic control zone?
2. Nationally 700 to 1000 fatalities are recorded in roadway work zones each year. What proven techniques can be applied to reduce this statistic?
3. Describe the design considerations and elements for use in an internal traffic control plan to prevent injuries on construction sites.
4. In today's litigious society, plaintiffs' attorneys and citizens seek judgments with financial compensation from political subdivisions and contractors. What techniques can be employed by the safety professional to eliminate or mitigate and minimize liability exposures to create positive defenses against lawsuits?
5. Compressed schedules and attempts to minimize impacts to motorists are reasons for increased frequency of nighttime work projects. What special considerations and injury prevention techniques are required for these projects?

■■ **About the Author** ■■

CONSTRUCTION SAFETY IN
HEALTHCARE FACILITIES
by Ralph E. Estep, R.N., CIH, CSP

Ralph E. Estep, R.N., CIH, CSP, is a principal consultant for Health Systems Safety, Inc. Actively involved at many levels in the professions of occupational safety and industrial hygiene for over twenty years, he currently specializes in health and safety concerns of the healthcare industry. He is a past president of the Western New York section of the American Industrial Hygiene Association, a member of the American Industrial Hygiene Association Occupational Epidemiology committee, and an adjunct faculty member for Niagara County Community College where he teaches students enrolled in the Occupational Safety and Health degree program. In addition to being a registered professional nurse, he holds a Master of Science in Occupational Health and Safety Engineering from the West Virginia University College of Engineering and a Bachelor of Arts degree in Chemistry, also from WVU.

CHAPTER 28

Construction Safety in Healthcare Facilities

LEARNING OBJECTIVES

- Discuss the potential hazards to employees during construction activities in healthcare facilities.
- Explain the role of the safety professional as a team member in preventing healthcare clients from developing infections during construction activities.
- Discuss the common activities of healthcare facilities and the hazards that are present.
- Explain the use of an Infection Control Risk Assessment as a tool in preplanning healthcare construction.
- Identify the types of microbial contamination and the consequences that can be associated with construction work in healthcare facilities.

INTRODUCTION

Many construction projects potentially pose hazards that may affect members of the general public. These may involve exposure to physical, chemical, or biological hazards. In addition to the traditional employee health and safety concerns, safety professionals must also consider these risks when managing a construction project.

A particular business environment that presents unique challenges during construction is the healthcare facility. Healthcare is one of the fastest growing segments of our economy and has seen a substantial increase in construction projects, particularly renovations. This has been due in part to an increasing demand for assisted-living facilities and skilled nursing centers. Construction of assisted-living facilities in 1998 rose by 49 percent, with 138 projects completed at a cost of $703 million.[1] Hospitals and healthcare facilities are also aging, increasing the need for repair and remediation work, while the introduction of new technologies in patient care and advances in new information systems have also led to renovation projects.

Many of these construction activities can introduce or increase contamination of the air and water in patient-care environments. Environmental disturbances during healthcare facility construction projects can create airborne and waterborne hazards for patients who may be at risk for healthcare-associated opportunistic infections.

This chapter will discuss the challenges that may be faced when providing safety oversight of a healthcare construction project. It will provide a number of tools for reducing the risk of injury to healthcare facility clients and employees.

531

HEALTH AND SAFETY CONCERNS

Day-to-day work activities in healthcare facilities can present uncommon hazards to construction employees—biological contaminants, chemicals, physical hazards unique to such settings, radioactive materials and other sources of ionizing and nonionizing radiation, to name a few. In general, healthcare hazards can be divided into two categories: noninfectious hazards and infectious hazards.

Noninfectious Hazards

Construction workers may encounter many noninfectious hazards in healthcare settings. Chemicals present may include numerous agents used as disinfectants. Some of the commonly used substances are: isopropyl alcohol, sodium hypochlorite (chlorine), iodine, phenolics, quaternary ammonia, and glutaraldehyde. Ethylene oxide, a colorless gas with a distinct sweet etherlike odor is used to sterilize medical instruments. This compound is regulated by the Occupational Safety and Health Administration (OSHA) as a carcinogen (29 CFR 1910.1047). Formaldehyde is another toxic substance that is widely used in healthcare. It may be found in histology laboratories or pathology departments as a tissue preservative. It is also used as a sterilizing agent in dialysis units. NIOSH considers formaldehyde a carcinogen, and OSHA regulates employee exposure (29 CFR 1910.1048). Asbestos as an insulating or fire-proofing agent is also commonly found in older healthcare facilities. Lead is also frequently used as a shielding method for radioactive sources and X rays.

Medical gases are commonly used in the healthcare industry. Examples of gases used are oxygen, nitrogen, nitrous oxide, medical-grade compressed air, and anesthetics. They may be present as individual compressed gas cylinders or as centralized bulk storage systems that are plumbed to various locations in the facility. Wall-mounted outlets of any medical gas present the potential for leakage into the work area.

Oxygen is one of the most common medical gases found in healthcare settings. It is a colorless, tasteless, odorless gas that supports combustion and should not come into contact with sparks or an open flame. Cylinders of this gas are used for patients requiring supplemental oxygen during transportation within the facility. The cylinders will have a regulator and flow meter that may or may not be attached. Cylinders containing oxygen have safety devices of either the frangible disc or frangible disc backed up with a fusible metal type.

Many patient rooms will have a wall-mounted oxygen outlet. A rotameter-type flow meter may be attached to the wall to regulate the amount of oxygen delivered to the patient. Oxygen is noncorrosive; however, extreme caution should be taken to avoid contact with combustible materials such as grease or oil, especially during construction of new systems. Lines to contain oxygen should be pretested for leaks with an inert gas such as nitrogen.

Nitrogen is used as an inert gas in healthcare, for example, as a purging agent in some types of sterilizing systems. Nitrogen is a colorless, odorless, nontoxic gas, but can act as an asphyxiant by displacing oxygen. Leaks can be detected using a soapy water solution. A leak will be evident by the formation of bubbles.

Nitrous oxide is used as an inhalation anesthetic during surgical and dental procedures. Nitrous oxide is a colorless, nonflammable gas with a sweetish taste and odor. Cylinders will have frangible discs as safety devices. Leaks can be detected with a soapy water solution, observing for bubble formation. Other anesthetic gases used in hospitals include a wide variety of halogenated hydrocarbons such as halothane, desflurane, enflurane, isofluane, methoxyflurane and sevoflurane. All of these agents are nonflammable and produce a sedative effect on the central nervous system.

Physical hazards might also present problems. Healthcare locations that could create noise and heat-related concerns can include laundries, dietary operations, and boiler rooms.

Sources of ionizing radiation may be associated with X-ray equipment and patient treatment with radionuclides. Nonionizing radiation may consist of ultraviolet light from germicidal lamps and lasers used in surgical applications.

A list of major noninfectious hazards to be considered when working in a healthcare environment is presented in Table 1.

■ ■ **Table 1** ■ ■

Common Healthcare Facility Noninfectious Hazards

Hazard	Location
Chemical disinfectants	Nursing unit, central supply
Sterilizing agents	Central supply, dialysis unit
Waste anesthetic gases	Operating room, out-patient clinic
Chemical preservatives	Laboratory, pathology
Heat, noise	Laundry, dietary facility, boiler room
Ionizing radiation	X-ray department, nursing unit
Nonionizing radiation	Outdoor entranceway, operating room

Infectious Hazards

Healthcare facilities, by the very nature of their business, involve potentially infectious materials. These can include blood and other body fluids that may be present on contaminated surfaces or equipment such as syringes, needles, and scalpel blades. Bloodborne viruses such as hepatitis B, C, and HIV are common. Healthcare workers are at risk for occupational exposure to bloodborne pathogens, including hepatitis B virus (HBV), hepatitis C virus (HCV), and human immunodeficiency virus (HIV). Exposures may occur through an unintentional needlestick or cuts from other sharp instruments that are contaminated with an infected patient's blood. The risk for actually developing a disease following occupational exposure to a bloodborne pathogen depends on a number of factors. These include the number of infected individuals in the patient population, the chance of becoming infected after a single blood contact from an infected patient, and the type and number of blood contacts.

After a specific exposure, the risk of infection may vary with factors such as the following:

■ the pathogen involved
■ the type of exposure
■ the amount of blood involved in the exposure
■ the amount of virus in the patient's blood at the time of exposure.

Most exposures do not result in infection.[2]

Hepatitis B is a serious disease caused by a virus that attacks the liver. The virus can cause lifelong infection, cirrhosis (scarring) of the liver, liver cancer, liver failure, and death. Hepatitis C is a disease of the liver caused by the hepatitis C virus. Human immunodeficiency virus is a bloodborne pathogen that can result in AIDS.

Clients in the facilities may be there for treatment of any number of infectious diseases. Airborne infections could include contagious diseases such as tuberculosis (TB). TB is caused by the bacteria *Mycobacterium tuberculosis* and is spread by airborne droplets generated when a person with TB coughs, speaks, or sneezes. Infection occurs when a susceptible person inhales droplet nuclei containing the bacteria, which then become established in the body. Increases in the incidence of TB have been observed in some geographic areas; these increases are related partially to the high risk for TB among immunosuppressed persons, particularly those infected with human immunodeficiency virus. Other factors (e.g., socioeconomic) have also contributed to these increases. Outbreaks have occurred in hospitals, correctional institutions, homeless shelters, nursing homes, and residential care facilities for AIDS patients. Heightened awareness of the disease in the United States has led to increased efforts at prevention and control, including the implementation of TB control measures recommended by the Centers for Disease Control and Prevention (CDC) and required by OSHA.[3]

Patients who are being treated for tuberculosis will be isolated from the healthcare employees in some manner. This could include being in a private room with negative pressure ventilation or some type of precaution involving the use of special personal protective equipment.

Construction workers, as a general rule, should have minimal contact with clients that have infectious diseases. They should be made aware of any potential hazards and how to identify their presence.

INFECTION-CONTROL ISSUES

One of the principal issues during construction in any healthcare facility is the potential for infection to clients from the construction activity itself. This

■ ■ **Figure 1** ■ ■ **Biohazard sign** (Photo courtesy of Lab Safety Supply, Inc.)

can be thought of as a function of increased susceptibility of certain clients to airborne or waterborne sources that may be present during construction or after it has occurred. Healthy individuals can usually fight off potential infection from microorganisms that are normally present in a construction environment. Biological sources that can cause infection during construction include dust, molds, or soil that has been contaminated with fungal spores or bacteria.

In the healthcare setting, certain clients may be at increased risk. These clients all have one thing in common. Their body's ability to fight infection has been weakened. The technical term for this condition is called immunosuppression or immunocompromised. It can be a result of age, types of medical conditions, certain therapies they are receiving such as cancer-fighting drugs, or medications used to help fight rejection during organ transplantation.

Outbreaks of disease due to infection from the microorganisms *Aspergillus*[4,5,6], *Mucoraceae*[7], and *Penicillium* have been documented in healthcare facilities when environmental controls were absent during periods of construction.[8] These infections can lead to increased length of recovery time or even death for clients who contract them.

Construction, renovation, repair, and demolition activities in healthcare facilities require substantial planning and coordination to minimize the risks of infection both during projects and after their completion. Traditionally, construction in healthcare has been a responsibility of the maintenance or physical plant departments. To prevent infection of susceptible clients, a multidisciplinary team approach—construction management, safety professional, infection control practitioner—should be considered in order to coordinate specific evaluation and control methods during the various stages of construction.

Examples of suggested team members and some of their responsibilities are presented in Table 2.

Preconstruction Phase

Establishment of a plan to prevent the potential for client infection should be considered prior to any

■ ■ **Table 2** ■ ■

**Construction Site
Infection Control Team Members**

Team Member	Responsibilities
Infection control practitioner	Identify at-risk clients, planning, inspections
Project manager	Planning, contract specifications
ES&H professional	Planning, education, inspections, air sampling
Effected department manager	Coordination, processing feedback
Administrator	Establish facility policy, conflict resolution
Architect	Planning, design, contract specifications
Contractor	Program implementation

actual construction work beginning. An integral part of this plan is an infection control risk assessment (ICRA). A good assessment will identify any at-risk clients that are in the construction area. This should be done by someone with the medical knowledge of diseases, procedures, treatments, and an understanding of how these relate to infection control. Most healthcare facilities will have an individual or team assigned the responsibility of infection control data collection and implementation of control measures. These professionals are becoming increasingly aware of their role in the prevention of infection during construction.[9] Infection control risk assessments are now required by numerous states for construction and renovation and by the Environment of Care standards of the Joint Commission for Accreditation of Healthcare Organizations (JCAHO).

The Guidelines for Design and Construction of Hospital and Healthcare Facilities issued by the American Institute of Architects' (AIA) Academy of Architecture for Health states that completing the ICRA should be the result of a "consultative" process. The AIA guidelines do not specify a particular ICRA tool or instrument. Many include matrices that classify construction projects by type and activities, since construction activities have varying degrees of risk associated with the potential for causing airborne or waterborne sources of microorganisms. Examples can range from removal of ceiling tiles to major renovation or new building construction. Once cli-

ents have been identified, they can also be grouped into risk categories. Figure 2 is a representation of an ICRA process and Figure 3 is an example of an infection control construction permit that addresses these issues.

A description of required infection control precautions both during construction and upon completion of the project and forms that ask for the signatures of contractors and subcontractors as a way of communicating construction project requirements can be included. Ideally, this information should become part of the project specifications and stop-work requirements.

Some examples of control measures that should be considered prior to the start of construction are:

- *Dust Control* – plastic barriers, floor mats, sealing windows, wetting, scheduled cleaning, enclosed scaffolding, sticky mats.
- *Ventilation* – flow patterns, negative pressure, filtration methods, sealing intakes.
- *Disinfection* – flushing and treating water lines per local and state codes.
- *Relocation* – moving clients to other areas of the facility, rerouting client traffic.

Consideration should also be given to construction activities that are external to the facility. Migration of dust from soil excavation or the building of new additions can also be potential sources for client infection. Examples of prevention strategies are presented in Table 3.

■■ **Table 3** ■■

**Methods to Reduce Dust Contamination During
External Demolition and Construction**

Location	Control Method
Demolition site	Dust control can be managed by misting the dirt and debris
Adjacent air intakes	Seal off/shut down/relocate all affected intakes
HVAC system	Maintain facility air pressure positive relative to outside air
Air intake filters	Change roughing filters frequently
Windows	Seal and caulk
Doors	Caulk or weather-strip frames
Pedestrian walkway	Relocate to decrease dust tracking

Infection Control Risk Assessment
Matrix of Precautions for Construction & Renovation

Step One:
Using the following table, *identify* the <u>Type</u> of Construction Project Activity (Type A-D)

TYPE A	**Inspection and Non-Invasive Activities**. Includes, but is not limited to: ▪ removal of ceiling tiles for visual inspection limited to 1 tile per 50 square feet ▪ painting (but not sanding) ▪ wallcovering, electrical trim work, minor plumbing, and activities which do not generate dust or require cutting of walls or access to ceilings other than for visual inspection.
TYPE B	**Small scale, short duration activities which create minimal dust** Includes, but is not limited to: ▪ installation of telephone and computer cabling ▪ access to chase spaces ▪ cutting of walls or ceiling where dust migration can be controlled.
TYPE C	**Work that generates a moderate to high level of dust or requires demolition or removal of any fixed building components or assemblies** Includes, but is not limited to: ▪ sanding of walls for painting or wall covering ▪ removal of floorcoverings, ceiling tiles and casework ▪ new wall construction ▪ minor duct work or electrical work above ceilings ▪ major cabling activities ▪ any activity which cannot be completed within a single workshift.
TYPE D	**Major demolition and construction projects** Includes, but is not limited to: ▪ activities which require consecutive work shifts ▪ requires heavy demolition or removal of a complete cabling system ▪ new construction.

Step 1_____

Steps 1–3 adapted with permission, V. Kennedy, B. Barnard, St. Luke Episcopal Hospital, Houston TX; C. Fine, CA.

■ ■ **Figure 2** ■ ■ **Precautions for new construction or renovations**

Step Two:

Using the following table, *identify* the <u>Patient Risk</u> Groups that will be affected.
If more than one risk group will be affected, select the higher risk group:

Low Risk	Medium Risk	High Risk	Highest Risk
▪ Office areas	▪ Cardiology ▪ Echocardiography ▪ Endoscopy ▪ Nuclear Medicine ▪ Physical Therapy ▪ Radiology/MRI ▪ Respiratory Therapy	▪ CCU ▪ Emergency Room ▪ Labor & Delivery ▪ Laboratories (specimen) ▪ Newborn Nursery ▪ Outpatient Surgery ▪ Pediatrics ▪ Pharmacy ▪ Post Anesthesia Care Unit ▪ Surgical Units	▪ Any area caring for immunocompromised patients ▪ Burn Unit ▪ Cardiac Cath Lab ▪ Central Sterile Supply ▪ Intensive Care Units ▪ Medical Unit ▪ Negative pressure isolation rooms ▪ Oncology ▪ Operating rooms including C-section rooms

Step 2_____

Step Three: <u>Match</u> the

Patient Risk Group (*Low, Medium, High, Highest*) with the planned ...
Construction Project Type (*A, B, C, D*) on the following matrix, to find the ...
Class of Precautions (*I, II, III or IV*) or level of infection control activities required.

Class I-IV or Color-Coded Precautions are delineated on the following page.

IC Matrix - Class of Precautions: Construction Project by Patient Risk

Patient Risk Group	Construction Project Type			
	TYPE A	TYPE B	TYPE C	TYPE D
LOW Risk Group	I	II	II	III/IV
MEDIUM Risk Group	I	II	III	IV
HIGH Risk Group	I	II	III/IV	IV
HIGHEST Risk Group	II	III/IV	III/IV	IV

Note: Infection Control approval will be required when the Construction Activity and Risk Level indicate that Class III or Class IV control procedures are necessary.

Step 3_____

Steps 1–3 adapted with permission, V. Kennedy, B. Barnard, St. Luke Episcopal Hospital, Houston TX; C. Fine, CA.

▪▪ **Figure 2** ▪▪ (cont.)

Description of Required Infection Control Precautions by Class

	During Construction Project	Upon Completion of Project
CLASS I	1. Execute work by methods to minimize raising dust from construction operations. 2. Immediately replace a ceiling tile displaced for visual inspection.	1. Clean work area upon completion of task.
CLASS II	1. Provide active means to prevent airborne dust from dispersing into atmosphere. 2. Water mist work surfaces to control dust while cutting. 3. Seal unused doors with duct tape. 4. Block off and seal air vents. 5. Place dust mat at entrance and exit of work area 6. Remove or isolate HVAC system in areas where work is being performed.	1. Wipe work surfaces with disinfectant. 2. Contain construction waste before transport in tightly covered containers. 3. Wet mop and/or vacuum with HEPA filtered vacuum before leaving work area. 4. Remove isolation of HVAC system in areas where work is being performed.
CLASS III	1. Remove or Isolate HVAC system in area where work is being done to prevent contamination of duct system. 2. Complete all critical barriers i.e. sheetrock, plywood, plastic, to seal area from non work area or implement control cube method (cart with plastic covering and sealed connection to work site with HEPA vacuum for vacuuming prior to exit) before construction begins. 3. Maintain negative air pressure within work site utilizing HEPA equipped air filtration units. 4. Contain construction waste before transport in tightly covered containers. 5. Cover transport receptacles or carts. Tape covering unless solid lid.	1. Do not remove barriers from work area until completed project is inspected by the owner's Safety Department and Infection Control Department and thoroughly cleaned by the owner's Environmental Services Department. 2. Remove barrier materials carefully to minimize spreading of dirt and debris associated with construction. 3. Vacuum work area with HEPA filtered vacuums. 4. Wet mop area with disinfectant. 5. Remove isolation of HVAC system in areas where work is being performed.
CLASS IV	1. Isolate HVAC system in area where work is being done to prevent contamination of duct system. 2. Complete all critical barriers i.e. sheetrock, plywood, plastic, to seal area from non work area or implement control cube method (cart with plastic covering and sealed connection to work site with HEPA vacuum for vacuuming prior to exit) before construction begins. 3. Maintain negative air pressure within work site utilizing HEPA equipped air filtration units. 4. Seal holes, pipes, conduits, and punctures appropriately. 5. Construct anteroom and require all personnel to pass through this room so they can be vacuumed using a HEPA vacuum cleaner before leaving work site or they can wear cloth or paper coveralls that are removed each time they leave the work site. 6. All personnel entering work site are required to wear shoe covers. Shoe covers must be changed each time the worker exits the work area. 7. Do not remove barriers from work area until completed project is inspected by the owner's Safety Department and Infection Control Department and thoroughly cleaned by the owner's Environmental Services Department.	1. Remove barrier material carefully to minimize spreading of dirt and debris associated with construction. 2. Contain construction waste before transport in tightly covered containers. 3. Cover transport receptacles or carts. Tape covering unless solid lid. 4. Vacuum work area with HEPA filtered vacuums. 5. Wet mop area with disinfectant. 6. Remove isolation of HVAC system in areas where work is being performed.

■■ Figure 2 ■■ (cont.)

Step 4. **Identify the areas surrounding the project area, assessing potential impact**

Unit Below	Unit Above	Lateral	Lateral	Behind	Front
Risk Group	Risk Group	Risk Group	Risk Group	Risk Group	Risk Group

Step 5. **Identify specific site of activity eg, patient rooms, medication room, etc.**

_____

Step 6. **Identify issues related to: ventilation, plumbing, electrical in terms of the occurrence of probable outages.**

_____

Step 7. **Identify containment measures, using prior assessment. What types of barriers? (Eg, solids wall barriers); Will HEPA filtration be required?**

_____

(Note: Renovation/construction area shall be isolated from the occupied areas during construction and shall be negative with respect to surrounding areas)

Step 8. **Consider potential risk of water damage. Is there a risk due to compromising structural integrity? (eg, wall, ceiling, roof)**

Step 9. **Work hours: Can or will the work be done during non-patient care hours?**

Step 10. **Do plans allow for adequate number of isolation/negative airflow rooms?**

Step 11. **Do the plans allow for the required number & type of handwashing sinks?**

Step 12. **Does the infection control staff agree with the minimum number of sinks for this project?**
(Verify against AIA Guidelines for types and area)

Step 13. **Does the infection control staff agree with the plans relative to clean and soiled utility rooms?**

Step 14. **Plan to discuss the following containment issues with the project team. Eg, traffic flow, housekeeping, debris removal (how and when),**

_____

_____

_____

Appendix: Identify and communicate the responsibility for project monitoring that includes infection control concerns and risks. The ICRA may be modified throughout the project.

Revisions must be communicated to the Project Manager.

Steps 4–14 adapted with permission, Fairview University Medical Center, Minneapolis MN.
Forms modified and provided courtesy of Judene Bartley, Epidemiology Consulting Services (ECS), Inc., Beverly Hills MI.

■ ■ **Figure 2** ■ ■ (cont.)

Infection Control Construction Permit

			Permit No:			
Location of Construction:			**Project Start Date:**			
Project Coordinator:			**Estimated Duration:**			
Contractor Performing Work			**Permit Expiration Date:**			
Supervisor:			**Telephone:**			

YES	NO	CONSTRUCTION ACTIVITY	YES	NO	INFECTION CONTROL RISK GROUP
		TYPE A: Inspection, non-invasive activity			GROUP 1: Low Risk
		TYPE B: Small scale, short duration, moderate to high levels			GROUP 2: Medium Risk
		TYPE C: Activity generates moderate to high levels of dust, requires greater than 1 work shift for completion			GROUP 3: Medium/High Risk
		TYPE D: Major duration and construction activities Requiring consecutive work shifts			GROUP 4: Highest Risk

CLASS I
1. Execute work by methods to minimize raising dust from construction operations.
2. Immediately replace any ceiling tile displaced for visual inspection.

3. Minor Demolition for Remodeling.

CLASS II
1. Provides active means to prevent airborne dust from dispersing into atmosphere.
2. Water mist work surfaces to control dust while cutting.
3. Seal unused doors with duct tape.
4. Block off and seal air vents.
5. Wipe surfaces with disinfectant.

6. Contain construction waste before transport in tightly covered containers.
7. Wet mop and/or vacuum with HEPA filtered vacuum before leaving work area.
8. Place dust mat at entrance and exit of work area.
9. Remove or isolate HVAC system in areas where work is being performed.

CLASS III — Date / Initial
1. Obtain infection control permit before construction begins.
2. Isolate HVAC system in area where work is being done to prevent contamination of the duct system.
3. Complete all critical barriers or implement control cube method before construction begins.
4. Maintain negative air pressure within work site utilizing HEPA equipped air filtration units.
5. Do not remove barriers from work area until complete project is thoroughly cleaned by Env. Services Dept.

6. Vacuum work with HEPA filtered vacuums.
7. Wet mop with disinfectant.
8. Remove barrier materials carefully to minimize spreading of dirt and debris associated with construction.
9. Contain construction waste before transport in tightly covered containers.
10. Cover transport receptacles or carts. Tape covering.
11. Remove or isolate HVAC system in areas where work is being performed.

CLASS IV — Date / Initial
1. Obtain infection control permit before construction begins.
2. Isolate HVAC system in area where work is being done to prevent contamination of duct system.
3. Complete all critical barriers or implement control cube method before construction begins.
4. Maintain negative air pressure within work site utilizing HEPA equipped air filtration units.
5. Seal holes, pipes, conduits, and punctures appropriately.
6. Construct anteroom and require all personnel to pass through this room so they can be vacuumed using a HEPA vacuum cleaner before leaving work site or they can wear cloth or paper coveralls that are removed each time they leave the work site.

7. All personnel entering work site are required to wear shoe covers.
8. Do not remove barriers from work area until completed project is thoroughly cleaned by the Environmental Service Dept.
9. Vacuum work area with HEPA filtered vacuums.
10. Wet mop with disinfectant.
11. Remove barrier materials carefully to minimize spreading of dirt and debris associated with construction.
12. Contain construction waste before transport in tightly covered containers.
13. Cover transport receptacles or carts. Tape covering.
14. Remove or isolate HVAC system in areas where work is being performed.

Additional Requirements:

Date Initials	_____ Exceptions/Additions to this permit Date Initials are noted by attached memoranda
Permit Request By:	Permit Authorized By:
Date:	Date:

■ ■ **Figure 3** ■ ■ **Sample Infection Control Construction Permit** (Source: ECS, Inc.)

Construction Phase

Once construction or demolition has begun, it is important to ensure that the preventative measures developed are actually being implemented. Key to this determination is a schedule of regular inspections of the work area. Checklists can be valuable tools during this phase of the project (see Figure 4).

The safety professional and infection control practitioner should visit the construction site with the project manager on a regular basis until the project is completed to ensure that controls are being implemented. If concerns are identified, they should be corrected immediately.

Another useful tool is monitoring the air conditions in and around the construction zones. This monitoring can be done in one of three ways: detection of particles, detection of microorganisms, and measurement of flow.

Microbiological sampling of air in healthcare facilities remains a controversial subject because of numerous unresolved technical limitations and the

Infection Control Checklist During Construction/Renovation			
Inspector:	*Location:*	*Date:*	*Time:*
Barriers		**Air Handling**	
Construction signs posted		All windows behind barrier closed	
Doors properly closed and sealed		Negative air pressure at barrier entrance	
Holes, pipes, conduits, punctures, etc. sealed		Portable air flow units used to maintain negative pressure running	
Dust barriers intact and sealed			
Floor and horizontal surfaces free of dust		**Trash and Debris**	
Ceiling tiles free of moisture		No visible evidence of insects (flies)	
Traffic Control		Trash placed in appropriate containers	
All doors and exits free of debris		Routine cleaning performed in work area	
Restricted to construction workers and essential staff		"Sticky" dust mats appropriately placed/clean	
		No evidence of dust outside the construction area	
Personal Protective Equipment (PPE)		Debris removed in covered container daily	
Workers wearing appropriate PPE		Regulated medical waste containers removed from work area before work is started	

■ ■ **Figure 4** ■ ■ **Construction infection control checklist** (Courtesy of U.S. Army Industrial Hygiene and Medical Safety Management Program)

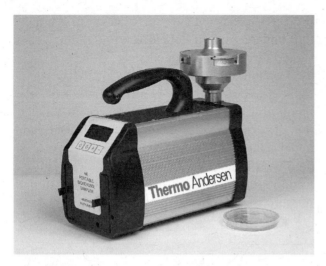

■■ Figure 5 ■■ Microbial air-sampling equipment (Photo courtesy Thermo-Andersen, Inc.)

■■ Table 4 ■■

Unresolved Issues Associated with Microbiological Air Sampling

Lack of standards linking fungal spore levels with infection rates (i.e., no safe level of exposure)

Lack of standard protocols for testing (e.g., sampling intervals, number/location of samples)

Need for substantial laboratory support

Unknown incubation period for Aspergillus infection

Variability of sampler readings

Sensitivity of the sampler used (i.e., the volumes of air sampled)

Lack of details in the literature about describing sampling circumstances (expected fungal concentrations, rate of outdoor air penetration)

Lack of correlation between fungal species and strains from the environment and clinical specimens

Confounding variables with high-risk patients (e.g., visitors)

need for substantial laboratory support. *Guidelines for Environmental Infection Control in Healthcare Facilities, 2001* from the CDC offers a compilation of recommendations for the prevention and control of infectious diseases that have been associated with healthcare environments. A number of the unresolved technical issues identified with microbial air sampling, as presented in the document, are listed in Table 4.

The most significant technical limitation of air sampling for airborne fungal agents is the lack of standards linking fungal spore levels with infection rates. Despite this limitation, several healthcare institutions have opted to use microbiologic sampling when construction projects are anticipated and/or underway in an effort to assess the safety of the environment for the immunocompromised client.[10]

Microbial air sampling should only be utilized with the help of a professional experienced in the collection methods involved and a clearcut strategy, including the interpretation and meaning of results when they are obtained. Laboratories that analyze results may vary in experience and proficiency. Accreditation for the type of analysis being requested should be considered when choosing a laboratory.

■■ Figure 6 ■■ Airborne dust sampling (Photo courtesy of Health Systems Safety, Inc.)

These considerations also apply for construction activities that are performed for the purpose of mold abatement when visible mold is present. In order for mold to be present, a source of food (carbon-based material such as paper on dry wall) and water must be present. Currently, there is much debate surrounding the potential health effects of mold exposure to people who are in good health. Resource documents are available that address mold abatement, such as the New York City Department of Health's *Guidelines on Assessment and Remediation of Fungi in Indoor Environments,* as well as from the Environmental Protection Agency.

Detection of particulates can help determine whether barriers and other efforts to control dust dispersion from construction are effective. This type of monitoring is more straightforward. Dust is collected on a filter and weighed on a scale in a laboratory. The results can be meaningful if they are compared to preconstruction baseline samples. Sampling should be performed at various times during the project and at barrier perimeter locations. Gaps or breaks in the barriers or seals can then be identified and subsequently repaired. Particulate sampling does not require microbiology laboratory services for the reporting of results.

Anemometers and smoke-trail visualization tests are used in airflow evaluation. The anemometer measures airflow velocity, which can then be used to determine air volumes. Smoke tubes can be used to determine airflow direction and the effectiveness of negative pressure ventilation.

Postconstruction Phase

No construction job is ever done until the final punch list is completed. Following are some items that should be considered before allowing entry, or reentry, of clients into a facility:

- A thorough cleaning, including all horizontal surfaces, before and after barriers have been removed. A period of time should be allowed to let dust settle before cleaning.
- Water lines should be flushed and disinfected per code.
- Hot water temperature should be assessed before use.
- Ventilation systems should be assessed (balanced).
- A final walk-through with the Infection Control Practitioner.

Lessons learned are a valuable tool by which to evaluate any project. A review of the effectiveness of the preventative controls implemented should be assessed by the multidisciplinary team, including both positive and negative outcomes.

Using a team approach to implement proper controls during construction will help to reduce risks to clients and workers alike. Establishing clear lines of communication and expectations is very important to controlling these potential hazards.

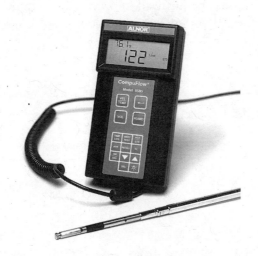

■ ■ **Figure 7** ■ ■ **Thermal anemometer** (Photo courtesy of Alnor™, Inc.)

REFERENCES

Abrutyn, E., D. A. Goldmann, and W. E. Scheckler. *Saunders Infection Control Reference Service,* 2nd ed. Philadelphia, PA: WB Saunders, 2000.

American Conference of Governmental Industrial Hygienists (ACGIH). *Bioaerosols: Assessment and Control,* Janet Macher (ed.). Cincinnati, OH: ACGIH, 1999.

American Industrial Hygiene Association (AIHA). *Field Guide for the Determination of Biological Contamination in Environmental Samples.* AIHA Biosafety Committee, H. Kenneth Dillon, Patricia A. Fleinsohn, and J. David Miller (eds.). Fairfax, VA: AIHA, 1996.

American Institute of Architects (AIA). *Guidelines for Design and Construction of Hospital and Health Care Facilities.* Washington DC: AIA Press; 2001.

Association for Professionals in Infection Control and Epidemiology, Inc. *Text of Infection Control and Epidemiology.* Washington DC: APIC, 2000. (Two binder volumes or CD-ROM).

Centers for Disease Control and Prevention, Healthcare Infection Control Practices Advisory Committee (HICPAC). *Draft Guideline for Environmental Infection Control in Healthcare Facilities, 2001.*

Environmental Protection Agency. *Mold Remediation in Schools and Commercial Buildings* (EPA 402-K-01-001). Washington, DC: EPA, March 2001.

Garvey, D., S. Sobczak, and A. Streifel. "Medical Facility Renovation—Safety and Health Considerations for Construction Safety Professionals," *Professional Safety Journal of the American Society of Safety Engineers*, Des Plaines, IL: ASSE, June 2001.

Institute of Inspection, Cleaning and Restoration Certification (IICRC). *Standard and Reference Guide for Professional Water Damage Restoration* (IICRC S500), www.iicrc.org

International Organization for Standardization (ISO). *Sterilization of medical devices—Microbiological methods, Part 1* (ISO standard 11737-1). Paramus NJ: ISO, 1995.

Jensen, P. A. and M. P. Schafer. "Sampling and characterization of bioaerosols," *NIOSH Manual of Analytical Methods* (revised 6/99). http://www.cdc.gov/niosh/nmam/pdfs/chapter-j.

Joint Commission on Accreditation of Healthcare Organizations (JCAHO). *Environment of Care Guidelines: 2000 Hospital Accreditation Standards.* Oak Brook, IL: JCAHO, 2000.

Occupational Health and Safety Administration, *Framework for a Comprehensive Health and Safety Program in the Hospital Environment..* Washington, DC: U.S. Department of Labor, OSHA, 1993.

REVIEW EXERCISES

1. What are some of the potential chemical and physical hazards that can be encountered in the healthcare environment?
2. What special precautions must be taken to protect healthcare clients when performing renovation and construction?
3. Give the location and types of potentially infectious materials that may be found in a healthcare setting.
4. List the agencies and organizations that could provide additional resources for addressing construction issues in healthcare.
5. Before construction work begins, who should be involved in preventing employee and client injury or illness?

Chapter Notes

[1] C. L. Croswell, "Better, not bigger: Construction costs soar on wings of patient demand, Construction and Design survey finds," *Modern Healthcare,* 1999, 29(12): 23–26, 28–34, 36–38.

[2] CDC Pamphlet "Exposure to Blood—What Healthcare Workers Need to Know." (http://www.cdc.gov)

[3] OSHA CPL 2.106 – Enforcement Procedures and Scheduling for Occupational Exposure to Tuberculosis. Washington, DC: Office of Health Compliance Assistance, February 9, 1996.

[4] F. A. Sarubbi, Jr., B. B. Kopf, N. O. Wilson, M. R. McGinnis, and W. A. Rutala, "Increased recovery of *Aspergillus flavus* from respiratory specimens during hospital construction," *American Review of Respiratory Disease,* 1982, 125:33–38.

[5] P. M. Arnow, M. C. Sadigh, D. Weil, and R. Chudy, "Endemic and epidemic aspergillosis associated with inhospital replication of *Aspergillus* organisms," *Journal of Infectious Disease,* 1991, 164:998–1002.

[6] P. M. Flynn, B. G. Williams, S. V. Hethrington, B. F. Williams, M. A. Giannini, and T. A. Pearson, "*Aspergillus terreus* during hospital renovation," *Infection Control and Hosp Epidemiology,* 1993, 14:363–365.

[7] J. J. Weems, Jr., B. J. Davis, O. C. Tablan, L. Kaufman, and W. J. Martone, "Construction activity: An independent risk factor for invasive aspergillosis and zygomycosis in patients with hematologic malignancy," *Infection Control,* 1987, 8:71–75.

[8] A. J. Streifel, P. P. Stevens, and F. S. Rhame, "In-hospital source of airborne *Penicillium* species spores," *J Clin Microbiol,* 25:1–4

[9] J. M. Bartley et al., "APIC State-of-the-Art Report: The Role of infection control during construction in health care facilities," *American Journal of Infection Control,* April 2000, Vol 28, No. 2, pp. 156–169.

[10] A. J. Streifel, "Air cultures for fungi," *Clinical Microbiology Procedures Handbook,* 2nd ed. M. Gilcrist (Washington DC: American Society for Microbiology Press), 1193:11.8.1–11.8.7.

■ ■ **About the Author** ■ ■

MSHA CONSIDERATIONS FOR CONTRACTORS
by Adele L. Abrams, Esq., CMSP, P.C.

Adele L. Abrams, is an attorney and president of the Law Office of
Adele L. Abrams, P.C., in Beltsville, MD. Ms. Abrams is a Certified Mine
Safety Professional and MSHA-approved trainer. She is a professional
member of the American Society of Safety Engineers' Construction and
Mining practice specialty groups (2000 "SPY" Award recipient, Mining
practice specialty). She is an active member of the International Society
of Mine Safety Professionals, the National Safety Council (past chairman
of the Cement, Quarry & Mineral Aggregates section), the Energy and
Mineral Law Foundation, Women in Mining (past president, DC chapter), the Holmes Safety Association, and the Washington Metropolitan
Area Construction Safety Association. She earned a Juris Doctor degree
from the George Washington University National Law Center in Washington, DC, and a Bachelor of Science degree from the University of
Maryland, College Park. She is admitted to the Maryland and DC Bars,
as well as to practice before the U.S. District Courts of Maryland and
DC, and the U.S. Court of Appeals, DC Circuit.

CHAPTER 29

MSHA Considerations for Contractors

LEARNING OBJECTIVES

■ Discuss distinctions between the regulatory authority of the Mine Safety and Health Administration (MSHA) and the Occupational Safety and Health Administration (OSHA).

■ Discuss MSHA's jurisdictional authority and multiemployer work-site policies and their significance for construction industry employers who perform contract work at mines.

■ Describe mandatory training and reporting requirements for contractors at surface and underground mining operations.

■ Discuss state regulatory requirements that are applicable to mine safety programs and the interface with federal enforcement authority.

■ Explain MSHA's inspection, investigative, and adjudicative process, and understand potential civil and criminal liability under the Mine Act for companies and individuals.

OVERVIEW OF THE MINE SAFETY AND HEALTH ADMINISTRATION

In 1891, the United States Congress passed the first federal statute governing mine safety. The 1891 law applied only to mines in U.S. territories and, among other things, established minimum ventilation requirements at underground coal mines and prohibited operators from employing children under twelve years of age. Initial regulation of mining work environments began in 1910 when the U.S. Bureau of Mines was established.

The first legislation that empowered the government to write notices of violations at mines was the 1952 Federal Coal Mine Safety Act (Pub. L. 552). In 1969, the Federal Coal Mine Health and Safety Act established mandatory safety and health regulations, and coal mine inspections, but did not cover metal/nonmetal mines.

The Federal Mine Safety & Health Act of 1977 (Mine Act), enacted March 9, 1978,[1] extended protections and regulatory/enforcement authority to all types of mines. This is the statute presently in effect. The Mine Act created the Mine Safety and Health Administration, which is part of the U.S. Department of Labor and has equal status with the Occupational Safety and Health Administration within that executive branch agency.[2] The Mine Act authorizes the Secretary of Labor to promulgate and enforce safety and health standards regarding working conditions of employees engaged in underground and surface mineral extraction (mining), related operations, and preparation and milling of the minerals extracted. The head of MSHA, who also serves as Assistant Secretary of Labor, is appointed by the President and must be confirmed by the U.S. Senate.

Key Definitions

MSHA has broad jurisdiction over all activities that occur on mine sites and significant latitude in determining what constitutes a "mine," what employers are "mine operators," and what workers are "miners" for purposes of enforcement under the Mine Act. As discussed in more detail below, any contractor who performs work on a mine site—regardless of whether it is related to the extraction and production processes—likely will be considered a miner and be subject to compliance with all MSHA standards as well as applicable reporting and training requirements.

MSHA classifies "Any owner, lessee, or other person who operates, controls, or supervises a coal or other mine, or any independent contractor performing services or construction at such mine" as an operator for purposes of MSHA compliance obligations. A miner is defined as "any individual working in a coal or other mine."

The Mine Act defines an independent contractor as "any person, partnership, corporation, subsidiary of a corporation, firm, association or other organization that contracts to perform services or construction at a mine." If the "person, partnership, . . . or other organization" contracts for the production of a mineral, the "person, partnership, . . . or other organization" is then classified as a mine operator, and it is required to file a Legal Identity Report with MSHA and keep the information up to date. In addition, it will be assigned a mine identification number, and will be subject to all requirements that apply to a mine operator.

MSHA's Enhanced Enforcement Powers

Because Congress deemed mining work sites to be more hazardous than those of other industries (as they were at the time the Mine Act was enacted), MSHA has powers that go well beyond those of OSHA. Even today, nearly 100 miners die each year at surface and underground coal and metal/non-

■ ■ Table 1 ■ ■

Coal Fatalities (1999–2002)

Accident Type	1999 UG	S	2000 UG	S	2001 UG	S	2002 UG	S
Electrical	0	0	0	1	0	1	1	0
Exp vessels u/pressure	0	0	0	0	0	0	0	1
Explosives	0	1	0	0	0	0	0	0
Fall/slide material	0	1	0	0	0	0	1	0
Hand tools	0	0	0	0	0	0	0	0
Handling material	0	0	0	0	0	0	0	0
Hoisting	0	0	0	0	0	0	0	0
Powered haulage	1	1	0	1	1	0	1	1
Ignition/exp of gas	0	0	0	0	0	0	0	0
Machinery	0	0	1	1	0	0	0	0
Rib/high wall	1	1	0	1	0	0	1	0
Roof fall	2	0	0	0	1	0	2	0
Slip/fall of person	0	0	0	1	0	1	0	0
Other	0	0	0	0	0	0	0	0
TOTALS	4	4	1	5	2	2	6	2
COMBINED TOTALS	8		6		4		8	
END OF YEAR TOTALS	35		38		42		27	

Source: MSHA, U.S. Department of Labor (March 2003)

metal mines across the United States (see Tables 1 and 2). Therefore, despite MSHA's extensive enforcement powers, there is still progress to be made in reducing fatal and nonfatal accident rates.

This type of incentive is relatively easy to implement and therefore common among construction companies. The employer simply needs to establish the performance objective (e.g., number of labor-hours without an injury) and the benefit (e.g., a baseball cap), and then monitor when an injury occurs and the length of time between injuries. The simple format helps when communicating the incentive to employees. A clear understanding of the incentive minimizes confusion and discouragement regarding employee participation and motivation to attain the level of performance.

MSHA has warrantless search authority at all mines. MSHA also has limited subpoena power (related to public hearings arising from MSHA investigations). The Mine Act is also a strict liability statute, which means that there are no affirmative defenses that can be offered to counter a "fact of violation"; such defenses can, of course, be offered with respect to the negligence assigned to the employer.

■ ■ **Table 2** ■ ■

Metal/Nonmetal Fatalities (1999–2002)

Accident Type	1999 UG	1999 S	2000 UG	2000 S	2001 UG	2001 S	2002 UG	2002 S
Electrical	0	0	0	0	0	0	0	0
Exp vessels u/pressure	1	0	0	0	0	0	0	0
Exp & breaking agents	0	0	0	0	0	0	0	0
Fall/slide material	0	0	0	0	0	0	0	1
Fall of face/high wall	0	1	1	0	0	0	0	0
Fall of roof/back/rib	0	0	0	0	0	0	0	0
Fire	0	0	0	0	0	0	0	0
Handling material	0	1	0	0	0	0	0	0
Hand tools	1	0	0	0	0	0	0	0
Nonpowered haulage	0	0	0	0	0	0	0	0
Powered haulage	1	3	2	3	1	2	0	3
Hoisting	0	0	0	0	0	0	0	1
Ignition/exp of gas	0	0	0	0	0	0	0	0
Inundation	0	0	0	0	0	0	0	0
Machinery	0	2	0	2	0	0	0	3
Slip/fall of person	0	0	0	0	0	0	0	1
Step/kneel on object	0	0	0	0	0	2	0	0
Striking or bumping	0	0	0	0	0	0	0	0
Other	0	1	0	0	0	0	0	0
TOTALS	**3**	**8**	**3**	**5**	**1**	**4**	**0**	**9**
COMBINED TOTALS	**11**		**8**		**5**		**9**	
END OF YEAR TOTALS	**55**		**47**		**30**		**40**	

Source: MSHA, U.S. Department of Labor (March 2003)

Perhaps most significant, MSHA is required to inspect each surface mine twice per year, and to inspect each underground mine four times per year. Intermittent operations, such as portable crushers, are inspected on an annual basis, and those operators must notify MSHA promptly when the operation is shut down and restarted. Therefore, a contractor's chances of being inspected while performing construction or other work at a mine are much higher than the odds of encountering an OSHA compliance officer.

Citation Procedures

MSHA compliance specialists (inspectors) are required by Section 104 of the Mine Act to issue citations or orders when they believe that a mandatory standard or regulation has been violated. A MSHA inspector does *not* have to witness a violation in progress in order to issue a citation. MSHA issues both "significant and substantial" citations, which are analogous to OSHA's "serious" citations (hazards reasonably likely to result in reasonably serious injury or illness), and "non-S&S" citations, akin to OSHA's "other than serious" category. However, MSHA does impose a civil penalty for every citation, and both S&S and non-S&S citations count toward a mine operator or contractor's "history of violations." This history is one of six statutory criteria used to determine the appropriate amount of a civil penalty. There are no *de minimis* citations permitted under the Mine Act.

Although there is no statute of limitations defining when MSHA must issue a citation or impose a civil penalty for alleged violations, MSHA strives to issue citations and penalties within 18 months of the occurrence of a violation, or the agency's discovery of the circumstances that constitute a violation. However, cases involving fatalities, special investigations, or criminal referrals may involve special circumstances that result in MSHA litigation lasting for several years.

NIOSH's Role in Mine Safety and Health

Representatives of the National Institute for Occupational Safety and Health, which is part of the U.S. Department of Health and Human Services, Centers for Disease Control, also have the right to enter mines without a warrant to perform safety and health research and to conduct Health Hazard Evaluations at the request of miners. Failure to admit authorized representatives of MSHA or NIOSH to the mine site is a violation of Section 103(a) of the Mine Act, and currently carries a maximum $60,000 civil penalty per violation.

MSHA/OSHA/STATE JURISDICTIONAL CONSIDERATIONS

MSHA/OSHA Jurisdictional Distinctions

MSHA has enforcement authority over all mine operators and owners, as well as independent contractors who perform work at mine sites. All persons performing work at mines, who are exposed to mine hazards, are considered miners under the Act. The term mine is broadly construed to include coal, metal, and nonmetal mines, aggregate operations, cement plants, portable crushing operations, and even off-site maintenance shops that repair mine equipment. MSHA and OSHA have an interagency agreement that defines the limits of each agency's jurisdiction. It can be found on MSHA's Web site at www.msha.gov.

The general principle embodied in the interagency agreement is that, in dealing with unsafe and unhealthful working conditions on mine sites and in milling operations, the Secretary will apply the provision of the Mine Act and MSHA standards to eliminate those conditions. If the Mine Act either does not cover or does not otherwise apply to occupational safety and health hazards on mine or mill sites (e.g., hospitals on mine sites) or where there is statutory coverage under the Mine Act, but where no MSHA standards applicable to particular working conditions on such sites exist, then the Occupational Safety and Health Act of 1970 (and its implementing regulations codified by OSHA in 29 CFR Parts 1910, 1915, and 1926) will be applied to those working conditions.

The interagency agreement further clarifies: "If an employer has control of the working conditions on the mine site or milling operation, and such employer is neither a mine operator nor an independent contractor subject to the Mine Act, the OSH Act may be applied to such an employer where the application of the OSH Act would, in such a case, provide a more effective remedy than citing a mine operator or an independent contractor subject to the Mine Act who does not, in such circumstances, have direct control over the working conditions."[3]

The interagency agreement specifically declares that MSHA jurisdiction includes: salt processing facilities on mine property; electrolytic plants, where the plants are an integral part of milling operations; stone cutting and stone sawing operations on mine property, where such operations do not occur in a stone polishing or finishing plant; and alumina and cement plants. OSHA jurisdiction includes the following, whether or not located on mine property: brick, clay pipe, and refractory plants; ceramic plants; fertilizer product operations; concrete batch, asphalt batch, and hot mix plants; smelters and re-

fineries. OSHA jurisdiction also includes salt and cement distribution terminals not located on mine property, and milling operations associated with gypsum board plants not located on mine property.[4]

A good rule of thumb is that MSHA-regulated milling and crushing operations generally involve the processing of raw materials, while OSHA-regulated milling and crushing sites involve recycled materials, asphalt, or concrete. However, MSHA has been known to attempt to assert jurisdiction over work sites that were previously inspected by OSHA, so jurisdictional challenges should be clearly articulated at once if there is any question about whether a work site is within the scope of MSHA's enforcement authority.

Of specific applicability to construction companies are MSHA's jurisdictional rules concerning "Borrow Pits." As used by MSHA, borrow pit means an area of land where the overburden, consisting of unconsolidated rock, glacial debris, or other earth material overlying bedrock is extracted from the surface. Extraction occurs on a one-time-only basis, or only intermittently as need occurs, for use as fill materials by the extracting party in the form in which it is extracted. No milling is involved, except for the use of a scalping screen to remove large rocks, wood, and trash. The material is used by the extracting party more for its bulk than its intrinsic qualities on land that is relatively near the borrow pit. MSHA has determined that borrow pits meeting this definition are subject to OSHA jurisdiction unless they are located on mine property or related to mining. Therefore, a borrow pit used to build a road or construct a surface facility on mine property would be subject to MSHA jurisdiction. Because MSHA and OSHA rules vary substantially—and employee training and documentation requirements are significantly different—contractors must clearly determine the scope and site of such work before commencing any job where agency jurisdiction may be at issue.

Also of relevance to construction is the guidance provided in MSHA's *Program Policy Manual* (*PPM*), which states that the agency lacks jurisdiction where a mineral is extracted incidental to the primary purpose of the activity. Under this circumstance, a mineral may be processed and disposed of and MSHA will not have jurisdiction since the company is not functioning for the purpose of producing a mineral. Operations not functioning for the purpose of producing a mineral include, but are not limited to, the following: (1) key cuts in dam construction (not on mining property or used in mining); (2) public road and highway cuts; (3) tunnels (railroad, highway, water diversion, etc.); and (4) storage areas (gas, petroleum reserves, high- and low-level radioactive waste). The question of jurisdiction in these and similar types of operations is contingent on the purpose and intent for which the facility is being developed.[5]

Another problematic jurisdictional issue concerns milling, which MSHA defines as "crushing, grinding, pulverizing, sizing, concentrating, washing, drying, roasting, pelletizing, sintering, evaporating, calcining, kiln treatment, sawing and cutting stone, heat expansion, retorting (mercury), leaching, and briquetting." Under prevailing interpretations of Section 3(h)(1) of the Mine Act, the scope of the term *milling* has been expanded to apply to mineral product manufacturing processes, where these processes are related, technologically or geographically, to milling.

Conversely, MSHA generally excludes from its jurisdiction the processes listed in Appendix A of the interagency agreement (included as Appendix A of this chapter), provided that such processes are unrelated, technologically or geographically, to mineral milling. Jurisdictional determinations often are made by agreement between MSHA and OSHA after the inspected employer files suit challenging the respective agency's enforcement authority.

There are some state agencies that also regulate mine safety and health issues (e.g., Cal-OSHA). Unlike the situation at OSHA-regulated workplaces, in these states employers who fall within the definition of mine operator will have to comply with both MSHA rules and more stringent state standards, and be subject to citations and penalties imposed by both federal and state agencies arising from the same violation.

State Agencies that Address Mine Safety

Although MSHA has primary authority to regulate safety and health at mine sites across the United

States, contractors should be aware that some state agencies have dual jurisdiction with MSHA and may promulgate safety and health standards that are stricter than those at the federal level. In some states, the mining-related agencies simply assist with education and consultative activities and provide training through the MSHA state grants program. Other states have personnel who regularly inspect mines and have authority to issue citations. Therefore, mine operators and contractors may, in some circumstances, be cited twice for the same violative condition—by federal inspectors and again by state enforcement personnel. Finally, some states have agencies that regulate environmental conditions at mines and special activities, such as blasting, but they work cooperatively with MSHA and frequently contact federal inspectors if they observe an unsafe or unhealthful condition while visiting the mine for other purposes.

The following is a summary of the key state agencies that contractors should be aware of when conducting business on mine property:

- *Alabama:* The Department of Industrial Relations has programs that address surface mining of nonfuel materials and abandoned mine land reclamation programs.
- *Alaska:* The Department of Natural Resources, Division of Mining and Water supplies public notices, special reports, and training information for mining operations. Important statutes include the Alaska Surface Coal Mining Control & Reclamation Act, which includes a mine permit program and 65 separate performance standards for coal-mining activities. State personnel inspect each operating coal mine an average of once each month and can impose criminal and civil penalties for violation of the Act.
- *Arizona:* The State Mine Inspector's office has an inspection team that visits all active mines; investigates fatalities, other serious accidents, and employee complaints at mines; provides training and educational services; performs mine site evaluations; and regulates reclamation of mined land. Mine inspectors look at compliance with the

Arizona Mine Code—a comprehensive set of safety and health laws and rules.

- **Arkansas:** The Department of Labor Safety Services is responsible for enforcing and promoting worker safety in the state, but actual inspection is left to federal MSHA. The Department of Environmental Quality's Surface Mining and Reclamation Division enforces the state's Surface Coal Mining and Reclamation Code (Act 134, Regulation No. 20), which deals with permit applications, performance and mine reclamation standards, and discharge of stormwater and process water.

- **California:** The Division of Occupational Safety and Health has five Mining and Tunneling offices that conduct enforcement at mine sites throughout the state. The agency can promulgate standards that are enforced at mines and issue monetary citations for violations of state OSHA rules at mine sites, sometimes for the same conditions that are cited and fined by MSHA. This often creates confusion for mining companies and their contractors because, at times, the MSHA and Cal-OSHA regulatory requirements conflict. The Mining and Tunneling section publishes a *Policy and Procedures Manual* (P&P C-29) that specifically addresses its authority to inspect surface, underground, and portable mine and mill operations; shafts and tunnel-related projects; concrete batch plants; asphalt plants; hot mix and recycle plants located on mine property; and all contractors who work on mine sites. The California Resources Agency and the Department of Conservation Minerals and Mining promote mineral resources and produce maps that are used in land-use planning decisions.

- **Colorado:** The Department of Natural Resources, Mining, Mine Safety, and Mine Land Reclamation deals with coal and mineral mines, abandoned mines, safety training, and land reclamation procedures. The state has promulgated extensive coal mine health and safety rules and regulations that address mine conditions, training, and certification of personnel. Colorado has been nationally recog-

nized for development of a Part 46 training program that is used by mine operators and contractors across the country. In addition, state employees acting under authority of the Mined Land Reclamation Act issue mining and reclamation permits for all noncoal mines in the state, while the Surface Coal Mining Reclamation Act is a parallel program for coal operations, and includes monthly inspections of each active coal mine.

- **Illinois:** The Department of Natural Resources' Office of Mines and Minerals deals with abandoned mines, land reclamation, and oil and gas leases. The Mine Safety and Training Division conducts health and safety training. The Division's two offices regulate working conditions in both surface and underground coal mines through monthly inspection by state personnel, issuance of "Certificates of Competency" to coal mine employees, maintenance of mine rescue stations and mine rescue teams, and operation of a laboratory to sample coal and analyze mine atmospheres. The Division also regulates the storage and use of explosives at mine sites.

- **Kentucky:** The Department of Mines and Minerals extensively regulates the coal-mining industry, including mine licensing, blasting, and mine mapping. The state has a full-time training and education program that provides certification for coal miners. Kentucky's mine safety standards are published in the state's Administration Regulations, Title 805, and in Kentucky Revised Statutes 351 and 352. The state also frequently brings criminal prosecutions against mine personnel who are involved in severe negligence activities or in fraudulent training programs and has updated its penalty regulations in 2003. Certified miners or mine foremen who are found guilty of intentional violations of safety laws can have their certifications suspended or revoked, which can preclude future mining employment in the state; they may also face personal monetary penalties that are based on the individual's wages. Mine owners who are convicted of intentional violations can have their licenses suspended or revoked and face monetary penalties of up to $10,000.

- *Louisiana:* The Department of Natural Resources regulates the mining of coal and lignite under the Surface Mining and Reclamation Act of 1980. The state ensures that mine operations do not harm the public, and oversees land restoration efforts. The surface mining rules promulgated by the Office of Conservation are published in Title 43 of the Louisiana Administrative Code.

- *Maryland:* The Department of Environment's Bureau of Mines addresses environmental issues associated with mining, but does not get actively involved with mine safety and health programs.

- *Minnesota:* The Iron Range Resources and Rehabilitation Agency provides incentives for mineral exploration and helps the taconite mining industry improve safety and health. The state also addresses mine land reclamation efforts, natural resources research, and drilling incentive programs.

- *Mississippi:* The Department of Environmental Quality's Office of Geology is involved with coal-mining regulation, as well as reclamation and environmental issues.

- *Montana:* The Department of Environmental Quality, Mine Waste Cleanup Bureau, addresses the safe disposal of toxic substances used or created by mines, as well as site reclamation under the Opencut Mining Act and its implementing regulations (82-4-401 et seq., MCA).

- *Nevada:* The Commission on Mineral Resources' Division of Minerals is involved with mine safety and training, abandoned mine lands, environmental issues, mining exploration, reclamation issues, and mine mapping. The state also has a Bureau of Mines that is a research entity through the University of Nevada, charged with conducting geological surveys and publishing reports on mineral resources and environmental and engineering geology.

- *New Mexico:* The Energy, Mineral and Natural Resources Department's Mining

and Minerals Division has over 30 staff who regulate both coal and hard rock mines, provide educational incentives related to mining, and reclaim abandoned mines.

- *North Carolina:* The Department of Environment and Natural Resources, Division of Land Resources, addresses environmental issues associated with mining operations, while the North Carolina Department of Labor conducts mine safety and health training, presents frequent conferences, and is involved with mine inspection and accident investigation.

- *North Dakota:* The Land Department's Minerals Management Division monitors mine activities on over 2.5 million acres from an environmental perspective, controls leases for gravel and aggregate mining, oversees oil and gas extraction, and conducts geologic surveys.

- *Ohio:* The Department of Natural Resources, Division of Mines and Reclamation, conducts mine enforcement activities, training, and education on safety and health; develops regulatory programs; and addresses abandoned mine issues. The state has 20 officers that inspect more than 400 mine sites on a quarterly basis. Mine-dust samples are also collected and analyzed at state laboratories, and water samples are tested as part of the state's Acid Abatement program.

- *Oregon:* The Department of Geology and Mineral Industries has a mined-land reclamation program that issues permits for all extraction, processing, and reclamation activities where more than 5,000 cubic yards of material are removed or where mining affects more than one acre of land within a one-year period.

- *Pennsylvania:* The Department of Environmental Protection, Mining and Mineral Resources, promulgates mine regulations, conducts extensive training and safety/health outreach activities, regulates blasting and drilling operations, and deals with abandoned mine reclamation and mapping and mine rescue operations. The Bureau of Deep Mine Safety is involved with safety and environmental issues, as well as enforcement of laws such as the Bituminous Mine Act, the

Anthracite Coal Mine Act, the Certified Mine Officials Act, and the Emergency Medical Personnel Act. In addition to its coal rules, the state also has set standards for the industrial minerals' sector, covering topics such as mine excavations, tunnels, ladders, and explosives.

- **South Dakota:** The Department of Environment and Natural Resources, Minerals and Mining, has laws and regulations that apply to mineral operations. They address inactive/abandoned mines, reclamation, and other environmental issues associated with mining activities.

- **Texas:** The Railroad Commission's Surface Mining and Reclamation Division and the Bureau of Economic Geology regulate and develop mine safety and health training and programs and address abandoned mine lands, geotechnology, and oil/gas discovery. State laws governing mining include the Coal Mining Regulations (16 Texas Admin. Code §12.1 et seq.), the Surface Coal Mining and Reclamation Act (Tex. Nat. Res. Code Ann. §134.001 et seq.), Uranium Mining laws (16 Texas Admin. Code §11.71 et seq.), the Uranium Surface Mining and Reclamation Act (Tex. Nat. Res. Code Ann. §131 et seq.), the Iron Ore/Iron Ore Gravel laws (Tex. Nat. Res. Code Ann. §134.012 et seq.), Quarry and Pit safety regulations (16 Texas Admin. Code §11.1001 et seq.), and the Texas Aggregate Quarry and Pit Safety Act (Tex. Nat. Res. Code Ann. §133.0001 et seq.). Safety aspects of sand and gravel pits are regulated by the RCT if they are located within 200 feet of a public roadway, but mining and reclamation of these pits are not otherwise regulated under state law.

- **Utah:** The Department of Natural Resources Mining implements the state's 1997 coal regulatory program, blaster certification, and non-coal-mining rules. It also regulates abandoned mine reclamation. The Utah Minerals Reclamation Program is a state agency that enforces the 1975 Mined Land Reclamation Act, Utah Code, Title 40, Ch. 8. It deals primarily with mineral leases at both large and small operations and is not involved in mine safety and health activities.

- **Virginia:** The Department of Mines, Minerals and Energy promulgates its own mine safety and health standards, and its 24 enforcement personnel conduct semiannual inspections of all underground mine sites and annual inspections of surface mines, pursuant to the Coal Mine Safety Laws of Virginia, Section 45.1-161.81. Additional inspections may be triggered by *spot* visits, accident investigations, reopening of a mine, or inspections targeting smoking articles in underground mines. The state has authority to impose monetary penalties, in addition to those levied by MSHA for the same violations, and can issue closure orders for the mine. Willful violations of state mine safety laws are prosecuted as a Class I misdemeanor. The state also provides outreach on training and education and can assist in Part 46 and Part 48 compliance. Virginia has requirements for certification of certain mine foremen and supervisors. The Department also works to eliminate off-site environmental damage associated with mining and to ensure proper restoration of mined lands.

- **Washington:** The Department of Natural Resources primarily is involved with surface mine reclamation and does not regulate safety and health at mines. The Metal Mining and Milling Act (Ch. 78.56 RCW) establishes a regulatory scheme that permits quarterly inspections of mines, establishes criteria for tailings impoundment control, waste-rock management plans, and water sampling.

- **West Virginia:** The Office of Miners' Health, Safety and Training promulgates regulations, provides miner certification examinations and mine safety training, and maintains a highly trained mine rescue team. The Office has enforcement and penalty powers at approximately 840 coal mines and quarries and over 2,500 independent contractors who work at mines. Maximum state civil penalties are $10,000, in addition to any fines levied by MSHA. The state's mine safety laws are published at Ch. 22A, Code of West Virginia, Title 56. Significantly, in West Virginia, "knowing" violations of state or federal mine safety laws provides injured workers or their estates with the ability to go outside workers' compensation "exclusive remedies" and seek additional monetary damages through tort litigation for personal injury or wrongful death.

- *Wyoming:* The Department of Environmental Quality administers and enforces all statutes, rules, and regulations dealing with mining and reclamation, and has authority to require permits and licenses for activities at surface and underground mines.

SPECIFIC REQUIREMENTS FOR INDEPENDENT CONTRACTORS

MSHA's *Program Policy Manual* establishes that "Any independent contractor that requests an identification number will receive one from MSHA." However, all contractors do not have to get identification numbers as a prerequisite to performing work at a mine. Independent contractors whose activities include any of the nine types of services or construction listed below must obtain identification numbers prior to commencing work and file their own injury/illness and quarterly hour reports as required under 30 CFR Part 50.

1. Mine development, including shaft and slope sinking.
2. Construction or reconstruction of mine facilities, including building or rebuilding preparation plants and mining equipment, and building additions to existing facilities.
3. Demolition of mine facilities.
4. Construction of dams.
5. Excavation or earthmoving activities involving mobile equipment.
6. Equipment installation, such as crushers and mills.
7. Equipment service or repair of equipment on mine property for a period exceeding five consecutive days at a particular mine.
8. Material handling within mine property, including haulage of coal, ore, refuse, etc., unless for the sole purpose of direct removal from or delivery to mine property.
9. Drilling and blasting.

Regardless of whether an MSHA identification number is obtained at the outset, the independent contractor who employs individuals classified as *miners,* or who falls within the definition of *operator,* must comply with the applicable MSHA standards at all times while on site. The producer-operator (e.g., the contracting mining company) has overall compliance responsibility for assuring that each independent contractor complies with the Mine Act and MSHA's standards and regulations. But, as discussed below, MSHA will issue citations to both the mine producer-operators and independent contractors for violations arising from the contractor's activities on the mine site.

MSHA RULEMAKING AUTHORITY

MSHA is vested with power to develop standards and regulations that address safety and health issues at mines. Often, these differ from OSHA or state agency standards that address similar subjects and work conditions. MSHA is granted broad deference in interpreting its own regulations and standards. Those standards are published at 30 CFR Parts 1-199.

MSHA does not have any rules that specifically address construction activities, although it does have some standards that address issues such as fall protection, crane and heavy equipment use and inspection, and requirements for work platforms. Construction contractors should carefully review the relevant MSHA standards, as the substantive requirements may vary from those of OSHA rules,[7] and failure to comply with the appropriate MSHA requirements will result in enforcement action.

The mandatory safety and health standards for surface metal/nonmetal mines are found in 30 CFR Parts 56 and 58, while underground metal/nonmetal standards are located in Parts 57 and 58. Coal standards are set forth in Parts 70–90 of Title 30. Mandatory underground coal health standards are in Part 70, underground coal safety standards are in Part 75, mandatory surface coal mine safety standards are found at Parts 70 and 77 (surface areas of underground coal mines), and surface coal mine health standards are promulgated at Parts 71 and 90. Parts 41 and 45 cover independent contractor requirements with respect to registration with the agency. Part 46 and Part 48 address training requirements for miners and contractors (aggregate and cement operations are covered by Part 46, while other types of mining fall under Part 48). Part 47 is MSHA's Hazard Communication Standard, which has different requirements from its OSHA counterpart. Part 62 establishes noise standards for all types of mining operations. Part 100 sets forth MSHA's enforcement and penalty criteria, as well as standards relating to a mine's history of violations. Part 104 establishes criteria for placing an employer on a "pattern of violations" program, which carries far more draconian punishment than do routinely issued citations.

Recordkeeping/reporting requirements with respect to production, employment, injuries/illness data, and notification of MSHA following certain categories of accidents are promulgated at 30 CFR Part 50. Violations of any paperwork requirement will result in civil penalties ranging from $60 to $60,000 per citation, regardless of whether it involves any hazard related to safety or health.

The agency publishes a *Program Policy Manual*, and other interpretative memoranda (e.g., Program Policy Letters, Program Information Bulletins, and Best Practice Guides), that provide interpretation of how MSHA plans to implement its standards. These documents can be accessed on MSHA's Web site (www.msha.gov) or at MSHA field offices across the United States. Knowledge of the information contained in MSHA policy pronouncements is imputed to the employer. However, MSHA cannot add new requirements through such interpretative documents, and it must go through "notice and comment" rulemaking under the Administrative Procedure Act[8] before imposing new mandates on the mining community.

MANDATORY TRAINING PROGRAMS

MSHA training requirements, codified at 30 CFR Parts 46 and 48, apply to independent contractors and their workers as well as to mine operators. This is because any workers who are exposed to the "hazards of mining" are considered to be "miners" by MSHA. Contractors who are exposed to mine hazards on a frequent (periodic, regularly repeated) or extended (more than five consecutive days) basis must receive complete new miner (or newly employed, experienced miner) training as well as annual refresher training each year. Each mine operator and contractor who is required to have MSHA-trained employees is required to have a written training plan. Model training plans are available from MSHA and can be created on-line by linking to http://www.msha.gov/FORMS/pt48train.htm.

New miner training is quite extensive (24 hours for surface mines and 40 hours for underground mines) and must be provided pursuant to a written training plan and conducted by either a "competent person" (Part 46—surface stone, sand, gravel, and cement facilities) or by a MSHA-approved trainer (Part 48—other types of surface mines and all underground mines). The annual refresher training provided under both Parts 46 and 48 must total 8 hours and be documented for each individual employee. New task training must also be provided and documented for employees whenever a job involves a new process, activity, or equipment. Finally, all individuals working at a new mine site must receive site-specific hazard training (this must be repeated annually when working at mines covered by Part 48).

Construction workers at mine sites are considered "miners" under Part 46 training, but they are specifically excluded from the comprehensive training requirements under Part 48 if they fall within the following categories: shaft and slope workers, workers engaged in construction activities ancillary to shaft and slope sinking, and workers engaged in the construction of major additions to an existing mine that requires the mine to cease operations.

Contractors, construction workers, customers, and visitors who are not exposed to mining hazards may receive only site-specific hazard training prior to beginning work or entering active work areas. Such hazard training can be simple (posting signs or handing out information sheets) or more complex, depending upon the nature and extent of hazards to which contractors and mine visitors will be exposed. However, contractors should be advised that MSHA interprets "mining hazards" quite broadly, and a wrong guess as to the potential exposures can result in a civil penalty of up to $60,000 and/or possible criminal prosecution.

■■ Table 3 ■■

Key Sections of 30 CFR Parts 1–100
Mine Safety and Health Standards and Regulations

Part	Content	Applicability
Subchapter B (Parts 5, 7, 15, 18, 19, 20, 21, 22, 23, 24, 26, 27, 28, 29, 31, 32, 33, 35, and 36	Testing, evaluation, and approval of mining products	Covers fees for testing/evaluation and approval of various products, as well as establishes standards for the following products: Explosives; electric motor-driven mine equipment and accessories; electric cap lamps; electric mine lamps; flame safety lamps; portable methane detectors; telephones and signaling devices; single-shot blasting units; lighting equipment for underground workings; methane-monitoring systems; fuses for DC current in trailing cables; portable coal dust/rock dust analyzers; warning lights and methane detectors; diesel mine locomotives and diesel-powered equipment; dust collectors; fire-resistant hydraulic fluids; and mobile diesel-powered transportation equipment.
Subchapter G (Parts 40, 41, 43, 44, and 45)	Filing and administrative requirements	Covers representative of miners; notification of legal identity; procedures for processing hazardous condition complaints; petitions for modification; and independent contractor requirements.
Part 46	Mandatory training	Selected surface nonmetal mines (stone, sand, gravel, cement, colloidal clay).
Part 47	Hazard communication	All mines.
Part 48	Mandatory training	All underground coal and metal-nonmetal mines; all surface coal mines; all surface metal mines; all surface nonmetal mines that are not otherwise included under Part 46.
Part 49	Mine rescue teams	All underground mines.
Part 50	Accident, injury, illness, employment, and production reports	All surface and underground mines.
Part 56	Mandatory safety and health standards	All surface metal/nonmetal mines.
Part 57	Mandatory safety and health standards	All underground metal/nonmetal mines.
Part 58	Mandatory health standards (abrasive blasting and drill dust control)	All metal/nonmetal mines.
Part 62	Mandatory noise standards	All surface and underground mines (coal and metal/nonmetal).
Part 70	Mandatory health standards	All underground coal mines.
Part 71	Mandatory health standards	Surface coal mines and surface work areas of underground coal mines.
Part 72	Health standards for coal mines (abrasive blasting and drill dust control)	All coal mines.
Part 74	Coal mine dust personal sampler units	All coal mines.
Part 75	Mandatory safety standards	Underground coal mines.
Part 77	Mandatory safety standards	Surface coal mines and surface work areas of underground coal mines.
Part 90	Mandatory health standards	Applicable to coal miners who have evidence of the development of pneumoconiosis.
Part 100	Criteria and procedures for proposed assessment of civil penalties	All mines.
Part 104	Pattern of violations	All mines.

Independent contractors required to provide training are also required to promptly produce miner training records (which must be maintained on MSHA Form 5000-23, or equivalent) to show that all mandatory training has been provided. This is interpreted as before the end of the inspection day. In comparison, usually MSHA will give the contractor a full business day to present its written Part 46/48 training plan.[9] The location where the records are maintained, such as a mine site, or at the contractor's office, is up to the independent contractor.

MSHA will issue an order under Section 104(g) of the Act to the direct employer of any miner who has not received the required training under Part 46 or Part 48. This means that a Section 104(g) order will be issued to the independent contractor or construction company for any persons who are direct employees and who are not properly trained.

MSHA policy states that the independent contractor should be issued a 104(g) order for any of his/her employees who are not trained in accordance with a plan approved under Part 48. If MSHA cannot determine who employs the untrained person, the agency will issue the Section 104(g) order to the producer-operator.[10] In addition to the order, MSHA will issue a corresponding citation to the independent contractor or production operator for failure to provide the miner with the requisite training. Therefore, both a citation and the Section 104(g) withdrawal order will be issued. The mine operator-contractor generally will receive one penalty assessment of up to $60,000 that combines both the citation and order, but multiple penalties may be proposed under certain circumstances if numerous untrained miners are involved.

Independent contractors are not required to have an approved training plan under Part 48, but they must receive appropriate training. Independent contractors may comply with the training requirements by either making arrangements to have their employees trained under an existing approved training plan and program (e.g., through a state grants program agency, a third-party training company with an MSHA-approved trainer, or from the mine operator) or by filing and adopting their own approved training plan. MSHA's Educational Field Services staff can also provide assistance in reviewing draft training plans and explaining instructor competency requirements.

MSHA operates the National Mine Health and Safety Training Academy in Beaver, West Virginia, which produces many written and audiovisual training materials that can be used by mine operators and contractors to comply with mandatory training requirements. The Academy's mission is to reduce accidents and improve health conditions by conducting a variety of education and training programs in health and safety and related subjects for mine inspectors, other government personnel, and the public. The Academy contains a comprehensive library on mine safety and health-related literature, and sponsors short courses and conferences on mine safety and health subjects and industrial hygiene practices.

MULTIEMPLOYER WORK-SITE ISSUES

MSHA's policy is to issue citations and, where appropriate, orders to independent contractors for violations of applicable provisions of the Act, standards, or regulations. This policy is based on the Mine Act's definition of an *operator*, which includes "independent contractors performing services or construction" at mines. In addition, basic compliance responsibilities of a mine producer–operator include assuring that independent contractors comply with the Mine Act and with applicable MSHA standards and regulations.

As a result, both independent contractors and producer-operators have *overlapping* compliance responsibility: There may be circumstances in which it is appropriate for MSHA to issue citations or orders to both the independent contractor and to the production operator for a violation.

Enforcement action taken by MSHA against a producer-operator for violations involving an independent contractor is considered appropriate in the following situations:

■ when the producer-operator has contributed by either an act or an omission to the occurrence of a violation in the course of an independent contractor's work

- when the producer-operator has contributed by either an act or omission to the continued existence of a violation committed by an independent contractor
- when the producer-operator's miners are exposed to the hazard
- when the producer-operator has control over the condition that needs abatement.

Inspectors also will cite independent contractors for violations committed by the contractor or by its employees. It is not unusual for the mine operator to be cited with a higher degree of negligence than that attributed to the independent contractor because MSHA theorizes that the mine operator is more familiar with the legal requirements under the Mine Act and 30 CFR Parts 1–199.

Contractors should be aware, however, that some provisions of the Mine Act, standards, or regulations are not directly applicable to independent contractors or their work, and/or independent contractor compliance with certain standards or regulations may duplicate the producer-operator's compliance efforts. It is critical to be certain that contractors know their rights and responsibilities to defend against unwarranted enforcement actions.

IMMEDIATE REPORTING REQUIREMENTS

MSHA's accident reporting requirements vary significantly from those of OSHA. Under 30 CFR §50.10, the mine operator or independent contractor has a legal obligation to immediately notify MSHA when an "accident" of the type defined in Section 3(k) of the Mine Act and in 30 CFR §50.2(h) occurs. *Immediately* has been construed by the Federal Mine Safety and Health Review Commission (FMSHRC) to mean within two (2) hours of the occurrence. Failure to provide this notification can result in civil penalties of up to $60,000. When MSHA is notified of such accidents, it generally will send an investigative team. Normally, a complete mine inspection will also be conducted within 30 days of the accident.

Events requiring immediate notification are:

- a death of an individual at a mine
- an injury to an individual at a mine which has a reasonable potential to cause death
- an entrapment of an individual for more than 30 minutes
- an unplanned inundation of a mine by a liquid or gas
- an unplanned ignition or explosion of gas or dust
- an unplanned mine fire not extinguished within 30 minutes of discovery
- an unplanned ignition or explosion of a blasting agent or an explosive
- an unplanned roof fall at or above the anchorage zone in active workings where roof bolts are in use; or, an unplanned roof or rib fall in active workings that impairs ventilation or impedes passage
- a coal or rock outburst that causes withdrawal of miners, or which disrupts regular mining activity for more than one hour
- an unstable condition at an impoundment, refuse pile, or culm bank that requires emergency action in order to prevent failure, or that causes individuals to evacuate an area; or, failure of an impoundment, refuse pile, or culm bank
- damage to hoisting equipment in a shaft or slope that endangers an individual or that interferes with use of the equipment for more than 30 minutes
- an event at a mine that causes death or bodily injury to an individual not at the mine at the time the event occurs.

The operator also has an independent obligation under Part 50 to investigate and prepare a report on such accidents, to determine the cause and means to prevent a recurrence, and to provide this report to MSHA upon request. Care must be exercised in preparing such reports because of the possible admission of violations (some of which may not have occurred to MSHA) and the use of such reports in criminal prosecutions and/or in related tort actions arising from personal injury, wrongful death, or property-damage litigation.

MANAGING THE MSHA INSPECTION/INVESTIGATION PROCESS

The Mine Act prohibits advance notification of MSHA inspections. Moreover, inspections cannot be delayed by demanding a warrant. However, the employer at the mine site does have the right to have a representative participate in the inspection, provided such a representative is available within a "reasonable" amount of time (generally interpreted as 30 minutes).

Each MSHA inspection has three parts: (1) opening conference, (2) the walkaround, and (3) the closing conference. The opening conference provides the opportunity to determine what type of inspection is about to take place (a routine quarterly or semiannual inspection, a special-emphasis inspection, a complaint-based investigation, or an incident-triggered investigation). The inspector is required to present his/her credentials as an "authorized representative" of the Secretary of Labor.

The *opening conference* sets the tone of the inspection and provides an opportunity to let the agency know about safety initiatives and achievements that have been successful in reducing injuries, illnesses, and incidents. However, any written documents voluntarily produced that are not required by statute or regulation may contain information that could form the basis for a separate citation (e.g., disclosing information about a near-miss). The compliance specialist does not have to observe a violation in progress in order to issue a citation, and information contained in records, audit documents, or statements made by hourly or management personnel can trigger an enforcement action.

Many MSHA standards have associated record-keeping requirements, and these should be reviewed in advance of an inspection to ensure that the management on site has all necessary materials and can also differentiate between requests for nonstatutorily required documents (which can be refused) and those documents that must be produced "or else." When in doubt, the supervisor can request that the inspector reference the particular standard or statutory section containing such a requirement. Contractors are also advised to inspect their work sites frequently for the most common mine hazards—those often triggering MSHA citations. Some of the most frequently cited metal/nonmetal and coal standards are delineated in Tables 4 and 5.

During the *walkaround* phase of the inspection, the MSHA representative will likely interview many employees about their training, procedures, and accident record. There is no legal reason why a management representative cannot participate in such interviews with the workers' consent. Similarly, workers can decline entirely to participate in MSHA interviews, but management must not use threats or coercion to discourage participation. MSHA has "informant privilege" rules that it takes very seriously!

Some MSHA inspections will include a *health inspection* component, and mine operators and contractors may take side-by-side samples (e.g., noise, silica) so that their results can be used to independently verify or refute MSHA's enforcement actions. MSHA's air contaminant standards are quite old and reference the threshold limit values (TLVs) published by the American Conference of Governmental Industrial Hygienists (ACGIH) in 1972 (coal mines) and 1973 (metal/nonmetal mines). Because these TLVs are not published directly in 30 CFR, it will be necessary to obtain this information from ACGIH or an MSHA field office to determine the applicable limits for each operation.

Similarly, during a safety inspection, a management representative will be well-served if he/she takes copious notes of inspector comments, conducts independent measurements, or otherwise photographs or sketches the allegedly violative conditions, and is conversant with MSHA requirements. It is rarely fruitful to argue with an inspector, and the management personnel should avoid making any incriminating statements during the course of the inspection.

The *closeout conference* is an opportunity to get a citation downgraded or vacated, but because these conferences are "on the record," statements can also be used to increase the severity or heighten the negligence of a citation. Any abatement issues should also

■ ■ **Table 4** ■ ■

Frequently Cited Standards—Metal/Nonmetal Mines

Standard	Description
30 CFR §§ 56/57.14107	Moving machine parts
30 CFR §§ 56/57.14132	Horns, back-up alarms, and automatic warning devices
30 CFR §§ 56/57.14100	Safety defects, examination, correction and records
30 CFR §§ 56/57.11001	Safe access to working places
30 CFR §§ 56/57.14101	Brakes
30 CFR §§ 56/57.12008	Insulation and fittings for power wires and cables
30 CFR §§ 56/57.14112	Construction and maintenance of guards
30 CFR §§ 56/57.20003	Housekeeping
30 CFR §§ 56/57.12032	Inspection and cover plates on electrical equipment
30 CFR §§ 56/57.9300	Berms or guardrails
30 CFR §§ 56/57.11002	Handrails and toeboards
30 CFR § 57.12004	Electrical conductors
30 CFR § 57.3200	Correction of hazardous ground/roof conditions
30 CFR § 57.12030	Correction of dangerous electrical conditions
30 CFR § 57.22305	Approved equipment (Class III gassy mines)
30 CFR §§ 56/57.15005	Safety belts and lines
30 CFR §§ 56/57.16005	Securing gas cylinders
30 CFR § 62.130(a)	Noise permissible exposure limit
30 CFR § 57.4101	Warning signs
30 CFR § 57.4201	Inspections

■ ■ **Table 5** ■ ■

Frequently Cited Standards—Coal Mines

Standard	Description
30 CFR § 77.404	Machinery and equipment—operation and maintenance
30 CFR §§ 75.1722 and 77.400	Mechanical equipment guards
30 CFR §§ 77.410(a) and (c) and 77.1605(b)	Horns, back-up alarms, and automatic warning devices
30 CFR § 77.1606(c)	Loading and haulage equipment—inspection and maintenance
30 CFR §§ 75.400 and 77.1104	Accumulations of combustible materials
30 CFR § 75.503	Permissible electric face equipment—maintenance
30 CFR § 77.205	Safe access/travelways
30 CFR § 77.1605	Loading and haulage equipment—installation and brakes
30 CFR §§ 75.1100 and 77.1110	Firefighting equipment
30 CFR §§ 75.515, 75.517 and 77.505	Insulation and fittings for power wires and cables
30 CFR § 72.620	Drill dust control
30 CFR § 62.130	Noise permissible exposure limit
30 CFR § 77.512	Inspection and cover plates on electrical equipment
30 CFR § 77.1605(k)	Berms or guardrails
30 CFR § 77.502	Electrical equipment—examination, testing, and maintenance
30 CFR § 75.202	Protection from falls of roof, face, and ribs
30 CFR § 75.220	Roof control plan
30 CFR § 75.370	Mine ventilation plan
30 CFR § 75.333(h)	Ventilation controls
30 CFR § 75.4103	Other safeguards

be raised during the closeout conference, as failure to abate can trigger high civil penalties. Unlike OSHA proceedings, MSHA's abatement requirements are not stayed by contesting a citation. Therefore, a realistic abatement period is one that provides sufficient time to further conference or have an expedited hearing on a citation or order.

SPECIAL INVESTIGATIONS, SECTION 110(C) LIABILITY, AND CRIMINAL PROSECUTIONS

MSHA occasionally conducts "special investigations" pursuant to Section 110(c) of the Mine Act. These investigations target management agents, officers, and directors who engage in aggravated conduct, exceeding ordinary negligence, and may lay the groundwork for criminal prosecutions under the Mine Act. Such individual prosecutions occur where MSHA believes the person failed to prevent violations by his employees (or those of another company) or, if the workers were under a supervisor's direction or control, where MSHA alleges that the supervisor/agent knew or should have known of the violative behavior. These individual civil penalties can reach $60,000 per violation and may be adjudicated separately from any actions taken against the employer.

Typically, citations and orders issued under Section 104(d) and Section 107 of the Mine Act, characterized as "high negligence, unwarrantable failure to comply, reckless disregard," and/or "imminent danger" provide the trigger for special investigations and for civil and/or criminal penalties under Section 110 of the Mine Act. Therefore, any citation received bearing these allegations should promptly be reviewed with counsel to determine appropriate defensive action.

In addition to individual penalties under Section 110(c), unwarrantable failure allegations can lead to higher civil penalties because they are specially assessed, rather than being subject to the penalty formula under 30 CFR Part 100. They may also be used in "pattern of violation" determinations, and adversely affect the mine's compliance history (which lets future inspectors identify the mine as a "bad actor" before ever coming onto the mine site).

Certain types of violations of the Mine Act and its implementing regulations—and conduct of the mine operator and its agents during mining activities and MSHA inspections/investigations—can trigger criminal prosecutions. These prosecutions are made when MSHA makes a referral to the U.S. Department of Justice for prosecution. In addition, some state attorneys general are reviewing occupational fatalities and even exposure to health hazards as potential bases for state criminal actions, including murder, manslaughter, assault, and battery.

The Mine Act contains statutory provisions for criminal penalties. Unlike the OSHA statute, no fatality or serious injury need occur in order to trigger a criminal prosecution for a willful violation. The terms of potential imprisonment are set forth in the Mine Act, 30 U.S.C. §811. Additional actions can be brought under the criminal provisions of 18 U.S.C. §3571, which increases fines against individuals to $250,000, and fines against corporations to $500,000.

The most severe provision is in Section 110(f) of the Mine Act, which provides a prison sentence of up to five years (a Class E felony) for any person who falsifies or alters documents required to be kept under the Mine Act. Although this is most commonly invoked with respect to the Form 5000-23 (records of Part 46/Part 48 training, and training mandated by MSHA's Hazard Communication Standard), it also applies to such things as workplace inspection reports, examination of hoists and cranes, roof-support inspection in underground mines, and electrical testing records. In addition to incarceration, Section 110(f) provides for a criminal fine of up to $10,000, and it is subject to enhancement through application of 18 U.S.C. §3571, which increases fines against individuals to $250,000 and fines against corporations to $500,000.

Criminal prosecutions under Section 110(c) for "knowing" violations of MSHA standards provide for both civil fines of up to $60,000 against individuals and criminal penalties. Section 110(d) applies to "willful" violations of MSHA rules, and to "knowing" violations of orders issued under Section 104 and/or Section 107 of the Act. Both of these provisions provide for jail terms of up to one year (Class A misdemeanor) and fines of up to $25,000 for the first offense and $50,000 for the second offense (with an increase in the prison sentence to five years—a Class E felony—for second offenders). As with prosecutions under Section 110(f), the Justice Department can also invoke 18 U.S.C. §3571, and its significantly higher criminal fines.

Section 110(c) and Section 110(f) proceedings can be brought against both management personnel and hourly employees who assume "management-like" responsibilities (e.g., conducting and documenting workplace examinations or providing training to fellow workers). Therefore, all persons working at an MSHA-regulated site should be aware that they cannot be compelled to provide a statement to the MSHA investigator in the absence of a subpoena, and they have the right to have counsel present if they choose to be interrogated. Conversely, all statements made voluntarily can and will be used against the proponent in a court of law.

Further, management representatives can be prosecuted for directing employees not to speak with MSHA because such actions to *impede* the investigation constitute a violation of Section 103(a) of the Mine Act. Therefore, any decision to refrain from participating in an MSHA investigation must be voluntary on the part of the worker.

Special investigations are initiated at the MSHA District Manager's recommendation, or at the direction of MSHA headquarters' officials. Once the investigation begins, the investigator will receive the enforcement file and the accident file (if any). The investigator will request to interview all persons involved, contractors, witnesses (whether employed as miners or not), and potential defendants.

Many of the special investigators have police backgrounds, and all of them receive extensive training on how to gather evidence and interview people. However, MSHA investigators do not have the right to use force, carry firearms, or take individuals into custody in the absence of a warrant. In addition, MSHA investigators cannot conduct surveillance of witnesses except with the express authorization of the Department of Justice and MSHA headquarters' officials.

However, waiting for a prospective witness outside his/her home or workplace is not considered by MSHA to constitute *surveillance*, although ap-

proval of the District Manager is needed if such action will be taken after dark. It is also not uncommon for MSHA to contact individuals at their homes (in person or by telephone) outside of working hours in an attempt to conduct the investigation without the knowledge of the employer. Rarely will the investigator inform the interviewee that he/she has the right to decline to make a statement or to have a representative (or counsel) present. Therefore, it is important that all possible witnesses know their rights in advance.

MSHA investigators may, from time to time, assist other law enforcement agencies in conducting undercover operations provided that: (1) the undercover operation is directed at uncovering violations of law pertaining to mine safety and health, (2) the operation is under the direct and constant supervision of a federal law enforcement agency, (3) the operation has been approved by the U.S. Attorney's Office, and (4) the MSHA Assistant Secretary and the Solicitor of Labor have authorized the use of MSHA personnel in the undercover activity.

The investigator may demand access to the property to conduct a physical examination. The investigator has the same warrantless right of entry as other authorized representatives of the Secretary, and the employer has the right to have its representative accompany the investigator at all times while on mine property. The investigator may demand records and other documents, but only those materials mandated by the Mine Act or under MSHA standards need be presented in the absence of a subpoena.

The investigator need not notify the producer-operator or independent contractor that a special investigation is in progress, unless he/she conducts the investigation at the mine site. Therefore, the investigator theoretically can conduct all witness interviews without letting the employer know that a special investigation has been authorized. However, if the operator has already notified MSHA that it and its agents are represented by counsel (e.g., during an informal or closeout conference), generally MSHA will let the company's attorney know that a special investigation is planned, and will schedule the meeting so management personnel can have counsel present.

SECTION 105(C) DISCRIMINATION ACTIONS

Section 105(c) of the Mine Act prohibits discrimination against mine employees based on the exercise of safety and health rights under the Act. A company can be penalized by MSHA, and additional relief can be ordered (reinstatement, back pay, etc.) to the miner who was the subject of discriminatory action. Because MSHA penalties against the employer and/or individual supervisors and agents can reach $60,000, and back pay may equal or exceed this amount by the time litigation is concluded, discrimination actions must be taken seriously. Potential remedies include: (1) temporary reinstatement, (2) permanent rehire or reinstatement of the miner, (3) back pay and interest, (4) attorney's fees and costs, and (5) civil penalties by MSHA.

In addition to raising safety and health complaints with management, miners can become "protected" within the meaning of Section 105(c) if they make a complaint to MSHA, speak with MSHA inspectors during inspections, or give statements during investigations. Even where the employee has not made a complaint or statement, he/she can be protected if management believes that worker is an informant or a complainant. Moreover, a complainant is protected even if MSHA investigates and determines that no violation exists. Employees are also protected if they are the subject of a medical evaluation and potential transfer under a standard (e.g., because of overexposure to lead or arsenic). An employer cannot take action because a worker exercised any statutory rights under the Mine Act (e.g., the right to be compensated while serving as the miner's rep, or during Part 46 or Part 48 training).

Finally, an employer cannot discriminate against a miner who refuses to work under an alleged unsafe or unhealthful condition, or who refuses to operate equipment that he/she believes is unsafe. The four-part test for evaluating whether a work refusal is protected is whether the miner has: (a) a *good faith* (b) *reasonable belief* that a (c) *hazard* exists that is ordinarily (d) *communicated* to the mine operator. Courts have ruled that workers were "constructively discharged" where the operator maintained conditions so intolerable that a reasonable individual would be compelled to resign. In such cases,

the worker is entitled to back pay for the period in which he/she was off the job.

Employers should not hinder any Section 105(c) investigations or prevent employees or management representatives from talking to inspectors or investigators at any time. All complaints must be thoroughly investigated, and appropriate remedial action taken promptly. Any disciplinary action that could have Section 105(c) implications should be carefully considered, and undertaken only with witnesses present. A record of all disciplinary action should be maintained, even prior to the receipt of a complaint, so that an employee's performance can be documented in the event that the employer must support his nondiscriminatory decisions.

Section 105(c) protects all workers at a mine—including salaried supervisors. An employee who believes that he/she was discriminated against because of participation in a protected activity must file a complaint with MSHA within 60 days of the alleged discriminatory action's occurrence. Construction companies should note that this is twice as long as the statutory period provided for similar Section 11(c) actions under the OSH Act. MSHA assigns a special investigator to determine whether the complaint is meritorious, and he/she will employ the same techniques described above related to Section 110(c) actions.

If MSHA finds that a complaint has merit, the FMSHRC can order temporary reinstatement of the worker, pending a final resolution of the complaint. MSHA will file a formal complaint with the Commission and will represent the miner against the employer. If MSHA finds that the claim does not have merit, it will notify the worker, but the worker can still proceed against the mine operator by representing himself/herself or by obtaining outside counsel—another distinction from OSHA proceedings. If the employee subsequently prevails, he/she can recover attorney's fees and the costs of litigation from the employer as part of the award in addition to the other types of monetary awards and reinstatement.

In Section 105(c) cases, the burden of proof is divided between the employee and the employer. The miner bears the burden of proving that he/she engaged in a protected activity and that the adverse action was motivated in any way by the protected activity. If the complainant meets this burden, this constitutes a *prima facie* case.

The Federal Mine Safety and Health Review Commission has inferred discriminatory intent from the following:

1. Knowledge of the employer that the complainant was making safety complaints.
2. Hostility toward safety matters and complaints about safety.
3. Proximity between the time safety complaints were made and the time that adverse action occurred.
4. Disparate treatment.

The employer may rebut the finding by showing either that the worker was not engaged in any protected activity or that the adverse action was not motivated in any part by the protected activity. Even if the protected activity was a factor in the decision, the employer can still defend if it demonstrates that it would have taken the adverse action in any event for the unprotected activity alone.

PROCEDURES FOR CONTESTING CITATIONS AND ORDERS

The Federal Mine Safety and Health Review Commission, an independent agency, handles adjudication of citations and orders issued by MSHA enforcement personnel. Cases initially come before one of the FMSHRC's Administrative Law Judges (ALJs). A five-member panel of Commissioners hears appeals of ALJ decisions, if the appellant's Petition for Review is granted. These Commissioners are appointed by the President, confirmed by the U.S. Senate, and serve six-year terms. If the FMSHRC declines to review a case, or if either party wishes to further appeal the FMSHRC's decision, the case proceeds to the U.S. Court of Appeals—either the District of Columbia Circuit or the Circuit governing the state where the mine at issue is located.

If an operator or contractor wishes to contest a citation or order prior to proposal of a civil penalty assessment, a *Notice of Contest* must be filed within 30 calendar days of receiving the citation. The Notice

of Contest is filed with the FMSHRC in Washington, D.C., and with the Office of the Solicitor of Labor, Mine Safety and Health in Arlington, Virginia. The Notice of Contest can be filed before or after an informal conference is held with MSHA district or field office personnel.

The operator or contractor can challenge the following in a notice of contest:

- fact of violation
- gravity of the violation (whether a serious accident/injury is reasonably likely to occur)
- negligence of the violation
- abatement requirements.

An expedited hearing can be held to obtain a judge's ruling on the validity of a citation or order before the mine operator must expend money to abate the alleged violation. However, unless an expedited hearing is requested, the mine operator is normally required to correct the alleged violation even if a contest hearing is pending.

If the operator does not initially contest the citation or order, MSHA will issue a proposed assessment of civil penalty. Once the penalty is received, the company must mail back the assessment form to MSHA within 30 calendar days to contest the penalty and/or the underlying citation/order. At this stage, the fact of violation, gravity, negligence, and amount of penalty can be challenged, but not the abatement method or deadline.

Once MSHA receives the completed assessment form, it must file a Petition for Penalty Assessment with the Federal Mine Safety & Health Review Commission within 45 calendar days and send a copy of the Petition to the mine operator. The mine operator must then file an Answer within 30 calendar days of receiving the Petition. If settlement of the matter is not reached, it will proceed to litigation before the FMSHRC[11] and may be appealed to the U.S. Court of Appeals and the U.S. Supreme Court (if certiorari is granted).

Companies may be entitled to reasonable attorney fees if, after full adjudication of the merits, MSHA's position is found to lack "substantial justification," provided that the company otherwise satisfies the size and monetary restrictions in the Equal Access to Justice Act, 5 U.S.C. §504.

CONCLUSION

Although construction contractors may perform work at mine sites infrequently, the potential exposures for prosecution by MSHA are high in light of frequent inspections and the variability of MSHA standards from OSHA standards, with which the employer is more familiar. Recent rulings have affirmed that compliance with a similar OSHA standard will not serve to negate liability under the Mine Act. Moreover, even construction companies with outstanding safety training programs may find themselves in violation of the more arduous MSHA requirements in Parts 46 and 48, and they may lack critical documents required by MSHA's hazard communication standard.

Any construction employer planning to engage in long-term work at a mine (e.g., more than five consecutive days) may wish to consider including the costs of *ramping up* on these unique regulatory requirements in any proposal, or requiring advance assistance from the mine operator in reviewing the contractor's existing training programs and other written materials to avoid being caught in a compliance trap.

REVIEW EXERCISES

1. Explain MSHA's new miner training requirements and describe under what circumstances contractor employees would be required to receive this type of training.
2. What types of hazards are often found at mine sites that may cause fatalities and result in enforcement activities?
3. What steps should be taken to investigate a safety complaint made by a miner?
4. Under what circumstances will MSHA issue citations to both a mine operator and a contractor?
5. Describe circumstances where MSHA might refer a violation case for criminal prosecution.
6. Define who are considered *miners* and *mine operators* under the Mine Act.
7. Can state agencies assert dual jurisdiction with MSHA over mine safety and health and, if so, what are the scope of these state powers?

APPENDIX A: EXCERPTS FROM MSHA–OSHA INTERAGENCY AGREEMENT

Definitions

Coal or other mine is defined in the Mine Act as:

(A) an area of land from which minerals are extracted in nonliquid form or, if in liquid form, are extracted with workers underground, (B) private ways and roads appurtenant to such area, and (C) lands, excavations, underground passageways, shafts, slopes, tunnels and workings, structures, facilities, equipment, machines, tools, or other property including impoundments, retention dams, and tailing ponds, on the surface or underground, used in, or to be used in, or resulting from, the work of extracting such minerals from their natural deposits in nonliquid form, or if in liquid form with workers underground, or used in, or to be used in, the milling of such minerals, or the work of preparing coal or other minerals, and includes custom coal preparation facilities.

Miner is defined in the Mine Act as:

"Any individual working in a coal or other mine."

Operator is defined in the Mine Act as:

Any owner, lessee, or other person who operates, controls, or supervises a coal or other mine or any independent contractor performing services or construction at such mine.

Mining and Milling:

Mining has been defined as the science, technique, and business of mineral discovery and exploitation. It entails such work as directed to the severence of minerals from the natural deposits by methods of underground excavations, opencast work, quarrying, hydraulicking, and alluvial dredging. Minerals so excavated usually required upgrading processes to effect a separation of the valuable minerals from the gangue constituents of the material mined. This latter process is usually termed "milling" and is made up of numerous procedures which are accomplished with and through many types of equipment and techniques.

Milling is the art of treating the crude crust of the earth to produce therefrom the primary consumer derivatives. The essential operation in all such processes is separation of one or more valuable, desired constituents of the crude from the undesired contaminants with which it is associated.

A *CRUDE* is any mixture of minerals in the form in which it occurs in the earth's crust. An *ORE* is a solid crude containing valuable constituents in such amounts as to constitute a promise of possible profit in extraction, treatment, and sale. The valuable constituents of an ore are ordinarily called valuable minerals, or often just minerals; the associated worthless material is called gangue.

In some ores the mineral is in the chemical state in which it is desired by primary consumers, e.g., graphite, sulphur, asbestos, talc, garnet. In fact, this is true of the majority of nonmetallic minerals. In metallic ores, however, the valuable minerals in their natural state are rarely the product desired by the consumer, and chemical treatment of such minerals is a necessary step in the process of beneficiation. The end products are usually the result of concentration by the methods of ore dressing (milling) followed by further concentration through metallurgical processes. The valuable produce of the ore-dressing treatment is called Concentrate; the discarded waste is Tailing.

Specific Examples of MSHA Authority

MINING–MSHA

Following is a list indicating mining operations and minerals over which MSHA has regulatory authority.

Mining Operations
Underground Mining
Open Pit Mining
Quarrying
Solution Mining (Precipitate & Leaching)
Dredging (when the primary purpose of a dredging operation is to recover metal or nonmetallic minerals for milling and/or sale or use.)
Hydraulicking
Ponds—Brine Evaporation
Auger Mining
Minerals
Coal

Metals:
(Included but not limited to)

Alumina
Antimony
Bauxite
Beryl
Bismuth
Chrome
Cobalt
Copper
Gold
Iron
Lead
Manganese
Mercury
Molybdenum
Nickel
Rare Earths
Silver
Titanium
Tungsten
Uranium
Vanadium
Zinc
Zirconium
Magnesite
Shale
Sulfur
Pyrophllite
Wollastonite

Nonmetals:
(Included but not limited to)

Abrasives
Aplite
Asbestos
Barite
Baron
Bromine
Calcium chloride
Clay
Mica
Mineral pigments
Oil shale
Peat
Perlite
Potash
Pumice
Potash rock
Diatomite
Feldspar
Fluorspar
Gilsonite
Graphite
Gypsum
Kyanite
Salt
Sodium compounds
Talc, Soapstone, and
Vermiculite

Subgroups of Nonmetals
(Sand and Gravel, and Crushed and Dimension
Stone Industries)

Sand
Gravel
Cement

Gabbro
Gneiss
Lime
Limestone

Marble
Native Asphalt
  (impregnated stone &
  sand)
Quartizite
Schist
Slate
Traprock or Diabase

MILLING–MSHA AUTHORITY

Following is a list with general definitions of milling processes over which MSHA has regulatory authority, subject to Paragraph B6 of the agreement. Milling consists of one or more of the following processes: crushing, grinding, pulverizing, sizing, concentrating, washing, drying, roasting, pelletizing, sintering, evaporating, calcining, kiln treatment, sawing and cutting stone, heat expansion, retorting (mercury), leaching, and briquetting.

Crushing: the process used to reduce the size of mined materials into smaller, relatively coarse particles. Crushing may be done in one or more stages, usually preparatory for the sequential stage of grinding, when concentration of ore is involved.

Grinding: the process of reducing the size of a mined product into relatively fine particles.

Pulverizing: the process whereby mined products are reduced to fine particles, such as to dust or powder size.

Sizing: the process of separating particles of mixed sizes into groups of particles of the same size, or into groups in which particles range between maximum and minimum sizes.

Concentrating: the process of separating and accumulating economic minerals from gangue, or the upgrading of ore or minerals.

Washing: the process of cleaning mineral products by the buoyant action of flowing water.

Drying: the process of removing uncombined water from mineral products, ores, or concentrates; for example, by the application of heat, in air-actuated vacuum-type filters, or by pressure-type equipment.

Roasting: the process of applying heat to mineral products to change their physical or chemical qualities for the purpose of improving their amenability to other milling processes.

Pelletizing: the process in which finely divided material is rolled in a drum, cone, or an inclined disk so that the particles cling together and roll up into small spherical pellets. This process is applicable to milling only when accomplished in relation to, and as an integral part of, other milling processes.

Sintering: the process of agglomerating small particles to form larger particles, cakes, or masses, usually by bringing together constituents through the application of heat at temperatures below the melting point. This process is applicable to milling only when accomplished in relation to, and as an integral part of, other milling processes.

Evaporating: the process of upgrading or concentrating soluble salts from naturally occurring or other brines, by causing uncombined water to be removed by application of solar or other heat.

Calcining: the process of applying heat to mineral materials to upgrade them by driving off volatile chemically combined components and effecting physical changes. This process is applicable to milling only when accomplished in relation to, and as an integral part of, other milling processes.

Kiln Treatment: the process of roasting, calcining, drying, evaporating, and otherwise upgrading mineral products through the application of heat. This process is applicable to milling only when accomplished in relation to, and as an integral part of, other milling processes.

Sawing and Cutting Stone: the process of reducing quarried stone to smaller sizes at the quarry site when the sawing and cutting is not associated with polishing or finishing.

Heat Expansion: a process for upgrading material by sudden heating of the substance in a rotary kiln or sinter hearth to cause the material to bloat or expand to produce a lighter material per unit of volume.

Retorting: a process usually performed at certain mine sites, accomplished by heating the crushed material in a closed retort to volatilize the metal, material, or hydrocarbon which is then condensed and recovered as upgraded metal, material, or hydrocarbon.

Leaching: the process by which a soluble metallic compound is removed from a mineral by selectively dissolving it in a suitable solvent, such as water, sulfuric acid, hydrochloric acid, cyanide, or other solvent, to make the compound amenable to further milling processes.

Briquetting: a process by which iron ore, or other pulverized mineral commodities, are bound together into briquettes, under pressure, with or without a binding agent, and thus made conveniently available for further processing.

MSHA AUTHORITY ENDS–OSHA AUTHORITY BEGINS

The following are types of operations that may be on or contiguous to mining and/or milling operations listed above over which MSHA does not have authority to prescribe and enforce employee safety and health standards, and over which OSHA has full authority, under the Act, to prescribe and enforce safety and health standards, regarding working conditions of employees. OSHA regulatory authority commences as indicated in the following types of operations:

Gypsum Board Plant: If the plant is located on mine property, commences at the point when milling, as defined, is completed, and the gypsum and other materials are combined to enter the sequential processes necessary to produce gypsum board. If not located on mine property, OSHA has authority over the entire plant.

Brick, Clay Pipe, and Refractory Plants: Commences after arrival of raw materials at the plant stockpile.

Ceramic Plant: Commences after arrival of the clay and other additives at the plant stockpile.

Fertilizer Products: Commences at the point when milling, as defined, is completed, and two or more raw materials are combined to produce another product. Note that a "kiln," as it relates to these products for roasting and drying, is considered to be within the scope of the milling definition.

Asphalt-Mixing Plant: Commences after arrival of sand and gravel or aggregate at the plant stockpile.

Concrete Ready-Mix or Batch Plants: Commences after arrival of sand and gravel or aggregate at the plant stockpile.

Custom Stone Finishing: Commences at the point when milling, as defined, is completed, and the stone is polished, engraved, or otherwise processed to obtain a finished product and includes sawing and cutting when associated with polishing and finishing.

Smelting: Commences at the point when milling, as defined, is completed, and metallic ores or concentrates are blended with other materials and are thermally processed to produce metal.

Electrowinning: Commences at the point when milling, as defined, is completed, and metals are recovered by means of electrochemical processes. Salt and cement distribution terminals not located on mine property.

Refining: Commences at the point when milling, as defined, is completed, and material enters the sequential processes to produce a product of higher purity.

APPENDIX B: MSHA DISTRICT OFFICES

Metal/Nonmetal Offices

Northeastern District: Connecticut, Delaware, District of Columbia, Maine, Maryland, Massachusetts, New Hampshire, New York, New Jersey, Pennsylvania, Rhode Island, Vermont, Virginia, and West Virginia. *Address:* Thorn Hill Industrial Park, 547 Keystone Dr., Warrendale, PA 15086-7574. *Telephone:* (724) 772-2333.

Southeastern District: Alabama, Florida, Georgia, Kentucky, Mississippi, North Carolina, Puerto Rico, South Carolina, Tennessee, and Virgin Islands. *Address:* 135 Gemini Circle, Suite 212, Birmingham, AL 35209. *Telephone:* (205) 290-7294.

North Central District: Illinois, Indiana, Iowa, Michigan, Minnesota, Ohio and Wisconsin. *Address:* Federal Building U.S. Courthouse, 515 W. 1st Street #333, Duluth, MN 55802-1302. *Telephone:* (218) 720-5448.

South Central District: Arkansas, Louisiana, Missouri, New Mexico, Oklahoma, and Texas. *Address:* 1100 Commerce Street, Room 462, Dallas, TX 75242-0499. *Telephone:* (214) 767-8401.

Rocky Mountain District: Arizona, Colorado, Kansas, Montana, Nebraska, Nevada, North Dakota, South Dakota, Utah, and Wyoming. *Address:* Denver Federal Center 6th & Kipling, 2nd Street, Bldg. 25, E-16, Denver, CO 80225. *Telephone:* (303) 231-5465.

Western District: Alaska, California, Hawaii, Idaho, Oregon, and Washington. *Address:* 2060 Peabody Road, Suite 610, Vacaville, CA 95687. *Telephone:* (707) 447-9842.

Coal Offices

District 1: Anthracite coal mining regions in Pennsylvania. *Address:* The Stegmaier Bldg., Suite 034, 7 North Wilkes-Barre Blvd., Wilkes-Barre, PA 18702. *Telephone:* (570) 826-6321.

District 2: Bituminous coal mining regions in Pennsylvania. *Address:* 319 Paintersville Road, Hunker, PA 15639. *Telephone:* (724) 925-5150.

District 3: Maryland, Ohio, and Northern West Virginia. *Address:* 5012 Mountaineer Mall, Morgantown, WV 26505. *Telephone:* (304) 291-4277.

District 4: Southern West Virginia. *Address:* 100 Bluestone Road, Mt. Hope, WV 25880. *Telephone:* (304) 877-3900.

District 5: Virginia. *Address:* P.O. Box 560, Norton, VA 24273. *Telephone:* (276) 679-0230.

District 6: Eastern Kentucky. *Address:* 100 Fae Ramsey Lane, Pikeville, Kentucky 1501-3211. *Telephone:* (606) 432-0944.

District 7: Central Kentucky, North Carolina, South Carolina, and Tennessee. *Address:* 3837 S. U.S. Hwy. 25E, Barbourville, KY 40906. *Telephone:* (606) 546-5123.

District 8: Illinois, Indiana, Iowa, Michigan, Minnesota, Northern Missouri, Wisconsin. *Address:* 2300 Willow Street, Suite 200, Vincennes, IN 47591. *Telephone:* (812) 882-7617.

District 9: All states west of the Mississippi River, except for Minnesota, Iowa, and Northern Missouri. *Address:* P.O. Box 25367, DFC, Denver, CO 80225-0367. *Telephone:* (303) 231-5458.

District 10: Western Kentucky. *Address:* 100 YMCA Drive, Madisonville, KY 42431-9019. *Telephone:* (270) 821-4180.

District 11: Alabama, Georgia, Florida, Mississippi, Puerto Rico, Virgin Islands. *Address:* 135 Gemini Circle, Suite 213, Birmingham, AL 35209. *Telephone:* (205) 290-7300.

Chapter Notes

[1] Pub. L. 91-173, *as amended by* Pub. L. 95-164, 30 U.S.C. 801 et seq.

[2] Prior to the creation of MSHA, mine safety and health rules were enforced by the Mining Enforcement and Safety Administration (MESA), which was part of the U.S. Department of Interior.

[3] MSHA/OSHA Interagency Agreement, 44 Fed. Reg. 22827 et seq. (Apr. 17, 1979).

[4] *Id.*

[5] MSHA *Program Policy Manual*, Vol. I-1.4 (May 16, 1996).

[6] MSHA *Program Policy Manual*, Vol. III, 45-3 (May 16, 1996).

[7] For example, MSHA does not have a "distance requirement" above which fall protection must be worn. Rather, MSHA simply requires safety belts and lines to be worn "where there is a danger of falling. . . ." See 30 CFR §§56/57.15005.

[8] 5 U.S.C. §551 et seq.

[9] Independent contractors are not required to have an approved training plan under Part 48, and they may comply with the training requirements by either making arrangements to have their employees trained under an existing approved training plan and program (e.g., through a state grants program agency, a third-party training company with an MSHA-approved trainer, or from the mine operator), or by filing and adopting their own approved training plan. Part 46 training plans need not be approved by MSHA but they must be distributed to employees or posted for their approval at least two weeks prior to provision of any training under this standard.

[10] See, *generally*, MSHA *Program Policy Manual*, Vol. III (May 16, 1996).

[11] See, e.g., *Secretary of Labor v. Western Industrial, Inc.*, 24 FMSHRC 269 (ALJ, 2002).

■■ **About the Author** ■■

MANAGING AND COMMUNICATING IN A CRISIS
by Janine Reid

Janine Reid is the founder of the Janine Reid Group, Inc., in Denver, Colorado. She is the author of three books: *Saving Lives! Proven Methods to Eliminate Job-Site Fatalities, Crisis Management: Planning and Media Relations for the Design and Construction Industry*, and *What To Do When The Sky Starts Falling*. She has also produced four video-tapes on crisis management and working with the news media.

CHAPTER 30

Managing and Communicating in a Crisis

LEARNING OBJECTIVES

■ Explain the definition of a crisis and then identify critical events that would leave your company most vulnerable.

■ Discuss how a crisis can go from bad to worse, and how that progression can be avoided.

■ Explain why a crisis management team is needed and who should participate.

■ Discuss your rights and responsibilities when working with the news media.

■ Identify what critical-incident stress debriefing is and the importance of its use.

INTRODUCTION

The preceding chapters dealt with key issues in the field of construction safety management, focusing primarily on the prevention of accidents. In this chapter, we will shift gears slightly to look at what companies can do to prepare for and effectively communicate should an emergency situation occur. As the U.S. Bureau of Labor statistics clearly establish, construction is an inherently dangerous business. Regardless of how good a company's safety and risk

management program is, it is an unfortunate reality that incidents can and do happen in construction.

Experience has shown that organizations must have a highly coordinated management and communications plan in place to respond to emergency situations. Otherwise, chaos will result when a critical event does occur. Having a plan in place may also help your company to anticipate and prevent some emergency situations from erupting in the first place because of an established heightened awareness of risks. This type of plan is generally referred to as a crisis management plan, and the discipline achieved by responding to these types of situations is considered crisis management planning.

WHAT IS A CRISIS?

Before we go further, we need to define the term *crisis*. A crisis can be defined as any incident that can focus negative attention on a company and have an adverse effect on its overall financial condition, its relationships with its audiences, and/or its reputation within the marketplace. Granted, there will be situations that occur from time to time that will never be known about outside the company. In other words, they will remain internal; the outside world will never hear of them. Nevertheless, there are a whole host of situations that could become known outside the company and potentially spell trouble.

In the construction industry, we often associate the term crisis with accidents involving injury or death; yet, these are not the only situations that would be considered a crisis. Consider all the potential situations that could impact your company, ranging from, but not limited to, defective materials, equipment failure, financial fraud, environmental actions, and natural disasters (see Figure 1).

Potential Crisis Situations

Natural Disaster
- Lightning
- Earthquake
- Extended severe cold/heat
- Extreme snow/ice conditions
- Flood/drought
- Hurricane/Tornado/Tsunamis

Operations
- Accident involving a company vehicle
- Bomb threat
- Construction delay
- Cost overrun
- Data/telecommunications failure/loss of critical data
- Design error/issue
- Sabotage by extremist environmental group
- Explosion
- Fire
- Major utility failure
- Neighborhood/community group opposition to a project
- Structural/subsidence collapse

Environmental Accidents/Liabilities
- Groundwater contamination
- Air-quality problem
- Gas leak
- Long-term exposure of toxic chemicals to the community
- Release of toxic chemicals into the air or waterways
- Mold

Employee Safety and Health
- Chronic safety problem
- Exposure to carcinogens
- Injury/fatality of an employee or nonemployee
- Personal injury suit
- Regulatory citations

Labor Relations
- Negotiations
- Organizing drive

- Unfair labor practices
- Violent strike/work stoppage

Management Issues
- Bankruptcy
- Contractual dispute with client resulting in litigation
- Death of owner or key employee
- Employee raiding by a competitor
- Loss of key customer
- Crisis in the same industry
- Someone else's crisis on your property (guilt by association)
- Kidnap, ransom, extortion
- Hostile takeover attempt
- Key employee starts competing company
- Management succession
- Merger/acquisition
- Negative publicity due to rumors
- Negative publicity relating to political contributions
- Reorganization/downsizing
- Serious cash-flow problems
- Sudden market shift
- Suicide
- Terrorism

Employee/Management Misconduct
- Bribery/kickbacks
- Disgruntled employee
- Executive misconduct/fraud/embezzlement
- Lawsuits from discrimination, sexual/racial harassment
- Murder
- Price fixing
- Sabotage
- Scandal involving top management
- Slander
- Suicide
- Theft/vandalism
- Workplace violence

Government Affairs
- Legislation and/or regulations that could affect business

■ ■ **Figure 1** ■ ■ **Crisis audit**

Can You Prevent Crises from Occurring in the First Place?

Once you have identified the situations that have the greatest potential for occurring in your workplace, take steps to prevent them from arising in the first place. In the large majority of cases, applying an increased awareness can do this. The only exception might be a natural disaster (an Act of God); however, advanced technology can forewarn us of most impending dangers and allow us to plan for them and the subsequent havoc they create. In man-made construction crises, we often times get early signals, but most of the time those signals are ignored. Here is an example.

Let us suppose that a general contractor is in the structural steel phase of a 15-story commercial office building located in a major metropolitan area. There are twelve subcontractors on site for a total of approximately 95 employees. It is 7:00 AM on a Tuesday morning, and a crane is being used to lift steel beams, which are being set by a team of ironworkers. This is all of the information needed for the most senior person on site to ask, "What could go wrong on this project today?" The logical answer might be to focus on the obvious issues related to the equipment; however, if we conduct a crisis audit we may determine that a whole host of situations could present themselves:

1. Equipment failure causing structural damage, injury, and/or death
2. Injury/death caused by unsafe safety practices
3. Neighborhood opposition to the project
4. Accident involving a company vehicle
5. Labor unrest
6. Fire/explosion
7. Chemical spill/hazardous-material release
8. Disgruntled employee/workplace violence
9. Harassment/discrimination issues
10. Complaints to the media about the project
11. Loss of a key subcontractor or supplier
12. Bomb threat
13. Schedule issues impacting costs and the relationship with the owner

A project is fluid and changes daily because of its schedule, workforce, equipment, materials, chemicals, and so forth. Therefore, it should become a daily habit for the most senior person on site to ask, "What could go wrong on this job today, and what needs to be done to prevent it?" It is vital for the project team to work closely with safety personnel who are trained in risk identification. Consistent attention to this collaboration is the key to eliminating or minimizing a potential job-site accident or other crisis event. This collaboration must begin before dirt is first scratched and cannot be terminated until the keys are turned over to the owner.

Can all risks be identified? Although there may be some "surprises," for the most part all members of a project team, as well as upper management, should be so intimate with their respective areas of responsibilities that "surprises" should be minimal.

How Can a Crisis Go from Bad to Worse?

Every crisis is unique and takes on a life of its own; even so, it is possible to anticipate possible reactions or effects of a crisis. A chain reaction will most likely occur as a result of a crisis, and "spin-off" crises can sometimes create more damage than the initial crisis itself. When an accident occurs and it "hits the fan," one must be able to look further than the initial crisis and determine the chain of events that could occur. This foresight can be accomplished through a process called a "what-if analysis" (see Figure 2).

THE WHAT-IF ANALYSIS

In a what-if analysis, you assume the role of a negative creative thinker and take a look at the domino effect that could occur as a result of a crisis. Envision the events and actions that could occur as a consequence of a crisis. Remember, it is not necessarily the initial crisis that can deal the crippling blow to your company; it could very well be an event that occurs as a result of the initial crisis and that your company could have identified and avoided.

To anticipate all of the scenarios, construct a what-if diagram that takes the form of a branched tree where each branch documents the possible occurrence of an additional crisis. In developing the various scenarios, the main question to ask is, "What is the worst that can happen?" The what-if diagram

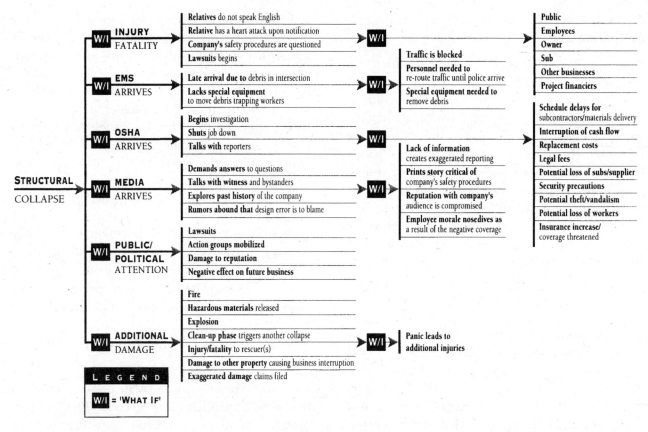

■■ Figure 2 ■■ What-if analysis (Source: Janine Reid Group, Inc.)

becomes a graphic presentation of a possible crisis progression. Because of its graphic nature, this diagram should help to track each of the many conceivable crisis developments. This exercise is best accomplished in a noncrisis environment when all possible spin-off crises can be anticipated without panic.

Do You Need to Examine the Impact of Your Crisis on Your Company and Various Audiences?

When a crisis occurs, it may have far-reaching consequences on your various audiences. An *audience* is defined as anyone who can have an effect on your business and/or reputation. The following list delineates some potential audiences for a construction company:

- shareholders/investors
- stock analysts
- board of directors
- employees and their families
- city/county officials
- site neighbors
- opinion leaders
- action groups
- the media in general and trade press
- governmental regulators
- financial community
- clients (current, past, potential)
- other contractors, architects, engineers
- unions
- suppliers
- insurance/bonding company.

Decisions, which are made rapidly during a crisis, can have tremendous impacts on your audiences. Determine what effect the crisis and your company's actions will have on these areas:

- your operations
- the general public
- the industry
- your financiers.

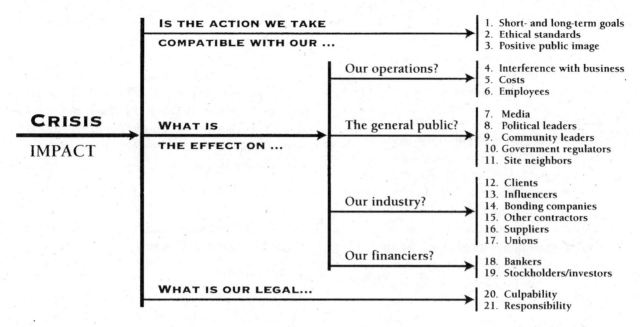

CRISIS
IMPACT

IS THE ACTION WE TAKE
COMPATIBLE WITH OUR ...
1. Short- and long-term goals
2. Ethical standards
3. Positive public image

WHAT IS
THE EFFECT ON ...

Our operations?
4. Interference with business
5. Costs
6. Employees

The general public?
7. Media
8. Political leaders
9. Community leaders
10. Government regulators
11. Site neighbors

Our industry?
12. Clients
13. Influencers
14. Bonding companies
15. Other contractors
16. Suppliers
17. Unions

Our financiers?
18. Bankers
19. Stockholders/investors

WHAT IS OUR LEGAL...
20. Culpability
21. Responsibility

■ ■ **Figure 3** ■ ■ **Crisis impact questions** (Source: Janine Reid Group, Inc.)

Communicating with your audiences is vitally important, especially if the news media becomes involved, which is discussed in more detail later in this chapter. Rest assured that all of your audiences will be paying very close attention to your actions and the impact they will have on them (see Figure 3).

Do You Need a Crisis Management Team?

There are two major goals to shoot for when a crisis occurs: *organize internally* to get the situation under control as quickly as possible and *communicate externally* your side of the story and what you are doing about the situation. A crisis management team is essential if you are to achieve these goals. Having worked on a variety of crisis situations, I have found that it is virtually impossible for *one* person to handle all of the tasks necessary to effectively manage a crisis. No matter what the size or scope of the crisis, it will require a variety of people with different areas of expertise to ensure that all facets of the problem are covered. Therefore, a team approach to crisis management is required.

Whether you are an owner/developer, program manager, construction manager, engineer, architect, general contractor, or subcontractor, your crisis management team should become a permanent unit

that can draw upon internal and external resources. The objective is to have a knowledgeable group that will work effectively and quickly in a crisis at both the corporate and project levels.

Who Are the Members of the Core Crisis Management Team?

It is recommended that you select your core crisis management team based on the largest potential crisis that could befall your company. This statement usually provokes the response, "What if I run a lean company and have only four people on my salaried payroll? I simply don't have enough people to fill a crisis management team!"

Good point. Nevertheless, it is important to understand that crises do not discriminate between small and large companies. Whether you have four salaried employees or six hundred, certain objectives need to be accomplished when a crisis hits. Regardless of whether you use internal or external sources, those objectives do not change because of the size of your company.

Your team should be built envisioning the most catastrophic crisis that could happen to your company. It should consist of employees and/or outside consultants who not only possess their respective areas of knowledge, but who also are willing to work on the team.

Let's start with the selection of your core crisis management team. The selection of a core team is based on two very important factors:

1. Each team member should possess an area of expertise that is useful to the team. For example, each member must have good organizational capabilities, have the trust and confidence of upper management, possess solid communications skills, and have the respect of the company's employees.
2. The team members should work well together in normal day-to-day situations. If they do not, whatever problems exist among them will be exacerbated in a crisis.

There are four members of your core crisis management team and two support members. The four core members are:

> Team Leader
> Spokesperson
> President
> Legal Counsel

These four members will be involved in any event classified as a crisis and will mobilize other team members as the situation warrants. Figure 4 shows the core crisis management team, along with support personnel and possible outside help that may be activated. Because the chain of command during a crisis may alter the company's usual organizational chart, all such changes must be clarified during the selection process *before* any crisis occurs.

What Are the Responsibilities of the Core Crisis Management Team?

TEAM LEADER

The team leader is the company's *internal organizer* during a crisis and will select and mobilize a team to control the crisis as rapidly as possible with the fewest complications. All team members will report to the team leader, who should be a member of mid-to-upper management and have a working knowledge of the company's short- and long-term goals. Making rapid decisions will be the norm for the team leader and will require that he or she has access to,

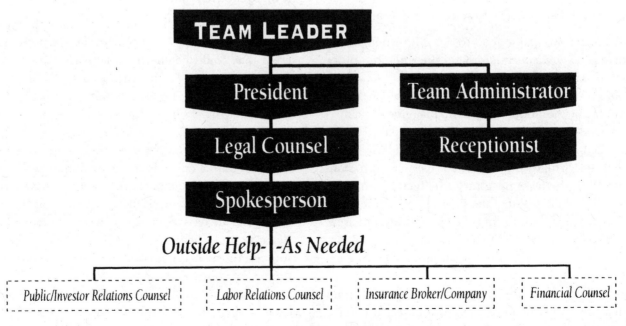

■ ■ **Figure 4** ■ ■ **Core crisis management team** (Source: Janine Reid Group, Inc.)

and the confidence of, the company's president and other top managers. Many companies will have a number of offices outside of corporate headquarters. In this case, there must be a team leader at each location and that team leader will communicate directly with the corporate team leader in the event of a crisis.

Select a backup team leader (in all offices) just in case the primary team leader is on vacation, at a remote job site, or is otherwise unavailable. Many companies will have two to three backups to make sure the coverage is there all the time.

SPOKESPERSON

The spokesperson is the company's *external communicator* to its various audiences, as described earlier. While each of a company's audiences must be contacted during a crisis, the primary responsibility of the spokesperson is working with the news media should they become involved in the crisis.

Working with the news media is an area where companies seem to get into trouble if they are not prepared. In today's business arena, any company that does not take the time to understand the media, learn its needs, and become comfortable dealing with media probably will get poor coverage or inaccurate, one-sided coverage. The penalty for mishandling a situation can be severe because there is no practical recourse to set the record straight. This penalty drives home the importance of selecting and training a spokesperson *before* your company finds itself in the spotlight. If you wait until a crisis occurs to make your selection, the results could be less than favorable.

The best candidate for the position of spokesperson is an individual who already speaks to reporters or outside groups on behalf of your firm. If you do not have a qualified candidate in your firm, now is the time to develop and train one.

As with the team leader, a backup spokesperson is highly recommended, not only to take on the responsibility if the primary spokesperson is out of town, but also to relieve the primary spokesperson if necessary. During a large crisis, for instance, the primary spokesperson can become overwhelmed and require some relief.

PRESIDENT

Appointing the company's president to the role of team leader or spokesperson is not recommended. A crisis will make heavy demands on time, and the president will need to be available for other duties. The president, or another company official, must make it his/her responsibility to allocate time to follow the crisis through to its conclusion. The responsibilities for the president or another member of upper management are:

- Work closely with the team leader and legal counsel to determine the company's direction and position on the incident.
- Review and approve all statements released to the media.
- Determine which of the company's audiences are affected by the incident and ensure that communication with these audiences is established and maintained throughout, and beyond, the crisis.
- Personally notify the next of kin in the event of a fatality of a company employee. This is discussed later in the chapter.
- Be prepared to make the initial statement to the news media (taking no questions) in the event of a tragic incident to show concern and personal involvement.
- Approve the crisis management plan and all additions and/or changes that occur during periodic updates.

LEGAL COUNSEL

Your legal counsel should *not* be used as a spokesperson because using a lawyer in that position may be interpreted as a sign of guilt. Legal counsel may also have a tendency to withhold information in an effort to "protect" the client; however, because the court of public opinion can be very unforgiving, the message intended by a lawyer may not be the message received by the media and then communicated to your audiences and the general public.

That said, your legal counsel should have a full understanding of your crisis management plan as well as the external communication demands that will become part and parcel of managing a crisis.

From an internal perspective, your legal counsel needs to be contacted at the first indication of a crisis and kept apprised of all decisions during a crisis because their guidance could prove invaluable. Your legal counsel, whether in-house or on retainer, should also have a current copy of your crisis management plan and become familiar with the players on your team.

As mentioned, there are two supporting members of your core crisis management team. They are the team administrator and the receptionist. The following will detail their responsibilities.

TEAM ADMINISTRATOR

The team administrator is the right hand of the team leader and is a critical support member of the crisis team. The responsibilities of the team administrator are as follows:

- Provide support to the crisis team [e.g., assisting in the screening of phone calls, organizing strategy meetings, maintaining communication with all team members, arranging for support services for the family (or families) of the injured, dispatching a critical-incidence stress counselor to witnesses and personnel, and performing clerical tasks].
- Under the direction of the team leader, maintain a timeline of the incident with a chronology of events. This chronology will prove to be invaluable during the debriefing of the crisis with affected stakeholders.
- Update and distribute changes and additions to the crisis management plan on a quarterly or as-needed basis.
- Work with the receptionist on how to route calls during a crisis.

RECEPTIONIST

The individual handling the phones during a crisis situation is critical and is often one of the last to be given instructions regarding what to do. It is the responsibility of the team leader and/or the team administrator to contact the receptionist at the first

warning of an impending crisis and to provide the receptionist with the following directives:

- Route all requests from the media to the individual who is screening calls for the spokesperson.
- Never release any information to outside callers—no matter how hard you are pushed.
- Do not allow a reporter to roam about the office, property, or job site unescorted—always notify the spokesperson of a reporter's presence as soon as possible. Determine, in advance, where media interviews will be held.

Who Are the Members of a Crisis Management Team for a Project?

From a project standpoint, the crisis management team should be selected prior to job start-up; however, it is never too late to develop a team. There should be no uncertainty about who is in control should a crisis occur. A good rule to follow is that *the highest entity on the project has the responsibility to select the crisis management team for the project* and then communicate the assigned responsibilities to the respective parties.

First, define who is in charge in the event of a project-related crisis. Among this group, whoever reports to the project owner/developer should be in charge in the event of a crisis. That rule does not absolve the other entities on the project from planning for a crisis. Every company involved should have a protocol to follow in the event of a crisis, but at the outset of the project it must be determined who has ultimate control in a crisis.

As previously discussed, it is important to establish a team based on the largest and most complex crisis that could potentially happen to your company and, in this case, on one of your projects. See Figure 5 for an example of a potential project crisis management team.

Why Is the Crisis Management Plan So Important?

Ugly headlines, loss of business, tarnished reputation, poor employee morale, these are a few of the negative effects that result from a poorly handled crisis.

A crisis management plan is an integral component of a company's safety and risk management program and should be part of strategic planning. I have worked on dozens upon dozens of construction-related accidents and other crises and have found there is absolutely no comparison between a company that has a written plan and employees trained to use it and a company that has not prepared in this way. The company that has included a crisis management plan as part of its risk and insurance management program stands a greater chance of maintaining control during a crisis, retaining customer and employee loyalty, minimizing its financial losses, and quickly restoring its operations than the company without a plan. The prepared company can react more quickly with an increased ability to think through action items from a number of different perspectives. The company that does not have a protocol to follow will compromise such matters and, perhaps, fall victim to half-baked decisions.

Still, a crisis management plan is not a panacea, nor is it a guarantee that no one will point an accusatory finger in your direction. Nevertheless, it will take the *panic* out of the situation and give you the presence of mind to be proactive rather than reactive during the rush of events.

Because no two companies are alike, the following information is presented to serve as a guideline for customizing your specific plan. These ideas are being offered to stimulate your critical thinking skills so you can make the best out of a difficult situation. Your job is to determine what should be included in your specific plan and what modifications are necessary to serve your particular needs.

No two crisis situations are alike; however, there are common patterns that surface in many crises. Instead of trying to develop a plan for every conceivable crisis, consider the commonalties that all of those crises may have and develop your plan around them.

What Should Be Included in Your Written Corporate Plan?

The secret to an effective crisis management plan—one that will *actually be used* during a crisis—is to keep it simple. Let me repeat—keep it simple! No one has time to read a philosophical thesis when the world is crashing down around them. Bullet-point statements using an easy to read (large) font and

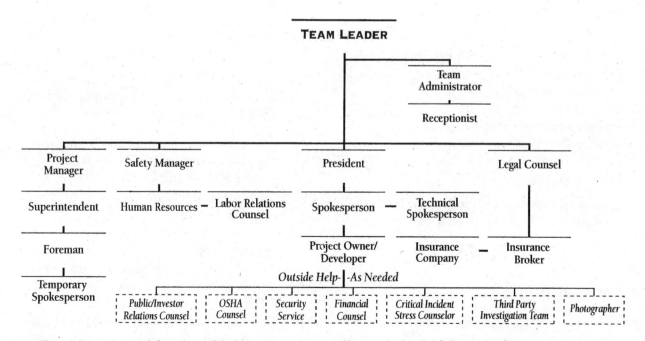

■ ■ **Figure 5** ■ ■ **Potential *project* crisis management team** (Source: Janine Reid Group, Inc.)

flowcharts can simplify plans, make them easy to read, and assist in accessing required information quickly. Remember, the reader of the plan is usually accessing this document during a period of stress, so the easier the plan's format, the more effective it will be.

With that in mind, here is a menu of items to consider for inclusion in your corporate crisis management plan. Use labeled tabs so that individual sections can be quickly accessed.

Cover sheet
Control sheet
Table of contents
First-hour response checklist
Contacts in the event of an emergency
Salaried-personnel contact list (office, home, cellular telephone, and pager numbers)
Internal notification procedures (who contacts who)
Notification procedures in the event of a serious employee injury
Notification procedures in the event of an employee fatality
Procedures for who will meet with governmental compliance officials
Procedures for identified crises
Responsibilities of the crisis management team
Community relations program
Tips on working with the news media
Safety history
Personal safety for international travel
Past emergencies
Company fact sheet
Project data sheets
Foreign language expertise
Emergency procedures contact list
Incident report form(s)
Evacuation plan

Choose the areas that would be most beneficial, and customize each section to your company and/or project.

WHEN A CRISIS HITS HOME

As an example, let's assume that a job-site accident has occurred on a project. This is what your *first-hour response checklist* might look like:

Most Senior Person on the Site

In the event of a job-site accident, it will be assumed that emergency services have been notified immediately. Subsequently, the most senior person on site has the following responsibilities:

- Contact and coordinate whatever emergency services are necessary.
- Secure the site and determine if it should be shut down.
- Contact the team leader and safety manager and request their assistance.
- Make certain that any and all evidence remains intact and unmoved.
- Work with subcontractors in obtaining a head count and identifying witnesses.
- Arrange for the witnesses to be debriefed.
- Determine, with the assistance of the team leader, who will make the notification to the next of kin in the event of an employee injury/fatality, and arrange for bilingual communication if needed.
- Post workers to restrict entry to the site until it is deemed safe.
- Notify the owner/developer of the project.
- Instruct the job-site receptionist on how to route phone calls.
- If a temporary spokesperson is needed, work with the team leader to determine the best candidate.

Team Leader

If a job-site accident occurs, the responsibilities of the team leader are as follows:

- Assess the situation quickly by asking the five famous questions of Who? What? When? Where? and How?
- Mobilize a team based on the answers to the above questions.
- Notify and work closely with upper management.
- Advise the team administrator and receptionist on how to route calls—both at the job site and corporate level.
- Determine what spin-off crisis could occur by using the what-if analysis technique described earlier in this chapter.

- Communicate with all employees via fax, e-mail, and/or voice mail if the crisis is getting outside attention. Tell them how they should handle requests for information and the person to whom any requests should be referred.
- Notify the human resources department if the situation calls for it.
- Notify the insurance broker/company if the situation warrants.

Safety

If a job-site accident occurs, it is the safety manager's responsibility to assist the project manager and/or superintendent/foreman in securing the site and ensuring that any further risks are eliminated. Additional safety manager responsibilities are as follows:

- Notify the necessary authorities.
- Begin the debriefing process with any witnesses and contact a critical-incident stress counselor (described later in this section) if necessary.
- Determine if the company should initiate a third-party investigation team to work with the authorities.
- Document the incident both in writing and on film.
- Liaison with the medical facilities should the incident involve an injury or fatality.

This checklist can be lengthened to include responsibilities for other members of your crisis management team.

What If a Reporter Shows Up On Site?

When a job-site incident or other crisis occurs, it tends to generate chaos, and this chaos will create an urge to stonewall the media. After all, the contractor has a site to secure and people to take care of, so the natural response would be to say, "No comment." However, this response screams "Guilty as charged!" Another thought that enters the minds of the crisis management team is, "To heck with reporters—they can wait!" Of course they can wait, but are a company's best interests served by this attitude?

From the media's perspective, they have a news story to deliver NOW. Wouldn't you prefer that your spokesperson provide whatever facts are available rather than have a bystander on the street tell your story? Case study after case study shows that reserving comment during those critical first hours of a crisis does not present a good company image in the media. Another option that the crisis management team may explore is to hold off on any comment until all of the facts of the incident have been discovered and verified. This course of action could present a problem from a communication standpoint because the investigation process in a crisis of this size could take considerable time.

Finally, some companies think that if their projects are located in remote areas, reporters will not find out about accidents. If, by chance, a reporter does find out about an incident, a company might mistakenly believe that the coverage would be restricted to the local media. Unfortunately, that line of thinking is not an option in today's technologically endowed societies. In a world of 24-hour news and instant Internet access, our goal in crisis communication is to be as open as possible as quickly as possible. To do otherwise will raise the red flag that information is being hidden. This action could also cause reporters to become very aggressive in their quest to discover that perceived, hidden information. Understand, however, that speed should not sacrifice organization and preparation.

Perception is reality, and the first message of a company in crisis to its audiences is the most crucial. A company must project the image of being proactive and of being in control of the flow of verifiable information. But how can you do that when you have precious little information? You deliver a statement that will "buy time."

The *buy-time statement* is one that a temporary spokesperson—either the senior person on site or the person assigned by the senior person—will deliver as the first statement about the incident to the media. It should accomplish the following goals:

- Acknowledge the incident.
- Buy time to gather more information and get the corporate spokesperson on site to take over the responsibility of working with the reporter(s).

A buy-time statement should sound like this:

My name is (_____) and I am (title) with (company name). The incident has just happened and I am not prepared to answer any questions at this time. Please stay in this safety area so we can do our job and take care of the situation. I need to return to the site, but either (spokesperson) or I will be back at (time) with an update. Thank you.

It is critically important not to take questions at this time. If badgered, simply state that you need to get back to the site and you will return at the stated time.

Let's fast-forward a few hours and assume that a buy-time statement has been delivered by your superintendent and now the media wants more information. Let's also assume that your spokesperson cannot get to the scene fast enough to deliver a second statement; however, he/she is available by cell phone and can help the superintendent craft this second buy-time statement.

After fully reviewing the latest *verified* information, the spokesperson and superintendent should identify the key points that the company wishes to communicate. Here are three key points that should be common in all statements:

- *Cause:* "At this time, we do not know the cause of the failure; however, we are working very closely with the authorities to find out exactly what happened."
- *Concern:* "We are so sorry to report that three employees were injured and are being transported to University Hospital. We are notifying the families and staying in close contact with the hospital."
- *Status:* "The site has been secured, and all employees are accounted for. Our corporate spokesperson, (name), is en route to the site. We will have an information update in one hour at (location)."

Now, based on the information that has been gathered and verified, as well as identifying our key points, our statement might read as follows:

My name is Justin Reed—that's J-u-s-t-i-n R-e-e-d— and I am the superintendent on this project for XYZ

Construction Company. Here is what I can confirm at this time.

At 3:30 PM this afternoon, while being lifted by the crane to the roof of the building, a cooling tower broke loose at the eighth floor. We do not know the cause of the failure and are sorry to report that it resulted in three employees being injured. Those employees are currently being transported to University Hospital, and our immediate focus is on notifying the families and determining the extent of the injuries. We will release the names of the injured workers as soon as the notification process has been completed.

Again, we do not know the cause of the failure, and we are working closely with the authorities to find out exactly what happened. The site has been secured, and all employees have been accounted for. John Smith is our corporate spokesperson, and he should be arriving shortly. We will both be your points of contact. If you elect to stay on site until the next update, we ask that you remain in this safety area. Right now, I need to return to the site to provide assistance and gather more information. John Smith or I will return in one hour to provide you with any updated and verified information.

Again, do not entertain any questions at this point.

In analyzing this statement you will note that the content is quite sketchy and merely verifies what the reporter can see. However, this statement will accomplish two things:

1. It will establish a working relationship with the reporter(s) and show that you will be forthcoming with new information as it becomes available and has been verified.
2. It will allow you to buy one hour of time to gather some additional information for the next interview. In large crises, you can get by delivering two buy-time statements without entertaining questions. After that, you need to be ready for Q&A.

One of the most critical parts of the buy-time statement is the close. When you have finished your statement, walk away and *do not* entertain any questions. Remember that the crisis has just happened, and verifiable information is almost nonexistent. As the temporary spokesperson, you will be vulnerable to potentially explosive questions that cannot be answered at this time. Clearly, reporters will not appreciate the fact that the spokesperson will not entertain their questions; however, a question-and-answer session at this juncture could create a very uncomfortable situation because of a real lack of information.

Whether you are the temporary or the corporate spokesperson, you can *read* the statement to avoid getting off track. This technique also helps to relieve some of the pressure caused by anxiety. The statement can then be copied and distributed to the reporter(s) to minimize the possibility of misunderstandings or misinterpretations that can result from a verbal presentation. Background information on your company and project can also be included.

You have probably noticed that one significant thread has been woven through the fabric of the first two statements, and it must remain within every statement when human life is involved. That is the thread of concern and compassion for those who have been affected by the crisis. This concern should continue to be the *primary focus* in all statements. Any company that avoids expressing concern and/or compassion for people hurt, killed, or inconvenienced will appear cold, or worse yet, calculating. This perception must be avoided at all costs.

That important issue considered, it is also important to be prepared to handle any questions a reporter may throw at you during the next interview. To prepare, think about the "ugly" questions, and develop responses to answer them. If possible, assemble a team of employees to rehearse the interview, having employees play the role of reporters. Certain interviewers will ask leading, confrontational questions that can lead inexperienced spokespersons into losing their focus—and sometimes their temper. The key is to stay focused during oral interviews, take plenty of time to respond, think about the answer, and try not to talk too fast. Without appearing misleading, you do not have to "take the bait" and answer the confrontational question with a direct answer. Instead, you can answer it with *facts* as you know them.

How Do You Communicate with Your Various Audiences When the News Is Bad?

As discussed earlier, an audience is defined as anyone who can have an effect on your business and/or

reputation. When a crisis strikes, insurance companies may cover the immediate damage, but their coverage does not include reclaiming your reputation and credibility in a competitive marketplace. This effort is your responsibility; it should be taken seriously and addressed quickly at the outset of a crisis. If you do not take the time to communicate your side of the story and explain what you are doing to your various audiences, they will have no choice but to believe what they read and hear from the news media and your competitors. Conversely, open and honest communications, executed quickly and consistently throughout the crisis, can maintain and perhaps enhance the company's reputation and credibility with its various audiences.

Take a moment to think about which audiences could have an effect on your business and/or reputation in the event of a crisis. Next, customize this list to your company by adding additional audiences or deleting ones that do not apply. A great resource to assist you in this process would be your marketing contact list—sometimes referred to as a marketing information database. If you do not have such a list, now would be a good time to develop one.

Once you have identified key companies within each audience, you should hold someone within your company responsible during a crisis for making contact with the key influencer from each company. The ideal match is someone who already has an existing relationship with that influencer.

It is a good practice to provide an update to your targeted audiences every time you have new information that has been released to the news media. That way, everyone is receiving the same message at the same time.

At some point, believe it or not, the media attention will begin to subside. Please do not believe for one instant that your various audiences are going to forget about your crisis just because it is not being reported in the papers or on local news stations. You must always be prepared to address any questions that may arise regarding the crisis. Also, make certain that your employees are always kept in the information loop about the status of the injured and the accident investigation because your employees are an important communication vehicle to the outside world.

There is never a reason to create external concern if an issue can be contained internally. However, news does have a way of "leaking," so it is wise to be prepared for information to wander outside your doors into the ears of your respective audiences. Should this leaking occur, be prepared to plug it as quickly as possible through direct and consistent communications with the interested parties.

CONCLUSION

The ultimate goal of an effective crisis management program is never having to use it. To ensure this outcome, an environment of risk-awareness must be part of a company's culture. No one is immune to risk; however, identification leads to prevention, and an alarm sounded early can prevent a possible nick in the company's armor. The key points addressed in this chapter can help a company fend off a disaster.

Develop a Crisis Management Plan Prior to a Crisis
Developing a crisis management plan prior to a crisis may sound like common sense; however, many companies feel they are immune to negative situations, so when the unthinkable happens, they fall victim to a corporate paralysis. In contrast, a company with a crisis management plan can avoid such paralysis because the plan will have prepared the company under siege. This is not to say that a plan will address every possible catastrophe that could befall a company. No plan can accomplish that task, but a well-thought-out crisis management plan will provide your company with the direction needed to make better and faster decisions during any crisis.

On the flip side, if your company has done something wrong, a plan will not make it right. However, a plan can help a company make the best out of a bad situation.

Keep Your Crisis Management Plan Current
You may spend weeks organizing your plan, researching information, and obtaining background materials. Do not become complacent and allow your plan to become outdated because nothing is more frustrating than needing information in the crunch of a crisis only to find out that the information is no longer valid. Instead, once developed, your crisis management plan should be reviewed and updated on a quarterly basis. Updating the plan should be a delegated assignment for an individual, specifically the team administrator.

Practice Your Plan
All salaried employees should receive training a minimum of once a year on what to do should catastrophe strike. Crisis communications training may be combined with other emergency response training as a "hands-on" or "table-top" exercise. This is an opportunity for the management team to remain familiar with the overall company emergency plans. During training and exercises, the company's actual written plans should be used to promote familiarity and identify discrepancies or shortcomings. All employees must understand that they have the ability to set the stage for a positive outcome or, conversely, to create so much negativity that the company will have an extremely difficult time just "breaking even." Training is critical for those times when an individual either cannot find the plan or elects not to use it. At such times, training is the only thing that stands between a company and chaotic disaster. The sports analogy that says, "You will play the way you have practiced," describes the way a well-trained crisis management team will rise to the occasion. Training will also reinforce the critical thinking skills of those on the team and help them remain focused during a crisis rather than default to emotional responses.

The company spokespersons should receive refresher training at least twice a year to keep skills sharp. The spokespersons must be fully prepared to stand in front of a group of reporters, with the hot lights in their eyes, microphones in their faces, and questions coming at them from all different directions. The feeling that this experience creates can be paralyzing; however, frequent media training will reduce much of the anxiety a reporter can create. Your worst day can be a reporter's best day, and an untrained spokesperson is juicy prey for a reporter.

Your Actions During a Crisis Will Be Closely Scrutinized by Your Various Audiences

In a crisis, always maintain open and honest communication with your various audiences because they will all keep a very close eye on your actions or lack of them.

> *In the event of an employee injury or fatality, notification of the employee's family must occur as soon as possible. Never release the names of an injured or deceased individual without first notifying the victim's family.*

I once worked with a company that experienced a job-site accident that resulted in a serious injury. As the Flight for Life Helicopter left the site, I approached the owner of the company and said that we needed to contact the employee's family immediately. The owner was not comfortable with making the call until he had a chance to speak with the attending physician. He felt as though he needed some solid information to communicate to the family before initiating the call. Unfortunately, my pleas for reconsideration went unheard, and the employee died two hours after admittance.

Again, practice common sense; put yourself in the position of the family and determine when you would want to receive notification. If the life of a member of my family were in peril, I would want to know as quickly as possible. To be notified any later is unacceptable.

Witnesses to an Accident Should Be Offered Critical-Incidence Stress Counseling

Research has shown that individuals who witness an accident can experience substantial physiological and psychological effects that can linger for months, or longer, if not addressed. Seek critical-incidence stress counseling for the witnesses and the project team within 24–48 hours of the incident. The goal is to return an individual to work, and a normal life, as quickly as possible.

Do Not Let a Crisis Overwhelm You

Easy to say, hard to do! A crisis can be an extremely overwhelming event to manage, and it will not hesitate to take control of your company—if you allow it to do so. When a crisis strikes, the team leader must take control and logically break the event down into manageable pieces. These "pieces" are then further reduced to a series of action items to be handled by the crisis management team and outside resources. This approach will reduce the anxiety that is caused by an otherwise unorganized approach to managing a crisis.

Expect Mistakes

If a crisis occurs, try to remember that no one is perfect under such duress. The only crime in making a mistake is failing to take action quickly and responsibly to correct the situation.

REVIEW EXERCISES

1. Is it possible to prevent a crisis? How?
2. Can the team leader and spokesperson roles be assigned to the same person? Why or why not?
3. If a crisis were to occur on a job site, what are the immediate responsibilities of the most senior person on site?
4. If a reporter walks on site during the first few hours of a crisis, what should happen and who is responsible for making that action happen?
5. If the media prints/airs a negative story relative to your crisis, what effect would this have on a company's internal and external audiences, and what action should be taken to minimize or eliminate that effect?

■ ■ **About the Author** ■ ■

CONSTRUCTION SAFETY AND HEALTH CERTIFICATIONS
by Delmar E. "Del" Tally, P.E., CSP

Del Tally, P.E., CSP, is a past president of the American Society of Engineers (ASSE) 1985–1986, and was a member of the ASSE Board of Directors until 1996. He served in the United States Air Force as a Lieutenant Colonel where he gained experience in aviation, nuclear, missile, and occupational safety. Subsequently, Mr. Tally was an OSHA Compliance Officer and a construction industry SHE manager. In addition to earning his CSP in Comprehensive Practice from the Board of Certified Safety Professionals (BCSP), he also earned a specialty in Construction Safety by BCSP examination.

Mr. Tally has served on the National Construction Advisory Committee to OSHA (ACOSH) and many other SHE committees and boards at local, state, and national levels. In 1997 Del Tally was elected an ASSE Fellow, the Society's highest lifetime achievement award.

CHAPTER 31

Construction Safety and Health Certifications

LEARNING OBJECTIVES

- Identify and distinguish the construction safety and health professional certifications that meet the nationally recognized requirements of the National Commission of Certifying Agencies and the Council of Engineering and Scientific Specialty Boards.
- Discuss the safety and health certification and the requirements for certification of construction craft supervisors, (i.e., Safety Trained Supervisor in Construction).
- Explain the safety and health certification and the requirements for certification of construction industry safety and health practitioners as a Construction Health and Safety Technician.
- Discuss the safety and health certification and the requirements for certification of construction safety and health professionals as a Certified Safety Professional.
- Discuss the safety and health certification and the requirements for certification of construction safety and health professionals in a construction safety specialty.
- Discuss other construction industry certifications that affect the overall safety, health, and environmental considerations that impact overall construction job-site safety.

INTRODUCTION

This chapter is written to give the construction industry safety, health, environmental (SHE) persons information on quality, professionally recognized, and accredited SHE certifications.

In the United States, there are approximately two hundred SHE certifications and designations of varying grades. Some only encourage or document training courses, and most of these have not met the national certification guidelines of the National Commission for Certifying Agencies (NCCA) or the Council of Engineering and Scientific Specialty Boards (CESB). These agencies are established to set standards for ensuring unbiased and quality certifications in many different work and specialty areas, including SHE certifications.

While researching materials for this chapter, the Internet produced some very interesting information concerning state and federal governmental agencies that have various authorities dealing with quality. It was interesting that most (maybe all) states have agencies that determine which universities offering various degrees are acceptable in their state. This means that Human Resource (HR) departments and employers can quickly check Internet sources to determine the validity of degrees earned and/or SHE certifications.

This is a major reason for professionals to ensure that the time, money, and effort put into professional development produces a quality and acceptable degree or SHE certification.

This chapter does not allow for discussion of every SHE certificate and/or certification program in the United States; consequently, it will address those which seem to have the greatest potential impact on the construction industry and the SHE

persons performing their duties in the construction industry. These certifications will also have met the standards of the National Commission for Certifying Agencies and/or the Council of Engineering and Scientific Specialty Boards (see Table 1).

CONSTRUCTION SAFETY PROFESSIONAL

Are you a construction safety person with no safety, health, or environmental degrees or certifications? If the answer is "yes," you are one of the many construction safety persons that have not been recognized for your training, your responsibilities, and your accomplishments.

This chapter will provide the information, procedures, and, hopefully, the inspiration to get the recognition, satisfaction, and certifications needed

■■ Table 1 ■■

Nationally Accredited Safety, Health, Environmental, and Ergonomics Certifications

Title	Operated By	Level	Accredited By
Safety Trained Supervisor (STS) in Construction	CCHEST*	Worker	CESB
Certified Crane Operator (CCO)	National Commission for Certification of Crane Operators	Worker	NCCA
Occupational Health and Safety Technologist (OSHT)	CCHEST*	Technician/ Technologist	CESB
Construction Health and Safety Technician (CHST)	CCHEST*	Technician/ Technologist	CESB
Certified Safety Professional (CSP)	Board of Certified Safety Professionals	Professional	NCCA & CESB
Certified Industrial Hygienist (CIH)	American Board of Industrial Hygiene	Professional	CESB
Certified Health Physicist (CHP)	American Board of Health Physicists	Professional	CESB
Qualified Environmental Professional (QEP)	Institute for Environmental Professionals	Professional	CESB
Certified Hazardous Materials Manager (CHMM)	Institute for Hazardous Materials Management	Professional	CESB
Diplomate in Environmental Engineering (DEE)	American Academy of Environmental Engineers	Professional	CESB
Certified Occupational Health Nurse (COHN)	American Board of Occupational Health Nurses	Professional	American Board of Nursing Specialties

*Council on Certification of Health, Environmental, and Safety Technologists

to prove to yourself, employers, and peers that you are truly a construction safety professional.

Every recognized profession (i.e., engineers, doctors, lawyers, pilots, electricians, pipefitters, etc.) has established formal training, apprenticeship requirements, or college degrees, plus internships or on-the-job (OTJ) training requirements. Many of those also require licensing or registration before a person begins public or private practice.

Safety has evolved as a profession. However, it is only recently that the industry has decided to establish professional requirements, college-degree accreditation procedures, certification, registration, and licensing requirements that distinguish qualifications for professional recognition.

This chapter will address those established certifications that have met the highest certification criteria. These criteria are mandated by recognized governmental agencies that act as oversight and approving entities for certifying organizations for recognized professions nationwide.

There are about two hundred (200+) safety-related certifications; however, there are only a few that meet the unbiased and strict requirements of the NCCA and/or the CESB. In addition, there are a number of schools offering certificate programs, which usually means that someone has completed a sequence of courses in a subject area that is *less* than what is required for a degree.

Certification from an organization not meeting the strict requirements can be compared to a college degree from a nonaccredited college or university—it could be a $3 certificate in a $100 frame. Before starting any certification preparation, verify that the issuing organization has achieved recognition by NCCA or CESB as an approved certifying agency. A training certificate is not a professional certification.

The certifications that follow merit serious consideration. As previously stated, there are numerous certificate programs, other certifications, and other organizations offering recognition programs. This chapter cannot address them all, so the following are discussed in some detail in order to offer guidance on construction safety, health, and environmental certifications that have met national certification requirements.

Safety Trained Supervisor (STS)

The Safety Trained Supervisor certification program is intended for individuals who are first-line supervisors, who have a safety responsibility for a work group that is part of other work duties, and who are not safety specialists or safety practitioners. Typical candidates have a safety responsibility that is adjunct, collateral, or ancillary to their regular duties. Their main job duties are in a craft or trade, in supervision or management, or in some technical specialty.

The typical STS helps an employer implement safety programs at the worker level. Safety tasks often include monitoring for job hazards, helping ensure regulatory compliance, training employees in safe practices, performing safety record-keeping tasks, and coordinating corrections for safety problems with other work groups, safety specialists, or management.

The STS safety responsibility is a part-time responsibility—usually less than one-third of the total job duties. If safety responsibilities involve a greater portion of job duties, the role is more likely to be at a safety technician level.

The STS certification establishes a minimum competency in general safety practices. Candidates must meet minimum safety training and work experience as well as demonstrate knowledge of safety fundamentals and standards by examination.

THE HISTORY OF STS IN CONSTRUCTION

The STS program, initiated in January of 1995, is operated by the Council on Certification of Health, Environmental, and Safety Technologists (CCHEST). It is intended for first-line construction supervisors, foremen, crew chiefs, and craftsmen who have a responsibility to maintain safe conditions and practices on job sites. This program emphasizes general job-site safety within and among work groups. It does not focus on specific safety knowledge and skills for particular crafts or trades.

STS BENEFITS

Construction supervisors, foremen, and crew chiefs play a very important role in making construction work safe. Safe construction practices add to both productivity and profit.

Construction contractors and owners benefit from the STS construction certification program through:

- evaluation of worker job-site safety knowledge
- demonstrated competency by examination
- increased safety awareness among employees
- an accredited, national credential
- reduced workers' compensation claims and premiums
- improved productivity as a result of better communication among work groups
- higher profits from safe work.

The STS supervisor can benefit from the STS construction certification through:

- demonstrated knowledge of fundamental safety practices
- having a recognized credential that may offer opportunities for increased job responsibilities or employment
- increased value to employers.

PROGRAM FEATURES

The STS certification program offers:

- a simple application and examination process
- convenient testing at more than 250 locations every business day
- distinctive wallet card and certificate
- a national database of STS construction certificate-holders available to employers
- assurance of current knowledge through recertification every three years
- group programs possible for employers and associations.

HOW THE CERTIFICATION WORKS

Qualifications
To gain eligibility for the Safety Trained Supervisor in Construction Examination, you must meet the following three requirements:

1. Have two years of experience in construction.
2. Have one year of experience as a construction foreman, supervisor, or crew chief. (This can be concurrent with general construction experience.) Two additional years of construction experience may be substituted for this requirement.
3. Have completed 30 hours of safety training through a single or multiple training courses.

To qualify for the Safety Trained Supervisor in Construction certification, you must pass the STS Construction Examination.

Examination
The STS Construction Examination consists of 75 multiple-choice questions. Candidates have 2 hours to complete the examination at a Prometric Testing Center. Before starting this computer-based, closed-book exam, all candidates receive a tutorial that explains the simple examination procedures. A clock on the computer screen shows the time remaining. Candidates can mark or skip questions and go back to them. Candidates receive pass or fail results at the testing center prior to leaving.

The STS Construction Examination covers the knowledge and skills related to the following tasks. (The portion of the examination devoted to each task appears in parenthesis):

- Conduct New Employee Safety Orientation (11%)
- Perform Basic Hazard Analysis (13%)
- Basic Hazard Recognition and Correction (18%)
- Issue and Monitor the Use of Personal Protective Equipment (9%)
- Conduct Safety Meetings (7%)
- Perform Hazard Prevention Analysis (11%)
- Inspect Tools and Equipment (7%)
- Apply Hazard Communication Standards (8%)
- Enforce Safety Standards on Job Sites (8%)
- Participate in Job Site Safety Inspections (4%)
- Respond to Accidents (4%).

THE CERTIFICATION PROCESS

The following is an outline of the STS construction certification process:

- You complete and submit your application form and fees to CCHEST.
- CCHEST staff reviews your application and determines if you are eligible for the examination.
- If you are eligible, CCHEST will send you an examination admission form good for 120 days and details about scheduling your examination at the location you select.
- You make your examination appointment with Prometric for a convenient time and location within the 120-day time limit.
- You sit for the examination at the appointed time and place.
- You receive examination results before leaving the examination center. If you pass the examination, CCHEST will send the STS wallet card and certificate by mail within a few days.

CONTACTING CCHEST

To receive additional information on being certified as a Safety Trained Supervisor or becoming a sponsoring organization, please contact CCHEST.

Council on Certification of Health,
Environmental, and Safety Technologists
208 Burwash Avenue
Savoy, IL 61874
Phone: 217-359-2686; Fax: 217-359-0055
Email: cchest@cchest.org; Web: www.cchest.org

Construction Health and Safety Technician (CHST)

Construction Health and Safety Technicians are persons who perform construction health and safety activities on a full-time or part-time basis as part of their job duties. Such duties may be ancillary to other job functions. Some examples of construction health and safety activities are safety inspections; job safety planning; organizing and conducting health and safety training; investigating and maintaining records of construction accidents, incidents, injuries and illnesses, and similar functions.

THE CHST CERTIFICATION

The Council on Certification of Health, Environmental and Safety Technologists (CCHEST) awards the Construction Health and Safety Technician certification to individuals who demonstrate competency and work part time or full time in health and safety activities devoted to the prevention of construction illnesses and injuries.

CHST HISTORY

The CHST certification began in October of 1994. Currently, 700 persons have achieved the CHST certification.

The CHST program recognized that many employers assign responsibility for construction safety and health functions to supervisors who may have other primary job functions, but have very important roles in protecting workers. The CHST may be a part-time or full-time supervisor assigned the duties of safety, health, or environmental tasks in the construction industry. This assignment may be at the general-contractor or specialty-contractor level. The responsibilities involve various SHE duties depending on the complexity of the construction activity or the craft function of the specialty contractor. Generally, they include safety inspections, job safety planning, organizing and conducting safety and health training, investigating and making recommendations for corrective actions for potential hazardous conditions, investigating any mishaps or injuries, and maintaining safety records.

There are several routes of entry into the CHST certification. The program accepts candidates who have gained knowledge of safety and health through academic preparation coupled with job experience in construction. It also accepts those who have developed expertise through construction job experience in various crafts and roles along with job safety and health training.

HOW THE CHST WORKS

Qualifications
To qualify for the CHST, an applicant must meet all of the following requirements:

- have good moral character and high ethical standards
- have experience in construction
- have health or safety training or education
- pass the CHST examination.

Experience and Education

Candidates from various backgrounds can qualify for the CHST certification. Some candidates may have experience in a construction craft; some may have served as a construction foreman, supervisor, crew chief, or job superintendent; they may have learned safety and health through job training courses. Other candidates may have studied safety and health in college and then gained their work experience in construction.

To qualify to sit for the CHST examination, a candidate must meet one of the three education/training requirements in Table 2 and satisfy one of the four experience requirements in Table 3.

■■ Table 2 ■■

Education and Training Options for CHST Candidates

Options	Education/Training	
1	High School Diploma **AND**	(a) Be qualified as an OSHA-authorized instructor for the OSHA Construction Safety and Health 10- and 30-hour courses **OR** (b) Have 40 hours of formal classroom training in construction safety and health courses **OR** (c) Have 3 years of experience in a construction position for which safety and health duties are 35% or more of the position duties
2	**OR** Have completed 9 semester-hours or 14 quarter-hours of college credit in safety and health courses	
3	**OR** Hold an associate or higher degree in safety and health	

■■ Table 3 ■■

Experience Options for CHST Candidates

Option	Construction Experience Required		Other Requirements
1	3 years	**AND**	2 years of work experience as a foreman, first-level supervisor, job superintendent, or manager in construction (can be concurrent with general construction experience)
2	3 years	**AND**	2 years in any position that has at least 35% of the job duties in safety and health
3	2 years	**AND**	12 semester-hours (or 20 quarter-hours) of completed college courses in safety and health
4	1 year	**AND**	holding an associate or higher degree in safety and health

STUDENT CANDIDATES

Students in the last semester or last two quarters of a safety and health degree program at the associate or higher level may take the CHST examination. However, CCHEST will not award the CHST certification until a candidate meets all requirements, including construction experience.

COMPUTING WORK EXPERIENCE

Because construction work often involves extended work weeks and concentrated work schedules, experience can be equated to years of experience by computing hours worked.

The reference period is a 40-hour week and 50 weeks per year (2000 hours per year). CHST candidates can convert the work experience in hours to years by dividing their total hours worked by 2000 hours. The number of months can be computed by dividing hours by 167 hours per month. To qualify for experience, a position must have at least 30 hours per week.

CREDIT FOR COLLEGE COURSES OR DEGREE

Candidates seeking credit for college courses or a degree must submit an original transcript from the school where the credit was earned. CCHEST accepts educational credit from colleges and universities accredited by a regional accrediting body recognized by the Commission on Higher Education Accreditation (CHEA) or the U.S. Department of Education. CCHEST evaluates foreign degrees for U.S. equivalency.

CHST EXAMINATION

The following describe the CHST Examination:

- It has 200 multiple-choice items with one correct answer and three incorrect answers.
- Candidates have four hours to complete the examination.
- The examination is closed-book, though a formula and conversion reference handout is provided.
- Candidates may bring a handheld calculator with scientific functions.

The CHST Examination tests knowledge used in eight safety and health tasks. Table 4 shows the tasks and the portion of the examination devoted to each, along with the supporting knowledge/subject areas for each task.

CHST RECIPIENT BENEFITS

Benefits of the CHST certification for construction health and safety practitioners include:

- satisfaction from knowing that you meet a standard of your professional peers
- self-esteem because you have respect of other health and safety professionals for your professionalism
- recognition from an employer or potential employer's health and safety qualifications
- potential recognition from an employer through increased job responsibility and/or pay
- improved ability to compete for construction health and safety positions
- an annual wallet card noting certification.

BENEFITS FOR EMPLOYERS

CHST certification benefits employers because it:

- is helpful when hiring qualified construction health and safety practitioners
- increases worker and public confidence in the employer's construction safety and health program
- enhances company profitability and quality by reducing construction accidents, illnesses, and insurance claims
- improves the company image for worker protection
- is a means for improving safety and health in the workplace through competence.

CHST TRAINING

CCHEST is not a training organization. To prepare for the CHST Examination, some candidates use national training programs, such as the 100-Hour Safety Technician training that is one part of the Construction Site Safety Program of the National

■ ■ Table 4 ■ ■

CHST Examination Areas

Subject/Knowledge Areas Required for Each Task (The bullets indicate which of the eight tasks listed at the right rely on knowledge of the subject areas listed below)	Safety and Health Inspections (30%)	General Safety Training and Safety Orientation (20%)	Safety and Health Record Keeping (5%)	Hazard Communication Compliance (10%)	Safety Analysis and Planning (10%)	Accident Investigations (10%)	Program Management and Administration (10%)	OSHA and Other Inspections (5%)
A. Working knowledge of applicable OSHA and other safety and health standards	•	•	•		•	•	•	•
1) OSHA Hazard Communication Standard				•				
2) Bloodborne pathogen standard		•		•				
3) Record keeping and safety program documentation				•				
4) First-aid practices								
B. Safe use of construction tools and equipment	•	•			•	•	•	
C. Safe construction practices	•	•			•	•	•	•
D. Selection and use of personal protective equipment	•	•			•	•	•	•
E. Use of safety test equipment	•	•			•	•		•
F. Job Hazard Analysis and other safety-analysis and job-planning methods	•	•			•		•	
G. Ergonomics	•	•			•		•	•
H. Knowledge of safety and health training and teaching methods		•			•			
I. Accident and incident investigation methods and use of investigative tools						•	•	
J. Inspection and appeal procedures of OSHA and other government agencies						•	•	
K. Basics of organizational and human behavior for safety							•	•
L. Program management and administration						•	•	•

Center for Construction Education and Research (1-888-NCCER-20, www.nccer.org) or the Construction Health and Safety Training Course offered periodically, or on a privately scheduled basis, through the American Society of Safety Engineers (ASSE) (www.asse.org, or contact ASSE customer service 847-768-2929).

CHST PROFESSIONAL DEVELOPMENT

A common question is: "Can the CHST help me qualify for the Certified Safety Professional (CSP)?" Yes, achieving the CHST will provide one year of experience for those who qualify for the CSP. For more information on the CSP, visit www.bcsp.org. Many use the CHST as a stepping stone to the CSP.

Certified Safety Professional (CSP)

A safety professional is one who applies the expertise gained from a study of safety science, principles, practices, and other subjects, and from professional safety experience to create or develop procedures, processes, standards, specifications, and systems to achieve optimal control or reduction of the hazards and exposures that may harm people, property, and/or the environment.

Professional safety work must be the primary function of a position and account for at least 50 percent of the position's responsibility. Professional safety experience involves analysis, synthesis, investigation, evaluation, research, planning, design, administration, and consultation to the satisfaction of peers, employers, and clients in the prevention of harm to people, property, and the environment. Professional safety experience differs from nonprofessional or paraprofessional safety experience in the degree of responsible charge and the ability to defend analytical approaches and recommendations for engineering or administrative controls.

HISTORY OF BCSP AND THE CSP

The Board of Certified Safety Professionals was organized in 1969 as a peer certification board. Its purpose is to certify practitioners in the safety profession. BCSP is not a membership organization. The specific functions of the Board are to:

- evaluate the academic and professional experience qualifications of safety professionals
- administer examinations
- issue certificates of qualification to those professionals who meet the Board's criteria and successfully pass its examinations.

In 1968, the American Society of Safety Engineers studied the issue of certification for safety professionals and recommended the formation of a certification program. In July 1969, this recommendation led to establishing BCSP as a not-for-profit corporation in Illinois. BCSP then began accepting applications for certification in 1970.

Thirteen people comprise the BCSP Board of Directors and represent a cross-section of the safety profession and the public. One person is a public director not involved with the safety profession and represents public interests.

BCSP's six membership organizations nominate six directors, while the other six directors are chosen from across the profession.

The six BCSP membership organizations are:

- American Society of Safety Engineers (since 1970)
- American Industrial Hygiene Association (since 1973)
- System Safety Society (since 1976)
- Society of Fire Protection Engineers (since 1983)
- National Safety Council (since 1994)
- Institute of Industrial Engineers (since 1994).

The Certified Safety Professional designation meets the highest national standards for certifications and is accredited by the Council of Engineering and Scientific Specialty Boards and the National Commission for Certifying Agencies. It is also recognized by the National Skill Standards Board (NSSB), which is an agency of the federal government.

Since BCSP was founded, more than 25,000 individuals have applied for, and nearly 17,000 individuals have achieved, the CSP. Currently, more than 10,000 people hold this designation and more than 4000 candidates are currently pursuing the CSP.

Over 93 percent of all current CSPs have a bachelor's degree or higher, and more than 44 percent have advanced degrees. About 65 percent of those certified have received their CSP since 1990.

CSPs work in manufacturing (19%), insurance (22%), petrochemicals (15%), consulting (15%), government (9%), and other industries (14%).

HOW THE CSP CERTIFICATION WORKS

Academic Requirements
The model educational background for a safety professional and a candidate for the Certified Safety Professional is a bachelor's degree in safety from a program accredited by the Accreditation Board for Engineering and Technology (ABET). Contact ABET

for accreditation standards and for a complete list of accredited degree programs (www.abet.org).

For U.S. degrees, BCSP requires that the educational institution hold accreditation from an accreditation body recognized by the Commission on Higher Education Accreditation (CHEA) or by the U.S. Department of Education. CHEA lists accredited schools on its Web site (www.chea.org).

Because so many people enter the safety profession from other educational backgrounds, candidates for the CSP may substitute other degrees plus professional safety experience for an accredited bachelor's degree in safety. All CSP candidates must meet one of the following minimum educational qualifications:

- a bachelor's degree in any field
- an associate degree in safety and health.

Experience Requirements

In addition to the academic requirement, CSP candidates must have four years of professional safety experience. The four years are in addition to any experience used to meet the academic requirement. Professional safety experience must meet all of the following criteria to be considered acceptable by BCSP:

- Professional safety must be the primary function of the position. Collateral duties in safety are not considered the primary function.
- The position's primary responsibility must be the prevention of harm to people, property, and the environment, rather than responsibility for responding to harmful events.
- Professional safety functions must be at least 50 percent of the position's duties. BCSP defines full-time as at least 35 hours per week. Part-time safety experience is allowed instead of full-time safety experience if the applicant has the equivalent of at least 900 hours of professional safety work during any year (75 hours per month or 18 hours per week) for which experience credit is sought.
- The position must be at the professional level. This is determined by evaluating the degree of responsible charge and reliance

of employers or clients on the person's ability to defend analytical approaches used in professional practice. This also encompasses their recommendations on how to control hazards through engineering and/or administrative approaches.

- The position must have breadth of professional safety duties. This is determined by evaluating the variety of hazards about which the candidate must advise and the range of skills involved in recognizing, evaluating, and controlling hazards. Examples of skill are analysis, synthesis, planning, administration, design, investigating, and communications.

Substitutions for Experience

Candidates may substitute advanced degrees and/or the Occupational Health and Safety Technologist (OHST) or Construction Health and Safety Technician (CHST) certifications from the Council on Certification of Health, Environmental and Safety Technologists for part of the experience requirement. Candidates may substitute advanced degrees from U.S. colleges and universities accredited by a CHEA-recognized accrediting body or one that is approved by the U.S. Department of Education. BCSP will evaluate degrees from schools outside the United States for U.S. equivalency.

Advanced degrees carry the following values, based on the fact that most master's degrees require about one-quarter of the course work of that required for a bachelor's degree:

Degree	Credit
Master's degree	one-quarter of the credit allowed by a bachelor's degree in the same field
Doctoral degree	one-half of the credit allowed for a bachelor's degree in the same field

If an applicant has more than one graduate degree, only the degree yielding the highest credit will be accepted.

Holding the OHST or CHST certification at the time of application will substitute for one year of professional safety experience (12 points). Even if someone holds both certifications, the maximum number of points one can receive is 12.

Experience Updates

Some candidates become eligible for and pass the Safety Fundamentals Examination, but must wait to achieve additional work experience to qualify for the Comprehensive Practice Examination (96 points). Eligibility to sit for the Comprehensive Practice Examination begins when a candidate achieves the 96 points in the CSP process and has passed, or has been granted a waiver of, the Safety Fundamentals Examination. When someone applies, BCSP will estimate when the applicant will achieve 96 points, assuming that the applicant remains employed in a position that will meet professional safety practice requirements. Near the date on which BCSP estimated that the applicant will achieve 96 points through the additional experience, BCSP will send instructions with an Experience Update Form. In order to demonstrate that an applicant actually met the 96 points, the applicant will have to provide details about employment since the time of the original application.

If BCSP accepts the additional experience and the applicant achieves 96 points, the person will be declared eligible for the Comprehensive Practice Examination and the three-year time limit to complete the CSP process will begin. If one has not yet gained 96 points after updating the experience, BCSP will again estimate what is needed. If an applicant fails to update experience when requested, or does not achieve the required experience within five years of the first request for an experience update, the applicant will risk being dropped from the CSP process.

EXAMINATION REQUIREMENTS

Achieving the Certified Safety Professional designation typically involves passing two examinations: Safety Fundamentals and Comprehensive Practice.

Safety Fundamentals Examination

The first examination is the Safety Fundamentals Examination. It covers basic knowledge appropriate to professional safety practice.

Candidates who meet the academic standard (achieve 48 points through an associate or bachelor's degree plus experience) may sit for the Safety Fundamentals Examination. Upon passing the Safety Fundamentals Examination, candidates receive the Associate Safety Professional (ASP) title to denote their progress toward the CSP. The ASP title is a temporary designation and can be held for only three years. Candidates must pass the Comprehensive Practice Examination within that time frame.

Students in ABET-accredited safety or safety-related degree programs at the bachelor's or master's level may sit for the Safety Fundamentals Examination during the last semester (or quarter) of their academic program—or any time thereafter. Like other candidates, students must apply to BCSP.

Some candidates, examined through other BCSP-accepted certification and licensing programs, and who currently hold such certifications or licenses, may be granted a waiver of the Safety Fundamentals Examination. BCSP accepts only the following certifications or licenses:

- Certified Industrial Hygienist (CIH) from the American Board of Industrial Hygiene
- Certified Health Physicist (CHP) from the American Board of Health Physics
- Professional Engineer (P.E.) from the engineering registration board of any U.S. state or territory
- National Diploma in Occupational Safety and Health from the British National Examination Board for Occupational Safety and Health (NEBOSH)
- Canadian Registered Safety Professional (CRSP) from the Board of Canadian Registered Safety Professionals
- Member, Singapore Institute of Safety Officers (SISO)
- Chartered Engineer (CE) from the Engineering Council (United Kingdom)

Those who receive a waiver of the Safety Fundamentals Examination do not receive the Associate Safety Professional (ASP) title.

Comprehensive Practice Examination

All CSP candidates must pass the second examination—the Comprehensive Practice Examination.

There are no waivers of this examination. To take this exam, a candidate must meet both the academic and experience requirements, acquire 96 points, and have passed or received a waiver of the Safety Fundamentals Examination. The total credit for academic degrees plus the months of professional safety experience must equal or exceed 96 points. After passing the Comprehensive Practice Examination and meeting all other requirements for the CSP, a candidate receives the Certified Safety Professional title. A certificate holder may continue to use the CSP title by paying an annual renewal fee and complying with Continuance of Certification requirements every five years.

CANDIDATE TIME LIMITS

Once candidates apply for the CSP, they must meet certain time limits as they progress toward the CSP. Failure to complete the CSP requirements within time limits may result in a terminated application. A candidate will then have to reapply and restart the process.

CSP Construction Safety Specialty

The CSP Construction Safety Specialty allows those holding the CSP to demonstrate knowledge in construction safety through a specialty examination. It does not lead to an additional certification, but does give recognition for the specialized knowledge. The unique safety, health, and environmental requirements for construction exposures create a need for construction-specific expertise to ensure more successful accident, injury, illness, and environmental prevention programs.

An employer can utilize the CSP Construction Safety Specialty Examination along with construction experience and construction-specific training to identify those safety professionals who have clearly developed construction safety expertise.

(Effective July 1, 2004, the BCSP will no longer offer the Construction Safety Specialty Examination.)

CONSTRUCTION SPECIALTY PROCESS

Qualifications
The applicant must be an active Certified Safety Professional in good standing, who applies for and successfully completes the separate Construction Safety Specialty Examination.

CONSTRUCTION SAFETY SPECIALTY EXAMINATION

The Construction Safety Specialty Examination contains four major subject areas. Each is listed below along with the portion of the examination devoted to the major subject area:

- Construction Means and Methods (25%)
- Construction Management (11%)
- Construction Safety Program Management (45%)
- Construction Safety Performance Measures (19%)

Each major subject area includes the major tasks required for the subject area and the supporting knowledge and/or skills appropriate to each major task. The tasks, knowledge, and skills are defined in the sections that follow.

Construction Means and Methods (25%)
(Construction Means and Methods refer to construction equipment, operations, practices, procedures, and related activities.)

Major Tasks:

- Develop construction risk management and safety programs applying knowledge of construction means and methods and engineering principles to ensure applicability and effectiveness of the programs to the work addressed.
- Evaluate the operation, inspection, and maintenance of construction equipment and tools using personal observation, document reviews, and testing to ensure that the equipment and tools are safe for use and that supervision, affected employees, and operators are competent and qualified.
- Develop hazard control measures for the different types of construction operations and environments using sound engineering principles and knowledge of construction industry practices to ensure the effectiveness and practicality of safety controls.

Supporting Knowledge and/or Skills

- engineering principles (e.g., civil, mechanical, electrical, materials, and chemical)
- construction safety programs (e.g., hazard evaluation, controls, training, enforcement, inspection, record keeping, measurement)
- construction safety and design codes, standards, and regulations
- risk management (e.g., risk identification, risk analysis, risk avoidance, risk transfer, and administering the risk management process)
- criteria and methodology for testing construction equipment and tools
- hazard control measures (e.g., elimination, engineering controls, administrative controls, and personal protective equipment)
- hazards and control measures related to the environment at the work area (e.g., operating equipment, existing and/or nearby facilities)
- construction practices (e.g., trade distinctions, employee skills) and construction conditions
- developing project-specific safety and health plans
- performing and interpreting construction safety engineering calculations
- interpreting construction standards, codes, regulations, contracts (e.g., OSHA 1926, ANSI)
- operating construction test equipment and interpreting results
- recognizing unsafe operations in the use and maintenance of construction equipment and tools
- selecting appropriate controls for construction industry hazards.

Construction Management (11%)

(Construction Management includes procedures and activities construction firms use to plan, organize, document, track, communicate, and evaluate costs for construction work.)

Major Tasks:

- Manage construction safety responsibilities and programs using knowledge of protocols and contractual relationships among owners, designers, construction managers, general contractors, and subcontractors to facilitate an effective safety program.
- Manage safety activities by applying knowledge and understanding of project scheduling methodologies (CPM, bar charts, etc.) to ensure effective and economic project planning and execution.
- Develop recommendations for risk management using knowledge of insurance data (workers' compensation and liability), potential litigation, and exposures to improve productivity, competitiveness, and safety.
- Advise employees and management using professional experience and knowledge in construction safety, health, and environmental principles, policies, and regulations to facilitate continuous improvement.
- Provide technical assistance during project conception, design, and estimating phases using professional experience and knowledge of construction safety to improve safety and constructability.
- Provide advice to marketing, purchasing, sales staff, and human resources using safety data and presentation materials to improve competitiveness and the quality of the workforce.
- Advise management on emergency preparedness by establishing policies and procedures to manage workplace violence, crisis situations, emergency responses, and environmental concerns.

Supporting Knowledge and/or Skills

- construction safety priorities
- construction contracts and relationships
- scheduling methods
- risk management principles
- construction safety knowledge (standards, codes, regulations, contracts, and means and methods)
- construction terminology
- construction hygiene
- environmental aspects of construction
- workplace violence in construction

- construction emergency response
- using critical path methodology and bar charts
- developing risk management programs

Construction Safety Program Management (45%) (Construction safety program management is recognition, evaluation, and administrative or engineering controls for hazards in construction work, and administration of safety programs for construction projects.)

Major Tasks:

- Implement effective safety controls to address identified project hazards by meeting or exceeding relevant standards, regulations, company policy, and contract requirements.
- Monitor project activities using personal observation and safety performance to ensure the effectiveness of planned safety controls and to recognize and control unforeseen hazards.
- Conduct safety training for management and employees using appropriate content, methods, and evaluation procedures to prevent incidents, foster a safety culture, and improve performance.
- Maintain construction safety data, records, and statistics by documenting incidents, injuries, illnesses, hazards, controls, disciplinary measures, training, inspections, safe practices, and other methods, to validate and improve safety performance.

Supporting Knowledge and/or Skills

- contract law and relationships
- construction terminology, symbols, and structure of drawings and specifications
- construction safety and design codes, standards, regulations, contracts, and means and methods
- safety controls used in construction and their limitations (e.g., fall protection methods, trench sloping, scaffolding)
- personal observation techniques and their limitations (i.e., complexity required in construction)

- human relations techniques (i.e., dynamic workforce)
- training methodology specific to construction
- construction safety culture (e.g., behaviors, psychology, sociology)
- visualizing components and their construction from pertinent drawings and documents
- prioritizing project hazards and organizing information related to effective safety controls (dynamics of the work site).

Construction Safety Performance Measures (19%) (Construction Safety Performance Measures are procedures and techniques for assessing the performance of safety programs.)

Major Tasks:

- Perform site safety inspections by walking through the project periodically to identify and record observable safe and/or unsafe acts and conditions.
- Using statistically sound procedures, review records, record-keeping procedure, available data bases, and damage reports to identify the number and types of injuries, illnesses, and incidents, and to identify possible trends.
- Evaluate program comprehensiveness (before construction) and/or effectiveness (during and after construction) through audits, documentation reviews, and/or inspections to ensure employee safety as well as regulatory and contractual compliance.
- Conduct baseline and periodic safety culture surveys using valid instruments and make appropriate interpretations of the results to identify true change.
- Perform economic analysis using workers' compensation and general-liability and property-loss data to improve competitiveness and performance.

Supporting Knowledge and/or Skills

- construction standards, codes, and regulations
- contract documents and relationships
- site-specific safety principles
- construction means and methods

- reading design and as-built drawings and specifications
- statistics (descriptive and inferential) meaningful to construction
- documentation and tabulations as required by construction dynamics
- significance of variableness in construction safety measurement
- program development and implementation pertinent to construction safety and workplace and workforce dynamics
- regional and social differences in work attitudes and accepted practices related to a mobile and dynamic workforce
- wrap-up insurance programs [e.g., owner-controlled (OCIPs), contractor-controlled (CCIPs), project (PIPs)]
- third-party lawsuits related to construction relationships
- evaluating performance from the safety records of contractors and from written program documents

PREPARING FOR THE CONSTRUCTION SAFETY SPECIALTY EXAMINATION

There are a number of sources that offer continuing education courses, seminars, symposia, conferences, and conference sessions to help candidates prepare for the specialty examination. The following references, while not exhaustive, provide a starting point to prepare for the Construction Safety Specialty Examination: The American Society of Safety Engineers at www.asse.org, the Construction Practice Specialty, or ASSE customer service 847-768-2929; or the National Safety Council (NSC) at www.nsc.org.

CSP CONSTRUCTION SAFETY SPECIALTY BENEFITS

For the Individual CSP with a Construction Safety Specialty

- recognized by SHE peers for their expertise in construction
- elation for their self-accomplishment
- potential for company recognition and raises
- potential for professional advancement

For the Employing Company

- additional credibility and acceptance of SHE programs
- reduced potential liability to SHE lawsuits due to SHE levels of expertise
- positive public relations and recognition of professionalism
- ISO certifications

OTHER CSP SPECIALTY EXAMS

- Ergonomics Specialty
- System Safety Specialty

Contact the Board of Certified Safety Professionals (www.bcsp.org) for additional information.

Certified Crane Operator (CCO)

A fairly new construction certification that is impacting one of the significant areas of construction accident prevention is the Certified Crane Operator (CCO).

Several major construction crane accidents during the past few years (the "Big Blue Crane" accident, the San Francisco tower crane accident, and the Texas A&M University stadium accident) cost millions of dollars, received widespread national attention, and caused major injuries and deaths. A coalition of many interested and affected organizations banded together to develop a national consensus for improving the safety of crane operations within the United States and worldwide.

A significant result was the formation of the National Commission for Certification of Crane Operators in January of 1995. That group (a not-for-profit organization) has developed a program and procedures for the certification of crane operators. Professional crane operators must meet strict crane-operator-experience requirements and pass medical requirements; then the crane operators may apply for a series of written examinations. The exams include a general crane knowledge exam and specialty knowledge exams based on the types and sizes of cranes they are qualified to operate. After successful completion of the written examinations, crane operators must then successfully complete a *hands-on, in-the-crane* proficiency exercise for the size and type of crane they operate.

CCOs are issued for different sizes and types of cranes, and the operator is only certified to operate the size and type of crane for which he has been certified.

This certification program has been a tremendous success. Major construction associations, construction insurance groups, and construction owners are endorsing it and requiring that only Certified Crane Operators are allowed to operate the cranes on their construction projects.

CCO CERTIFICATION POLICIES

Eligibility

Requirements for the crane operator certification include the following:

- be at least 18 years of age
- have 1000 hours crane-related experience in the last 5 years
- meet medical requirements
- comply with CCO's substance abuse policy
- pass written examinations (core and one specialty)
- pass practical examination(s).

Candidates must pass the practical exam within twelve months of passing the written examination.

Experience

CCO certification examinations are designed for operators who are trained and who currently work in crane operation.

Crane-related experience is defined as: operating, maintenance, inspection, or training. Candidates who do not meet the experience requirements are *NOT* eligible for certification.

Physical Evaluation

Candidates must submit one of the following:

- CCO physical examination form
- current DOT (Department of Transportation) Medical Examiner's Certificate.

Certified crane operators must continue to meet ASME B30.5 physical requirements throughout their certification period. The CCO certification card is valid only with a current medical certificate that meets the requirements of ASME B30.5.

CCO WRITTEN EXAMINATIONS

The written examination program consists of a core examination in crane operation, as well as four crane specialty examinations. The Core Examination has 90 multiple-choice questions. Candidates are allowed 90 minutes to complete the core exam.

All specialty examinations consist of 26 multiple-choice questions. Candidates are allowed 55 minutes to complete each specialty exam.

All candidates are required to take the Core Examination regardless of the specialties in which they wish to be certified.

CCO Specialty Examinations

- Lattice Boom Crawler Cranes
- Lattice Boom Truck Cranes
- Large Telescopic Boom Cranes (greater than 17.5 tons)
- Small Telescopic Boom Cranes (less than 17.5 tons)

Candidates must register for the Core and at least one of the specialty examinations. Certification requires competency in both the elements on the core exam and one or more of the specialty categories. Candidates meeting the eligibility requirements and passing the written examinations are eligible to take the Practical Examination.

CCO Practical Examination

The Practical Examination, demonstrating crane operation proficiency, is available for three crane types:

- Lattice Boom Cranes
- Large Telescopic Boom Cranes (greater than 17.5 tons)
- Small Telescopic Boom Cranes (less than 17.5 tons)

A candidate must pass both the written Core and at least one Specialty Examination as well as the corresponding Practical Examination in order to achieve the CCO and be certified for a 5-year period.

The training programs for the crane operators desiring update or refresher training in preparation for these examinations are absolutely fantastic. The author personally attended one of the Crane Operators Certification training courses presented by St. Paul Construction Insurance Group. Every construction safety person responsible for any aspect of crane operations should attend CCO training to attain the crane safety operating knowledge for integration into their construction safety and health programs. Additional information is available at www.cco.org or from the National Commission for the Certification of Crane Operators; 2750 Prosperity Avenue, Suite 505; Fairfax, VA 22031-4312 (phone: 703-560-2391; fax: 703-560-2392; email: info @nccco.org).

Certified Industrial Hygienist (CIH)

The health aspects of construction SHE exposures have continued to grow as additional emphasis has been placed on the effect of asbestos, lead, silica, noise, and other health hazards in the construction industry.

DEFINITIONS AND FUNCTIONS OF INDUSTRIAL HYGIENE

Industrial hygiene is the science and practice devoted to the anticipation, recognition, evaluation, and control of those environmental factors and stresses arising in or from the workplace that may cause sickness, impaired health and well-being, or significant discomfort among workers, and that may also impact the general community.

Industrial hygienist is a person having a baccalaureate or graduate degree from an accredited college or university in industrial hygiene, biology, chemistry, engineering, physics, or a closely related physical or biological science who, by virtue of special studies and training, has acquired competence in industrial hygiene. Such special studies and training must have been sufficient in the above cognate sciences to provide the ability and competency (1) to anticipate and recognize the environmental factors and stresses associated with work and work operations and to understand their effects on people and their well-being; (2) to evaluate, on the basis of training and experience, and with the aid of quantitative measurement techniques, the magnitude of these factors and stresses in terms of their ability to impair human health and well-being; and (3) to prescribe methods to prevent, eliminate, control, or reduce such factors and stresses and their effects.

FUNCTIONS OF THE INDUSTRIAL HYGIENIST

Industrial hygiene is a broad, multifaceted profession. To illustrate the nature of industrial hygiene practice, the American Board of Industrial Hygiene (ABIH) offers the following examples of *typical* functions performed by industrial hygienists. These are examples only. Performing one or more of these activities *DOES NOT* constitute certification eligibility.

Industrial hygienists may typically:

- Review projects, designs, and purchases to anticipate health hazards.
- Critically evaluate work environments, processes, material inventories, and worker demographics to recognize potential health risks to persons or communities.
- Assess human exposures to hazards through a combination of qualitative and quantitative methods to determine health risks, regulatory compliance, and legal liabilities.
- Recommend effective control measures to mitigate risks via engineering, administrative, or personnel protective methods.
- Communicate risks and control measures to workers, management, clients, customers, and/or communities.
- Provide specific training to workers about risks and control measures.
- Perform laboratory analysis of samples taken to assess worker exposure.
- Conduct research and development of industrial hygiene methods and tools.
- Integrate industrial hygiene programs with related health-risk management efforts, including safety, environmental, and medical.

■ Interface with regulatory, community, and professional organizations.

■ Manage, supervise, or advise other industrial hygiene staff.

■ Audit industrial hygiene programs.

■ Provide technical support to legal proceedings in matters related to industrial hygiene.

■ Provide academic training in industrial hygiene at the college or university level.

HISTORY OF THE AMERICAN BOARD OF INDUSTRIAL HYGIENE

The ABIH was incorporated in 1960 as a not-for-profit corporation. It is governed by a Board of Directors comprised of eighteen CIHs and one public member.

The American Board of Industrial Hygiene exists for the express purpose of improving the practice and educational standards of the industrial hygiene profession.

The Board achieves this purpose by: (1) offering certification examinations to industrial hygienists with the required educational background and professional experience; (2) acknowledging individuals who successfully complete the examination by issuing a certificate; (3) requiring diplomates to maintain their certification by submitting evidence of continued professional development.

The need for qualified industrial hygienists has never been greater. As concerns about health, safety, and environmental issues have grown, so has the demand for qualified professionals to evaluate working conditions and community exposures. Increasing importance is being attached to the studies and judgments of the industrial hygienist as the legal ramifications surrounding employee and community health continue to expand.

Through 1999, more than 7900 professionals had achieved the ABIH Certified Industrial Hygienist credential. This credential not only enhances the credibility and opportunities of the individual, but also promotes high standards of professional conduct for those serving the health interests of workers and the community. CIHs are employed throughout the industrial hygiene profession in industry; labor unions; state, provincial, and local governments; federal agencies; uniformed services; consulting firms; and academia in the United States, Canada, Australia, and many other countries.

Certification in the practice of industrial hygiene is a two-stage process. First, the individual needs to demonstrate his or her educational and experience qualifications and then must successfully complete the written certification examination required by the Board.

The certification and maintenance of certification processes provide mechanisms for advancing the profession, providing a frame of reference for identifying industrial hygiene professionals, and distinguishing them from others involved in industrial hygiene activities. Many employers see the CIH designation as one of their selection and promotion criteria. By participating in and supporting the ABIH program, qualified persons are helping to identify and maintain the professional stature of industrial hygiene.

CIH EXAMINATION PROCESS

To qualify for admission to the ABIH examination, an applicant must comply with all regulations of the Board in effect at the time the application is filed. An applicant must:

■ be practicing industrial hygiene full time

■ have good moral character and high ethical and professional standing

■ meet an academic requirement

■ meet a professional industrial hygiene experience requirement supported by references.

Academic Requirements

Academic Degree: Graduation from a college or university, acceptable to the Board, with a bachelor's degree in industrial hygiene; biology; chemistry; sanitary, chemical, or mechanical engineering; physics; or safety (ABET-accredited program) is required.

The Board will consider, and may accept, any other bachelor's degree from an acceptable college or university if the degree contains at least 60 semester-hours in undergraduate or graduate-level courses in science, mathematics, engineering, and science-based technology, with at least 15 of those hours at the upper (junior, senior, or graduate) level. A degree that is heavily comprised of only one of those subject areas may be judged not acceptable. The social sciences are not considered to be qualifying sciences.

Experience Requirement: Five years of full-time employment in the professional practice of industrial hygiene, acceptable to the Board, and subsequent to completion of an acceptable degree, is required to be eligible for the Comprehensive Practice Exam.

For additional information, contact: American Board of Industrial Hygiene; 6015 W. St. Joseph, Suite 102; Lansing, MI 48917; (phone: 517-321-2638; Web site: www.abih.org; email: abih@abih.org).

ADDITIONAL CERTIFICATIONS

Additional certifications that meet the NCCA or CESB certification criteria are listed below with contact information. They will be happy to provide their application criteria and procedures for all of their certifications.

Due to space restrictions for this chapter, details about these certifications have not been included.

1. Occupational Health and Safety Technologist (OHST)
   Contact: Council on Certification of Health, Environmental, and Safety Technologists (www.cchest.org)
2. Qualified Environmental Professional (QEP)
   Contact: Institute of Professional Environmental Practice (www.ipep.org)
3. Certified Hazardous Materials Manager (CHMM)
   Contact: Institute of Hazardous Materials Management (www.ihmm.org)
4. Diplomate in Environmental Engineering
   Contact: American Academy of Environmental Engineers (www.envior-engrs.org)
5. Certified Occupational Health Nurse (COHN)
   Contact: American Board for Occupational Health Nurses, Inc. (www.abohn.org)

The research for this chapter has been a labor of love. During the last forty years, there has been a tremendous evolution within the safety, health, and environmental profession. Accredited academic degrees associated with the SHE profession have proliferated. Public acceptance of SHE principles is now a reality. Employees are demanding a workplace free from SHE exposures that impact their lifestyles and their working environment. Management understands that safety, health, and environmental managers must have the accredited academic degrees, proper technical training, on-the-job experience, and the SHE certifications to positively impact the company's net annual profits.

SHE professionals in this business of safety, health, and environmental challenges must be aware of the ethical requirements of the profession and declare they will "do their best" to expect everyone in these areas in the construction field to improve through day-by-day professionalism and accredited certifications.

REFERENCES

American National Standards Institute. *ANSI Accreditation, Certification Bodies Operating a Personnel Certification Program.* New York: ANSI, 2002.

Board of Certified Safety Professionals. "A Brief History of the Board of Certified Safety Professionals, Advancing the Safety, Health and Environmental Professional for 30 Years," special supplement to the *1999 BCSP Annual Report.* Savoy, IL: BCSP, 1999.

_____. *Annual Report, 2001.* Savoy, IL: BCSP, 2001.

_____. *Career Guide to the Safety Profession.* Savoy, IL: ASSE Foundation and BCSP, 2000.

_____. *Certified Safety Professional (CSP) Application Guide,* 2nd ed. Savoy, IL: BCSP, September 2002.

_____. *Certified Safety Professional (CSP) Examination Guide.* Savoy, IL: BCSP, July 2001.

_____. *Certified Safety Professional, CSP-Specialty Examinations Handbook,* 3rd ed. Savoy, IL: BCSP, February 2000.

_____. *Code of Ethics and Professional Conduct.* Savoy, IL: BCSP, April 2001.

_____. *Certified Safety Professionals: Competent Management Team Members,* video. Savoy, IL: BCSP.

Brauer, Dr. Roger L., Ph.D., CSP, P.E., CPE, "Safety Competency and Certification," magazine articles. State of Oregon, Office of Degree Authorization, Unaccredited Colleges, Web site: www.osac.state.OR.US/oda/unaccreditated.ht.

Council on Certification of Health, Environmental and Safety Technologist. *Candidate Handbook, CHST-Construction Health and Safety Technician Certification*, 4th ed. Savoy, IL: CCHEST, June 1999.

_____. *Candidate Handbook, OHST-Occupational Health and Safety Technologist Certification*, 5th ed. Savoy, IL: CCHEST, April 1999.

_____. *Candidate Handbook, STS-Construction, Safety Trained Supervisor in Construction*, 6th ed. Savoy, IL: CCHEST, May 2001.

_____. *Code of Ethics for CHST and OHST*. Savoy, IL: CCHEST.

Freiberger, Fred, CSP, CIH, CBAE, P.E., and T. Spiers, CSP. Presentation paper, "Professional Certification, Registration, and Licensure: Benefits and Issues," given at the American Society of Safety Engineers Professional Development Conference, Anaheim, CA, June 2001.

Manuale, Fred A. *On the Practice of Safety*, 2nd ed. New York: John Wiley & Sons, 1997.

National Commission for the Certification of Crane Operators. *Candidate Handbook, CCO-Certified Crane Operator*. Fairfax, VA: NCCCO, 2003.

National Commission for Certifying Agencies, "Standards for the Accreditation of Certification Programs, National Organization for Competency Assurance," 2002, www.ncca.org.

National Skills Standards Board, "Certification Recognition," May 2001, www.nssb.org.

U.S. Department of Education, Office of Post Secondary Education, "Overview of Accreditation," October 2002, http://www.ed.gov/offices/OPE/accreditation/index.ht.

REVIEW EXERCISES

1. Is the STS (Safety Trained Supervisor) considered a safety professional? Explain the issues and the STS primary job function.
2. The CHST (Construction Health and Safety Technician) is assigned as a project safety person. Explain their job responsibilities and experience requirements.
3. What is the impact of the CCO (Certified Crane Operator) certification on construction project safety? Explain potential crane hazards and the financial impact of a crane accident or incident.
4. If the company safety manager is a CSP (Certified Safety Professional), what is the potential financial impact, and is liability enhanced?
5. Construction health hazards impact the long-term work life of construction employees. Discuss common construction health hazards.

■ ■ **About the Author** ■ ■

OSHA CONSTRUCTION PARTNERSHIPS/VPP
by T. J. Lyons, OHST, CSP

T. J. Lyons, OHST, CSP, is a safety coordinator with Turner Construction. Previously he was the VPP coordinator for a $30M highway project that received OSHA VPP approval—the first such site in New York State, and only the second in the United States. In recognition of this achievement, Mr. Lyons received the International Risk Management Institute's 2001 Best Practices in Construction Award.

Mr. Lyons is also a volunteer fire chief, an EMT, and heads a county hazmat team.

CHAPTER 32

OSHA Construction Partnerships/VPP

LEARNING OBJECTIVES

■ Explain the value of partnerships with regulators and trade organizations to improve safety and increase profits.

■ Discuss the implications of a safer work site in marketing potential and positive exposure for a firm.

■ Discuss the potential reduction in safety performance ratings based on real data from case histories.

■ Explain the Occupational Safety and Health Administration's Voluntary Protection Program (VPP) and its impact on work sites where it is in place.

■ Discuss how to "sell" the profit-making ability of a safe workplace to owners and the workforce.

PARTNERSHIPS

An urban legend during the Vietnam conflict describes an Army Jeep with a U.S. soldier driving a prisoner down a rutted, one-lane road—a sheer drop to one side, a rock wall on the other. As they speed around a bend, they see a Russian-built enemy tank heading toward them. Unable to turn, the U.S. sol-

dier frees his prisoner; they lift the jeep, spin it on its axis, and speed away. They later became best friends—but, first, they had to become partners.

This chapter details the advantages of working in a partnership with regulators or trade organizations. The value to your firm in market exposure, business contacts, and positive press are just a few of the advantages such partnerships engender. However, a large commitment is involved once you have chosen this route. But the payoffs in profit and a considerable increase in safety performance make the effort worth it. Several case studies are also given that prove safety success—the kind of success that changed the way organizations like General Electric do their business *and do safety.*

THE NEED FOR PARTNERSHIPS

From OSHA's Perspective

It is apparent that we are killing American workers at an embarrassing rate. Though we preach "safety first" the workplace still kills over 60,000 moms, dads, and kids every decade—and all they did was go to work. As safety professionals, our job is to make sure employees go home each night *uninjured.* In response to injury statistics, OSHA recognized the need to establish good-working, nonconfrontational relationships with employers, and in 1982 it organized the Voluntary Protection Program (VPP). About partnerships, John Henshaw Assistant Secretary of Labor for Occupational Safety and Health has written:

> The process is simple and straightforward. We start with a shared commitment and a common goal among partners: preventing injuries and illness in the workplace. We all stand behind a simple understanding: safety and health add value—to your business, to your people, and to your life. And each of us seeks out opportunities to work together to identify and control workplace hazards before an injury or illness occurs. America's workers and businesses deserve our best efforts to reduce injuries and illnesses on the job. To get these positive outcomes, we need to work in partnership (*Job Safety and Health*).

From the Construction Contractor's Perspective

The contractor must also recognize the value of keeping people healthy. A clear link has been established between profits and a safe work site. Taking a proven partnering model like the OSHA Voluntary Protection Program and applying it to a construction project will result in:

- establishing a nonconfrontational relationship with OSHA
- confirming to the client that safety is a fundamental part of the project
- significant, positive exposure for the firm
- potential for an increase in profit
- a steep reduction in accident and injury rates
- increases in worker productivity
- enviable returns on investments.

Looking at the cost of injuries across the country, current data indicate workplace injuries produced $40.1 billion in direct costs (payments to injured workers and their medical care providers) in 1999, the last year for which this data is available. Workplace accidents had a total financial impact of between $120 billion and $240 billion in 1999, the last year for which information is available (Liberty Mutual Workplace Safety Index 2002).

Direct and Indirect Costs

A simple formula to help contractors calculate these costs is offered below:

$_____	A. Direct cost of the injury
$_____	B. Indirect cost of the injury
$_____	C. Total cost of the injury (A + B)
_____% (as decimal)	D. Profit margin on job where injury occurred
$_____	E. Added revenue company must generate to recover cost (C/D)

Here is one example worked out using an injury that PPE typically would prevent:

$10,000	A. Direct cost of eye injury
$30,000	B. Indirect cost of eye injury
$40,000	C. Total cost of the injury (A + B)
.05	D. Profit margin on job where injury occurred
$800,000	E. Added revenue company must generate to recover cost (C/D)

Daniel K. Shipp, *Roads and Bridges*, July 2002.

■■ **Figure 1** ■■ **Estimating the effect of injuries on a specific project**

CONSTRUCTION PROFITS THROUGH SAFETY

Good construction contractors are aware of the impact of an accident on their projects. In the past, accidents and *safety* costs were often considered overhead, and those expenses were spread across all active projects. Recently, losses that occur during the work are taken directly off the project profit margin. A project's safety record has become a significant factor in the bottom line. Like buying concrete or gravel, project safety costs are now budgeted and reviewed as the job progresses. If someone gets hurt, that is "money out," and a cost to the project.

Direct and Indirect Costs of Accidents

The direct cost of the injuries, treatment, and rehabilitation is easy to add up; however, often overlooked are the indirect costs. The Liberty Mutual Workplace Safety Index 2002 shows these direct costs produced between $80 billion and $200 billion of indirect costs (lost productivity, overtime, etc.). American employers absorb all of these indirect costs, while the financial impact of direct costs on a company depends on its specific workers' compensation program (self-insurance, size of deductible, etc.). (See Figure 1.)

The cost might seem excessive since the indirect cost of the injury, such as replacing the worker for the project, is minimal. But the effect of the accident on fixed profit/loss structures like insurance and bonding are significant: the increase in an Experience Modifier Rating (EMR) will stay with the firm for three years; compensation costs will rise; and if an OSHA investigation was involved, this must be listed on all bid documents. That can translate into lost bids.

Once the full cost of the accident is known, (i.e., the true loss), next consider the time directly needed to complete the work that was not performed by the worker or crew; this replaces the loss—not indirect costs. However, additional work must be completed *to retain the profit*. Since this extra work was not part of the bid, again, a loss is sustained. Safety professionals should make this an argument that is part of their accident prevention plan. Getting a project manager to see the potential cost of an injury to his bottom line is selling safety.

In these cases, money saved by not paying expected compensation costs on a project are used to quantify success. Again, money saved is money earned. Indirect costs must also be included when summarizing any safety success. The direct and indirect costs must be weighed against total project revenue to determine the savings of accident prevention *and* prove the effectiveness of everyone's safety efforts.

These hidden costs further prove the need to eliminate injuries and accidents through effective safety programs. Programs that must exceed high levels of success and reduce or eliminate injuries will have a dramatic effect on profit.

Why Partnerships Work

In this chapter, partnerships between those doing the work and those regulating the workplace will be explored. It will become clear that any partnership, in particular programs like OSHA's Voluntary Protection Program, provides a proven model for reducing the costs of injuries in the workplace.

The idea of a construction site in the VPP is something new. Traditional VPP members were the widget makers of the world—fixed industrial facilities with longstanding safety records and considerable data. By design, once a site is "constructed," there is no site! This makes the fit between the OSHA VPP and construction a difficult one.

A Safety Return on Investment

Current safety philosophers quote a return on safety investment of $3–$7 saved for every $1 spent on a safety program.

Roger Hannay, of Hannay Reels in Eastern New York, states, "Money saved is money earned," in reference to the profitability of his safety programs.

The safety professional soon learns that his or her efforts rarely put a check in the bank. Safety success is related to quantifying reductions in injury and illness rates and, more often, a decline in the Workers' Compensation Experience Modifier Rating.

Participation in a partnering program such as OSHA's VPP is a recognized method of confirming safety success. The return on investment (ROI) as reported by many of the participants easily exceeds the current 7–1 ratio.

PARTNERSHIPS IN CONSTRUCTION

History of OSHA and Previous Safety Partnerships

Since the days of Pliny the Elder, the deaths and illnesses of workers were recognized as directly affecting production. The toll on the workforce, though significant, was simply overcome. There was plenty of asbestos to be mined and slaves to do it. One must still consider the effects on the mine managers of that time, since their exposure was similar to the miners. It is likely the toxicity of the asbestos was more readily apparent when supervisors and foremen took ill—since their absence affected profit, and supervisors were probably more difficult to replace than slaves.

Profit has always driven changes in the workplace, and today's safety successes must be weighed in both the reduction of workplace injuries and illnesses and the return on investment.

CASE STUDY

The author had the pleasure (and pain) of putting a construction project through the OSHA Voluntary Protection Program. The project was a $30 million highway expansion that entailed moving bridges and railroads and converting a two-lane country road into a seven-mile divided highway (see Figure 2).

While a quarter of a million tons of blacktop were installed and 300,000 yards of dirt moved, the crew achieved the OSHA VPP Merit level. Of note was the fact that the job went 200,000 hours with not one lost workday. A review of the project savings due to the VPP put the return on investment near $283,000. Not a bad return on an investment of about $14,000.

Partnerships and Alliances

Partnerships are arranged for specific construction projects or long-term agreements. These agreements range from the recovery efforts at the World Trade Center to an understanding with the National Park Services. The Park Service had one of the highest injury rates for the Department of the Interior before entering a partnership with OSHA. In fact, the driving factor behind many *partnerships* is a fatality or excessive injury rate.

The myriad of firms participating in partnering with OSHA indicates both the business sense of participating and the value of trade organizations like the Associated Builders and Contractors (ABC) and Associated General Contractors (AGC). Both are leading trade groups for the construction industry and heavily promote construction safety. These organizations, insurance companies, and similar trade or lobby groups have developed agreements across the country in an effort to protect their members and improve work-site conditions *and their profits*.

Though many formal agreements are established at the national level, local chapters also entice their members to participate using incentives like annual awards and recognition programs.

■ ■ **Figure 2** ■ ■ **VPP Merit Construction Site, Canandaigua, NY** (Photo courtesy of Norm Stahlman, Rifenburg Construction, Inc.)

ASSOCIATED BUILDERS AND CONTRACTORS (ABC)

OSHA and the Associated Builders and Contractors are partnering to reduce the number of injuries, illnesses, and fatalities affecting participant employers; reduce worker compensation costs and OSHA penalties; and identify general contractors and trade companies with effective safety and health programs. This is a national partnership with individual agreements and implementation at the local level.

ASSOCIATED GENERAL CONTRACTORS (AGC)

The Associated General Contractors of America and OSHA partnered to improve safety and health at construction sites across the country. Goals include annual reductions of 3 percent in injuries, illnesses, and fatalities; implementation of effective safety and health programs at partner sites; and the provision of training for management, supervisors, and all employees.

Construction insurance firms are also driving their insured to consider VPP and similar partnerships since they are a proven method of reducing injuries. The owner saves on renewal fees, and the insurer reduces overall risk on the targeted project.

TARGETED PARTNERSHIPS

When accident rates indicate a trend in injuries, like the poultry processing trade, targeted partnerships have been put in place to deal with the hazard.

ORGANIZED LABOR

Organized labor also recognizes the value of entering into a partnership with OSHA and construction firms to which they supply labor. In many cases, partnerships require cooperation of any on-site organized labor force. When the agreements are made, union employees are a critical part of the success of the program. Members often participate in the daily inspections and attend local safety meetings and national conferences like the VPP annual conference.

Examples

Some recent or ongoing construction and industry partnerships include:

- *National Park Service (NPS)* OSHA's partnership with the National Park Service (and others, depending on the park) aims to reduce injuries and illnesses after NPS was identified as having the highest injury rate in the Department of Interior.
- *Roofing Industry Partnership Program for Safety and Health* Partners are OSHA; National Roofing Contractors Association; United Union of Roofers, Waterproofers and Allied Workers; CNA Insurance companies; and National Safety Council. The partnership encourages roofing contractors to improve safety and health performance, and seeks to foster an environment of cooperation among partners.
- *Motor Company/UAW/OSHA* Partnership between OSHA; International Union, United Auto, Aerospace and Agricultural Implement Workers (UAW); and the Ford Motor Company to reduce injuries and illnesses at each Ford location through the creation of a proactive health and safety culture and a cooperative, nonadversarial relationship that optimizes the resources of all parties. Systematic anticipation, identification, evaluation, and control of health and safety hazards is an objective designed to continuously reduce worker injury and illness.
- *Tyson Foods* OSHA has entered into a partnership with Tyson Foods, covering two of their poultry processing plants (one in Arkansas and one in Missouri). The five-year agreement has a goal of improving and strengthening the company's safety and health programs, reducing injuries and illnesses, and serving as a model for improved worker protection throughout the company.
- *Maine Safe Logging 2000* Employers and employees (nonunion) and OSHA have partnered to reduce fatalities and serious injuries in the logging industry in Maine and to develop effective safety and health programs.
- *NPS—Cape Cod National Seashore* The National Park Service, OSHA, and AFGE

Local 3789 have joined as partners to reduce injury and illness, develop and implement an effective safety and health program, promote safety awareness and education, encourage employee participation in hazard identification and abatement, significantly reduce the lost-time case rate, and to establish a system of accountability for safety and health at Cape Cod National Seashore and its park units in Wellfleet, Massachusetts.

- *Maine Healthcare Association* This partnership between the Maine Healthcare Association, OSHA, and nursing homes and personal care facilities in the state of Maine has the goal of reducing injuries and illnesses as well as streamlining or simplifying the implementation of industry regulations and procedures.

- *NJ Pilot Silica Partnership [New Jersey]* OSHA, the NJ Department of Transportation, the New Jersey Department of Health and Senior Services, the New Jersey Department of Labor, the Utilities and Transportation Contractors Association—NJ (UTCA), the Laborers Intl. Health and Safety Fund, Laborers Locals 172 and 472, NIOSH, and several major highway construction contractors partnered to control silica hazards in the heavy highway construction industry.

- *NJ Highway Work Zone* Partnership with the New Jersey State Police, New Jersey Department of Transportation, UTCA and Laborers Unions, and Rutgers University to address the high rate of injuries and fatalities among highway construction workers. A major component of the agreement is safety hazard awareness training for state and local police and DOT staffers. The agreement calls for contract language requiring commitment to safe practices in order to bid on state-funded highway work. OSHA provides outreach to contractors and training to state personnel.

- *NPS—Fire Island Seashore* The National Park Service, OSHA, and AFGE Local 3432 have joined to reduce injury and illness, develop and implement an effective safety and health program, promote safety awareness and education, encourage employee

participation in hazard identification and abatement, significantly reduce the lost-time case rate, and establish a system of accountability for safety and health at Fire Island National Seashore. The partnership covers 100 employees at Fire Island.

- *World Trade Center* The WTC Emergency Project Partnership formalizes a commitment to safety and health among contractors, employees, employee representatives, and governmental agencies participating in the emergency response efforts at Ground Zero in lower Manhattan. The agreement outlines cooperative efforts to ensure a safe work environment, including a site-orientation training program to familiarize all participants with potential hazards and personal protective equipment requirements.

- *Staten Island Recovery Project World Trade Center* The WTC Project Staten Island Recovery Operation Partnership calls on the partners to exercise leadership in preventing occupational fatalities and serious injuries and illnesses for all workers involved in the WTC Staten Island Landfill Recovery Operation. Partners agree to work cooperatively with all organizations assisting in the operation to ensure implementation of and compliance with the WTC Staten Island Recovery Operation Emergency Project Environmental, Safety and Health Plan; to abate all serious hazards immediately; and to share safety hazard and exposure monitoring data. Joining OSHA in the partnership are the U.S. Army Corps of Engineers; the Environmental Protection Agency; New York Police Department; New York State Department of Environmental Conservation; New York City Department of Health; New York City Department of Sanitation; Hugo Neu Schnitzer East; Phillips and Jordan; Evans Environmental & Geosciences; Yanuzzi & Sons, Inc.; Mazzochi Wrecking; Taylor Recycling Facility LLC; International Union of Operating Engineers; Local 14-14B and Local 15; and Garner Environmental Services, Inc.

- *OSHA/Bechtel Delayed Coker Plant* The partnership agreement is specific to work being performed by Bechtel and its subcontractors during the construction of the Delayed Plant Project, located at the Hovensa, LLC site in St. Croix, U.S. Virgin Islands. Partners are OSHA, Bechtel, HOVENSA, the U.S. Virgin Islands Department of Labor, and the United Steel Workers of America (Local 8526). This partnership supports OSHA's strategic plan to reduce injuries, illnesses, and fatalities in the construction industry and has reduction of fall, electrical, struck-by and caught-in hazards as an objective. A 20 percent reduction in the OSHA Recordable Incident Rate, per year, is also a goal.

- *National Park Service—National Parks Central* OSHA, NPS, International Brotherhood of Painters and Allied Trades Local 1997, the Fraternal Order of Police, and the U.S. Park Police Labor Committee are partnering to improve the employee safety and health program, promote safety awareness, and reduce injury and illness in Washington, DC parks. The agreement provides for employee participation in hazard identification and abatement, training, and the implementation of an effective safety and health program. The partnership seeks to reduce the lost-time case rate from 15.56 in 1997 to 7.78 in 2003.
- *Baltimore Marine* This agreement between OSHA, the Baltimore Marine Industries, the Industrial Union of Marine and Shipbuilding Workers of America, and the International Association of Machinists and Aerospace Workers, Local Lodge S33, seeks to prevent fatalities, accidents, and injuries resulting from work related to ship disposal within the Baltimore/Washington OSHA Area Office jurisdiction.
- *Telecommunication Tower Safety & Health Partnership* The goal of this partnership between six Philadelphia broadcasting tower owners, I.B.E.W. Local #1241 and Ironworkers Local #401, and OSHA is to prevent all accidents, injuries, and fatalities resulting from work on or from telecommunication towers. Impetus for this partnership: 32

fatalities occurred in the five years prior to this partnership agreement.

- *NIST-AML/OSHA/Clark* A partnering arrangement between OSHA, the National Institute of Standards and Technology, and Clark/Guilford covers construction of NIST's Advanced Measurement Laboratory. The partnership has goals of encouraging and assisting Washington, DC construction contractors to improve their safety and health performance, to eliminate serious accidents, and to recognize those contractors with exemplary safety and health management systems.
- *Georgia Poultry Initiative* Nineteen local poultry plants, members of the Georgia Poultry Association, have partnered with the United Food and Commercial Workers and OSHA to focus on hazard identification, employee training, and implementation of effective safety and health programs. The partnership has distributed video training tapes to the poultry industry in the Atlanta area. More than 19 employers and 9964 employees are participating. Thus far, 11,051 hazards have been identified.
- *NATE* Partnership between OSHA and the National Association of Tower Erectors (NATE) to prevent serious accidents and fatalities in the tower erection industry through enhanced safety and health programs, increased training, and implementation of best work practices.
- *Oil and Gas Well Servicing* Association of Energy Service Companies, Mitchell Energy, the Energy Service Company of Bowie, Texas, and the Texas Workers' Compensation Commission, covering 31 employers and 1000+ employees, partnered to develop and utilize effective safety and health programs.
- *SCA Shipbuilding* The Shipbuilder Council of America (SCA), Houston SCA members, and OSHA have partnered to reduce injuries and illnesses at SCA member-sites located in and around Houston, Texas. These member-sites are involved in the building, repair, and cleaning of all types of vessels.
- *Grain Handling Industry* The purpose of this Omaha partnership is to conduct training

and outreach and to encourage voluntary compliance. Outreach packets (650) have been mailed to grain handling employers, and seven seminars have been held with approximately 375 attendees. Fatalities went from 3 to 0, grain elevator explosions went from 1 to 0, and severe injuries have been reduced by 50%.

- **SESAC** More than 30 small businesses in Colorado's steel erection industry are partnering with OSHA and both union and non-union workers to improve workplace safety and health. A unique outcome of this OSPP is the development of a training school (built through industry and county donations) that is shared among the 38 employers.

ESTABLISHING A PARTNERSHIP

Prepare the Foundation

As a safety professional, you must be prepared to deal with numerous issues before considering an alliance or partnership with a trade organization or OSHA. The foundation for this success is trust. Trust will be required of your workers, your boss, and the partners. It will not be an easy task. Unless the firm has a philosophy of "safety first" and means it, your battle will be to show everyone the value and the return of such a partnership. Remember, it is a balance of profit and safety that must be sold. Every successful safety professional knows you must sell safety. When you have sold your firm on the idea of a partnership and they are looking forward to the next step, you have won the battle.

Establish a Company Culture that Will Accept Working with "the Enemy"

One recognized obstacle to establishing a relationship with OSHA is an inherent distrust of the department. A common fear is that OSHA will come back after a partnering meeting to issue citations. This mindset of labor versus regulator is changing, but the philosophy needs to be tempered with hard successes on both sides.

Since its inception in 1970, OSHA has been looked upon by many as the *warden* watching over the workplace, while business owners have been seen as the *prisoner*. Surprise visits, anonymous tips, focused inspections, and tips by disgruntled employees often bring compliance officers into a facility or work site, unannounced or uninvited. These visits are rarely friendly. Owners respond with attorneys, tips to the employees on "What to do if OSHA visits," and an overall picture of the Department of Labor as safety cops. The press further expands on this by photographing OSHA inspectors at every catastrophic event they can—from a bridge collapse to the September 11th attack on New York's World Trade Center. OSHA does not supply articles to the newspapers claiming "Acme Industrial Undergoes OSHA Inspection—They Did Great!" Regrettably, it is the fatalities investigated and the citations issued that make the news.

Changes Are Taking Place at OSHA

An OSHA Compliance Assistance Specialist (CAS), noted to a group of newly trained construction safety professionals that one of the advantages of the OSHA VPP effort was, "You can tell your clients OSHA will stay away for three years." A member of the crowd then commented that accident and injury rates were also about half the industry average at VPP sites. "Oh, yeah, that too," the inspector replied. He then went on to speak about the value of recent emphasis programs, targeted inspections, focused inspections, enforcement, and increases in penalties to achieve compliance. After a 30-minute lecture on what OSHA was hard at work doing, someone raised his hand and asked why prevention was not once mentioned? In partnering, both sides have some learning to do.

OSHA's initiative toward working with those it regulated was the placement of a Compliance Assistance Specialist at each regional office. This staff is dedicated to helping employers understand, meet, and exceed the standards. They have no citation authority and are valued in the safety community for their insight on *best practices* on compliance and the ability to answer "I have this friend . . ." questions.

Establish a Relationship with OSHA

Getting inspected and cited in many cases may lead to the first heart-to-heart with your Regional OSHA Director across his conference table at an informal citation hearing. Safety professionals recognize the need to establish relationships long before citations are issued. But, as mentioned earlier, there is an inherent distrust to be overcome.

Your first and best step is to call OSHA and ask them to come to *your* conference room for a visit. This invitation will go a long way toward any further relations. If you are comfortable asking them to your office, then you have trust. You have also taken the lead. Trust and leadership will be needed throughout the partnership program.

A firm could also participate in a state-sponsored safety program. Often funded in part by OSHA, these programs—like NY state's SHARP (Safety and Health Recognition Program)—are a great step toward achieving and exceeding safety standards.

Another way to get comfortable with OSHA is to invite them to speak at one of your safety committee meetings or your annual corporate staff meeting. Find a topic that you and the committee are interested in and ask them to present.

You need to get your firm's name recognized in the regulator's world as a firm that is a high performer. And if you are the first among your competitors to give them a call—that's market advantage. Remember, safety is selling. If your first meeting with OSHA is at their table and there are no citations being discussed—that's a partnership!

Safety Committees Spell Partnership Success

The value of effective safety committees to the safety professional has been recognized for years. Having employees meet to improve safety at a construction site puts the power of the entire workforce behind the effort. Give them a little recognition for their safety efforts and you have one of the fundamentals in place for targeting a partnership or alliance—a healthy, working safety committee. To consider a partnership, alliance, or VPP recognition, a safety committee is the key.

OSHA'S VOLUNTARY PROTECTION PROGRAM

Program Overview

Though partnering and alliances are effective as tools, they fall short of the success of the OSHA VPP. This program is the Cadillac of partnerships. The application process is difficult, the inspection can be painful, but the reward is immeasurable. OSHA clearly states, ". . . applying for VPP is a major undertaking."

"Preparation of your application will involve a thorough and detailed review of your work-site safety and health program, first by you and then by OSHA" ("So You Want to Apply to VPP? Here's How to Do It!" OSHA, 1997).

Within the Federal VPP structure, there are also state-sponsored VPP programs that may have differing requirements for participation. Figures 3 and 4 show the continued growth in both the federal and state VPP plans.

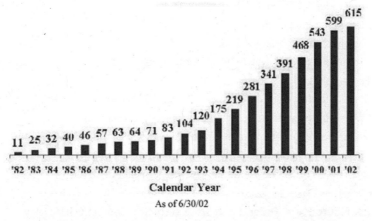

Calendar Year

As of 6/30/02

■ ■ **Figure 3** ■ ■ **Growth of companies participating in the federal VPP** (Source: OSHA, Division of Voluntary Programs)

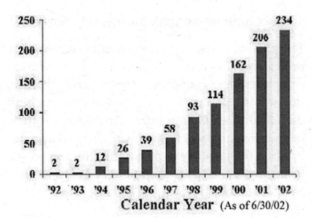

■ ■ Figure 4 ■ ■ Growth of state-plan VPP sites
(Source: OSHA, Division of Voluntary Programs)

The U.S. Department of Energy (DOE) has also had great success in VPPs at facilities across the United States. Figure 5 presents some interesting statistics on federal and state VPP sites.

LEVELS OF RECOGNITION: STAR, MERIT, AND DEMONSTRATION

In the early years of the VPP program there were three program levels: STAR, TRY, and PRAISE. Some years later, these were renamed: STAR, MERIT, and DEMONSTRATION.

You may be considering recognition for OSHA's premier program—*Star*—open to companies with comprehensive, successful safety and health programs that are in the forefront of employee protection. Or you may qualify for *Merit*, designed for firms with a strong commitment to health and safety that need a stepping stone to achieve Star performance. Or perhaps you have an alternative approach to safety and health that does not meet all of the requirements of VPP, but still protects your employees at the Star level of excellence. Then you would apply to the *Demonstration* VPP.

CONSTRUCTION AND VPP

One downside of the VPP is the difficulty of entering a construction project into the program. In many cases, construction work may require only one or two years for completion. You would put the effort into a project, obtain recognition, and be out of the program in a year or so. The ideal model for a construction site would be a project that would extend three years and be valued over $10 million.

One current construction partnership requires participants to have an EMR of 1.0. Rewarding safety achievement for average performance must be looked at closely. To drive down accident and injury rates, safety professionals must target exceptional rates.

One VPP firm had lost workday incidence rates (LWDIR) that dropped from 17.9 to 0.6 after initiating the VPP philosophy. To simply *meet* the standard

---

OSHA VPP SITES

Total federal and state VPP sites – 840

Sites in federal programs – 610

Sites in state programs – 230

Federal sites with fewer than 200 employees – 200

Federal sites with Star status – 547 (90%)

State with the most federal sites – Texas (127 sites)

State with the most state programs – North Carolina (48 sites)

Union/nonunion breakdown of federal sites –39% union, 73% nonunion

Employee breakdown of VPP – 39% union, 48% nonunion, 13% other

Industry with the most federal sites – Chemical (157)

■ ■ Figure 5 ■ ■ Federal and state statistics (Source: OSHA, May 31, 2002)

U.S. BUSINESSES	
3M	Marathon Oil
American Refuel	Mead Paper
Bestfoods	Milliken
BP	Mobil
Bridgestone	Monsanto
Delta Air Lines	Motorola
Dow	NASA Langley
Entergy	Nutrasweet
FMC	Occidental
General Electric	Pactiv
Georgia-Pacific	Potlach
Gillette	Rhom + Haas
Goodyear	Rockwell
Halliburton	Solutia
Honeywell	Solvay
IBM	Tropicana
International Paper	U.S. Postal Service
Johnson Space Center	Westinghouse
Kerr-McGee	Weyerhauser
Kiewit	Wheelabrator
Kraft Foods	White Sands
Lockheed Martin	

■ ■ **Figure 6** ■ ■ **Selected companies currently participating in Voluntary Protection Programs** (Source: OSHA, October 2002)

will result in average performance. OSHA is looking for *high-performance* firms to lead this partnership initiative.

From Frito-Lay to the Kennedy Space Center, from Tropicana to the White Sands Missile Base, these are just a few firms that enjoy VPP success. The selected companies listed in Figure 6 are a veritable "Who's Who" of U.S. businesses. In fact, John Henshaw, the Assistant Director of OSHA, was recruited from Monsanto—a VPP firm.

INTERNATIONAL PAPER AND GENERAL ELECTRIC

Many of the firms listed have multiple plants and projects accredited in the VPP Program. International Paper and General Electric are just two companies that have over 20 sites in the program—and that add several more each year. Targets for both firms are to have all of their sites and plants in the program.

They recognize VPP as good business, and as the list indicates, so do their competitors.

The Benefits of Participating in a VPP

VPP participants establish and maintain excellent safety and health programs in their workplaces that are recognized by OSHA as models for their industries. Cooperative interaction with OSHA gives companies the opportunity to provide OSHA with input on safety and health matters and to provide industry with models of effective means for accomplishing workplace safety and health objectives. While it certainly is necessary to maintain compliance activity, resources used to promote the Voluntary Protection Program will benefit both the worker and the bottom line. When a compliance officer cites a work site for unguarded machinery, the company pays the fine and provides appropriate guards. When a work site institutes the elements necessary for membership in the VPP, every person entering the work site is protected from occupational safety and health hazards, and the other improvements follow.

VPP participants are not subject to routine OSHA inspections because OSHA's VPP on-site reviews ensure that their safety and health programs provide superior protection. Establishing and maintaining safety and health programs on the VPP model are reflected in substantially lower than average worker injury rates at VPP work sites (OSHA, 2002):

- Injury Incidence Rates: In 1994, of the 178 companies in the program, 9 sites had no injuries at all. Overall, the sites had only 45 percent of the injuries expected, or were 55 percent below the expected average for similar industries.

- Lost Workday Injury Rates: In 1994, of the 178 companies in the program, 31 had no lost workday injuries. Overall, the sites had only 49 percent of the lost workdays expected, or were 51 percent below the expected average for similar industries.

While protecting workers from occupational safety and health hazards, companies following the management guidelines mandated for VPP membership also experience decreased costs in workers' compensation and lost work time, and often experience increased production and improved employee morale.

VPP Success Stories and Statistics

To sell your company on a VPP program, the first thing you will need are several partnering success stories from other companies in your industry—maybe competitors or your suppliers. Then take one of your examples and get it out to the guys in the field. Perhaps put an article in a newsletter or general posting. Just like a politician "leaks" an idea out to get a feel for the response, you need to do that as well. Here is one example cited in the August 2002 issue of *Occupational Health and Safety.*

"BRB Contractors, Inc., was a key player in a partnering program between OSHA and the Kansas City Contractors Association (KCA). As a result of these efforts BRB, which has over 1000 employees working throughout the United States, had zero worker compensation claims and zero lost time."

The following OSHA "success stories" highlight the results of a focused safety program. Implementing effective safety and health programs reduced their injury rates.

- The lost workday case rate at Thrall Car Manufacturing Company in Winder, Georgia decreased from 17.9 in 1989 when the facility began implementing a quality safety and health program to 4.6 in 1992 when the plant was ready to qualify for the VPP Star Program. In 1994 the rate was 0.6.
- At Monsanto Chemical Company's Plant in Pensacola, Florida, which employs 1600 workers, the lost workday case rates have steadily declined during the period the work site was implementing effective safety and health programs and in the four years since approval to the VPP. The rates fell from 2.7 in 1986 to 0.1 in 1994.
- Mr. Robert Brant testified to the following experience for Mobil Chemical Company on August 30, 1988. Between 1983 and 1987, Mobil Chemical Company brought all of its then-existing plants (plastics production

and chemical plants) into the VPP. During this period, recordable injuries for the company were reduced by 32 percent. Lost workday cases were reduced by 39 percent. The severity of cases was reduced by 24 percent. In subsequent years through 1994, recordable injuries and lost workday cases continued at the low rates experienced at all these Mobil Chemical Company plants.

- Mobil Oil Company's Joliet, Illinois refinery experienced a reduction in its lost workday case rate from 3.8 in 1987, the year before it began implementing VPP quality safety and health programs, to 0.2 in 1994, three years after approved to the Star Program.
- Occidental Chemical Company determined that, as their Safety Process Systems Implementation percentage increased company-wide, their injury/illness rate decreased from 6.84 in 1987 to 1.84 in 1993—a 73% decline.
- Testimony from Ron Amerson at Georgia Power provided this information. In the construction industry, Georgia Power Company brought two large power plant construction sites into the VPP in 1983 and 1984. By 1986, one site had reduced its total recordables by 24 percent and its lost workday cases by one-third. The other site reduced recordables by 56 percent and its lost workday cases by 62 percent.
- Between 1989, when Thrall Car's plant in Winder, Georgia, began implementing its programs to qualify for the VPP, and 1992, workers' compensation costs dramatically declined by 85% from $1,376,000 to $204,000.
- Mobil Chemical Company reduced its workers' compensation costs by 70 percent, or more than $1.6 million, from 1983 to 1986, during the years it was qualifying its plants for the VPP. This reduction has been sustained through 1993.
- At Georgia Power's two power plant construction sites, the direct cost savings from accidents prevented was $4.14 million at one site and $.5 million at the other for 1986 alone.

- Mobil Oil Company's Joliet, Illinois refinery experienced a drop of 89 percent in its workers' compensation costs between 1987 and 1993.
- During three years in the VPP, the Ford New Holland Plant noted a 13 percent increase in productivity and a 16 percent decrease in scrapped product that had to be reworked.
- During one evaluation of the Kerr-McGee Chemical Corporation's Mobile, Alabama plant in July 1991, the VPP team found that, at the same time work-related injuries continued to decline, production hit an all-time high that exceeded their goal by 35 percent.
- One Mobil Chemical Company plant manager has testified that the adoption of a single work-practice change at his 44-employee chemical plant during the first three years of VPP participation resulted in increased volume of product and a savings of $265,000 per year.
- In the three years since its approval to the VPP, Mobil Oil Company's Joliet refinery reports a 25 percent decrease in absenteeism and the highest employee morale ever experienced; in the same period productivity and quality remained high.

Introduction and maintenance of a quality worksite safety and health program often leads to improvements in employee morale, in productivity, and in product quality. OSHA has received considerable information on such improvements. Although anecdotal in nature, these improvements are referred to frequently enough by participants in the VPP to indicate there is a good possibility of a direct relationship between improved management of safety and health protection and these benefits (OSHA data, 2002).

Usefulness of the VPP to OSHA

Participants provide OSHA with examples of the most effective way to protect workers in their industries, often in ways which exceed the requirements of OSHA standards.

Participants provide effective input into OSHA's standard-setting process.

- CIBA Inc.'s McIntosh, Alabama site and DOW Chemical's Freeport, Texas facility both provided OSHA standards-setters with demonstrations of effective 100 percent fall protection.
- Many participants in the petrochemical industry provided OSHA standards-setters with models of effective process safety management.

Participants provide training for OSHA staff.

- Mobil Oil Company's Joliet, Illinois refinery provides OSHA compliance officers with hands-on training in process safety management.
- Participants from several companies help provide OSHA compliance and program personnel with training in safety and health program management.

Participation in VPP on-site evaluations provides OSHA with model plants that can be shown to others. As one compliance officer said at an evaluation closing conference: "I've always felt it could be done this way, and now I've seen that it can be done right, so no one will ever be able to tell me again that it can't be done." (OSHA data, 2002)

The application for participating in a voluntary protection program is a breath of fresh air in the world of regulations. Words like "delighted" and "we are glad" are used in the introduction. The application is available from the OSHA.gov Web page or in booklet form ("So You Want to Apply to VPP? Here's How to Do It!" U.S. Department of Labor, Occupational Safety and Health Administration, 1997).

OSHA's commitment to the program is clearly stated in the introduction: "We are glad you are seriously considering VPP, and we look forward to reviewing your application. OSHA's VPP is a strong component of the 'New OSHA's' commitment to partnership with companies that want to do the right thing—improve workers' safety and health."

You have a choice of downloading the application or contacting your regional OSHA office for a copy. The latter is recommended since the request will notify them that you are taking a serious look at the program. Take the time to meet and speak with a regional OSHA Compliance Assistance Specialist (CAS), express your interest and ask for a contact in the area. Do not reinvent the wheel; over 800 successful firms have already done it!

A Review of the VPP Application

The application for VPP runs 23 pages and is prepared more like a worksheet than an application. Initially, OSHA is looking for information on your firm—the kind of work you do and how many people do it. Data must also be provided on the site, organized labor, and company contact information. Your firm's safety record is also needed, specifically injury rates and lost workday cases and comparisons against the Bureau of Labor Statistics for your trade. The balance of the application mirrors the philosophy of the VPP program:

- management leadership
- employee involvement
- work-site analysis
- hazard prevention and control
- training.

Preparation of the application may take several dedicated weeks to several months, depending on the project scope. Honesty in preparation is most important. During the inspection of the project, OSHA will verify everything you stated your firm did in the application. Again, trust is an underlying theme of this partnership. Should an error or misstatement be uncovered during the early hours of the inspection, it will likely tarnish the balance of your efforts.

The following are some tips for completing the VPP application:

- Do not state something you cannot prove you did or do.
- Dedicate one person to oversee the coordination of the information.
- Spread the completion of the application around; use your safety committee.

- Prepare the text in the first person for easy reading: "At Acme Supplies we . . .".
- Use a successful VPP application as an example (ask your mentor).
- Keep it accurate and short—if you do accident investigations, say so. There is no need to send the policy. They will ask for it during the inspection.
- Never lie.

Before starting the work on the application, get familiar with the requirements. Obtain a copy of the *Training Education Document (TED) 8.1a—Revised Voluntary Protection Programs (VPP) Policies and Procedures Manual*, commonly referred to as the TED. This is a comprehensive, 241-page manual for the entire program, including the inspection team requirements. A review of the TED and application booklet is critical for what OSHA needs and what you are expected to supply. For example, if the application asks for employee involvement in accident investigations, and you have no such policy, get some people trained, let them get some experience—now you have both formed a team and satisfied another component of the application.

Some of the major benefits of a close look at the VPP application are:

- It will highlight areas you do well.
- Programs that you likely need will be put together.
- Training needs that might have been put off become a priority.
- Small but important items—like your annual respirator written review—get done.

Finally, if you put together all the items needed for the VPP, try them out in the field, and prove they are effective, but then decide not to apply; you have made some significant, recognized steps toward tightening up your safety program.

THE INSPECTION

Preparing for the Inspection Team

Suppose you submitted the VPP application a few months ago and you just received a phone call from your regional VPP inspection team requesting a date for the inspection. This is where the rubber meets the road. OSHA will be at your construction site for a week or so with a team of professionals, looking at everything. Will they write citations? No. Can you fix little things they might see that need correction? Maybe. Will the inspection disrupt your construction site? Sure. OSHA is on your site and you asked them there!

Length of the Inspection

The length of the inspection depends on many factors. Should the inspection go poorly, the team might hand back the application and suggest the site attempt recognition at a later date. This is a rare occurrence. However, how much time the team will spend on your construction site is important to the success of the effort.

The site must be prepared for the OSHA team to visit all operations, pull employees out of productive work for interviews, and other activities—all disruptive and distracting.

The duration of the visit will depend upon the size and complexity of the site (refer to TED 8.1a):

- Planning for the duration of the visit should take into account time needed to hold the opening conference, conduct the on-site review, prepare the draft preapproval report, and carry out the closing conference.
- On-site visits should average 4 to 5 days, including travel time, unless the site has some unusual characteristics (e.g., the site is large and/or the processes are complex).
- Chemical plants producing, using, or storing one or more "highly hazardous chemicals" may require additional time for the on-site inspection.

Preparing for the Inspection

If your application has been approved and the inspection scheduled, you are ready. Perhaps some little things will be found, but if you have had an effective safety program and the accident rates to show it, you're already doing a lot of things right. VPP is simply the next step in your safety progression. Many safety professionals say that getting VPP on their site was the biggest accomplishment of their careers. This is the time to show off your project site and your company's commitments to safety.

The Inspection Team

Your inspection team will be composed mainly of OSHA employees. The staff will be represented by a team leader, compliance specialist, and industrial hygienist. Depending on the type of construction project, the team might also include a special government employee (SGE). These are private industry volunteers that work at a VPP site (similar to the one being inspected) and function as volunteer inspectors for OSHA. They are OSHA-trained, and offer specific insight into your project due to their experience in the trade. The SGE is a valuable tool for you and the inspection team. He or she understands why you might have to perform a task a certain way, whereas OSHA might look at it strictly from a compliance perspective. The SGE brings common sense to the team.

WHAT OSHA WILL NEED ON SITE

Before the inspection, the OSHA team leader will send a letter confirming the inspection date to the VPP contact. Expect that some additional things will be requested. They might include:

- a computer and printer for the entry of field notes and the draft inspection report
- dedicated space for the team to work
- private areas for interviews of staff and employees
- suggestions on lodging for the team if needed
- a request for the inspectors to view company documents and records
- OSHA logs for the past several years.

A common question is "Do we have to feed these guys?" Well, yes and no. This team was invited to your site, but since they do represent a federal agency, free lunches might be considered a gift. It is recommended that during the inspection, the on-site safety committee meet with the inspection team at lunch to review the morning's progress. Have lunch brought in for the crowd and simply ask the inspection team to join you.

WHAT OSHA WANTS TO SEE

Evidence of Effective Safety and Health Programs
OSHA will form an impression on the effectiveness of your safety efforts through the inspection and interviews. As a safety professional, you will need all your skills as you accompany the inspection team, especially your listening skills. It is also important that you take charge of the inspection. Take the time to tell the inspection team what *you* expect of them.

They should be accompanied at all times by one of your staff. However, they may take someone aside for an informal interview. Respect this and step away, out of earshot.

If your site requires safety vests, hard hats, and work shoes, make sure the team wears them. Keep in mind that you asked them to your site as guests. The inspectors spend the majority of their time writing citations. Listen to them, and perhaps teach them. Everyone has something to learn, and you have everything to gain. Now is not the time to get defensive. Cooperation and trust will be the keystone of the inspection's success.

TIPS FOR THE INSPECTION PROCESS

Always be courteous. The inspection team is your guest, and they are devoting time away from their work and families to complete the inspection. To make it easier:

- If they note something, record it yourself.
- Try to lead the tour and limit free-ranging.
- Practice mock interviews with site employees before the inspection.
- Put up a bulletin board and post your safety alerts, near-misses, atta-boys, etc.
- If they have a question, explain why it's done that way. They may not understand the task.
- If you can fix something they find, correct it right away.

WHAT YOU DON'T WANT OSHA TO SEE

The key to a successful VPP inspection is the site safety committee. Long before the OSHA VPP team arrives, they should be doing weekly full-site inspections and making corrections as needed. Expect the OSHA VPP team to find some minor items during their visit; for example, antibiotic ointment might be expired in first-aid kits. Correct these items immediately and put a process in place so it does not reoccur.

SUBCONTRACTORS

Subcontracted trades are fair game during the on-site inspection. Though they were involved during the application process and during the subcommittee inspections, new employees and new vehicles might be inspected for the first time. The team will look at these vehicles in addition to yours for adequate maintenance, the condition of the tools, MSDS-matching for covered materials, lead cord condition, condition of lifting devices, and other things.

After the On-site Inspection

During the latter days of the inspection, your on-site team will indicate your level of success in the VPP—in most cases, Merit or Star. In either case, a celebration is in order. Put together a party to thank everyone on the VPP team, your mentors, and the regional OSHA staff, and remember to invite in your local Senators, Congressmen, and any other VIPs. This is your time to shine. Plus, you get a cool VPP recognition flag to fly!

Maintaining your VPP Status

If you qualify for VPP Merit recognition, the OSHA team will go over specific goals you need to address and will schedule another inspection in a year. Please keep in mind, Merit is still a fantastic distinction and everyone should be proud.

If you qualify as a Star site, you will be inspected again in three years. There are certain requirements to be met in both cases before the next inspection. The most significant is an annual report on progress at the site and examples of how the program has helped your business, your people, and your industry. Outreach in the safety community is encouraged (and expected) among VPP participants.

The Voluntary Protection Program Participants' Association

The Voluntary Protection Program Participants' Association (VPPPA), a nonprofit 501-C(3) charitable organization, is a leader in safety, health, and environmental excellence through cooperative efforts among labor, management, and government.

As part of its efforts to share the benefits of cooperative programs, the VPPPA works closely with the Occupational Safety and Health Administration, OSHA state-plan states, the Department of Energy, and the Environmental Protection Agency in the development and implementation of cooperative programs within the agencies. The Association provides expertise to these groups in the form of comments and stakeholder feedback on agency rulemakings and policies. The VPPPA also offers comments and testimony to members of Congress regarding legislative bills on safety and health issues.

VPPPA members include almost 900 companies and work sites that are involved in, or are in the process of applying to, OSHA's or DOE's Voluntary Protection Programs (VPP) and government agencies that are developing or implementing coopera-

tive recognition programs. For more information, contact the Voluntary Protection Participants' Association; 7600 E Leesburg Pike, Ste. 440; Falls Church, VA 22043; (703) 761-1146; Fax: (703) 761-1148.

REVIEW EXERCISES

1. Prepare a one-page summary for your facility manager to "sell" him on the idea of a partnership. However, the company already has an excellent safety record. You are the safety director for the firm and report to the manager.
2. Prepare a similar summary for your manager, but you have one of the worst lost workday averages for your region and morale is in the dumpster.
3. Think ahead five years. You have just joined the ranks of a biopharmaceutical firm as safety manager. The firm is two years old and they have never assigned safety responsibility in the plant. How would you approach getting a partnership like the VPP established within your five-year plan?

■■ **About the Author** ■■

SAFETY AND HEALTH ISSUES
FOR WOMEN IN CONSTRUCTION
by Carol Schmeidler, CSP

Carol Schmeidler, CSP, is the manager of General Safety and Industrial Hygiene Programs for the Department of Environment, Health & Safety Services at the State University of New York at Buffalo. Her professional experience includes training, safety program design, and hazard evaluation and control. She holds a Bachelor of Arts degree, cum laude, in psychology from the State University of New York at Albany. She is a member of the Western New York Federal Safety and Health Council, and a professional member of the Niagara Frontier chapter of the ASSE, and has served in all elected offices; currently, she is the chapter historian.

CHAPTER 33

Safety and Health Issues for Women in Construction

INTRODUCTION

The foreman walked toward me from across the cavernous, dark industrial floor. I was dressed for work in navy coveralls, steel-toed work boots, safety glasses, and hard hat. I saw his eyes widen when he saw my large, pink gear bag. "You're a girl!" he said, looking me up and down. "We don't get too many girls here; they get dirty."

"I clean up pretty good," I replied, and smiled at him. It wasn't the worst comment I'd ever heard; it wasn't like the whistles, gestures, sexually offensive comments—or like this response I got to instructions I'd given as a safety professional, ignored, because I "couldn't know what I was talking about. I was only a girl. . . ."

This chapter is for the women who get dirty—and those who don't—and all of the men who work with us.

In volume IX of *History of Women in Industry in the United States*, Helen Sumner writes:

The story of women's work in gainful employments is a story of constant changes or shiftings of work and workshop, accompanied by long hours, low wages, insanitary [*sic*] conditions, overwork, and the want on the part of the woman of training, skill, and vital interest in her work.

While there is a rich vein of literature on the history of women in the workforce, little has been written on the history of women in the construction workforce. Early histories of craftswomen or tradeswomen describe women working out of their homes, in manufacturing settings, during economic depressions, or in wartime, and early reports about safety and health list injuries sustained and illnesses contracted while working in factories and businesses. These reports do not include information about attempts to predict, control, or eliminate hazards.

Women's roles in construction are still being defined. Groundbreaking women who entered the construction industry as laborers or tradeswomen less than 30 years ago did so, in most cases, to make a better life for themselves and their children. Most did not want to be groundbreakers; they wanted to be wage earners. They didn't want to be injured or killed on the job; they wanted to be trained to do their jobs well—and safely. They didn't want to be abused, threatened, and humiliated; they wanted to go to work and go home feeling they had done a good job and had contributed to their projects. Their stories have been documented in books and studies. The lessons they learned and the impact they made on the industry have implications for today's workers—both men and women.

This chapter provides background on women in the trades, who led the way for the female construction "pioneers" of the mid-1970s and later. It examines issues that include differences between women and men, as well as hazards and protection that apply to all workers. To a much lesser degree, barriers to success for women will also be considered, including discrimination by coworkers and in equipment and machinery, training, and legislation. Finally, this chapter will examine the potential for improvement, and how making workplaces safer for women can benefit the construction industry as a whole.

While the information provided is applicable to other countries, the focal point of this chapter is the United States, due to differences in legislation, history, and customs elsewhere in the world.

The thrust of this chapter is that women are capable of contributing much to the construction industry despite differences in background, physiology, and training. Recognizing and addressing issues that affect women may decrease hazards for both genders.

HISTORY

The Way It Was

"The scarcity of labor supply in particular places or at particular times has often been responsible for the use of women's work. . . ." (Sumner, H., see the reference section)

The 5.3 million women working in the United States in 1900 comprised less than 20 percent of the total workforce. By 1950 18.4 million female workers were almost 30 percent of the total labor force. In 2001, 66 million working women counted as 46.6 percent of all workers.

Economic depressions, war, and the development of technology often took men out of the workforce, causing labor shortages that had to be (and ultimately were) filled by women. Men traditionally took newer, more skilled, and better-paying jobs, leaving the lower-skilled, lower-wage jobs to women. Depressions and economic downturns often took the primary wage earner, traditionally the male, out of the labor market, forcing women to take whatever job they could find to help their families, and at whatever wage they could get. Wars not only required the presence of men on the battlefield, but also created jobs in industries serving the war effort that men were no longer at home to fill.

The editors of *Working Women: Past, Present, Future* noted that since the early years when women were needed to fill the void in factories and businesses, work has been dirty and unsafe, with long hours, and often with unequal pay when compared to that of men. Workplace safety practices and procedures were virtually nonexistent. Hiring, job assignments, and salaries reflected the norms of the times. It was customary to employ women in less-skilled positions; company policy often required women's salaries to be, at best, less than that of the lowest-paid male.

In time, women formed many of their own unions, and were brought into the male-only unions during the Great Depression and World War II. As Helen Sumner tells us, male-only unions, which barred women from membership for decades, forced the entry of women into certain professions—often as strikebreakers—while keeping them out of other disciplines for a time. Also, in some cases, union support for women in the workplace originated because of policies stipulating that women be paid wages that were often one-half to one-third of men's salaries for comparable work. The United States government legislated this one-half to one-third policy during the Civil War for female clerks (Koziara et al., *Working Women: Past, Present, Future*). Private industry soon recognized this precedent.

Contemporary unions soon promised higher wages, more or less, to protect men's wages. The other main platform of activity in early union history was the reduction of work-hours in many businesses and industries. These initiatives met with mixed success for many years in several areas. Eventually, reducing the working hours was successfully accomplished.

The progression of the economy from farm-based to industrial, along with advances in technology and increases in population, produced new jobs. These and other factors led to a gradual increase in the number of women in the workforce—and in a wider variety of occupations. During World War I and World War II, women performed jobs vacated by men who went to war, or jobs created for the war effort. These jobs ranged from the production of aircraft and ordinance to skilled trade work in both government and industry. Women performed manual labor during World War II and again continued to learn skilled trades. In these cases, women were often at high risk from poor training and unsafe conditions in factories and plants throughout the country (AFL-CIO Fact Sheet).

The Huntsville Arsenal in Huntsville, Alabama, offers us a good example. It manufactured colored smoke-munitions, gel-type incendiaries, and toxic agents, including mustard gas, phosgene, lewisite, white phosphorous, and tear gas. The Redstone Arsenal, adjacent to Huntsville, produced burster charges, chemical artillery ammunition, rifle grenades, demolition blocks, and bombs of varying sizes and weights. At peak employment, Huntsville's labor force was 37 percent female. At Redstone, women comprised 54 percent of the labor force in 1944 and peaked at 62 percent in 1945.

Salary differentials were equalized at both Arsenals in 1942 when the National War Labor Board issued General Order 16, a voluntary order that allowed companies to provide "equal pay for equal work." *Equal pay acts* were introduced in a number of states, and some were actually enacted. Federal legislation, including the Gender Pay Equity Act introduced in 1945, and its successor, the Equal Rights Act, never passed. (AFL-CIO Fact Sheet)

During this time, three women died at Huntsville and two at Redstone from work-related causes.

There were also numerous serious injuries; many of the injured women were able to return to work.

In an article on the Redstone Arsenal Web site, Dr. Kaylene Hughes reports that all of the female employees at Redstone Arsenal were laid off by the end of October 1945 when all of the ordinance lines were shut down at the end of the war. Huntsville Arsenal's female employees were furloughed at the end of the war, and then laid off by September 1945.

During World War II the Vocational Education for National Defense (VEND) program trained men and women in defense work. Social services were supplied by a variety of organizations, including the National Travelers' Aid Association, the Family Welfare Association of America, the YWCA, the YMCA, labor unions, religious groups such as the Salvation Army, civic clubs, and multiagency coalitions. Day care was provided through a federal program at over 2000 centers, for at least 12 hours a day. The Works Progress Administration (WPA) also had nurseries located in schools. Although American women petitioned to keep the benefit of the day-care program after the war, it was terminated in February 1946. Several states continued the day-care programs for a few years after the war (Knapp).

The first woman to be certified as an aircraft welder in New Jersey received the same training—and pay—as her male counterparts. Her job ended after Hiroshima was bombed, and she did not work as a welder again until 1966, approximately the same time the Vietnam War started (*Hard-Hatted Women*).

The 1970s and Later

By the mid-1970s, about 18 percent of the 29 million blue-collar workers in America were women. An 80 percent increase of women working in skilled trades (from 277,000 in 1960 to 495,000 in 1970) was at a rate twice that of women in all occupations, and was eight times the rate of increase for men in skilled trades. Significant increases were seen in the number of women working in construction trades, such as carpenters, electricians, plumbers, painters, tool and die makers, and machinists. As high as these increases were, the absolute numbers and percentages of women in each occupation remained low

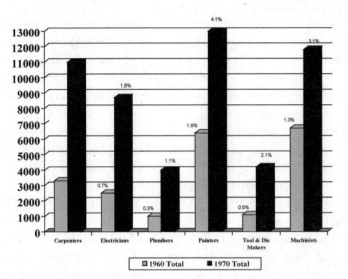

■ ■ **Figure 1** ■ ■ **Number of women in skilled trades, 1960–1970**
(Source: *Blue-Collar Women*, © 1981 by Mary Lindenstein Walshock.
Used by permission of Doubleday, a division of Random House, Inc.

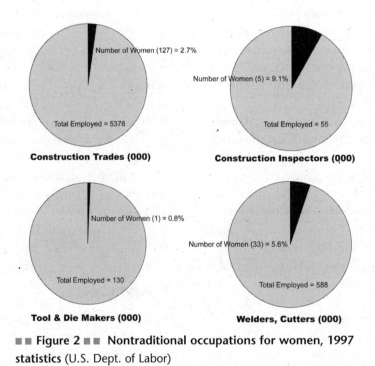

■ ■ **Figure 2** ■ ■ **Nontraditional occupations for women, 1997
statistics** (U.S. Dept. of Labor)

compared to males in these same occupations—
from 1.1–4.1 percent of each total (from *Blue-Collar
Women* by Mary Lindenstein Walshock). Refer to
Figures 1–3 for more details.

Although most unions welcomed women as
members, Koziara et al. point out that the following
unions were still all male in 1978:

■ United Slate, Tile and Composition Roofers
■ Damp and Waterproof Workers Association
■ Brotherhood of Maintenance of Way
Employees

In 2000, a total of 877,000 women were employed in the construction industry. The majority

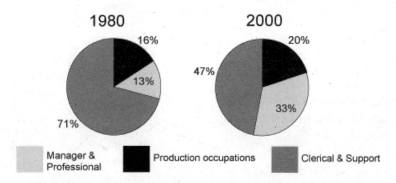

■ ■ **Figure 3** ■ ■ **Distribution of female construction workers among occupations** (Reprinted wtih permission from *The Construction Chart Book*, 3rd ed., the Center to Protect Workers' Rights, 2002, chart 19d)

of women are still in nonproduction occupations; however, the number of women in production jobs as well as those in managerial and professional jobs has risen in the last 20 years. Figure 3 is a chart illustrating the percentage of women in construction and other industries.

PROFILES AND PERCEPTIONS

A study entitled "Women in the Construction Workplace: Providing Equitable Safety and Health Protection" was submitted to OSHA in June 1999 by the Health and Safety of Women in Construction (HASWIC) Workgroup. The study and recommendations summarized a survey of tradeswomen by the Chicago Women in Trades (CWIT) as well as research studies conducted by the National Institute of Occupational Safety and Health (NIOSH).

Concerns were noted in these studies that were believed to significantly impact respondents' safe job performance. They included a hostile workplace, poor sanitary facilities, ill-fitting personal protective equipment and clothing, and on-the-job training.

■ Workplace hostility can take many forms, including overt violence and threats, sabotaged work, poor training, and isolation of the female worker. Although technically prohibited by law, it is often not formally reported, and stories of sexual harassment are heard repeatedly from women in construction. This includes exposure to explicit photos, unwelcome remarks, stares, gestures,

and unwanted physical contact. Stress resulting from dealing with these situations can lead to illness, or distractions, which may cause injuries.

■ Sanitary-facility concerns included no toilet facilities for women or for the entire job site, poorly maintained facilities, and lack of privacy when using facilities. Infections could result from not urinating when needed, or from using dirty facilities. Restricting water intake to delay the need to use facilities could put workers at risk for heat stress and other health problems.

■ Personal protective equipment includes clothing and equipment such as safety glasses, shoes, gloves, earplugs, and fall protection harnesses. Ill-fitting clothing and equipment does not properly protect the wearer, and can be a hazard by itself. Tools designed for larger hands may not work well with smaller hands, and techniques developed to use the larger tools may cause ergonomic problems for the smaller-handed user. Lifting guidelines and techniques developed using data from male workers may be inappropriate for female workers.

■ Training includes job-specific training and safety training. Due to hostility and attitudes of male co-workers, women may feel that they are not provided with opportunities to learn and practice job skills. There is also concern expressed that on-the-job safety training and education are not supported or encouraged by co-workers or bosses.

Response from the Questionnaire

One survey respondent from Massachusetts said, ". . . the subcontractor told me he wished I was a man so he could pop me one."

To gather additional information for this chapter, a questionnaire was distributed to the National Association for Women in Construction (NAWIC). Members were also encouraged to offer surveys to nonmembers.

This survey was not meant to be a scientific study; rather, it was to be a gathering of information, thoughts, and ideas of women in all areas of the construction industry. NAWIC has 5800 members in chapters in the United States and Canada, and has international affiliation agreements with groups in Australia, New Zealand, and South Africa (NAWIC Facts).

The questionnaire asked about respondents' occupations, safety training, on-the-job injuries and illnesses, working while pregnant, satisfaction with personal protective equipment, adequacy of bathroom facilities, exposure to inappropriate behavior, the biggest safety challenge, and the best and worst things about their job.

Almost 200 surveys were returned from women in 37 states and two countries. Many similarities with the previously mentioned study were noted. Results included the following:

- 10% of respondents were tradeswomen— plumbers, pipefitters, roofers, tile setters, electricians, truckers, road construction workers, hardware installers, communications installers, environmental/asbestos workers, and carpenters
- 10% owned companies
- 75% considered themselves management, clerical, or other; some women checked more than one job function. Professions covered a wide range, including safety professionals, general mangers, project managers, estimators, technical writers, marketing coordina-

tors, technicians, closet designers, trainers, saleswomen, and employment counselors

- 6% of respondents are union members; 45% of tradeswomen are union members
- 62% have been doing their jobs 0–10 years; 48% 11 or more
- 83% were trained to do their jobs safely; 95% of responding tradeswomen were trained to do their jobs safely
- 18% were injured or became ill on the job; 56% of tradeswomen responded yes to this question
- 30% of respondents worked while pregnant; 22% of tradeswomen
- 32% of respondents who answered the question about the fit/function of safety equipment were dissatisfied
- 88% of respondents said that there were adequate bathroom facilities at their work sites, although some split the question and answered "yes" for the office and "no" for the field; 39% of tradeswomen answered "sometimes" or "no." One respondent from Louisiana said, ". . . even the men complain."
- 21% of all respondents said they had been a target of inappropriate behavior, comments, or suggestions; 44% of tradeswomen answered "yes." Comments included reports of physical threats and humiliation, a sexual proposition that prevented a respondent from completing a college degree, and support from male co-workers in response to an incident
- 2% of respondents stated that they had done an apprenticeship (electrician and carpenter)
- 77% of respondents were 31–60 years old
- 84% said their employers promote safety, manage safety effectively, and address safety violations quickly.

Comments from the questionnaire about dealing with inappropriate behavior included:

- company had policies to follow and recourse for women who felt they had been harassed
- lawsuit filed with EEOC

- threats from union because company is nonunion
- felt very uncomfortable and out of place at site, not unsafe
- did not feel unsafe, but violated, angry, ashamed, and hostile.

Comments from the questionnaire concerning the biggest health and safety challenge included:

- 3% of those who commented said harassment or harassment-related issues, including the threat of physical assault and pornography on job sites
- 24% said physical or health hazards, including clean air, dust, temperature extremes, mold, lifting, scaffolding, trenching, chemicals, ladders, and ergonomics
- 12% said the actions of others, including selling management on safety; on-the-job concerns, including traffic; general concern for the safety of others; company involvement in maintaining and repairing equipment; drugs in the workplace; and communication difficulties due to language barriers
- 4% expressed a concern specific to women's issues, including proper fit of personal protective equipment, sanitary facilities and issues, equipment design, and advancement of women in the safety field.

Comments on the best thing about the job included:

- 1% of those who answered this question mentioned salary, benefits, and position
- 1% mentioned women's issues, including breaking stereotypes and employers who are supportive and aware of women's needs
- 3% mentioned safety, including employer support of safety and knowing that they have made a difference in the safety conditions of the workplace
- all other respondents mentioned more intangible qualities, including satisfaction, competency, variety, people, fascination, personal growth, sense of accomplishment and pride,

independence, creativity, flexibility, learning, and co-workers.

Comments on the worst thing about the job included:

- 7% of those who answered this question mentioned hazards, including weather, noise, dust, rollovers, and the threat of physical violence
- 17% mentioned difficulties with male co-workers, including having to prove themselves, not being taken seriously, harassment and hostility, and unrealistic expectations
- 4% mentioned stress, including time pressure and budget restraints
- other respondents mentioned excessive computer use, government red tape, managing growth, lack of protocol and supervision, other women, long hours, and illegal workers who have no safety training and poor English-language skills.

ISSUES TO CONSIDER

Physical and Anthropometric Comparisons

Anthropometrics is the study of human body dimensions. Information gleaned from such study is used to design equipment, materials, workspaces and buildings. Much of the early data was collected by the military; later data came mainly from the automotive and clothing industries. After collection, results were graphed, and each measurement analyzed. Ranges from smallest to largest, lightest to heaviest, and others were calculated. Data from one measurement may be analyzed in relationship to another measurement. Designs are usually based on the fifth to ninety-fifth percentiles (i.e., the smallest and the largest measures of any variable are excluded).

The graphs in Figure 4 show the 50 percentile, approximately the average for white men and women aged 20–65.

SIZE DIFFERENCES

There is great variation in sizes among racial and ethnic groups. For instance, Alvin Tilley in *The*

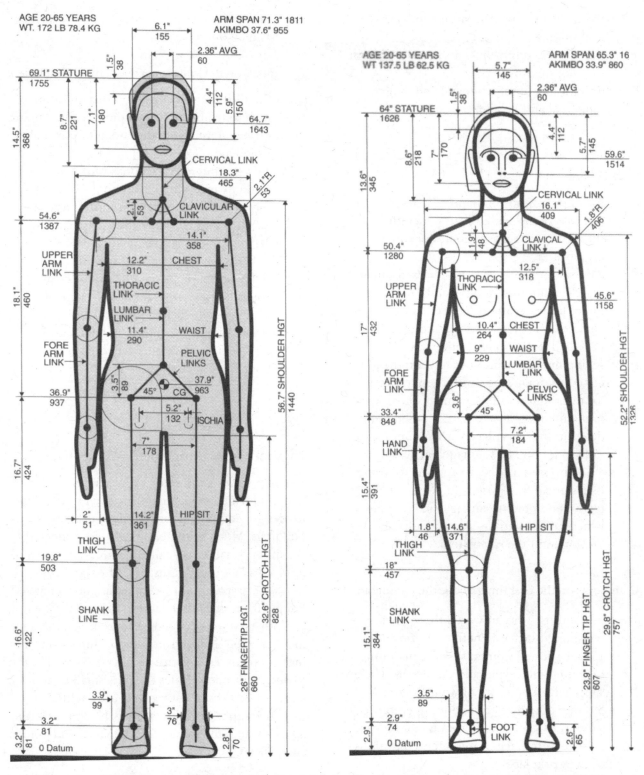

■ ■ **Figure 4** ■ ■ ■ **Fiftieth percentile anatomical characteristics of men and women** (Source: *The Measure of Man and Woman*, by Alvin R. Tilley, Henry Dreyfuss Associates, published by John Wiley & Sons, New York, 1993. Reprinted with permission.)

Measure of Man and Woman tells us that the 1 percentile for a Japanese woman is 2.3 inches shorter than the 1 percentile for an American woman.

The Cornell University Ergonomics Web page states: ". . . different ethnic groups have different physical characteristics. Black Africans have proportionally longer legs than Europeans. Eastern people (Asians) have proportionally shorter lower limbs; this is most pronounced for Japanese, and less so for Chinese, Koreans, Thais and Vietnamese."

This information is not provided to justify not hiring people; rather, it is given as another reminder that our diverse workforce is comprised of many people, male and female, who may be outside the traditional design criteria of fifth to ninety-fifth percentile. If we consider human factors and ergonomics in job design, we must also consider the needs of these men and women.

GENDER DIFFERENCES

Men and women are proportioned differently.

- Women's hips and thighs are, on average, larger than those of the average man. This results in the average woman's center of gravity being lower, women's hip joints being slightly different, legs proportionately longer and arms shorter than men's (Messing et al., *Tradeswoman*, January 31, 1990).
- Men's upper and lower limbs are greater, both proportionally and absolutely, than women's, except for buttock-knee length (Alan Hedge on the Cornell University Ergonomics Web page).
- In a study of 100 men and women, women were found to have significantly more mobility than men in a total of 24 measurements. Men showed more mobility in only two areas, the ankle and the wrist (Pulat, M. in *Fundamentals of Industrial Ergonomics*).
- On the average, women have two-thirds the strength of men; however, exercise can increase strength and endurance by as much as 50% (Pulat, M.)
- Men have twice the grip strength of women, on average, and there are large variations within each gender (Pulat, M.).

- In general, women experience more discomfort than men from the same exposure to and intensity of vibration (Pulat, M.).
- Average height begins to decline with age after 20, and by 40 most people are shrinking. Women shrink more than men, and the shrinkage increases with age. The shrinkage occurs in the spinal discs; some may be from shrinkage of lower limbs around the joints (Hedge, A.).

What can we conclude from this data? If we are to design work sites and equipment to make work safer, the size differences among men and women and between men and women must be taken into account. If job sites and equipment are designed for the fifth to ninety-fifth percentile, there is the potential that a number of members of our diverse workforce are being left out. Have modifications been made for the tallest, largest worker that could also be made for the shortest, smallest worker?

Physical and Health Hazards

Construction sites can be dangerous places. The Department of Labor's *Bureau of Labor Statistics 2001* gives information on workplaces in 2000:

- A total of 194,410 injuries and illnesses were recorded; 4382 of those were to women.
- The most common injury to women and men was a sprain or strain. Women suffered more bruises and contusions; men more cuts, lacerations, and punctures. The trunk was the part of the body most often affected.
- A higher percentage of women were victims of assaults or violent acts (2.5%) than men (0.9%). A higher percentage of women also suffered repetitive motion injuries (7.9%) than men (2.2%).

Fatality figures showed an increase of 6% in construction fatalities, despite a decline in overall fatalities and a decline in total employment.

A study of selected construction fatalities from 1980–1992 showed that approximately 1%, or 139, of the 14,257 fatalities were women. An article on this study by Timothy Ore in the *American Journal of*

■■ **Table 1** ■■

Physical and Health Hazards at Construction Sites

Potential Hazard	Contributing Equipment/Condition	Potential Cause
Falls	Scaffolding	Under construction, lack of fall protection
	Ladders	Positioning, poor equipment maintenance
	Roofs, floors	Unprotected openings in roofs and floors
Struck by/crushed	Excavations	Shoring/trenching deficiencies, unprotected edges, unmarked areas
	Buildings	Under construction/demolition, poor barrier protection
	Falling objects	No toe boards on scaffolding; poor housekeeping; lack of storage facilities; improper hoisting and rigging
	Vehicles	Automobiles at general construction sites or road construction sites, by construction vehicles or passing traffic
	Machinery	Inadequate barriers; improper repairs; inadequate or no lockout
Caught in/pinched	Equipment	Inadequate or no lockout; inadequate training; inadequate maintenance; improper guarding, improper fit of personal protective equipment; personal protective equipment being drawn into equipment
	Tools	Improper use; poor fit; improper body position; poor tool maintenance
Electrocution		Inadequate or no lockout; contact with energized equipment/lines; damaged or no insulation
Eye injuries	Foreign objects, dust, projectiles	Lack of personal protective equipment; poorly maintained personal protective equipment; lack of guards; not wetting down work
Temperature	Hot/cold	Inadequate or poorly fitting personal protective equipment; inadequate work/rest regimen for weather conditions; lack of water/cool, shaded break area or warm area
Noise	Equipment	Lack of hearing protection; lack of training; engineering controls not possible or not used
Vibration	Equipment	Pneumatic tools; inadequate or no personal protective equipment; no insulation
Musculo-skeletal disorders	Sprains/strains	Lifting technique; unbalanced loads; too much weight; Awkward positioning; repetitive motion; lack of training in proper technique; not using aids such as carts, levers, stools
	Carpal Tunnel	Hand position; tools; lack of assistive equipment
	Bursitis	Kneeling; concrete work, floor or carpet laying
	Other repetitive motion injuries	Tools; overwork; lack of training; lack of assistive equipment
Cancer, respiratory disease	Particulate from cement, lead, asbestos, wood, fiber board	Inhalation while welding, sanding, sandblasting, pouring, demolition, removal; dry work; inadequate local or area ventilation; inadequate respiratory protection and clothing; lack of proper washing facilities
Neurological difficulties, sensitizers, dermatitis, reproductive difficulties	Solvents, nickel; hexavalent chromium	Inhalation of or skin contact with paints, varnishes, lacquers, adhesives; grinding; welding; cutting
	Pesticides Fire retardants	Lawn or wood treatments
Biological hazards	Bacteria	Inadequate hygiene facilites, contaminated water; inadequate hazard control in healthcare facilities

Industrial Medicine (1998) stated that, although more men were killed in all categories, a higher percentage of women were killed in motor vehicle accidents, followed by machinery accidents; the most common causes of death for men were falls, followed by electrocutions. Overall fatality rates for this period were calculated to be 1.80 per 100,000 female and 17.29 per 100,000 male construction workers.

Each trade and each occupation has numerous physical and health hazards, some specific to that trade, and some in common with all workers. Although more men are employed in construction and are killed and injured, the number of women employed in all areas of construction (except for clerical positions) continues to increase; with that increase, will corresponding increases in fatalities and injuries be far behind?

There has been little research of construction hazards specifically for women, partly because the number of women on jobs has been so small. Table 1 lists physical and health hazards found on construction sites. It is not meant to be an all-inclusive table.

Other hazards and areas of particular concern to women include:

- Strains and sprains – due to differences in size and strength, and often lack of experience and training; many women, particularly those new to their jobs, are vulnerable to injury from moving and carrying loads. Healthcare workers often lift very heavy loads numerous times during a day, but construction workers tend to lift loads of varying weights at a different frequency.
- Hygiene facilities – not urinating when a need is felt is associated with a higher incidence of bladder infections. While this can happen to both men and women, women are more likely to contract infections from sitting on dirty seats and using dirty toilet paper. Inadequate washing facilities, particularly for menstruating women, pose an awkward and often difficult problem. Adequate washing facilities are essential for all workers to clean chemicals and dust at

the job site, to decrease the probability that they will harm themselves or their families by bringing workplace contaminants home.

- Personal protective equipment – although designed to protect the wearer, it is often not designed for the wearer (i.e., women who are smaller than the average man). Poorly fitting personal protective equipment may not only not protect women from hazardous conditions, it can cause injuries and illnesses by getting in the way, being stepped on, or being drawn into equipment or machinery. Safety equipment manufacturers in designing safety footwear, eyewear, gloves, respiratory protection, earplugs, and fall protection harnesses in different sizes have made many advances in recent years.
- Pregnancy – physical changes associated with pregnancy may affect the pregnant worker and how work is done. There is an additional risk of carpal tunnel syndrome due to increased fluid retention. The major changes to body structure affect movement, agility, and balance. Different personal protective equipment (i.e., larger) is required if the worker remains in the field during pregnancy, including footwear if feet and legs are swollen. Endurance may also be affected, especially in the later stages of pregnancy. The availability of sanitary facilities is also critical during pregnancy. Exposure to chemical substances is discussed in the next section of this chapter.

Reproductive Hazards

Reproductive hazards are often cited as reasons why women should not be in certain occupations, but reproductive hazards also affect men. The issue of reproductive hazards is complex and still not fully understood.

Reproductive hazards are substances, agents, or conditions that affect the reproductive health of men and women or the ability of couples to have healthy children. This includes couples trying to conceive and pregnant women.

Workplace exposures or stress can affect the sex drive of partners or affect menstrual cycles, which will affect conception, damage eggs or sperm, cause changes in genetic material called *mutations*, which can be passed on to future generations, or cause cancer or other diseases in the reproductive organs of men and women. Effects are dependent on time and exposure, and individuals may react differently to exposures. (See the module "Male and Female Reproductive Health Hazards in the Workplace" on the International Labour Organization's Web site.)

- Changes to an individual's sex drive or to a woman's menstrual cycle can be caused by exposure to chemicals, including antimony and benzene, and those that have depressant effects, such as some solvents; or rotating shifts or work stress.
- Damage to eggs or sperm, or the number of sperm produced, is associated with exposure to radiation on work sites and with numerous work-site chemicals, including lead, welding fumes, arsenic, benzene, cadmium, nickel, and mercury. Sperm count can also be affected by heat (ILO Web site).
- Numerous chemicals are known or suspected to cause mutations, which can result in birth defects, stillbirth, or miscarriages, and can also affect future children. Chemicals include antimony, cadmium, carbon dioxide, chlorinated hydrocarbons, lead, mercury, nitrous oxides, and polychlorinated biphenyls (PCBs). Radiation can also act as a mutagen.
- Cancer of reproductive organs is associated with exposure to arsenic, cadmium, and radiation.

When a woman is pregnant, exposure to different substances at different times can have different effects. It is thought that a fetus is most at risk early in the pregnancy, through the first trimester when internal organs and limbs are formed. About 1 in 6 pregnancies ends in miscarriage, often before the woman knows she is pregnant.

The Centers for Disease Control publishes a NIOSH Fact Sheet, *The Effects of Workplace Hazards on Female Reproductive Health*, which addresses the many things that can cause miscarriages. Maternal health issues such as diet, smoking, and alcohol con-

sumption will not be discussed in this chapter, but can have an effect on the fetus. Teratogens prevent the normal development of a fetus. They can pass from the mother's blood across the placenta to the fetus. Teratogens can cause low birth weight, premature birth, and physical or developmental abnormalities.

- Exposure to lead, mercury, organic solvents, carbon monoxide, chemicals including PCBs, and ionizing radiation are associated with teratogenic effects (ILO Web site).
- Increased stress and hard physical work are also associated with premature birth and low birth weight.

Although not technically a reproductive hazard, babies and young children can also be harmed by workplace exposures. Spouses/partners can bring contaminants home on their bodies and in their hair and clothes and expose family members. Substances can also enter breast milk and be passed along to nursing infants.

Stress

While stress is not a uniquely female issue, the entrance of women into the traditionally male construction world has created many kinds of stressful situations. Susan Eisenberg, in her book *We'll Call You If We Need You* (see the reference section), relates one such situation.

> Some jobs had the feel of trench warfare. Men who wanted to drive women out; women who were determined to stay. Knowledge of tools and experience at the trade did not prevent an "accident" that broke . . . nose, when a journeyman did not want her—not only a woman, but a Cherokee Indian—working with him.

The NIOSH Fact Sheet, *Women's Safety and Health Issues at Work*, reported on one survey's findings that 60 percent of employed women considered stress their biggest problem at work. In all fields of employment, almost twice as many women as men suffered from stress-related illnesses. NIOSH reiterated that job stress has been associated with cardiovascular disease, musculoskeletal disorders, depression, and burnout. Stress can also distract workers from the task being performed or conditions around them, leaving them more likely to commit errors and in more danger of being hurt.

The NIOSH Fact Sheet listed the conditions that contribute to stress:

- heavy workloads
- shift work
- lack of control over work
- conflict and ambiguity of roles and expectations, unrealistic expectations
- poor relationships with co-workers and supervisors
- boring, repetitive work
- harassment
- family issues or balancing work and family.

The women pioneers in construction often got their start through early women's initiatives, or by going to union halls or employers until they were hired; a small number reported having friends or relatives in a position to recommend or hire them. Often little training was provided before going to job sites; training was usually on the job, in classrooms taught by experienced tradesmen, or in apprenticeship programs. In many cases, women were hired only to meet quotas on government-funded jobs.

Trying to quantify the number of women who were "harassed" while training for work or while at work would be futile because of differences in the definition of harassment, perceptions of the receivers, and the number of unreported incidents. Many women also reported that once on the job, they were ignored, verbally and/or physically harassed, threatened, and in some cases, assaulted.

The stress of these situations was enough to drive many women out of their jobs, or to send them from job to job and partner to partner. The number of injuries and illnesses resulting directly or indirectly from this harassment are unknown.

BUILDING A BETTER WORKFORCE

For *Blue-Collar Women, Pioneers on the Male Frontier* (see the reference section), Mary Lindenstein Walshok interviewed women entering "nontraditional" jobs and followed them for a number of years. She iden-

tified some traits and characteristics of these women that were essential to finding, learning, and keeping jobs, and situational characteristics of the jobs that lead to success. Below are some of her conclusions.

- Finding a job is a function of needing employment; the jobs available in the geographical area and demand for workers; the formal and informal information available about a job and how to act on it; access to employers or agencies to provide the information and do the hiring; and training and experience that lead to commitment to a job.
- Learning a job is a function of learning and developing new skills, both physical and mental; getting used to the "process," or how work is done, including familiarity with tools and equipment; getting used to formal and informal work norms, including hours, work pacing, and the process of learning the job; and adjusting to the work setting, the environment, nature of the work, and the product.
- Keeping a job is a function of having good job skills; understanding how to behave as a worker; understanding the norms and culture of the workplace; and being committed to the job, for the tangibles of money and benefits, or for the satisfaction brought by the work.

In addition to things that workers can do, employers can facilitate worker development by providing stable jobs; good compensation and benefits; mechanisms for employee representation that work; providing familiarity with work and workers prior to assignment; facilitating the development of job competencies, including mentors; communicating a clear understanding of job requirements and expectations; and giving feedback on performance.

Needs

Although more women today take shop and technical training in secondary school than in the past, most still do not have the background many men obtain to prepare them for construction work. Safety training as part of secondary education is almost nonexistent. The following must occur if women are to continue as productive—and safe—construction employees:

- Recruitment – identify the best possible places to find good recruits; present opportunities honestly to recruits; examine hiring preferences and any restrictions for validity and legality.
- Training, including safety training – most training is offered through a partnership of unions and employers, usually as apprenticeships. Some training involves a combination of classroom and on-the-job experience. On-the-job training may be most critical when classroom training is limited or not tailored to certain jobs. Training styles and communication skills can differ greatly; even the most skilled workers may be unwilling or unable to convey job-specific knowledge and techniques. Requirements and expectations for students must be clear, and opportunities must be available to practice existing skills and learn new ones. Assigning new people to "go-fer" or other nonparticipatory jobs will not provide opportunities to learn about the tools, equipment, machinery, and techniques needed to work safely or gain proficiency. Opportunities for feedback and follow-up must be provided if training is to be improved.
- Providing proper equipment and facilities – personal protective equipment that fits, is appropriate for the job, and is properly maintained. Training is needed to ensure that equipment is used properly. Sanitary facilities must also be provided, including adequate water for washing.
- Enforcing laws and standards – enforce hiring standards where they exist. Hire and train women to be workers, not numbers to meet quotas. Enforce health and safety standards on job sites. Enforce antidiscrimination and antiharassment policies. Unions and management must show their support.
- Efforts to retain employees – communicate and enforce policies of job assignment and promotion. Address safety concerns. Studies have shown a higher retention rate of female workers when they have been provided with more information about what a job entails during the hiring process, particularly information about job demands and the potential for difficult working conditions. (Refer to *Sex Segregation in the Workplace—Trends, Explanations, Remedies,* edited by Barbara Reskin.)
- Studies – more information is needed about the injuries and illnesses suffered by women in construction jobs. Specific information is needed in order to eliminate hazards.

CONCLUSION

While women have been in the workforce for many years, their entry into construction trades highlighted many needs and issues. The key safety and health issues are important for all workers.

Training and proper equipment are essential, as are support from employers and unions. Sanitary facilities, poor-fitting personal protective equipment and reproductive hazards are often cited as "women's issues," but their hazards are not limited to women. Looking out for one group will protect everyone.

REFERENCES

AFL-CIO. *Fact Sheet,* "Facts About Working Women," at http://www.aflcio.org/women/wwfacts.htm.

Centers for Disease Control, National Institute of Occupational Safety and Health. *The Effects of Workplace Hazards on Female Reproductive Health.* DDHS NIOSH Publication Number 99-104, NIOSH, 1999.

_____. *Women's Safety and Health Issues at Work.* DHHS NIOSH Publication Number 2001-123, NIOSH, 2001.

Center to Protect Workers' Rights. *The Construction Chart Book,* 3rd ed. (calculations by Xiuwen Dong). Silver Spring, MD: Center to Protect Workers' Rights, 2002. Also see eLCOSH, Electronic Library of Construction Safety and Health Web page at http://www.cdc.gov/niosh/elcosh/docs/d0100/d00038/contents.html.

Eisenberg, Susan. *We'll Call You If We Need You.* Ithaca and London: Cornell University Press, 1998.

Hard-Hatted Women: Stories of Struggle and Success in the Trades, Molly Martin (ed.). Seattle, Washington: Seal Press, 1988.

Health and Safety of Women in Construction (HASWIC) Workgroup, Advisory Committee on Construction Safety and Health (ACCSH). "Women in the Construction Workplace: Providing Equitable Safety and Health Protection." Washington, DC: Department of Labor, OSHA, June 1999.

Hedge, Alan. DEA 325/651, Class Notes, Anthropometrics and Design. Cornell University Ergonomics Web page at http://ergo.human.cornell.edu/DEA325notes/anthrodesign.html.

Hughes, Dr. Kaylene. "Women at War," article from Redstone Arsenal Web page at http://www.redstone.army.mil/history/women/welcome.html/, 1992.

International Labour Organization. "Male and Female Reproductive Health Hazards in the Workplace," *Your Health and Safety at Work: A Collection of Modules.* Available at http://www.itcilo.it/english/actrav/telelearn/osh/rep/prod.htm.

Knapp, Gretchen E. *Home Front Maneuvers: Civilian Mobilization and Social Problem-solving in Western New York during World War II.* Ann Arbor, MI: University Microfilms, 1994.

Messing, Karen, Julie Courville, and Nicole Vezina. "Research Report: When Women Work at "Men's" Jobs; What are the Differences?" *Tradeswoman,* vol. 9; no.1, 31 January 1990.

National Association of Women in Construction. "NAWIC Facts." Available at http://www.nawic.org.

Ore, Timothy. "Women in the United States Construction Industry: An analysis of Fatal Occupational Injury Experience, 1980 to 1992," *American Journal of Industrial Medicine.* 33:256–262, 1998.

Pulat, B. Mustafa. *Fundamentals of Industrial Ergonomics.* Englewood Cliffs, New Jersey: Prentice Hall, 1993.

Sex, Segregation in the Workplace—Trends, Explanations, Remedies, Barbara F. Reskin (ed.). Washington DC: National Academy Press, 1984.

Sumner, Helen L., Ph.D. "Report on Condition of Woman and Child Wage-Earners in the United States," *Volume IX: History of Women in Industry in the United States.* New York: Arno Press, 1974.

Tilley, Alvin R. *The Measure of Man and Woman.* New York: John Wiley & Sons, 1993.

United States Department of Labor. Bureau of Labor Statistics, Figures for 2001.

_____. Women's Bureau. "Nontraditional Occupations for Employed Women in 1997." Available on the Web page at http://www.dol.gov/wb/wb_pubs/nolntrad.1997.htm.

Walshok, Mary Lindenstein. *Blue-Collar Women, Pioneers on the Male Frontier.* Garden City, New York: Anchor Books, 1981.

Working Women Past, Present, Future, Karen Shallcross Koziara, Michael H. Moskow, and Lucretia Dewey Tanner (eds.). Washington, DC: The Bureau of National Affairs, 1987.

REVIEW EXERCISES

1. How did corporate policies and social norms affect women's employment?
2. In what ways can a workplace perceived to be hostile affect a worker's performance?
3. How can differences in size and strength be compensated for?
4. Why can personal protective equipment be a hazard?
5. Which reproductive hazards can affect both men and women?
6. What contributes to job stress?
7. What can employers do to facilitate worker development?

■■ **About the Author** ■■

HOW TO COMMUNICATE WITH YOUR SPANISH-SPEAKING WORKFORCE

by Hector M. Escarcega, ARM, CSP

Hector M. Escarcega, ARM, CSP, is President of Bilingual Solutions International/Esteem Ahead Training & Seminars, specializing in providing consulting and training to the Spanish-speaking workforce.

Mr. Escarcega received a B.S. in Public Administration and Occupational Health and Safety and an M.S. in Industrial Hygiene from the University of Southern California. He conducts seminars in both English and Spanish in the areas of construction and general industry safety training for the OSHA Training Institute. He is an adjunct faculty member at West Los Angeles College, Culver City, California, conducting training in team building, communications, and problem solving. He is also an instructor for the Insurance Education Association (IEA), and is an adjunct faculty member at UCLA in the Labor/Occupational Safety and Health Program in Hazardous Materials Management.

Mr. Escarcega, is active in many professional organizations, including the National Speakers Association, the Latin Business Association, and the American Industrial Hygiene Association. He is a Professional Member of ASSE and a past president of the Los Angeles chapter of the American Society of Safety Engineers.

CHAPTER 34

How to Communicate with Your Spanish-Speaking Workforce

LEARNING OBJECTIVES

- Provide a general overview of the Spanish-speaking Latino workforce.
- Describe how to improve communication skills between the employer and the company's Latino workers. This will allow the employer to work with greater efficiency and decrease serious accidents and fatalities.
- Discuss Latino values, traditions, and culture, and their influence in the workplace.

INTRODUCTION

According to census statistics, Latinos totaled 35.3 million, or about 13 percent, of the *total population* of the United States in the year 2000. By 2050, Latinos are expected to comprise 25 percent, or 96 million residents. Today's Latino worker contributes significantly to our nation's workforce, and, statistically, in states like California, Florida, Texas, Illinois, Colorado, and Kentucky, Latino workers make up a large percentage of the workforce. The 2000 Census shows that well over 30 percent of the total workforce in Southern California is Latino.

Our Spanish-speaking workforce experiences many difficulties on the job. Many come from coun-

tries where the majority of the people live at a low socioeconomic level. They are seeking a better life by coming to the United States to work, get paid, and hopefully save some money for their families. Because of their desperation and desire to make things work for themselves, they take on whatever jobs their bodies can physically manage. Even without any formal safety training or education in the English language, these hard-working Latinos are willing to take dangerous risks in order to rid themselves of the personal family problems associated with their poverty.

The purpose of this chapter is to provide upper management, supervisors, SHE professionals, consultants, and trainers with suggestions for working effectively with Latino workers. Employers must learn how to communicate with and develop their Spanish-speaking workforces so that, in return, Latino workers will have fewer serious accidents and fatalities and acheive greater efficiency and productivity.

BACKGROUND ON THE SPANISH-SPEAKING WORKFORCE

Latinos: Who Are They? Where Do They Come From? Why Are They Here?

Many Latinos are of mixed race. Many are "Mestizo" or *mixed* between indigenous Indians (Aztecs, Mayans,

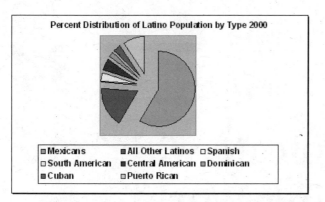

■■ **Figure 1** ■■ **Percent distribution of Latino population in the United States by type in 2000** (Source: U.S. Census Bureau)

Incas) and Spaniards. There is also European influence from Germany, France, and other European countries. The words Hispanic and Latino are synonymous. The term "Hispanic" was created by the U.S. Census Bureau for the purpose of categorizing different ethnic groups with similarities under one category. For the purposes of this chapter, the term "Latino" is used.

WHERE DO LATINOS COME FROM?

When we talk about Latinos, we are talking about people who come from some of these countries:

Chile
Bolivia
Ecuador
Argentina
El Salvador
Guatemala
Nicaragua
Cuba
Puerto Rico
Mexico
Spain
Venezuela
Columbia

The U.S Census estimates that:
- people from Mexico (Mexicans) make up 65% of the Latinos in the United States
- Puerto Ricans make up 12%

- Cubans are 5%
- Central Americans are 6%.

As managers, supervisors, trainers, safety directors and coordinators, human resource professionals, and consultants, we need to understand this powerful—and growing—population of workers. Why? Because they are one of the fastest-growing ethnic groups in this country; they are a high-revenue-generating population for our country; and they have one of the highest fatality rates in both the general and construction industries. This is both a human loss to our society and an economic loss to our nation.

WHY ARE THEY HERE?

A Mexican's Journey to "El Norte"
Many Mexicans leave their country in order to have a better way of life; more importantly, they leave in order to survive. The jobs these hard-working foreigners will take on may be: dishwasher, cook, janitor, painter, street-corner handy man, manual laborer, subcontractor in construction work, weed hauler, rubbish and debris remover, tree trimmer, gardener, concrete or brick layer, roofer, machine-shop worker, plating worker, farm laborer, and any other kind of work that will yield them greenbacks or "dolares" at the end of the day.

A Quick View of Our Nation's Accident Rate

In December of 2002, U.S. Secretary of Labor Elaine L. Chao made this statement about the 2001 workplace injury and illness rates: "Workplace injuries and illnesses have dropped for the ninth consecutive year, reaching an all-time low. Construction injuries are down. And there has been a 10 percent reduction in repeated trauma injuries, such as carpal tunnel syndrome and other musculoskeletal disorders. There were 500,000 fewer job-related injuries in 2001 than the year before, according to the Bureau of Labor Statistics. This year's rate of 5.7 injuries per 100 full-time employees reflects an 8 percent drop in cases from the previous year."

ACCIDENT STATISTICS AND THE LATINO WORKFORCE

While the Bureau of Labor Statistics (BLS) indicates that there has been a significant improvement in safety and health conditions in America, there are some very disturbing and sad statistics relating to Latino workers in the United States. Latinos make up 11 percent of the U.S. workforce. Unfortunately, Latinos make up 14 percent of the fatality rate.

The BLS report on fatalities for the year 2000 shows that 815 Latino workers, including 494 foreign-born Latino workers, died as a result of job-related injuries—an 11.6 percent increase from the previous year. Latinos accounted for a disproportionate number of workplace fatalities in 2000 (13.8 percent), compared with their proportion of employment, which was 10.7 percent. This appears to be largely due to the fact that Latinos are disproportionately employed in the more dangerous industries. For example, the construction industry accounts for about 7 percent of all employment, but 20 percent of fatalities. Latinos comprise almost 15 percent of construction employment, well above their representation in the workforce overall.

UNDERSTANDING THE PROBLEM WITH WORK AND ACCIDENTS

Now that we have some understanding of who Latinos are, and why they are here, it is now time to take a look at why Latinos are having such high accident rates. We will take a look at how the issues of Mexicans living in the United States creates a safety problem for them, their fellow co-workers, and the companies and organizations they work for. Understanding the lives of this Latino population will provide us with valuable information on why certain behaviors and risks are taken, why current safety practices are not having a positive effect on this hard-working population, and, lastly, what ideas can be implemented to control and prevent serious accidents, injuries, and unfortunate fatalities.

Let's take a quick look at their history, traditions, customs, and values to see how these come into play in the American workforce and how under-

standing them or not can either reduce or increase accidents in the workplace.

Crossing the Border

Mexican people, primarily males, leave home at a young age—between 16 and 19. In making their way north into the United States, they must cross the border, "la frontera." This is literally a life or death situation. Unfortunately, since one of the primary reasons they come to the United Staes is to make and save money, more than likely they don't have the necessary resources nor the time to come into the United States legally. If they have any money at all, it will be the money they will need to live on while crossing the border from Mexico into the United States.

As a Mexican-American who has worked with Spanish-speaking employees for over twenty years, it has been my experience that, in general, Mexican immigrants, whether they entered the United States legally or illegally, have a sense of economic deprivation and scarcity. They are desperate to make as much money as they can. This is why we find these workers taking on the high-risk jobs. To them any job is better than no job. They will take shortcuts because the sooner they finish the sooner they'll get paid. If they are involved in an accident, their attitude is something like the following: "So what is the big deal? Nothing major has ever happened to me before. This is the way I have always done it in my country and nothing major has ever happened before. If others can do it this way, then so can I. Accidents have happened before and they will happen again. This is part of the job and it's part of life. I leave it up to God. When it's time to go, it's time to go. There is nothing we can do about it!"

Why Traditions and Values Are Hard to Leave Behind

Unfortunately, even those Mexicans who begin to have a better understanding of the American way, its language, traditions, and holidays, will still find it very challenging to acclimate to the American culture. They still cling to their own food, customs, music, traditions, and values. And they should. This is what makes the United States so unique and rich. However, the problem lies in not understanding that they can learn American customs and traditions while at the same time keeping their rich Mexican culture. Remember, their way of thinking is to live in the moment.

What are some of the values, beliefs, attitudes, traditions, and customs of the Mexican people that at times may get in the way of working safely and that may be contradictory of a proactive approach to safety in the workplace? They provide some additional reasons, in the form of values and traditions, that may contribute to the high risks that these Latino, Mexican workers take while on the job:

- low self-esteem
- need to survive
- loyalty to "la familia"
- envy, jealousy
- machismo, pride, ego
- lack of education and formal training.

LOW SELF-ESTEEM, DESPERATION, NEED TO SURVIVE

Self-esteem is defined as "having value for oneself. The capability of knowing how to value oneself and how to love oneself. To see that one receives the best and to have the capability to be open to being loved as well as to love." Therefore, *low* self-esteem would consist of valuing oneself very little or not at all. Feelings of low self-esteem can be very detrimental to one's spirit, and one's way of thinking and living in the world. Employees at work may not care about their future. Their low self-esteem and beliefs lead them to live for today and not be concerned about matters dealing with health, safety, and the future. People with low self-esteem will not care about team-building or the sense of belonging outside of their family. Feelings of low self-esteem may show up as depression or feelings of low self-worth. Some of these feelings of low self-esteem come from their native country's history of being attacked, suppressed, and controlled by other countries— conquered by the Spanish, French, and Americans.

The feeling of being less than who they really are. The feeling of not having the freedom to live their lives and to do what they would like to do. Adding to this is the realization that they don't have much education and many do not know how to read and write. Unfortunately, there are many employers who take advantage of this situation and pay them less than minimum wage and never give them proper safety training while at the same time putting them into high-risk jobs. Many other people look down on them, treating them as inferiors.

Latinos work in high-risk jobs such as construction, manufacturing, and agriculture. Unfortunately, it has been reported that they are two to three times more likely to suffer work-related injuries. This is due to a number of different factors, including lack of appropriate training and cultural and language barriers (Agnvall; Halcraz).

Many Mexican immigrants don't know how to communicate their needs and feelings of frustration. For males to show emotions, cry, or share their personal problems and feelings with anyone, is taking away the only bit of self-respect they have as a man. This is called their *machismo*. As soon as they begin to subscribe to the "American" way, then there is nothing left for them. Women also are expected not to say anything at all. Ultimately, this leads to a chain of effects (similar in concept to Heinrich's Domino Theory) that will result in serious accidents, and possibly death.

As you can see, this is a recurring pattern that needs to stop if we are to have these employees work in a safe manner while at the same time keeping up and eventually increasing production levels. They need help in learning how to value themselves. If you as an employer understand this and take action to turn it around, the results will be a win/win situation. Let's take a look at how we might possibly motivate our employees to move ahead at work and, as a result, move ahead in their personal lives.

SAFETY AND SURVIVAL

Maslow's Hierarchy of Needs is a good model and tool to use in understanding the needs of our Latino workforce. Abraham Maslow was an in-

Self-actualized ⟹

Esteem/Recognition ⟹

Social/Sense of belonging ⟹

Safety/Security ⟹

Physiological needs ⟹

■ ■ **Figure 2** ■ ■ Maslow's Hierarchy of Needs

dustrial psychologist during the early to mid 1930s. His theory on motivating workers was based on the five basic human needs shown in Figure 2. He felt that motivating an employee takes place every time one of these needs is fulfilled, in ascending order from physiological to self-actualization, with those listed on the bottom of the model having to be met first.

On average, the blue-collar working Latino or Mexican worker can be found at the physiological and safety/security levels. Immigrant workers who come directly from Mexico with no money, no food, and very little clothing or belongings find themselves at this first level. Desperate for food, shelter, or clothing, they are open to working in almost any situation for whatever pay they may receive, as the photos in Figures 3–6 illustrate. These workers often will not hesitate to take shortcuts in the workplace in order to produce more output, which in turn will produce more money for them to take home. This is why we say that the Latino worker can be found at this first level on the hierarchy of needs. Filling those needs leaves little time for socializing, improving oneself, or for self-actualization.

■ ■ **Figure 3** ■ ■ Construction workers in Mexico City manually handle concrete in 5-gallon buckets

■ ■ **Figure 4** ■ ■ Construction employee climbing a ladder in Mexico while carrying a 5-gallon bucket filled with cement. Notice the lack of PPE, improper ladder safety procedures, and a fall hazard created by only a two-point contact.

■ ■ **Figure 5** ■ ■ Although safety standards exist in Mexico, no agency enforces them. Latino workers are in the habit of not using safety protection. Can you guess what is wrong in this picture?

■ ■ **Figure 6** ■ ■ **Latinos on a construction site without PPE**

■ ■ **Figure 7** ■ ■ **Latinos during lunch break; co-workers play dominos in the break room**

LOYALTY TO LA FAMILIA—"FAMILISMO"

The Latino culture is collective and cohesive. For them the family and group closeness (such as the group photographed in Figure 7), is their most important priority. Latinos have incredibly strong ties and loyalties to family and friends. If one person loses a job, the whole family may quit. As a result, disputes can have a more complex quality than Euro-Americans are accustomed to. The net effect is for Latinos to hold back from speaking up until they can stand it no longer and then to strike out. Since they may go to great lengths to avoid disputes, using a third party to intercede or mediate a dispute might be productive. The problem arises from their cultural tendency to not make any waves and keep peace within the family, since this is what they are supposed to do. However, this may ultimately have long-lasting negative effects both at home and at work.

LACK OF EDUCATION

The majority of the Mexican blue-collar workers unfortunately have an average education no higher than a sixth-grade level. The reason for this is simple. Mexico's economy is poor with very few good-paying jobs. Mexican children start to work at a very young age (6 to 8) because their parents need their help in making ends meet. The children must sacrifice their opportunities to go to school and enjoy their childhood in order to support their family,

which is such an important element of the Latino group. As a result of not going beyond the sixth grade, their fundamental reading, writing, and communication skills are limited. The majority of their jobs will involve manual labor. And they have very little to no exposure to any type of formal training. If or when they do receive formal training or education, it will take some time for them to adjust to this new experience, plus the issue of trust will come up: They will ask, "What is the catch?" or "What do they want from me?" Because they are not familiar with the concept of formal training, it can be a bit confusing and intimidating if not presented in an appropriate manner. It will take time, but eventually they will appreciate the efforts to improve themselves.

The "appropriate manner" may consist of presenting the training in their native language, providing them with the objectives of the training, not talking down to them, and making sure you or someone from the organization follows up with them to see if there are any questions and to evaluate the level of comprehension.

ENVY AND JEALOUSY

There is a joke, or saying, among the Mexican people, which basically points out that as fellow Mexicans they are always keeping each other from advancing both in the workplace and in their personal lives. While this may be a joke among fellow Latinos and

Mexicans, it is a sad reality, because it means keeping themselves from growing both as individuals and as a group.

Unfortunately, I believe this may be another way of dealing with their personal frustrations and not knowing how to get ahead in life in this new country. The feelings of frustration eventually turn into envy and jealousy. This jealousy turns into feelings of anger and despair. While this occurs within the Mexican population, it doesn't mean that all working Mexicans react this way. Nevertheless, let's take a look at how this envy and jealousy might manifest itself.

Some hard-working Mexicans come here to make money, but really resent the fact that they have to be in a foreign country, and are very angry at their financial situation. They don't like the fact that someone else is telling them what to do, and resent the fact that they can't seem to get ahead in life. In other words, they come here just to work and make their money. They are not really interested in understanding the American way, nor do they want to attempt to learn English. Their attitude is one of pride, ego, and machismo (discussed in the next section). Their way of thinking may be something like this. "I am here to work. I am a hard worker. I know how to work and I will do things the way I know how to do them. These silly American rules and policies of arriving on time and working safely don't work for me. I have never done this before, so why should I do it now? "

While this type of attitude is very negative and may have a negative effect on both safety and productivity for themselves as well as others, you may be surprised to learn that at times this behavior is tolerated by management. Management many times may be tempted to ignore this behavior because it results in higher productivity—a direct result of the strong Latino work ethic. For some American employers this is too good a deal to pass up, regardless of the legal liability. This sends the message to other workers that it is okay to behave in this manner, which at the same time discourages others from following company policies and procedures.

Another problem that arises from the negativity of envy and jealousy is that these employees make it very difficult for Latino co-workers to advance both at work and in their personal lives—and it impacts the rest of their population. Unfortunately, when a Mexican employee shows skill and the capabilities for getting the job done and is recognized by upper management through a promotion, some fellow Mexican workers, who feel insecure, will be jealous and envious. They will do all they can to sabotage that promoted employee. They will not respect the new position of the promoted employee, his instructions to them, nor his decision to move ahead. They will talk about him behind his back and call him a traitor. They will call him names and make fun of him. Many times they will disassociate from him. Why? Because they are envious that someone like themselves got the promotion, but they didn't.

They are frustrated to see that they are working just as hard as the individual who got promoted, yet feel like they are not being recognized at all. They are disappointed and angry that they don't know how to do the things that are necessary to get ahead. They are too proud and stubborn to ask the promoted Mexican worker how he did it and if he could show them how to succeed.

The first step for employers is to recognize some of the important personal characteristscs and values (i.e., lack of education, machismo, pride) of the average Mexican blue-collar worker, which may impede acceptance of safe work practices. Employers can then incorporate the techniques discussed in this chapter to assist their Latino employees in creating new cultural attitudes that put safety first.

MACHISMO, PRIDE, EGO

We have often heard the term "machismo" used for men in general, but especially for Latino men. So what is machismo, where does it come from, and what does it have to do with safety in the workplace?

Webster's Dictionary defines *machismo* as "a strong sense of masculine pride. An exaggerated awareness and assertion of masculinity." *Pride* is defined as "a reasonable or justifiable self-respect. Excessive self-esteem." *Ego* is defined as "focused on the self."

According to the above definitions, machismo is a strong sense and assertion of masculinity. It is an innate feeling that men are the providers and leaders of their families. Many Latino men will use machismo to compensate for their lack of formal education, their low self-worth, and their frustration at not knowing how to get ahead in life. It is important for the Latino man to show that he is a leader, a winner, and that he knows what he is doing. He needs to show that he is dependable and capable of taking care of his family.

Often they will assert their masculinity through their muscular strength, their voice, and their anger. Then machismo becomes an over-exaggerated awareness and assertion of masculinity, acting as a defense mechanism so that others around them will not notice that this particular male is feeling frustrated, embarrassed, sad, and is emotionally upset. But to show these feelings is interpreted as a sign of weakness. They identify the male as a loser and not a leader. If a Latino man doesn't know something and does not want to appear stupid or inept, he will use machismo to mask that insecurity. Machismo can also cover feelings of inadequacy from not having a sense of his own identity.

The only identity they can relate to is that of being a provider, a leader of some kind. Many times they will use this type of behavior to establish themselves in the world as someone who belongs. This characteristic of machismo allows them to save face when they make a mistake or don't know something. Machismo emerges from wanting to be right when they know they are wrong.

There is a way to get around this issue—to give them a way out. Let them save face by offering some training. Exercise some group dynamics so they are not singled out and focused on. Maybe the trainer will need to have them make fun of themselves first in order to show that it's okay to lighten up and still learn.

SOLUTIONS

Research shows that Americans born in the United States tend to be more focused on their individual accomplishments and goals. As we have seen and heard many times before, Americans are always striving to be "Numer One," striving for the gold medal. This is where much of our identity comes from.

And it is totally contrary to about 70 percent of the rest of the world, including foreigners who come here from other countries where people tend to be more group-oriented. They are more concerned about others within their immediate family and cultural group and are accustomed to living their lives and accomplishing their goals and projects as a group. For them, the aspect of family plays a very powerful role in their lives. These two key differences are vital in understanding the cultural differences in other ethnic groups as well. Understanding ourselves as Americans and some of the differences among other cultures, will give us the necessary edge for communicating better and motivating our employees. In turn, communicating better and knowing how to motivate our employees will assist us as managers, consultants, and business owners in decreasing accidents in the workplace and improving quality and productivity.

Communicating with Your Spanish-speaking Workforce

The following list offers some tips for achieving better communication with your Spanish-speaking workers.

1. Take the time to understand some of the most important aspects of the customs, values, and traditions of your Spanish-speaking workforce. Be aware of their value for the family and their religious practices. For example: Mexican people place a high value on family and cohesiveness. When they sense this concept of cohesiveness at work they are more at ease and feel like they are at home. They are readily open to new concepts and change. Many organizations with Spanish-speaking employees understand this idea of cohesiveness and will periodically have a company picnic, a pizza party during safety meetings, or some kind of raffle with a safety incentive program. They enjoy having fun as a group and with others. It does take time for the employees to warm up to this idea; however, once accepted, the benefits are tremendous.
2. Take the time to get to know your employees on an individual basis. When managers,

supervisors, and safety consultants take the time to get to know employees on a one-on-one basis, it gives them a sense of belonging. It improves their self-esteem and confidence and opens communication channels.

3. When you go to meet your employees one on one, make sure to have a sincere smile on your face. Make it easy and okay for them to communicate and open up to you. Remember, Latino workers respect position-power and hierarchal status within an organization. For Latinos, a person who is in a position of power must be looked up to and they should agree with the beliefs of those in authority. They have learned from their parents that they are not to make direct eye contact with their supervisors or upper management. To do so is a sign of disrespect, which may result in a variety of negative consequences. It will take time for them to open up and trust you. Your Latino employees may have *respect* for you because of your position and power, but this doesn't mean that they *trust* you. That is a whole different ball game. Trust must be earned. This will take continuous effort by you and the organization.

4. Offer your employees positive feedback. Let them know some positive things about their work ethic and their skills in getting the job done. Let them know specifically what you like about how they work and what you like about them. Research shows that on the average constructive criticism is given out ten times more than positive feedback. The irony is that as humans we like and enjoy positive feedback, which, in turn, has the potential to make your employees happier as well as increase employee morale and productivity.

5. If appropriate, compliment your Latino employee with a pat on the shoulder or a firm, sincere handshake. Remember Latinos in general enjoy the idea of being friendly and sharing what they have. They enjoy being close as a group. If you decide to use this idea, make sure your organization reviews its polices on sexual harassment.

6. In addition to positive feedback, consider using Maslow's hierarchy of needs to get a better understanding of who your employees are and what their needs are. This gives managers and supervisors a kind of evaluation form when going through the different levels of Maslow's hierarchy (refer to Figure 2). This tool can be used as a checklist of questions to ask your employees in order to get to know them.

7. Provide opportunities to have an exchange of ideas and concerns between your workforce and management. Make sure you share your organization's mission statement and companywide goals. When you include them, this gives your Latino workers a sense of belonging and being valued.

8. Make sure employees understand the policies and procedures required of them when it comes to safety and during safety meetings. This will require communication in Spanish when possible. The concept of presenting a safety meeting in English with your Spanish-speaking workforce does not lend itself to successful results. Some organizations use other bilingual employees to translate from English to Spanish. While this is a slight step above presenting training in English only to your Spanish-speaking employees, it is still undesirable because of the loss of ideas during the interpretation. In addition, it simply takes more time.

9. The best step to take is to use a formal in-house bilingual trainer or outside training and safety consultant. Using an outside bilingual consultant/trainer will provide effective benefits in the form of better understanding, comprehension, and an increase in morale.

10. Treat all of your employees fairly and make sure they are familiar with the company's policies and procedures. This is especially true with your Spanish-speaking workforce. Make sure to provide all of them with formal training, tools, and personal protective equipment. Stay away from the idea of working only with your favorite employees. This kind of favoritism will make other employees resentful and angry. These emotions can eventually turn into negative behavior, such as sabotage of certain employees, operations,

and production. Resentment can even turn into physical altercations. As mentioned earlier, many times Latino workers harbor this characteristic of jealousy and envy toward fellow employees who tend to move up within the organization. This comes from the frustration of not knowing how to achieve their workplace and personal goals, while at the same time seeing their fellow employees succeed through promotions.

11. Offer ESL (English as a Second Language) classes. If possible offer some kind of incentive such as certificates of completion and maybe a potluck luncheon, or offer the classes during the last few hours of work. Make sure all of management and the company overall is in support of this program.

12. Provide basic and formal training in the following areas:
    – New employee orientation
    – OSHA-required safety training
    – How to set and reach your goals.

    Again, it is highly recommended that your training be conducted in Spanish. If you are going to provide employees the time to be trained, it makes sense to do it the right way.

13. Be aware of the holidays that Latinos and Mexicans celebrate. Many of their religious days of celebration are deeply rooted in their customs and traditions. If as an organization you can work something out to give your Latino workers some of these days off, perhaps letting them take comp time, this will show management's commitment to supporting and showing interest in their employees. In return, the organization will receive the employees' appreciation, and they will move in the direction of trust. Examples of some of these celebrated holidays are:

    • Christmas
    • New Year's eve and day
    • Easter
    • Dia de Los Meurtos (Day of the Dead) festivities, spanning November 1, and November 2. In most localities November 1 (All Saints Day) is set aside for remembrance of deceased infants and children; those who have died as adults are honored November 2 (All Souls Day).
    • Day to honor the Virgin Guadalupe (La Virgen de Guadalupe), December 12
    • Quincenera (coming of age for a girl at age 15)
    • Mother's Day, celebrated May 10
    • Father's Day
    • The 16th of September (Mexican Independence Day)
    • Cinco de Mayo—celebrating a significant victory over French troops. (Celebrated much more in the United States than in Mexico.)

Additional Quick Tips for Working with Your Latino Workforce

1. Acknowledge who they are as individuals.
2. Show them respect.
3. Watch to see how you are using your body language.
4. Smile. It feels good for them and you.
5. Keep your communication at the same level. Do not be condescending.
6. Don't put them down, especially in front of others.
7. Provide social functions.
8. Use kinesthetic and visual communication.
9. Make sure to remember family is important to them. Use this to your advantage.

Tips for Upper Management

1. As a company, acknowledge and respect any cultural differences.
2. Correctly interpret behavior by providing training to managers and supervisors.
3. Explain your expectations to each and every employee (in Spanish and English).
4. Employ specific motivational actions.

Training

Providing training to your employees is so important in preventing accidents that I feel it is important to provide some tips and ideas pertaining to training.

Make sure to use previous information discussed with regard to traditions, values, and customs when tailoring your training for your Spanish-speaking workforce. Each organization will have unique situations, which will also require your consideration prior to providing your training.

Consider some important aspects of how adults learn. We fall into one of three categories, or a combination of all three: visual, kinesthetic, and auditory. Here we will primarily focus on visual and kinesthetic learners.

VISUAL LEARNERS

A visual learner will draw and sketch out concepts and ideas for greater clarification, and will look at charts, graphs, and diagrams, and then read about the diagrams in order to gain a better understanding. Often a visual learner will close his or her eyes and picture concepts and ideas of how they think things should be in their mind. They use their imagination and are good at visualizing goals, projects, or specific things to try. Pictures, drawings, and images help a visual learner understand ideas and information better than explanations. A visual learner will say phrases like, "I *see* your point." Because English is a big issue for Spanish-speaking employees, consider using these aids during training. Use pictures, slides, flipcharts, PowerPoint presentations, and video tapes of your employees. How can you apply this concept in your workplace?

KINESTHETIC LEARNERS

A kinesthetic person learns by touch and body movements. This kind of person will take a highlighter and pencil to mark notes. They want to sense the position and movement of what they are working on. Kinesthetic learners want to handle things. They like to roll up their sleeves and get into action. This kind of learner will say something like, "I feel we're *moving* in the right direction." Some suggestions for using this aspect of learning are: make sure your participants have a note pad and pencil or pen to take notes, and when possible bring in props (e.g., for fall protection training definitely bring in the harness, lanyards, self-retracting lines, hooks, anchor

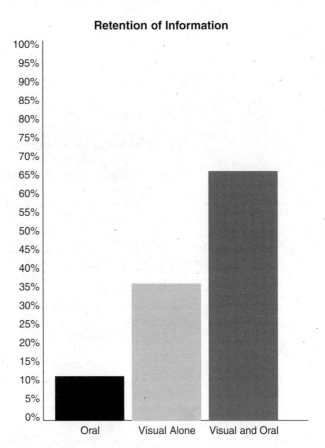

■ ■ **Figure 8** ■ ■ **Effectiveness of Visual Aids**

points etc.). Let them get a realistic feel for what you are teaching.

Visual Aids and Other Techniques

Visual aids add impact and interest to a presentation. They enable you to appeal to more than one sense at the same time, thereby increasing the audience's understanding and retention level. With pictures, the concepts or ideas you present are no longer simply words—but words plus images. The chart in Figure 8 depicts the effectiveness of visual aids on an audience's retention. If using words, flipcharts, PowerPoint, or other elements, remember to have everything written and illustrated in Spanish.

Here are some visual aids that we want to be aware of: flip chart, overhead projector, posters, slide projector, video tape, PowerPoint presentations, props, and the use of a VCR and monitor. Table 1 takes a look at each one of these and outlines the pros and cons of using each.

■ ■ **Table 1** ■ ■

Pros and Cons of Visual Aids

Visual Aid	Pros	Cons
Flip chart	• Are quick, inexpensive for briefing small groups • Conveys information • Can be prepared before and during presentation • Shows speaker has given thought to presentation • Can be used to record audience comments	• Really not suitable in large audience • May be difficult to transport • Penmanship is important
Overhead projector and transparencies	• Transparencies can be produced quickly, easily, and inexpensively on a computer • Use black and white or color • Use PowerPoint slides • Use charts and graphs • Use digital pictures, clip art	• May be difficult to write on • Projector head gets in way • Need overhead projector • Cumbersome to transport • Projector glass may be scratched • Older projectors not bright • May have difficulty with bulb
Posters	• Are permanent and portable • Can be simple or elaborate • Can be used alone or with others	• May contain too much information • Elaborate posters may be costly • Difficult to secure
Video tape	• Creates good picture and sound • Holds audience attention • Can create your own tailor-made videos • Relatively inexpensive for value	• Need some basic training • In large audience, may be difficult to see monitor • May require additional equipment
Slide projector	• Good for any size audience • Good quality of information	• Requires projector and screen • Lighting must be good • Requires much preparation

Communication—A Common Thread

A common thread among the previously mentioned solutions is communication. Unfortunately, people underestimate the power of good communication and it's value. In addition, many people think that communication is just talking. While the concept may sound simple and basic, effective communication that renders positive results can actually be rather complicated.

The basic communication model consists of:

1. A sender to send a message
2. A message
3. How the message is sent (verbal, email, letter, PowerPoint)
4. Background noise
5. A receiver of the message
6. The receiver's interpretation of the sender's message.

We may think that there is really nothing to communication. However, there are many possibilities where the sender's message can be misunderstood or even be incorrect. For example, there is background noise, there is the language barrier, and there is body language, which includes tone of

Verbal Communication vs. Body Language		
7%	38%	55%
Words	Voice tone	Body language

■■ Figure 9 ■■ Importance of body language in communication

voice. Also, realize that the receiver may respond by communicating in a poor manner as well.

As you can see from information presented in Figure 9, over 50 percent of our communication is accomplished through body language. You may say one thing, but your body is saying something different. People can sense this. And they will know when you are being truthful or insincere. Sending contradictory messages, can be a very big mistake to make with any person or ethnic group. Make sure your words are congruent with your body language. Otherwise you may end up wasting your resources and creating even bigger problems.

SEND THE RIGHT NONVERBAL MESSAGES

Your body talks...

- eyes
- face
- posture.

Nonverbal signals can leak out whether or not we are conscious of them. What are some evident body language messages that you send to others? Are you communicating one message to your Latino workers in words and then saying something totally different with your body language?

REVIEW

Use the ideas and information in this chapter to discuss the following case studies.

Family—"Familismo"
Three types of value orientation are involved in the Latino concept of family—"familismo":

1. Feeling obligated to provide material and emotional support to extended family members.
2. Relying on relatives for help and support.
3. Constantly checking with relatives about the way they see various behaviors and attitudes and being influenced by their perceptions and feelings.

Case Study: "The Flu"

JOAN: I was hoping we could have that safety committee meeting with our managers and the supervisors' team tomorrow morning.

MARIA: Actually, my daughter has some kind of flu and I was going to take her to the doctor tomorrow morning.

JOAN: I see. Well, let me check with Bob and see if he can sit in for you. Shouldn't be any problem. I'll let you know.

MARIA: Thank you.

JOAN: Don't mention it.

REVIEW QUESTIONS

1. Knowing what you now know about family in the Latino culture, what do you think about the scenario above?
2. If you were in Joan's shoes and you knew a bit more than she did, how would you handle this situation?

3. Take the time to think about your Latino workforce and write down instances where family issues may become a work issue and how you might handle those situations.

Hierarchy and Status

A sense of hierarchy and status is a strong element in virtually all Latino countries. People are born into an upper or lower class, and the middle class in most countries is small but growing. Traditionally, the masses live in destitute poverty and the elite live with great wealth. The social conventions of the upper classes are more formal and elaborate than in the United States. The work you do is directly related to your social class; therefore, if a manager or person with position-power were to do manual labor, such as helping out with tasks at a construction work site or personally helping a blue-collar worker with some task, it would be viewed as undignified and inappropriate.

Case Study: "Don't Worry About It. I Am Your Boss"

How would you handle this situation?
Your employees receive formal, in-house safety training, which instructs them in the proper way to wear fall protection, including a harness, safety lanyard, and anchor point and hook. The next day, the employee takes the time to put on the fall protection safety equipment and follows the proper safety procedures. His supervisor, who is concerned about production, sees this and tells him that he doesn't need to put this safety gear on. Contrary to the training procedures, the employee makes the decision to follow his supervisor's directions.

REVIEW QUESTIONS

1. What do you think about this scenario?
2. What are your suggestions to remedy this?
3. If you were a safety manager and witnessed this, how would you handle it?

Respect—"Respeto"

Latino people show respect for someone of superior status through their tone of voice and manner. The traditional belief is that the reason people are poor or rich, have power or don't have power, is because it is God's will. A "jefe" (boss) is a man of power or wealth who receives loyalty from people of lesser status. He may be the employer, a politician, a landowner, or a businessman. The jefe makes the decisions, and others don't question him. While Americans attempt to minimize differences (status, age, sex, etc.) between persons, Latinos tend to stress them. They have been taught to do this at a very young age.

Case Study: "The Worker Speaks"

MR. BILL: Efficiency is falling off in the quality control division. What can we do?
MS. RAMIREZ: The workers may have some ideas.
MR. BILL: Good. Why don't we call a meeting and ask them?
MS. RAMIREZ: A meeting?
MR. BILL: Yes, I'll run it myself and let them know how much we value their input.
MS. RAMIREZ: You'll go to the meeting?

REVIEW QUESTIONS

1. Knowing what you now know about hierarchy, status, and respect in the Latino culture, what do you think about this scenario? What may be some other alternatives for handling this situation?
2. Take the time to think about your Latino workforce, and write down instances where hierarchy, status, and respect may become a work issue. How would you handle these situations?

Relating in a Personal Way—"Personalismo"

To relate well to people of Latino culture usually means relating everything to them on a personal level. Instead of talking in generalities, you would talk in terms of them as persons, their families, their town, and, most of all, their personal pride. Especially for Latino workers, the more the communication is personalized, the more successful it tends to be. Latinos tend to trust only those whom they have a personal relationship with, for only those people can appreciate their soul, or inner self, and therefore only those people can be trusted. This is why it is difficult, or impossible, to do business with a Latino if you don't first establish some type of personal relationship.

The need for personalismo means that Latinos are less likely than Euro-Americans to open up and self-disclose. They are reluctant to share their thoughts and feelings with mere acquaintances. When people reveal personal information, they become vulnerable to how the listener will use that information, and their personal honor could be damaged.

Case Study: "A Helping Hand"

CARL: Hey, Juan, is everything okay?

JUAN: Yes, sir. I was just explaining to Raul here about the new drill press. Some of the men are not sure about it yet.

CARL: I know. Actually, I overheard you. What you were telling Raul isn't exactly right.

JUAN: No?

CARL: No. You have to turn on the fan BEFORE you switch on the water jet, not after. Now try it.

RAUL: (PAUSE) Yes. That is it. Any more problems with this, Juan, just come and ask me. That is what I am here for.

JUAN: Thank you, sir.

REVIEW QUESTIONS

1. Knowing what you now know about personalismo, trust, and hierarchy with the Latino culture, what do you think about the scenario above?
2. What may be some other alternatives for handling this situation that deals with personalismo, status, and trust?

3. Take the time to think about your Latino workforce, and write down instances where personlismo, status, and trust or a lack of it, got in the way of work and productivity. How can this become an issue in the workplace?

CONCLUSION

Nearly everyone in the United States can probably trace their heritage to the culture of another country. For Latino immigrants, work is an important aspect of their lives, especially as a way to become part of the United States. These people left their homelands to avoid political oppression and economic hardship.

The chapter explored specific areas of Latino customs, traditions, and values that could create obstacles for communication, productivity, and safety in the workplace:

- Where the Latino workforce comes from and why they come to the United States.
- Their values, traditions, and customs in order to gain a better understanding of how we as Americans can narrow the communication gap.
- Problems that exist as a result of the differences between both cultures, and suggestions and solutions for closing the gap.
- How social and cultural conditioning influences communication, and how we must learn about and implement effective tools in order to have effective cross-cultural communication.

Because they encounter so many diverse ethnic backgrounds, today's managers and supervisors are realizing that there is a gap between them and these hard-working employees, and they are asking themselves, "How do I motivate and communicate with these employees who have different values, needs, and traditions?" Safety managers, supervisors, and safety consultants need to try out some of the tools recommended and explained in this chapter in dealing with their Spanish-speaking workforces. When implemented appropriately, the positive results will far outweigh any investment of resources. And don't forget, sincerity and treating them as you would like to be treated goes a very long way.

REFERENCES

Agnvall, Elizabeth. "Working to Improve Lives of Hispanics," *Safety & Health*, December 12, 2000.

Billings-Harris, Lenora. *The Diversity Advantage. A Guide to Making Diversity Work.* Greensboro, NC: Oakhill Press, 1998.

Branden, Nathaniel. *The Six Pillars of Self-Esteem.* New York: Bantam Publishing Group, Inc., 1994.

Canfield, Jack and Jacqueline Miller. *Heart at Work.* New York: McGraw-Hill, 1996.

Conejo, Carlos A. *Motivating Hispanic Employees. A Practical Guide to Understanding and Managing Hispanic Employees.* Thousand Oaks, CA: Multi-Cultural Press, 2001.

Civitello, Andrew, Jr. *Construction Safety and Loss Control Program.* Armonk, NY: M.E. Sharpe, Inc., 1998.

Escarcega, Hector M. *High Impact Communication, Presentation and Training Techniques for the Health & Safety Professional.* Los Angeles, CA: Bilingual Solutions Int'l, 2002.

Halcraz, Joseph F., Sr. "Safety Training for the Diverse Working Populations," *Occupational Hazards*, February 2003.

Holladay, Neal T. *Hispanics and the American Workplace: An Employer's Guide.* Thomasville, NC: Holladay Management Services, Inc., 2000.

Morales, Rebecca and Frank Bonilla. *Latinos in a Changing U.S. Economy.* Thousand Oaks, CA: Sage, 1993.

Pike, Robert W. *Creative Training Techniques Handbook,* 2nd ed. Minneapolis, MN: Lakewood Books, 1994.

U.S. Census Bureau. *Census 2000.* Washington, DC: Government Printing Office, 2000.

U.S. Department of Labor, Bureau of Labor and Statistics. Annual Average Data, "Characteristics of the Employed." Washington, DC: Government Printing Office, 2002.

Williams, J. Clifton. *Human Behavior in Organizations.* Cincinnati, OH: South-Western Publishing Co., 1978.

Zurcher, A. L. et al. "Value Orientation, Role Conflict, and Alienation from Work," *American Sociological Review* 30:530–560, 1965.

REVIEW EXERCISES

1. List and discuss why traditional safety management techniques will not work alone in preventing serious accidents with the Spanish-speaking workforce.
2. List some major reasons why accidents and fatalities are on the rise with the Spanish-speaking workforce.
3. Discuss other factors relevant to cultural issues that can cause serious accidents and fatalities among the Spanish-speaking workforce.
4. List at least three specific accident prevention techniques or other tools that you can apply to prevent accidents and fatalities with your Spanish-speaking workforce.

■■ **About the Author** ■■

FUTURE TRENDS IN CONSTRUCTION SAFETY
by Jim E. Lapping, P.E., CSP

Jim E. Lapping, P.E., CSP is Vice-President and Director of Safety for Power Maintenance & Constructors LLC. He is also a senior editor with *Construction & Engineering Safety Magazine.* He has served four years as special assistant for construction and engineering to the Assistant Secretary of Labor for Occupational Safety and Health, for twenty-five years as Director of Safety and Health for the Building and Construction Trades Department with the AFL-CIO, and for twelve years on the Department of Labor OSHA Advisory Committee on Construction Safety and Health. He is coauthor of the *Handbook of OSHA Construction Safety and Health.*

Future Trends in Construction Safety

TRENDS IMPACTING THE CONSTRUCTION INDUSTRY

An overall understanding of the current state of the construction industry is needed before one can address future trends in construction safety. Several factors are exerting major influence on the construction profession and on the safety, health, and environmental process.

At its core, owners are requesting that construction companies demonstrate superior product and service value before awarding contracts. As a result, service and maintenance are the emerging capabilities for value creation among construction firms. Service, maintenance, commissioning, and renovation comprise a significant percentage of the total industry revenue and are seen as primary components to added value for customers and users of construction services.

Life-cycle costing, including increased use of risk-management techniques and loss control, continues to reinforce this shift in value creation away from pure construction. Therefore, value is being built through alternative combinations of services, not just construction alone.

Specialization continues to provide advantages for those entities that have committed to developing truly distinctive and unique competencies in specific areas. Market strategies and delivery performance also offer leverageable advantages. Public and private users of construction are now expecting the industry to provide delivery of nontraditional services. Many times this could include the use of outsourced services. Specialization now allows the service provider to select those services that offer value to the end-user.

Project delivery and the "just-in-time" concept is gaining increased attention. Owners' desires for "better, faster, safer, cheaper" are causing evolutionary activities to redefine work procurement and delivery methods. Owners are the driving force for the rapid evolution of alternative project delivery methods and processes. Partnering allows for cooperation, collaboration, and innovation that ultimately will impact the productivity, quality, and safety at the job site. Design-build is one of many processes that is redefining the traditional delivery methods. Collaboration among construction parties involved in the overall process (design, engineer, construction, service) contributes to the skills needed to participate and succeed in certain markets.

Competition will continue to be a major influence. The construction industry in the United States is relatively mature, and the number of companies entering the marketplace to deliver these products and services will force existing companies to be on the *cutting edge*. Core competencies are improving bottom-line results and customer satisfaction for the more efficient. Companies that continue to compete on price alone will face difficulties in the future global market.

Many construction industry managers are technically competent, but lack the leadership qualities needed in the early twenty-first century. This fact represents a major opportunity for project managers to acquire the tools and techniques to become effective and productive leaders. Companies are increasingly utilizing employee participation to assist in laying out the strategic direction for the company. There is a shortage of experienced project managers, which is raising pressure for devising innovative organizational structures to increase employee participation, feedback, and involvement.

The owner or construction services' user is becoming the primary driver for procurement and delivery methods. The owner will be more involved in research and development, technological advancement, and project development and coordination. This trend will ultimately lead to individually cus-

tomized products and services based upon specific and unique requirements. These owners will continue to evaluate these new or expanded procurement and delivery systems. Performance metrics that are based upon leading indicators will be important tools. These metrics along with benchmarking and historical data will allow the user and provider of construction services to improve performance through self-evaluation and comparison to industry leaders.

The Growing Need for Construction Safety Professionals

The Construction Chart Book, 3rd edition, published in 2002 by The Center To Protect Workers' Rights, documents that there are 656,000 contractors employing 5,700,000 construction workers across the United States. The Board of Certified Safety Professionals reports that there are only 177 Certified Safety Professionals in the Construction Specialty, or one CSP for every 3700 contractors and 32,200 construction employees. The need for qualified safety professionals seems overwhelming. However, with new safety planning techniques and intensive supervisor safety training, the expertise of the Construction CSP can best be focused on the high-hazard workplaces and projects where unique situations exist that require planning and monitoring.

The Changing Role of the Construction Safety Professional

The status of the safety professional has improved dramatically during the past five years due to the increase in insurance costs and stringent criteria set by owners for contractors to qualify to bid on their projects. The days when safety was viewed as first aid, PPE selection and inventory, and basic orientation training for new-hires are in the past. Craft foremen and tool-room attendants can handle those duties. The changing roles are often seen as a threat to those who have not advanced in their management skills and have not continued to acquire the latest knowledge in planning and communication techniques.

The safety professional of tomorrow will spend his or her time developing project safety plans that are blueprints for senior project supervisors to run their projects. The immediate implementation of the safety plan will be at the foreman and general foreman level. Hazard recognition and prevention are front-line responsibilities that require continuous and detailed attention. Given that many projects have hundreds of employees with tens of supervisors working at remote and ever-changing locations, it is impossible for a safety person to provide the required oversight. Front-line supervisors must assume the safety duties and responsibilities that have been traditionally viewed as the sole duties of the safety professional. Many owners will not pay for a safety person unless there is a minimum of twenty-five employees on the project. In these cases, which are more the norm than not, safety is often *more critical* to the overall success and profitability of the project. Poor or inadequate safety planning can result in losses that far exceed the anticipated profit margin for the project. When a contractor has ten projects under way, all with fewer than twenty-five employees, safety must be handled by the on-site supervisor with as much help as possible from the safety staff. The safety staff often consists of one person with several collateral duties and little time to get out to the job sites.

As a result of the changing role and increased importance of safety, contractors are looking for a higher level of competence and leadership from their safety staff. Safety at the job-site level will, in the future, be planned and implemented by foremen who have been trained as safety technicians, with the senior job-site supervisor having the final responsibility and accountability for safety.

Safety professionals will provide the road map, but developing the job-site safety plan and implementing the plan will be the responsibility of line supervisors. The safety professional who believes that his or her job is to be "out in the field and staying in touch with all levels of the company" has not established an effective reporting and accountability structure that will enable the safety activities and results of each project and supervisor to be monitored and analyzed.

The future safety professional in construction will need to rely on front-line supervisors to identify situations and conditions that require the expertise of specialty experts in such areas as fall prevention, rigging, critical crane lifts, complex scaffolding, and deep excavations. Regardless of how experienced and educated the staff safety professional is, he or she will not possess all the knowledge necessary to address every situation that exists in the construction industry. The corporate safety director will be a key manager with access to a cadre of specialty experts who provide assistance to project supervisors.

Computerized Planning and Scheduling for Construction Projects

Many current projects use computer software programs to plan and track work schedules and project costs. More projects in the future will use similar software. The natural next step is the inclusion of safety in the project-management and cost-control programs. One such initiative is the Salus CPM Program being developed by the Center to Protect Workers' Rights with help from the University of Florida at Gainesville. The program, funded by the National Institute for Occupational Safety and Health, is a plug-in software program that identifies project activities that require safety and health planning and safe work procedures. A drop-down menu allows the supervisor to select Job Safety Analysis from a data bank of analysis conducted on similar activities. The supervisor can either modify an existing analysis or open a new analysis format and conduct an analysis from scratch that then can be added to the database.

The software program can generate checklists, OSHA standards, and training programs specific to the activity that can be accessed and printed for use in inspections and craft briefings. An additional value of the safety software program is the ability to document the contractor's hazard identification and control efforts to ensure that supervisors are carrying out their responsibilities and that the OSHA regulations are being followed.

The safety software program could be monitored and reviewed from a central location, such as the corporate office or a regional office. The corporate safety director could have access to all current projects and provide assistance and guidance whenever the existing safety planning was lacking in specificity or detail.

Construction phases, activities, or tasks could be marked for extra planning and supervision based on past experience of the contractor with similar projects. At some point, the contractor associations could pool their expertise and make available to all of their members the experience and techniques that improved employee safety and reduced equipment damage. As in the past, where contractors refused to share their basic safety manuals, the sharing of safety planning, evaluation techniques, and information may be an ideal situation that will not be possible to achieve.

The Role of Project Managers and Supervisors

Accountability for safety at the supervisor level has been a difficult task for managers as there has been a lack of measurable criteria to evaluate the safety efforts of supervisors. Most contractors and owners continue to use the number of injuries and lost-time days as a measure of the success or failure of contractor safety programs. Future managers will rely on the safety activities undertaken by supervisors to prevent injuries rather than measuring the random, uncontrollable incidence of injuries. The computerized hazard identification and tracking software will enable managers to monitor each level of job-site supervision to assess their planning, training, and workplace and equipment inspections, and documentation. The adage that "what gets measured gets done" can be a powerful tool to accomplish safety program implementation and OSHA compliance.

In addition to the ability of senior managers to monitor safety activities at the job-site level, the responsibility for safety can be shifted from the safety professional to the project and front-line supervisors. Safety professionals will be in demand to provide assistance and expertise in the computerized safety planning and safety analysis activities rather than for the current duties of training and job-site inspections. Using the job safety analysis and checklists generated by the software, supervisors will be more capable and inclined to conduct inspections and training themselves, thereby improving project safety by having the individuals with the

direct responsibility conducting the activities and then documenting their performance so that they can be held accountable for the safety program implementation rather than relying on the number of injuries that may or may not occur.

Prequalification of the Contractor Safety Program

Owner and contractor safety professionals are becoming more involved in the prequalification of contractors and subcontractors prior to the bid process. This process usually happens in one of two ways. Either contractors submit information to qualify to be included on the approved-to-bid list, which requires that contractors demonstrate they have adequate resources and experience to perform the work safely, or contractors are required to demonstrate their safety capabilities after being placed on the short-list of bidders. In both processes, the success of the contractor in documenting past safety accomplishments can make the difference between winning the contract award or losing work that would have a meaningful impact on the firm's bottom line.

Even when contractors can show that they have excellent safety histories, many owners require safety programs that go well beyond the industry norms and OSHA standards. The potential for high-visibility incidents that are covered by local and national press prompts many owners to allow only the highest qualified contractors to perform work in their facilities. An equal concern of owners is the potential for third-party lawsuits. They believe that the exposure to damaging publicity and the cost of legal battles can override any savings that could have been gained by ignoring contractor safety records and capabilities.

On previous projects owners may have experienced that better safety planning and control results in a more efficient and time-saving overall project that ends on schedule and within budget guidelines. This experience and acquired expertise may enable an owner to allow a contractor with a minimum safety program and record to bid and perform work because the owner will provide guidance and closely monitor the contractor's safety activities. The owner will often require contractors with poor records to closely adhere to the owner's safety program for contractors. With the appropriate oversight conducted by qualified safety staff, the owner may need to hire contractors that he or she normally would not hire because of the availability of specialty contractors.

Previous knowledge and experience with the contractor is a major factor in allowing contractors with high injury rates to work in an owner's facility. By careful monitoring of the contractor, the owner can provide assistance in areas where the contractor lacks expertise or resources. The liability for the owner is dependent on how closely the contractor on the job is monitored and supervised. Relying on a contractor who makes a strong presentation and provides low injury data may be a mistake for the owner because performance is dependent on the contractor's commitment to the specific project and the capabilities of the assigned supervision. Not all projects share equally in a contractor's safety record. The injury statistics are an *average* of the overall contractor performance and do not guarantee that the supervision assigned to the project will not be one with an unacceptable past safety record that is obscured by the excellent work of counterpart supervisors on other projects.

Design and Engineering

There is continuing consolidation toward a two-tiered market. Industry consolidation is accelerating, especially with the availability of capital in the markets. Large design firms keep getting larger, and smaller design firms continue to specialize in selective practices and markets. There has also been a steady trend toward increased design-firm risk-taking in the delivery process. In the past, design professionals have passed along project risks to contractors. But, as contractors have become more involved and responsive to owners and purchasers of construction services, designers have been resigned to a lesser role in the overall process. In an attempt to enhance and expand their role, designers have responded by assuming additional risk with the anticipation of generating higher rewards.

A convergence of construction and design is increasingly found in the industry. It is estimated that approximately 50 percent of construction will be performed via design-build. Construction and design firms see a comingling of resources as a way of becoming more competitive. Construction companies and design firms are merging to provide full in-house services to owners desiring to consolidate their project administration. This shift is viewed positively by many safety professionals due to the design-in-safety concept and its importance to safe job sites.

The construction market in recent years has experienced outstanding performance in the United States and in the international arena. Despite economic difficulties in Asia, design and construction firms are proceeding with construction and investment in international markets. Europe, China, and South America still foresee growth in their markets, a trend that will draw the highly capitalized design-build firms to work in these countries. Cross-border acquisitions, mergers, and joint ventures are viewed as creating an extension for their base of operations. Engineering firms will accelerate their level of risk-taking, including their approaches to delivery, contractual liability, and cost control. The convergence of engineering and construction will continue, exerting an even heavier influence on construction safety.

Certifying Craftsmen Safety Training and Capabilities

The ever-growing shortage of trained and skilled craftsmen will continue for many years, significantly impacting safety performance. With 86,000 union construction apprentices and 36,000 nonunion apprentices, the rate of replacement for retiring and injured journeyman is wholly inadequate. Because many construction workers do not complete formal apprenticeship training, they lack the safety skills and knowledge necessary to perform their work.

Owners and contractors have recognized the impact of untrained workers and have required, through negotiated contracts, that a minimal level of training be obtained prior to being allowed to work on their projects. In response to these mandates, several union–contractor associations have established training and certification programs that enable craftsmen to attend training and be placed on a master list of eligible workers that can easily be verified by email or fax-on-demand.

An example of preemployment qualification is the MOST Program, established by the construction boilermaker industry, where a labor–management trust has been established to provide safety training and substance-abuse screening and education. Qualified craftsmen carry identification cards that can be read like a credit card when swiped through a scanner attached to a job-site computer. The card provides documentation on a craftsman's most recent welding tests, physical examinations for respirator use, and fit-testing results. The card also verifies OSHA 10-hour safety training and foreman training.

Owner-funded training and drug screening has also been tried in the Southwest and Northwest. Local unions and contractor associations, such as International Brotherhood of Electrical Workers, Local 176, and the National Electrical Contractors Association, Eastern Illinois chapter, have established programs that provide for drug screening of all craftsmen referred to job sites by the local union. The drug screening is also conducted on a random annual basis so that all participants are tested at least once every two years.

Readily available and reliable documentation of craftsmen qualifications, training, and substance-abuse testing will be utilized more in the future as unions and contractors search for economical and efficient ways of providing trained, safety-conscious craftsmen.

Recognition Programs Initiated by Labor–Management Agreements

The entire construction industry needs to place more importance on safety and provide recognition for those who have performed at the highest levels. One such innovative program is the "Safety Incentive Program" (SIP) being developed by the Chicago area Laborers–Employers Cooperation and Education Trust (LECET). LECET describes their program as follows: "The purpose of the Safety Incentive Program is to encourage union laborers to be aware of working safely on their job sites. Reducing accidents is not only good economic sense; it is a moral obligation for all of us. The Safety Incentive Program is not intended to replace an individual company's safety policy or culture but to enhance it."

Eligible participants are union laborers who are field construction workers and members in good standing with any participating labor local affiliated with the Construction & General Laborers' District Council of Chicago and Vicinity. These participants must have no lost-time or OSHA-recordable events due to work-related accidents or illness and must have worked a stipulated number of hours in a calendar quarter in order to qualify for a quarterly award. On a quarterly basis, a contractor will be randomly selected to nominate a laborer who has worked safely for that contractor.

LECET SIP holds Annual Safety Awards Luncheons where contractors nominate laborers who have worked safely for the year. One of the nominees will be selected to win a new pickup truck.

Manufacturers and Suppliers

Electronic commerce is becoming more common throughout the distribution channel and will grow at a faster rate over the next few decades. The consequences of this will benefit both the manufacturer and its partners with greater distribution efficiencies. All parties will need this technology just to be a key player in the construction industry. Many manufacturers sell to fewer, larger clients and are concerned about the shift in power to these large clients. Manufacturers will continue to evaluate the benefits of selling direct versus selling through distribution to alter the balance of power.

The prosperous 1990s have fueled expansions that will result in an excess-capacity situation in many industry segments. This excess capacity will cause prices to soften as manufacturers struggle to maintain share when the market inevitably turns. Creating highly differentiated, value-added services continues to be an objective throughout the distribution channel. Manufacturers selling to retailers, distributors, and contractors search for unique methods to build greater loyalty and to create exit barriers for their preferred clients. Manufacturers will participate in shared knowledge networks to maintain a competitive advantage and become a strategic partner within the industry.

The Owner's Role in Safe Work Performance

In addition to requiring contractors to document their safety program and injury record to be eligible to bid on construction contracts, owners are beginning to take a direct role in the safety activities of contractors performing work on their property. Owners require that each new-hire attend a site-specific orientation that covers the safety and health policies and practices of the owner. The orientation is conducted by a facility safety official and includes emergency response procedures and hazardous materials present that require special protective measures. Owners often issue confined space permits after conducting air testing, and they require that their emergency rescue teams be notified of all confined space entries. Hot-work permits and lockout/tagout permits are controlled by the owner with routine audits and monitoring of high-hazard contractor activities to assure conformance with owner procedures.

The owner mandates safety committees and the safety meetings to be attended by all contractors and craft representatives; the meetings are conducted by an owner safety official. The owner performs walk-around safety inspections, and all safety violations are brought to the contractor's attention. When an imminent hazard is encountered, the owner will issue a "stop work" order and not allow work to proceed until corrective action is taken. The role of the owner has increased dramatically in the past few years, and based on draft policy procedures of organizations like the Construction User's Roundtable, contractors can look forward to even more involvement and oversight in the construction process by owners in the future.

General Contractors

General contractors continue to explore ways to separate themselves from competitors. One trend that is emerging is contractors who increase self-performing work in trades, such as framing and masonry. Also, many general contractors are looking for additional or broader project management roles, including construction management and design-build. The design-build option will increase in popularity and become the preferred method of procurement.

Training and manpower issues will continue to be an important factor for general contractors. Multiskilling is becoming an option for contractors who are willing to teach open-shop craft workers a new trade, allowing them to perform multiple tasks on the same project. This will reduce the total labor requirements on a specific project and provide employment continuity for the workers. Projects that utilize the multiskilling concept usually experience less rework and significantly reduced hiring requirements. Contractors must continue to meet the challenge in finding qualified craft workers. Creative hiring and recruiting practices are being used to identify new-hires throughout the construction industry. Several contractors are paying "bounties" to current employees who refer new-hires who accept a position and stay with the company for a minimum specified period. There are numerous opportunities for local contractors to create alliances with large national and international contractors on local projects, thus expanding and sharing safe work practices and improving job-site safety performance.

Many general contractors are improving communications and reducing paperwork at job sites by using handheld technology to enter time and cost code information. Many superintendents and foremen are using laptop computers to enter daily field requirements and track performance against the estimate and schedule. These activities then leave supervisors with additional time to focus on safety.

Contractors are measuring more than just downstream results, such as job costs, accident rates, and productivity. Additional emphasis is being placed on identifying and correcting system deficiencies that lead to improved quality, increased productivity, and fewer incidents.

Government Policies that Impact Construction Safety

The current OSHA administration has a goal of increasing partnerships with industry leaders to 174 agreements and 12 alliances. The partnerships and alliances enable OSHA to build coalitions that expand the limited government resources and bring responsible employers into the process of providing services that in the past were carefully restricted to internal OSHA functions.

The message that OSHA wants to communicate is that, along with their enforcement responsibilities, the agency is a *resource* for business. A specific application to the construction industry is the OSHA offer to develop a program that allows compliance officers to receive training from contractors and unions so that compliance officers will have real-world experience in climbing steel and installing fall protection and prevention equipment. This approach to industry training of OSHA compliance officers was conceived during the exchange of ideas within the process of writing the new Subpart R Steel Erection Standard. Future standards for cranes and platforms may present possible areas for cooperation in compliance officer training as well.

Another project that OSHA is working on is the revamping of the Construction Safety Excellence Program that would enable owners and contractors to participate in a cooperative initiative in a type of voluntary protection program; this initiative is more specific to the short-term high-hazard projects in the construction industry. The concept is to focus on project qualification and also empower employees to become involved. A company/project can progress through the CSE process to VPP Star status. The Directorate of Construction would provide an initial contractor review to decide who gets into the program, which consists of various stages (ranging from "Commitment Agreement Compacts" through "Covenants" and "OSHA Strategic Partnerships"). Ultimately, a company would reach VPP Merit or VPP Star status. It would give recognition to the company, and the company would pick the project(s) it wished to have participate.

Partnerships will continue to be a critical link between OSHA and the construction industry. The Associated General Contractors (AGC) and Associated Builders and Contractors (ABC) were partnerships that provided a national template. There are approximately 100 total partnerships. Other partnerships that will contribute to construction safety include: National Commission for Certification of Crane Operators (NCCCO), NEA steel erection training partnership, and Board of Certified Safety Professionals (BCSP). This last alliance will revolve around the Safety Trained Supervisor (STS) and Construction Health and Safety Technician (CHST) certifications in the construction industry. Also, the Industrial and Electrical Contractors alliance will promote education and outreach for electrical contractors in the industrial sector. All of these partnerships and alliances will contribute greatly to construction safety and health.

HEARING CONSERVATION

Hearing conservation is a safety issue that will gain more attention in the construction industry. OSHA estimates that 750,000 construction workers are exposed to noise levels at or above the TLV-TWA. Engineering administrative controls are often not used, or not feasible, in construction to bring the noise level to its required PEL, which is set at 90 decibels. Also, currently the use of hearing protectors is low. Many construction workers lose their hearing at an early age. Tinnitus (ringing in the ears) is a common experience among construction workers. OSHA's construction standard as compared to the general industry standard is not specific.

OSHA's work in nonregulatory arenas can help bring about solutions in this safety and health area. For example, OSHA is working with a contractor at a concrete project to enhance existing toolbox talks. This approach can result in a train-the-trainer option, deliverable in terms of a program and a training component.

EMPLOYMENT TRENDS IN THE CONSTRUCTION INDUSTRY

Perhaps more than any other major industry, construction is particularly sensitive to shifting economic trends. When looking at the overall health of the industry, three factors must be addressed. First, the overall level of economic activity is volatile. The construction industry suffered from the recession of the early nineties. The slowdown in the economy and high unemployment reverberated through the construction sector. The stronger economy from 1992 until the end of the century gave a needed boost to construction. Due to a stagnant economy, the industry is experiencing a "slow down," except in areas such as heavy construction (roads, sewers) and commercial construction. Second, the sector is particularly sensitive to changes in government spending. This is especially true in heavy construction where governments are the dominant source of funding. As governments at all levels struggled to reduce debt and hold the line on taxes, investment in infrastructure was reduced. Finally, and of longer-term significance, is the changing demographic situation. The population is aging. While slower growth is a negative for the industry, the changing make-up of the population may bring benefits to certain parts of the construction industry. This may be especially so for those involved in household renovation work. An aging population may be less inclined to tackle projects on their own and more interested (and financially better prepared) to hire others to do the work.

Stagnant economy aside, there will continue to be a high demand for skilled tradespeople in many areas and there may be a potential shortage in several of the key construction trades. While there is a promise of increased demand for labor in the construction sector, the volatility of the industry makes forecasting hiring needs rather difficult. Moreover, there are several other features of the labor force in the construction industry that complicate the process of predicting employment needs.

One factor is the high level of mobility among workers in the industry. Steady employment often requires that workers must commute or relocate outside their local or state region. This is important because it suggests that local economic trends may have more to do with *where* people work rather than *whether* they work. By the same token, a surge in demand locally may be met in part by the movement of skilled workers from other regions. Both companies and unions have become accustomed to a situation in which mobility is common. However, while the practice of traveling in search of work may allow some individuals to avoid unemployment, it may also encourage others to abandon construction for work in other industries. This may be particularly true for those who are older, have family commitments, or for whom constant travel is undesirable. Related to this is the lure for many in the

skilled trades of regular employment in manufacturing or related sectors. There is also a growing concern about the loss of experienced workers who preferred the greater stability of employment in other sectors of the economy. Another factor is the strenuous nature of the work in many trades, which may produce a different pattern of retirement or job change than in many other industries. Some jobs, such as roofing, are so physically demanding that relatively few workers may be likely to spend their whole careers at this type of work. These factors should be considered when examining the age distribution of workers in various trades. If early retirement or movement into work in other industries is common, the potential for labor shortages may be greater than would be apparent from data on the age distribution of current workers.

CONCLUSION

Safety professionals will be in high demand as owners and contractors implement complex and detailed computer-based safety planning and monitoring programs. Owners will continue to screen contractors to ensure that they have the experience and resources to implement effective safety programs. Owners will increase their influence over the planning and monitoring of contractor safety procedures and control high-hazard activities through an owner-managed permit system. The role of OSHA in construction safety will be to identify and target contractors with high injury rates while at the same time offering to partner with contractors who exhibit a commitment to job-site safety. Contractors and unions will develop safety training that is documented and verifiable that will improve job-site safety and minimize duplicative costs related to redundant testing and record keeping.

The future for construction safety professionals, owners, contractors, and workers is promising in that the industry is beginning to recognize the value of qualified safety staff that can assist in the planning and implementation of injury prevention programs that prove to be both cost effective and easily managed.

REVIEW EXERCISES

1. List two factors that are impacting the construction industry and discuss how each may have an influence on job-site safety.

2. Has the construction safety profession kept pace with other industries (i.e., manufacturing, healthcare) as it relates to protecting people, property, and the environment? Support your answer.

3. Explain the importance of education as it relates to the changing role of the construction safety professional.

4. Who will have a greater influence on job-site safety, owners or general contractors? Support your answer.

5. Discuss how the employment outlook in the construction industry will have an impact on construction injuries.

Safety and Health Regulations for Construction

- 1926 – Table of Contents
- 1926 Subpart A – General
  - 1926.1 – Purpose and Scope
  - 1926.2 – Variances from safety and health standards
  - 1926.3 – Inspections – right of entry
  - 1926.4 – Rules of practice for administrative adjudications for enforcement of safety and health standards
  - 1926.5 – OMB control numbers under the Paperwork Reduction Act

- 1926 Subpart B – General Interpretations
  - 1926.10 – Scope of subpart
  - 1926.11 – Coverage under section 103 of the act distinguished
  - 1926.12 – Reorganization Plan No. 14 of 1950
  - 1926.13 – Interpretation of statutory terms
  - 1926.14 – Federal contract for "mixed" types of performance
  - 1926.15 – Relationship to the Service Contract Act; Walsh-Healey Public Contracts Act
  - 1926.16 – Rules of construction

- 1926 Subpart C – General Safety and Health Provisions
  - 1926.20 – General safety and health provisions
  - 1926.21 – Safety training and education
  - 1926.22 – Recording and reporting of injuries
  - 1926.23 – First aid and medical attention
  - 1926.24 – Fire protection and prevention
  - 1926.25 – Housekeeping
  - 1926.27 – Sanitation
  - 1926.28 – Personal protective equipment

- 1926.29 – Acceptable certifications
- 1926.30 – Ship building and ship repairing
- 1926.31 – Incorporation by reference
- 1926.32 – Definitions
- 1926.33 – Access to employee exposure and medical records
- 1926.34 – Means of egress
- 1926.35 – Employee emergency action plans

▪ 1926 Subpart D – Occupational Health and Environmental Controls
  - 1926.50 – Medical services and first aid
    • 1926.50 Appendix A – Medical services and first aid
  - 1926.51 – Sanitation
  - 1926.52 – Occupational noise exposure
  - 1926.53 – Ionizing radiation
  - 1926.54 – Nonionizing radiation
  - 1926.55 – Gases, vapors, fumes, dusts, and mists
    • 1926.55 Appendix A – Gases, vapors, fumes, dusts, and mists
  - 1926.56 – Illumination
  - 1926.57 – Ventilation
  - 1926.58 – [Reserved]
  - 1926.59 – Hazard communication
  - 1926.60 – Methylenedianiline
    • 1926.60 Appendix A – Substance data sheet, for 4-4' METHYLENEDIANILINE
  - 1926.60 – Methylenedianiline
  - 1926.60 – Methylenedianiline
    • 1926.60 Appendix D – Sampling and analytical methods for MDA monitoring and measurement procedures
    • 1926.60 Appendix E – Qualitative and quantitative fit-testing procedures
  - 1926.61 – Retention of DOT markings, placards, and labels
    • Qualitative and quantitative 1926.62 – Appendix A – Substance data sheet for occupational exposure to lead
    • 1926.62 Appendix B – Employee standard summary
    • 1926.62 Appendix C – Medical surveillance guidelines
    • 1926.62 Appendix D – Qualitative and quantitative fit-test protocols
  - 1926.64 – Process safety management of highly hazardous chemicals
    • 1926.64 Appendix A – List of highly hazardous chemicals, toxics, and reactives (mandatory)
    • 1926.64 Appendix B – Block flow diagram and simplified process flow diagram (nonmandatory)
    • 1926.64 Appendix C – Compliance guidelines and recommendations for process safety management (nonmandatory)
  - 1926.65 – Hazardous waste operations and emergency response
  - 1926.65 Appendix A – Personal protective equipment test methods
  - 1926.65 Appendix B – General description and discussion of the levels of protection and protective gear
  - 1926.65 Appendix C – Compliance guidelines
  - 1926.65 Appendix D – References
  - 1926.65 Appendix E – Training curriculum guidelines nonmandatory
  - 1926.66 – Criteria for design and construction of spray booths

■ 1926 Subpart E – Personal Protective and Life-Saving Equipment
  - 1926.95 – Criteria for personal protective equipment
  - 1926.96 – Occupational foot protection
  - 1926.97 – [Reserved]
  - 1926.98 – [Reserved]
  - 1926.99 – [Reserved]
  - 1926.100 – Head protection
  - 1926.101 – Hearing protection
  - 1926.102 – Eye and face protection
  - 1026.103 – Respiratory protection
  - 1926.104 – Safety belts, lifelines, and lanyards
  - 1926.105 – Safety nets
  - 1926.106 – Working over or near water
  - 1026.107 – Definitions applicable to this subpart

■ 1926 Subpart F – Fire Protection and Prevention
  - 1926.150 – Fire protection
  - 1926.151 – Fire prevention
  - 1926.152 – Flammable and combustible liquids
  - 1926.153 – Liquefied petroleum gas (LP – gas)
  - 1926.154 – Temporary heating devices
  - 1926.155 – Definitions applicable to this subpart
  - 1926.156 – Fixed extinguishing systems, general
  - 1926.157 – Fixed extinguishing systems, gaseous agent
  - 1926.158 – Fire detection systems
  - 1926.159 – Employer alarm systems

■ 1926 Subpart G – Signs, Signals, and Barricades
  - 1926.200 – Accident prevention signs and tags
  - 1926.201 – Signaling
  - 1926.202 – Barricades
  - 1026.203 – Definitions applicable to this subpart

■ 1926 Subpart H – Material Handling, Storage, Use, and Disposal
  - 1926.250 – General requirements for storage
  - 1926.251 – Rigging equipment for material handling
  - 1926.252 – Disposal of waste materials

■ 1926 Subpart I – Tools, Hand and Power
  - 1926.300 – General requirements
  - 1926.301 – Hand tools
  - 1926.302 – Power-operated hand tools
  - 1926.303 – Abrasive wheels and tools
  - 1926.304 – Woodworking tools
  - 1926.305 – Jacks, lever and ratchet, screw, and hydraulic
  - 1926.306 – Air receivers
  - 1926.307 – Mechanical power-transmission apparatus

■ 1926 Subpart J – Welding and Cutting
  - 1926.350 – Gas welding and cutting
  - 1926.351 – Arc welding and cutting
  - 1926.352 – Fire prevention
  - 1926.353 – Ventilation and protection in welding, cutting, and heating
  - 1926.354 – Welding, cutting, and heating in any way of preservative coatings

■ 1926 Subpart K – Electrical
  - 1926.400 – Introduction
  - 1926.401 – [Reserved]
  - 1926.402 – Applicability
  - 1926.403 – General requirements
  - 1926.404 – Wiring design and protection
  - 1926.405 – Wiring methods, components, and equipment for general use
  - 1926.406 – Specific-purpose equipment and installations
  - 1926.407 – Hazardous (classified) locations
  - 1926.408 – Special systems
  - 1926.409 – [Reserved]
  - 1926.410 – [Reserved]
  - 1926.411 – [Reserved]
  - 1926.412 – [Reserved]
  - 1926.413 – [Reserved]
  - 1926.414 – [Reserved]
  - 1926.415 – [Reserved]
  - 1926.416 – General requirements
  - 1926.417 – Lockout and tagging of circuits
  - 1926.418 – [Reserved]
  - 1926.419 – [Reserved]
  - 1926.420 – [Reserved]
  - 1926.421 – [Reserved]
  - 1926.422 – [Reserved]
  - 1926.423 – [Reserved]
  - 1926.424 – [Reserved]
  - 1926.425 – [Reserved]
  - 1926.426 – [Reserved]
  - 1926.427 – [Reserved]
  - 1926.428 – [Reserved]
  - 1926.429 – [Reserved]
  - 1926.430 – [Reserved]
  - 1926.431 – Maintenance of equipment
  - 1926.432 – Environmental deterioration of equipment
  - 1926.433 – [Reserved]
  - 1926.434 – [Reserved]
  - 1926.435 – [Reserved]
  - 1926.436 – [Reserved]
  - 1926.437 – [Reserved]
  - 1926.438 – [Reserved]

- 1926.555 – Conveyors
- 1925.556 – Aerial lifts

■ 1926 Subpart O – Motor Vehicles, Mechanized Equipment, and Marine Operations
  - 1926.600 – Equipment
  - 1926.601 – Motor vehicles
  - 1926.602 – Material-handling equipment
  - 1926.603 – Pile-driving equipment
  - 1926.604 – Site clearing
  - 1926.605 – Marine operations and equipment
  - 1926.606 – Definitions applicable to this subpart

■ 1926 Subpart P – Excavations
  - 1926.650 – Scope, application, and definitions applicable to this subpart
  - 1926.651 – Specific excavation requirements
  - 1926.652 – Requirements for protective systems
  - 1926 Subpart P Appendix A – Soil classification
  - 1926 Subpart P Appendix B – Sloping and benching
  - 1926 Subpart P Appendix C – Timber shoring for trenches
  - 1926 Subpart P Appendix D – Aluminum hydraulic shoring for trenches
  - 1926 Subpart P Appendix E – Alternatives to timber shoring
  - 1926 Subpart P Appendix F – Selection of protective systems

■ 1926 Subpart Q – Concrete and Masonry Construction
  - 1926.700 – Scope, application, and definitions applicable to this subpart
  - 1926.701 – General requirements
  - 1926.702 – Requirements for equipment and tools
  - 1926.703 – Requirements for cast-in-place concrete
  - 1926.703 Appendix – General requirements for formwork
  - 1926.704 – Requirements for precast concrete
  - 1926.705 – Requirements for lift-slab operations
  - 1926.705 Appendix – Lift-slab operations
  - 1926.706 – Requirements for masonry construction
  - 1926 Subpart Q Appendix A – References to Subpart Q of Part 1926

■ 1926 Subpart R – Steel Erection
  - 1926.750 – Scope
  - 1926.751 – Definitions
  - 1926.752 – Site layout, site-specific erection plan, and construction sequence
  - 1926.753 – Hoisting and rigging
  - 1926.754 – Structural steel assembly
  - 1926.755 – Column anchorage
  - 1926.756 – Beams and columns
  - 1926.757 – Open web steel joists
  - 1926.758 – Systems-engineered metal buildings
  - 1926.759 – Falling-object protection
  - 1926.760 – Fall protection

- 1926.761 – Training
- 1926 Subpart R Appendix A – Guidelines for establishing the components of a site-specific erection plan: nonmandatory guidelines for complying with 1926.752(e)
- 1926 Subpart R Appendix B – Acceptable test methods for testing slip-resistance of walking/working surfaces: nonmandatory guidelines for complying with 1926.754(c)(3)
- 1926 Subpart R Appendix C – Illustrations of bridging terminus points: nonmandatory guidelines for complying with 1926.757(a)(10) and 1926.757(c)(5)
- 1926 Subpart R Appendix D – Illustration of the use of control lines to demarcate controlled decking zones (CDZs): nonmandatory guidelines for complying with 1926.760(c)(3)
- 1926 Subpart R Appendix E – Training: nonmandatory guidelines for complying with 1926.761
- 1926 Subpart R Appendix F – Perimeter columns: nonmandatory guidelines for complying with 1926.756(e) to protect the unprotected side or edge of a walking/working surface
- 1926 Subpart R Appendix G – 1926.502(b)-(e) Fall protection systems criteria and practices
- 1926 Subpart R Appendix H – Double connections: illustration of a clipped-end connection and a staggered connection: nonmandatory guidelines for complying with 1926.7560(c)(1)

■ 1926 Subpart S – Underground Construction, Caissons, Cofferdams, and Compressed Air
- 1926.800 – Underground construction
- 1926.801 – Caissons
- 1926.802 – Cofferdams
- 1926.803 – Compressed air
- 1926.804 – Definitions applicable to this subpart
- 1926 Subpart S Appendix A – Decompression tanks

■ 1926 Subpart T – Demolition
- 1926.850 – Preparatory operations
- 1926.851 – Stairs, passageways, and ladders
- 1926.852 – Chutes
- 1926.853 – Removal of materials through floor openings
- 1926.854 – Removal of walls, masonry sections, and chimneys
- 1926.855 – Manual removal of doors
- 1926.856 – Removal of walls, floors, and material with equipment
- 1926.857 – Storage
- 1926.858 – Removal of steel construction
- 1926.859 – Mechanical demolition
- 1926.860 – Selective demolition by explosives

■ 1926 Subpart U – Blasting and the Use of Explosives
- 1926.900 – General provisions
- 1926.901 – Blaster qualifications
- 1926.902 – Surface transportation of explosives
- 1926.903 – Underground transportation of explosives
- 1926.904 – Storage of explosives and blasting agents
- 1926.905 – Loading of explosives or blasting agents
- 1926.906 – Initiation of explosive charges—electric blasting
- 1926.907 – Use of safety fuse
- 1926.908 – Use of detonating cord

- 1926.909 – Firing the blast
- 1926.910 – Inspection after blasting
- 1926.911 – Misfires
- 1926.912 – Underwater blasting
- 1926.913 – Blasting in excavation work under compressed air
- 1926.914 – Definitions applicable to this subpart

- 1926 Subpart V – Power Transmission and Distribution
  - 1926.950 – General requirements
  - 1926.951 – Tools and protective equipment
  - 1926.952 – Mechanical equipment
  - 1926.953 – Material handling
  - 1926.954 – Grounding for protection of employees
  - 1926.955 – Overhead lines
  - 1926.956 – Underground lines
  - 1926.957 – Construction in energized substations
  - 1926.958 – External load helicopters
  - 1926.959 – Lineman's body belts, safety straps, and lanyards
  - 1926.960 – Definitions applicable to this subpart

- 1926 Subpart W – Rollover Protective Structures; Overhead Protection
  - 1926.1000 – Rollover protective structures (ROPS) for material-handling equipment
  - 1926.1001 – Minimum performance criteria for rollover protective structures for designated scrapers, loaders, dozers, graders, and crawler tractors
  - 1926.1002 – Protective frames (rollover protective structures, known as ROPS) for wheel-type agricultural and industrial tractors used in construction
  - 1926-1003 – Overhead protection for operators of agricultural and industrial tractors

- 1926 Subpart X – Ladders
  - 1926.1050 – Scope, application, and definitions applicable to this subpart
  - 1926.1051 – General requirements
  - 1926.1052 – Stairways
  - 1926.1053 – Ladders
  - 1926.1054 – [Reserved]
  - 1926.1055 – [Reserved]
  - 1926.1056 – [Reserved]
  - 1926.1057 – [Reserved]
  - 1926.1058 – [Reserved]
  - 1926.1059 – [Reserved]
  - 1926.1060 – Training requirements
  - 1926 Subpart X Appendix A – Ladders

- 1926 Subpart Y – Commercial Diving Operations
  - 1926.1071 – Scope and application
  - 1926.1072 – Definitions
  - 1926.1076 – Qualifications of dive team
  - 1926.1080 – Safe practices manual

- 1926.1081 – Predive procedures
- 1926.1082 – Procedures during dive
- 1926.1083 – Postdive procedures
- 1926.1084 – SCUBA diving
- 1926.1085 – Surface-supplied air diving
- 1926.1086 – Mixed-gas diving
- 1926.1087 – Liveboating
- 1926.1090 – Equipment
- 1926.1091 – Record-keeping requirements
- 1926.1092 – Effective date
- 1926 Subpart Y Appendix A – Examples of conditions which may restrict or limit exposure to hyperbaric conditions
- 1926 Subpart Y Appendix B – Guidelines for scientific diving

■ 1926 Subpart Z – Toxic and Hazardous Substances
- 1926.1100 – [Reserved]
- 1926.1101 – Asbestos
- 1926.1101 Appendix A – OSHA reference method, mandatory
- 1926.1101 Appendix B – Sampling and analysis, nonmandatory
- 1926.1101 Appendix C – Qualitative and quantitative fit-testing procedures, mandatory
- 1926.1101 Appendix D – Medical questionnaires, mandatory
- 1926.1101 Appendix E – Interpretation and classification of chest-roentgenograms, mandatory
- 1926.1101 Appendix F – Work practices and engineering controls for Class I asbestos operations, nonmandatory
- 1926.1101 Appendix G – [Reserved]
- 1926.1101 Appendix H – Substance technical information for asbestos, nonmandatory
- 1926.1101 Appendix I – Medical surveillance guidelines for asbestos, nonmandatory
- 1926.1101 Appendix J – Smoking cessation program information for asbestos, nonmandatory
- 1926.1101 Appendix K – Polarized light microscopy of asbestos, nonmandatory
- 1926.1102 – Coal tar pitch volatiles; interpretation of term
- 1926.1103 – 13 Carcinogens (4-Nitrobiphenyl, etc.)
- 1926.1104 – Alpha-naphthylamine
- 1926.1105 – [Reserved]
- 1926.1106 – Methyl chloromethyl ether
- 1926.1107 – 3,3′-Dichlorobenzidine (and its salts)
- 1926.1108 – Bis-chloromethyl ether
- 1926.1109 – Beta-naphthylamine
- 1926.1110 – Benzidine
- 1926.1111 – 4 Aminodiphenyl
- 1926.1112 – Ethyleneimine
- 1926.1113 – Beta-propiolactone
- 1926.1114 – 2 Acetylaminofluorene
- 1926.1115 – 4 Dimethylaminoazobenzene
- 1926.1116 – N-Nitrosodimethylamine
- 1926.1117 – Vinyl chloride
- 1926.1118 – Inorganic arsenic
- 1926.1127 – Cadmium

- 1926.1127 Appendix A – Substance safety data sheet – cadmium
- 1926.1127 Appendix B – Substance technical guidelines for cadmium
- 1926.1127 Appendix C – Qualitative and quantitative fit-testing procedures
- 1926.1127 Appendix D – Occupational health history interview with reference to cadmium exposure
- 1926.1127 Appendix E – Cadmium in workplace atmospheres
- 1926.1127 Appendix F – Nonmandatory protocol for biological monitoring
- 1926.1128 – Benzene
- 1926.1129 – Coke oven emissions
- 1926.1144 – 1,2-dibromo-3-chloropropane
- 1926.1145 – Acrylonitrile
- 1926.1147 – Ethylene oxide
- 1926.1148 – Formaldehyde
- 1926.1152 – Methylene chloride
- 1926 Subpart Z Appendix A – Designations for general industry standards incorporated into body of construction standards

OSHA Position Letter– Press Release (HHS)

OSHA's Position on Providing a Drug-free Workplace

May 2, 1998

Mr. Patrick J. Robinson
Safety Coordinator
Starline Manufacturing Co., Inc.
6060 West Douglas Avenue
Milwaukee, WI 53218-1561

Dear Mr. Robinson:

This is in response to your letter dated April 9, to Gerald Cunningham, Area Director, Milwaukee Area Office. You requested clarification of the Occupational Safety and Health Administration's (OSHA's) viewpoint of providing a safe work environment, specifically related to workers performing work under the influence of alcohol or illicit drugs. Your letter was forwarded to us from the Milwaukee Area Office to answer.

In your letter you asked the following questions:

1. **Does OSHA believe that an employer has a duty to provide a workplace free of employees performing assigned duties with mechanical machinery under the intoxicating influence of alcohol or under the influence of illicit drugs?**

Response: OSHA strongly supports measures that contribute to a drug-free environment and reasonable programs of drug testing within a comprehensive workplace program for certain workplace environments, such as those involving safety-sensitive duties like operating machinery. Such programs, however, need to also take into consideration employee rights to privacy.

2. If so, has OSHA been in a position to enforce such a viewpoint?

<u>Response</u>: Although OSHA supports workplace drug and alcohol programs, at this time OSHA does not have a standard. In some situations, however, OSHA's General Duty Clause, Section 5(a)(1) of the OSHA Act, may be applicable where a particular hazard is not addressed by any OSHA standard.

Citations for violation of the General Duty Clause are issued to employers when the four components of this provision are present, and when no specific OSHA standard has been promulgated to address the recognized hazard. The four components are: (1) the employer failed to keep its workplace free of "hazard"; (2) the hazard was "recognized" either by the cited employer individually or by the employer's industry generally; (3) the recognized hazard was causing or was likely to cause death or serious physical harm; and (4) there was a feasible means available that would eliminate or materially reduce the hazard. An employer's duty will arise only when all four elements are present.

3. Generally speaking, what steps would be deemed reasonable in attempting to provide an alcohol/ drug-free workplace?

<u>Response</u>: On April 12, 1996, the U.S. Department of Health and Human Services (DHHS) released a study of drug abuse among U.S. workers. An announcement of this study may be found on the DHHS Internet web site (www.dhhs.gov/search) as item #15 when "drug abuse" is entered as the search term. The DHHS announcement (copy enclosed) gives an 800 number for employers to call for guidance and technical assistance in setting up a substance abuse prevention program. This service is free and available to all employers during regular working hours in both English and Spanish languages. The number is 800-WORKPLACE.

An employer's trade association, or workers' compensation insurance company may also be able to give helpful advice. Any education/training activity that helps employers and employees become aware of the dangers of working under the influence of alcohol, illicit drugs, and even some over-the-counter and prescription medications would be a good first step.

Thank you for your interest in occupational safety and health. If we can be of further assistance, please contact Helen Rogers of my staff at (202) 219-8031 x106.

Sincerely,

John B. Miles, Jr., Director
Directorate of Compliance Programs

Enclosure

Date: Friday, April 12, 1996
FOR IMMEDIATE RELEASE
Contact: Terri Gates, SAMHSA, (301)443-8956

HHS RELEASES STUDY OF DRUG ABUSE AMONG U.S. WORKERS

New Initiative Will Target Industries Needing Prevention Efforts

The Department of Health and Human Services today released a first-time report examining illicit drug use by U.S. workers. The report estimates prevalence and trends in drug abuse, as well as alcohol abuse, by occupation and industry categories.

The report, based on a new detailed analysis of the National Household Survey on Drug Abuse, finds that use of illicit drugs has declined by more than half among American workers since the mid-1980s. However, when broken down by industry and occupation, the analysis shows wide variation in the extent of illicit drug use and alcohol abuse among workers.

HHS Secretary Donna E. Shalala released the new report at a business-labor-government meeting, saying the report "helps show us where our next steps must be taken." She announced a new partnership initiative to target drug- and alcohol-abuse prevention efforts, especially in industries where abuse rates are still high.

"This report is a milestone in our efforts against <substance abuse>, Secretary Shalala said at the Teaming Up for Prevention Forum in Washington, D.C. "It tells us that we've made very real and substantial progress since the mid-1980s. It tells us that prevention strategies are reaching employees, especially those in positions of public trust and public safety. But it also tells us we need to do more, and we need to team up with labor and management, in a number of industries."

Shalala said a new "targeted, worker-oriented initiative will be part of the Clinton administration's broad-scale efforts to reduce and prevent <substance abuse> in America."

The new report, "Drug Use Among U.S. Workers: Prevalence and Trends by Occupation and Industry Categories," was produced by HHS' <substance abuse> and Mental Health Services Administration. The new analysis of the Household Survey covers data from 1991 to 1993 on drug and alcohol use among full- and part-time U.S. workers, aged 18–49. It was undertaken to provide information to help tailor prevention, intervention, and treatment efforts for occupations and industries most affected by drug use.

The Household Survey data show that among full-time employees, the portion reporting illicit drug use in the month prior to interview decreased from 16.7 percent in 1985 to 7.0 percent in 1992. The lower levels have remained steady since 1992, according to SAMHSA data.

The report finds that:

- Workers in occupations that require a considerable amount of public trust, such as police officers, teachers, and child-care workers, report the lowest rates of illicit drug use.

- The highest rates of illicit drug use were reported by workers in the following occupations: construction, food preparation, and waiters and waitresses.

- Workers in occupations that impact public safety, including truck drivers, fire fighters and police, report the highest rate of participation in drug testing.

Other highlights include:

- Heavy alcohol use rates were highest among construction workers, as well as auto mechanics, food preparation workers, light truck drivers, and laborers. The lowest rates of heavy alcohol use were reported by data clerks, personnel specialists, and secretaries.

- Broken down by sex of the worker, males in the following occupations reported the highest rates of illicit drug use: entertainers, food preparation and service, cleaning services, and construction. Among females, the highest rates of illicit drug use were reported by workers involved in food preparation, social work, and the legal professions, including lawyers and legal assistants. (However, in each of these occupations, males reported similar levels of illicit drug use).

- In general, unmarried workers reported about twice the rate of illicit drug and heavy alcohol use as married workers. In occupations such as food preparation, transportation drivers, and mechanics, and in industries such as construction and machinery (not electrical), the discrepancy between married and unmarried workers was especially notable.

- Workers who reported having three or more jobs in the previous five years were about twice as likely to be current or past-year illicit drug users as those who had two or fewer jobs.

- About 13 percent of full-time workers, ages 18–49, reported past-year involvement in a mandatory drug test at work.

Secretary Shalala also announced the new partnership initiative to help employers implement <substance abuse> prevention programs and policies. "This is an area where new efforts are needed—as part of President Clinton's comprehensive drug strategy that includes interdiction, law enforcement, prevention, treatment, drug testing, and research," Shalala said.

The initiative will include special efforts to help small businesses protect their employees from <substance abuse>, and HHS plans to work with the Department of Labor and the Small Business Administration on the partnership initiatives.

"Small businesses are the engine of a growing economy, so it is critical that all of us work to help prevent drug abuse among their workers. The fact is, small businesses are less likely to provide <substance abuse> information for employees, to have a written policy on <substance abuse>, and to provide access to employee assistance programs for drug and alcohol use. Many smaller businesses often don't have the resources to develop <substance abuse> prevention programs. This is where our efforts are needed most."

Secretary Shalala said employers can call 1-800-WORKPLACE for guidance and technical assistance. The service is free and available during regular working hours in both English and Spanish languages.

According to Nelba Chavez, Ph.D., administrator of SAMHSA, "Reaching out to employers is an essential part of our overall strategy to prevent <substance abuse>. Employers are much more than an economic force in their communities—they influence families and community values. In addition, most Americans access their health care through their employers. Business and labor can ensure that prevention and early intervention remain options for their employees and families."

RADIO STATION'S NOTE: Actualities from HHS Secretary Donna E. Shalala available April 12 on HHS Radio Hotline at 1-800-621-2984 or 202-690-8317.

OSHA Partnerships

How to Propose a Partnership

Step 1: Become Familiar with the Strategic Program Requirements
- Review the Strategic Partnership Policy Directive and the Partnership Overview and Summary document included in this packet to understand the requirements of the program. These documents are available on OSHA's Web page (www.osha.gov), or contact your Regional Partnership Coordinator. A Partnership Contact List is included in this package.

Step 2: Identify and Contact Companies and Organizations that Might Benefit from this Partnership
- One of the primary benefits of a Partnership with OSHA is that the more people we can bring under the Partnership umbrella the greater impact we have to the benefit of all. Bringing together groups and organizations with similar issues and similar problems provides a basis for efficient leverage of limited resources. Examples include: trade associations, professional associations, unions, insurance firms, universities, state and local governments, companies in industry with successful programs.

Step 3: Develop Your Preliminary Partnership Goal and Strategy
- What is the purpose of the Partnership? Of course it is designed to prevent injuries and illnesses, but the question you need to ask yourself and partners is what strategy are you planning on utilizing to accomplish that goal? The strategy should be incorporated into the goal to focus it, and make it more meaningful. Additionally, OSHA expects that your goal(s) be measurable. For example:
  - Reduce the likelihood of injuries and fatalities in our association by implementing 100% fall protection when working at heights over 6'; or
  - Develop and implement a cost-effective small business safety and health program utilizing contracted and OSHA safety and health experts for hazard awareness training, compliance assistance, and hazard correction, and using peer-to-peer safety and health audits and inspections in our industry.

Fortunately, OSHA's partners are extremely creative in finding new and effective means of involving more employers and employees and reducing injuries and illnesses.

Step 4: Consider What Type of Partnership (Limited or Comprehensive) Will Work Best

■ Deciding upon your specific goals and objectives will help you determine which type of partnership is best for you. For example: if you want to implement a successful and comprehensive safety and health program within your association, industry, or organization, then it is likely that a comprehensive partnership will serve you best.

However, if you want to focus on a particular area of safety and health management, such as: employee involvement, hazard analysis and identification, or a particular hazard control strategy for falls, then a limited partnership may be a better choice.

Step 5: Contact a Regional OSHA Partnership Representative

■ Contact your OSHA Partnership representative (see included list) and discuss your preliminary goals and strategy for your proposed Partnership efforts. Involving the OSHA Partnership representative early in the process can save a significant amount of time and energy.

Step 6: Develop and Refine Your Partnership Strategy

■ Your strategy must include several key elements to be considered for a Partnership. If it is a Comprehensive Partnership, it will require the implementation of an effective safety and health program. Your strategy should have the following elements as a bare minimum:

- Well-documented Strategic Goal(s)
- Partner Identification, Their Roles, Responsibilities, and Resources Required
- Employee Involvement
- List of Incentives and Benefits
- Performance Measurement System
- Terms of the Partnership and Circumstances/Rules for Termination of the Partnership
- A Written Plan Detailing the Proposed Partnership

Step 6a: Strategic Goals

■ Although not required in your Partnership, OSHA is very interested in addressing and impacting those hazards and industries described in its Strategic Plan. These include but are not limited to: increasing employer and worker awareness of, commitment to, and involvement in safety and health; reducing injuries and illnesses that are a result of silica, lead, and amputations; reducing the top four causes of injury, illnesses and fatalities in the construction, meatpacking, nursing home, logging, and shipbuilding industries. Regardless of what goal(s) are developed, they should demonstrate how they support the overall strategy of the Partnership and be measurable.

Step 6b: Partner Identification, Their Roles, Responsibilities, and Resources Required

■ Just as in any good safety and health system, responsibilities need to be defined. Who will track and maintain the data from the partnership? Who will do verification? Who will ensure programs are developed and distributed? Who will be the partnership's point of contact? Do each of the partners have certain responsibilities? How will resources (financial, time, expertise, etc.) be utilized and for who? How often will the partners meet? These questions need to be addressed in any Partnership. The final Partnership proposal should outline and describe with sufficient information the signatories of the primary partners and the roles and responsibilities of the key partners.

Step 6c: Employee Involvement

■ Businesses can benefit through OSHA Partnerships through a variety of ways, but the ultimate purpose of any Partnership is to ensure worker safety and health through better ways of working. Successes in OSHA's Voluntary Protection Programs (VPP) have demonstrated the value of involving employees in the safety and health management system. There is no single correct way to involve employees in safety and health. OSHA expects all employees under a Partnership to be meaningfully involved and aware of their rights under the OSHA act.

Examples of employee involvement that have been used successfully in the past, once trained in the appropriate skills, include but are not limited to: labor–management safety and health committees, performing hazard analyses, safety and health inspections, peer-to-peer audits, policy development, program reviews, and topical safety and health training.

Step 6d: Incentives and Benefits

■ There are many tangible and intangible benefits with Partnerships. Ideally a primary benefit is reduced injury and illness. However, what OSHA is looking for in this element of the Partnership is a discussion of how the Partnership is utilizing incentives and benefits that allow all of the partners (including OSHA) to leverage their limited resources. OSHA can offer (with some restrictions) recognition, outreach training, safety and health expertise (and other forms of technical assistance), consultation, focused inspections on a site's most serious hazards, and in some cases reduced penalty reductions as a result of citations. However, all OSHA Partnerships must stipulate that partnering employers remain subject to OSHA inspections and investigations in accordance with established agency procedures.

Step 6e: Performance Measurement System

■ It is very important that Partnership elements be measurable. Measurement is especially important in order to ensure that the approaches utilized are effective. Measures of success are important for demonstrating a cause and effect relationship based upon the incentives utilized, hazard controls implemented and overall success and failure of the Partnership. If the Partnership cannot show a definitive relationship, how can an accurate measure of its effectiveness be made? As was mentioned earlier, one of the main benefits of the Partnership is leveraging resources. Neither OSHA nor its partners can afford to expend valuable resources in a Partnership program without being reasonably assured of their investment return. Measures are the best way to validate the Partnership.

Obviously, one of the prime measures OSHA expects its partner to keep is injury and illness data as required under the record-keeping portion of the OSHA Act. Injury and illness statistics are an important indirect measure of the quality of the safety and health system being implemented. There are many measures that can and should also be considered. OSHA is looking for both quantitative and qualitative measures.

For example, if training is a significant aspect of the Partnership, then it is likely three measures would be needed. These might include: 1) how many were trained and to what level; 2) what was the quality of the training; and 3) how effective is the training at actually changing skill or behavior/or minimizing the severity and/or likelihood of the hazard(s)?

In the example above, the measures for the quality can be accomplished easily with test scores, and the measures of effectiveness can utilize a tailored inspection and/or audit form. Specific names of individuals do not need to be shared, and OSHA recognizes the need for confidentiality. However any measures should be verifiable when OSHA conducts its verification visits.

Lastly, all Comprehensive Partnerships require a thorough evaluation be performed and submitted to OSHA on an annual basis. Appendix C of the Partnership Directive provided guidance on the minimum information necessary for this evaluation. Essentially, this report will help document the effectiveness of the Partnership, identify strengths and weaknesses, and update OSHA with any significant changes to the Partnership.

Step 6f: Terms of the Partnership
■ The terms for continuation and termination of the Partnership should be clearly spelled out. The conditions and any sunset provisions should be documented in case any of the partners decide to voluntarily withdraw or, as a result of not meeting specific goals of the Partnership, it needs to be ended. Withdrawal of one of the primary partners from the agreement will always result in the Partnership discontinuation.

Step 6g: Develop Draft Partnership Proposal
■ Once the above elements have taken shape, document the proposal and contact your OSHA Partnership Representative. Submit your draft proposal so that it can be fine-tuned and modified if necessary. This will help ensure that the proposal is ready and that it is feasible for OSHA to act upon.

The description of the Partnership needs to be clear and concise. It needs to address all of the elements in the directive and should demonstrate how all the partners benefit from undertaking this proposal. Well-thought-out, descriptive summaries and rationale will help the administrators and Partnership evaluators make a timely decision on the viability of the proposal. Example proposals are available to use as a guide.

Step 7: Submit Final Proposal to OSHA for Acceptance
■ Submit your final proposal to your Regional Partnership Coordinator for approval. The timeframe for approval is variable, depending upon available resources and other factors. Not all proposals can be accepted, that is why it is important to have up-front talks with OSHA to avoid developing a Partnership proposal that is not viable. Good luck with your submission, and please do not hesitate to contact your Partnership Representative with any questions.

Construction Partnership: Florida CARE Program

CONSTRUCTION ACCIDENT REDUCTION EMPHASIS (CARE) PROGRAM

The mission of the Florida Construction Accident Reduction Emphasis (CARE) Program is to reduce costruction accidents and fatalities in Florida by focusing resources on Enforcement, Outreach, and Partnerships. On March 22, 1999, the OSHA Florida "CARE" inspection program began in the Fort Lauderdale area. Since then, the Jacksonville and Tampa OSHA Area Offices have begun conducting inspections under the "CARE" program guidelines in their areas.

Additional "volunteer" compliance officers from within Region IV, as well as from outside our region, have volunteered to do temporary duty and have been conducting inspections in the three-county area around Fort Lauderdale and Miami. As of August 1, 1999, the volunteer compliance officers have conducted approximately 174 additional construction inspections under the "CARE" program. This number is in addition to the construction inspections conducted by the permanent staff of compliance officers located in the Fort Lauderdale area office.

Approximately 700 construction inspections have been conducted by staff in the Jacksonville, Tampa, and Fort Lauderdale area offices in Florida under the "CARE" program.

CONTENTS

Outreach Activities

Scaffolding Hazards

Focus on Construction

Partnerships

Spanish Construction Course

CARE Posters

FLORIDA "CARE" VOLUNTEERS

Volunteers keep coming out of the woodwork and indicating their desire to assist in the Florida CARE program. Some have indicated a willingness to do a repeat performance if we still need their help. We do intend on putting out the call again. To achieve the impact what we all want regarding this effort, this will not be a short-term patch—and then move on to something new. This effort is one that we will be focusing on for some time to come.

Since March 22, 1999, volunteer compliance officers have each spent two to three weeks in the Fort Lauderdale area conducting inspections. These inspections are in addition to the inspections conducted by the permanent staff in the office.

We thank the volunteers that have served so far: Jim Critopoulos, Mobile AO; Harlan Browning, Birmingham AO; Mike Shea, Atlanta Regional Office; Floyd Gattis, Atlanta-E AO; Mike Leek, Birmingham AO; Ed Keith, Birmingham AO; Eric Harbin, Birmingham AO; Scott Brooks, Milwaukee AO; Ron Hynes, Birmingham AO; Robin Cathey, Atlanta-E AO; Patrick Whavers, Jackson AO; Luis Ramirez, Savannah AO; Marianne Bonito, Atlanta-E AO; Patrick Sharp, Birmingham AO; Billie Kizer, Atlanta-E AO; Xavier Aponte and Ron Byrd, Savannah AO; Patrick Sharp, Birmingham AO; and Scott Maloney, Kansas City AO.

There have been many very fines cases developed during these past few months by the volunteers as well as the staff in the Fort Lauderdale office. Some of the comments that have been passed on to the region after a volunteer has returned home have included:

"This was a good experience for me."

"Many of the employers had heard of the "CARE" program prior to the CSHOs arrival and were eager to learn more about it."

"I could not believe the amount of construction going on down there … it's going on everywhere you look."

We are still looking for a few good volunteers for September, as well as the beginning of the next fiscal year. If anyone wants to repeat a tour in Florida or if there are others that have yet to offer their services to this effort, we would like to hear from you. Anyone wishing to volunteer should get approval from their supervisor and then call Ron McGill in the Regional office at (404)562-2277 to set up a schedule for their visit.

We are encouraging volunteers to be available for three-week tours in order to be more cost effective and increase productivity.

OUTREACH ACTIVITIES

There are many different efforts going on throughout Florida with regard to training and education relating to the CARE effort.

The local chapters of the *National Safety Council* are offering various safety and health classes relating to the CARE program. Many of these are being presented in conjunction with the *Florida Department of Labor and Employment Security's Division of Safety* and OSHA.

To obtain information on classes being presented in your area, please contact your local chapter of the *National Safety Council* or the Florida Department of Labor and Employment Security or the local OSHA area office in Jacksonville (904)232-2985, Tampa (813)626-1177, or Fort Lauderdale (954)424-0242. Many of these classes are free of charge.

The *Florida Division of Safety* has conducted sixteen 10-hour OSHA outreach courses throughout Florida, providing training to approximately 352 employers and employees in the construction industry.

The *Florida Safety Division* has also developed a partnership with the *National Safety Council's* ten Florida chapters, which are also providing further training for the construction industry.

SCAFFOLDING HAZARDS

The photo below illustrates several hazards associated with scaffolding on construction job sites. The picture was taken during a recent compliance inspection at a job site in the Jacksonville, Florida, area.

How many hazards can you find?

1) Failure to provide proper guardrails
2) Failure to provide a ladder for safe access to the scaffolding (climbing up the lanyard rope of the hoisting device)
3) Missing X-bracing at inside of scaffolds
4) Not solid decking
5) No hard hats with overhead hazards
6) Planing spans between scaffold stages are too long
7) Single board over to roof

FOCUS ON CONSTRUCTION

The CARE program is an extensive effort aimed at reducing accidents, injuries, and fatalities in the construction industry throughout the state of Florida.

This program will continue with that goal in mind, and OSHA has stated that over the next three years our goal is to reduce fatalities by at least 15%. In 1998 total construction fatalities in Florida were 68.

On March 22, 1999, the CARE program was started with emphasis placed on getting to as many construction sites as possible. The three area offices in Florida have concentrated their efforts in the construction industry through compliance officer self-referrals and increased emphasis on screening of available information regarding construction sites that are in progress.

The first two tables show comparative data for the period from the beginning of the CARE program (March 22, 1999) to the present time (August 20, 1999) compared to the same period a year ago. The third table shows the data for this fiscal year, to date (FY-99 – YTD).

■■ Table 1 ■■

Construction Fatalities
March 22, 1999–August 20, 1999

Area Office	Number Construction FAT/CAT	Number Construction Inspections
Jacksonville	8	259
Tampa	5	98
Fort Lauderdale	10	387
Totals	**23**	**744**

■■ Table 2 ■■

Construction Fatalities
March 22, 1998–August 20, 1998

Area Office	Number Construction FAT/CAT	Number Construction Inspections
Jacksonville	2	165
Tampa	10	91
Fort Lauderdale	13	48
Totals	**25**	**304**

■■ Table 3 ■■

Florida Fatality Report (FY–99–YTD)
October 1, 1998–August 20, 1999

Area Office	Number Construction FAT/CAT	Number General Industry	Total FAT/CAT
Jacksonville	14	10	24
Tampa	19	33	52
Fort Lauderdale	22	29	51
Totals	**55**	**72**	**127**

PARTNERSHIPS

In an effort to further one of the three major components of the "CARE" program in Florida, a group of interested parties in the south Florida area came together at the outset of the program to discuss the formation of a partnership. This group included employers, employee representatives, and academic representatives from the *State of Florida Division of Safety* and representatives from OSHA.

The "South Florida Construction Safety and Health Partnership" is taking shape with the working draft of the agreement being reviewed by all interested parties to the partnership. A subcommittee of the group's steering committee has been meeting on a regular basis and is working on development of the agreement. Their stated goal is "the elimination of serious injuries and fatalities in the construction industry." Their secondary goal is "to obtain recognition from OSHA for those contractors that have developed and implemented effective safety and health programs and site-specific safety and health plans that assure compliance with OSHA standards, policies and procedures."

This agreement will establish a Partnership Management Committee which will have specific responsibilities for review and approval of all applicants to the partnership, notification to OSHA of accepted new partners, termination of a partner's participation in the program for cause, and preparation and submission of an annual report.

Specific methods of measurement will be developed and implemented to ensure that the appropriate evaluation of the programs and the participating partners will be properly evaluated on a periodic basis. Other partnerships are currently being discussed in the south Florida area. In addition, there are partnerships being discussed and developed in the central and north Florida areas. Each of the partnerships are unique in themselves and involve various organizations and employer/employee groups.

The Tampa OSHA office has had initial discussions with NASA safety officials regarding the development of an alliance with OSHA and CARE.

CONSTRUCTION SAFETY COURSES TO BE GIVEN IN SPANISH

As part of the CARE outreach program, *Florida's Division of Safety* (you can reach them on the Web at http://www.safety.fdles.state.fl.us/) will be offering safety courses in Spanish at Miacon and Miami '99, to be held at the Miami Beach Convention Center on September 16 and 17.

Course topics will include fall protection, electrical safety, excavations and trenching, and scaffolding. The courses are provided without charge, and registration to Miacon & Miami '99 is free to the trade. The convention's Web site address is http://www.miamin.com/.

In addition to the Division of Safety's presentations, the *National Safety Council* will present a construction safety/hazmat course during the show.

CARE Newsletter Staff

Editor – Ron McGill, Atlanta RO Clerk – Frank Marques
Fort Lauderdale, Florida, A.O.

CARE POSTERS STILL AVAILABLE

All three Florida OSHA offices have CARE posters available, in English and Spanish, for those employers who wish to visit their local area office. Posters can also be mailed, on request. Several varieties of the CARE poster are available. Posters are available without charge.

Construction Safety Resources

American Society of Safety Engineers (ASSE)
1800 E. Oakton Street
Des Plaines, IL 60018-2187
(847) 699-2929
www.asse.org

Founded in 1911, with currently over 30,000 members, ASSE is this nation's oldest and largest global safety society, serving the safety, health, and environmental professional. ASSE speaks up for safety . . . in industry, government, standards, education, training, and professional ethics. ASSE provides such services as education, public affairs, government affairs, and national and international safety standards development, technical publications, and timely ongoing communications on safety advancements worldwide. ASSE has 150 chapters, 56 sections, and 65 student sections with eight U.S. regions. There are members in 64 countries, including Saudi Arabia, Kuwait, the United Kingdom, and Egypt. Interested construction safety professionals may wish to join the ASSE Construction Practice Specialty, one of 13 practice specialty groups, to supplement their membership. Other practice specialty groups include: Academics, Consultants, Engineering, Environmental, Healthcare, Industrial Hygiene, International, Management, Mining, Public Sector, Risk Management/Insurance, and Transportation.

American Society for Testing & Materials (ASTM) International
100 Barr Harbor Drive
West Conshohocken, PA 19428
(610) 832-9585
www.astm.org

Organized in 1898, ASTM International is one of the largest voluntary standards development organizations in the world. ASTM Intl. is a not-for-profit organization that provides a forum for the development and publication of voluntary consensus standards for materials, products, systems, and services. More than

20,000 members representing producers, users, ultimate consumers, and representatives of government and academia develop documents that serve as a basis for manufacturing, procurement, and regulatory activities.

American National Standards Institute (ASNI)

1819 L. Street, NW, Suite 600
Washington, DC 20036
(202) 293-8020
www.ansi.org

The American National Standards Institute (ANSI) is a private, nonprofit organization that administers and coordinates the U.S. voluntary standardization and conformity assessment system. The Institute's mission is to enhance both the global competitiveness of U.S. business and the U.S. quality of life by promoting and facilitating voluntary consensus standards and conformity assessment systems, and safeguarding their integrity.

Associated Builders and Contractors (ABC), Inc.

4250 N. Fairfax Dr., 9th Floor
Arlington, VA 22203
(703) 812-2000
www.abc.org

The organization consists of construction contractors, subcontractors, suppliers, and associates. Their purpose is to foster the principles of rewarding construction workers and management on the basis of merit. They sponsor management education programs, craft training, and apprenticeship and skill training programs. A publications catalog is available.

Associated General Contractors (AGC) of America

333 John Carlyle Street, Suite 200
Alexandria, VA 22314
(703) 548-3118
www.agc.org

Located in the metropolitan, DC area, the AGC has more than 80 years of experience in the construction industry. Comprised of over 33,000 members, including general contractors and specialty contractors as well as suppliers and service providers, AGC has 100 chapters throughout the nation. Both at the national level and the state level, the AGC goal is to work in the best interest of its membership.

Board of Certified Safety Professionals (BCSP)

208 Burwash Avenue
Savoy, Illinois 61874
(217) 359-9263
www.bcsp.org

The Board of Certified Safety Professionals was organized as a peer certification board with the purpose of certifying practitioners in the safety profession. The specific functions of the Board, as outlined in its charter, are to evaluate the academic and professional experience qualifications of safety professionals, to administer examinations, and to issue certificates of qualification to those professionals who meet the Board's criteria and successfully pass its examinations. BCSP holds national accreditation.

Construction Industry Institute (CII)
3925 West Braker Lane
Austin, TX 78759-5316
(512) 232-3000
www.construction-institute.org

The Construction Industry Institute (CII), established in 1983, is a consortium of leading owners, contractors, and suppliers who have joined with academia, intent on finding better ways of planning and executing capital construction programs.

Construction Safety Council (CSC)
4100 Madison St.
Hillside, IL 60162
(708) 544-2082
www.buildsafe.org

The Construction Safety Council was founded in 1989 as a nonprofit organization dedicated to the advancement of safety and health interests in the field of construction throughout the world. Chartered by a board of directors composed mostly of large construction company owners and operators, the organization is a professional construction consortium with associations that span the globe.

Dynamic Scientific Controls, Inc.
P.O. Box 415
Wilmington, DE 19808
(800) 376-7775
www.fallsafety.com

Dynamic Scientific Controls—DSC—provides nationwide local fall protection services to industry, including assessment, engineering, supervision of contractor's installation of engineered system, and training.

Janine Reid Group, Inc.
1770 Kearney Street
Denver, CO 80220-1547
(303) 322-3211
www.janinereid.com

The Janine Reid Group offers the following services: crisis management planning, crisis simulation workshop, spokesperson training, and coaching during/after a crisis. These services equip employees in construction and other areas with the skills necessary to respond quickly and appropriately during a crisis.

Mechanical Contractors Association of America (MCAA), Inc.
1385 Piccard Drive
Rockville, MD 20850
(301) 869-5800
www.MCAA.org

MCAA is an association of more than 2200 mechanical, plumbing, and service contractors. Services include: education, safety, legislative assistance, business services, and a publications catalog.

National Fire Protection Association (NFPA)
1 Batterymarch Park
Quincy, MA 02169-7471
(617) 770-3000
www.nfpa.org

The mission of the international nonprofit NFPA is to reduce the worldwide burden of fire and other hazards on the quality of life by providing and advocating scientifically based consensus codes and standards, research, training, and education.

National Safety Council (NSC)
1121 Spring Lake Drive
Itasca, IL 60143-3201
(630) 285-1121
www.nsc.org

NSC is a not-for-profit public service organization dedicated to improving the safety and health of all people. The mission of NSC is to educate and influence society to adopt safety, health, and environmental policies, practices, and procedures that prevent and mitigate suffering economic losses arising from preventable causes.

NEA—The Association of Union Constructors
1501 Lee Hwy.
Arlington, VA 22209
(703) 524-3336
www.nea-online.org

NEA is a nationwide network of over 5000 union contractors, local union contractor trade associations, industrial maintenance contractors, and industry suppliers. The NEA is dedicated to providing its members with the highest level of labor relations and safety services, as well as the promotion of positive labor–management programs in the construction industry.

NESTI
8951 Treeland Lane
Dayton, OH 45458
(937) 434-8951
Hayslip@aol.com

An organization that offers straightforward safety training and litigation support with an emphasis in hands-on excavation work, concrete placement, and project scheduling.

Occupational Safety and Health Administration (OSHA)
200 Constitution Avenue, NW
Washington, DC 20210
(202) 693-1999
www.osha.gov

The mission of the Occupational Safety and Health Administration is to save lives, prevent injuries, and protect the health of America's workers. OSHA partners with the more than 100 million working men and women and their six-and-a-half-million employers who are covered by the Occupational Safety and Health Act of 1970.

Scaffold Industry Association
20335 Ventura Blvd.
Suite 420
Woodland Hills, CA 91364
(818) 610-0320
www.Scaffold.org

The Association promotes safety by developing and providing educational seminars and training courses, providing audio-visual programs and codes of safety practices as well as other training materials and safety aids.

Scaffold Training Institute (STI)
311 East Walker
League City, TX 77573
(281) 332-1613
www.scaffoldtraining.com

A comprehensive scaffold training company.

Sheet Metal and Air Conditioning Contractors National Association (SMACNA)
4201 Lafayette Center Drive
Chantilly, VA 20151-1209
(703) 803-2980
www.SMACNA.org

SMACNA offers contractors professional assistance in labor relations, legislative assistance, research and technical standards development, safety, marketing, business management, industry issues, and insurance and retirement plans.

The Center to Protect Workers' Rights (CPWR)
8484 Georgia Avenue, Suite 1000
Silver Spring, MD 20910
(301) 578-8500
www.cpwr.com

CPWR's main focus is to develop practical ways to improve safety and health for construction workers and their families. As the research, development, and training arm of the Building and Construction Trades Department and the construction unions in the AFL-CIO, CPWR works with more than 30 organizations nationwide.

Underwriters Laboratory (UL), Inc.
333 Pfingsten Road
Northbrook, IL 60062
(847) 272-8800
www.ul.com

UL is an independent, not-for-profit product-safety testing and certification organization. UL offers services to consumers, manufacturers, and regulators.

Index

Bryant Electric, 358

Builders' risk insurance, 83–84

Bullard, Edward D., 6

Bureau of Labor Statistics (BLS) 3, 68, 81, 85, 93, 116, 147, 157, 158, 179, 272, 353, 430, 516, 575, 626, 648, 649

   Bureau of Labor Statistics 2001, 639

Business Roundtable, 7, 18, 68, 98–100, 104–107, 245

Canadian Registered Safety Professional (CRSP), 601

Census of Fatal Occupational Injuries, 158

Centers for Disease Control and Prevention (CDC), 533, 542, 549, 642

Center to Protect Workers' Rights, 68, 70, 635, 666, 667, 706

Certified Crane Operator (CCO), 605–607

Certified Hazardous Materials Manager (CHMM), 609

Certified Health and Safety Technician (CHST), 14

Certified Health Physicist (CHP), 601

Certified Industrial Hygienist (CIH), 601, 607–609

Certified Occupational Health Nurse (COHN), 609

Certified Safety Professional (CSP), 14, 599–602

   Comprehensive Practice Examination, 601–602

   Construction Safety Specialty, 602–605

   Safety Fundamentals Examination, 601

Chain-of-events Theory, 51–52

Changing role of construction safety professional, 666–667

Chartered Engineer (CE), 601

Chicago Women in Trades (CWIT), 635

Choice of law principles, 211

Christmas treeing, *see* multiple-lift rigging

Chromium, *see* hexavalent chromium

CIBA Inc., 625

Clark/Guilford, 619

Clean Air Act (CAA), 417

Clean Water Act (CWA), 415, 416, 417

Cleary, Timothy, 271

CNA Insurance companies, 617

Coble, R., 75

Code of Federal Regulations (CFR), 20

Colorado Contractors Association (CCA), 517

   Instructor's Manual for Traffic Control Supervisor, 517

Colorado Department of Transportation (CDOT), 520, 523

Colorado Steel Erection Partnership, 620

Commission on Higher Education Accreditation (CHEA), 597, 600

Common law "right to control" test, 209, 211, 213

Comparative negligence, 208

Competent person, 19, 103, 144, 146, 147, 178, 221, 236–237, 238, 256, 258, 260, 261, 262, 264, 274, 308, 316, 334, 335, 373, 376, 377, 378, 380, 383, 384, 385, 433, 438, 439, 441

Compliance Safety and Health Officer (CSHO), 190, 192, 196, 197

Comprehensive Environmental Response, Compensation and Liability Act (CERCLA), 415, 417

Comprehensive Practice Examination, *see* Certified Safety Professional

Computer Assisted Design (CAD), 347

Computerized planning and scheduling for construction projects, 667–668

Concrete work

   common hazards, 445, 447

   ergonomic considerations, 451–452

   fall protection for, 450

   flammable chemicals used in, 452

   formwork, 452–453

   handling precast units, 455–456

   mixer hazards, 454

   prestress operations, 454–455

   protective equipment for, 447–450

   pumping ready-mix concrete, 456–457

   regulations governing, 446–447

   respirator requirements, 449

   safety during mixing and batching, 453–454

   toxicity issues, 450–451

   training requirements, 447

Confined space

   alternate-entry classification, 473

   Anatomy of Confined Spaces in Construction, 478

   attendant, 462, 474–475

   atmospheric conditions, 467–468, 471–472

   definition, 169, 407, 462

   electrical work in, 403–405

   entrant, 462, 475

   entry procedure, 475

   entry supervisor, 462, 473–474

   equipment/tool usage in, 476

   glossary of terms, 462–464

   hazard control equipment/measures, 471

   hazards, 465–469

      equipment to identify, 469

   nonpermit-required classification, 463, 472

   owner/employer/contractor responsibilities, 479

   permit-required classification, 464–466, 473

      permit cancellation, 475

      task-specific system, 473

   rescue, 476–478

   retrieval system requirements, 477–478

   testing instruments, 470–471

   testing methods, 469–470

   training topics, 476

   unauthorized entry, 472

   unusual conditions, 478–479

   ventilating, 472

Consolidated insurance program (CIP), 79–88

   project safety management, 88–90

Construction & General Laborers' District Council of Chicago and Vicinity, 670